Advanced Lectures in Mathematics (ALM)

ALM 1: Superstring Theory
ALM 2: Asymptotic Theory in Probability and Statistics with Applications
ALM 3: Computational Conformal Geometry
ALM 4: Variational Principles for Discrete Surfaces
ALM 6: Geometry, Analysis and Topology of Discrete Groups
ALM 7: Handbook of Geometric Analysis, No. 1
ALM 8: Recent Developments in Algebra and Related Areas
ALM 9: Automorphic Forms and the Langlands Program
ALM 10: Trends in Partial Differential Equations
ALM 11: Recent Advances in Geometric Analysis
ALM 12: Cohomology of Groups and Algebraic K-theory
ALM 13: Handbook of Geometric Analysis, No. 2
ALM 14: Handbook of Geometric Analysis, No. 3
ALM 15: An Introduction to Groups and Lattices: Finite Groups and Positive Definite Rational Lattices
ALM 16: Transformation Groups and Moduli Spaces of Curves
ALM 17: Geometry and Analysis, No. 1
ALM 18: Geometry and Analysis, No. 2
ALM 19: Arithmetic Geometry and Automorphic Forms
ALM 20: Surveys in Geometric Analysis and Relativity
ALM 21: Advances in Geometric Analysis
ALM 22: Differential Geometry: Under the Influence of S.-S. Chern
ALM 23: Recent Developments in Geometry and Analysis
ALM 24: Handbook of Moduli, Volume I
ALM 25: Handbook of Moduli, Volume II
ALM 26: Handbook of Moduli, Volume III
ALM 27: Number Theory and Related Areas
ALM 28: Selected Expository Works of Shing-Tung Yau with Commentary, Volume I
ALM 29: Selected Expository Works of Shing-Tung Yau with Commentary, Volume II
ALM 30: Automorphic Forms and *L*-functions
ALM 31: Handbook of Group Actions, Volume I
ALM 32: Handbook of Group Actions, Volume II
ALM 33: Introduction to Modern Mathematics
ALM 34: Some Topics in Harmonic Analysis and Applications

Advanced Lectures in Mathematics
Volume 34

Some Topics in Harmonic Analysis and Applications

edited by

Junfeng Li · Xiaochun Li · Guozhen Lu

International Press
www.intlpress.com

HIGHER EDUCATION PRESS

Advanced Lectures in Mathematics, Volume 34
Some Topics in Harmonic Analysis and Applications

Volume Editors:
Junfeng Li (Beijing Normal University)
Xiaochun Li (University of Illinois at Urbana-Champaign)
Guozhen Lu (Wayne State University)

ISBN: 978-1-57146-315-9

Printed in the United States of America.

20 19 18 17 16 1 2 3 4 5 6 7 8 9

Advanced Lectures in Mathematics

This special volume is dedicated to

Professor Shanzhen Lu

Preface

This is a special volume dedicated to Professor Shanzhen Lu on the occasion of his 75th birthday. As friends and students of Professor Lu, it is our great pleasure to edit this volume to celebrate the mathematical life and career of Professor Lu.

Professor Lu has made important mathematical contributions in harmonic analysis and applications. In addition, Professor Lu has educated and trained many younger mathematicians in China. Professor Lu supervised around 50 M.S. and Ph.D. students, post-doctors and visiting scholars over the years. We refer the reader to "A List of Ph.D. Students, Post-doctors and Visiting Scholars Supervised by Professor Shanzhen Lu and Foreign Collaborators Who Worked with Professor Shanzhen Lu". We also refer the reader to "An Introduction of Professor Shanzhen Lu" for more information about his personal and mathematical life.

This volume describes some recent development on some topics in harmonic analysis and applications to partial differential equations and geometric analysis. Experts in these areas are invited to contribute works which address recent development in some of the topics. Subjects covered in this volume include local Tb theorems in non-homogeneous or vector-valued setting, multilinear embedding theorems, norm estimates of quasi-modes on Riemannian manifolds, Calderón-Zygmund singular integral operators and commutators, Dirichlet-to-Neumann maps, multilinear and multi-parameter Fourier multipliers, weighted normed inequalities, nonlinear Schrödinger equations, function space theory, etc. It will be an excellent reference book for mathematicians working in analysis and PDEs and advanced graduate students, post-doctors in these and related areas.

We are indebted to many people who have contributed to the publication of this volume. First of all, we would like to take this opportunity to sincerely thank all contributors of articles to this volume. This volume could not have been completed without their hard work and mathematical contributions. In particular, we like to thank them for their patience in waiting for this volume to be finally published. We also like to thank Guido Weiss, Elinor Anheuser Professor of Mathematics at Washington University in St. Louis, for his short tribute to Professor Shanzhen Lu. Professor Lu spent two years at Washington University working with Professors Weiss and Mitchell Taibleson in early 1980's. We also thank Editors of the series of Advanced Lectures in Mathematics Professors S.-T. Yau, K. Liu, L. Ji for their support in the process of editing of this volume. In particular Lizhen has always been a good person to talk to when we have questions. Most importantly, special thanks are due to Ms. Huaying Li and Ms. Liping Wang from the Higher Education Press in China for their tireless support and hard work in editing this volume. Without the tremendous support and devotion of Ms. Huaying Li and Ms. Liping Wang and many other staff members at the Higher Education Press, we could not have had completed this well organized volume. In particular, this is the second time for the third Editor (Guozhen Lu) of this special volume to collaborate with both of Ms. Huaying Li and Ms. Liping Wang. It is indeed a pleasure for him to work with them through this long process.

Lastly, it is our great pleasure and honor to be in charge of editing this volume in celebration of the mathematical life of Professor Lu. All of us have benefited from Professor Lu in many aspects over the years. As former students and/or friends of Professor Lu, we would like to take this opportunity to thank Professor Lu for his teaching, continuous support to our mathematical careers and his long-lasting friendship.

Junfeng Li, Beijing Normal University, China
Xiaochun Li, University of Illinois at Urbana-Champaign, USA
Guozhen Lu, Wayne State University, USA
May, 2015

A Tribute to Professor Shanzhen Lu

Professor Shanzhen Lu is a dear friend who has contributed considerably to the mathematical world. I first met Professor Lu when he came to the Department of Mathematics at Washington University in St. Louis to work with me in the fall of 1980. Professor Lu remained in St. Louis until the end of the spring semester in 1982. During that time we developed a wonderful friendship and collaboration. This friendship continues to this day in the year 2015.

I am proud to say that I have many close Chinese friends as well as having had some excellent Chinese Ph.D. students. These rich connections have permitted me to visit China often and I always made times to see Professor Lu and reestablish our old ties. Professor Lu and all of my other substantial relationships with Chinese mathematicians have helped make my career a wonderful experience.

Professor Lu and I began collaborating in harmonic analysis soon after his arrival in St. Loius. It is important to state that the late Professor Mitchell Taibleson joined this collaboration. On several occasions Professor Taibleson expressed his pleasure with being a part of this joint work.

Professor Lu and I studied and developed many properties of important operators in harmonic analysis in the study of certain function spaces that we described as being generated by blocks. "Blocks" are generalizations of "atoms". The latter is an important notion in the study and extension of Hardy spaces. In 1989, Professor Lu wrote a book on the theory of spaces generated by blocks. He very generously included my name and that of Professor Taibleson as co-authors of this book.

In this short note I cannot describe in great detail Professor Lu's research. However, it is quite extensive and deals with many topics in analysis. Let me name some of the areas in which he has worked: Hardy spaces and their extensions to higher dimensions, applications of Fourier integrals, almost everywhere convergence of important means of multiple Fourier series and the boundedness of several important operators.

Professor Lu was born in November 1939. In 1961 he graduated from East China Normal University. In 1961–1980 he was an Assistant Professor in Beijing Normal University and was promoted in 1983 to a Full Professor in this institution. At that time, Professor Lu was the Director of the Institute of Mathematical

Sciences and Education. He was Chancellor of Beijing Normal University during the period 1995–1999. He also was a member of the Editorial Board of Acta Mathematica Sinica and the International Journal of Applied Mathematical Sciences. In addition, he has received a number of important prizes in both education and research.

From this short history it is clear that Professor Lu has had a most distinguished, varied and impressive career. I am very proud to have been his friend and collaborator over these many years.

Guido Weiss
Washington University in St. Louis
February 19, 2015

An Introduction of Professor Shanzhen Lu

Professor Shanzhen Lu was born in 1939 in Wenzhou, a coastal city in the east of China. Lu was the second of an eight-children family. It was very tough to take care of such a big family only with the salary of an office worker during the war time. But his parents still tried their best to let him finish his high school in 1957. Wenzhou, now a famous city by its rapid commercial development, is also thought to be the hometown of mathematicians in China. Many renowned mathematicians in modern China call Wenzhou as their hometown. As one of the early opened cities in China after the Opium War in 1840, its education in science was better than most other parts of China at that time, especially in the mathematical education. Some excellent mathematicians were teaching in high schools those years. Lu was lucky to have the opportunity to access to some advanced mathematics starting his years during high school study. The beauty of mathematics attracted this young man. He made up his mind to study mathematics further. From 1957 to 1961, he studied at the Department of Mathematics of East China Normal University. He graduated in 1961 and moved to Beijing. From that on, he has worked at Beijing Normal University for more than 50 years as a faculty member.

As an outstanding mathematician who specializes in harmonic analysis, Professor Shanzhen Lu has made significant contributions in a number of subjects in harmonic analysis and their applications. His works involve many important fields, such as the boundedness of the Bochner-Riesz means operators, which was motivated to set up the pointwise convergence of Fourier series in high dimension spaces. He has also done important works on the boundedness of singular integral operators with rough kernels, the oscillatory integrals and their commutators. Function space theory is also one of his research interests. He obtained many interesting results in the theory of Hardy spaces and Herz type spaces.

Professor Lu is one of the pioneers helping the development of modern harmonic analysis in China. Together with a few others in the older generation, he introduced the most updated development of the real variable method on Euclidean spaces, an important branch of harmonic analysis, to the mathematical community in China. In the beginning of 1980's, the research of harmonic analysis in China was very weak. After Lu and his colleague's forty-year long effort and contribution, most of major universities in China now have faculty members working in this important area. Professor Lu himself has also trained many students and visiting scholars. Among 29 people who obtained their Ph.D under Professor Lu's supervision, most of them now are active harmonic analysts and are leading members of faculty in their own universities in China.

In addition to his achievement in mathematics and education, he is also a successful administrator. In 1994, he became the first elected president of Beijing Normal University, a top research university focusing on liberal arts and sciences in China. He was also one of the first two elected presidents of Chinese universities since the founding of the People's Republic of China. As a president of a university in China, he realized that the most important thing is the quality of the faculty, especially the healthy growth of young faculty members. He paid a considerable attention to support young faculty members by implementing some new promotion policy that was designed for helping young faculty members to improve their research and teaching quality and to grow up to a leader as soon as possible. More than 15 years have passed since he was the president of Beijing Normal University, Professor Lu has been making substantial contributions to education in China. He was invited in 2012 to become the founding president of Wenzhou-Kean University, a joint adventure between Kean University in US and municipal government of the city of Wenzhou.

Professor Lu is a passionate person who loves Chinese 5000-year history and culture. He is a great teacher holding Chinese traditional wisdom and modern visions. All of his students have benefited from and been influenced by those precious characters through his teaching.

At his leisure time, Professor Lu enjoys the classical music and traditional Chinese opera. He is also good at swimming and he swam almost everyday when he was younger.

Contents

Some Topics in Harmonic Analysis and Applications
ALM 34, pp. 1–26

Multilinear Embedding and Hardy's Inequality

William Beckner *

Abstract

Multilinear trace restriction inequalities are obtained for Hardy's inequality. More generally, detailed development is given for new multilinear forms for Young's convolution inequality, and a new proof for the multilinear Hardy-Littlewood-Sobolev inequality, and its realization on hyperbolic space.

2000 Mathematics Subject Classification: 58J79, 42B10, 35A15.

Keywords and Phrases: Hardy's inequality, multilinear embedding, Hardy-Littlewood-Sobolev inequality.

Hardy's inequality

$$f \rightsquigarrow (Tf)(x) = \frac{1}{x}\int_0^x f(t)\,dt\ , \quad x > 0,$$

$$\int_0^\infty \left|\frac{1}{x}\int_0^x f(t)\,dt\right|^p dx < \left(\frac{p}{p-1}\right)^p \int_0^\infty |f(x)|^p\,dx\ , \quad 1 < p < \infty \qquad (1)$$

lies at the core of understanding for many parts of modern Fourier analysis — embedding estimates, maximal functions, singular integrals, weighted inequalities and measures of uncertainty. This operator plays an intrinsic role for understanding how dilation invariance and representation for convolution integrals on a group determine optimal estimates. Recent work by S. Lu and his collaborators has led to new multilinear and higher-dimensional forms for Hardy's inequality. Our purpose here is to consider such operators in the context of classical convolution inequalities.

*Department of Mathematics, The University of Texas at Austin, 1 University Station C1200, Austin TX 78712-0257 USA. *E-mail:* beckner@math.utexas.edu

1 Multilinear convolution inequalities

Consider the convolution integral with multi-component decomposition

$$F(w) = \int_{\mathbb{R}^n \times \cdots \times \mathbb{R}^n} G(w-y)\, H(y)\, dy \ , \qquad w \in \mathbb{R}^{mn}.$$

(1) "diagonal trace restriction"

$$F(w) \rightsquigarrow F(\underbrace{x, \ldots, x}_{m \text{ slots}}) \equiv F(x) \ , \qquad x \in \mathbb{R}^n.$$

(2) "multilinear products"

$$F(x) = \int_{\mathbb{R}^{mn}} \prod g_k(x-y_k)\, H(y_1, \ldots, y_m)\, dy \ ;$$

the objective is to determine how the components of G and the nature of H control the size of F. Here the g_k's will be taken as inputs, so the multilinear map is given by

$$H \in L^p(\mathbb{R}^{mn}) \rightsquigarrow F \in L^q(\mathbb{R}^n) \ , \qquad 1 < p < q. \tag{2}$$

A relatively simple characterization for this framework can be obtained from the classical Young's inequality using a varied set of techniques.

Theorem 1. *For $1 < s_k < 2 \leq q < \infty$, $k = 1, \ldots, m$ and $m/2 + 1/q = \sum 1/s_k$,*

$$\|F\|_{L^q(\mathbb{R}^n)} \leq C \prod \|g_k\|_{L^{s_k}(\mathbb{R}^n)} \|H\|_{L^2(\mathbb{R}^{mn})}, \tag{3}$$

C is a constant less than or equal to 1.

Proof. Consider the dual problem (primes denote dual exponents, $1/q + 1/q' = 1$)

$$\left[\int_{\mathbb{R}^{nm}} \left| \int_{\mathbb{R}^n} \prod_k g_k(x-y_k)\, h(x)\, dx \right|^2 dy \right]^{1/2} \leq C \prod \|g_k\|_{L^{s_k}(\mathbb{R}^n)} \|h\|_{L^{q'}(\mathbb{R}^n)}, \tag{4}$$

then the left-hand side becomes

$$\int_{\mathbb{R}^n \times \mathbb{R}^n} \overline{h(w)} \prod (g_k * \tilde{g}_k)(w-x) h(x)\, dx\, dw$$

with $\tilde{f}(x) = \bar{f}(-x)$. Here the problem can be equivalently stated as finding the norm for the operator

$$h \in L^{q'}(\mathbb{R}^n) \rightsquigarrow k * h \in L^q(\mathbb{R}^n)$$

with $k = \prod_k (g_k * \tilde{g}_k)$ — a map from a space to its dual. So applying Young's inequality, it will suffice to show that $k \in L^r(\mathbb{R}^n)$ with $r = q/2$. To complete the proof, we need to exhibit a sequence of indices $\{t_k\}$ with $\sum 1/t_k = 1$ and such that $g_k * g_k \in L^r(\mathbb{R}^n)$ with $r = qt_k/2$. Then $1/t_k = q[1/s_k - 1/2]$ and $\sum 1/t_k = q[\sum 1/s_k - m/2] = 1$ by hypothesis. By this argument, the classical value of $C = 1$ can be used. □

The structure of inequality (4) allows application of the Brascamp-Lieb-Luttinger rearrangement inequality [10] so that the functions h, $\{g_k\}$ can be replaced by their radial decreasing rearrangements in order to have an improved inequality. Then methods from [9] determine the sharp constant C by calculating the optimal value when the functions are gaussian. From this explicit calculation, one can see that the analysis does not reduce to iterative processes or product function states. For example, when $m = 2$ extremal functions for (3) will have the form

$$e^{-\alpha_1 y_1^2} e^{-\alpha_2 y_2^2} e^{-\beta(y_1-y_2)^2}, \qquad \alpha_1, \alpha_2, \beta > 0$$

which is not a product function in the variables y_1, y_2.

Corollary 1. *The optimal value for the constant C in Theorem 1 is given by*

$$C = \left[2^{-m} (q')^{2/q'} \prod_{k=1}^{m} (s_k)^{2/s_k} \prod_{\ell=0}^{m} (\lambda_\ell)^{\lambda_\ell} \right]^{n/4} \tag{5}$$

where $\lambda_0 = 1 - 2/q$ and $\lambda_k = 1 - 2/s_k'$ with $\sum \lambda_\ell = 1$.

Proof. As observed above, application of the Brascamp-Lieb-Luttinger rearrangement inequality allows determination of the best constant to be attained when all functions are gaussian. Consider positive parameters $\{a_k\}$, $k = 0, 1, \ldots, m$ with

$$h(x) = (a_0)^{n/2} e^{-\pi a_0 x^2}, \quad g_k(x) = (a_k)^{n/2} e^{-\pi a_k x^2}, \qquad k = 1, \ldots, m,$$

$$\|h\|_{L^{q'}(\mathbb{R}^n)} = (a_0)^{n/2q} (q')^{-n/2q'}\,; \quad \|g_k\|_{L^{s_k}(\mathbb{R}^n)} = (a_k)^{n/2s_k'} (s_k)^{-n/2s_k}.$$

Then

$$(g_k \times g_k)(x) = (a_k/2)^{n/2} e^{-\pi a_k x^2/2}$$

and

$$C = 2^{-mn/4} (q')^{n/2q'} \prod_k (s_k)^{n/2s_k} \max \left[\frac{\prod a_\ell^{\lambda_\ell}}{\sum a_\ell} \right]^{n/4}.$$

Observe by variation argument for $a_\ell > 0$ and $\sum \lambda_\ell = 1$,

$$\max \left[\frac{\prod a_\ell^{\lambda_\ell}}{\sum a_\ell} \right] = \prod \lambda_\ell^{\lambda_\ell}. \tag{6}$$

□

The symmetry in the role of the functions h and $\{g_k\}$ for the computation of the optimal constant suggests that there should be an equivalent formulation of the inequality in Theorem 1 that displays this symmetry. Such a result can be given using the Fourier transform and shows the connection with the Hausdorff-Young inequality as suggested by the role of a map from a space to its dual. And a closer look at the left-hand side in (4) shows that the role of h can be interchanged with any single g_k which also displays this symmetry.

Corollary 2. *With constraints as given in Theorem* 1, *an equivalent inequality is*

$$\Big(|\hat{h}|^2 \times |\hat{g}_1|^2 * \cdots * |\hat{g}_m|^2\Big)(0) \le C^2 \Big[\|h\|_{L^{q'}(\mathbb{R}^n)} \prod \|g_k\|_{L^{s_k}(\mathbb{R}^n)}\Big]^2 \tag{7}$$

with the value of C given in (5).

A more general result, though with restricted conditions for an elementary proof, can be given for the multilinear map

$$H \in L^p(\mathbb{R}^{nm}) \rightsquigarrow F \in L^q(\mathbb{R}^n), \qquad 1 < p \le q < \infty.$$

Theorem 2. *For $1 < p < q < \infty$, $1 < s_k < p'/q'$, $k = 1, \dots, m$ and $1/q + m/p' = \sum 1/s_k$, $m < p'/q'$,*

$$\|F\|_{L^q(\mathbb{R}^n)} \le C \prod \|g_k\|_{L^{s_k}(\mathbb{R}^n)} \|H\|_{L^p(\mathbb{R}^{nm})}, \tag{8}$$

C is a constant less than or equal to 1.

Proof. Consider the dual problem (primes denote dual exponents, $1/q + 1/q' = 1$)

$$\left[\int_{\mathbb{R}^{nm}} \left| \int_{\mathbb{R}^n} \prod g_k(x - y_k) h(x)\, dx \right|^{p'} dy\right]^{1/p'} \le C \prod \|g_k\|_{L^{s_k}(\mathbb{R}^n)} \|h\|_{L^{q'}(\mathbb{R}^n)}. \tag{9}$$

Choose the sequence $\beta_k = 1/t_k = q(1/s_k - 1/p')$; then $\sum \beta_k = 1$ and $1 < t_k < s_k$. Apply Hölder's inequality to the inner integral to give

$$\left| \int \prod g_k(x - y_k) h(x)\, dx \right| \le \prod_k \Big[\big(u_k * |h|\big)(-y_k)\Big]^{\beta_k}$$

for $u_k = |g_k|^{t_k}$. Then

$$\left[\int_{\mathbb{R}^{nm}} \left| \int_{\mathbb{R}^n} \prod g_k(x - y_k) h(x)\, dx \right|^{p'} dy\right]^{1/p'} \le \prod \Big[\big\|u_k * |h|\,\big\|_{L^{p'/t_k}(\mathbb{R}^n)}\Big]^{\beta_k}.$$

Observe that $p'/t_k > 1$ since $t_k < s_k < p'/q' < p'$. Then using Young's inequality

$$\big\|u_k * |h|\big\|_{L^{p'/t_k}(\mathbb{R}^n)} \le \Big[\|g_k\|_{L^{s_k}(\mathbb{R}^n)}\Big]^{t_k} \|h\|_{L^{q'}(\mathbb{R}^n)}$$

and

$$\prod \Big[\big\|u_k * |h|\big\|_{L^{p'/t_k}(\mathbb{R}^n)}\Big]^{\beta_k} \le \Big[\prod \|g_k\|_{L^{s_k}(\mathbb{R}^n)}\Big] \|h\|_{L^{p'}(\mathbb{R}^n)}$$

which demonstrates inequality (8) with the classical value of $C = 1$ from Young's inequality. □

There is no overlap between these two theorems since for $p = 2$ the second theorem requires $m < 2/q' < 2$. More generally, the method of radial-decreasing equimeasurable rearrangements and reduction to gaussian functions (see [9]) can be applied for the case where p' is a positive integer at least two. For $m = 2$ an iterative argument for multidimensional estimates allows a full determination in terms of allowed parameters for this multilinear extension of Young's inequality.

Theorem 3. *For $1 < p < q < \infty$, $1 < s_1, s_2 < p' < \infty$ and $2/p' + 1/q = 1/s_1 + 1/s_2$,*

$$\|F\|_{L^q(\mathbb{R}^n)} \le C\|g_1\|_{L^{s_1}(\mathbb{R}^n)}\|g_2\|_{L^{s_2}(\mathbb{R}^n)}\|H\|_{L^p(\mathbb{R}^n\times\mathbb{R}^n)}, \tag{10}$$

C is a constant less than or equal to 1.

Proof. Consider the dual problem

$$\begin{aligned}&\left[\int_{\mathbb{R}^n\times\mathbb{R}^n}\left|\int_{\mathbb{R}^n} g_1(x-y_1)g_2(x-y_2)h(x)\,dx\right|^{p'} dy_1\,dy_2\right]^{1/p'}\\ &\le C\|g_1\|_{L^{s_1}(\mathbb{R}^n)}\|g_2\|_{L^{s_2}(\mathbb{R}^n)}\|h\|_{L^{q'}(\mathbb{R}^n)}.\end{aligned} \tag{11}$$

Write the left-hand side as

$$\begin{aligned}&\left[\int_{\mathbb{R}^n}\left[\int_{\mathbb{R}^n}\left|\int_{\mathbb{R}^n} g_1(x-y_1)w(x)\,dx\right|^{p'} dy_1\right]^{1/p\cdot p'} dy_2\right]^{1/p'}\\ &=\left[\int_{\mathbb{R}^n}\left[\int_{\mathbb{R}^n}|(\tilde g_1 * w)(y_1)|^{p'}dy_1\right]^{1/p'\cdot p'} dy_1\right]^{1/p'}\\ &\le \|g_1\|_{L^{s_1}(\mathbb{R}^n)}\left[\int_{\mathbb{R}^n}\left[\|w\|_{L^r(\mathbb{R}^n)}\right]^{p'} dy_2\right]^{1/p'}\\ &=\|g_1\|_{L^{s_1}(\mathbb{R}^n)}\left[\int_{\mathbb{R}^n}\left[\left(|\tilde g_2|^r * |h|^r\right)(y_2)\right]^{p'/r} dy_2\right]^{r/p'\cdot 1/r}\end{aligned}$$

where $w(x) \equiv g_2(x-y_2)h(x)$ and $1/s_1 + 1/r - 1 = 1/p'$.

It is necessary to show that $1 < r < s_2$ and $r/s_2 + r/q' - 1 = r/p'$. Then $1/r = 1/p' + 1/s_1' < 1/s_1 + 1/s_1' = 1$ since $s_1 < p'$; and $1/r = 1/p' + 1/s_1' > 1/s_2$ since $1/p' + 1 = 2/p' + 1/p > 2/p' + 1/q = 1/s_1 + 1/s_2$ as $p < q$. Finally $r/s_2 + r/q' - 1 = r/p'$ is equivalent to $1/s_2 + 1/q' - 1/r = 1/p'$ or $1/s_2 + 1/q' - 1/p' = 1/r = 1/p' + 1/s_1'$ or $1/s_1 + 1/s_2 = 2/p' + 1 - 1/q' = 2/p' + 1/q$ (initial condition required by dilation invariance).

Under these constraints

$$\left[\int_{\mathbb{R}^n}\left[\left(|\tilde g_2|^r * |h|^r\right)(y_2)\right]^{p'/r} dy_2\right]^{r/p'\cdot 1/r} \le \|g_2\|_{L^{s_2}(\mathbb{R}^n)}\|h\|_{L^{q'}(\mathbb{R}^n)}$$

which completes the demonstration of inequality (11) and the proof of the theorem. □

By extending the argument given above, one obtains a proof by induction for the allowed range of parameters for convolution inequalities associated with (2) for all values of m.

Theorem 4 (Multilinear Young's Inequality). *For $1 < p < q < \infty$, $1 < s_k < p'$, $k = 1, \dots, m$ and $1/q + m/p' = \sum 1/s_k$,*

$$\|F\|_{L^q(\mathbb{R}^n)} \le \prod \|g_k\|_{L^{s_k}(\mathbb{R}^n)}\|H\|_{L^p(\mathbb{R}^{nm})}. \tag{12}$$

Proof. Consider the dual problem

$$\left[\int_{\mathbb{R}^{nm}}\left|\int_{\mathbb{R}^n}\prod g_k(x-y_k)h(x)dx\right|^{p'}dy\right]^{1/p'} \le \prod \|g_k\|_{L^{s_k}(\mathbb{R}^n)}\|h\|_{L^{q'}(\mathbb{R}^n)}. \qquad (13)$$

The method used in Theorem 3 allows the construction of an argument so that the inequality for m copies of $\mathbb{R}^n$ is reduced to proving the inequality for the case of $m-1$ copies of $\mathbb{R}^n$. Then the theorem is established from using Theorem 3 where the case $m=2$ is proved. Order the values of $\{s_k\}$ so that $1 < s_1 \le s_2 \le \cdots \le s_m < p'$. Start with the integral

$$\left[\int\left|\int_{\mathbb{R}^n}\prod g_k(x-y_k)h(x)dx\right|^{p'}dy_1\right]^{1/p'} = \left[\int\left|\int_{\mathbb{R}^n} g_1(x-y_1)w(x)dx\right|^{p'}dy_1\right]^{1/p'}$$
$$\le \|g_1\|_{L^{s_1}(\mathbb{R}^n)}\|w\|_{L^r(\mathbb{R}^n)}$$

where

$$w(x) = \prod_{k\ge 2} g_k(x-y_k)h(x)\ , \qquad \frac{1}{s_1}+\frac{1}{r}-1=\frac{1}{p'}\ .$$

Now it is necessary to show that $1 < r < s_2 \le s_k < p'$. First, $1/r = 1/p' + 1/s_1' < 1/s_1 + 1/s_1' = 1$; second, $m/p' + 1/q = \sum 1/s_k$ and

$$\frac{2}{p'}+\frac{1}{q} = \frac{1}{s_1}+\frac{1}{s_2}+\sum\left(\frac{1}{s_k}-\frac{1}{p'}\right) > \frac{1}{s_1}+\frac{1}{s_2},$$

$$\frac{1}{r} = \frac{1}{p'}+\frac{1}{s_1'} = \frac{2}{p'}+\frac{1}{p}-\frac{1}{s_1} > \frac{2}{p'}+\frac{1}{q}-\frac{1}{s_1} > \frac{1}{s_2},$$

so $r < s_2 \le s_k < p'$. So writing out this representation

$$\left[\int_{\mathbb{R}^{(m-1)n}}\left[\|w\|_{L^r(\mathbb{R}^n)}\right]^{p'}dy\right]^{1/p'} = \left[\int_{\mathbb{R}^{(m-1)n}}\left|\int{\prod}' u_k(x-y_k)v(x)dx\right|^{p'/r}dy\right]^{1/p'}$$

where $u_k = |g_k|^r$ and $v = |h|^r$, and for $t_k = s_k/r$, $\alpha = p'/r$, $\sigma' = q'/r$, then the corresponding constraint becomes $(m-1)/\alpha + 1/\sigma = \sum' 1/t_k$ which by using $1/r = 1/p' + 1/s_1'$ becomes the mth-level constraint $m/p' + 1/q = \sum 1/s_k$ (here $\sum'$ indicates that some terms are not included). Since the number of y-integrations has now been reduced by one, this process can be repeated until one reaches $m=2$ which is already demonstrated in Theorem 3. □

Because of the iterative nature of the proof of Theorem 4, it would be natural to consider whether this product structure could determine the nature of the optimal constant. But since the extremal functions will not be simple product functions, such an argument would not be sufficient. However this method of iterative reduction does enable an extension to include Riesz potentials with a resulting multilinear Hardy-Littlewood-Sobolev inequality. For the case $p=2$, this problem was treated in [6]. Still a different approach is given in [3] (see page 48) where conformal invariance is used.

Theorem 5 (Multilinear Hardy-Littlewood-Sobolev Inequality). *Let*

$$F(x) = \int_{\mathbb{R}^{nm}} \prod |x - y_k|^{-\lambda_k} H(y_1, \dots, y_m)\, dy \ , \qquad \lambda_k = n/s_k \tag{14}$$

for $1 < p < q < \infty$, $1 < s_k < p'$, $k = 1, \dots, m$, $1/q + m/p' = \sum 1/s_k$; *then*

$$\|F\|_{L^q(\mathbb{R}^n)} \le C\|H\|_{L^p(\mathbb{R}^{nm})}. \tag{15}$$

Proof. Look at the dual inequality and use the same induction argument as for Theorems 3 and 4 while applying the standard inequality for bounds for the Riesz potentials on $\mathbb{R}^n$. □

Corollary. *Let*

$$F(x) = \int_{\mathbb{R}^{nm}} \prod g_k(x - y_k) H(y_1, \dots, y_m)\, dy$$

for $1 < p < q < \infty$, $1 < s_k < p'$, $k = 1, \dots, m$ *and* $1/q + m/q' = \sum 1/s_k$; *then*

$$\|F\|_{L^q(\mathbb{R}^n)} \le C \prod \|g_k\|_{L_{s_k,\infty}(\mathbb{R}^n)} \|H\|_{L^p(\mathbb{R}^{nm})}.$$

Proof. Return to the proof of Theorem 4, and the inductive step

$$\begin{aligned}
&\left[\int \left| \int_{\mathbb{R}^n} \prod g_k(x - y_k) h(x)\, dx \right|^{p'} dy_1 \right]^{1/p'} \\
&= \left[\int \left| \int_{\mathbb{R}^n} g_1(x - y_1) w(x)\, dx \right|^{p'} dy_1 \right]^{1/p'} \\
&\le \left[\int \left| \int_{\mathbb{R}^n} g_1^*(x - y_1) w^*(x)\, dx \right|^{p'} dy_1 \right]^{1/p'} \\
&\le \|g_1\|_{L_{s_1,\infty}(\mathbb{R}^n)} \left[\int \left| \int_{\mathbb{R}^n} |x - y_1|^{-\lambda_1} w^*(x) dx \right|^{p'} dy_1 \right]^{1/p'} \\
&\le C\|g_1\|_{L_{s_1,\infty}} \|w^*\|_{L^r(\mathbb{R}^n)} = C\|g_1\|_{L_{s_1,\infty}(\mathbb{R}^n)} \|w\|_{L^r(\mathbb{R}^n)}.
\end{aligned}$$

Repeat this inductive step for g_2 and continue the iterative steps. Here g^* denotes the non-negative equimeasurable radial decreasing rearrangement of g, and

$$g^*(x) \le C\|g\|_{L_{s,\infty}(\mathbb{R}^n)} |x|^{-n/s}$$

for g in the weak-Lebesgue space $L_{s,\infty}(\mathbb{R}^n)$. □

2 Diagonal trace restriction for Hardy's inequality

The multilinear convolution structure described above can be applied to Hardy's inequality:

$$(Tf)(x) = \frac{1}{x}\int_0^x f(t)\,dt\ , \qquad x > 0,$$

$$\|Tf\|_{L^p(0,\infty)} < \frac{p}{p-1}\|f\|_{L^p(0,\infty)}\ , \qquad 1 < p < \infty, \tag{16}$$

$$\|x^{1/r}Tf\|_{L^q(0,\infty)} \le C_{p,q}\|f\|_{L^p(0,\infty)}\ , \qquad 1 < p < q < \infty, \tag{17}$$

$$\frac{1}{r} = \frac{1}{p} - \frac{1}{q}\ , \quad C_{p,q} = (p'/q)^{1/q}\left[\frac{q/r\ \Gamma(r)}{\Gamma(r/q)\Gamma(r/q')}\right]^{1/r}.$$

The constants are best possible (see [14, 15, 8, 16]). Equality is attained for functions of the form

$$f(t) = A(1 + ct^{q/r})^{-q/(q-p)}.$$

These operators extend by tensor product to the manifold $[0,\infty)^m$: $T\otimes T\otimes\cdots\otimes T$ acts on functions defined on $[0,\infty)^m$ and $S\otimes S\otimes\cdots\otimes S$ with $S = x^{1/r}T$ similarly acts on such functions. By iteration one obtains:

Theorem 6. *For* $1 < p < q < \infty$,

$$\|(T\otimes\cdots\otimes T)H\|_{L^p[0,\infty)^m} < \left(\frac{p}{p-1}\right)^m \|H\|_{L^p[0,\infty)^m}, \tag{18}$$

$$\|(S\otimes\cdots\otimes S)H\|_{L^q[0,\infty)^m} \le [C_{p,q}]^m\|H\|_{L^p[0,\infty)^m}. \tag{19}$$

Proof. This extension to higher dimensions is obtained by iteration of one-dimensional estimates (see [1]). The constants are best possible from considering product functions. □

Let Λ_m denote the first quadrant in $\mathbb{R}^m$; namely, $[0,\infty)^m$. For a non-negative function f defined on Λ_m, let $f_\#$ denote the equimeasurable rearrangement of f as a radial decreasing function on Λ_m. Then

Lemma.

$$\int_{\Lambda_m} f(x)g(x)\,dx \le \int_{\Lambda_m} f_\#(x)g_\#(x)\,dx. \tag{20}$$

Proof. Extend f and g to all of $\mathbb{R}^m$ by copying the functions to all equivalent sectors; then

$$\begin{aligned}\int_{\Lambda_m} f(x)g(x)\,dx = 2^{-m}\int_{\mathbb{R}^m} f(x)g(x)\,dx &\le 2^{-m}\int_{\mathbb{R}^m} f^*(x)g^*(x)\,dx \\ &= \int_{\Lambda_m} f_\#(x)g_\#(x)\,dx.\end{aligned}$$

□

Such a rearrangement is optimal for any sector in $\mathbb{R}^m$ having vertex at the origin which you can replicate to pave $\mathbb{R}^m$.

Theorem 7. *For $1 < p \le q < \infty$ and $\gamma = m/p - 1/q$ with $H \in L^p(\Lambda_m)$ and $\psi(x) = x^\gamma [(T \otimes \cdots \otimes T)H](x, \ldots, x)$,*

$$\|\psi\|_{L^q[0,\infty)} \le C \|H\|_{L^p(\Lambda_m)}. \tag{21}$$

Proof. By applying the rearrangement lemma for reduction to radial functions and rescaling, a simple bound for this diagonal trace restriction inequality is obtained from the weighted Hardy inequality (17).

$$\psi(x) \le x^{-m/p'-1/q} \int_{B_x \cap \Lambda_m} f^*(t)\, dt = 2^{-m} \mathrm{vol}(B) \int_0^{ax} f^*(u) u^{m-1}\, du \tag{22}$$

where $u = |t|$ and B_x is a ball centered at the origin with $\mathrm{vol}(B_x) = 2^m x^m$; that is, the ball has radius ax so that $a^m = 2^m/\mathrm{vol}(B)$ with $g(v) = f^*(av^{1/m})$, $v = |t|^m$ and B being the unit ball of radius 1. Then after rescaling to remove the factor a, changing variables from $t \in \Lambda_m$ to $v = |t|^m$, and clearing away factors of spherical volume, inequality (21) is equivalent to the weighted one-dimensional Hardy inequality

$$\left[\int_0^\infty \left(x^{-1/p'-1/q} \int_0^x g(v)\, dv\right)^q dx\right]^{1/q} \le C_{p,q} \left[\int_0^\infty |g(v)|^p\, dv\right]^{1/p}. \tag{23}$$

□

To have a better sense of the method used in this argument, it will be useful to look at an n-dimensional variant of Hardy's inequality which was suggested in [13]. On $\mathbb{R}^n$ let

$$W(x) = \frac{1}{\mathrm{vol}(B)} \frac{1}{x^n} \int_{|t|<x} f(t)\, dt\,, \qquad x > 0$$

where now $W(x)$ is considered as a function on $\mathbb{R}^n$ with x being the radial variable and B is the unit ball in $\mathbb{R}^n$.

Theorem 8. *For $1 < p \le q < \infty$ and $\gamma = n/p - n/q = n/r$,*

$$\|x^\gamma W\|_{L^q(\mathbb{R}^n)} \le C \|f\|_{L^p(\mathbb{R}^n)}. \tag{24}$$

The optimal constant is given by $[\mathrm{vol}(B)]^{-1/r} C_{p,q}$ (see inequality (17)).

Proof. Observe that $W(x, f) \le W(x, f^*)$ and choose new variables in terms of $|t|^n$; then simplify to obtain (24). □

Because of dilation invariance and reduction to one dimensional estimates, inequalities in this section have equivalent representations as convolution inequalities on the multiplicative group $\mathbb{R}_+$ where Haar measure is given by $1/t\, dt$ and convolution on the group is realized as

$$(g * h)(x) = \int_G g(y) h(y^{-1}x)\, dy$$

with dy denoting Haar measure on G. Then bounds are obtained using Young's inequality though for $p \neq q$ the constants will not be optimal.

Theorem 9. *For $1 < p \leq q < \infty$ with $1/r = 1/p - 1/q$ the following two inequalities are equivalent:*

$$\|x^{1/r} Tf\|_{L^q(0,\infty)} \leq C_{p,q} \|f\|_{L^p(0,\infty)} \tag{17}$$

and

$$\|g * h\|_{L^q(\mathbb{R}_+)} \leq C_{p,q} \|h\|_{L^p(\mathbb{R}_+)} \tag{25}$$

with $g(y) = y^{1/p'} \chi_{[0,1]}(y)$.

Proof. In (17) set $f(t) = t^{-1/p} h(1/t)$ and simplify terms to obtain (25). □

The sharp constants for convolution on the multiplicative group $\mathbb{R}_+$ transfer directly from the additive translation group $\mathbb{R}$ (see [1]).

Lemma (Young's inequality on $\mathbb{R}_t$). *For $1 < p \leq q < \infty$ and $1/q = 1/p - 1/s'$,*

$$\|f * g\|_{L^q(\mathbb{R}_+)} \leq D_{p,q} \|f\|_{L^p(\mathbb{R}_+)} \|g\|_{L^s(\mathbb{R}_+)}. \tag{26}$$

But one can reverse this map from $\mathbb{R}$ to $\mathbb{R}_+$ and examine the nature of the convolution inequality (25) and the corresponding Hardy inequality in terms of analysis on the entire line. A question of particular interest is to determine sharp estimates in Young's inequality when one of the functions is specified and the map is from a space to its dual:

$$f \rightsquigarrow \varphi * f \,, \quad |\varphi * f\|_{L^{p'}(\mathbb{R}^n)} \leq C_p \|f\|_{L^p(\mathbb{R}^n)} \,, \qquad 1 < p < 2. \tag{27}$$

Two intrinsic cases on $\mathbb{R}^n$ are: (a) gaussian $\varphi(x) = e^{-|x|^2}$ (see [1]); (b) Riesz potential $\varphi(x) = |x|^{-2n/p'}$ (see [18]). Hardy's inequality allows the extension of this program to include the exponential density $\varphi(x) = e^{-|x|}$ for one dimension.

Theorem 10 (Young's Inequality for an Exponential Density). *For $f \in L^p(\mathbb{R})$, $1 < p < 2$, $1/p + 1/p' = 1$ and $\varphi(x) = e^{-|x|}$,*

$$\|\varphi * f\|_{L^{p'}(\mathbb{R})} \leq C_p \|f\|_{L^p(\mathbb{R})}, \tag{28}$$

$$C_p = \left(\frac{p'}{2}\right)^{2/p} \left[\frac{\Gamma(\frac{2p}{2-p})}{\Gamma(\frac{2}{2-p})\Gamma(\frac{p}{2-p})}\right]^{2/p \,-\, 1}.$$

Proof. From the Hardy inequality for $q = 2$:

$$\int_0^\infty \left| \int_0^s g(t)\, dt \right|^2 s^{-1 \,-\, 2/p'}\, ds \leq K^{2/p} \left[\int_0^\infty |g|^p\, ds \right]^{2/p}, \qquad K = [C_{p,2}]^p.$$

First, set $g(t) = h(t)t^{-1/p}$, and then set $t = e^x$,

$$\int_0^\infty \left| \int_0^s h(t)(t/s)^{1/p'} \frac{1}{t}\, dt \right|^2 \frac{1}{s}\, ds \le K^{2/p} \left[\int_0^\infty |h|^p \frac{1}{s}\, ds \right]^{2/p},$$

$$\int_{-\infty}^\infty \left| \int_{-\infty}^x h(y) e^{-(x-y)/p'}\, dy \right|^2 dx \le K^{2/p} \left[\int_{-\infty}^\infty |h|^p\, dx \right]^{2/p},$$

$$\left| \int_{\mathbb{R}\times\mathbb{R}} h(u) e^{-2/p'|u-v|} h(v)\, du\, dv \right| \le \frac{2}{p'} K^{2/p} \left[\int_{\mathbb{R}} |h|^p\, dx \right]^{2/p},$$

$$C_p = \left(\frac{p'}{2}\right)^{2/p\ -\ 1} K^{2/p} = \left(\frac{p'}{2}\right)^{2/p} \left[\frac{\Gamma(\frac{2p}{2-p})}{\Gamma(\frac{2}{2-p})\Gamma(\frac{p}{2-p})} \right]^{2/p\ -\ 1}.$$

An extremal function for inequality (28) is given by

$$f(x) = \Big[\cosh\big[p'x/(4p\delta)\big]\Big]^{-\delta}, \qquad \delta = \frac{2}{2-p}.$$

□

Following the direction defined by the calculation for the optimal constant in Theorem 1 and the principal results in [6], the question of finding the optimal constant for $p = 2$ in Theorem 7 is examined. The importance of such estimates lies with providing insight into the structure of "maps from a space to its dual". Here there are two independent calculations corresponding to the cases $p = 2 = q'$, $m \ge 2$ and $p = 2 > q'$. The results for the Hardy operator are analogous to the cases that use Riesz potentials and extend Pitt's inequality and the Hardy-Littlewood-Sobolev inequality for multilinear embedding in [6].

Theorem 11. *For $m \ge 2$, $2 \le q < \infty$ and $\gamma = m/2 - 1/q$ with $H \in L^2(\Lambda_m)$ and $\psi(x) = x^\gamma[(T \otimes \cdots \otimes T)H](x, \ldots, x)$,*

$$\|\psi\|_{L^q(0,\infty)} \le C\|H\|_{L^2(\Lambda_m)} \tag{29}$$

where for $q = 2$,

$$C = 2/\sqrt{m} \tag{30}$$

and for $q > 2$, C is determined by the optimal constant for the weighted Hardy inequality.

Proof. As in the calculations in the previous section, consider the dual problem where $\alpha = m/2 + 1/q = \gamma + 2/q$,

$$\int_{\Lambda_m} \left| \int_{x>\{t_1,\ldots,t_m\}} x^{-\alpha} h(x)\, dx \right|^2 dt_1 \cdots dt_m$$

$$= m! \int_{\Lambda_m} \left| \int_{x>t_1>\cdots>t_m} x^{-\alpha} h(x)\, dx \right|^2 dt_1 \cdots dt_m$$

$$= 2m! \int_{\Lambda_m} \iint_{w>x>t_1>\cdots>t_m} x^{-\alpha} w^{-\alpha} h(x)h(w)\, dx\, dw\, dt_1 \cdots dt_m$$

$$= 2 \int_0^\infty h(w) w^{-\alpha} \left[\int_0^w x^\gamma h(x)\, dx \right] dw \le C^2 \left[\|h\|_{L^{q'}[0,\infty)} \right]^2$$

which is equivalent to the inequality

$$\|k * g\|_{L^q(\mathbb{R}_+)} \le C^2/2 \; \|g\|_{L^{q'}(\mathbb{R}_+)} \tag{31}$$

where $k(t) = t^{-m/2} \chi_{[1,\infty)}(t)$. Then for $q = 2$, the best constant is given by the $L^1(\mathbb{R}_+)$ norm of k which has the value $2/m$ with $C = 2/\sqrt{m}$. To determine the optimal constant for the case $q > 2$, one must relate bounds for the operator

$$h \in L^{q'}(0,\infty) \rightsquigarrow w^{-\sigma-1/q} \int_0^w x^{\sigma-1/q} h(x)\, dx \in L^q(0,\infty) \tag{32}$$

to those for the weighted Hardy inequality (17) in the case of dual exponents and $\sigma = m/2$ by making a simple change of variables. Note that the dilation character of the symmetric form

$$\int_0^\infty h(w) w^{-\sigma-1/q} \left[\int_0^w x^{\sigma-1/q} h(x)\, dx \right] dw$$

is measured by the index $2/q$ which corresponds to the example of Riesz potentials and the Hardy-Littlewood-Sobolev inequality on $\mathbb{R}^n$ where the index would be $2n/q$. Set

$$h(x) = x^\beta u(x^\delta)\ , \qquad q'\beta = \delta - 1 .$$

Then

$$\|h\|_{L^{q'}(0,\infty)} = \delta^{-1/q'} \|u\|_{L^{q'}(0,\infty)}$$

and

$$\int_0^\infty h(w) w^{-\sigma-1/q} \left[\int_0^2 x^{\sigma-1/q} h(x)\, dx \right] dw = \delta^{-2} \int_0^\infty u(w) w^{-2/q} \left[\int_0^w u(x)\, dx \right] dw$$

$$= \delta^{-2} \int_0^\infty u(w) w^{1/q' - 1/q} (Tu)(w)\, dw$$

for the choice $\delta = q\sigma$. Then C^2 for the constant appearing in the inequality (29) when $q > 2$ is given by

$$2(2/mq)^{2/q} \sup \left\{ \|x^{1/r} Tu\|_{L^q(0,\infty)} \Big/ \|u\|_{L^{q'}(0,\infty)} \right\} = 2(2/mq)^{2/q} C_{q',q}$$

with $1/r = 1/q' - 1/q$ and $2 < q < \infty$. Then the constant C for inequality (29) is given by

$$C = \sqrt{2}\ (2/mq)^{1/q} \sqrt{C_{q',q}} . \tag{33}$$

□

A functional integral that has similar properties to the Hardy functional is the Hilbert integral:

$$f \rightsquigarrow I(f)(x) = \int_0^\infty \frac{1}{x+y} f(y)\, dy \ , \qquad x > 0, \tag{34}$$

$$\|I(f)\|_{L^p(0,\infty)} \le \Gamma(1/p)\Gamma(1/p')\|f\|_{L^p(0,\infty)} \ , \qquad 1 < p < \infty.$$

Observe that the Hilbert integral is controlled by the Hardy operator: $I(f) \le T(f) + T^*(f)$ where T^* denotes the adjoint for T. Such simple estimates allow one to recover elementary inequalities such as $\theta(1-\theta)\Gamma(\theta)\Gamma(1-\theta) \le 1$ and provide a check that "calculated constants make sense". Because of the associated role of exponential densities when dilation-invariant operators are transferred from the real half-line to the multiplicative group $\mathbb{R}_+$ and then to the additive group $\mathbb{R}$ (see proof of Theorem 10), it is instructive to apply the methods used here and in [6] to calculate the precise norm for the diagonal trace map corresponding to the Hilbert integral from $L^2(\Lambda_m)$ to $L^2(0,\infty)$.

Theorem 12. *For $m \ge 2$ and $\gamma = (m-1)/2$ with $H \in L^2(\Lambda_m)$ and $\psi(x) = x^\gamma[(I \otimes \cdots \otimes I)H](x,\ldots,x)$,*

$$\|\psi\|_{L^2(0,\infty)} \le C\|H\|_{L^2(\Lambda_m)}, \tag{35}$$

$$C = 2\left[\int_0^\infty \left[\frac{x}{\sinh x}\right]^m dx\right]^{1/2} .$$

Proof. As with the proof of the preceding theorem, consider the dual problem

$$\int_{\Lambda_m} \left| \int_0^\infty x^\gamma \prod \frac{1}{x+y_k}\, h(x)\, dx \right|^2 dy_1 \cdots dy_m$$

$$= \int_0^\infty \int_0^\infty h(x) x^\gamma \left[\int_0^\infty \frac{1}{x+y}\,\frac{1}{w+y}\, dy\right]^m w^\gamma h(w)\, dx\, dw$$

$$= \int_0^\infty \int_0^\infty h(x) \left[\frac{(xw)^{(m-1)/2}}{|x-w|^m} \left|\ln \frac{x}{w}\right|^m\right] h(w)\, dw$$

$$\le C^2 \Big[\|h\|_{L^2(0,\infty)}\Big]^2 .$$

First set $h(x) = x^{-1/2} g(x)$ and then change variables $x = e^t$ and rescale by a factor of 2 to obtain a convolution inequality on the real line

$$\|\varphi * f\|_{L^2(\mathbb{R})} \le C^2 \|f\|_{L^2(\mathbb{R})} \ , \qquad \varphi(x) = 2[x/\sinh x]^m$$

with

$$C^2 = \|\varphi\|_{L^1(\mathbb{R})} = 4\int_0^\infty [x/\sinh x]^m\, dx \ .$$

□

3 Diagonal trace restriction for a multilinear fractional integral

Modeled on results such as (2) above plus ideas from [6, 17] (see Theorem 1 and Lemma 7) and [20], a diagonal trace restriction inequality is obtained for a multilinear fractional integral.

Theorem 13. *For $H \in L^p(\mathbb{R}^{mn})$, $\rho(x_1, \ldots, x_m) = \sum |x_k|$, $x_k \in \mathbb{R}^n$ and*

$$F(x) = \int_{\mathbb{R}^{mn}} \rho(x - y_1, \ldots, x - y_m)^{-(mn-\alpha)} H(y_1, \ldots, y_m)\, dy$$

with $\alpha/n = m/p - 1/q$; then for $q \geq p$,

$$\|F\|_{L^q(\mathbb{R}^n)} \leq C \|H\|_{L^p(\mathbb{R}^{mn})}. \tag{36}$$

For $p = 2$, the optimal constant is

$$C = \sqrt{B}\ \pi^{n/2q} \left[\frac{\Gamma(\frac{n}{2} - \frac{n}{q})}{\Gamma(n)} \right]^{1/2} \left[\frac{\Gamma(n)}{\Gamma(\frac{n}{q'})} \right]^{1/q'}, \tag{37}$$

$$B = \int_{\mathbb{R}^{mn}} \Big(\sum |\hat{\eta} - y_k| \Big)^{-(mn-\alpha)} \Big(\sum |y_k| \Big)^{-(mn-\alpha)} dy\ , \qquad |\hat{\eta}| = 1.$$

Proof. By using a simple comparison between sums and products, the first part of this theorem follows from the multilinear Hardy-Littlewood-Sobolev inequality (Theorem 5 above). For the case $p = 2$, use duality to consider

$$\left[\int_{\mathbb{R}^{nm}} \left| \int_{\mathbb{R}^n} \rho(x - y_1, \ldots, x - y_m)^{-(mn-\alpha)} h(x)\, dx \right|^2 dy \right]^{1/2} \leq C \|h\|_{L^{q'}(\mathbb{R}^n)}. \tag{38}$$

Then

$$\int_{\mathbb{R}^{mn}} \rho(x - y_1, \ldots, x - y_m)^{-(mn-\alpha)} \rho(w - y_1, \ldots, w - y_m)^{-(mn-\alpha)} dy$$

$$= \int_{\mathbb{R}^{mn}} \rho(x - w - y_1, \ldots, x - w - y_m)^{-(mn-\alpha)} \rho(y_1, \ldots, y_m)^{-(mn-\alpha)} dy$$

$$= |x - w|^{-(mn-2\alpha)} B = |x - w|^{-2n/q} B$$

where

$$B = \int_{\mathbb{R}^{mn}} \rho(\hat{\eta} - y_1, \ldots, \hat{\eta} - y_m)^{-(mn-\alpha)} \rho(y_1, \ldots, y_m)^{-(mn-\alpha)} dy\ , \qquad \hat{\eta} \in S^{n-1}.$$

Observe that B does not depend on the value of the unit vector $\hat{\eta}$. So the dual expression becomes

$$B \int_{\mathbb{R}^n \times \mathbb{R}^n} h(x) |x - w|^{-2n/q} h(w)\, dx\, dw \leq C^2 \Big[\|h\|_{L^{q'}(\mathbb{R}^n)} \Big]^2.$$

Using the sharp constant for the classical Hardy-Littlewood-Sobolev inequality

$$C^2 = B\,\pi^{n/q}\,\frac{\Gamma(\frac{n}{2}-\frac{n}{q})}{\Gamma(\frac{n}{q'})}\left[\frac{\Gamma(n)}{\Gamma(\frac{n}{q'})}\right]^{2/q'-1}.$$

□

The dilation character of the above argument easily transfers to a more general statement. Consider a non-negative function $G(x_1,\dots,x_m)$ defined on $\mathbb{R}^n\times\cdots\times\mathbb{R}^n \simeq \mathbb{R}^{mn}$ that is invariant under a uniform rotation:

$$G(Rx_1,\dots,Rx_m) = G(x_1,\dots,x_m)\ , \qquad R\in SO(n)$$

and with the dilation property — there exists a sequence $(\beta_1,\dots,\beta_m)$, $\beta_k\geq 0$ and $\sum\beta_k = mn-\alpha = mn/p'+n/q$, $q>p$ such that for $\delta_k>0$,

$$G(\delta_1x_1,\dots,\delta_mx_m) = \prod \delta_k^{-\beta_k}G(x_1,\dots,x_m).$$

Corollary. *For $H\in L^2(\mathbb{R}^{mn})$ and*

$$F(x) = \int_{\mathbb{R}^{mn}} G(x-y_1,\dots,x-y_m)H(y_1,\dots,y_m)\,dy$$

with $\alpha = m/2-1/q$; then for $q>2$,

$$\|F\|_{L^q(\mathbb{R}^n)} \leq C\|H\|_{L^2(\mathbb{R}^{mn})} \tag{39}$$

with the optimal constant

$$C = \sqrt{B}\,\pi^{n/2q}\left[\frac{\Gamma(\frac{n}{2}-\frac{n}{q})}{\Gamma(n)}\right]^{1/2}\left[\frac{\Gamma(n)}{\Gamma(\frac{n}{q'})}\right]^{1/q'},$$

$$B = \int_{\mathbb{R}^{mn}} G(\hat\eta-y_1,\dots,\hat\eta-y_m)G(y_1,\dots,y_m)\,dy\ , \qquad |\eta|=1$$

if this value for B is finite.

Proof. Follow directly the previous argument given above for Theorem 13. □

4 Multilinear integrals and rearrangement

A central focus for the methods developed here has been: (1) demonstration of estimates for multilinear multidimensional forms that arise from duality arguments, and (2) calculation of optimal constants or norms in select cases. Classic examples beyond Hardy's inequality involve:

Young's inequality: $f \rightsquigarrow f*g$.

Hardy-Littlewood-Sobolev inequality: $f \rightsquigarrow |x|^{-\lambda} * f$.

Stein-Weiss fractional integral: $f \rightsquigarrow |x|^{-\beta}(|x|^{-\lambda} * |x|^{-\alpha} f)$.

Representation of norm estimates in terms of multilinear integral functionals allows application of equimeasurable rearrangement techniques to obtain improved inequalities. Insight was first gained from the Riesz-Sobolev Rearrangement Inequality:

$$\int_{\mathbb{R}^n \times \mathbb{R}^n} f(x) g(x-y) h(y)\, dx\, dy \leq \int_{\mathbb{R}^n \times \mathbb{R}^n} f^*(x) g^*(x-y) h^*(y)\, dx\, dy \tag{40}$$

where $f, g, h \geq 0$ and f^* denotes the equimeasurable radial decreasing rearrangement of f on $\mathbb{R}^n$. The proof based on the Brunn-Minkowski inequality allowed a much more general formulation given by Brascamp, Lieb and Luttinger [10]: for non-negative functions f_k, $k = 1, \ldots, M$,

$$\begin{aligned} &\int_{\mathbb{R}^n \times \cdots \times \mathbb{R}^n} \prod f_k \Big(\sum a_{kj} x_j \Big)\, dx_1 \cdots dx_m \\ &\leq \int_{\mathbb{R}^n \times \cdots \times \mathbb{R}^n} \prod f_k^* \Big(\sum a_{kj} x_j \Big)\, dx_1 \cdots dx_m . \end{aligned} \tag{41}$$

Some restrictions will be necessary to ensure finite values for the integrals. Brascamp and Lieb used this rearrangement improvement combined with an iterative high-dimensional asymptotic argument to show that the optimal bound for this integral under constraints given by Lebesgue space norms for the $\{f_k\}$ will either be attained for gaussian functions or by limits obtained using gaussian functions if the value is finite. By applying this technique to a characteristic example for the integral given by (41) and suggested in part by calculations from [12], a multilinear generalization is obtained for the classic Young's inequality

$$f \rightsquigarrow \varphi * f\ , \quad \|\varphi * f\|_{L^p(\mathbb{R}^n)} \leq \|\varphi\|_{L^1(\mathbb{R}^n)} \|f\|_{L^p(\mathbb{R}^n)}$$

where equality for this estimate is never attained for $1 < p < \infty$.

Theorem 14 (Multilinear Young's Inequality). *For $f \in L^p(\mathbb{R}^n)$ and $g \in L^q(\mathbb{R}^n)$, $f, g \geq 0$ with $1 < p \leq m$ and $1/p' = (m-1)/2q$,*

$$\int_{\mathbb{R}^{mn}} \prod f(x_k) \prod_{i<j} g(x_i - x_j)\, dx_1 \cdots dx_m \leq C \Big[\|f\|_{L^p(\mathbb{R}^n)} \Big]^m \Big[\|g\|_{L^q(\mathbb{R}^n)} \Big]^{m(m-1)/2}, \tag{42}$$

$$C = \sup_{0<\rho} \left[\frac{\rho^{m/p'}}{(1+m\rho)^{m-1}} \right]^{n/2} \left[p^{m/p} q^{m(m-1)/2q} \right]^{n/2} .$$

If $p = m$, then $C = [m/2^{m-1}]^{n/2}$ and equality is never attained.

Proof. Let $f(x) = e^{-\pi\alpha x^2}$ and $g(x) = e^{-\pi\beta x^2}$, $\alpha, \beta > 0$; then to calculate the value of the integral on the left-hand side for gaussian functions, we need to find the determinant of the matrix of coefficients corresponding to the gaussian function

$$\exp\Big[-\pi(\alpha + (m-1)\beta) \sum x_k^2 + 2\pi\beta \sum_{i<j} x_i x_j \Big] .$$

This only requires computing the determinant of the matrix since a rotation applied to the matrix would then display eigenvalues on the diagonal corresponding to a symmetric matrix and the determinant is invariant under a rotation

$$\begin{bmatrix} \alpha + (m-1)\beta & -\beta & \cdots & -\beta \\ -\beta & \alpha + (m-1)\beta & \cdots & -\beta \\ \vdots & \vdots & & \vdots \\ -\beta & -\beta & \cdots & \alpha + (m-1)\beta \end{bmatrix},$$

m variables $x_k \in \mathbb{R}^n$ results in an $m \times m$ matrix. To simplify the calculation, consider the matrix

$$\begin{bmatrix} \delta & -\beta & \cdots & -\beta \\ -\beta & \delta & \cdots & -\beta \\ \vdots & \vdots & & \vdots \\ -\beta & -\beta & \cdots & \delta \end{bmatrix}.$$

Add all the columns ($k = 2, \ldots, m$) to the first column, then subtract the first row from each successive row:

$$\begin{bmatrix} \delta - (m-1)\beta & -\beta & \cdots & -\beta \\ \delta - (m-1)\beta & \delta & \cdots & -\beta \\ \vdots & \vdots & & \vdots \\ \delta - (m-1)\beta & -\beta & \cdots & \delta \end{bmatrix} \rightsquigarrow \begin{bmatrix} \delta - (m-1)\beta & -\beta & \cdots & -\beta \\ 0 & \delta + \beta & \cdots & 0 \\ \vdots & \vdots & & \vdots \\ 0 & 0 & \cdots & \delta + \beta \end{bmatrix},$$

the determinant equals $[\delta - (m-1)\beta](\delta + \beta)^{m-1}$ which for the variables (α, β) becomes $\alpha(\alpha + m\beta)^{m-1}$. Then the value of the integral becomes

$$\Big[\alpha(\alpha + m\beta)^{m-1}\Big]^{-n/2} .$$

So the ratio of this integral to the norms on the right-hand side becomes

$$\left[\frac{\alpha^{m/p}\beta^{m(m-1)/2q}}{\alpha(\alpha + m\beta)^{m-1}}\right]^{n/2} \Big[p^{m/p} q^{m(m-1)/2q}\Big]^{n/2}$$

$$= \left[\frac{(\beta/\alpha)^{m/p'}}{(1 + m\,\beta/\alpha)^{m-1}}\right]^{n/2} \Big[p^{m/p} q^{m(m-1)/2q}\Big]^{n/2} \leq C.$$

This calculation makes explicit the constraint that $1 < p \leq m$ or equivalently $m/2q \leq 1$. For $m/2q < 1$, the optimal constant is attained for $\beta/\alpha = (2q - m)^{-1}$. □

This theorem is the formative step in demonstrating the existence of bounds that lead to the proof of a multilinear Hardy-Littlewood-Sobolev inequality that is conformally invariant. In extending Theorem 14 to allow the input function g to be replaced by the Riesz potential $|x|^{-\lambda}$ with $\lambda = n/q$, then p is restricted to lie below the critical index $1 < p < m$, as suggested by the lack of extremals at that index in the multilinear Young's inequality. It is striking that such a natural extension would preserve conformal symmetry.

Theorem 15 (Multilinear Hardy-Littlewood-Sobolev Inequality). *For $f \in L^p(\mathbb{R}^n)$, $f \geq 0$, with $1 < p < m$ and $2n/p' = (m-1)\gamma$, $\gamma = n/q$,*

$$\int_{\mathbb{R}^{mn}} \prod f(x_k) \prod_{i<j} |x_i - x_j|^{-\gamma} dx_1 \cdots dx_m \leq C \Big[\|f\|_{L^p(\mathbb{R}^n)} \Big]^m, \tag{43}$$

$$C = \Big[(4\pi)^{n/2} \Gamma(n/2)/\Gamma(n) \Big]^{m/p'} \int_{S^n \times \cdots \times S^n} \prod_{i<j} |\xi_i - \xi_j|^{-\gamma} d\xi_1 \cdots d\xi_m$$

and extremal functions are given by $f(x) = A(1+|x|^2)^{-n/p}$ up to conformal automorphism. Here $d\xi$ is normalized surface measure.

Proof. This result without the sharp constant and in one dimension is given in [12] (see Section 2) using an interpolation argument. Observe that the condition $0 < \gamma < 2n/m$ follows from the restriction on the index $1 < p < m$. The intertwined singularity structure can be viewed as an obstruction in determining conditions for the integral to be bounded. Here rearrangement, dilation invariance, conformal symmetry and axial representation are applied explicitly to obtain the inequality, and the background role of the result from Theorem 14 guarantees integrability of the singularity.

The nature of the proof follows the classic steps from the original arguments given to obtain sharp constants for the Hardy-Littlewood-Sobolev inequality and the Moser-Trudinger inequality (see [2, 3, 4, 18]): existence of bounds, realization with inversion symmetry on the multiplicative group $\mathbb{R}_+$, existence of extremal functions, determination of optimal constants using spherical symmetry, and $SL(2,\mathbb{R})$ invariance on hyperbolic space.

For determination of bounds, the proof will be developed in two parts — first for $n = 1$, and then $n > 1$ where integration over the sphere S^{n-1} comes into play. By applying the Brascamp-Lieb-Luttinger rearrangement inequality, the estimate is improved when f is a radial decreasing function. Then set $h(y) = |x|^{n/p} f(x)$ where $y = |x|$; notice that h is bounded so that $h \in L^p(\mathbb{R}_+) \cap L^\infty(\mathbb{R}_+)$. The resulting inequality becomes

$$\int_{(\mathbb{R}_+)^m \times (S^{n-1})^m} \prod h(y_k) \prod_{i<j} \left[\frac{y_i}{y_j} + \frac{y_j}{y_i} - 2\xi_i \cdot \xi_j \right]^{-\gamma/2} \frac{dy_1}{y_1} \cdots \frac{dy_m}{y_m} \, d\xi_1 \cdots d\xi_m$$

$$\leq C\Big[\|h\|_{L^p(\mathbb{R}_+)}\Big]^m . \tag{44}$$

Because the "potentials" are now symmetric decreasing away from the origin $\{y = 1\}$, the Brascamp-Lieb-Luttinger inequality can be applied to improve the inequality so that $h(1/y) = h(y)$ and $h(y)$ is monotone decreasing for $y > 1$. This step then implies that inequality (43) is improved if (1) f is radial decreasing, (2) $|x|^{n/p} f(x)$ is decreasing for $|x| > 1$, and (3) $f(|x|^{-1}) = |x|^{2n/p} f(|x|)$. These conditions are precisely what is meant by saying that f possesses an "inversion symmetry". Set $y = e^t$; then (44) becomes

$$2^{-mn/2p'}\int_{\mathbb{R}^m\times(S^{n-1})^m} \prod h(t_k)\prod_{i<j}\Big[\cosh(t_i-t_j)-\xi_i\cdot\xi_j\Big]^{-\gamma/2} dt_1\cdots dt_m\, d\xi_1\cdots d\xi_m$$

$$\leq C\Big[\|h\|_{L^p(\mathbb{R})}\Big]^m . \tag{45}$$

Here the objective first is to show existence of a bound and second to show existence of extremals. This theme — to show first bounds and then second existence of limits — in many essential ways characterizes the objectives in Stein's classic text on *Singular integrals* ([21]). Observe that the function h on $\mathbb{R}$ inherits the bounds of h on $\mathbb{R}_+$ (a simple change of variables) and these bounds for $h \in L^p(\mathbb{R}) \cap L^\infty(\mathbb{R})$ are controlled by the value $\|f\|_{L^p(\mathbb{R}^n)}$. For the one-dimensional setting where $S^{n-1} = \{\pm 1\}$, the simplest estimate is that

$$\Big[\cosh(t_i - t_j) - \xi_i\xi_j\Big]^{-\gamma/2} \leq \Big[\cosh(t_i - t_j) - 1\Big]^{-1/(p'(m-1))}$$

so the inequality requires an estimate for

$$2^{m(1-1/2p')}\int_{\mathbb{R}^m} \prod h(t_k)\prod_{i<j}\Big[\cosh(t_i - t_j) - 1\Big]^{-\gamma/2} dt_1\cdots dt_m . \tag{46}$$

Since $1 < p < m$, choose $p < p_* < m$; since $h \in L^p(\mathbb{R}) \cap L^\infty(\mathbb{R})$, $h \in L^{p_*}(\mathbb{R})$. Define q_* by $1/p'_* = (m-1)/2q_*$; since $p'_* < p'$, then $q_* < q$. Now claim that $g(t) = (\cosh t - 1)^{-\gamma/2} \in L^{q_*}(\mathbb{R})$ since $\gamma q_* = p'_*/p' < 1$ so that by the Multilinear Young's Inequality (Theorem 14), the integral given by (46) is bounded in terms of $[\|f\|_{L^p(\mathbb{R})}]^m$.

To address the question of boundedness in higher dimensions for

$$\Lambda(h) = \int_{\mathbb{R}^m\times(S^{n-1})^m} \prod h(t_k)\prod_{i<j}\Big[\cosh(t_i - t_j) - \xi_i\cdot\xi_j\Big]^{-\gamma/2} dt_1\cdots dt_m\, d\xi_1\cdots d\xi_m$$

observe that

$$\Big[\cosh(t_i - t_j) - \xi_i\cdot\xi_j\Big]^{-n/p'(m-1)}$$
$$\leq A\Big[\cos(t_i - t_j) - 1\Big]^{-1/p'(m-1)}\Big[1 - \xi_i\cdot\xi_j\Big]^{-(n-1)/p'(m-1)} .$$

Then the integration splits and the "t-integration" is handled as in the one-dimensional case above. The integral over the $(n-1)$-dimensional sphere is bounded by again applying Theorem 14. □

Lemma. *For* $1 < p < m$ *and* $2N/p' = (m-1)\gamma$, $\gamma = N/q$, $N \geq 1$ *the following integral is bounded*

$$\int_{S^N \times \cdots \times S^N} \prod_{i<j} |\xi_i - \xi_j|^{-\gamma} d\xi_1 \cdots d\xi_m \leq C. \tag{47}$$

Proof. Apply Theorem 14 in the case where f is constant with bounded support for the index p_*, $p < p_* < m$. Then $|\xi - \eta|^{-\gamma}$ as a function of ξ is in $L^{q^*}(S^N)$ where $2/p'_* = (m-1)/q_*$ and $\gamma q^* < N$. □

Now having determined that the integral $\Lambda(h)$ is bounded for $\|h\|_{L^p(\mathbb{R}^*)} = 1$, it follows directly that an extremal function exists for this functional where the "best constant" C is attained (see the nature of the argument given on page 40 in [3]). Choose a sequence $\{h_k\}$ with $\|h_k\| = 1$ so that $\Lambda(h_k) \rightsquigarrow C$; the h_k will be symmetric decreasing and uniformly bounded from the earlier application of the Brascamp-Lieb-Luttinger rearrangement inequality. Since the functions $\{h_k\}$ are decreasing, the Helly selection principle can be applied to choose a subsequence that converges almost everywhere to a function $h \in L^p(\mathbb{R})$ with $\|h\|_p \leq 1$ and $\Lambda(h) \leq C$. Since h_k has a uniform functional bound

$$h_k(t) \leq \Lambda\left(1 + |t|^{n/p}\right)^{-1} = u(t)$$

and the argument used above to show that the integral given by (46) is bounded in terms of L^p norms now shows that $\Lambda(u)$ is bounded and hence the integrand for the integrals $\Lambda(h_k)$ has a uniform $L^1(\mathbb{R}^m \times (S^{n-1}))$ majorant so that the dominated convergence theorem implies that

$$\Lambda(h_k) \longrightarrow \Lambda(h) , \qquad k \to \infty$$

and hence $\Lambda(h) = C$ and h is an extremal function which is symmetric decreasing. Tracing back through successive changes of variables, this exhibits a radial decreasing extremal function f for inequality (43) which possesses the inversion symmetry

$$f(|x|^{-1}) = |x|^{2n/p} f(|x|) . \tag{48}$$

The next objective is to characterize the extremal functions, and show that up to conformal automorphism an extremal function for inequality (43) is given by conformal factors

$$f(x) = \Lambda(1 + |x|^2)^{-n/p} .$$

Conformal invariance allows the inequality to be transferred to the sphere S^n. The equivalence between $\mathbb{R}^n$ and $S^n - \{\text{pole}\}$ is given by

$$\xi = \left(\frac{1-|x|^2}{1+|x|^2}, \frac{2x}{1+|x|^2}\right) , \qquad d\xi = \pi^{-n/2}\left[\frac{\Gamma(n)}{\Gamma(n/2)}\right](1+|x|^2)^{-n}\, dx,$$

$$|x-y| = \frac{1}{2}|\xi - \eta|\Big[(1+|x|^2)(1+|y|^2)\Big]^{1/2}$$

where $d\xi$ corresponds to normalized surface measure on the sphere. By setting $F(\xi) = (1+|x|^2)^{n/p} f(x)$, then an inequality on S^n equivalent to (43) becomes

$$\int_{S^n \times S^n} \prod F(\xi_k) \prod_{i<j} |\xi_i - \xi_j|^{-\gamma}\, d\xi_1 \cdots d\xi_m \le D\Big[\|F\|_{L^p(S^n)}\Big]^m , \tag{49}$$

$$D = C\Big[(4\pi)^{-n/2}\Gamma(n)\,/\,\Gamma(n/2)\Big]^{m/p'} .$$

So if extremal functions are given by $(1+|x|^2)^{-n/p}$, then the constant C for Theorem 15 is given by

$$C = \Big[(4\pi)^{n/2}\Gamma(n/2)\,/\,\Gamma(n)\Big]^{m/p'} \int_{S^n \times \cdots \times S^n} \prod_{i<j} |\xi_i - \xi_j|^{-\gamma} d\xi_1 \cdots d\xi_m \,.$$

To demonstrate that extremal functions for $n \ge 2$ are given by $A(1+|x|^2)^{-n/p}$ up to conformal automorphism, an equivalent realization for the multilinear Hardy-Littlewood-Sobolev inequality on hyperbolic space is used. It's very natural to recognize the conformal equivalence for the classical geometries — the plane, the sphere and the two-sheeted hyperboloid, the latter corresponding to the homogeneous space for the group $SO(n,1)$. But the equivalence for the standard Liouville-Beltrami upper half-space model for hyperbolic space represents a more subtle dilation invariance which utilizes axial symmetry. More striking is that the Hardy-Littlewood-Sobolev inequality on $\mathbb{R}^n$ for $n \ge 2$ (Theorem 15) has an equivalent formulation on the hyperbolic space $\mathbb{H}^\ell$ for $2 \le \ell \le n$ which depends on the dimension of the submanifold used for the axial symmetry.

The Liouville-Beltrami upper half-space model includes the Lie group structure of affine mappings and incorporates "axial symmetry" from the Euclidean picture. Here

$$w = (x,y) \in \mathbb{R}^\ell_+ \simeq \mathbb{R}^{\ell-1} \times \mathbb{R}^+ \simeq SO(\ell,1)\,/\,SO(\ell) \simeq \mathbb{H}^\ell$$

with the Poincaré distance

$$d(w,w') = \frac{|w-w'|}{2\sqrt{yy'}}$$

and left-invariant Haar measure $d\nu = y^{-\ell}\, dy\, dx$. The Riemannian metric on $\mathbb{H}^\ell$ is defined by

$$ds^2 = y^{-2}(dx^2 + dy^2)$$

and the invariant gradient is given by $D = y\nabla$. The group structure of hyperbolic space corresponds to a non-unimodular Lie group that is an extension of the affine "$ax+b$ group". Hyperbolic space $\mathbb{H}^\ell$ is identified with the subgroup of $SL(\ell,\mathbb{R})$ given by all matrices of the form

$$\sqrt[\ell]{y}\begin{pmatrix} \mathbf{1} & x/y \\ 0 & 1/y \end{pmatrix}$$

where $x \in \mathbb{R}^{\ell-1}$ is represented by a column vector and $y > 0$. Such matrices can act via fractional linear transformation on $\mathbb{R}^\ell_+ \simeq \mathbb{H}^\ell$,

$$w = x + iy\xi \in \mathbb{R}^\ell_+ \longrightarrow \frac{Aw+B}{Cw+D}$$

for a matrix $\begin{pmatrix} A & B \\ C & D \end{pmatrix}$ where $A = (\ell-1)\times(\ell-1)$ matrix, $B = (\ell-1)\times 1$ matrix, $C = 1\times(\ell-1)$ matrix, $D = 1\times 1$ matrix and fixed non-zero $\xi \in \mathbb{R}^{\ell-1}$. The group action here corresponds to the multiplication rule

$$(x,y)(u,v) = (x+yu, yv)$$

for $x, u \in \mathbb{R}^{\ell-1}$ and $y, v > 0$ so this is again a non-unimodular group with the modular function $\Delta(x,y) = y^{-(\ell-1)}$. The group identity is $\widehat{0} = (0,1)$.

Following the argument given in [4], consider non-negative radial functions f in equation (43) and set $y = |x'|$ where $x \in \mathbb{R}^n = (u, x') \rightsquigarrow w = (u,y) \in \mathbb{R}^{\ell-1}\times\mathbb{R}_+ \simeq \mathbb{H}^\ell$. Let $F(u,y) = y^{n/p} f(u,x')$; then

$$\|f\|_{L^p(\mathbb{R}^n)} = \left[\frac{2\pi^{n-\ell+1}}{\Gamma[(n-\ell+1)/2]}\right]^{1/p} \|F\|_{L^p(\mathbb{H}^\ell)}$$

and equation (43) results in the family of estimates for $2 \le \ell \le n$;

$$\int_{(\mathbb{H}^\ell)^m} \prod F(w_k) \int_{(S^{n-\ell})^m} \prod_{i<j} \left[d^2(w_i,w_j) + (1-\xi_i\cdot\xi_j)/2\right]^{-\gamma/2} d\xi_1\cdots d\xi_m\, d\nu_1\cdots d\nu_m$$
$$\le C\left[\frac{\pi^{n-\ell+1}}{2^{n-1}\Gamma[(n-\ell+1)/2]}\right]^{-m/p'} \left[\|F\|_{L^p(\mathbb{H}^\ell)}\right]^m \tag{50}$$

where C is the optimal constant for equation (43). The multiplicity of "fractional integral" inequalities with varied parameters corresponds to earlier results obtained for embedding estimates and Sobolev inequalities on hyperbolic space (see [4, 5]).

The Brascamp-Lieb-Luttinger rearrangement inequality and the nature of the multiplication rule for group action on $\mathbb{H}^\ell$ result in a Riesz-Sobolev rearrangement inequality on hyperbolic space. This general result is presented here in the reduced form that is needed for determining the form of the extremal functions for Theorem 15.

Lemma (Riesz-Sobolev Inequality on Hyperbolic Space). *Let $\{F_k\}$ be a sequence of non-negative functions on $\mathbb{H}^n$ and $\{g_{ij}\}$ a sequence of non-negative decreasing functions on $\mathbb{R}_+$. Then*

$$\int_{(\mathbb{H}^n)^m} \prod F_k(w_k) \prod_{i<j} g_{ij}\left[d(w_i,w_j)\right] dw_1\cdots dw_m$$
$$\le \int_{(\mathbb{H}^n)^m} \prod F_k^*(w-k) \prod_{i<j} g_{ij}\left[d(w_i,w_j)\right] dw_1\cdots dw_m \tag{51}$$

where F^* corresponds to the equimeasurable radial decreasing rearrangement on $\mathbb{H}^n$ of F with respect to the distance function $d(w,\widehat{0})$ where for $w = (u,y)$,

$$1 + 2d^2(w,\widehat{0}) = \frac{y^2 + u^2 + 1}{2y} .$$

Proof. Apply the Brascamp-Lieb-Luttinger rearrangement inequality to the integrations over the Euclidean variables $\{u_k\}$. Use the Lusternik technique to take a sequence of successive rearrangements followed by a change of variables so that the sequence of rearranged functions converges to the resulting equimeasurable radial decreasing rearrangements of the functions $\{F_k\}$. □

Apply this lemma to the left side of equation (50) and follow the argument back to equation (43). Then there must exist an extremal $f_\#$ of the form

$$f_\#(u,y) = y^{-n/p} F^* \left(\frac{y^2 + u^2 + 1}{2y} \right) , \qquad y = |x'|$$

up to conformal action, the only possible form for $f_\#$ is to be radial. Hence

$$f_\#(x) = f_\#(u,y) = y^{-n/p} F^* \left(\frac{y^2 + u^2 + 1}{2y} \right)$$

and

$$f_\#(x) = f_\#(u,y) = A(1 + u^2 + y^2)^{-n/p} = A(1 + |x|^2)^{-n/p} .$$

This completes proof of Theorem 15 when the dimension n is at least two. The argument is important because it illustrates how hyperbolic symmetry is embedded in the conformal structure of the Riesz functional and the Hardy-Littlewood-Sobolev inequality corresponding to Theorem 15 and equation (43). The competing radial and cylindrical symmetry force the extremal to be of the form $f(x) = A(1 + |x|^2)^{-n/p}$ up to conformal automorphism. For the one-dimensional case where "axial symmetry" is not available, the observation by Carlen and Loss that conformal transformations can be chosen that break geodesic symmetry so that a sequence of successive rearrangements followed by conformal automorphisms will generate an extremal function in the limit (see [11] and page 42 in [3]) can be used to fully complete the proof of Theorem 15. While simple and elegant, such an argument does not bring out the geometric symmetry that is encompassed by the conformal structure. Additionally, the arguments here can be used for the case of multiple functions $\{f_k\}$, $k = 1, \ldots, m$ in Theorem 15 rather than a single function f. □

Remark. It's perhaps useful in terms of possible application to quantum dynamical processes to consider Coulomb potentials and many body interactions. For low dimensions, interesting cases are: $n = 3$, $\gamma = 1$, $m = 4$ and $p = 2$; $n = 4$, $\gamma = 2$, $m = 3$ and $p = 2$.

The framework for these arguments can be extended to include multilinear Stein-Weiss fractional integrals though without the capability to compute sharp constants and extremal functions. The argument including existence for extremals follows from adapting the mechanism of the proof used in Theorem 15. A variant of this result in one dimension is given in [12].

Theorem 16 (Multilinear Stein-Weiss Inequality). *For $f \in L^p(\mathbb{R}^n)$, $f \geq 0$, $1 < p \leq m$, $\beta > 0$ and $2n/p' = 2\beta + (m-1)\gamma$, $\gamma = n/q > 0$,*

$$\int_{\mathbb{R}^{mn}} \prod \left[f(x_k)|x_k|^{-\beta}\right] \prod_{i<j} |x_i - x_j|^{-\gamma} dx_1 \cdots dx_m \leq \left[\|f\|_{L^p(\mathbb{R}^n)}\right]^m . \tag{52}$$

In addition, extremal functions exist for $1 < p < m$.

Corollary. *For $2 \leq p \leq m$, $1/p + 1/p' = 1$,*

$$\left| \int_{\mathbb{R}^{mn}} \prod \widehat{f}(x_k) \prod_{i<j} |x_i - x_j|^{-\gamma} dx_1 \cdots dx_m \right| \leq C\left[\|(-\Delta/4\pi^2)^{\beta/2} f\|_{L^{p'}(\mathbb{R}^n)}\right]^m \tag{53}$$

where $[(-\Delta/4\pi^2)^{\beta/2} f]^\wedge = |x|^\beta \widehat{f}$.

Proof. Apply the Brascamp-Lieb-Luttinger rearrangement inequality for the integral that appears in equation (52) so that the function f may be taken to be radial decreasing. Then follow the development of the proof as for Theorem 15 and the inequality (45). Here one obtains the similar functional integral

$$\int_{\mathbb{R}^m \times (S^{n-1})^m} \prod h(t_k) \prod_{i<j} \left[\cosh(t_i - t_j) - \xi_i \cdot \xi_j\right]^{-\gamma/2} dt_1 \cdots dt_m \, d\xi_1 \cdots d\xi_m$$

and since $(m-1)\gamma < 2n/p'$ due to $\beta > 0$, application of the multilinear Young's inequality (Theorem 14) will show that the functional is bounded for $1 < p \leq m$ and allow the Helly selection principle to show the existence of an extremal function where the optimal constant is attained for the index range $1 < p < m$. The corollary is a direct restatement of the theorem to demonstrate the connection with embedding. □

Acknowledgements

I would like to thank Guozhen Lu for his warm encouragement.

References

[1] W. Beckner, Inequalities in Fourier analysis, Annals Math. **102** (1975), 159–182.

[2] W. Beckner, Sharp Sobolev inequalities on the sphere and the Moser-Trudinger inequality, Annals Math. **138** (1993), 213–242.

[3] W. Beckner, Geometric inequalities in Fourier analysis, Essays on Fourier Analysis in Honor of Elias M. Stein, Princeton University Press, Princeton, NJ, 1995, 36–68.

[4] W. Beckner, Sharp inequalities and geometric manifolds, J. Fourier Anal. Appl. **3** (1997), 825–836.

[5] W. Beckner, On the Grushin operator and hyperbolic symmetry, Proc. Amer. Math. Soc. **129** (2001), 1233–1246.

[6] W. Beckner, Multilinear embedding estimates for the fractional Laplacian, Math. Res. Lett. **19** (2012), 1–15.

[7] W. Beckner, Embedding estimates and fractional smoothness, Int. Math. Res. Notices **3** (2014), 390–417.

[8] G. A. Bliss, An integral inequality, J. London Math. Soc. **5** (1930), 40–46.

[9] H. J. Brascamp and E. H. Lieb, Best constants in Young's inequality, its converse, and its generalization to more than three functions, Adv. Math. **20** (1976), 151–173.

[10] H. J. Brascamp, E. H. Lieb, and J. M. Luttinger, A general rearrangement inequality for multiple integrals, J. Funct. Anal. **17** (1974), 227–237.

[11] E. A. Carlen and M. Loss, Extremals of functionals with competing symmetries, J. Funct. Anal. **88** (1990), 437–456.

[12] M. Christ, On the restriction of the Fourier transform to curves: endpoint results and the degenerate case, Trans. Amer. Math. Soc. **287** (1985), 223–238.

[13] M. Christ and L. Grafakos, Best constants for two nonconvolution inequalities, Proc. Amer. Math. Soc. **123** (1995), 1687–1693.

[14] G. H. Hardy, Note on a theorem of Hilbert, Math. Z. **6** (1920), 314–317.

[15] G. H. Hardy and J. E. Littlewood, Notes on the theory of series (XII): on certain inequalities connected with the calculus of variations, J. London Math. Soc. **5** (1930), 34–39.

[16] G. H. Hardy, J. E. Littlewood, and G. Pólya, Inequalties, Cambridge University Press, 1967.

[17] C. E. Kenig and E. M. Stein, Multilinear estimates and fractional integration, Math. Res. Lett. **6** (1999), 1–15.

[18] E. H. Lieb, Sharp constants in the Hardy-Littlewood-Sobolev and related inequalities, Annals Math. **118** (1983), 349–374.

[19] S. Z. Lu, D. Yan, and F. Zhao, Sharp bounds for Hardy type operators on higher-dimensional product spaces, J. Inequalities and Appl. **148** (2013), 1–11.

[20] Z. Y. Si and S. Z. Lu, Weighted estimates for iterated commutators of multilinear fractional operators, Acta. Math. Sinica **28** (2012), 1769–1778.

[21] E. M. Stein, Singular Integrals and Differentiability Properties of Functions, Princeton University Press, 1970.

Some Topics in Harmonic Analysis
and Applications
ALM 34, pp. 27–70

Real-variable Theory of Orlicz-type Function Spaces Associated with Operators — A Survey

Der-Chen Chang*, Dachun Yang† and Sibei Yang‡

Abstract

This article is devoted to presenting a survey for the real-variable theory of Orlicz-type function spaces, associated with several linear operators, developed in recent years. More precisely, the real-variable theory for the (Musielak-)Orlicz-Hardy space, the (Musielak-)Orlicz BMO-type space and the (Musielak-)Orlicz VMO-type space associated with the linear operator in $L^2(\mathbb{R}^n)$ satisfying the pointwise Poisson estimate, the nonnegative self-adjoint operator in $L^2(\mathcal{X})$ satisfying the Davies-Gaffney estimate, and the linear operator of type ω, with $\omega \in [0, \frac{\pi}{2})$, having a bounded H_∞-functional calculus in $L^2(\mathcal{X})$ and satisfying some off-diagonal estimates on balls, where $\mathcal{X}$ is a metric measure space with doubling measure, is described.

2010 Mathematics Subject Classification: Primary 42B35; Secondary 42B30, 42B20, 42B25, 46E30.

Keywords and Phrases: nonnegative self-adjoint operator, elliptic operator, Schrödinger operator, operator of type ω, Hardy space, BMO-type space, VMO-type space, (Musielak-)Orlicz function, Davies-Gaffney estimate, Poisson estimate, off-diagonal estimate, Lusin area function, atom, molecule, Riesz transform.

*Department of Mathematics and Department of Computer Science, Georgetown University, Washington D.C. 20057, USA; Department of Mathematics, Fu Jen Catholic University, Taipei 242, China; *E-mail:* `chang@georgetown.edu`

†School of Mathematical Sciences, Beijing Normal University, Laboratory of Mathematics and Complex Systems, Ministry of Education, Beijing 100875, China; *E-mail:* `dcyang@bnu.edu.cn`

‡School of Mathematics and Statistics, Lanzhou University, Lanzhou 730000, China; *E-mail:* `yangsb@lzu.edu.cn`

1 Introduction

The real-variable theory of Hardy spaces on the n-dimensional Euclidean space $\mathbb{R}^n$ was initiated by Stein and Weiss [81] and then systematically developed by Fefferman and Stein [40]. As we have seen in many examples, Hardy space $H^p(\mathbb{R}^n)$ with $p \in (0,1]$ is a suitable substitute of the Lebesgue space $L^p(\mathbb{R}^n)$, especially when we study estimates for singular integral operators of convolution type. It plays an important role in various fields of analysis and partial differential equations (see, for example, [27, 78, 80]). In order to apply the real-variable theory of $H^p(\mathbb{R}^n)$ with $p \in (0,1]$, their several equivalent characterizations were established one by one (see, for example, [26, 40, 64, 84, 80]). Among these, a very important and useful characterization of Hardy spaces is their atomic characterizations, which were established by Coifman [26] when $n = 1$ and Latter [64] when $n > 1$. Later, as an extension of this characterization, the molecular characterization of Hardy spaces was obtained by Taibleson and Weiss [84]. Moreover, García-Cuerva [42] and Strömberg and Torchinsky [83] systematically studied the weighted Hardy space $H^p_\omega(\mathbb{R}^n)$ with $p \in (0,1]$, where $\omega \in A_q(\mathbb{R}^n)$ and $q \in [1,\infty]$. Hereafter, $A_q(\mathbb{R}^n)$ denotes the class of *Muckenhoupt weights* (see, for example, [44]).

Furthermore, due to the need for more inclusive classes of function spaces than the $L^p(\mathbb{R}^n)$ space from applications, the Orlicz space was introduced by Birnbaum and Orlicz [9] and Orlicz [74], which is widely used in various branches of mathematics (see, for example, [52, 76, 77]). Moreover, as an extension of the theory of Orlicz spaces, Orlicz-Hardy spaces and their dual spaces were studied by Strömberg [82] and Janson [53] on $\mathbb{R}^n$ and Viviani [85] on spaces of homogeneous type in the sense of Coifman and Weiss [28].

In the meanwhile, the classical BMO space (the *space of functions with bounded mean oscillation*), originally introduced by John and Nirenberg [62], plays an important role in the study of partial differential equations and harmonic analysis (see, for example, [40]). In particular, Fefferman and Stein [40] obtained the Carleson measure characterization of $\mathrm{BMO}(\mathbb{R}^n)$ and showed that it is the dual space of $H^1(\mathbb{R}^n)$. Later, the generalized BMO-type space $\mathrm{BMO}_\rho(\mathbb{R}^n)$ was studied in [46, 53, 82, 85] and it was proved therein to be the dual space of the Orlicz-Hardy space $H_\Phi(\mathbb{R}^n)$, where Φ denotes an *Orlicz function* on $[0,\infty)$, namely, Φ is nondecreasing, $\Phi(0) = 0$, $\Phi(t) > 0$ for $t \in (0,\infty)$ and $\lim_{t\to\infty} \Phi(t) = \infty$, and

$$\rho(t) := t^{-1}/\Phi^{-1}(t^{-1}) \tag{1.1}$$

for all $t \in (0,\infty)$. Here and in what follows, Φ^{-1} denotes the *inverse function* of Φ. It is worth pointing out that, different from the classical definition of Orlicz functions, the *Orlicz function Φ in this survey may not be convex.* Meanwhile, the Carleson measure characterization of $\mathrm{BMO}_\rho(\mathbb{R}^n)$ was obtained in [46].

For a function $\varphi : \mathbb{R}^n \times [0,\infty) \to [0,\infty)$ such that, for any $x \in \mathbb{R}^n$, $\varphi(x,\cdot)$ is an Orlicz function, φ is said to be of *uniformly upper type p* (resp. *uniformly lower type p*) for some $p \in [0,\infty)$, if there exists a positive constant C such that, for all $x \in \mathbb{R}^n$, $t \in [1,\infty)$ (resp. $t \in [0,1]$) and $s \in [0,\infty)$, $\varphi(x,st) \le Ct^p\varphi(x,s)$. Assume that $\varphi : \mathbb{R}^n \times [0,\infty) \to [0,\infty)$ is a function such that $\varphi(x,\cdot)$ is an Orlicz function with uniformly upper type 1 and lower type p for some

$p \in (0,1]$, and $\varphi(\cdot,t)$ is a Muckenhoupt $A_\infty(\mathbb{R}^n)$ weight uniformly in t. Such a φ is called a *growth function*. Recently, the Musielak-Orlicz-Hardy space $H_\varphi(\mathbb{R}^n)$ was introduced by Ky [70], which includes Hardy spaces $H^p(\mathbb{R}^n)$ with $p \in (0,1]$, Orlicz-Hardy spaces $H_\Phi(\mathbb{R}^n)$ and weighted Hardy spaces $H^p_\omega(\mathbb{R}^n)$ with $p \in (0,1]$ and $\omega \in A_\infty(\mathbb{R}^n)$. Moreover, Ky [70] also further introduced the Musielak-Orlicz BMO-type space $\mathrm{BMO}_\varphi(\mathbb{R}^n)$ and proved that it is the dual space of $H_\varphi(\mathbb{R}^n)$. Furthermore, some interesting applications of these spaces were also presented in [10, 12, 13, 47, 70, 68, 69, 90]. Recall that Musielak-Orlicz functions are the natural generalization of Orlicz functions that may vary in the spatial variables (see, for example, [70, 72]). Moreover, the motivation to study function spaces of Musielak-Orlicz type is due to their wide applications in several branches of physics and mathematics (see, for example, [11, 12, 13, 70]).

In the last decade, there appear many papers which pay attention to the problem of function spaces associated to different operators (see, for example, [4, 6, 7, 14, 19, 22, 36, 37, 48, 49, 50, 58, 55, 59, 61, 67, 79, 86]). In this article, due to limitation of length, we present a survey of the real-variable theory *only on Orlicz-type function spaces, associated with several linear operators*, developed in recent years. More precisely, let $\mathcal{X}$ be a metric measure space endowed with a doubling measure. We describe some real-variable theory for the (Musielak-) Orlicz-Hardy space, the (Musielak-) Orlicz BMO-type space and the (Musielak-) Orlicz VMO-type space associated with a linear operator in $L^2(\mathcal{X})$ which satisfying the pointwise Poisson estimate, the nonnegative self-adjoint operator in $L^2(\mathcal{X})$ satisfying the Davies-Gaffney estimate, and the operator of type ω, with $\omega \in [0,\pi/2)$, having a bounded H_∞-functional calculus in $L^2(\mathcal{X})$ and satisfying some off-diagonal estimates on balls.

The layout of this article is as follows.

Let Φ be an Orlicz function and ρ as in (1.1). Assume that L is a linear operator in $L^2(\mathbb{R}^n)$ which generates an analytic semigroup $\{e^{-tL}\}_{t>0}$ with kernel satisfying a pointwise upper bound estimate of Poisson type. In Section 2, we state the John-Nirenberg inequality and the ρ-Carleson measure characterization for the Orlicz BMO-type space $\mathrm{BMO}_{\rho,\,L}(\mathbb{R}^n)$, the molecular characterization for the Orlicz-Hardy space $H_{\Phi,\,L}(\mathbb{R}^n)$, the duality between $\mathrm{BMO}_{\rho,\,L}(\mathbb{R}^n)$ and $H_{\Phi,\,L}(\mathbb{R}^n)$, and the duality between the Orlicz VMO-type space $\mathrm{VMO}_{\rho,\,L}(\mathbb{R}^n)$ and the Banach completion space of $H_{\Phi,\,L}(\mathbb{R}^n)$, established in [58, 61, 65].

Let $\mathcal{X}$ be a metric measure space endowed with a doubling measure and $\varphi : \mathcal{X}\times[0,\infty) \to [0,\infty)$ be a growth function. Assume that L is a nonnegative self-adjoint operator in $L^2(\mathcal{X})$ satisfying the Davies-Gaffney estimate. In Section 3, we recall the φ-Carleson characterization for the Musielak-Orlicz BMO-type space $\mathrm{BMO}_{\varphi,\,L}(\mathcal{X})$, the atomic and the molecular characterizations for the Musielak-Orlicz-Hardy space $H_{\varphi,\,L}(\mathcal{X})$ and the duality between $\mathrm{BMO}_{\varphi,\,L}(\mathcal{X})$ and $H_{\varphi,\,L}(\mathcal{X})$ obtained in [15, 91]. Moreover, as a special case of this setting, several maximal function characterizations for the Musielak-Orlicz-Hardy space $H_{\varphi,\,L}(\mathbb{R}^n)$, associated with the Schrödinger operator $L := -\Delta + V$ on $\mathbb{R}^n$ with $0 \le V \in L^1_{\mathrm{loc}}(\mathbb{R}^n)$, and the boundedness of the Riesz transform $\nabla L^{-1/2}$ on $H_{\varphi,\,L}(\mathbb{R}^n)$ are also presented.

Let L satisfy the reinforced (p_L, q_L, k) off-diagonal estimates on balls, where

$p_L \in [1, 2)$, $q_L \in (2, \infty]$ and $k \in \mathbb{N}$. In Section 4, we recall the molecular characterization for the Musielak-Orlicz-Hardy space $H_{\varphi,\,L}(\mathcal{X})$ obtained in [15]. Moreover, as a special case of this setting, we also state several maximal function characterizations of the (Musielak-)Orlicz-Hardy space $H_{\varphi,\,L}(\mathbb{R}^n)$, associated with the divergence form elliptic operator L on $\mathbb{R}^n$ with complex bounded measurable coefficients, and the boundedness of the Riesz transform $\nabla L^{-1/2}$ on $H_{\varphi,\,L}(\mathbb{R}^n)$ obtained in [16, 55, 56].

Finally we make some conventions on notation. Throughout the whole article, we denote by C a *positive constant* which is independent of the main parameters, but it may vary from line to line. We also use $C_{(\gamma,\beta,\dots)}$ to denote a *positive constant depending on the indicated parameters* γ, β, ... The *symbol* $A \lesssim B$ means that $A \le CB$. If $A \lesssim B$ and $B \lesssim A$, then we write $A \sim B$. The *symbol* $\lfloor s \rfloor$ for $s \in \mathbb{R}$ denotes the maximal integer not more than s. Also, for any given $\lambda > 0$, we write λB for the λ-*dilated ball*, which is the ball with the same center as B and with radius $r_{\lambda B} := \lambda r_B$. We also set $\mathbb{N} := \{1, 2, \dots\}$ and $\mathbb{Z}_+ := \{0\} \cup \mathbb{N}$. For any ball $B \subset \mathcal{X}$, we set

$$U_0(B) := B \text{ and } U_j(B) := 2^j B \setminus 2^{j-1} B$$

for $j \in \mathbb{N}$. For any measurable subset E of $\mathcal{X}$, we denote by $E^{\complement}$ the *set* $\mathcal{X} \setminus E$ and by χ_E its *characteristic function*. For any subsets $E, F \subset \mathcal{X}$, let $d(E, F) := \inf_{x\in E,\, y\in F} d(x, y)$. For $1 \le q \le \infty$, we denote by q' the *conjugate exponent* of q, namely, $1/q + 1/q' = 1$.

2 Orlicz type spaces associated with operators satisfying Poisson estimates

Let L be a linear operator in $L^2(\mathbb{R}^n)$ and generate an analytic semigroup $\{e^{-tL}\}_{t>0}$ with kernel satisfying a pointwise upper bound estimate of Poisson type, whose decay is measured by $\theta(L) \in (0, \infty)$. Auscher, Duong and McIntosh [4] initially studied the Hardy space $H^1_L(\mathbb{R}^n)$ associated with the operator L. Based on this, Duong and Yan [36, 37] introduced the BMO-type space $\mathrm{BMO}_L(\mathbb{R}^n)$ associated with L and proved that the dual space of $H^1_L(\mathbb{R}^n)$ is just $\mathrm{BMO}_{L^*}(\mathbb{R}^n)$, where L^* denotes the *adjoint operator* of L in $L^2(\mathbb{R}^n)$. Moreover, Yan [86] further generalized these results to the Hardy space $H^p_L(\mathbb{R}^n)$ with $p \in (n/(n + \theta(L)), 1]$ and its dual space.

Recall that a function $\Phi : [0, \infty) \to [0, \infty)$ is called an *Orlicz function* if it is nondecreasing, $\Phi(0) = 0$, $\Phi(t) > 0$ for $t \in (0, \infty)$ and $\lim_{t\to\infty} \Phi(t) = \infty$ (see, for example, [72, 76, 77]). The function Φ is said to be of *upper type* p (resp. *lower type* p) for some $p \in [0, \infty)$, if there exists a positive constant C such that, for all $t \in [1, \infty)$ (resp. $t \in (0, 1]$) and $s \in (0, \infty)$,

$$\Phi(st) \le C t^p \Phi(s). \tag{2.1}$$

If Φ is of both upper type p_1 and lower type p_0, then Φ is said to be of *type* (p_0, p_1). Let

$$p^+_\Phi := \inf\{p \in (0, \infty) : \text{ there exists a positive constant } C$$

$$\text{such that (2.1) holds for all } t \in [1,\infty) \text{ and } s \in (0,\infty)\} \tag{2.2}$$

and

$$p_\Phi^- := \sup\{p \in (0,\infty) : \text{ there exists a positive constant } C \text{ such that (2.1) holds for all } t \in (0,1] \text{ and } s \in (0,\infty)\}. \tag{2.3}$$

The function Φ is said to be of *strictly lower type* p if, for all $t \in (0,1)$ and $s \in (0,\infty)$, $\Phi(st) \le t^p\Phi(s)$, and we let

$$p_\Phi := \sup\{p \in (0,\infty) : \ \Phi(st) \le t^p\Phi(s) \text{ holds for all } t \in (0,1) \text{ and } s \in (0,\infty)\}. \tag{2.4}$$

It is easy to see that $p_\Phi \le p_\Phi^- \le p_\Phi^+$ for all Φ. In what follows, p_Φ, p_Φ^- and p_Φ^+ are called the *strictly critical lower type index*, the *critical lower type index* and the *critical upper type index* of Φ, respectively. We point out that, if p_Φ is defined as in (2.4), then Φ is also of strictly critical lower type p_Φ; namely, p_Φ is attainable. (see [55] for the proof).

Let Φ satisfy the following assumptions.

Assumption $(\Phi)_1$. The function Φ is a positive, continuous and strictly increasing function on $(0,\infty)$ whose critical lower type index $p_\Phi^- \in (n/(n+\theta(L)),1]$. Also assume that Φ is of upper type 1 and subadditive. □

A typical example of Φ satisfying Assumption $(\Phi)_1$ is $\Phi(t) := t^p$, with $p \in (n/(n+\theta(L)),1]$, for all $t \in (0,\infty)$. Jiang et al. [61] introduced the Orlicz-Hardy space $H_{\Phi,L}(\mathbb{R}^n)$ and the Orlicz BMO-type space $\mathrm{BMO}_{\rho,L}(\mathbb{R}^n)$ and proved that the dual space of $H_{\Phi,L}(\mathbb{R}^n)$ is the space $\mathrm{BMO}_{\rho,L^*}(\mathbb{R}^n)$, where ρ is as in (1.1). Moreover, Jiang et al. [61] also characterized $H_{\Phi,L}(\mathbb{R}^n)$ in terms of the molecule. As an application, the boundedness of the generalized fractional operator $L_\rho^{-\gamma}$ with some $\gamma \in [0,\infty)$ from $H_{\Phi,L}(\mathbb{R}^n)$ to $H_L^1(\mathbb{R}^n)$ (when $\gamma = 0$) or $L^q(\mathbb{R}^n)$ with some $q \in (1,\infty)$ was obtained. Moreover, Jiang and Yang [58] introduced the Orlicz VMO-type space $\mathrm{VMO}_{\rho,L}(\mathbb{R}^n)$ and showed that it is just the predual space of $B_{\Phi,L^*}(\mathbb{R}^n)$, where $B_{\Phi,L^*}(\mathbb{R}^n)$ denotes the *Banach completion* of the Orlicz-Hardy space $H_{\Phi,L^*}(\mathbb{R}^n)$.

Assume further that Φ satisfies the following stronger Assumption $(\Phi)_2$.

Assumption $(\Phi)_2$. The function Φ is a positive, continuous and strictly increasing function on $(0,\infty)$ whose strictly critical lower type index

$$p_\Phi \in (n/(n+\theta(L)),1].$$

Also assume that Φ is concave. □

Recall that the assumption that Φ is concave implies that Φ is of upper type 1 and subadditive (see [55, p. 1175]). Liang et al. [65] established a new molecular characterization of $H_{\Phi,L}(\mathbb{R}^n)$ and, as applications, proved that the corresponding fractional integral $L^{-\gamma}$ for some $\gamma \in (0,\infty)$ maps $H_{\Phi,L}(\mathbb{R}^n)$ continuously into

$H_{\widetilde{\Phi},\,L}(\mathbb{R}^n)$, where $\widetilde{\Phi}$ has the same properties as Φ and is determined by Φ and γ, and also that L has a bounded holomorphic functional calculus in $H_{\Phi,\,L}(\mathbb{R}^n)$. Moreover, Liang et al. [65] proved that a generalized Riesz transform associated with L is bounded from $H_{\Phi,\,L}(\mathbb{R}^n)$ to the Orlicz space $L^{\Phi}(\mathbb{R}^n)$ or the Orlicz-Hardy space $H_{\Phi}(\mathbb{R}^n)$. Furthermore, under the assumption that Φ satisfies Assumption $(\Phi)_1$ with $p_{\Phi}^{-} \in (n/(n+\theta(L)), 1]$ replaced by $p_{\Phi} \in (n/(n+\theta(L)), 1]$, Liang et al. [65] proved that the Littlewood-Paley g-function g_L maps $H_{\Phi,\,L}(\mathbb{R}^n)$ continuously into the Orlicz space $L^{\Phi}(\mathbb{R}^n)$, and further characterized $H_{\Phi,\,L}(\mathbb{R}^n)$ in terms of the Littlewood-Paley g_{λ}^{*}-function with $\lambda \in (n(2/p_{\Phi}+1), \infty)$. Moreover, the local Hardy space $h_L^1(\mathbb{R}^n)$ associated with L was studied in [60].

In this section, we state some results for the BMO-type space $\mathrm{BMO}_{\rho,\,L}(\mathbb{R}^n)$, the Orlicz-Hardy space $H_{\Phi,\,L}(\mathbb{R}^n)$ and the VMO-type space $\mathrm{VMO}_{\rho,\,L}(\mathbb{R}^n)$ established in [61, 58, 65].

2.1 Some notions and two assumptions on operators

We begin with some notions of bounded holomorphic functional calculi which were introduced by McIntosh [71].

Let $0 < \nu < \gamma < \pi$. Define the *closed sector* S_ν in the complex plane $\mathbb{C}$ by setting $S_\nu := \{z \in \mathbb{C} : |\arg z| \le \nu\} \cup \{0\}$ and denote by S_ν^0 its *interior*. We employ the following subspaces of the space $H(S_\nu^0)$ of *all holomorphic functions on* S_ν^0:

$$H_\infty(S_\nu^0) := \left\{ b \in H(S_\nu^0) : \ \|b\|_{L^\infty(S_\nu^0)} := \sup_{z \in S_\nu^0} |b(z)| < \infty \right\}$$

and

$$\Psi(S_\nu^0) := \{\psi \in H(S_\nu^0) : \text{ there exist } s \in (0,\infty) \text{ and } C \in (0,\infty) \text{ such that, for all } z \in S_\nu^0,\ |\psi(z)| \le C|z|^s(1+|z|^{2s})^{-1}\}.$$

For any given $\nu \in (0,\pi)$, a closed operator L in $L^2(\mathbb{R}^n)$ is said to be of *type* ν if $\sigma(L) \subset S_\nu$, where $\sigma(L)$ denotes the *spectra* of L and, for all $\gamma > \nu$, there exists a positive constant C_γ such that, for all $\lambda \notin S_\gamma$, $\|(L-\lambda I)^{-1}\|_{L^2(\mathbb{R}^n)\to L^2(\mathbb{R}^n)} \le C_\gamma |\lambda|^{-1}$. Let $\mathscr{X}$ and $\mathscr{Y}$ be two normed linear spaces and T a continuous linear operator from $\mathscr{X}$ to $\mathscr{Y}$. Here and hereafter, $\|T\|_{\mathscr{X}\to\mathscr{Y}}$ denotes the *operator norm of* T *from* $\mathscr{X}$ *to* $\mathscr{Y}$. Let $\theta \in (\nu,\gamma)$ and Γ be the *contour* $\{\xi = re^{\pm i\theta} : r \ge 0\}$ parameterized clockwise around S_ν. Then, if L is of type ν and $\psi \in \Psi(S_\nu^0)$, the *operator* $\psi(L)$ is defined by

$$\psi(L) := \frac{1}{2\pi i}\int_\Gamma (L-\lambda I)^{-1}\psi(\lambda)\,d\lambda,$$

where the integral is absolutely convergent in $\mathfrak{L}(L^2(\mathbb{R}^n), L^2(\mathbb{R}^n))$ (*the class of all bounded linear operators in* $L^2(\mathbb{R}^n)$). By the Cauchy theorem, we conclude that the definition of $\psi(L)$ is independent of the choices of ν and γ satisfying $\theta \in (\nu,\gamma)$. Moreover, if L is one-to-one and has dense range, and $b \in H_\infty(S_\gamma^0)$, then $b(L)$ is defined by setting $b(L) := [\psi(L)]^{-1}(b\psi)(L)$, where $\psi(z) := z(1+z)^{-2}$ for all $z \in S_\gamma^0$.

It was proved that $b(L)$ is a well-defined linear operator in $L^2(\mathbb{R}^n)$ by McIntosh [71]. Moreover, the operator L is said to have a *bounded H_∞-calculus* in $L^2(\mathbb{R}^n)$ provided that, for all $\gamma \in (\nu, \pi)$, there exists a positive constant $\widetilde{C}_\gamma$ such that, for all $b \in H_\infty(S_\gamma^0)$, $b(L) \in \mathfrak{L}(L^2(\mathbb{R}^n), L^2(\mathbb{R}^n))$ and $\|b\|_{L^2(\mathbb{R}^n)\to L^2(\mathbb{R}^n)} \le \widetilde{C}_\gamma \|b\|_{L^\infty(S_\nu^0)}$.

Let L be a linear operator of type ν in $L^2(\mathbb{R}^n)$ with $\nu \in [0, \frac{\pi}{2})$. Then L generates a holomorphic semigroup $\{e^{-zL}\}_z$ with z satisfying $0 \le |\arg(z)| < \frac{\pi}{2}$, where $\arg(z) \in (-\pi, \pi]$ is the *argument* of z. We make the following two assumptions on L.

Assumption $(L)_1$. Assume that, for all $t \in (0, \infty)$, the *distribution kernel* p_t of e^{-tL} belongs to $L^\infty(\mathbb{R}^n \times \mathbb{R}^n)$ and satisfies the estimate $|p_t(x, y)| \le h_t(x, y)$ for all $x, y \in \mathbb{R}^n$, where h_t is given by

$$h_t(x, y) := t^{-\frac{n}{m}} g\left(\frac{|x - y|}{t^{\frac{1}{m}}}\right), \tag{2.5}$$

in which m is a positive constant and g is a positive, bounded and decreasing function satisfying

$$\lim_{r\to\infty} r^{n+\epsilon} g(r) = 0 \tag{2.6}$$

for some $\epsilon \in (0, \infty)$. □

Assumption $(L)_2$. Assume that the operator L is one-to-one, has dense range in $L^2(\mathbb{R}^n)$ and a bounded H_∞-functional calculus in $L^2(\mathbb{R}^n)$. □

From Assumptions $(L)_1$ and $(L)_2$, we deduce some useful estimates related to the operator L as follows.

First, if $\{e^{-tL}\}_{t>0}$ is a bounded analytic semigroup in $L^2(\mathbb{R}^n)$ whose kernels $\{p_t\}_{t>0}$ satisfy the estimates (2.5) and (2.6), then, for any $k \in \mathbb{N}$, there exists a positive constant C, depending on k, such that the time derivatives of p_t satisfy

$$\left| t^k \frac{\partial^k p_t(x, y)}{\partial t^k} \right| \le \frac{C}{t^{\frac{n}{m}}} g\left(\frac{|x - y|}{t^{\frac{1}{m}}}\right) \tag{2.7}$$

for all $t \in (0, \infty)$ and almost every $x, y \in \mathbb{R}^n$. It should be pointed out that, for all $k \in \mathbb{N}$, the function g may depend on k but it always satisfies (2.6) (see, for example, [75, Theorem 6.17]).

Second, L has a bounded H_∞-functional calculus in $L^2(\mathbb{R}^n)$ if and only if, for all $\gamma \in (\nu, \pi]$ and non-zero functions $\psi \in \Psi(S_\gamma^0)$, L satisfies the square function estimate and its reverse, namely, there exists a positive constant C such that, for all $f \in L^2(\mathbb{R}^n)$,

$$C^{-1}\|f\|_{L^2(\mathbb{R}^n)} \le \left\{\int_0^\infty \|\psi_t(L) f\|_{L^2(\mathbb{R}^n)}^2 \frac{dt}{t}\right\}^{1/2} \le C\|f\|_{L^2(\mathbb{R}^n)},$$

where $\psi_t(x) := \psi(tx)$ for all $t \in (0, \infty)$ and $x \in \mathbb{C}$. Notice that different choices of $\gamma > \nu$ and $\psi \in \Psi(S_\gamma^0)$ lead to equivalent quadratic norms of f (see [71] for the details).

Moreover, Duong and Yan [36] introduced the class of functions that the operators $\{e^{-tL}\}_{t>0}$ act upon. More precisely, for all $\beta \in (0, \infty)$, let $\mathcal{M}_\beta(\mathbb{R}^n)$ be the *set of all functions* $f \in L^2_{\mathrm{loc}}(\mathbb{R}^n)$ such that

$$\|f\|_{\mathcal{M}_\beta(\mathbb{R}^n)} := \left\{\int_{\mathbb{R}^n} \frac{|f(x)|^2}{1+|x|^{n+\beta}}\,dx\right\}^{1/2} < \infty.$$

Then $\mathcal{M}_\beta(\mathbb{R}^n)$ is a Banach space under the norm $\|\cdot\|_{\mathcal{M}_\beta(\mathbb{R}^n)}$. For any given operator L, let

$$\theta(L) := \sup\{\epsilon \in (0,\infty) : \text{ (2.6) holds true}\} \tag{2.8}$$

and define

$$\mathcal{M}(\mathbb{R}^n) := \begin{cases} \mathcal{M}_{\theta(L)}(\mathbb{R}^n), & \text{if } \theta(L) < \infty, \\ \bigcup_{0<\beta<\infty} \mathcal{M}_\beta(\mathbb{R}^n), & \text{if } \theta(L) = \infty. \end{cases}$$

Assume that $s \in \mathbb{Z}_+$. For all $(x,t) \in \mathbb{R}^{n+1}_+ := \mathbb{R}^n \times (0,\infty)$ and $f \in \mathcal{M}(\mathbb{R}^n)$, let

$$P_{s,t}f(x) := f(x) - (I - e^{-tL})^{s+1} f(x) \text{ and } Q_{s,t}f(x) := t^{s+1}L^{s+1}e^{-tL}f(x). \tag{2.9}$$

If $s = 0$, write

$$P_t f(x) := P_{0,t}f(x) = e^{-tL}f(x) \text{ and } Q_t f(x) := Q_{0,t}f(x) = tLe^{-tL}f(x). \tag{2.10}$$

For any $f \in \mathcal{M}(\mathbb{R}^n)$, by (2.7), we see that $P_{s,t}f$ and $Q_{s,t}f$ are well defined. Moreover, by (2.7) again, we know that the kernels $p_{s,t}$ and $q_{s,t}$, respectively, of $P_{s,t}$ and $Q_{s,t}$ satisfy that, for all $t \in (0,\infty)$ and $x,\ y \in \mathbb{R}^n$,

$$|p_{s,t^m}(x,y)| + |q_{s,t^m}(x,y)| \lesssim t^{-n} g\left(\frac{|x-y|}{t}\right), \tag{2.11}$$

where g satisfies the condition (2.6).

Some examples satisfy Assumptions $(L)_1$ and $(L)_2$ are given in the following remark.

Remark 2.1. (i) Let $\Delta := \sum_{k=1}^n \frac{\partial^2}{\partial x_k^2}$ be the Laplace operator on $\mathbb{R}^n$.

(i)$_1$ If $L := -\Delta$, then $\theta(L) = \infty$ and L satisfies Assumptions $(L)_1$ and $(L)_2$; more generally, if the kernels $\{p_t\}_{t>0}$ of $\{e^{-tL}\}_{t>0}$ satisfy a Gaussian upper bound, namely, there exists a positive constant C such that, for all $x,\ y \in \mathbb{R}^n$ and $t \in (0,\infty)$, $|p_t(x,y)| \le \frac{C}{t^{n/2}} e^{-\frac{|x-y|^2}{t}}$, then $\theta(L) = \infty$.

(i)$_2$ If $L := \sqrt{-\Delta}$, then $\theta(L) = 1$ and L also satisfies Assumptions $(L)_1$ and $(L)_2$.

(ii) Let $A := \{a_{ij}(x)\}_{1\le i,\,j\le n}$ be an $n \times n$ matrix with entries $a_{ij} \in L^\infty(\mathbb{R}^n, \mathbb{C})$ satisfying $\Re(\sum_{i,\,j=1}^n a_{ij}(x)\xi_i\xi_j) \ge \lambda|\xi|^2$ for all $x \in \mathbb{R}^n$, $\xi = (\xi_1, \xi_2, \ldots, \xi_n) \in \mathbb{C}^n$ and some $\lambda \in (0,\infty)$. Define a divergence form operator $Lf := -\mathrm{div}(A\nabla f)$, which is interpreted in the usual weak sense via a sesquilinear form. Notice that the Gaussian bound of the heat kernels of $\{e^{-tL}\}_{t>0}$ holds true when A has real entries or, when $n \in \{1, 2\}$, in the case of complex entries (see, for example, [8]). In this case, $\theta(L) = \infty$ and L satisfies Assumptions $(L)_1$ and $(L)_2$.

(iii) Let $V \in L^1_{\mathrm{loc}}(\mathbb{R}^n)$ be a nonnegative function. The Schrödinger operator with potential V is defined by $L := -\Delta + V$ on $\mathbb{R}^n$. From the Trotter-Kato product formula, it follows that the kernels $\{p_t\}_{t>0}$ of the semigroup $\{e^{-tL}\}_{t>0}$ satisfy the estimate that, for all $x, y \in \mathbb{R}^n$ and $t \in (0, \infty)$,

$$0 \le p_t(x, y) \le \frac{C}{(4\pi t)^{n/2}} e^{-\frac{|x-y|^2}{4t}},$$

where C is a positive constant independent of x, y and t. In this case, $\theta(L) = \infty$ and L satisfies Assumptions $(L)_1$ and $(L)_2$. However, unless V satisfies additional conditions, the heat kernel can be a discontinuous function of the space variables and Hölder's continuous estimates may fail to holds true (see, for example, [31]).

2.2 Orlicz BMO-type spaces associated with L

In this subsection, we recall the Orlicz BMO-type space $\mathrm{BMO}_{\rho, L}(\mathbb{R}^n)$ introduced in [61] and, the John-Nirenberg inequality and the ρ-Carleson measure characterization for $\mathrm{BMO}_{\rho, L}(\mathbb{R}^n)$ obtained in [61].

Let ρ be an Orlicz function on $(0, \infty)$. To state the definition of the Orlicz BMO-type space $\mathrm{BMO}^{q,\,s}_{\rho,\,L}(\mathbb{R}^n)$, we assume that ρ satisfies the following assumptions.

Convention. Let p^+_ρ and p^-_ρ be, respectively, as in (2.2) and (2.3) with Φ replaced by ρ. Assume that ρ is of lower type 0 and $0 \le p^+_\rho < \theta(L)/n$, which implies that $0 \le p^-_\rho \le p^+_\rho < \theta(L)/n$. In what follows, if (2.1) holds true for p^+_ρ and all $t \in [0, \infty)$ with Φ replaced by ρ, then we choose $\widetilde{p}^+_\rho := p^+_\rho$; otherwise we choose $\widetilde{p}^+_\rho \in (p^+_\rho, \theta(L)/n)$ such that $\lfloor \frac{n}{m}\widetilde{p}^+_\rho \rfloor = \lfloor \frac{n}{m} p^+_\rho \rfloor$. Similarly, if (2.1) holds true for p^-_ρ and all $t \in (0, 1)$ with Φ replaced by ρ, then we choose $\widetilde{p}^-_\rho := p^-_\rho$; otherwise we choose $\widetilde{p}^-_\rho \in (0, p^-_\rho)$. We also call $\widetilde{p}^+_\rho$ and $\widetilde{p}^-_\rho$ the *critical lower type index* and the *critical upper type index* of ρ, respectively. □

Definition 2.2. Let L satisfy Assumptions $(L)_1$ and $(L)_2$, ρ be of lower type 0 and $p^+_\rho \in [0, \frac{\theta(L)}{n})$, $q \in [1, \infty]$ and $s \in \mathbb{Z}^n_+$ with $s \ge \lfloor \frac{n}{m} p^+_\rho \rfloor$, where p^+_ρ is as in (2.2) with Φ replaced by ρ. A function $f \in \mathcal{M}(\mathbb{R}^n)$ is said to be in the *space* $\mathrm{BMO}^{q,\,s}_{\rho,\,L}(\mathbb{R}^n)$ if, when $q \in [1, \infty)$,

$$\|f\|_{\mathrm{BMO}^{q,\,s}_{\rho,\,L}(\mathbb{R}^n)} := \sup_{B \subset \mathbb{R}^n} \frac{1}{\rho(|B|)} \left\{ \frac{1}{|B|} \int_B |f(x) - P_{s,\,r_B^m} f(x)|^q \, dx \right\}^{1/q} < \infty$$

and, when $q = \infty$,

$$\|f\|_{\mathrm{BMO}^{\infty,\,s}_{\rho,\,L}(\mathbb{R}^n)} := \sup_{B \subset \mathbb{R}^n} \operatorname*{ess\,sup}_{x \in B} \frac{1}{\rho(|B|)} |f(x) - P_{s,\,r_B^m} f(x)| < \infty,$$

where the supremum is taken over all balls $B \subset \mathbb{R}^n$.

Remark 2.3. Let q, s, ρ and L be as in Definition 2.2.

(i) The space $(\mathrm{BMO}^{q,\,s}_{\rho,\,L}(\mathbb{R}^n), \|\cdot\|_{\mathrm{BMO}^{q,\,s}_{\rho,\,L}(\mathbb{R}^n)})$ is a semi-normed vector space with the seminorm vanishing on the space $\mathcal{K}_{(L,\,s)}(\mathbb{R}^n)$ defined by

$$\mathcal{K}_{(L,\,s)}(\mathbb{R}^n) := \{f \in \mathcal{M}(\mathbb{R}^n) :\ P_{s,\,t}f(x) = f(x) \text{ for almost every } x \in \mathbb{R}^n \text{ and all } t \in (0,\infty)\}.$$

The space $\mathrm{BMO}^{q,\,s}_{\rho,\,L}(\mathbb{R}^n)$ is understood to be modulo $\mathcal{K}_{(L,\,s)}(\mathbb{R}^n)$.

(ii) If $\rho = 1$, $q \in (1,\infty)$ and $s = 0$, then $\mathrm{BMO}^{q,\,s}_{\rho,\,L}(\mathbb{R}^n)$ is just the space $\mathrm{BMO}_L(\mathbb{R}^n)$ introduced by Duong and Yan [36, 37]. If $\alpha \in (0, \theta(L)/n)$ and $\rho(t) := t^\alpha$ for all $t \in (0,\infty)$, then $\mathrm{BMO}^{q,\,s}_{\rho,\,L}(\mathbb{R}^n)$ is just the space $\mathcal{L}_L(\alpha,\,q,\,s)$, introduced in [86], for all $q \in [1,\infty)$ and $s \in \mathbb{Z}^n_+$ with $s \geq \lfloor \frac{n}{m}\alpha \rfloor$. Observe that, if $\widetilde{p}^-_\rho = \widetilde{p}^+_\rho = p^+_\rho = p^+_\rho$, then ρ is equivalent to $t^{\widetilde{p}^-_\rho}$ and hence, for $q \in [1,\infty)$ and $s \geq \lfloor \frac{n}{m}\widetilde{p}^-_\rho \rfloor$, $\mathrm{BMO}^{q,\,s}_{\rho,\,L}(\mathbb{R}^n)$ coincides with $\mathrm{BMO}_L(\mathbb{R}^n)$ when $\widetilde{p}^-_\rho = 0$ and with $\mathcal{L}_L(\widetilde{p}^-_\rho,\,q,\,s)$ when $\widetilde{p}^-_\rho \in (0, \theta(L)/n)$.

For the space $\mathrm{BMO}^{1,\,s}_{\rho,\,L}(\mathbb{R}^n)$, we have the following John-Nirenberg inequality, which is just [61, Theorem 3.1].

Theorem 2.4. *Let L, ρ and s be as in Definition 2.2. Then there exist positive constants C_1, C_2 and C_3 such that, for all $f \in \mathrm{BMO}^{1,\,s}_{\rho,\,L}(\mathbb{R}^n)$, $\lambda \in (0,\infty)$ and $B := B(x_B, r_B)$ with $x_B \in \mathbb{R}^n$ and $r_B \in (0,\infty)$,*

(i) *when $p^-_\rho = 0$,*

$$\left|\left\{x \in B :\ \frac{1}{\rho(|B|)}|f(x) - P_{s,\,r_B^m}f(x)| > \lambda\right\}\right| \leq C_1|B|\exp\left\{-\frac{C_2\lambda}{\|f\|_{\mathrm{BMO}^{1,\,s}_{\rho,\,L}(\mathbb{R}^n)}}\right\};$$

(ii) *when $p^-_\rho \in (0,\infty)$, for almost every $x \in B$,*

$$\frac{1}{\rho(|B|)}|f(x) - P_{s,\,r_B^m}f(x)| \leq C_3\|f\|_{\mathrm{BMO}^{1,\,s}_{\rho,\,L}(\mathbb{R}^n)}.$$

As a corollary of Theorem 2.4, we have the following conclusion, which is just [61, Corollary 3.1].

Corollary 2.5. *Let L, ρ, q and s be as in Definition 2.2.*

(i) *If $p^-_\rho = 0$, then the spaces $\mathrm{BMO}^{q,\,s}_{\rho,\,L}(\mathbb{R}^n)$ coincide with equivalent norms for different $q \in [1,\infty)$.*

(ii) *If $p^-_\rho \in (0,\infty)$, then the spaces $\mathrm{BMO}^{q,\,s}_{\rho,\,L}(\mathbb{R}^n)$ coincide with equivalent norms for different $q \in [1,\infty]$.*

Remark 2.6. (i) We remark that, if $\alpha \in (0, \theta(L)/n)$ and $\rho(t) := t^\alpha$ for all $t \in (0,\infty)$, then, by Corollary 2.5, we know that $\mathcal{L}_L(\alpha,\,q,\,s) = \mathcal{L}_L(\alpha,\,1,\,s)$ with equivalent norms for all $q \in (1,\infty]$.

(ii) Based on Corollary 2.5, in what follows, we denote the space $\mathrm{BMO}^{q,\,s}_{\rho,\,L}(\mathbb{R}^n)$, when $p^-_\rho = 0$ and $q \in [1,\infty)$ or when $p^-_\rho \in (0,\infty)$ and $q \in [1,\infty]$, simply by $\mathrm{BMO}^{s}_{\rho,\,L}(\mathbb{R}^n)$.

To state the ρ-Carleson measure characterization of the space $\mathrm{BMO}^s_{\rho,\,L}(\mathbb{R}^n)$, which is presented in [61, Theorem 4.5], we recall the ρ-Carleson measure introduced in [46]. A measure $d\mu$ on $\mathbb{R}^{n+1}_+$ is called a *ρ-Carleson measure* if

$$\|d\mu\|_\rho := \sup_{B\subset\mathbb{R}^n}\left\{\frac{1}{|B|[\rho(|B|)]^2}\int_{\widehat{B}}|d\mu|\right\}^{1/2}<\infty,$$

where the supremum is taken over all balls $B\subset\mathbb{R}^n$ and $\widehat{B}$ denotes the *tent* over B, namely, $\widehat{B}:=\{(x,t):\ |x-x_B|<r_B-t\}$.

Theorem 2.7. *Let L and Φ satisfy, respectively, Assumptions $(L)_1$, $(L)_2$ and $(\Phi)_1$, $s\in\mathbb{Z}_+$ with $s\geq s_0:=\lfloor\frac{n}{m}(\frac{1}{p_\Phi^-}-1)\rfloor$, where p_Φ^- is as in (2.3), and ρ be as in (1.1). Then the following are equivalent:*

(i) *$g\in\mathrm{BMO}^{s_0}_{\rho,\,L}(\mathbb{R}^n)$;*

(ii) *$g\in\mathcal{M}(\mathbb{R}^n)$ and $d\mu_g(x,t):=|Q_{s,\,t^m}(I-P_{s_0,\,t^m})g(x)|^2\frac{dx\,dt}{t}$ is a ρ-Carleson measure on $\mathbb{R}^{n+1}_+$.*

Moreover, $\|d\mu_g\|_\rho$ is equivalent to $\|g\|_{\mathrm{BMO}^{s_0}_{\rho,\,L}(\mathbb{R}^n)}$.

Remark 2.8. Let L satisfy Assumptions $(L)_1$ and $(L)_2$ and φ be as in Definition 3.2. It is worth pointing out that the φ-Carleson measure characterization and the John-Nirenberg inequality for the Musielak-Orlicz BMO-type space associated with L is not established.

2.3 Orlicz-Hardy spaces associated with L

In this subsection, we recall the Orlicz-Hardy space $H_{\Phi,\,L}(\mathbb{R}^n)$ introduced in [61], its molecular characterization, and the duality between the Orlicz BMO-type space $\mathrm{BMO}_{\rho,\,L}(\mathbb{R}^n)$ and $H_{\Phi,\,L}(\mathbb{R}^n)$ obtained in [61, 65]. In what follows, we *always assume* that the operator L satisfies Assumptions $(L)_1$ and $(L)_2$, and that the Orlicz function Φ satisfies Assumption $(\Phi)_1$. We first recall the Orlicz-Hardy space $H_{\Phi,\,L}(\mathbb{R}^n)$ introduced in [61].

For all $f\in L^2(\mathbb{R}^n)$, the Lusin area function $S_L(f)$ associated with L is defined by setting, for all $x\in\mathbb{R}^n$,

$$S_L(f)(x):=\left\{\int_{\Gamma(x)}|Q_{t^m}f(y)|^2\frac{dy\,dt}{t^{n+1}}\right\}^{1/2},\tag{2.12}$$

where $\Gamma(x):=\{(y,t)\in\mathbb{R}^n\times(0,\infty):\ |x-y|<t\}$ and Q_{t^m} is as in (2.10).

Let Φ be an Orlicz function on $[0,\infty)$. Recall that a measurable function f on $\mathbb{R}^n$ is said to be in the *space* $L^\Phi(\mathbb{R}^n)$ if

$$\int_{\mathbb{R}^n}\Phi(|f(x)|)\,dx<\infty.$$

Moreover, for any $f\in L^\Phi(\mathbb{R}^n)$, define

$$\|f\|_{L^\Phi(\mathbb{R}^n)}:=\inf\left\{\lambda\in(0,\infty):\ \int_{\mathbb{R}^n}\Phi\left(\frac{|f(x)|}{\lambda}\right)dx\leq1\right\}.$$

Jiang, Yang and Zhou [61] introduced the Orlicz-Hardy space $H_{\Phi,\,L}(\mathbb{R}^n)$ associated with L as follows.

Definition 2.9. Let L satisfy Assumptions $(L)_1$ and $(L)_2$ and Φ Assumption $(\Phi)_1$. A function $f \in L^2(\mathbb{R}^n)$ is said to be in the set $\widetilde{H}_{\Phi,\,L}(\mathbb{R}^n)$ if $S_L(f) \in L^\Phi(\mathbb{R}^n)$ and, moreover, define

$$\|f\|_{H_{\Phi,\,L}(\mathbb{R}^n)} := \|S_L(f)\|_{L^\Phi(\mathbb{R}^n)} = \inf\left\{\lambda \in (0,\infty) :\ \int_{\mathbb{R}^n} \Phi\left(\frac{S_L(f)(x)}{\lambda}\right) dx \le 1\right\}.$$

The *Orlicz-Hardy space* $H_{\Phi,\,L}(\mathbb{R}^n)$ associated with L is defined to be the completion of $\widetilde{H}_{\Phi,\,L}(\mathbb{R}^n)$ in the norm $\|\cdot\|_{H_{\Phi,\,L}(\mathbb{R}^n)}$.

Now we recall the definition of the $(\Phi,\,s)$-molecule introduced in [61]. We begin with the notion of $T_\Phi(\mathbb{R}^{n+1}_+)$-atoms, here and hereafter,

$$\mathbb{R}^{n+1}_+ := \mathbb{R}^n \times (0,\infty).$$

Let $q \in (1,\infty)$. A function a on $\mathbb{R}^{n+1}_+$ is called a $T_\Phi(\mathbb{R}^{n+1}_+)$-*atom* if

(i) there exists a ball $B \subset \mathbb{R}^n$ such that $\operatorname{supp} a \subset \widehat{B}$, here and hereafter,

$$\widehat{B} := \{(y,t) \in \mathbb{R}^n \times (0,\infty) :\ |y - x_B| < r_B - t\};$$

(ii) $\|a\|_{T^2_2(\mathbb{R}^{n+1}_+)} := \left\{\int_{\widehat{B}} |a(x,t)|^2 \,\frac{dy\,dt}{t}\right\}^{1/2} \le |B|^{-1/2}[\rho(|B|)]^{-1}.$

Moreover, for a sequence $\{a_j\}_j$ of $T_\Phi(\mathbb{R}^{n+1}_+)$-atoms and $\{\lambda_j\} \subset \mathbb{C}$, let

$$\Lambda_2(\{\lambda_j a_j\}_j) := \inf\left\{\lambda \in (0,\infty) :\ \sum_j |B_j|\Phi\left(\frac{|\lambda_j|\|a_j\|_{T^2_2(\mathbb{R}^{n+1}_+)}}{\lambda|B_j|^{1/2}}\right) \le 1\right\}. \tag{2.13}$$

Definition 2.10. Let Φ satisfy Assumption $(\Phi)_1$ and $s \in \mathbb{Z}^n_+$ with $s \ge s_0 := \lfloor \frac{n}{m}(\frac{1}{p_\Phi^-} - 1)\rfloor$. A function α on $\mathbb{R}^n$ is called a $(\Phi,\,s)$-*molecule* if there exists a $T_\Phi(\mathbb{R}^{n+1}_+)$-atom a such that, for all $x \in \mathbb{R}^n$,

$$\alpha(x) := \pi_L(a)(x) := C_{m,\,s}\int_0^\infty Q_{s,\,t^m}(I - P_{s_0,\,t^m})(a(\cdot,t))(x)\frac{dt}{t}, \tag{2.14}$$

where $C_{m,\,s}$ is a positive constant such that

$$C_{m,\,s}\int_0^\infty t^{m(s+2)}e^{-2t^m}(1 - e^{-t^m})^{s_0+1}\frac{dt}{t} = 1.$$

The following molecular characterization of $H_{\Phi,\,L}(\mathbb{R}^n)$ is just [61, Theorems 4.3 and 4.6].

Theorem 2.11. *Let Φ satisfy Assumption $(\Phi)_1$ and $s \in \mathbb{N}$ with*

$$s \ge s_0 := \lfloor \frac{n}{m}(\frac{1}{p_\Phi^-} - 1)\rfloor.$$

(i) *For all* $f \in H_{\Phi,L}(\mathbb{R}^n)$, *there exist a sequence* $\{\alpha_k\}_k$ *of* (Φ, s)*-molecules and* $\{\lambda_k\}_k \subset \mathbb{C}$ *such that* $f = \sum_k \lambda_k \alpha_k$ *in* $H_{\Phi,L}(\mathbb{R}^n)$. *Furthermore, define* $\widetilde{\Lambda}_2(\{\lambda_k \alpha_k\}_k) := \inf \Lambda_2(\{\mu_k a_k\}_k)$, *where* $\Lambda_2(\{\mu_k a_k\}_k)$ *is as in* (2.13) *and the infimum is taken over all possible* $T_\Phi(\mathbb{R}^{n+1}_+)$*-atoms* $\{a_k\}_k$ *and* $\{\mu_k\}_k \subset \mathbb{C}$ *such that* $\sum_k \lambda_k \alpha_k = \sum_k \mu_k \pi_L(a_k)$ *in* $H_{\Phi,L}(\mathbb{R}^n)$. *Then there exists a positive constant* C, *independent of* f, *such that*

$$\widetilde{\Lambda}_2(\{\lambda_k \alpha_k\}_k) \le C\|f\|_{H_{\Phi,L}(\mathbb{R}^n)}.$$

(ii) *Suppose that* $\{\alpha_k\}_k$ *is a sequence of* (Φ, s)*-molecules and* $\{\lambda_k\}_k \subset \mathbb{C}$ *satisfying* $\Lambda_2(\{\mu_k a_k\}_k) < \infty$ *with* $\alpha_k := \pi_L(a_k)$ *for all* k. *Then* $\sum_k \lambda_k \alpha_k \in H_{\Phi,L}(\mathbb{R}^n)$ *and*

$$\left\|\sum_k \lambda_k \alpha_k\right\|_{H_{\Phi,L}(\mathbb{R}^n)} \le C\widetilde{\Lambda}_2(\{\lambda_k \alpha_k\}_k),$$

where C *is a positive constant independent of* $\{\lambda_k \alpha_k\}_k$.

Moreover, Liang et al. [65] introduced another kind of molecules and further established the equivalent characterization of $H_{\Phi,L}(\mathbb{R}^n)$ in terms of those molecules under assuming that Φ satisfies Assumption $(\Phi)_2$.

Definition 2.12. Let Φ satisfy Assumption $(\Phi)_2$, $q \in (1,\infty)$, $s \in \mathbb{N}$ and $\eta \in (0,\infty)$. A function $\alpha \in L^q(\mathbb{R}^n)$ is called a (Φ, q, s, η)*-molecule*, associated with the ball B, if there exists a ball $B \subset \mathbb{R}^n$ such that

(i) $\|\alpha\|_{L^q(U_j(B))} \le 2^{-j\eta}|2^jB|^{1/q-1}[\rho(|2^jB|)]^{-1}$, $j \in \mathbb{Z}_+$;

(ii) for every $k \in \{1, \ldots, s\}$ and $j \in \mathbb{Z}_+$, it holds true that

$$\left\|\left(r_B^{-m}L^{-1}\right)^k \alpha\right\|_{L^q(U_j(B))} \le 2^{-j\eta}\left|2^jB\right|^{1/q-1}\left[\rho\left(\left|2^jB\right|\right)\right]^{-1}.$$

If α is a (Φ, q, s, η)-molecule for all $q \in (1,\infty)$, then α is called a (Φ, ∞, s, η)-*molecule.*

Definition 2.13. Let L satisfy Assumptions $(L)_1$ and $(L)_2$ and Φ Assumption $(\Phi)_2$. Let $q \in (1,\infty)$, $s \in \mathbb{N}$ with

$$s \ge s_0 := \lfloor \frac{n}{m}(\frac{1}{p_\Phi} - 1)\rfloor \quad \text{and} \quad \eta \in (n[1/p_\Phi - 1/p_\Phi^+], \infty),$$

where p_Φ and p_Φ^+ are, respectively, as in (2.4) and (2.2). A function $f \in L^q(\mathbb{R}^n)$ is said to be in $\widetilde{H}^{s,q,\eta}_{\Phi,L}(\mathbb{R}^n)$ if there exist $\{\lambda_j\}_{j=1}^\infty \subset \mathbb{C}$ and a sequence $\{\alpha_j\}_{j=1}^\infty$ of (Φ, q, s, η)-molecules such that $f = \sum_{j=1}^\infty \lambda_j \alpha_j$ in $L^2(\mathbb{R}^n)$ and

$$\Lambda(\{\lambda_j \alpha_j\}_{j=1}^\infty) := \inf\left\{\lambda \in (0,\infty) : \sum_{j=1}^\infty |B_j|\Phi\left(\frac{|\lambda_j|}{\lambda|B_j|\rho(|B_j|)}\right) \le 1\right\} < \infty,$$

where, for each $j \in \mathbb{Z}_+$, α_j is associated with the ball B_j. Moreover, define

$$\|f\|_{\widetilde{H}^{s,q,\eta}_{\Phi,L}(\mathbb{R}^n)} := \inf \Lambda(\{\lambda_j \alpha_j\}_{j=1}^\infty),$$

where the infimum is taken over all decompositions of f as above. The *molecular Orlicz-Hardy space* $H^{s,\,q,\,\eta}_{\Phi,\,L}(\mathbb{R}^n)$ is defined to be the completion of $\widetilde{H}^{s,\,q,\,\eta}_{\Phi,\,L}(\mathbb{R}^n)$ in the norm $\|\cdot\|_{H^{s,\,q,\,\eta}_{\Phi,\,L}(\mathbb{R}^n)}$.

The following molecular characterization of $H_{\Phi,\,L}(\mathbb{R}^n)$ is obtained in [65, Theorem 3.1].

Theorem 2.14. *Let L satisfy Assumptions $(L)_1$ and $(L)_2$ and Φ Assumption $(\Phi)_2$. Let $q\in(1,\infty)$, $s\in\mathbb{N}$ with $s\ge s_0:=\lfloor\frac{n}{m}(\frac{1}{p_\Phi}-1)\rfloor$ and $\eta\in(n[1/p_\Phi-1/p_\Phi^+],\infty)$, where p_Φ and p_Φ^+ are, respectively, as in (2.4) and (2.2). If*

$$\frac{n}{n+\theta(L)}<\frac{p_\Phi}{2-p_\Phi/p_\Phi^+},$$

then the spaces $H_{\Phi,\,L}(\mathbb{R}^n)$ and $H^{s,\,q,\,\eta}_{\Phi,\,L}(\mathbb{R}^n)$ coincide with equivalent norms.

Let $H_{\Phi,\,L,\,\mathrm{fin}}(\mathbb{R}^n)$ denote the *set of all finite linear combinations of* $(\Phi,\,s)$-*molecules*. Then we have the following duality theorem, which is just [61, Theorem 4.4]. Recall that L^* denotes the *adjoint operator* of L in $L^2(\mathbb{R}^n)$.

Theorem 2.15. *Let L satisfy Assumptions $(L)_1$ and $(L)_2$ and Φ Assumption $(\Phi)_1$, $s\in\mathbb{N}$ with $s\ge s_0:=\lfloor\frac{n}{m}(\frac{1}{p_\Phi^-}-1)\rfloor$ and ρ be as in (1.1). Then the dual space of $H_{\Phi,\,L}(\mathbb{R}^n)$, $(H_{\Phi,\,L}(\mathbb{R}^n))^*$, coincides with $\mathrm{BMO}^{s_0}_{\rho,\,L^*}(\mathbb{R}^n)$ in the following sense:*

(i) *Let $g\in\mathrm{BMO}^{s_0}_{\rho,\,L^*}(\mathbb{R}^n)$. Then the linear functional ℓ, which is initially defined on $H_{\Phi,\,L,\,\mathrm{fin}}(\mathbb{R}^n)$ by*

$$\ell(f):=\int_{\mathbb{R}^n}f(x)g(x)\,dx,\tag{2.15}$$

has a unique extension to $H_{\Phi,\,L}(\mathbb{R}^n)$ with $\|\ell\|_{(H_{\Phi,\,L}(\mathbb{R}^n))^*}\le C\|g\|_{\mathrm{BMO}^{s_0}_{\rho,\,L^*}(\mathbb{R}^n)}$, where C is a positive constant independent of g.

(ii) *Conversely, for any $\ell\in(H_{\Phi,\,L}(\mathbb{R}^n))^*$, there exists $g\in\mathrm{BMO}^{s_0}_{\rho,\,L^*}(\mathbb{R}^n)$ such that (2.15) holds true for all $f\in H_{\Phi,\,L}(\mathbb{R}^n)\cap L^2(\mathbb{R}^n)$ and $\|g\|_{\mathrm{BMO}^{s_0}_{\rho,\,L^*}(\mathbb{R}^n)}\le C\|\ell\|_{(H_{\Phi,\,L}(\mathbb{R}^n))^*}$, where C is a positive constant independent of ℓ.*

Remark 2.16. (i) By using the molecular characterization of $H_{\Phi,\,L}(\mathbb{R}^n)$ in Theorem 2.11, the boundedness of the fractional integral associated with L from $H_{\Phi,\,L}(\mathbb{R}^n)$ to the Lebesgue space $L^q(\mathbb{R}^n)$ with some $q\in(1,\infty)$ or the Hardy space $H^1_L(\mathbb{R}^n)$ was obtained in [61, Theorem 5.1].

(ii) Assume further that Φ satisfies Assumption $(\Phi)_2$. The boundedness of the fractional integral associated with L from $H_{\Phi,\,L}(\mathbb{R}^n)$ to the Orlicz-Hardy space $H_{\widetilde{\Phi},\,L}(\mathbb{R}^n)$ was established in [65, Theorem 4.1] by applying Theorem 2.14, where the Orlicz function $\widetilde{\Phi}$ is determined by Φ. By using the molecular characterization of $H_{\Phi,\,L}(\mathbb{R}^n)$ in Theorem 2.14, the boundedness of the holomorphic functional calculus, the Littlewood-Paley g-function and g_λ^*-function, and the Riesz transform associated with L on $H_{\varphi,\,L}(\mathbb{R}^n)$ was, respectively, studied in [65, Theorems 4.2, 4.3, 4.4 and 4.5].

(iii) The Musielak-Orlicz-Hardy space associated with L is not introduced and studied.

2.4 Predual spaces of $H_{\Phi,\,L}(\mathbb{R}^n)$

In this subsection, we recall the space $\mathrm{VMO}^s_{\rho,\,L}(\mathbb{R}^n)$ introduced in [58], which was proved to be the predual space of the Banach completion of $H_{\Phi,\,L}(\mathbb{R}^n)$.

Definition 2.17. Let L satisfy Assumptions $(L)_1$ and $(L)_2$ and Φ Assumption $(\Phi)_1$. Assume that ρ is as in (1.1) and $s\in\mathbb{N}$ with $s\ge\lfloor\frac{n}{m}(\frac{1}{p_\Phi^-}-1)\rfloor$. A function $f\in\mathrm{BMO}_{\rho,\,L}(\mathbb{R}^n)$ is said to be in the *space* $\mathrm{VMO}^s_{\rho,\,L}(\mathbb{R}^n)$, if it satisfies the limiting conditions $\gamma_1(f)=\gamma_2(f)=\gamma_3(f)=0$, where

$$\gamma_1(f):=\lim_{c\to 0}\sup_{\text{ball }B:\ r_B\le c}\left\{\frac{1}{|B|[\rho(|B|)]^2}\int_B|f(x)-P_{s,\,r_B^m}f(x)|^2\,dx\right\}^{1/2},$$

$$\gamma_2(f):=\lim_{c\to\infty}\sup_{\text{ball }B:\ r_B\ge c}\left\{\frac{1}{|B|[\rho(|B|)]^2}\int_B|f(x)-P_{s,\,r_B^m}f(x)|^2\,dx\right\}^{1/2}$$

and

$$\gamma_3(f):=\lim_{c\to\infty}\sup_{\text{ball }B\subset[B(0,c)]^\complement}\left\{\frac{1}{|B|[\rho(|B|)]^2}\int_B|f(x)-P_{s,\,r_B^m}f(x)|^2\,dx\right\}^{1/2}.$$

Moreover, for $f\in\mathrm{VMO}^s_{\rho,\,L}(\mathbb{R}^n)$, its norm is defined by

$$\|f\|_{\mathrm{VMO}^s_{\rho,\,L}(\mathbb{R}^n)}:=\|f\|_{\mathrm{BMO}^s_{\rho,\,L}(\mathbb{R}^n)}.$$

For all measurable functions g on $\mathbb{R}^{n+1}_+$ and all $x\in\mathbb{R}^n$, define

$$\mathcal{C}_\rho(g)(x):=\sup_{B\ni x}\frac{1}{\rho(|B|)}\left\{\frac{1}{|B|}\int_{\widehat{B}}|g(x,t)|^2\frac{dy\,dt}{t}\right\}.$$

The space $T^\infty_\Phi(\mathbb{R}^{n+1}_+)$ is defined to be the set of all measurable functions g on $\mathbb{R}^{n+1}_+$ such that $\|g\|_{T^\infty_\Phi(\mathbb{R}^{n+1}_+)}:=\|\mathcal{C}_\rho(g)\|_{L^\infty(\mathbb{R}^n)}<\infty$.

In what follows, let $T^\infty_{\Phi,\,0}(\mathbb{R}^{n+1}_+)$ be the set of all $f\in T^\infty_\Phi(\mathbb{R}^{n+1}_+)$ satisfying $\eta_1(f)=\eta_2(f)=\eta_3(f)=0$, where

$$\eta_1(f):=\lim_{c\to 0}\sup_{\text{ball }B:\ r_B\le c}\frac{1}{\rho(|B|)}\left\{\frac{1}{|B|}\int_{\widehat{B}}|g(x,t)|^2\frac{dy\,dt}{t}\right\},$$

$$\eta_2(f):=\lim_{c\to\infty}\sup_{\text{ball }B:\ r_B\ge c}\frac{1}{\rho(|B|)}\left\{\frac{1}{|B|}\int_{\widehat{B}}|g(x,t)|^2\frac{dy\,dt}{t}\right\}$$

and

$$\eta_3(f):=\lim_{c\to\infty}\sup_{\text{ball }B\subset[B(0,c)]^\complement}\frac{1}{\rho(|B|)}\left\{\frac{1}{|B|}\int_{\widehat{B}}|g(x,t)|^2\frac{dy\,dt}{t}\right\}.$$

It is easy to see that $T^\infty_{\Phi,\,0}(\mathbb{R}^{n+1}_+)$ is a closed linear subspace of $T^\infty_\Phi(\mathbb{R}^{n+1}_+)$.

Then we have the following characterization for the space $\mathrm{VMO}^s_{\rho,\,L}(\mathbb{R}^n)$, which is just [58, Theorem 3.2].

Theorem 2.18. *Let L, Φ and ρ be as in Definition* 2.17, *s, $s_1 \in \mathbb{N}$ with $s_1 \geq s_0 := \lfloor \frac{n}{m}(\frac{1}{p_\Phi^-} - 1)\rfloor$ and $s \geq 2s_1$. Then the following conditions are equivalent:*

(i) *$f \in \mathrm{VMO}^{s_0}_{\rho,\,L}(\mathbb{R}^n)$;*

(ii) *$f \in \mathcal{M}(\mathbb{R}^n)$ and $Q_{s,\,t^m}(I - P_{2s_1,\,t^m})f \in T^{\infty}_{\Phi,0}(\mathbb{R}^{n+1}_+)$.*

Remark 2.19. We remark that Theorem 2.18 completely covers [32, Proposition 3.3] by taking $\Phi(t) := t$ for all $t \in (0, \infty)$.

Now we recall the Banach completion of the space $H_{\Phi,\,L}(\mathbb{R}^n)$. Let s, $s_1 \in \mathbb{Z}_+$ with $s \geq s_1 \geq \lfloor \frac{n}{m}(\frac{1}{p_\Phi^-} - 1)\rfloor$. The $(\Phi,\, s,\, s_1)$-molecule is defined via replacing s_0 in (2.14) by s_1.

Definition 2.20. Let L satisfy Assumptions $(L)_1$ and $(L)_2$ and Φ Assumption $(\Phi)_1$. Assume that ρ is as in (1.1) and s, $s_1 \in \mathbb{N}$ with $s \geq s_1 \geq \lfloor \frac{n}{m}(\frac{1}{p_\Phi^-} - 1)\rfloor$. The *space* $B^{s,\,s_1}_{\Phi,\,L}(\mathbb{R}^n)$ is defined to be the set of all $f = \sum_j \lambda_j \alpha_j$, where the series converges in $(\mathrm{BMO}_{\rho,\,L^*}(\mathbb{R}^n))^*$, $\{\lambda_j\}_j \in \ell^1$ and $\{\alpha_j\}_j$ is a sequence of $(\Phi,\, s,\, s_1)$-molecules. Moreover, for all $f \in B^{s,\,s_1}_{\Phi,\,L}(\mathbb{R}^n)$, define

$$\|f\|_{B^{s,\,s_1}_{\Phi,\,L}(\mathbb{R}^n)} := \inf\left\{\sum_j |\lambda_j|\right\},$$

where the infimum is taken over all the possible decompositions of f as above.

Then we have the following conclusion that the predual space of the space $B_{\Phi,\,L}(\mathbb{R}^n)$ is $\mathrm{VMO}_{\rho,\,L^*}(\mathbb{R}^n)$, which is just [58, Theorem 4.4].

Theorem 2.21. *Assume that L, Φ and ρ are as in Definition* 2.17. *Then*

$$(\mathrm{VMO}_{\rho,\,L^*}(\mathbb{R}^n))^* = B_{\Phi,\,L}(\mathbb{R}^n)$$

in the following sense:

(i) *For any given $g \in B_{\Phi,\,L}(\mathbb{R}^n)$, define the linear functional ℓ by setting, for all $f \in \mathrm{VMO}_{\rho,\,L^*}(\mathbb{R}^n)$,*

$$\ell(f) := \langle f, g\rangle. \tag{2.16}$$

Then there exists a positive constant C, independent of g, such that

$$\|\ell\|_{(\mathrm{VMO}_{\rho,\,L^*}(\mathbb{R}^n))^*} \leq C\|g\|_{B_{\Phi,\,L}(\mathbb{R}^n)}.$$

(ii) *Conversely, for any $\ell \in (\mathrm{VMO}_{\rho,\,L^*}(\mathbb{R}^n))^*$, there exists $g \in B_{\Phi,\,L}(\mathbb{R}^n)$ such that* (2.16) *holds true and there exists a positive constant C, independent of ℓ, such that*

$$\|g\|_{B_{\Phi,\,L}(\mathbb{R}^n)} \leq C\|\ell\|_{(\mathrm{VMO}_{\rho,\,L^*}(\mathbb{R}^n))^*}.$$

Remark 2.22. (i) It was proved in [32, Theorem 4.1] that the predual space of $H^1_L(\mathbb{R}^n)$ is $\mathrm{VMO}_{L^*}(\mathbb{R}^n)$ in the sense as in Theorem 2.21. From the definition of $H^1_L(\mathbb{R}^n)$ and its molecular characterization, we deduce that the Banach completion of $H^1_L(\mathbb{R}^n)$ is just itself. Thus, Theorem 2.21 completely covers [32, Theorem 4.1] by taking $\Phi(t) := t$ for all $t \in (0, \infty)$.

(ii) The Musielak-Orlicz VMO-type space is not introduced associated with L and hence whether the results in Theorems 2.18 and 2.21 are holds true or not for such space is not clear.

3 Musielak-Orlicz type spaces associated with nonnegative self-adjoint operators satisfying Davies-Gaffney estimates

Let $\mathcal{X}$ be a *set*, d a *metric* on $\mathcal{X}$ and μ a *nonnegative Borel regular measure* on $\mathcal{X}$. For all $x \in \mathcal{X}$ and $r \in (0, \infty)$, let $B(x, r) := \{y \in \mathcal{X} : d(x, y) < r\}$ and $V(x, r) := \mu(B(x, r))$. Moreover, we assume that there exists a constant $C \in [1, \infty)$ such that, for all $x \in \mathcal{X}$ and $r \in (0, \infty)$,

$$V(x, 2r) \leq CV(x, r) < \infty. \tag{3.1}$$

Observe that $(\mathcal{X}, d, \mu)$ is a *space of homogeneous type* in the sense of Coifman and Weiss [28]. Recall that, in the definition of spaces of homogeneous type in [28, Chapter 3], d is assumed to be a quasi-metric. However, for simplicity, we *always assume* that d is a metric. Notice that the doubling property (3.1) implies that the following strong homogeneity property that, for some positive constants C and n,

$$V(x, \lambda r) \leq C\lambda^n V(x, r) \tag{3.2}$$

uniformly for all $\lambda \in [1, \infty)$, $x \in \mathcal{X}$ and $r \in (0, \infty)$. There also exist constants $C \in (0, \infty)$ and $N \in [0, n]$ such that, for all $x, y \in \mathcal{X}$ and $r \in (0, \infty)$,

$$V(x, r) \leq C\left\{1 + \frac{d(x, y)}{r}\right\}^N V(y, r). \tag{3.3}$$

Indeed, the property (3.3) with $N = n$ is a simple corollary of the triangle inequality for the metric d and the strong homogeneity property (3.2). In the cases of Euclidean spaces and Lie groups of polynomial growth, N can be chosen to be 0.

Let L be a nonnegative self-adjoint operator in $L^2(\mathcal{X})$ satisfying the Davies-Gaffney estimate, where $\mathcal{X}$ is a metric measure space endowed with a doubling measure. Hofmann et al. [48] introduced the Hardy space $H^1_L(\mathcal{X})$ via the Lusin area function associated with the semigroup $\{e^{-tL}\}_{t>0}$ and the BMO space $\mathrm{BMO}_L(\mathcal{X})$, which was proved to be the dual space of $H^1_L(\mathcal{X})$. Hofmann et al. [48] also obtained two characterizations of $H^1_L(\mathcal{X})$ in terms of atoms and molecules.

Let Φ satisfy the following assumptions.

Assumption $(\Phi)_3$. Let Φ be a positive, continuous and strictly increasing function on $(0, \infty)$ whose strictly critical lower type index $p_\Phi \in (0, 1]$. Also assume that Φ is concave. □

Let ρ be as in (1.1). Jiang and Yang [58] introduced the Orlicz-Hardy space $H_{\Phi, L}(\mathcal{X})$ via the Lusin area function associated with the heat semigroup of L, and the BMO-type space $\mathrm{BMO}_{\rho, L}(\mathcal{X})$ and proved that the dual space of $H_{\Phi, L}(\mathcal{X})$ is the space $\mathrm{BMO}_{\rho, L}(\mathcal{X})$. Moreover, Jiang and Yang [58] characterized the space $H_{\Phi, L}(\mathcal{X})$ in terms of the atom, the molecule and the Lusin area function associated with the Poisson semigroup of L. As a special case of this setting, several equivalent characterizations and some applications of the Hardy space $H^1_L(\mathbb{R}^n)$ and the Orlicz-Hardy space $H_{\Phi, L}(\mathbb{R}^n)$ associated with the Schrödinger operator $L := -\Delta + V$

were, respectively, obtained in [48] and [58], where $0 \le V \in L^1_{\mathrm{loc}}(\mathbb{R}^n)$. Moreover, Song and Yan [79] studied the weighted Hardy space $H^1_{\omega,L}(\mathbb{R}^n)$ associated with the Schrödinger operator L, where $\omega \in A_1(\mathbb{R}^n)$. Furthermore, the Hardy space $H^p_L(G)$ with $p \in (0,1]$ associated with the Schrödinger operator L on the nilpotent Lie group G was studied in [54]. Moreover, Hardy spaces associated with (magnetic) Schrödinger operators on the strongly Lipschitz domain of $\mathbb{R}^n$ were studied in [51, 87].

For a given function $\varphi : \mathcal{X} \times [0,\infty) \to [0,\infty)$ such that, for any $x \in \mathcal{X}$, $\varphi(x,\cdot)$ is an Orlicz function, φ is said to be of *uniformly upper type p* (resp. *uniformly lower type p*) for some $p \in [0,\infty)$, if there exists a positive constant C such that, for all $x \in \mathcal{X}$, $t \in [1,\infty)$ (resp. $t \in [0,1]$) and $s \in [0,\infty)$,

$$\varphi(x, st) \le Ct^p \varphi(x,s); \tag{3.4}$$

φ is said to be of *positive uniformly upper type* (resp. *uniformly lower type*) if it is of uniformly upper type (resp. uniformly lower type) p for some $p \in (0,\infty)$. Moreover, let

$$I(\varphi) := \inf\{p \in (0,\infty) : \ \varphi \text{ is of uniformly upper type } p\} \tag{3.5}$$

and

$$i(\varphi) := \sup\{p \in (0,\infty) : \ \varphi \text{ is of uniformly lower type } p\}. \tag{3.6}$$

In what follows, $I(\varphi)$ and $i(\varphi)$ are, respectively, called the *uniformly critical upper type index* and the *uniformly critical lower type index* of φ. Observe that $I(\varphi)$ and $i(\varphi)$ may not be attainable, namely, φ may not be of uniformly upper type $I(\varphi)$ and uniformly lower type $i(\varphi)$ (see below for some examples).

Definition 3.1. Let $\varphi : \mathcal{X} \times [0,\infty) \to [0,\infty)$ satisfy that $x \mapsto \varphi(x,t)$ is measurable for all $t \in [0,\infty)$. The function $\varphi(\cdot,t)$ is said to satisfy the *uniformly Muckenhoupt condition for some $q \in [1,\infty)$*, denoted by $\varphi \in \mathbb{A}_q(\mathcal{X})$, if, when $q \in (1,\infty)$,

$$\begin{aligned}\mathbb{A}_q(\varphi) :&= \sup_{t\in(0,\infty)} \sup_{B\subset\mathcal{X}} \frac{1}{\mu(B)} \int_B \varphi(x,t)\, d\mu(x) \\ &\quad \times \left\{\frac{1}{\mu(B)} \int_B [\varphi(y,t)]^{-q'/q}\, d\mu(y)\right\}^{q/q'} < \infty,\end{aligned}$$

where $1/q + 1/q' = 1$, or, when $q = 1$,

$$\mathbb{A}_1(\varphi) := \sup_{t\in(0,\infty)} \sup_{B\subset\mathcal{X}} \frac{1}{\mu(B)} \int_B \varphi(x,t)\, d\mu(x) \left\{\operatorname*{ess\,sup}_{y\in B} [\varphi(y,t)]^{-1}\right\} < \infty.$$

Here the first suprema are taken over all $t \in (0,\infty)$ and the second ones over all balls $B \subset \mathcal{X}$.

The function $\varphi(\cdot,t)$ is said to satisfy the *uniformly reverse Hölder condition for some $q \in (1,\infty]$*, denoted by $\varphi \in \mathbb{RH}_q(\mathcal{X})$, if, when $q \in (1,\infty)$,

$$\mathbb{RH}_q(\varphi) : = \sup_{t\in(0,\infty)} \sup_{B\subset\mathcal{X}} \left\{\frac{1}{\mu(B)} \int_B [\varphi(x,t)]^q\, d\mu(x)\right\}^{1/q}$$

$$\times \left\{ \frac{1}{\mu(B)} \int_B \varphi(x,t)\, d\mu(x) \right\}^{-1} < \infty,$$

or, when $q = \infty$,

$$\mathbb{RH}_\infty(\varphi) := \sup_{t\in(0,\infty)} \sup_{B\subset\mathcal{X}} \left\{ \operatorname*{ess\,sup}_{y\in B} \varphi(y,t) \right\} \left\{ \frac{1}{\mu(B)} \int_B \varphi(x,t)\, d\mu(x) \right\}^{-1} < \infty.$$

Here the first suprema are taken over all $t \in (0,\infty)$ and the second ones over all balls $B \subset \mathcal{X}$.

Recall that, in Definition 3.1, when $\mathcal{X} = \mathbb{R}^n$, $\mathbb{A}_q(\mathbb{R}^n)$ with $q \in [1,\infty)$ was introduced by Ky [70].

Let $\mathbb{A}_\infty(\mathcal{X}) := \cup_{q\in[1,\infty)} \mathbb{A}_q(\mathcal{X})$ and the *critical indices* of φ be defined as follows:

$$q(\varphi) := \inf\{q \in [1,\infty) :\ \varphi \in \mathbb{A}_q(\mathcal{X})\} \tag{3.7}$$

and

$$r(\varphi) := \sup\{q \in (1,\infty] :\ \varphi \in \mathbb{RH}_q(\mathcal{X})\}. \tag{3.8}$$

Observe that, if $q(\varphi) \in (1,\infty)$, then $\varphi \notin \mathbb{A}_{q(\varphi)}(\mathcal{X})$, and there exists $\varphi \notin \mathbb{A}_1(\mathcal{X})$ such that $q(\varphi) = 1$ (see, for example, [63]). Similarly, if $r(\varphi) \in (1,\infty)$, then $\varphi \notin \mathbb{RH}_{r(\varphi)}(\mathcal{X})$, and there exists $\varphi \notin \mathbb{RH}_\infty(\mathcal{X})$ such that $r(\varphi) = \infty$ (see, for example, [29]).

Now we introduce the notion of growth functions.

Definition 3.2. A function $\varphi :\ \mathcal{X} \times [0,\infty) \to [0,\infty)$ is called a *growth function* if the following hold true:

(i) φ is a *Musielak-Orlicz function*, namely,

(i)$_1$ the function $\varphi(x,\cdot) :\ [0,\infty) \to [0,\infty)$ is an Orlicz function for all $x \in \mathcal{X}$;

(i)$_2$ the function $\varphi(\cdot,t)$ is a measurable function for all $t \in [0,\infty)$.

(ii) $\varphi \in \mathbb{A}_\infty(\mathcal{X})$.

(iii) The function φ is of uniformly upper type 1 and of uniformly lower type p for some $p \in (0,1]$.

Clearly, for all $x \in \mathcal{X}$ and $t \in [0,\infty)$,

$$\varphi(x,t) := \omega(x)\Phi(t) \tag{3.9}$$

is a growth function if $\omega \in A_\infty(\mathcal{X})$ and Φ is an Orlicz function of upper type 1 and lower type p for some $p \in (0,1]$. It is known that, for $p \in (0,1]$, if $\Phi(t) := t^p$ for all $t \in [0,\infty)$, then Φ is an Orlicz function of type (p,p); for $p \in [\frac{1}{2},1]$, if $\Phi(t) := t^p/\ln(e+t)$ for all $t \in [0,\infty)$, then Φ is an Orlicz function of lower type q for $q \in (0,p)$ and of upper type p; for $p \in (0,1]$, if $\Phi(t) := t^p \ln(e+t)$ for all $t \in [0,\infty)$, then Φ is an Orlicz function of lower type p and of upper type q for $q \in (p,1]$. Recall that, if an Orlicz function is of upper type $p \in (0,1)$, then it is also of upper type 1. We also remark that these Orlicz functions satisfy Assumption $(\Phi)_3$.

Let $x_0 \in \mathcal{X}$. Another typical and useful example of the growth function φ is

$$\varphi(x,t) := \frac{t^\alpha}{[\ln(e+d(x,x_0))]^\beta + [\ln(e+t)]^\gamma}$$

for all $x \in \mathcal{X}$ and $t \in [0,\infty)$ with some $\alpha \in (0,1]$, $\beta \in [0,n)$ and $\gamma \in [0, 2\alpha(1+\ln 2)]$. It is easy to show that $\varphi \in \mathbb{A}_1(\mathcal{X})$, φ is of uniformly upper type α, $I(\varphi) = i(\varphi) = \alpha$, $i(\varphi)$ is not attainable, but $I(\varphi)$ is attainable. It is worth pointing out that such a function φ naturally appears in the study of the pointwise multiplier characterization for the BMO-type space on the metric space endowed with a doubling measure (see [73]).

Let L be a nonnegative self-adjoint operator in $L^2(\mathcal{X})$ satisfying the Davies-Gaffney estimate, and $\varphi : \mathcal{X} \times [0,\infty) \to [0,\infty)$ a growth function as in Definition 3.2. Moreover, we always assume that $\varphi \in \mathbb{RH}_{2/[2-I(\varphi)]}(\mathcal{X})$. The Musielak-Orlicz-Hardy space $H_{\varphi,L}(\mathcal{X})$ associated with L was introduced in [91] via the Lusin area function. Moreover, the space $H_{\varphi,L}(\mathcal{X})$ was characterized in [91] in terms of the atom, the molecule, the Lusin area function associated with the Poisson semigroup of L and the Littlewood-Paley g_λ^*-function. Furthermore, the Musielak-Orlicz BMO-type space $\mathrm{BMO}_{\varphi,L}(\mathcal{X})$ was also introduced in [91], which was proved to be the dual space of $H_{\varphi,L}(\mathcal{X})$ and further characterized by the φ-Carleson measure. As a special case of this setting, several equivalent characterizations of the Musielak-Orlicz-Hardy space $H_{\varphi,L}(\mathbb{R}^n)$ associated with the Schrödinger operator $L := -\Delta + V$, where $0 \le V \in L^1_{\mathrm{loc}}(\mathbb{R}^n)$, in terms of the Lusin-area function, the non-tangential and the radial maximal functions associated with L, were presented in [91]. Moreover, the boundedness of the corresponding Riesz transform $\nabla L^{-1/2}$ on $H_{\varphi,L}(\mathbb{R}^n)$ was studied in [16, 91]. Furthermore, some special Musielak-Orlicz-Hardy spaces associated with the Schrödinger operator $L := -\Delta+V$ on $\mathbb{R}^n$, where the nonnegative potential V satisfies the reverse Hölder inequality of order $n/2$, were studied by Ky [69] and further applied to the study of commutators of singular integral operators associated with the operator L. Especially, the Hardy space associated with the Schrödinger operator $-\Delta + V$ was studied in [38, 39], where the nonnegative potential V satisfies the reverse Hölder inequality (see, for example, [43, 44] for the definition of the reverse Hölder inequality).

Let L be a nonnegative adjoint operator in $L^2(\mathcal{X})$ satisfying the Davies-Gaffney estimate. In this section, we state the results for the Musielak-Orlicz BMO-type space $\mathrm{BMO}_{\varphi,L}(\mathcal{X})$ and the Musielak-Orlicz-Hardy space $H_{\varphi,L}(\mathcal{X})$ established in [15, 91].

3.1 Musielak-Orlicz BMO-type spaces associated with L

In this subsection, we recall some results for the space $\mathrm{BMO}_{\varphi,L}(\mathcal{X})$ obtained in [91]. Let $(\mathcal{X}, d, \mu)$ be a metric measure space with μ being doubling. For $p \in (0,\infty]$, the *space of p-integrable functions on* $\mathcal{X}$ is denoted by $L^p(\mathcal{X})$ and the *(quasi-)norm* of $f \in L^p(\mathcal{X})$ by $\|f\|_{L^p(\mathcal{X})}$. We first recall the following assumptions for the operator L.

Assumption $(L)_3$. L is a nonnegative self-adjoint operator in $L^2(\mathcal{X})$. □

Assumption $(L)_4$. The operator L generates an analytic semigroup $\{e^{-tL}\}_{t>0}$ which satisfies the Davies-Gaffney estimates, namely, there exist positive constants C_2 and C_3 such that, for all closed sets E and F in $\mathcal{X}$, $t \in (0,\infty)$ and $f \in L^2(E)$,

$$\left\|e^{-tL}f\right\|_{L^2(F)} \le C_2 \exp\left\{-\frac{[\mathrm{dist}\,(E,F)]^2}{C_3 t}\right\} \|f\|_{L^2(E)}, \tag{3.10}$$

here and hereafter, $L^2(E)$ denotes the *set of all μ-measurable functions supported in E* such that

$$\|f\|_{L^2(E)} := \left\{\int_E |f(x)|^2\,d\mu(x)\right\}^{1/2} < \infty$$

and $\mathrm{dist}\,(E,F) := \inf_{x\in E,\,y\in F} d(x,y)$. □

Examples of operators satisfying Assumptions $(L)_3$ and $(L)_4$ include second order elliptic self-adjoint operators in divergence form on $\mathbb{R}^n$ with bounded measurable coefficients, (degenerate) Schrödinger operators with nonnegative potential or with magnetic field, and Laplace-Beltrami operators on all complete Riemannian manifolds (see, for example, [30, 41]).

Let φ be as in Definition 3.2, $M \in \mathbb{N}$ and $\phi = L^M \nu$ be a function in $L^2(\mathcal{X})$, where $\nu \in \mathcal{D}(L^M)$, $\mathcal{D}(L^M)$ denotes the domain of the operator L^M and L^M the M-fold composition of L with itself. Following [48, 49, 58, 91], for $\epsilon \in (0,\infty)$, $M \in \mathbb{N}$ and fixed $x_0 \in \mathcal{X}$, we introduce the space

$$\mathcal{M}_\varphi^{M,\epsilon}(L) := \left\{\phi = L^M\nu \in L^2(\mathcal{X}) : \|\phi\|_{\mathcal{M}_\varphi^{M,\epsilon}(L)} < \infty\right\},$$

where

$$\|\phi\|_{\mathcal{M}_\varphi^{M,\epsilon}(L)} := \sup_{j\in\mathbb{Z}_+}\left\{2^{j\epsilon}[V(x_0,1)]^{-1/2}\|\chi_{B(x_0,1)}\|_{L^\varphi(\mathcal{X})}\sum_{k=0}^{M}\|L^k\nu\|_{L^2(U_j(B(x_0,1)))}\right\}.$$

Let A_t denote either $(I+t^2L)^{-1}$ or e^{-t^2L} and f belong to the dual space of $\mathcal{M}_\varphi^{M,\epsilon}(L)$, $(\mathcal{M}_\varphi^{M,\epsilon}(L))^*$. It was proved in Section 4.2 of [91] that $(I-A_t)^M f \in L^2_{\mathrm{loc}}(\mathcal{X})$ in the sense of distributions. Moreover, for any $M \in \mathbb{N}$, define

$$\mathcal{M}_\varphi^M(\mathcal{X}) := \bigcap_{\epsilon > n[q(\varphi)/i(\varphi)-1/2]} (\mathcal{M}_\varphi^{M,\epsilon}(L))^*,$$

where n, $q(\varphi)$ and $i(\varphi)$ are, respectively, as in (3.6), (3.7) and (3.8).

Now we recall the Musielak-Orlicz BMO-type space associated with L introduced in [91].

Definition 3.3. Let φ be as in Definition 3.2, L satisfy Assumptions $(L)_3$ and $(L)_4$, and $M \in \mathbb{N}$ with $M > \frac{n}{2}[\frac{q(\varphi)}{i(\varphi)} - \frac{1}{2}]$, where n, $q(\varphi)$ and $i(\varphi)$ are, respectively, as in (3.2), (3.6) and (3.5). A functional $f \in \mathcal{M}_\varphi^M(\mathcal{X})$ is said to be in the *space* $\mathrm{BMO}_{\varphi,L}^M(\mathcal{X})$ if

$$\|f\|_{\mathrm{BMO}_{\varphi,L}^M(\mathcal{X})} := \sup_{B\subset\mathcal{X}} \frac{[\mu(B)]^{1/2}}{\|\chi_B\|_{L^\varphi(\mathcal{X})}}\left\{\int_B \left|(I-e^{-r_B^2 L})^M f(x)\right|^2 d\mu(x)\right\}^{1/2} < \infty,$$

where the supremum is taken over all balls B of $\mathcal{X}$.

For the space $\mathrm{BMO}^M_{\varphi,\,L}(\mathcal{X})$, we have the following proposition, which is just [91, Proposition 4.11].

Proposition 3.4. *Let φ, L and M be as in Definition 3.3. Then $f \in \mathrm{BMO}^M_{\varphi,\,L}(\mathcal{X})$ if and only if $f \in \mathcal{M}^M_\varphi(\mathcal{X})$ and*

$$\sup_{B\subset\mathcal{X}} \frac{[\mu(B)]^{1/2}}{\|\chi_B\|_{L^\varphi(\mathcal{X})}} \left\{\int_B \left|[I-(I+r_B^2L)^{-1}]^M f(x)\right|^2 \,d\mu(x)\right\}^{1/2} < \infty,$$

where the supremum is taken over all balls B of $\mathcal{X}$. Moreover, the quantity appeared in the left-hand side of the above formula is equivalent to $\|f\|_{\mathrm{BMO}^M_{\varphi,\,L}(\mathcal{X})}$.

Now we recall the notion of the φ-Carleson measure associated with φ introduced in [91, Definition 4.18].

Definition 3.5. A measure $d\mu$ on $\mathcal{X}\times(0,\infty)$ is called a *φ-Carleson measure* if

$$\|d\mu\|_\varphi := \sup_{B\subset\mathcal{X}} \frac{[\mu(B)]^{1/2}}{\|\chi_B\|_{L^\varphi(\mathcal{X})}} \left\{\int_{\widehat{B}} |d\mu(x,t)|\right\}^{1/2} < \infty,$$

where the supremum is taken over all balls $B\subset\mathcal{X}$ and $\widehat{B}$ denotes the tent over B.

Then, for the space $\mathrm{BMO}^M_{\varphi,\,L}(\mathcal{X})$, we have the following φ-Carleson measure characterization, which is just [91, Theorem 4.19].

Theorem 3.6. *Let L satisfy Assumptions $(L)_3$ and $(L)_4$, φ be as in Definition 3.2 with $\varphi \in \mathbb{RH}_{2/[2-I(\varphi)]}(\mathcal{X})$ and $I(\varphi)$ as in (3.5), and $M\in\mathbb{N}$ with $M > \frac{n}{2}[\frac{q(\varphi)}{i(\varphi)}-\frac{1}{2}]$, where n, $q(\varphi)$ and $i(\varphi)$ are, respectively, as in (3.2), (3.7) and (3.6). Then the following are equivalent:*

(i) *$f\in \mathrm{BMO}^M_{\varphi,\,L}(\mathcal{X})$;*

(ii) *$f\in\mathcal{M}^M_\varphi(\mathcal{X})$ satisfies*

$$\int_{\mathcal{X}} \frac{|[I-(I+L)^{-1}]^M f(x)|^2}{1+[d(x,x_0)]^{N+\epsilon+2nq_0/p_2}}\,d\mu(x) < \infty$$

for some $x_0\in\mathcal{X}$, $q_0\in(q(\varphi),\infty)$, $p_2\in(0,i(\varphi))$ and $\epsilon\in(0,\infty)$, and $d\mu_f$ is a φ-Carleson measure, where $d\mu_f$ is defined by setting, for all $x\in\mathcal{X}$,

$$d\mu_f(x) := \left|(t^2L)^M e^{-t^2L} f(x)\right|^2 \frac{d\mu(x)\,dt}{t}.$$

Moreover, $\|f\|_{\mathrm{BMO}^M_{\varphi,\,L}(\mathcal{X})}$ and $\|d\mu_f\|_\varphi$ are comparable.

Remark 3.7. We remark that, when φ is as in (3.9) with $\omega\equiv 1$ and Φ concave, the Musielak-Orlicz BMO-type space $\mathrm{BMO}^M_{\varphi,\,L}(\mathcal{X})$ is just the BMO-type space $\mathrm{BMO}^M_{\rho,\,L}(\mathcal{X})$ introduced in [57]. Moreover, Proposition 3.4 and Theorem 3.6, respectively, completely cover [57, Proposition 4.4 and Theorem 4.2] by taking φ as in (3.9) with $\omega\equiv 1$ and Φ concave.

3.2 Musielak-Orlicz-Hardy spaces associated with L

In this subsection, we state some results for Musielak-Orlicz-Hardy spaces $H_{\varphi,\,L}(\mathcal{X})$ obtained in [91]. We begin with the notion of the space $H_{\varphi,\,L}(\mathcal{X})$. For all $f \in L^2(\mathcal{X})$ and $x \in \mathcal{X}$, define

$$S_L(f)(x) := \left\{\int_{\Gamma(x)} \left|t^2 L e^{-t^2 L} f(y)\right|^2 \frac{d\mu(y)\,dt}{V(x,t)t}\right\}^{1/2},$$

here and hereafter, $\Gamma(x) := \{(y,t) \in \mathcal{X} \times (0,\infty) : \ d(x,y) < t\}$.

Let φ be a growth function on $\mathcal{X} \times [0,\infty)$. A measurable function f on $\mathcal{X}$ is said to be in the *space* $L^{\varphi}(\mathcal{X})$ if

$$\int_{\mathcal{X}} \varphi(x, |f(x)|)\,d\mu(x) < \infty.$$

Moreover, for any $f \in L^{\varphi}(\mathcal{X})$, define

$$\|f\|_{L^{\varphi}(\mathcal{X})} := \inf\left\{\lambda \in (0,\infty) : \ \int_{\mathcal{X}} \varphi\left(x, \frac{|f(x)|}{\lambda}\right) d\mu(x) \le 1\right\}.$$

We now recall the notion of the Musielak-Orlicz-Hardy space $H_{\varphi,\,L}(\mathcal{X})$ introduced in [91, Definition 4.1].

Definition 3.8. Let L satisfy Assumptions $(L)_3$ and $(L)_4$, and φ be as in Definition 3.2. Assume that $H^2(\mathcal{X}) := \overline{R(L)}$, where $\overline{R(L)}$ denotes the *closure of the range of* L *in* $L^2(\mathcal{X})$. A function $f \in H^2(\mathcal{X})$ is said to be in $\widetilde{H}_{\varphi,\,L}(\mathcal{X})$ if $S_L(f) \in L^{\varphi}(\mathcal{X})$; moreover, define

$$\begin{aligned}\|f\|_{H_{\varphi,\,L}(\mathcal{X})} &:= \|S_L(f)\|_{L^{\varphi}(\mathcal{X})}\\ &:= \inf\left\{\lambda \in (0,\infty) : \ \int_{\mathcal{X}} \varphi\left(x, \frac{S_L(f)(x)}{\lambda}\right) d\mu(x) \le 1\right\}.\end{aligned}$$

The *Musielak-Orlicz-Hardy space* $H_{\varphi,\,L}(\mathcal{X})$ is defined to be the completion of $\widetilde{H}_{\varphi,\,L}(\mathcal{X})$ in the quasi-norm $\|\cdot\|_{H_{\varphi,\,L}(\mathcal{X})}$.

Moreover, the atomic and the molecular Musielak-Orlicz-Hardy spaces were also introduced in [91, Definitions 5.1 and 5.2].

Definition 3.9. (I) Let $M \in \mathbb{N}$. A function $\alpha \in L^2(\mathcal{X})$ is called a (φ, M)*-atom* associated with the operator L, if there exist a function $b \in \mathcal{D}(L^M)$ and a ball $B \subset \mathcal{X}$ such that

(i) $\alpha = L^M b$;

(ii) $\operatorname{supp}(L^k b) \subset B$, $k \in \{0, \ldots, M\}$;

(iii) $\|(r_B^2 L)^k b\|_{L^2(\mathcal{X})} \le r_B^{2M}[\mu(B)]^{1/2}\|\chi_B\|_{L^{\varphi}(\mathcal{X})}^{-1}$, $k \in \{0, \ldots, M\}$.

(II) Let $M \in \mathbb{N}$ with $M > \frac{n}{2}[\frac{q(\varphi)}{i(\varphi)} - \frac{1}{2}]$, where n, $q(\varphi)$ and $i(\varphi)$ are, respectively, as in (3.2), (3.7) and (3.6). A distribution $f \in (\mathrm{BMO}_{\varphi,\,L}(\mathcal{X}))^*$ is said to be in the *space* $H^M_{\varphi,\,\mathrm{at}}(\mathcal{X})$ if there exist $\{\lambda_j\}_j \subset \mathbb{C}$ and a sequence $\{\alpha_j\}_j$ of $(\varphi,\,M)$-atoms such that $f = \sum_j \lambda_j \alpha_j$ in $(\mathrm{BMO}_{\varphi,\,L}(\mathcal{X}))^*$ and

$$\sum_j \varphi\left(B_j, \frac{|\lambda_j|}{\|\chi_{B_j}\|_{L^\varphi(\mathcal{X})}}\right) < \infty,$$

where, for each j, $\operatorname{supp}\alpha_j \subset B_j$. Moreover, for any $f \in H^M_{\varphi,\,\mathrm{at}}(\mathcal{X})$, its *quasi-norm* is defined by $\|f\|_{H^M_{\varphi,\,\mathrm{at}}(\mathcal{X})} := \inf\{\Lambda(\{\lambda_j\alpha_j\}_j)\}$, where

$$\Lambda(\{\lambda_j\alpha_j\}_j) := \inf\left\{\lambda \in (0,\infty) : \sum_j \varphi\left(B_j, \frac{|\lambda_j|}{\lambda\|\chi_{B_j}\|_{L^\varphi(\mathcal{X})}}\right) \le 1\right\} \tag{3.11}$$

and the infimum is taken over all possible decompositions of f as above.

Definition 3.10. (I) Let $M \in \mathbb{N}$ and $\epsilon \in (0,\infty)$. A function $\beta \in L^2(\mathcal{X})$ is called a $(\varphi,\,M,\,\epsilon)$-*molecule* associated with the operator L, if there exist a function $b \in \mathcal{D}(L^M)$ and a ball $B \subset \mathcal{X}$ such that

(i) $\beta = L^M b$;

(ii) for each $k \in \{0,\,\ldots,\,M\}$ and $j \in \mathbb{Z}_+$, it holds true that

$$\|(r_B^2 L)^k b\|_{L^2(U_j(B))} \le 2^{-j\epsilon} r_B^{2M} [\mu(B)]^{1/2} \|\chi_B\|^{-1}_{L^\varphi(\mathcal{X})},$$

where $U_j(B)$ with $j \in \mathbb{Z}_+$ is as in (3.4).

(II) Let $M \in \mathbb{N}$ with $M > \frac{n}{2}[\frac{q(\varphi)}{i(\varphi)} - \frac{1}{2}]$ and $\epsilon \in (n[\frac{q(\varphi)}{i(\varphi)} - \frac{1}{2}], \infty)$, where n, $q(\varphi)$ and $i(\varphi)$ are, respectively, as in (3.2), (3.7) and (3.6). A distribution $f \in (\mathrm{BMO}_{\varphi,\,L}(\mathcal{X}))^*$ is said to be in the *space* $H^{M,\,\epsilon}_{\varphi,\,\mathrm{mol}}(\mathcal{X})$ if there exist $\{\lambda_j\}_j \subset \mathbb{C}$ and a sequence $\{\beta_j\}_j$ of $(\varphi,\,M,\,\epsilon)$-molecules such that $f = \sum_j \lambda_j \beta_j$ in $(\mathrm{BMO}_{\varphi,\,L}(\mathcal{X}))^*$ and

$$\sum_j \varphi\left(B_j, \frac{|\lambda_j|}{\|\chi_{B_j}\|_{L^\varphi(\mathcal{X})}}\right) < \infty,$$

where, for each j, β_j is associated with the ball B_j. Moreover, for any $f \in H^{M,\,\epsilon}_{\varphi,\,\mathrm{mol}}(\mathcal{X})$, its *quasi-norm* is defined by $\|f\|_{H^{M,\,\epsilon}_{\varphi,\,\mathrm{mol}}(\mathcal{X})} := \inf\{\Lambda(\{\lambda_j\beta_j\}_j)\}$, where $\Lambda(\{\lambda_j\beta_j\}_j)$ is as in (3.11) and the infimum is taken over all possible decompositions of f as above.

For all $f \in L^2(\mathcal{X})$ and $x \in \mathcal{X}$, define the Lusin area function associated with the Poisson semigroup of L by

$$S_P f(x) := \left\{\int_{\Gamma(x)} \left|t\sqrt{L}e^{-t\sqrt{L}} f(y)\right|^2 \frac{d\mu(y)\,dt}{V(x,t)t}\right\}^{1/2}. \tag{3.12}$$

Similar to Definition 3.8, the space $H_{\varphi,\,S_P}(\mathcal{X})$ can be defined by the following way.

Definition 3.11. Let L satisfy Assumptions $(L)_3$ and $(L)_4$, and φ be as in Definition 3.2. A function $f \in H^2(\mathcal{X})$ is said to be in $\widetilde{H}_{\varphi,\,S_P}(\mathcal{X})$ if $S_P(f) \in L^{\varphi}(\mathcal{X})$; moreover, define

$$\|f\|_{H_{\varphi,\,S_P}(\mathcal{X})} := \|S_P(f)\|_{L^{\varphi}(\mathcal{X})} := \inf\left\{\lambda \in (0,\infty):\ \int_{\mathcal{X}} \varphi\left(x, \frac{S_P(f)(x)}{\lambda}\right) d\mu(x) \le 1\right\}.$$

The *Musielak-Orlicz-Hardy space* $H_{\varphi,\,S_P}(\mathcal{X})$ is defined to be the completion of $\widetilde{H}_{\varphi,\,S_P}(\mathcal{X})$ in the quasi-norm $\|\cdot\|_{H_{\varphi,\,S_P}(\mathcal{X})}$.

Then we have the following equivalent characterizations for $H_{\varphi,\,L}(\mathcal{X})$, which is just [91, Theorems 5.5 and 5.7].

Theorem 3.12. *Let L satisfy Assumptions $(L)_3$ and $(L)_4$, φ be as in Definition 3.2 with $\varphi \in \mathbb{RH}_{2/[2-I(\varphi)]}(\mathcal{X})$ and $I(\varphi)$ as in (3.5), $M \in \mathbb{N}$ with $M > \frac{n}{2}[\frac{q(\varphi)}{i(\varphi)} - \frac{1}{2}]$ and $\epsilon \in (n[\frac{q(\varphi)}{i(\varphi)} - \frac{1}{2}], \infty)$, where n, $q(\varphi)$ and $i(\varphi)$ are, respectively, as in (3.2), (3.7) and (3.6). Then the spaces $H_{\varphi,\,L}(\mathcal{X})$, $H^M_{\varphi,\,\mathrm{at}}(\mathcal{X})$, $H^{M,\,\epsilon}_{\varphi,\,\mathrm{mol}}(\mathcal{X})$ and $H_{\varphi,\,S_P}(\mathcal{X})$ coincide with equivalent quasi-norms.*

Remark 3.13. If L satisfies Assumption $(L)_3$ and the reinforced $(p_L,\, p'_L,\, 1)$ off-diagonal estimates on balls with $p_L \in [1,2)$ (see Assumptions $(L)_6$ below for its definition), the atomic characterization of the space $H_{\varphi,\,L}(\mathcal{X})$ was also obtained in [15] via replacing the assumption that $\varphi \in \mathbb{RH}_{2/[2-I(\varphi)]}(\mathcal{X})$ in Theorem 3.12 by the weaker assumption that $\varphi \in \mathbb{RH}_{(p'_L/I(\varphi))'}(\mathcal{X})$ (see [15, Theorem 5.4]).

Let $H^{\widetilde{M}}_{\varphi,\,\mathrm{at,\,fin}}(\mathcal{X})$ and $H^{M,\,\epsilon}_{\varphi,\,\mathrm{mol,\,fin}}(\mathcal{X})$, respectively, denote the *set of all finite linear combinations of $(\varphi,\, \widetilde{M})$-atoms and $(\varphi,\, M,\, \epsilon)$-molecules*. Then we have the following dual theorem for the spaces $H_{\varphi,\,L}(\mathcal{X})$ and $\mathrm{BMO}^M_{\varphi,\,L}(\mathcal{X})$, which is just [91, Theorem 4.16].

Theorem 3.14. *Let L satisfy Assumptions $(L)_3$ and $(L)_4$, φ be as in Definition 3.2 with $\varphi \in \mathbb{RH}_{2/[2-I(\varphi)]}(\mathcal{X})$ and $I(\varphi)$ as in (3.5), $M \in \mathbb{N}$ with $M > \frac{n}{2}[\frac{q(\varphi)}{i(\varphi)} - \frac{1}{2}]$ and $\widetilde{M} \in \mathbb{N}$ with $\widetilde{M} > M + \frac{N}{4} + \frac{nq(\varphi)}{2i(\varphi)}$, where n, N, $q(\varphi)$ and $i(\varphi)$ are, respectively, as in (3.2), (3.3), (3.7) and (3.6). Then the dual space of $H_{\varphi,\,L}(\mathcal{X})$, $(H_{\varphi,\,L}(\mathcal{X}))^*$, coincides with the space $\mathrm{BMO}^M_{\varphi,\,L}(\mathcal{X})$ in the following sense:*

(i) Let $g \in \mathrm{BMO}^M_{\varphi,\,L}(\mathcal{X})$. Then the linear functional ℓ, which is initially defined on $H^{\widetilde{M}}_{\varphi,\,\mathrm{at,\,fin}}(\mathcal{X})$ by

$$\ell(f) := \langle g, f\rangle, \tag{3.13}$$

has a unique extension to $H_{\varphi,\,L}(\mathcal{X})$ with $\|\ell\|_{(H_{\varphi,\,L}(\mathcal{X}))^*} \le C\|g\|_{\mathrm{BMO}^M_{\varphi,\,L}(\mathcal{X})}$, where C is a positive constant independent of g.

(ii) Conversely, let $\epsilon \in (n[q(\varphi)/i(\varphi) - 1/2], \infty)$. Then, for any $\ell \in (H_{\varphi,\,L}(\mathcal{X}))^*$, there exists $g \in \mathrm{BMO}^M_{\varphi,\,L}(\mathcal{X})$ such that (3.13) holds true for all $f \in H^{M,\,\epsilon}_{\varphi,\,\mathrm{mol,\,fin}}(\mathcal{X})$ and

$$\|g\|_{\mathrm{BMO}^M_{\varphi,\,L}(\mathcal{X})} \le C\|\ell\|_{(H_{\varphi,\,L}(\mathcal{X}))^*},$$

where C is a positive constant independent of ℓ.

Remark 3.15. When φ is as in (3.9) with $\omega \equiv 1$ and Φ concave, the Musielak-Orlicz-Hardy space $H_{\varphi,\,L}(\mathcal{X})$ is just the Orlicz-Hardy space $H_{\Phi,\,L}(\mathcal{X})$ introduced in [57]. Moreover, Theorems 3.12 and 3.14, respectively, completely cover [57, Theorems 4.1, 5.1 and 5.2] by taking φ as in (3.9) with $\omega \equiv 1$ and Φ being concave.

3.3 Musielak-Orlicz-Hardy spaces associated with Schrödinger operators

In this subsection, we recall some results for the Musielak-Orlicz-Hardy space $H_{\varphi,\,L}(\mathbb{R}^n)$ associated with the Schrödinger operator L, established in [15, 91]. Let $\mathcal{X} := \mathbb{R}^n$ and

$$L := -\Delta + V \tag{3.14}$$

be a Schrödinger operator, where $0 \le V \in L^1_{\mathrm{loc}}(\mathbb{R}^n)$.

Since V is a nonnegative function, from the Feynman-Kac formula, we deduce that the kernel of the semigroup e^{-tL}, h_t, satisfies that, for all $x,\, y \in \mathbb{R}^n$ and $t \in (0,\infty)$,

$$0 \le h_t(x,y) \le (4\pi t)^{-n/2} \exp\left\{-\frac{|x-y|^2}{4t}\right\}.$$

It is easy to see that L satisfies Assumptions $(L)_3$ and $(L)_4$.

For all $f \in L^2(\mathbb{R}^n)$ and $x \in \mathbb{R}^n$, let

$$\mathcal{N}_h(f)(x) := \sup_{y\in B(x,t),\, t\in(0,\infty)} \left|e^{-t^2L}(f)(y)\right|,$$

$$\mathcal{N}_P(f)(x) := \sup_{y\in B(x,t),\, t\in(0,\infty)} \left|e^{-t\sqrt{L}}(f)(y)\right|,$$

$\mathcal{R}_h(f)(x) := \sup_{t\in(0,\infty)} |e^{-t^2L}(f)(x)|$ and $\mathcal{R}_P(f)(x) := \sup_{t\in(0,\infty)} |e^{-t\sqrt{L}}(f)(x)|$.

Then we have the following Musielak-Orlicz Hardy spaces associated with L, which were introduced in [91].

Definition 3.16. Let L be as in (3.14) and φ as in Definition 3.2. A function $f \in H^2(\mathbb{R}^n)$ is said to be in $\widetilde{H}_{\varphi,\mathcal{N}_h}(\mathbb{R}^n)$ if $\mathcal{N}_h(f) \in L^{\varphi}(\mathbb{R}^n)$; moreover, let $\|f\|_{H_{\varphi,\mathcal{N}_h}(\mathbb{R}^n)} := \|\mathcal{N}_h(f)\|_{L^{\varphi}(\mathbb{R}^n)}$. The *Musielak-Orlicz-Hardy space* $H_{\varphi,\mathcal{N}_h}(\mathbb{R}^n)$ is defined to be the completion of $\widetilde{H}_{\varphi,\mathcal{N}_h}(\mathbb{R}^n)$ with respect to the quasi-norm $\|\cdot\|_{H_{\varphi,\mathcal{N}_h}(\mathbb{R}^n)}$.

The *spaces* $H_{\varphi,\mathcal{N}_P}(\mathbb{R}^n)$, $H_{\varphi,\mathcal{R}_h}(\mathbb{R}^n)$ and $H_{\varphi,\mathcal{R}_P}(\mathbb{R}^n)$ are defined in a similar way.

For the space $H_{\varphi,\,L}(\mathbb{R}^n)$, we have the following equivalent characterizations, which were obtained in [15, Theorem 8.3].

Theorem 3.17. *Let L be as in* (3.14) *and φ as in Definition* 3.2. *Then the spaces $H_{\varphi,\,L}(\mathbb{R}^n)$, $H_{\varphi,\mathcal{N}_h}(\mathbb{R}^n)$, $H_{\varphi,\mathcal{N}_P}(\mathbb{R}^n)$, $H_{\varphi,\mathcal{R}_h}(\mathbb{R}^n)$, $H_{\varphi,\mathcal{R}_P}(\mathbb{R}^n)$ and $H_{\varphi,\,S_P}(\mathbb{R}^n)$ coincide with equivalent quasi-norms.*

Now we recall the definition of the Musielak-Orlicz-Hardy space $H_\varphi(\mathbb{R}^n)$ introduced by Ky [70] and the weak Musielak-Orlicz-Hardy space $WH_\varphi(\mathbb{R}^n)$ introduced in [15].

Definition 3.18. Let φ be as in Definition 3.2 and $\psi \in \mathcal{S}(\mathbb{R}^n)$ satisfy $\operatorname{supp}\psi \subset B(0,1)$ and

$$\int_{\mathbb{R}^n} \psi(x)\,dx = 1.$$

(i) The *Musielak-Orlicz-Hardy space* $H_\varphi(\mathbb{R}^n)$ is defined to be the space of all distributions $f \in \mathcal{S}'(\mathbb{R}^n)$ such that $\mathcal{G}(f) \in L^\varphi(\mathbb{R}^n)$ with $\|f\|_{H_\varphi(\mathbb{R}^n)} := \|\mathcal{G}(f)\|_{L^\varphi(\mathbb{R}^n)}$, where $\mathcal{S}'(\mathbb{R}^n)$ and $\mathcal{G}(f)$ denote, respectively, the *dual space* of the space of all Schwartz functions (namely, the *space of tempered distributions*) and the *grand maximal function of* f.

(ii) The *weak Musielak-Orlicz-Hardy space* $WH_\varphi(\mathbb{R}^n)$ is defined to be the set $\{f \in S'(\mathbb{R}^n) : \|f\|_{WH_\varphi(\mathbb{R}^n)} < \infty\}$, where

$$\begin{aligned}
&\|f\|_{WH_\varphi(\mathbb{R}^n)}\\
&:= \left\| \sup_{t\in(0,\infty)} \psi_t * f \right\|_{WL^\varphi(\mathbb{R}^n)}\\
&:= \inf\left\{\lambda \in (0,\infty) : \sup_{\eta\in(0,\infty)} \varphi\left(\left\{x \in \mathbb{R}^n : \sup_{t\in(0,\infty)} |(\psi_t * f)(x)| > \eta\right\}, \frac{\eta}{\lambda}\right) \le 1\right\}
\end{aligned}$$

and, for all $t \in (0,\infty)$ and $x \in \mathbb{R}^n$, $\psi_t(x) := t^{-n}\psi(\frac{x}{t})$.

Then we have the following boundedness of the Riesz transforms $\nabla L^{-1/2}$, which was obtained in [15, Theorems 8.5 and 8.6].

Theorem 3.19. *Let φ and L be as in Definition 3.2 and (3.14), respectively. Assume that $\nabla L^{-1/2}$ is bounded on $L^r(\mathbb{R}^n)$ for all $r \in (1, p_0)$ with some $p_0 \in (2,\infty)$. Let $i(\varphi)$, $q(\varphi)$ and $r(\varphi)$ be, respectively, as in (3.5), (3.7) and (3.8).*

(i) If $r(\varphi) > (p_0/I(\varphi))'$, then $\nabla L^{-1/2}$ is bounded from $H_{\varphi,L}(\mathbb{R}^n)$ into $L^\varphi(\mathbb{R}^n)$.

(ii) If $i(\varphi) \in (\frac{n}{n+1}, 1]$, $\frac{q(\varphi)}{i(\varphi)} \in (1, \frac{n+1}{n})$ and $r(\varphi) > (p_0/q(\varphi))'$, then $\nabla L^{-1/2}$ is bounded from $H_{\varphi,L}(\mathbb{R}^n)$ into $H_\varphi(\mathbb{R}^n)$.

(iii) Assume further that $\varphi \in \mathbb{A}_1(\mathbb{R}^n) \cap \mathbb{RH}_{(p_0/q(\varphi))'}(\mathbb{R}^n)$. If $i(\varphi) = \frac{n}{n+1}$ is attainable and $I(\varphi) \in (0,1)$, then $\nabla L^{-1/2}$ is bounded from $H_{\varphi,L}(\mathbb{R}^n)$ to $WH_\varphi(\mathbb{R}^n)$.

Remark 3.20. (i) When φ is as in Definition 3.2 and satisfies the additional assumption $\varphi \in \mathbb{RH}_{2/[2-I(\varphi)]}(\mathbb{R}^n)$, then Theorem 3.17 was obtained in [91, Theorem 7.4]. Moreover, when φ is as in (3.9) with $\omega \equiv 1$ and Φ satisfying that Φ is concave on $(0,\infty)$ and there exist $q_1, q_2 \in (0,\infty)$ such that $q_1 < 1 < q_2$ and $[\Phi(t^{q_2})]^{q_1}$ is a convex function on $(0,\infty)$, then Theorem 3.17 was obtained in [57, Theorem 6.4].

(ii) Theorem 3.19(i) completely covers [57, Theorem 6.2] by taking φ as in (3.9) with $\omega \equiv 1$ and Φ being concave, and Theorem 3.19(ii) completely covers [57, Theorem 6.3] by taking φ as in (3.9) with $\omega \equiv 1$, Φ being concave and $p_\Phi \in (\frac{n}{n+1}, 1]$, where p_Φ is as in (3.8). Furthermore, Theorem 3.19(i) completely covers [48, Theorem 8.6] by taking $\varphi(x,t) := t$ for all $x \in \mathbb{R}^n$ and $t \in [0,\infty)$.

(iii) Theorem 3.19(iii) completely covers [21, Corollary 1.1] by taking $\varphi(x,t) := t^{n/(n+1)}$ for all $x \in \mathbb{R}^n$ and $t \in [0,\infty)$.

(iv) The Musielak-Orlicz VMO-type space associated with L is not introduced and studied so far.

4 Musielak-Orlicz type spaces associated with operators satisfying bounded H_∞-functional calculus

Let L be the second order divergence form elliptic operator on $\mathbb{R}^n$ with complex bounded measurable coefficients. Hofmann and Mayboroda [49] introduced the Hardy space $H_L^1(\mathbb{R}^n)$ associated with L via molecules and the BMO-type space $\mathrm{BMO}_L(\mathbb{R}^n)$, and further proved that the dual space of $H_L^1(\mathbb{R}^n)$ is just the space $\mathrm{BMO}_{L^*}(\mathbb{R}^n)$, where L^* denotes the *adjoint operator* of L in $L^2(\mathbb{R}^n)$. Moreover, Hofmann and Mayboroda [49] established several equivalent characterizations of $H_L^1(\mathbb{R}^n)$ in terms of the maximal functions and the Lusin area function associated with the heat semigroup $\{e^{-tL}\}_{t>0}$ and the Poisson semigroup $\{e^{-t\sqrt{L}}\}_{t>0}$ of L. Based on this, Hofmann et al. [50] further developed a systematical theory of Sobolev and Hardy spaces associated with L including their Riesz transform characterization.

Let Φ satisfy Assumption $(\Phi)_3$ and ρ be as in (1.1). Jiang and Yang [55] introduced the Orlicz-Hardy space $H_{\Phi,\,L}(\mathbb{R}^n)$ via the Lusin area function associated with L and the BMO-type space $\mathrm{BMO}_{\rho,\,L}(\mathbb{R}^n)$, and proved that the dual space of $H_{\Phi,\,L}(\mathbb{R}^n)$ is $\mathrm{BMO}_{\rho,\,L^*}(\mathbb{R}^n)$. Furthermore, the ρ-Carleson measure characterization and the John-Nirenberg inequality for the space $\mathrm{BMO}_{\rho,\,L}(\mathbb{R}^n)$ were established in [55]. The characterizations of $H_{\Phi,\,L}(\mathbb{R}^n)$ in terms of molecules, various maximal functions associated with the heat semigroup and the Poisson semigroup of L, and the Lusin area function associated with the Poisson semigroup of L were also established in [55]. As applications, Jiang and Yang [55] proved that the Riesz transform $\nabla L^{-1/2}$ and the Littlewood-Paley g-function g_L are bounded from $H_{\Phi,\,L}(\mathbb{R}^n)$ to $L^\Phi(\mathbb{R}^n)$, and further proved that $\nabla L^{-1/2}$ is bounded from $H_{\Phi,\,L}(\mathbb{R}^n)$ to the Orlicz-Hardy space $H_\Phi(\mathbb{R}^n)$ introduced in [85] when $p_\Phi \in (n/(n+1), 1]$ and the corresponding fractional integral $L^{-\gamma}$ for some $\gamma \in (0,\infty)$ is bounded from $H_{\Phi,\,L}(\mathbb{R}^n)$ to $H_{\widetilde{\Phi},\,L}(\mathbb{R}^n)$, where $\widetilde{\Phi}$ satisfies the same property as Φ and is determined by Φ and γ. Moreover, Jiang and Yang [56] introduced the Orlicz VMO-type space $\mathrm{VMO}_{\rho,\,L}(\mathbb{R}^n)$ and showed that it is just the predual space of $B_{\Phi,\,L^*}(\mathbb{R}^n)$, where $B_{\Phi,\,L^*}(\mathbb{R}^n)$ denotes the Banach completion of the Orlicz-Hardy space $H_{\Phi,\,L^*}(\mathbb{R}^n)$. We remark that, as special cases, a theory of Hardy spaces associated with L was also obtained in [55, 56], which partly overlaps the results for $p \in (0,1)$ in [50], but was obtained independently. Furthermore, (weighted local) Orlicz-Hardy spaces associated with a second order divergence form elliptic operator having some boundary conditions on the strongly Lipschitz domain of $\mathbb{R}^n$ were studied in [17, 88, 89]. It is worth pointing out that the strongly Lipschitz domain of $\mathbb{R}^n$ is a special space of homogeneous type in the sense of Coifman and

Weiss [28]. Recall that the Hardy spaces on strongly Lipschitz domains associated with the Laplace operator having some boundary conditions were originally and systematically studied by Chang et al. in [23, 24, 25], Auscher and Russ [7] and Duong et al. [34]. Moreover, the theory of Hardy spaces associated with the higher order elliptic operator on $\mathbb{R}^n$ was studied in [20, 21, 33].

Let L be an operator of type ω having a bounded H_∞ functional calculus on $L^2(\mathcal{X})$ and satisfying the Davies-Gaffney estimates, where $\mathcal{X}$ is a metric measure space endowed with a doubling measure. Duong and Li [35] introduced the Hardy space $H_L^p(\mathcal{X})$, and the Lipschitz space $\Lambda_L^\alpha(\mathcal{X})$ which was further proved to be the dual space of $H_{L^*}^p(\mathcal{X})$. Furthermore, the atomic and the molecular characterizations of $H_L^p(\mathcal{X})$ were obtained in [35]. It is worth pointing out that one needs an additional assumption for L when establishing the atomic characterization of $H_L^p(\mathcal{X})$. Moreover, the Orlicz-Hardy space $H_{\Phi,\,L}(\mathcal{X})$, which is characterized via the molecule, and its dual space were studied in [1], where the Orlicz function Φ satisfies Assumption $(\Phi)_3$. Based on this, in [66], the VMO-type space $\mathrm{VMO}_{\rho,\,L}(\mathcal{X})$ was studied and further proved to be the predual space of $B_{\Phi,\,L^*}(\mathcal{X})$, where ρ is as (1.1), L^* denotes the adjoint operator of L in $L^2(\mathcal{X})$ and $B_{\Phi,\,L^*}(\mathcal{X})$ the Banach completion of the Orlicz-Hardy space $H_{\Phi,\,L^*}(\mathcal{X})$.

Assume further that L satisfies the reinforced (p_L, q_L, k) off-diagonal estimates on balls, where $p_L \in [1,2)$, $q_L \in (2,\infty]$ and $k \in \mathbb{N}$ (see Assumption $(L)_6$ in Section 4.1 below). Let $\varphi : \mathcal{X} \times [0,\infty) \to [0,\infty)$ be a growth function and $\varphi \in \mathbb{RH}_{(q_L/I(\varphi))'}(\mathcal{X})$, where $(q_L/I(\varphi))'$ denotes the conjugate exponent of $q_L/I(\varphi)$. Bui et al. [15] introduced the Musielak-Orlicz-Hardy space $H_{\varphi,\,L}(\mathcal{X})$, via the Lusin-area function associated with L, and further established its molecular characterization. In particular, when L is nonnegative self-adjoint and satisfies the Davies-Gaffney estimates, the atomic characterization of $H_{\varphi,\,L}(\mathcal{X})$ was also presented in [15]. Moreover, as a special case of this setting, for the Musielak-Orlicz-Hardy space $H_{\varphi,\,L}(\mathbb{R}^n)$ associated with the second order elliptic operator in divergence form on $\mathbb{R}^n$ or the Schrödinger operator $L := -\Delta + V$ with $0 \le V \in L^1_{\mathrm{loc}}(\mathbb{R}^n)$, its several equivalent characterizations, in terms of several non-tangential and radial maximal functions, were also established in [15]. Furthermore, the boundedness of the Riesz transform $\nabla L^{-1/2}$ on $H_{\varphi,\,L}(\mathbb{R}^n)$ was also obtained in [15]. It is worth pointing out that all the results in [15] for the space $H_{\varphi,\,L}(\mathbb{R}^n)$ with L being a second order elliptic operator in divergence form on $\mathbb{R}^n$ generalized the corresponding results in [55] by taking φ as (3.9) with $\omega \equiv 1$ and Φ being concave. Moreover, all the results in [15] for $H_{\varphi,\,L}(\mathcal{X})$ with L being an nonnegative self-adjoint operator and satisfying the Davies-Gaffney estimates generalized the corresponding results in [91] by enlarging the range of the weights.

Let L be an operator of type ω, with $\omega \in [0, \frac{\pi}{2})$, having a bounded H_∞ functional calculus in $L^2(\mathcal{X})$ and satisfying some off-diagonal estimates on balls. In this section, we state the results for the Musielak-Orlicz-Hardy space $H_{\varphi,\,L}(\mathcal{X})$ established in [15].

4.1 Assumptions on operators

Before giving the assumptions on the considered operators L, we first recall some notions of bounded holomorphic functional calculus introduced by McIntosh [71].

Let $\omega \in [0, \pi)$. A closed operator T in $L^2(\mathcal{X})$ is said to be of *type* ω if

(i) the *spectrum* of T, $\sigma(T)$, is contained in S_ω;

(ii) for each $\theta \in (\omega, \pi)$, there exists a nonnegative constant C such that, for all $z \in \mathbb{C} \setminus S_\theta$,

$$\|(T - zI)^{-1}\|_{\mathcal{L}(L^2(\mathcal{X}))} \le C|z|^{-1}.$$

For $\mu \in [0, \pi)$ and $\sigma, \tau \in (0, \infty)$, let

$$H(S_\mu^0) := \{f : f \text{ is a holomorphic function on } S_\mu^0\},$$

$$H_\infty(S_\mu^0) := \left\{f \in H(S_\mu^0) : \|f\|_{L^\infty(S_\mu^0)} < \infty\right\}$$

and

$$\Psi_{\sigma,\tau}(S_\mu^0) := \{f \in H(S_\mu^0) : \text{there exists a positive constant } C \text{ such that, for all } \xi \in S_\mu^0, |f(\xi)| \le C\inf\{|\xi|^\sigma, |\xi|^{-\tau}\}\}.$$

It is known that every one-to-one operator T of type ω in $L^2(\mathcal{X})$ has a unique holomorphic functional calculus (see, for example, [71]). More precisely, let T be a one-to-one operator of type ω, with $\omega \in [0, \pi)$, $\mu \in (\omega, \pi)$, $\sigma, \tau \in (0, \infty)$ and $f \in \Psi_{\sigma,\tau}(S_\mu^0)$. The *function of the operator* T, $f(T)$, can be defined by the H_∞-functional calculus in the following way,

$$f(T) := \frac{1}{2\pi i}\int_\Gamma (\xi I - T)^{-1} f(\xi)\, d\xi, \tag{4.1}$$

where $\Gamma := \{re^{i\nu} : \infty > r > 0\} \cup \{re^{-i\nu} : 0 < r < \infty\}$, $\nu \in (\omega, \mu)$, is a curve consisting of two rays parameterized anti-clockwise. It is known that $f(T)$ in (4.1) is independent of the choice of $\nu \in (\omega, \mu)$ and the integral in (4.1) is absolutely convergent in $\|\cdot\|_{\mathcal{L}(L^2(\mathcal{X}))}$ (see [71, 45]).

In what follows, we *always assume* $\omega \in [0, \pi/2)$. Then, it follows, from [45, Proposition 7.1.1], that, for every operator T of type ω in $L^2(\mathcal{X})$, $-T$ generates a holomorphic C_0-semigroup $\{e^{-zT}\}_{z\in S^0_{\pi/2-\omega}}$ on the open sector $S^0_{\pi/2-\omega}$ such that $\|e^{-zT}\|_{\mathcal{L}(L^2(\mathcal{X}))} \le 1$ for all $z \in S^0_{\pi/2-\omega}$ and, moreover, every *nonnegative self-adjoint operator* is of type 0.

Let $\Psi(S_\mu^0) := \cup_{\sigma,\tau>0}\Psi_{\sigma,\tau}(S_\mu^0)$. It is well known that the above holomorphic functional calculus defined on $\Psi(S_\mu^0)$ can be extended to $H_\infty(S_\mu^0)$ via a limit process (see [71]). Recall that, for $\mu \in (0, \pi)$, the operator T is said to have a *bounded* $H_\infty(S_\mu^0)$ *functional calculus* in the Hilbert space $\mathcal{H}$, if there exists a positive constant C such that, for all $\psi \in H_\infty(S_\mu^0)$, $\|\psi(T)\|_{\mathcal{L}(\mathcal{H})} \le C\|\psi\|_{L^\infty(S_\mu^0)}$ and T is said to have a *bounded* H_∞ *functional calculus* in the Hilbert space $\mathcal{H}$ if there exists $\mu \in (0, \pi)$ such that T has a bounded $H_\infty(S_\mu^0)$ functional calculus.

For any given $f \in L^1_{\mathrm{loc}}(\mathcal{X})$, each ball $B \subset \mathcal{X}$ and $j \in \mathbb{Z}_+$, let

$$\fint_{U_j(B)} |f(x)|d\mu(x) := \frac{1}{\mu(2^j B)} \int_{U_j(B)} |f(x)|d\mu(x).$$

Now we recall the notion of $L^p - L^q$ off-diagonal estimates on balls, which was first introduced in [5].

Definition 4.1. Let $k \in \mathbb{N}$, p, $q \in [1,\infty]$ with $p \le q$, and $\{A_t\}_{t>0}$ be a family of sublinear operators. The family $\{A_t\}_{t>0}$ is said to satisfy *$L^p - L^q$ off-diagonal estimates on balls of order k*, denoted by $A_t \in \mathcal{O}_k(L^p - L^q)$, if there exist constants θ_1, $\theta_2 \in [0,\infty)$ and C, $c \in (0,\infty)$ such that, for all $t \in (0,\infty)$ and all balls $B \subset \mathcal{X}$ and $f \in L^p_{\mathrm{loc}}(\mathcal{X})$,

$$\left\{\fint_B |A_t(\chi_B f)(x)|^q d\mu(x)\right\}^{1/q} \le C\left\{\Upsilon\left(\frac{r_B}{t^{1/2k}}\right)\right\}^{\theta_2}\left\{\fint_B |f(x)|^p\,d\mu(x)\right\}^{1/p}$$

and, for all $j \in \mathbb{N}$ with $j \ge 3$,

$$\begin{aligned}&\left\{\fint_{U_j(B)} |A_t(\chi_B f)(x)|^q\,d\mu(x)\right\}^{1/q}\\ &\le C2^{j\theta_1}\left\{\Upsilon\left(\frac{2^j r_B}{t^{1/2k}}\right)\right\}^{\theta_2} e^{-c\frac{(2^j r_B)^{2k/(2k-1)}}{t^{1/(2k-1)}}}\left\{\fint_B |f(x)|^p\,d\mu(x)\right\}^{1/p}\end{aligned}$$

and

$$\begin{aligned}&\left\{\fint_B |A_t(\chi_{U_j(B)} f)(x)|^q\,d\mu(x)\right\}^{1/q}\\ &\le C2^{j\theta_1}\left\{\Upsilon\left(\frac{2^j r_B}{t^{1/2k}}\right)\right\}^{\theta_2} e^{-c\frac{(2^j r_B)^{2k/(2k-1)}}{t^{1/(2k-1)}}}\left\{\fint_{U_j(B)} |f(x)|^p\,d\mu(x)\right\}^{1/p},\end{aligned}$$

where $\Upsilon(s) := \max\{s, \frac{1}{s}\}$ for all $s \in (0,\infty)$.

Now, we make the following two assumptions on operators L.

Assumption $(L)_5$. Assume that the operator L is a one-to-one operator of type ω in $L^2(\mathcal{X})$ with $\omega \in [0,\pi/2)$, has dense range in $L^2(\mathcal{X})$ and a bounded H_∞-functional calculus in $L^2(\mathcal{X})$. □

Assumption $(L)_6$. Let $k \in \mathbb{N}$. Assume that there exist $p_L \in [1, 2)$ and $q_L \in (2,\infty]$, depending on L, such that the family $\{(tL)^\kappa e^{-tL}\}_{t>0}$, with $\kappa \in \mathbb{Z}_+$, satisfies the *reinforced (p_L, q_L, k) off-diagonal estimates on balls*, namely, for all $t \in (0,\infty)$ and p, $q \in (p_L, q_L)$ with $p \le q$, $(tL)^\kappa e^{-tL} \in \mathcal{O}_k(L^p - L^q)$. □

Now we recall some examples of operators satisfying Assumptions $(L)_5$ and $(L)_6$ as follows (see also [15, Remark 2.9(c)]).

(i) The second order divergence form elliptic operators with complex bounded coefficients as in [49] (see also (4.2) below).

(ii) The $2k$-*order homogeneous divergence form elliptic operators*

$$(-1)^k \sum_{|\alpha|=k=|\beta|} \partial^\beta \left(a_{\alpha,\beta}\partial^\alpha\right)$$

interpreted in the usual weak sense via a sesquilinear form, with complex bounded measurable coefficients $a_{\alpha,\,\beta}$ for all multi-indices α and β (see, for example, [20]).

(iii) The *Schrödinger operator* $-\Delta + V$ on $\mathbb{R}^n$ with $0 \le V \in L^1_{\mathrm{loc}}(\mathbb{R}^n)$ (see, for example, [48, 58]).

(iv) The *Schrödinger operator* $-\Delta + V$ on $\mathbb{R}^n$ with the suitable real potential V as in [2].

(v) The nonnegative self-adjoint operators satisfying *Gaussian upper bounds*, namely, there exist positive constants C and c such that, for all $x,\, y \in \mathcal{X}$ and $t \in (0, \infty)$,

$$|p_t(x,y)| \le \frac{C}{V(x, t^{1/2k})} \exp\left\{-c\frac{[d(x,y)]^{2k/(2k-1)}}{t^{1/(2k-1)}}\right\},$$

where p_t is the associated kernel of e^{-tL} and $k \in \mathbb{N}$.

4.2 Musielak-Orlicz-Hardy spaces associated with L

In this subsection, we recall some results for the Musielak-Orlicz-Hardy space $H_{\varphi,\,L}(\mathcal{X})$ associated with L obtained in [15].

Definition 4.2. Let φ be as in Definition 3.2 and L satisfy Assumptions $(L)_5$ and $(L)_6$. Assume that p_L and q_L are as in Assumption $(L)_6$. A function $f \in L^p(\mathcal{X})$ with $p \in (p_L, q_L)$ is said to be in $\widetilde{H}_{\varphi,\,L,\,p}(\mathcal{X})$ if $S_L(f) \in L^\varphi(\mathcal{X})$ and, moreover, define

$$\begin{aligned}\|f\|_{H_{\varphi,\,L,\,p}(\mathcal{X})} &:= \|S_L f\|_{L^\varphi(\mathcal{X})}\\ &:= \inf\left\{\lambda \in (0,\infty) : \int_{\mathcal{X}} \varphi\left(x, \frac{S_L(f)(x)}{\lambda}\right) d\mu(x) \le 1\right\}.\end{aligned}$$

The *Musielak-Orlicz-Hardy space* $H_{\varphi,\,L,\,p}(\mathcal{X})$ is defined to be the completion of $\widetilde{H}_{\varphi,\,L,\,p}(\mathcal{X})$ with respect to the quasi-norm $\|\cdot\|_{H_{\varphi,\,L,\,p}(\mathcal{X})}$.

In what follows, for the simplicity of the notation, we *write* $H_{\varphi,\,L}(\mathcal{X}) := H_{\varphi,\,L,\,2}(\mathcal{X})$.

Definition 4.3. Let φ be as in Definition 3.2, L satisfy Assumptions $(L)_5$ and $(L)_6$, p_L and q_L be as in Assumption $(L)_6$. Let $q \in (p_L, q_L)$, $M \in \mathbb{N}$ and $\epsilon \in (0, \infty)$.

(I) A function $\alpha \in L^q(\mathcal{X})$ is called a $(\varphi,\, q,\, M,\, \epsilon)_L$-*molecule* associated with the ball $B \subset \mathcal{X}$ if, for each $k \in \{0,\, \ldots,\, M\}$ and $j \in \mathbb{Z}_+$, it holds true that

$$\left\|\left(r_B^{-2m}L^{-1}\right)^k \alpha\right\|_{L^q(S_j(B))} \le 2^{-j\epsilon}[\mu(2^j B)]^{1/q}\|\chi_B\|^{-1}_{L^\varphi(\mathcal{X})}.$$

Moreover, if α is a $(\varphi, q, M, \epsilon)_L$-molecule for all $q \in (p_L, q_L)$, then α is called a $(\varphi, M, \epsilon)_L$*-molecule.*

(II) The equality $f = \sum_j \lambda_j \alpha_j$ is said to be a *molecular* $(\varphi, r, q, M, \epsilon)$-*representation* of f for some $r \in (p_L, q_L)$, if each α_j is a $(\varphi, q, M, \epsilon)_L$-molecule associated to the ball $B_j \subset \mathcal{X}$, the summation converges in $L^r(\mathcal{X})$ and $\{\lambda_j\}_j \subset \mathbb{C}$ satisfies that $\sum_j \varphi(B_j, |\lambda_j| \|\chi_{B_j}\|^{-1}_{L^\varphi(\mathcal{X})}) < \infty$. Let

$$\widetilde{H}^{M,q,\epsilon}_{\varphi,L}(\mathcal{X}) := \{f : f \text{ has a molecular } (\varphi, r, q, M, \epsilon)\text{-representation for some } r \in (p_L, q_L)\}$$

with the *quasi-norm* $\|\cdot\|_{H^{M,q,\epsilon}_{\varphi,L}(\mathcal{X})}$ given by setting, for all $f \in \widetilde{H}^{M,q,\epsilon}_{\varphi,L}(\mathcal{X})$,

$$\|f\|_{H^{M,q,\epsilon}_{\varphi,L}(\mathcal{X})} := \inf\left\{\Lambda(\{\lambda_j\alpha_j\}_j) : f = \sum_j \lambda_j\alpha_j \text{ is a molecular } (\varphi, r, q, M, \epsilon)\text{-representation}\right\},$$

where $\Lambda(\{\lambda_j\alpha_j\}_j)$ is as in (3.11).

The *molecular Musielak-Orlicz-Hardy space* $H^{M,q,\epsilon}_{\varphi,L}(\mathcal{X})$ is then defined as the completion of $\widetilde{H}^{M,q,\epsilon}_{\varphi,L}(\mathcal{X})$ with respect to the quasi-norm $\|\cdot\|_{H^{M,q,\epsilon}_{\varphi,L}(\mathcal{X})}$.

For the space $H_{\varphi,L}(\mathcal{X})$, we have the following molecular characterization, which is just [15, Theorems 4.8 and 4.9].

Theorem 4.4. *Let* φ *be as in Definition* 3.2 *and* L *satisfy Assumptions* $(L)_5$ *and* $(L)_6$.

(i) *Assume that* $q \in [2, q_L) \cap ([r(\varphi)]' I(\varphi), q_L)$, $M \in \mathbb{N}$ *with*

$$M > \frac{n}{2k}\left[\frac{q(\varphi)}{i(\varphi)} + \frac{\theta_1}{n} - \frac{2}{nq}\right],$$

$\epsilon \in (nq(\varphi)/i(\varphi), \infty)$, *where* q_L, $r(\varphi)$, $I(\varphi)$, $q(\varphi)$ *and* $i(\varphi)$ *are, respectively, as in Assumption* $(L)_6$, (3.8), (3.4), (3.7) *and* (3.6). *Then* $H_{\varphi,L}(\mathcal{X})$ *and* $H^{M,q,\epsilon}_{\varphi,L}(\mathcal{X})$ *coincide with equivalent quasi-norms.*

(ii) *Assume that* $\varphi \in \mathbb{RH}_{(q_L/I(\varphi))'}(\mathcal{X})$. *Then, for any* $s \in (p_L, q_L)$, *the spaces* $H_{\varphi,L}(\mathcal{X})$ *and* $H_{\varphi,L,s}(\mathcal{X})$ *coincide with equivalent quasi-norms.*

Remark 4.5. Assume that the operator L satisfies Assumptions $(L)_4$ and $(L)_5$. When the Orlicz function Φ satisfies Assumption $(\Phi)_3$, the Orlicz-Hardy space $H_{\Phi,L}(\mathcal{X})$ was introduced in [1] and its molecular characterization was also established in [1, Theorem 3.5]. Moreover, Anh and Li [1] also introduced the BMO-type space $\mathrm{BMO}_{\rho,L}(\mathcal{X})$ and proved that $\mathrm{BMO}_{\rho,L^*}(\mathcal{X})$ is the dual space of $H_{\Phi,L}(\mathcal{X})$ (see [1, Theorems 3.13 and 3.16]), where ρ is as in (1.1) and L^* denotes the adjoint operator of L in $L^2(\mathcal{X})$.

Specially, when $\Phi(t) := t^p$, with $p \in (0, 1]$, for all $t \in (0, \infty)$, the Hardy space $H^p_L(\mathcal{X})$ was introduced in [35] and its molecular characterization (see [35, Theorem 3.15]) and dual theory (see [35, Theorem 3.28]) were also established.

Remark 4.6. (i) For the BMO-type space and VMO-type space associated with L, we also have some results similar to those in Proposition 3.4, Theorems 3.6, 3.14, 2.18 and 2.21.

More precisely, assume that L and Φ, respectively, satisfy Assumptions $(L)_4$, $(L)_5$ and $(\Phi)_3$, and ρ is as in (1.1). The BMO-type space $\mathrm{BMO}_{\rho,\,L}(\mathcal{X})$ was introduced in [1] with replacing $\|\chi_B\|_{L^\varphi(\mathcal{X})}$ and the nonnegative self-adjoint operator in Definition 3.3 by $\rho(\mu(B))\mu(B)$ and L, respectively. In [1, Proposition 3.11], a property for the space $\mathrm{BMO}_{\rho,\,L}(\mathcal{X})$ similar to Proposition 3.4 was established. Similar to Theorem 3.6, a ρ-Carleson measure characterization for the space $\mathrm{BMO}_{\rho,\,L}(\mathcal{X})$ was obtained in [66, Theorem 3.3]. Moreover, in [1, Theorems 3.13 and 3.16], it was proved that $\mathrm{BMO}_{\rho,\,L^*}(\mathcal{X})$ is the dual space of $H_{\Phi,\,L}(\mathcal{X})$ (see also [66, Theorem 3.2]), where ρ is as in (1.1) and L^* denotes the adjoint operator of L in $L^2(\mathcal{X})$.

Similar to Definition 2.17, the VMO-type space $\mathrm{VMO}_{\rho,\,L}(\mathcal{X})$ was introduced in [66] and proved to be the predual space of $B_{\Phi,\,L^*}(\mathcal{X})$ (see [66, Theorem 4.2]), where $B_{\Phi,\,L^*}(\mathcal{X})$ denotes the Banach completion of the Orlicz-Hardy space $H_{\Phi,\,L^*}(\mathcal{X})$ and is defined by a way similar to Definition 2.20 (see also [66, Definition 4.3]).

(ii) The Musielak-Orlicz BMO-type and VMO-type spaces associated with L are not introduced and studied so far.

4.3 Applications to second order elliptic operators in divergence form

We begin this subsection with recalling some necessary notions and notation. Let A be an $n\times n$ matrix with entries $\{a_{i,\,j}\}_{i,\,j=1}^n \subset L^\infty(\mathbb{R}^n,\,\mathbb{C})$ satisfying the *ellipticity condition*, namely, there exist constants $0<\lambda_A\le\Lambda_A<\infty$ such that, for all $\xi,\,\zeta\in\mathbb{C}$ and almost every $x\in\mathbb{R}^n$,

$$\lambda_A|\xi|^2\le \Re e\langle A(x)\xi,\xi\rangle \quad\text{and}\quad |\langle A(x)\xi,\zeta\rangle|\le\Lambda_A|\xi||\zeta|,$$

where $\langle\cdot,\,\cdot\rangle$ denotes the *inner product* in $\mathbb{C}$ and $\Re e\,\xi$ denotes the *real part* of the complex number ξ. Then the *second order elliptic operator L in divergence form* is defined by

$$Lf := -\mathrm{div}(A\nabla f), \tag{4.2}$$

interpreted in the weak sense via a sesquilinear form. It is well known that there exists a positive constant $\omega\in[0,\,\pi/2)$ such that the operator L is of type ω on $L^2(\mathbb{R}^n)$ and L has a bounded H_∞-functional calculus on $L^2(\mathbb{R}^n)$ (see, for example, [50]). Moreover, let $(p_-(L),\,p_+(L))$ be the *interior of the maximal interval of exponents $p\in[1,\,\infty]$ for which the semigroup $\{e^{-tL}\}_{t>0}$, generated by L, is $L^p(\mathbb{R}^n)$ bounded.* By [3, Proposition 3.2] (see also [50, Lemma 2.25]), we conclude that, for all $p_-(L)<p\le q<p_+(L)$, $\{e^{-tL}\}_{t>0}$ satisfy the L^p-L^q off-diagonal estimates. Thus, L satisfies Assumptions $(L)_5$ and $(L)_6$ with $k=1$.

We also recall the definitions of some maximal functions associated with L from [49]. Let $f\in L^2(\mathbb{R}^n)$ and $x\in\mathbb{R}^n$. The *radial maximal functions* $\mathcal{R}_h$ and $\mathcal{R}_P$, respectively, associated with the heat semigroup and Poisson semigroup generated by L, are defined by setting, for all $f \in L^2(\mathbb{R}^n)$ and $x \in \mathbb{R}^n$,

$$\mathcal{R}_h(f)(x) := \sup_{t>0} \left\{ \frac{1}{t^n} \int_{B(x,t)} \left| e^{-t^2 L}(f)(y) \right|^2 dy \right\}^{\frac{1}{2}} \tag{4.3}$$

and

$$\mathcal{R}_P(f)(x) := \sup_{t>0} \left\{ \frac{1}{t^n} \int_{B(x,t)} \left| e^{-t\sqrt{L}}(f)(y) \right|^2 dy \right\}^{\frac{1}{2}}. \tag{4.4}$$

Similarly, the *non-tangential maximal functions* $\mathcal{N}_h$ and $\mathcal{N}_P$, respectively, associated with the heat semigroup and Poisson semigroup generated by L, are defined by setting, for all $f \in L^2(\mathbb{R}^n)$ and $x \in \mathbb{R}^n$,

$$\mathcal{N}_h(f)(x) := \sup_{(y,t)\in\Gamma(x)} \left\{ \frac{1}{t^n} \int_{B(y,t)} \left| e^{-t^2 L}(f)(z) \right|^2 dz \right\}^{\frac{1}{2}} \tag{4.5}$$

and

$$\mathcal{N}_P(f)(x) := \sup_{(y,t)\in\Gamma(x)} \left\{ \frac{1}{t^n} \int_{B(y,t)} \left| e^{-t\sqrt{L}}(f)(z) \right|^2 dz \right\}^{\frac{1}{2}}, \tag{4.6}$$

where $\Gamma(x) := \{(y,t) \in \mathbb{R}^n \times (0,\infty) : |x-y| < t\}$.

We also define the *Lusin-area functions*, S_h and S_P, associated respectively to the heat semigroup and Poisson semigroup, by setting, for all $f \in L^2(\mathbb{R}^n)$ and $x \in \mathbb{R}^n$,

$$S_h(f)(x) := \left\{ \int_{\Gamma(x)} \left| t\nabla e^{-t^2 L}(f)(y) \right|^2 \frac{dy\,dt}{t} \right\}^{\frac{1}{2}} \tag{4.7}$$

and

$$S_P(f)(x) := \left\{ \int_{\Gamma(x)} \left| t\nabla e^{-t\sqrt{L}}(f)(y) \right|^2 \frac{dy\,dt}{t} \right\}^{\frac{1}{2}}. \tag{4.8}$$

We now recall the Musielak-Orlicz-Hardy space defined via the above maximal functions as follows.

Definition 4.7. Let φ and L be respectively as in Definition 3.2 and (4.2), and S_P as in (4.8). The *S_P-adapted Musielak-Orlicz-Hardy space* $H_{\varphi,\,S_P}(\mathbb{R}^n)$ is defined to be the completion of the *set*

$$\left\{ f \in L^2(\mathbb{R}^n) : \ \|f\|_{H_{\varphi,\,S_P}(\mathbb{R}^n)} := \|S_P(f)\|_{L^\varphi(\mathbb{R}^n)} < \infty \right\}$$

with respect to the *quasi-norm* $\|\cdot\|_{H_{\varphi,\,S_P}(\mathbb{R}^n)}$.

In a similar way, the *S_h-adapted, $\mathcal{R}_h$-adapted, $\mathcal{R}_P$-adapted, $\mathcal{N}_h$-adapted* and *$\mathcal{N}_P$-adapted Musielak-Orlicz-Hardy spaces,* $H_{\varphi,\,S_h}(\mathbb{R}^n)$, $H_{\varphi,\,\mathcal{R}_h}(\mathbb{R}^n)$, $H_{\varphi,\,\mathcal{R}_P}(\mathbb{R}^n)$, $H_{\varphi,\mathcal{N}_h}(\mathbb{R}^n)$ and $H_{\varphi,\mathcal{N}_P}(\mathbb{R}^n)$ are also defined.

The following maximal function characterizations of the space $H_{\varphi, L}(\mathbb{R}^n)$ were obtained in [15, Proposition 7.4 and Theorem 7.5].

Theorem 4.8. *Let φ and L be, respectively, as in Definition 3.2 and (4.2), $\mathcal{R}_h$, $\mathcal{R}_P$, $\mathcal{N}_h$, $\mathcal{N}_P$, S_h and S_P, respectively, as in (4.3), (4.4), (4.5), (4.6), (4.7) and (4.8). Assume further that $\varphi \in \mathbb{RH}_{(\min\{p_+(L), q_+(L)\}/I(\varphi))'}(\mathcal{X})$, where $I(\varphi)$ is as in (3.5). Then $H_{\varphi,\mathcal{R}_h}(\mathbb{R}^n)$, $H_{\varphi,\mathcal{R}_P}(\mathbb{R}^n)$, $H_{\varphi,\mathcal{N}_h}(\mathbb{R}^n)$, $H_{\varphi,\mathcal{N}_P}(\mathbb{R}^n)$, $H_{\varphi, S_P}(\mathbb{R}^n)$, $H_{\varphi, S_h}(\mathbb{R}^n)$ and $H_{\varphi, L}(\mathbb{R}^n)$ coincide with equivalent quasi-norms.*

Moreover, the boundedness of the Riesz transform $\nabla L^{-1/2}$ associated with L on the space $H_{\varphi, L}(\mathbb{R}^n)$ was also obtained in [15, Theorems 7.8 and 7.11].

Theorem 4.9. *Let φ and L be, respectively, as in Definition 3.2 and (4.2). Let $I(\varphi)$, $i(\varphi)$, $q(\varphi)$ and $r(\varphi)$ be, respectively, as in (3.5), (3.6), (3.7) and (3.8).*

(i) *If $r(\varphi) > (\frac{\min\{p_+(L), q_+(L)\}}{I(\varphi)})'$, then $\nabla L^{-1/2}$ is bounded from $H_{\varphi, L}(\mathbb{R}^n)$ to $L^\varphi(\mathbb{R}^n)$.*

(ii) *If $i(\varphi) \in (\frac{n}{n+1}, 1]$, $q(\varphi) < \frac{(n+1)i(\varphi)}{n}$ and $r(\varphi) > (\frac{\min\{p_+(L), q_+(L)\}}{q(\varphi)})'$, then $\nabla L^{-1/2}$ is bounded from $H_{\varphi, L}(\mathbb{R}^n)$ to $H_\varphi(\mathbb{R}^n)$.*

(iii) *If $i(\varphi) = \frac{n}{n+1}$ is attainable, $\varphi \in \mathbb{A}_1(\mathbb{R}^n)$, $I(\varphi) \in (0, 1)$ and*

$$r(\varphi) > \left(\frac{\min\{p_+(L), q_+(L)\}}{q(\varphi)}\right)',$$

then $\nabla L^{-1/2}$ is bounded from $H_{\varphi, L}(\mathbb{R}^n)$ to $WH_\varphi(\mathbb{R}^n)$.

Remark 4.10. (i) When φ is as in (3.9) with $w \equiv 1$ and Φ being concave, Theorem 4.8 was obtained in [55, Theorem 5.2 and Corollary 5.1].

(ii) When φ is as in (3.9) with $w \equiv 1$ and Φ being concave, (i) and (ii) of Theorem 4.9 were obtained in [55, Theorems 7.1 and 7.4]. Moreover, when $\varphi(x,t) := t^{n/(n+1)}$ for all $x \in \mathbb{R}^n$ and $t \in [0,\infty)$, Theorem 4.9(iii) was obtained in [21, Theorem 1.2].

(iii) Recall that the strongly Lipschitz domain of $\mathbb{R}^n$ is a special space of homogeneous type in the sense of Coifman and Weiss [28]. Orlicz-Hardy spaces associated with a second order divergence form elliptic operator on the strongly Lipschitz domain of $\mathbb{R}^n$ with the boundary condition were studied in [17, 88, 89]. More precisely, let Ω be a strongly Lipschitz domain of $\mathbb{R}^n$, L a second order divergence form elliptic operator on $L^2(\Omega)$ with the Dirichlet or Neumann boundary condition, and the heat semigroup generated by L has the Gaussian property $(G_{\mathrm{diam}(\Omega)})$ with the regularity of their kernels measured by $\mu \in (0,1]$, where $\mathrm{diam}(\Omega)$ denotes the diameter of Ω. When Φ satisfies Assumption $(\Phi)_1$ with $p_\Phi^- \in (n/(n+\theta(L)), 1]$ replaced by $p_\Phi \in (n/(n+\mu), 1]$, the Orlicz-Hardy space $H_{\Phi, L}(\Omega)$ associated with L was introduced in [88, 89] in terms of the Lusin area function and its nontangential maximal function characterization was also obtained in [88, Theorems 1.6 and 1.12] and [89, Theorem 1.9]. Moreover, the similar results were obtained in [17] for the weighted local Orlicz-Hardy space associated with L on Ω. However, the (local) Musielak-Orlicz-Hardy spaces associated with L on Ω are not introduced and hence whether the above results hold true or not for such spaces is not clear so far. Moreover, the Musielak-Orlicz-Hardy space associated with the Schrödinger operator on the strongly Lipschitz domain of $\mathbb{R}^n$ is also not introduced and studied.

(iv) Assume that Φ satisfies Assumption $(\Phi)_3$ and ρ is as in (1.1). The BMO-type space $\mathrm{BMO}_{\rho,\,L}(\mathbb{R}^n)$ and the VMO-type space $\mathrm{VMO}_{\rho,\,L}(\mathbb{R}^n)$ associated with L were, respectively, introduced in [55, Definition 4.3] and [56, Definition 3.3]. The ρ-Carleson measure characterization and the John-Nirenberg inequality for the space $\mathrm{BMO}_{\rho,\,L}(\mathbb{R}^n)$, which are, respectively, similar to those in Theorem 2.7 (or Theorem 3.6) and Corollary 2.5, were established in [55, Theorems 6.1 and 6.2]. It was also proved that $\mathrm{BMO}_{\rho,\,L}(\mathbb{R}^n)$ is the dual space of the Orlicz-Hardy pace $H_{\Phi,\,L^*}(\mathbb{R}^n)$ in [55, Theorem 4.1]. Moreover, it was proved in [56, Theorem 4.10] that the VMO-type space $\mathrm{VMO}_{\rho,\,L}(\mathbb{R}^n)$ is the predual space of $B_{\Phi,\,L^*}(\mathbb{R}^n)$, where $B_{\Phi,\,L^*}(\mathbb{R}^n)$ denotes the Banach completion of $H_{\Phi,\,L^*}(\mathbb{R}^n)$ and L^* the adjoint operator of L in $L^2(\mathbb{R}^n)$.

Remark 4.11. Let $k \in \mathbb{N}$. Other typical examples of operators satisfying Assumptions $(L)_5$ and $(L)_6$ include the *$2k$-order divergence form homogeneous elliptic operator*

$$L_1 := (-1)^k \sum_{|\alpha|=k=|\beta|} \partial^\beta(a_{\alpha,\beta}\partial^\alpha)$$

interpreted in the usual weak sense via a sesquilinear form, with complex bounded measurable coefficients $a_{\alpha,\,\beta}$ satisfying the elliptic condition for all multi-indices α and β with $|\alpha| = k = |\beta|$, and the *$2k$-order Schrödinger-type operator*

$$L_2 := (-\Delta)^k + V^k,$$

where $0 \le V \in L^k_{\mathrm{loc}}(\mathbb{R}^n)$. The Hardy space $H^p_{L_i}(\mathbb{R}^n)$ with $i \in \{1,\,2\}$ was studied in [20, 21]. More precisely, the molecular characterization of the space $H^p_{L_i}(\mathbb{R}^n)$ with $p \in (0,1]$ and $i \in \{1,\,2\}$ was established in [20, Theorem 4.5], which is a special case of Theorem 4.4(i). Furthermore, it was proved that the associated Riesz transform $\nabla^k L_i^{-1/2}$ for $i \in \{1,2\}$ is bounded from $H^p_{L_i}(\mathbb{R}^n)$ to the classical Hardy space $H^p(\mathbb{R}^n)$ with $p \in (\frac{n}{n+k},\,1]$ in [20, Theorem 6.2], and bounded from $H^{n/(n+k)}_{L_i}(\mathbb{R}^n)$ to the weak Hardy space $WH^{n/(n+k)}(\mathbb{R}^n)$ in [21, Theorem 4.2].

Let Φ satisfy Assumption $(\Phi)_3$ with $p_\Phi \in (0,1]$ replaced by $p_\Phi^- \in (0,1]$. The Orlicz-Hardy space $H_{\Phi,\,L_i}(\mathbb{R}^n)$ with $i \in \{1,\,2\}$ was studied in [18]. More precisely, the molecular characterization for the Orlicz-Hardy space $H_{\Phi,\,L_i}(\mathbb{R}^n)$ with $i \in \{1,\,2\}$ was obtained in [18, Theorem 2.23]. Let $i \in \{1,\,2\}$, p_Φ be as in (2.4), $\lfloor\gamma\rfloor$ the maximal integer not more than γ and $(p_-(L_i),\,p_+(L_i))$ the range of exponents $p \in [1,\,\infty]$ for which the semigroup $\{e^{-tL_i}\}_{t>0}$ is bounded on $L^p(\mathbb{R}^n)$. As applications of the molecular characterization for $H_{\Phi,\,L_i}(\mathbb{R}^n)$, it was proved in [18, Proposition 3.10] that the generalized Riesz transform $\nabla^\gamma L_i^{-\delta/(2k)}$ is bounded from the Orlicz-Hardy space $H_{\Phi,\,L_i}(\mathbb{R}^n)$ to the Orlicz space $L^{\widetilde{\Phi}}(\mathbb{R}^n)$ when $p_\Phi^- \in (0,\,\frac{n}{n+\delta-\gamma}]$, $0 < \gamma \le \delta < \infty$ and $\delta-\gamma < n(\frac{1}{p_-(L_i)} - \frac{1}{p_+(L_i)})$, or from $H_{\Phi,\,L_i}(\mathbb{R}^n)$ to the Orlicz-Hardy space $H^{\widetilde{\Phi}}(\mathbb{R}^n)$ when $p_\Phi^- \in (\frac{n}{n+\delta+\lfloor\gamma\rfloor-\gamma},\,\frac{n}{n+\delta-\gamma}]$, $1 \le \gamma \le \delta < \infty$ and $\delta-\gamma < n(\frac{1}{p_-(L_i)} - \frac{1}{p_+(L_i)})$, or from $H_{\Phi,\,L_i}(\mathbb{R}^n)$ to the weak Orlicz-Hardy space $WH^{\Phi}(\mathbb{R}^n)$ when $\gamma = \delta$ and $p_\Phi = n/(n+\lfloor\gamma\rfloor)$ or $p_\Phi^- = n/(n+\lfloor\gamma\rfloor)$ with p_Φ^-

attainable, where $\widetilde{\Phi}$ is an Orlicz function whose inverse function $\widetilde{\Phi}^{-1}$ is defined by $\widetilde{\Phi}^{-1}(t) := \Phi^{-1}(t)t^{\frac{1}{n}(\gamma-\delta)}$ for all $t \in (0, \infty)$.

Acknowledgements

Der-Chen Chang is partially supported by an NSF grant DMS-1203845 and Hong Kong RGC competitive earmarked research grant #601410. Dachun Yang is supported by the National Natural Science Foundation of China (Grant Nos. 11171027 and 11361020), the Specialized Research Fund for the Doctoral Program of Higher Education of China (Grant No. 20120003110003) and the Fundamental Research Funds for Central Universities of China (Grant No. 2014KJJCA10).

References

[1] B. T. Anh and J. Li, Orlicz-Hardy spaces associated to operators satisfying bounded H_∞ functional calculus and Davies-Gaffney estimates, J. Math. Anal. Appl. **373** (2011), 485–501.

[2] J. Assaad and E. M. Ouhabaz, Riesz transforms of Schrödinger operators on manifolds, J. Geom. Anal. **22** (2012), 1108–1136.

[3] P. Auscher, On necessary and sufficient conditions for L^p-estimates of Riesz transforms associated to elliptic operators on $\mathbb{R}^n$ and related estimates, Mem. Amer. Math. Soc. **186** (2007), no. 871.

[4] P. Auscher, X. T. Duong, and A. McIntosh, Boundedness of Banach space valued singular integral operators and Hardy spaces, Unpublished Manuscript, 2005.

[5] P. Auscher and J. Martell, Weighted norm inequalities, off-diagonal estimates and elliptic operators. II. Off-diagonal estimates on spaces of homogeneous type, J. Evol. Equ. **7** (2007), 265–316.

[6] P. Auscher, A. McIntosh, and E. Russ, Hardy spaces of differential forms on Riemannian manifolds, J. Geom. Anal. **18** (2008), 192–248.

[7] P. Auscher and E. Russ, Hardy spaces and divergence operators on strongly Lipschitz domains of $\mathbb{R}^n$, J. Funct. Anal. **201** (2003), 148–184.

[8] P. Auscher and P. Tchamitchian, Square Root Problem for Divergence Operators and Related Topics, Astérisque **249**, Society of Mathmatics, France, 1998.

[9] Z. Birnbaum and W. Orlicz, Über die verallgemeinerung des begriffes der zueinander konjugierten potenzen, Studia Math. **3** (1931), 1–67.

[10] A. Bonami, J. Feuto, and S. Grellier, Endpoint for the DIV-CURL lemma in Hardy spaces, Publ. Mat. **54** (2010), 341–358.

[11] A. Bonami and S. Grellier, Hankel operators and weak factorization for Hardy-Orlicz spaces, Colloq. Math. **118** (2010), 107–132.

[12] A. Bonami, S. Grellier, and L. D. Ky, Paraproducts and products of functions in BMO($\mathbb{R}^n$) and $H^1(\mathbb{R}^n)$ through wavelets, J. Math. Pures Appl. (9) **97** (2012), 230–241.

[13] A. Bonami, T. Iwaniec, P. Jones, and M. Zinsmeister, On the product of functions in BMO and H^1, Ann. Inst. Fourier (Grenoble) **57** (2007), 1405–1439.

[14] H.-Q. Bui, X. T. Duong, and L. Yan, Calderón reproducing formulas and new Besov spaces associated with operators, Adv. Math. **229** (2012), 2449–2502.

[15] T. A. Bui, J. Cao, L. Ky, D. Yang, and S. Yang, Musielak-Orlicz-Hardy spaces associated with operators satisfying reinforced off-diagonal estimates, Anal. Geom. Metr. Spaces **1** (2013), 69–129.

[16] T. A. Bui, J. Cao, L. Ky, D. Yang, and S. Yang, Weighted Hardy spaces associated with operators satisfying reinforced off-diagonal estimates, Taiwanese J. Math. **17** (2013), 1127–1166.

[17] J. Cao, D.-C. Chang, D. Yang, and S. Yang, Weighted local Orlicz-Hardy spaces on domains and their applications in inhomogeneous Dirichlet and Neumann problems, Trans. Amer. Math. Soc. **365** (2013), 4729–4809.

[18] J. Cao, D.-C. Chang, D. Yang, and S. Yang, Boundedness of generalized Riesz transforms on Orlicz-Hardy spaces associated to operators, Integral Equations Operator Theory **76** (2013), 225–283.

[19] J. Cao, Y. Liu, and D. Yang, Hardy spaces $H^1_L(\mathbb{R}^n)$ associated to Schrödinger type operators $(-\Delta)^2 + V^2$, Houston J. Math. **36** (2010), 1067–1095.

[20] J. Cao and D. Yang, Hardy spaces $H^p_L(\mathbb{R}^n)$ associated with operators satisfying k-Davies-Gaffney estimates, Sci. China Math. **55** (2012), 1403–1440.

[21] J. Cao, D. Yang, and S. Yang, Endpoint boundedness of Riesz transforms on Hardy spaces associated with operators, Rev. Mat. Complut. **26** (2013), 99–114.

[22] A. Carbonaro, A. McIntosh, and A. J. Morris, Local Hardy spaces of differential forms on Riemannian manifolds, J. Geom. Anal. **23** (2013), 106–169.

[23] D.-C. Chang, G. Dafni, and E. M. Stein, Hardy spaces, BMO and boundary value problems for the Laplacian on a smooth domain in $\mathbb{R}^n$, Trans. Amer. Math. Soc. **351** (1999), 1605–1661.

[24] D.-C. Chang, S. G. Krantz, and E. M. Stein, Hardy spaces and elliptic boundary value problems, Contemporary Math. **137** (1992), 119–131.

[25] D.-C. Chang, S. G. Krantz, and E. M. Stein, H^p theory on a smooth domain in $\mathbb{R}^N$ and elliptic boundary value problems, J. Funct. Anal. **114** (1993), 286–347.

[26] R. R. Coifman, A real variable characterization of H^p, Studia Math. **51** (1974), 269–274.

[27] R. R. Coifman, P.-L. Lions, Y. Meyer, and S. Semmes, Compensated compactness and Hardy spaces, J. Math. Pures Appl. (9) **72** (1993), 247–286.

[28] R. R. Coifman and G. Weiss, Analyse Harmonique Non-commutative sur Certains Espaces Homogènes, Lecture Notes in Math. **242**, Springer, Berlin, 1971.

[29] D. Cruz-Uribe and C. J. Neugebauer, The structure of the reverse Hölder classes, Trans. Amer. Math. Soc. **347** (1995), 2941–2960.

[30] E. B. Davies, Heat kernel bounds, conservation of probability and the Feller property, J. Anal. Math. **58** (1992), 99–119.

[31] E. B. Davies, Heat Kernels and Spectral Theory, Cambrige University Press, 1989.

[32] D. Deng, X. T. Duong, L. Song, C. Tan, and L. Yan, Functions of vanishing mean oscillation associated with operators and applications, Michigan Math. J. **56** (2008), 529–550.

[33] Q. Deng, Y. Ding, and X. Yao, Characterizations of Hardy spaces associated to higher order elliptic operators, J. Funct. Anal. **263** (2012), 604–674.

[34] X. T. Duong, S. Hofmann, D. Mitrea, M. Mitrea, and L. Yan, Hardy spaces and regularity for the inhomogeneous Dirichlet and Neumann problems, Rev. Mat. Iberoam. **29** (2013), 183–236.

[35] X. T. Duong and J. Li, Hardy spaces associated to operators satisfying Davies-Gaffney estimates and bounded holomorphic functional calculus, J. Funct. Anal. **264** (2013), 1409–1437.

[36] X. T. Duong and L. Yan, New function spaces of BMO type, the John-Nirenberg inequality, interpolation, and applications, Comm. Pure Appl. Math. **58** (2005), 1375–1420.

[37] X. T. Duong and L. Yan, Duality of Hardy and BMO spaces associated with operators with heat kernel bounds, J. Amer. Math. Soc. **18** (2005), 943–973.

[38] J. Dziubański and J. Zienkiewicz, Hardy space H^1 associated to Schrödinger operator with potential satisfying reverse Hölder inequality, Rev. Mat. Iberoam. **15** (1999), 279–296.

[39] J. Dziubański and J. Zienkiewicz, H^p spaces associated with Schrödinger operators with potentials from reverse Hölder classes, Colloq. Math. **98** (2003), 5–38.

[40] C. Fefferman and E. M. Stein, H^p spaces of several variables, Acta Math. **129** (1972), 137–193.

[41] M. P. Gaffney, The conservation property of the heat equation on Riemannian manifolds, Comm. Pure Appl. Math. **12** (1959), 1–11.

[42] J. García-Cuerva, Weighted H^p spaces, Dissertationes Math. (Rozprawy Mat.) **162** (1979), 1–63.

[43] J. García-Cuerva and J. Rubio de Francia, Weighted Norm Inequalities and Related Topics, Amsterdam, North-Holland, 1985.

[44] L. Grafakos, Modern Fourier Analysis, Second edition, Graduate Texts in Mathematics **250**, Springer, New York, 2009.

[45] M. Haase, The Functional Calculus for Sectorial Operators, Operator Theory: Advances and Applications **169**, Birkhäser Verlag, Basel, 2006.

[46] E. Harboure, O. Salinas, and B. Viviani, A look at $\mathrm{BMO}_\phi(\omega)$ through Carleson measures, J. Fourier Anal. Appl. **13** (2007), 267–284.

[47] S. Hou, D. Yang, and S. Yang, Lusin area function and molecular characterizations of Musielak-Orlicz Hardy spaces and their applications, Commun. Contemp. Math. **15** (2013), no. 6, 1350029.

[48] S. Hofmann, G. Lu, D. Mitrea, M. Mitrea, and L. Yan, Hardy spaces associated to non-negative self-adjoint operators satisfying Davies-Gaffney estimates, Mem. Amer. Math. Soc. **214** (2011), no. 1007.

[49] S. Hofmann and S. Mayboroda, Hardy and BMO spaces associated to divergence form elliptic operators, Math. Ann. **344** (2009), 37–116.

[50] S. Hofmann, S. Mayboroda, and A. McIntosh, Second order elliptic operators with complex bounded measurable coefficients in L^p, Sobolev and Hardy spaces, Ann. Sci. École Norm. Sup. (4) **44** (2011), 723–800.

[51] J. Huang, Hardy spaces associated to the Schrödinger operator on strongly Lipschitz domains of $\mathbb{R}^d$, Math. Z. **266** (2010), 141–168.

[52] T. Iwaniec and J. Onninen, $\mathcal{H}^1$-estimates of Jacobians by subdeterminants, Math. Ann. **324** (2002), 341–358.

[53] S. Janson, Generalizations of Lipschitz spaces and an application to Hardy spaces and bounded mean oscillation, Duke Math. J. **47** (1980), 959–982.

[54] R. Jiang, X. Jiang, and D. Yang, Maximal function characterizations of Hardy spaces associated with Schrödinger operators on nilpotent Lie groups, Rev. Mat. Complut. **24** (2011), 251–275.

[55] R. Jiang and D. Yang, New Orlicz-Hardy spaces associated with divergence form elliptic operators, J. Funct. Anal. **258** (2010), 1167–1224.

[56] R. Jiang and D. Yang, Generalized vanishing mean oscillation spaces associated with divergence form elliptic operators, Integral Equations Operator Theory **67** (2010), 123–149.

[57] R. Jiang and D. Yang, Orlicz-Hardy spaces associated with operators satisfying Davies-Gaffney estimates, Commun. Contemp. Math. **13** (2011), 331–373.

[58] R. Jiang and D. Yang, Predual spaces of Banach completions of Orlicz-Hardy spaces associated with operators, J. Fourier Anal. Appl. **17** (2011), 1–35.

[59] R. Jiang, Da. Yang, and Do. Yang, Maximal function characterizations of Hardy spaces associated with magnetic Schrödinger operators, Forum Math. **24** (2012), 471–494.

[60] R. Jiang, D. Yang, and Y. Zhou, Localized Hardy spaces associated with operators, Appl. Anal. **88** (2009), 1409–1427.

[61] R. Jiang, D. Yang, and Y. Zhou, Orlicz-Hardy spaces associated with operators, Sci. China Ser. A **52** (2009), 1042–1080.

[62] F. John and L. Nirenberg, On functions of bounded mean oscillation, Comm. Pure. Appl. Math. **14** (1961), 415–426.

[63] R. Johnson and C. J. Neugebauer, Homeomorphisms preserving A_p, Rev. Mat. Iberoam. **3** (1987), 249–273.

[64] R. H. Latter, A characterization of $H^p(\mathbb{R}^n)$ in terms of atoms, Studia Math. **62** (1978), 93–101.

[65] Y. Liang, D. Yang, and S. Yang, Applications of Orlicz-Hardy spaces associated with operators satisfying Poisson estimates, Sci. China Math. **54** (2011), 2395–2426.

[66] Y. Liang, D. Yang, and W. Yuan, Vanishing mean oscillation spaces associated with operators satisfying Davies-Gaffney estimates, Kyoto J. Math. **52** (2012), 205–247.

[67] S. Liu and L. Song, An atomic decomposition of weighted Hardy spaces associated to self-adjoint operators, J. Funct. Anal. **265** (2013), 2709–2723.

[68] L. D. Ky, Bilinear decompositions and commutators of singular integral operators, Trans. Amer. Math. Soc. **365** (2013), 2931–2958.

[69] L. D. Ky, Endpoint estimates for commutators of singular integrals related to Schrödinger operators, arXiv:1203.6335 or Rev. Mat. Iberoam. (to appear).

[70] L. D. Ky, New Hardy spaces of Musielak-Orlicz type and boundedness of sublinear operators, Integral Equations Operator Theory **78** (2014), 115–150.

[71] A. McIntosh, Operators which have an H_∞ functional calculus, Miniconference on operator theory and partial differential equations (Canberra), Proc. Centre Math. Anal. Austral. Nat. Univ. **14**, Austral. Nat. Univ., Canberra, 1986, pp.210–231.

[72] J. Musielak, Orlicz Spaces and Modular Spaces, Lecture Notes in Math. **1034**, Springer-Verlag, Berlin, 1983.

[73] E. Nakai and K. Yabuta, Pointwise multipliers for functions of weighted bounded mean oscillation on spaces of homogeneous type, Math. Japon. **46** (1997), 15–28.

[74] W. Orlicz, Über eine gewisse Klasse von Räumen vom Typus B, Bull. Int. Acad. Pol. Ser. A **8** (1932), 207–220.

[75] E. M. Ouhabaz, Analysis of Heat Equations on Domains, Princeton University Press, Princeton, N. J., 2005.

[76] M. Rao and Z. Ren, Theory of Orlicz Spaces, Marcel Dekker, Inc., New York, 1991.

[77] M. Rao and Z. Ren, Applications of Orlicz Spaces, Marcel Dekker, Inc., New York, 2002.

[78] S. Semmes, A primer on Hardy spaces, and some remarks on a theorem of Evans and Müller, Comm. Partial Differential Equations **19** (1994), 277–319.

[79] L. Song and L. Yan, Riesz transforms associated to Schrödinger operators on weighted Hardy spaces, J. Funct. Anal. **259** (2010), 1466–1490.

[80] E. M. Stein, Harmonic Analysis: Real-variable Methods, Orthogonality, and Oscillatory integrals, Princeton University Press, Princeton, New Jersey, 1993.

[81] E. M. Stein and G. Weiss, On the theory of harmonic functions of several variables. I. The theory of H^p-spaces, Acta Math. **103** (1960), 25–62.

[82] J.-O. Strömberg, Bounded mean oscillation with Orlicz norms and duality of Hardy spaces, Indiana Univ. Math. J. **28** (1979), 511–544.

[83] J.-O. Strömberg and A. Torchinsky, Weighted Hardy Spaces, Lecture Notes in Math. **1381**, Springer-Verlag, Berlin, 1989.

[84] M. H. Taibleson and G. Weiss, The molecular characterization of certain Hardy spaces, in: Representation Theorems for Hardy Spaces, pp. 67–149, Astérisque **77**, Soc. Math. France, Paris, 1980.

[85] B. E. Viviani, An atomic decomposition of the predual of BMO(ρ), Rev. Mat. Iberoam. **3** (1987), 401–425.

[86] L. Yan, Classes of Hardy spaces associated with operators, duality theorem and applications, Trans. Amer. Math. Soc. **360** (2008), 4383–4408.

[87] Da. Yang and Do. Yang, Hardy spaces associated with magnetic Schrödinger operators on strongly Lipschitz domains, Nonlinear Anal. **75** (2012), 6433–6447.

[88] D. Yang and S. Yang, Orlicz-Hardy spaces associated with divergence operators on unbounded strongly Lipschitz domains of $\mathbb{R}^n$, Indiana Univ. Math. J. **61** (2012), 81–129.

[89] D. Yang and S. Yang, Real-variable characterizations of Orlicz-Hardy spaces on strongly Lipschitz domains of $\mathbb{R}^n$, Rev. Mat. Iberoam. **29** (2013), 233–288.

[90] D. Yang and S. Yang, Local Hardy spaces of Musielak-Orlicz type and their applications, Sci. China Math. **55** (2012), 1677–1720.

[91] D. Yang and S. Yang, Musielak-Orlicz Hardy spaces associated with operators and their applications, J. Geom. Anal. **24** (2014), 495–570.

Some Topics in Harmonic Analysis and Applications
ALM 34, pp. 71–91

Boundedness of Rough Strongly Singular Integral Operators*

Jiecheng Chen[†], Dashan Fan[‡] and Meng Wang[§]

Abstract

In this paper, we study the operator

$$\mathcal{T}_{\Omega,\alpha,\beta}f(x) = p.v. \int_{\mathbf{R}^n} e^{i|y|^{-\beta}}|y|^{-n-\alpha}\Omega(y')f(x-y)dy.$$

We obtain that when $\Omega \in L^q$ or Ω satisfies some smooth condition, $\mathcal{T}_{\Omega,\alpha,\beta}$ is bounded from L^p to L^r, in which $p \le r$ satisfies some condition.

2010 Mathematics Subject Classification: 42B20.

Keywords and Phrases: strongly singular integral operator, rough kernel, Lipschitz condition.

*The subject is supported by NSF of China (Grant No. 11271330, 10931001, 11201103, 11371316).

†Department of Mathematics, Zhejiang Normal University, Jinhua, 321004, China; *E-mail:* `jcchen@zjnu.edu.cn`

‡Department of Mathematics, University of Wisconsin-Milwaukee, Milwaukee, WI53201, U.S.A.; *E-mail:* `fan@uwm.edu`

§Department of Mathematics, Zhejiang University, Hangzhou, 310027, China; *E-mail:* `mathdreamcn@zju.edu.cn`

1 $L^p \to L^q$ boundedness on rough operators

Let $\beta > \alpha/n \geq 0$. The strongly singular integral operator $T_{\alpha,\beta}$ is defined, initially on the test space $\mathcal{S}(\mathbb{R}^n)$, by

$$T_{\alpha,\beta}f(x) = \int_{\mathbf{R}^n} e^{i|y|^{-\beta}} |y|^{-n-\alpha} f(x-y)dy. \tag{1.1}$$

Such operator was studied by Wainger in [21] and by Fefferman-Stein in [8]. The following theorem can be found in [8].

Theorem 1.1. *The operator $T_{\alpha,\beta}$ is bounded on $L^p(\mathbb{R}^n)$ if and only if $|1/2-1/p| \leq (n\beta-\alpha)/n\beta$.*

On the other hand, for an L^∞ function b on the interval $(0,\infty)$ and an L^1 function Ω on the unit sphere satisfying the mean zero property

$$\int_{S^{n-1}} \Omega(y')d\sigma(y') = 0, \tag{1.2}$$

the rough singular integral

$$S_{\Omega,\alpha}f(x) = p.v. \int_{\mathbf{R}^n} b(|y|)|y|^{-n-\alpha}\Omega(y')f(x-y)dy \tag{1.3}$$

was well-studied by Calderón and Zygmund and by many authors. The reader can see [2, 3, 4, 6, 7, 9, 11, 12, 17, 18], et al., among numerous papers in the case $\alpha = 0$.

When $\alpha > 0$, the kernel of $S_{\Omega,\alpha}$ has a higher singularity at zero. This case was studied by Wheeden [22] about 40 years ago. Recently, it attracted some new studies. (See [1, 15, 16], et al.) As a natural question, one will ask what happens if the operator in (1.1) carries some rough things on its kernel. Namely, it is worth to clarify what is the boundedness property of the operator

$$\mathcal{T}_{\Omega,\alpha,\beta}f(x) = p.v. \int_{\mathbf{R}^n} e^{i|y|^{-\beta}} |y|^{-n-\alpha}\Omega(y')f(x-y)dy \tag{1.4}$$

when Ω is some L^1 function on S^{n-1}.

We note that, when $\alpha > 0$, the kernel

$$K_{\alpha,\beta}(y) = e^{i|y|^{-\beta}} |y|^{-n-\alpha}\Omega(y')$$

of $\mathcal{T}_{\Omega,\alpha,\beta}$ has no singularity at the infinity. Thus the L^p boundedness of $\mathcal{T}_{\Omega,\alpha,\beta}$ is equivalent to those of its local operator

$$L_{\Omega,\alpha,\beta}f(x) = p.v. \int_{|y|\leq 1} e^{i|y|^{-\beta}} |y|^{-n-\alpha}\Omega(y')f(x-y)dy.$$

We have the following sharp result.

Theorem 1.2. *For* $\alpha > 0$,

$$\|T_{\Omega,\alpha,\beta}(f)\|_{L^p(\mathbf{R}^n)} \leq C\|\Omega\|_{L^1(S^{n-1})}\|f\|_{L^p(\mathbf{R}^n)}$$

if and only if $|1/p - 1/2| \leq (\beta - 2\alpha)/2\beta$, *where the constant* C *is independent of* f *and* Ω.

Proof. To prove the sufficient part, clearly we only need to prove the same boundedness for the local operator $L_{\Omega,\alpha,\beta}$. We note that $\alpha = 0$ is also allowed in our arguments on the operator $L_{\Omega,\alpha,\beta}$.

By duality, it suffices to show the case $1 < p \leq 2$. Also, using the polar coordinates, we have

$$L_{\Omega,\alpha,\beta}f(x) = \int_{S^{n-1}} \Omega(y')\left(\int_0^1 e^{it^{-\beta}}t^{-1-\alpha}f(x-ty')dt\right)d\sigma(y').$$

We write the inside integral of $L_{\Omega,\alpha,\beta}$ as

$$H_{y'}f(x) = \int_0^1 e^{it^{-\beta}}t^{-1-\alpha}f(x-ty')dt.$$

By Minkowski's inequality, we have

$$\|L_{\Omega,\alpha,\beta}(f)\|_{L^p(\mathbf{R}^n)} \leq \int_{S^{n-1}} \|H_{y'}(f)\|_{L^p(\mathbf{R}^n)}|\Omega(y')|d\sigma(y').$$

Thus, to prove the sufficient part of the theorem, it suffices to show that there is a constant C independent of f and y' such that

$$\|H_{y'}(f)\|_{L^p(\mathbf{R}^n)} \leq C\|f\|_{L^p(\mathbf{R}^n)} \tag{1.5}$$

whenever $|1/p - 1/2| \leq (\beta - 2\alpha)/2\beta$. Denote $SO(n)$ the rotation group of $\mathbb{R}^n$. For fixed y', pick $A \in SO(n)$ such that $Ay' = \mathbf{1} = (1, 0, 0, ..., 0)$ is the north pole of S^{n-1}. We define a function g by $g(Ay) = f(y)$. Thus

$$\|g\|_{L^p(\mathbf{R}^n)} = \|f\|_{L^p(\mathbf{R}^n)}$$

and

$$f(x - ty') = g(Ax - t\mathbf{1}).$$

After changing variables,

$$\begin{aligned}\int_{\mathbf{R}^n} |H_{y'}f(x)|^p dx &= \int_{\mathbf{R}^n}\left|\int_0^1 e^{it^{-\beta}}t^{-1-\alpha}f(x-ty')dt\right|^p dx\\ &= \int_{\mathbf{R}^n}\left|\int_0^1 e^{it^{-\beta}}t^{-1-\alpha}g(x-t\mathbf{1}dt)\right|^p dx.\end{aligned}$$

Therefore, by iterated integration and Theorem 1.1 we obtain the "if" part of the theorem.

To prove "only if" part, for simplicity in the argument, we work the case $n=2$. The general case follows by the same argument. Let $f_2(x_2)$ be a positive continuous even function whose support is inside the interval $[-10,10]$ and whose value is identically one in the interval $[-5,5]$. For a very small $\varepsilon>0$, let $f_1(x_1)=f_2(10x_1)$ and $f_\varepsilon(x)=f_1(x_1/\varepsilon^{\beta+1})f_2(x_2)$. Let $\Omega_\varepsilon(\cos\theta,\sin\theta)=\tilde{\Omega}_\varepsilon(\theta)$ be defined by $\tilde{\Omega}_\varepsilon(\theta)=\varepsilon^{-2}$ if $0\le\theta\le\varepsilon^2$ and $\tilde{\Omega}_\varepsilon(\theta)=0$ elsewhere. It is easy to see that

$$\|f_\varepsilon\|^p_{L^p(\mathbf{R}^2)}\cong\varepsilon^{\beta+1},\quad \|\Omega_\varepsilon\|_{L^1(S^1)}\cong 1.$$

The inequality

$$\|T_{\Omega_{\varepsilon,\alpha,\beta}}(f)\|_{L^p(\mathbf{R}^n)}\le C\|\Omega\|_{L^1(S^{n-1})}\|f\|_{L^p(\mathbf{R}^n)} \tag{1.6}$$

implies that there is a constant C independent of $\varepsilon>0$ such that

$$\|T_{\Omega_\varepsilon,\alpha,\beta}(f_\varepsilon)\|^p_{L^p(\mathbf{R}^2)}\le C\|\Omega_\varepsilon\|^p_{L^1(S^1)}\|f_\varepsilon\|^p_{L^p(\mathbf{R}^2)}\cong\varepsilon^{\beta+1}.$$

On the other hand,

$$\|T_{\Omega_\varepsilon,\alpha,\beta}(f_\varepsilon)\|^p_{L^p(\mathbf{R}^2)}$$

$$\ge\int_{\mathbf{R}^2}\left|\int_0^\infty\int_{-\pi}^{\pi}\Omega_\varepsilon(\theta)f_1((x_1-t\cos\theta)\varepsilon^{-\beta-1})f_2(x_2-t\sin\theta)e^{-it^{-\beta}}t^{-1-\alpha}dtd\theta\right|^p dx$$
$$\ge\int_0^1\left|\int_0^\varepsilon\varepsilon^{-2}\int_0^2 f_1((x_1-t\cos\theta)\varepsilon^{-\beta-1})f_2(x_2-t\sin\theta)e^{-it^{-\beta}}t^{-1-\alpha}dtd\theta\right|^p dx$$
$$\ge\int_0^1\left|\int_0^\varepsilon\varepsilon^{-2}\int_0^2 f_1((x_1-t\cos\theta)\varepsilon^{-\beta-1})\cos(t^{-\beta})t^{-1-\alpha}dtd\theta\right|^p dx_1.$$

We construct intervals $\{J_k\}$ by setting

$$J_k=[(2\pi k+\pi/4)^{-1/\beta},(2\pi k-\pi/4)^{-1/\beta}].$$

It is easy to check that the length $|J_k|$ of J_k is

$$(2\pi k-\pi/4)^{-1/\beta}-(2\pi k+\pi/4)^{-1/\beta}\cong k^{-1-1/\beta}.$$

Without loss of generality, we assume $|J_k|=10k^{-1-1/\beta}$. Thus for $0<\varepsilon\le k^{-1/\beta}$, we construct disjoint intervals $\{I_k\}$ by letting

$$I_k=[(2\pi k+\pi/4)^{-1/\beta}+2\varepsilon^{\beta+1},(2\pi k-\pi/4)^{-1/\beta}-2\varepsilon^{\beta+1}].$$

We have

$$|I_k|\cong|J_k|\cong k^{-1-1/\beta}.$$

Note that the support condition of f_1 implies $f_1((x_1 - t\cos\theta)/\varepsilon^{\beta+1}) = 0$ if $t\cos\theta$ is outside the interval $[x_1 - \varepsilon^{\beta+1}, x_1 + \varepsilon^{\beta+1}]$. Thus, if $x_1 \in I_k$, the function $f_1((x_1 - t\cos\theta)/\varepsilon^{\beta+1}) = 0$, except $t\cos\theta$ is inside the interval

$$[(2\pi k + \pi/4)^{-1/\beta} + \varepsilon^{\beta+1}, (2\pi k - \pi/3)^{-1/\beta} - \varepsilon^{\beta+1}].$$

When $0 \le \theta \le \varepsilon^2$, it is easy to see that the function $f_1((x_1 - t\cos\theta)/\varepsilon^{\beta+1}) = 0$ except $t \in [(2\pi k + \pi/3)^{-1/\beta}, (2\pi k - \pi/3)^{-1/\beta}]$. In this interval, $\cos t^{-\beta} \ge 1/2$ and $t^{-1-\alpha} \cong k^{(1+\alpha)/\beta}$. Therefore, we obtain

$$\int_0^1 \left| \int_0^{\varepsilon^2} \varepsilon^{-2} \int_0^2 f_1((x_1 - t\cos\theta)\varepsilon^{-\beta-1}) \cos(t^{-\beta}) t^{-1-\alpha} dt d\theta \right|^p dx_1$$

$$\ge \sum_{k=4}^{\varepsilon^{-\beta}} \int_{I_k} \left| \int_0^{\varepsilon^2} \varepsilon^{-2} \int_0^2 f_1((x_1 - t\cos\theta)\varepsilon^{-\beta-1}) \cos(t^{-\beta}) t^{-1-\alpha} dt d\theta \right|^p dx_1$$

$$\ge C(1/2)^p \sum_{4 \le k < 1/\varepsilon^\beta} k^{p(\alpha+1)/\beta} \int_{I_k} \left| \int_{\{|t-x_1| < \varepsilon^{\beta+1}/2\}} dt \right|^p dx_1$$

$$\ge C(1/2)^p \varepsilon^{p(\beta+1)} \sum_{4 \le k < 1/\varepsilon^\beta} k^{p(1+\alpha)/\beta} |I_k|$$

$$\cong \varepsilon^{p(\beta+1)} \sum_{4 \le k < 1/\varepsilon^\beta} k^{p(1+\alpha)/\beta - 1/\beta - 1} \cong \varepsilon^{p(\beta-\alpha)+1}.$$

As a conclusion, the inequality

$$\|\mathcal{T}_{\Omega,\alpha,\beta}(f)\|_{L^p(\mathbf{R}^n)} \le C \|\Omega\|_{L^1(S^{n-1})} \|f\|_{L^p(\mathbf{R}^n)}$$

implies that for all small $\varepsilon > 0$,

$$\varepsilon^{\beta+1} \ge C \varepsilon^{p(\beta+1)} \sum_{4 \le k < 1/\varepsilon^\beta} k^{p(1+\alpha)/\beta - 1/\beta - 1}. \tag{1.7}$$

This means that there is a C independent of ε such that

$$\varepsilon^{p(\beta-\alpha)} \le C \varepsilon^\beta$$

for any $\varepsilon > 0$.

We now obtain that (1.6) implies $p(\beta - \alpha) \ge \beta$, which is equivalent to $1/p - 1/2 \le (\beta - 2\alpha)/2\beta$ in the case $1 \le p \le 2$. By duality, we must have

$$|1/p - 1/2| \le (\beta - 2\alpha)/2\beta \tag{1.8}$$

if the inequality in (1.6) holds. The theorem is proved. □

If $p = 1$, from (1.8) we find that α must be zero. Thus (1.7) gives us an impossible inequality

$$\varepsilon^{\beta+1} \geq C\varepsilon^{(\beta+1)} \sum_{4 \leq k < 1/\varepsilon^\beta} k^{-1}$$

for all $\varepsilon > 0$. This leads to a corollary:

Corollary 1.1. *Let $1 \leq p < \infty$. For any $\beta > 0$,*

$$\|L_{\Omega,0,\beta}\|_{L^p(\mathbf{R}^n)} \leq C\|\Omega\|_{L^1(S^{n-1})}\|f\|_{L^p(\mathbf{R}^n)}$$

if and only if $p > 1$.

We note that, when $\alpha = 0$, the operator $\mathcal{T}_{\Omega,\alpha,\beta}$ has worse boundedness properties than those of $L_{\Omega,\alpha,\beta}$. To see this fact, checking the Fourier transform and using the Plancherel Theorem, one can see that $\mathcal{T}_{\Omega,0,\beta}$ is not bounded on $L^2(\mathbb{R}^n)$. The reason of causing such a damage is that its kernel

$$K_{0,\beta}(y) = e^{i|y|^{-\beta}}|y|^{-n}\Omega(y')$$

has singularities at both zero and infinity, while the oscillating factor $e^{i|y|^{-\beta}} \cong 1$ as y in a neighborhood of the infinity. Such a roadblock can be removed by assuming the cancellation condition (1.2) on Ω together with a better size condition on Ω (actually, these are very natural conditions on many applications). For an example, we have the following results.

Theorem 1.3 ([3, 4]). *Let $\beta \geq 0$ and $1 < p < \infty$. If $\Omega \in L^q(S^{n-1})$ for some $q > 1$, and satisfies the cancellation condition (1.2), then the operator $\mathcal{T}_{\Omega,0,\beta}$ is bounded on $L^p(\mathbb{R}^n)$. Also, the operator $S_{\Omega,0}$ is weak type $(1,1)$ if $b(t) = 1$.*

Theorem 1.4 ([6]). *Let $1 < p < \infty$. If $b \in L^\infty(0,\infty)$, $\Omega \in L^q(S^{n-1})$ for some $q > 1$, and satisfies the cancellation condition (1.2), then the operator $S_{\Omega,0}$ is bounded on $L^p(\mathbb{R}^n)$.*

Now we claim that we can gain a little more than those in Theorem 1.4. More precisely, we expect to obtain some better estimates for $S_{\Omega,0}$ if $b(t) = e^{it^{-\beta}}$. We have the following proposition on $p = 2$.

Proposition 1.1. *If $\Omega \in L^q(S^{n-1})$. Then*

(i) *if $1/q' > \beta/(2\beta+2)$, we have, for any $0 < \gamma \leq \beta/(2\beta+2)$,*

$$\|\mathcal{T}_{\Omega,0,\beta}f\|_{L^2(\mathbf{R}^n)} \leq C\|f\|_{L^2_{-\gamma}(\mathbf{R}^n)}.$$

(ii) *If* $1/q' \le \beta/(2\beta+2)$, *we have, for any* $0 < \gamma \le 1/q'$,

$$\|\mathcal{T}_{\Omega,0,\beta} f\|_{L^2(\mathbf{R}^n)} \le C\|f\|_{L^2_{-\gamma}(\mathbf{R}^n)}.$$

Proof. For $\Omega \in L^q(S^{n-1})$, we define the function F_Ω by

$$F_\Omega(s) = (1-s^2)^{(n-3)/2}\chi_{(-1,1)}(s)\int_{S^{n-2}}\Omega(s,(1-s^2)^{1/2}\tilde{y})d\sigma(\tilde{y}),$$

if $n \ge 3$, and

$$F_\Omega(s) = (1-s^2)^{-1/2}\chi_{(-1,1)}(s)\left(\Omega(s,(1-s^2)^{1/2}) + \Omega(s,-(1-s^2)^{1/2})\right),$$

if $n = 2$. We decompose the operator $\mathcal{T}_{\Omega,0,\beta}$ as

$$\begin{aligned}\mathcal{T}_{\Omega,0,\beta} f(x) &= \sum_k \int_{2^k \le |y| < 2^{k+1}} e^{i|y|^{-\beta}}|y|^{-n}\Omega(y')f(x-y)dy \\ &= \sum_k \sigma_k * f(x),\end{aligned}$$

where the Fourier transform of σ_k is

$$\hat{\sigma}_k(\xi) = m_k(\xi) \cong \int_{S^{n-1}}\Omega(y')\left(\int_{2^k}^{2^{k+1}} e^{it^{-\beta}}t^{-1}e^{-2\pi it<y',\xi>}dt\right)d\sigma(y').$$

By the rotation method of Calderón and Zygmund, we have

$$\begin{aligned}m_k(\xi) &\cong \int_{2^k}^{2^{k+1}} e^{it^{-\beta}}t^{-1}\int_{S^{n-1}}\Omega(y')e^{-2\pi it<y',\xi>}d\sigma(y')dt \\ &\cong \int_{\mathbf{R}} F_\Omega(s)N_k(s|\xi|)ds,\end{aligned}$$

where

$$\begin{aligned}N_k(s|\xi|) &= \int_{2^k}^{2^{k+1}} e^{it^{-\beta}}t^{-1}e^{-2\pi ts|\xi|}dt \\ &\cong \int_1^2 e^{i\theta_k(t,s|\xi|)}t^{-1}dt\end{aligned}$$

and

$$\theta_k(t,s|\xi|) = 2^{k\beta}t^{-\beta} - 2^{1-k}\pi ts|\xi|.$$

Taking the derivative on t variable, we have

$$\theta_k'(t,s|\xi|) = -\beta 2^{k\beta}t^{-\beta-1} - 2^{1-k}\pi s|\xi|.$$

Let

$$E_1 = \{k \in \mathbf{Z} : |\beta 2^{k\beta}| \ge 2^{\beta-k+4}\pi|s\xi|\},$$

$$E_2 = \{k \in \mathbf{Z} : |\beta 2^{k\beta}| \le 2^{-\beta-k-4}\pi|s\xi|\},$$

and

$$E_3 = \{k \in \mathbf{Z} : 2^{-\beta-k-4}\pi|s\xi| < |\beta 2^{k\beta}| < 2^{\beta-k+4}\pi|s\xi|\}.$$

Then we easily see that $\mathbf{Z} = E_1 \cup E_2 \cup E_3$ and E_3 is a finite set and the total number of E_3 is independent of s, ξ and k. Also, it is easy to see that there exist positive numbers c_1, c_2 and c such that for $t \in [1/2, 2]$ we have

$$|\theta_k'(t, |s\xi|)| \ge c2^{k\beta} \text{ and } 2^k \ge c_2|s\xi|^{1/(\beta+1)} \text{ if } k \in E_1;$$

$$|\theta_k'(t, |s\xi|)| \ge c2^{-k}|s\xi| \text{ and } 2^k \le c_1|s\xi|^{1/(\beta+1)} \text{ if } k \in E_2.$$

Thus, an easy integration by parts shows that

$$|N_k(s|\xi|)| \le C2^{-k\beta} \text{ if } k \in E_1$$

and

$$|N_k(s|\xi|)| \le C2^k|s\xi|^{-1} + C2^{k(\beta+2)}|s\xi|^{-2} \text{ if } k \in E_2.$$

We have

$$\sum_k m_k(\xi) = \int_{\mathbf{R}} F_\Omega(s) \sum_k N_k(s|\xi|)ds.$$

We write

$$\sum_k N_k(s\xi) = \sum_{k\in E_1} N_k(s|\xi|) + \sum_{k\in E_2} N_k(s|\xi|) + \sum_{k\in E_3} N_k(s|\xi|).$$

We now consider that the case $\beta/2(\beta+1) < 1/q'$.

Note that $|N_k(s|\xi|)| \le C$ uniformly on $|s\xi|$. We may have for any $0 < \theta \le 1$,

$$|N_k(s|\xi|)| \le C2^{-\theta k\beta/2} \text{ if } k \in E_1.$$

Thus,

$$\left|\sum_{k\in E_1} N_k(s|\xi|)\right| \le C \sum_{2^{-k}\le|s\xi|^{-1/(\beta+1)}c_2^{-1}} 2^{-\theta k\beta/2}$$
$$\le C|s\xi|^{-\theta\beta/(2\beta+2)}.$$

Similarly, we have

$$\left|\sum_{k\in E_2} N_k(s|\xi|)\right| \le C|s\xi|^{-\theta} \sum_{2^k\le C|s\xi|^{1/(\beta+1)}} 2^{\theta k} \le C|s\xi|^{-\theta\beta/(2\beta+2)}.$$

E_3 contains only finitely many terms. Using the Van der Corput Lemma, we have

$$\left|\sum_{k\in E_3} N_k(s|\xi|)\right| \le C2^{-\theta k\beta/2} \cong C|s\xi|^{-\theta\beta/(2\beta+2)}.$$

Therefore, for any $0 < \theta \le 1$,

$$\begin{aligned}\left|\sum_k m_k(\xi)\right| &\le C\int_{-1}^{1} |F_\Omega(s)||s\xi|^{-\theta\beta/(2\beta+2)}ds \\ &\le C\|\Omega\|_{L^q(S^{n-1})}|\xi|^{-\beta\theta/(2\beta+2)}\int_{-1}^{1}|s|^{-\theta\beta q'/(2\beta+2)}ds \\ &\cong |\xi|^{-\theta\beta/(2\beta+2)},\end{aligned}$$

since $\theta\beta q'/(2\beta+2) < 1$. This shows that

$$\widehat{\mathcal{T}_{\Omega,0,\beta}(f)}(\xi) = \hat{f}(\xi)|\xi|^{-\theta\beta/(2\beta+2)}.$$

Therefore, by Plancherel Theorem, we have

$$\|\mathcal{T}_{\Omega,0,\beta}f\|^2_{L^2(\mathbf{R}^n)} \le C\int_{\mathbf{R}^n} |\hat{f}(\xi)|^2|\xi|^{-\theta\beta/(\beta+1)}d\xi.$$

(i) of the proposition is proved. The proof of (ii) is the same as that of (i) with only a minor modification. We omit the proof. □

Now by Theorem 1.3 and Proposition 1.1 with the Sobolev imbedding theorem, we have the following (L^r, L^p) result.

Theorem 1.5. *Let $1 < p < \infty$ and $\Omega \in L^q(S^{n-1})$ satisfy the cancellation condition* (1.2). *Then*

(i) *if $1/q' > \beta/(2\beta+2)$, we have*

$$\|\mathcal{T}_{\Omega,0\,\beta}f\|_{L^p(\mathbf{R}^n)} \le C\|f\|_{L^r(\mathbf{R}^n)}$$

for any $p_0 \le r \le p$ with $1/p_0 = 1/p + \beta/n(\beta+1)\max\{p,p'\}$,

(ii) *If $1/q' \le \beta/2(\beta+2)$, we have*

$$\|\mathcal{T}_{\Omega,0,\beta}f\|_{L^p(\mathbf{R}^n)} \le C\|f\|_{L^r(\mathbf{R}^n)}$$

for any $p_1 \le r \le p$ with $1/p_1 = 1/p + 2/nq'\max\{p,p'\}$.

Proof. Modifying the proof of Theorem 1.3, one can check that $\mathcal{T}_{\Omega,0,\beta}$ is also of weak type (1,1). Interpolating the results in Proposition 1.1 and weak (1,1) bound, we obtain if $1/q' > \beta/(2\beta+2)$, $1 < p \le 2$ and $0 < \gamma \le \beta/(\beta+1)p'$,

$$\|\mathcal{T}_{\Omega,0,\beta}f\|_{L^p(\mathbf{R}^n)} \le C\|f\|_{L^p_{-\gamma}(\mathbf{R}^n)}. \tag{1.9}$$

If $1/q' > \beta/(2\beta+2)$, $2 < p < \infty$ and $0 < \gamma < \beta/(\beta+1)p$,

$$\|T_{\Omega,0,\beta}f\|_{L^p(\mathbf{R}^n)} \leq C\|f\|_{L^p_{-\gamma}(\mathbf{R}^n)}, \tag{1.10}$$

where $s > -2\gamma/p$. Thus the (i) of the theorem follows by the Sobolev imbedding theorem.

The proof of (ii) is the same as (i). Here we omit it. □

We now easily extend our results to the case $\alpha > 0$. As we discussed before, in this case the cancellation condition (1.2) on Ω is not needed. Similar to Proposition 1.1, we have

Proposition 1.2. *Let $\Omega \in L^q(S^{n-1})$ with $q > 1$ and $\beta \geq 2\alpha > 0$. Then*

(i) *If $1/q' > (\beta - 2\alpha)/(2\beta+2)$, we have, for any $0 < \gamma \leq (\beta-2\alpha)/(2\beta+2)$,*

$$\|T_{\Omega,\alpha,\beta}f\|_{L^2(\mathbf{R}^n)} \leq C\|f\|_{L^2_{-\gamma}(\mathbf{R}^n)}.$$

(ii) *If $1/q' \leq (\beta-2\alpha)/(2\beta+2)$, we have, for any $0 < \gamma \leq 1/q'$,*

$$\|T_{\Omega,\alpha,\beta}f\|_{L^2(\mathbf{R}^n)} \leq C\|f\|_{L^2_{-\gamma}(\mathbf{R}^n)}.$$

In order to apply an interpolation, we recall a boundedness result at the end point $|1/p - 1/2| = (\beta - 2\alpha)/2\beta$ in Theorem 1.2. In fact, we have

$$\|T_{\Omega,\alpha,\beta}f\|_{L^{\beta/(\beta-\alpha)}} \leq C\|f\|_{L^{\beta/(\beta-\alpha)}}, \tag{1.11}$$

and

$$\|T_{\Omega,\alpha,\beta}f\|_{L^{\beta/\alpha}} \leq C\|f\|_{L^{\beta/\alpha}}. \tag{1.12}$$

Interpolating between (1.11), (1.12) with the results in Proposition 1.2 and applying the Sobolev imbedding Theorem, we now obtain

Theorem 1.6. *Let $\beta/(\beta-\alpha) \leq p \leq \beta/\alpha$, $\Omega \in L^q(S^{n-1})$ and $\beta \geq 2\alpha > 0$. Then*

(i) *if $1/q' > (\beta-2\alpha)/(2\beta+2)$, we have*

$$\|T_{\Omega,\alpha,\beta}f\|_{L^p(\mathbf{R}^n)} \leq C\|f\|_{L^r(\mathbf{R}^n)}$$

for any $p_0 \leq r \leq p$ with $1/p_0 = 1/p + (\beta - \alpha\max\{p,p'\})/n(\beta+1)\max\{p,p'\}$.

(ii) *If $1/q' \leq (\beta-2\alpha)/(2\beta+2)$, we have*

$$\|T_{\Omega,\alpha,\beta}\|_{L^p(\mathbf{R}^n)} \leq C\|f\|_{L^p(\mathbf{R}^n)}$$

for any $p_1 \leq r \leq p$ with $1/p_1 = 1/p + (1/\max\{p,p'\})/nq'(1/2 - \alpha/\beta)$.

2 The phase function is not radial

In the previous discussion, the oscillating factor $e^{i|y|^{-\beta}}$ is a radial function. This condition also can be weakened if $\Omega \in L^q(S^{n-1})$. For simplicity we consider the case $\alpha = 0$, and a local operator

$$R_\Omega f(x) = \int_{|y|\leq 1} e^{i\Psi(y)}|y|^{-n}\Omega(y')f(x-y)dy.$$

We assume that the phase function Ψ satisfies

$$\Psi(ty') = t^{-\beta}P(y'), \quad \beta > 0. \tag{2.1}$$

We have the following theorem.

Theorem 2.1. *Let $\Omega \in L^q(S^{n-1})$, $q > 1$ (no cancellation condition is needed), and Ψ is defined in (2.1). If there is a $\delta > 0$ such that*

$$\int_{S^{n-1}} |P(y')|^{-\delta}d\sigma(y') < \infty,$$

then the operator R_Ω is bounded on $L^p(\mathbb{R}^n)$ for any $1 < p < \infty$.

Proof. Let $\{\Phi_k\}_{-\infty}^{\infty}$ be a smooth partition of the unity in $(0,\infty)$ adapted to the intervals $(2^{k-1}, 2^{k+1})$. To be precise, we choose a radial function $\Phi \in C^\infty(\mathbb{R}^n)$ satisfying $\text{supp}(\Phi) \subset \{x : 1/2 < |x| \leq 2\}$ and $0 \leq \Phi(x) \leq 1$. We let $\Phi_k(x) = \Phi(2^k x)$ and require that Φ satisfies, for all $t \in (0,\infty)$,

$$\sum_{k=-\infty}^{\infty} \Phi_k(t) = 1.$$

It is easy to see $\text{supp}(\Phi_k) \subset (2^{-k-1}, 2^{-k+1})$. Thus we have

$$R_\Omega f(x) = \sum_{k=0}^{\infty} R_{\Omega,k}f(x),$$

where

$$R_{\Omega,k}f(x) = \int_{\mathbf{R}^n} e^{i\Psi(y)}|y|^{-n}\Omega(y')\Phi_k(y)f(x-y)dy.$$

By Fubini's theorem, we have

$$\|R_{\Omega,k}f\|_{L^1(\mathbf{R}^n)} \leq C\|f\|_{L^1(\mathbf{R}^n)}$$

uniformly on all k. The Fourier transform of $R_{\Omega,k}$ is $\hat{f}(\xi)m_k(\xi)$, where

$$m_k(\xi) \cong \int_{S^{n-1}} \Omega(y')\left(\int_0^\infty I_k(ty',\xi)dt\right)d\sigma(y')$$

and

$$\int_0^\infty I_k(ty', \xi)dt = \int_0^\infty \Phi(t)t^{-1}e^{\left(i\Psi(2^{-k}ty')-2^{-k}t<y',\xi>\right)}dt.$$

By the definition of Ψ,

$$\Psi(2^{-k}ty') = 2^{k\beta}t^{-\beta}P(y').$$

By the Van der Corput's lemma, we now have

$$\left|\int_0^\infty I_k(ty', \xi)dt\right| \le C|2^{k\beta}P(y')|^{-1/2}.$$

On the other hand, it is trivial to see

$$\left|\int_0^\infty I_k(ty', \xi)\right| \le C.$$

Thus we further have

$$\left|\int_0^\infty I_k(ty', \xi)dt\right| \le C|2^{k\beta}P(y')|^{-\delta/q'}.$$

By Hölder's inequality, we have

$$\begin{aligned}|m_k(\xi)| &\le C\int_{S^{n-1}} |\Omega(y')||2^{k\beta}P(y')|^{-\delta/q'}d\sigma(y')\\ &\le C2^{-k\theta}\|\Omega\|_{L^q(S^{n-1})}\end{aligned}$$

for some $\theta > 0$. By Plancherel theorem, one has

$$\|R_{\Omega,k}f\|_{L^2(\mathbf{R}^n)} \le C2^{-k\theta}\|f\|_{L^2(\mathbf{R}^n)}.$$

Thus, there is another $0 < \varepsilon \le \theta$ such that

$$\|R_{\Omega,k}f\|_{L^p(\mathbf{R}^n)} \le C2^{-k\varepsilon}\|f\|_{L^p(\mathbf{R}^n)}.$$

Thus

$$\|R_\Omega f\|_{L^p(\mathbf{R}^n)} \le C\sum_k \|R_{\Omega,k}f\|_{L^p(\mathbf{R}^n)} \le C\sum_{k\ge 0} 2^{-k\theta}\|f\|_{L^p(\mathbf{R}^n)}. \qquad \square$$

3 The kernel satisfies a Lipschitz condition

In the previous section, we study the case that Ω is a rough function on the unit sphere. Now we study the case that Ω has some smoothness condition. For simplicity in the discussion, we study the case $\alpha = 0$. Thus we focus on the local operator $L_{\Omega,0,\beta}$. Our results can be easily extended to case $\alpha > 0$.

Theorem 3.1. *Suppose* $\beta > 0$. *If* F_Ω *satisfies a Lipschitz condition* ρ, *then for any* $0 \leq \nu \leq \beta(\rho+1)/2(\beta+1)$, *we have*

$$\| L_{\Omega,0,\beta}(f) \|_{L^2(R^n)} \leq C \| f \|_{L^2_{-\nu}(R^n)} .$$

Proof. Recall that the Fourier transform of $L_{\Omega,0,\beta}$ is $\hat{f}(\xi)m(\xi)$, where

$$m(\xi) \cong \int_{S^{n-1}} \Omega(y') \int_0^1 e^{it^{-\beta}} t^{-1} e^{-2\pi i t<y',\xi>} dt \, d\sigma(y').$$

By the Plancherel theorem, it suffices to show that

$$|m(\xi)| \precsim |\xi|^{-\nu}$$

for any $0 \leq \nu \leq \beta(\rho+1)/2(\beta+1)$.

Following the same argument as in Section 1, for each fixed ξ, let $A \in SO(n)$ satisfy $A^{-1}(\xi) = |\xi|(1,0,0,...,0)$ and write $\Omega_1(y') = \Omega(A(y'))$. By the rotation method of Calderón-Zygmund, we have

$$m(\xi) \cong \int_R F_{\Omega_1}(s) N(s|\xi|) ds,$$

where

$$N(s|\xi|) = \int_0^1 e^{it^{-\beta}} t^{-1} e^{-2\pi i t s|\xi|} dt.$$

It is easy to check that $\Omega(y')$ and $\Omega(A(y'))$ have the same size and smoothness conditions uniformly on $A \in SO(n)$. Thus, without loss of generality, again we write $\Omega_1(y') = \Omega(y')$. Also, checking the proof of Proposition 1.1, we easily see that the absolute value of $N(s|\xi|)$ is uniformly bounded. On the other hand, by a well-known estimate (see Stein's book[19], p.357), we know that if $|s\xi|$ is sufficiently large then

$$\int_0^1 e^{it^{-\beta}} t^{-1} e^{-2\pi i t s|\xi|} dt \cong e^{ic_1|s\xi|^{\frac{\beta}{\beta+1}}} |s\xi|^{\frac{-\beta}{2(\beta+1)}} + O(|s\xi|^{\frac{-\beta}{(\beta+1)}}),$$

where c_1 is a constant independent of all essential variables. Thus we may assume $c_1 = 1$ in our argument. By the boundedness of $m(\xi)$, clearly for any $|\xi| \leq 20$ and any $0 \leq \theta \leq 1$,

$$|m(\xi)| \precsim (1+|\xi|)^{-\theta(\frac{\beta}{\beta+1})}.$$

Assume $|\xi| \geq 20$. Let $c = \frac{(100\pi)^{\beta+1/\beta}}{200}$. We write

$$|m(\xi)| = \left| \int_{|s\xi| \geq 200c} F_\Omega(s) N(s|\xi|) ds + \int_{|s\xi| < 200c} F_\Omega(s) N(s|\xi|) ds \right| .$$

Clearly, the second integral above satisfies

$$\left|\int_{|s\xi|<200c} F_\Omega(s)N(s|\xi|)ds\right| \precsim \|\Omega\|_{L^\infty}\int_{|s|\le 200c/|\xi|} ds \approx |\xi|^{-\theta}$$

for any $0\le\theta\le 1$.

For the first integral, we write

$$\begin{aligned}\int_{|s\xi|\ge 200c} F_\Omega(s)N(s|\xi|)ds &\approx \int_{|s\xi|\ge 200c} F_\Omega(s)e^{ic_1|s\xi|^{\frac{\beta}{\beta+1}}}|s\xi|^{\frac{-\beta}{2(\beta+1)}})ds\\ &\quad + O\left(\int_{|s\xi|\ge 200c}|F_\Omega(s)||s\xi|^{\frac{-\beta}{\beta+1}}ds\right).\end{aligned}$$

It is trivial to check that the second term above is bounded by

$$O\left(\int_{1\ge|s|\ge 200c/|\xi|}|s\xi|^{\frac{-\beta}{\beta+1}}(1-s^2)^{1/2}ds\right) = O(|\xi|^{\frac{-\beta}{\beta+1}}).$$

For the first term, (remember we assume c_1 in the phase function to be 1) changing variables $s=t^{\frac{\beta+1}{\beta}}$, we have

$$\begin{aligned}&\int_{200c/|\xi|}^{1} F_\Omega(s)e^{i|s\xi|^{\frac{\beta}{\beta+1}}}|s\xi|^{\frac{-\beta}{2(\beta+1)}}ds\\ &\cong |\xi|^{\frac{-\beta}{2(\beta+1)}}\int_{(200c/|\xi|)^{\frac{\beta}{\beta+1}}}^{1} F_\Omega(t^{\frac{\beta+1}{\beta}})e^{it|\xi|^{\frac{\beta}{\beta+1}}}t^{\frac{2-\beta}{2\beta}}dt\\ &= |\xi|^{\frac{-\beta}{2(\beta+1)}}\sum_{k=100}^{N}\int_{\pi k/|\xi|^{\frac{\beta}{\beta+1}}}^{\pi(k+1)/|\xi|^{\frac{\beta}{\beta+1}}} F_\Omega(t^{\frac{\beta+1}{\beta}})e^{it|\xi|^{\frac{\beta}{\beta+1}}}t^{\frac{2-\beta}{2\beta}}dt,\end{aligned}$$

where $N\approx|\xi|^{\frac{\beta}{(\beta+1)}}$, without loss of generality we assume N is even. By changing variables, we see that

$$\begin{aligned}&|\xi|^{\frac{-\beta}{2(\beta+1)}}\left|\sum_{k=100}^{N}\int_{\pi k/|\xi|^{\frac{\beta}{\beta+1}}}^{\pi(k+1)/|\xi|^{\frac{\beta}{\beta+1}}} F_\Omega(t^{\frac{\beta+1}{\beta}})e^{it|\xi|^{\frac{\beta}{\beta+1}}}t^{\frac{2-\beta}{2\beta}}dt\right|\\ &\le |\xi|^{\frac{-\beta}{2(\beta+1)}}\sum_{k=50}^{N/2}\int_{\pi 2k/|\xi|^{\frac{\beta}{\beta+1}}}^{\pi(2k+1)/|\xi|^{\frac{\beta}{\beta+1}}} |(t+\pi/|\xi|^{\frac{\beta}{\beta+1}})^{\frac{2-\beta}{2\beta}}F_\Omega((t+\pi/|\xi|^{\frac{\beta}{\beta+1}})^{\frac{\beta+1}{\beta}})\\ &\quad -t^{\frac{2-\beta}{2\beta}}F_\Omega(t^{\frac{\beta+1}{\beta}})|dt\\ &\le |\xi|^{\frac{-\beta}{2(\beta+1)}}\sum_{k=50}^{N/2}\int_{\pi 2k/|\xi|^{\frac{\beta}{\beta+1}}}^{\pi(2k+1)/|\xi|^{\frac{\beta}{\beta+1}}} |\{(t+\pi/|\xi|^{\frac{\beta}{\beta+1}})^{\frac{2-\beta}{2\beta}}-t^{\frac{2-\beta}{2\beta}}\}\\ &\quad \cdot F_\Omega((t+\pi/|\xi|^{\frac{\beta}{\beta+1}})^{\frac{\beta+1}{\beta}})|dt\end{aligned}$$

$$
\begin{aligned}
&+|\xi|^{\frac{-\beta}{2(\beta+1)}} \sum_{k=50}^{N/2} \int_{\pi 2k/|\xi|^{\frac{\beta}{\beta+1}}}^{\pi(2k+1)/|\xi|^{\frac{\beta}{\beta+1}}} |t^{\frac{2-\beta}{2\beta}} \{F_\Omega((t+\pi/|\xi|^{\frac{\beta}{\beta+1}})^{\frac{\beta+1}{\beta}}) \\
&-F_\Omega(t^{\frac{\beta+1}{\beta}})\}|dt \\
&= I_1 + I_2.
\end{aligned}
$$

It is easy to compute that

$$
I_1 \precsim |\xi|^{\frac{-\beta}{(\beta+1)}} \int_{100\pi/|\xi|^{\frac{\beta}{\beta+1}}}^{1} |F_\Omega(t^{\frac{\beta+1}{\beta}})| t^{\frac{2-\beta}{2\beta}} \cdot dt \precsim |\xi|^{\frac{-\beta}{(\beta+1)}},
$$

and

$$
\begin{aligned}
I_2 &\precsim |\xi|^{\frac{-\beta}{2(\beta+1)}} \sum_{k=50}^{N/2} \int_{\pi 2k/|\xi|^{\frac{\beta}{\beta+1}}}^{\pi(2k+1)/|\xi|^{\frac{\beta}{\beta+1}}} t^{\frac{2-\beta}{2\beta}} \parallel F_\Omega \parallel_{lip_\rho} |(t+\pi/|\xi|^{\frac{\beta}{\beta+1}})^{\frac{\beta+1}{\beta}} - t^{\frac{\beta+1}{\beta}}|^\rho dt \\
&\precsim |\xi|^{\frac{-\beta-\rho\beta}{2(\beta+1)}}.
\end{aligned}
$$

We prove the theorem by combining the estimates on $|\xi| > 20$ and $|\xi| \leq 20$. □

By Theorem 3.1, we can easily obtain $L^q \to L^p$ estimate by using the Sobolev imbedding theorem. We also can follow the same idea as in the previous section to extend our results to the case $\alpha > 0$.

4 The kernel is C^∞

In this section, we assume that Ω is a C^∞ function on S^{n-1}. We view the operator $L_{\Omega,\alpha,\beta}$ as a bilinear operator

$$
T_{\alpha,\beta}(f,\Omega)(x) = \int_{|y|\leq 1} \frac{e^{i|y|^{-\beta}}}{|y|^{n+\alpha}} \Omega(y) f(x-y) dy,
$$

when $\Omega \in C^\infty(S^{n-1})$.

Then, we can obtain the following sharp $L^q \to L^p$ boundedness theorem:

Theorem 4.1. *Let $n\beta > 2\alpha \geq 0$ and $p > 1$. Then*

$$
\|T_{\alpha,\beta} f\|_{L^p(\mathbf{R}^n)} \leq C \|\Omega\|_{C^{2n}} \|f\|_{L^q(\mathbf{R}^n)}
$$

for all $q_0 \leq q \leq p$ if and only if

$$
\begin{cases}
\left|\dfrac{1}{p} - \dfrac{1}{2}\right| \leq \dfrac{n\beta - 2\alpha}{2n\beta}, \\
q_0 = \dfrac{\beta+1}{\beta + (1/p - \alpha/n)},
\end{cases}
$$

where the C^k norm of Ω is defined by

$$\|\Omega\|_{C^k} = \sum_{0\le|\alpha|\le k} \|D^\alpha\Omega\|_{L^\infty(S^{n-1})}.$$

Proof. We first proof the "only if" part.

Let g be a test function supported in $|y| \le 2$ and identically equal one in the set $|y| \le 1$. We also assume that $0 \le g(y) \le 1$. For small $\varepsilon > 0$, let $f_\varepsilon(x) = g(x/\varepsilon^{\beta+1})$ and take $\Omega \equiv 1$. We have

$$\|\Omega\|_{C^{2n}}\|f_\varepsilon\|_{L^q(\mathbf{R}^n)} \cong \varepsilon^{n(\beta+1)/q}.$$

Let $\{J_k\}$ be a family of sets defined by

$$J_k = \{y \in \mathbb{R}^n : \ (2\pi k + \pi/4)^{-1/\beta} \le |y| \le (2\pi k - \pi/4)^{-1/\beta}\}.$$

We have

$$\text{Volume}(J_k) \cong k^{-1-n/\beta}$$

and

$$\text{diam}(J_k) \cong k^{-1-1/\beta}.$$

Without loss of generality, we may assume $\text{diam}(J_k) \cong 10k^{-1-1/\beta}$. Let $0 < \varepsilon \le k^{-1/\beta}$, we define another sequence of sets $\{I_k\}$ by

$$I_k = \{y \in \mathbb{R}^n : (2\pi k + \pi/4)^{-1/\beta} + 2\varepsilon^{\beta+1} \le |y| \le (2\pi k - \pi/4)^{-1/\beta} - 2\varepsilon^{\beta+1}\}.$$

Clearly, we have $I_k \subset J_k$ and $\text{Volume}(J_k) \cong \text{Volume}(I_k)$, $\text{diam}(J_k) \cong \text{diam}(I_k)$.

Now we have

$$\|T_{\alpha,\beta}(f_\varepsilon, 1)\|_{L^p(\mathbf{R}^n)} \ge C\sum_{k=0}^{\varepsilon^{-\beta}} \int_{I_k} \left| \int_{|y|\le 1} \frac{\cos|y|^{-\beta}}{|y|^{n+\alpha}} f\left(\frac{x-y}{\varepsilon^{\beta+1}}\right) dy\right|^p dx.$$

Note that if $x \in I_k$, $f(x - y/\varepsilon^{\beta+1}) = 0$ unless $y \in J_k$. Therefore we have

$$\|T_{\alpha,\beta}(f_\varepsilon, 1)\|_{L^p(\mathbf{R}^n)} \gtrsim \sum_{k=0}^{\varepsilon^{-\beta}} \int_{I_k} \left| \int_{J_k} \frac{f((x-y)/\varepsilon^{\beta+1})}{|y|^{n+\alpha}} dy\right|^p dx$$

$$\gtrsim \varepsilon^{np(\beta+1)} \sum_{k=0}^{\varepsilon^{-\beta}} k^{p(n+\alpha)/\beta-1-n/\beta}.$$

Thus, we must have $\varepsilon^{n(\beta+1)/q} \gtrsim \varepsilon^{n\beta-\alpha+n/p}$ for all small $\varepsilon > 0$. It implies that $q \ge \frac{n(\beta+1)}{n\beta-\alpha+n/p} = q_0$. The "only if" part of the theorem is proved.

To prove the "if" part, we recall the following lemma, which is an easy modification of results in Littman [14] and Domar [5].

Lemma 4.1. *Let $A \subseteq \mathbb{R}^n$ denote an open set and $\Phi \in C_0^\infty(A)$. If $\Psi \in C^\infty(A)$ satisfies*

$$\left|\det\left(\frac{\partial^2}{\partial x_i \partial x_j}\Psi(x)\right)\right| \geq C > 0$$

for all $x \in \operatorname{supp}(\Phi)$, then for large $|\lambda|$,

$$\left|\int_A e^{i(\lambda\Psi(x)+<\xi,x>)\Phi(x)}dx\right| \lesssim |\lambda|^{-n/2}\|\Phi\|_{C^{2n}}.$$

Taking the Fourier transform on $T_{\alpha,\beta}(f,\Omega)(x)$, we have

$$\widehat{T_{\alpha,\beta}(f,\Omega)}(\xi) \cong \hat{f}(\xi)m(\xi),$$

and

$$m(\xi) = \int_{|y|\leq 1} \frac{\Omega(y')}{|y|^{n+\alpha}} e^{i(|y|^{-\beta}-<y,\xi>)}dy.$$

We have the following estimate on $m(\xi)$.

Lemma 4.2. $|m(\xi)| \lesssim (1+|\xi|)^{-\beta n - 2\alpha/2(\beta+1)}\|\Omega\|_{C^{2n}}$.

Proof. Let $\{\Phi_j\}_{-\infty}^{\infty} = \{\Phi(2^j|y|)\}_{-\infty}^{\infty}$ be a smooth partition of unity in $(0,\infty)$ adapted to the intervals $(2^{-j-1}, 2^{-j+1})$. Thus Φ is supported in the set $\{y : 1/2 \leq |y| \leq 2\}$. We also can require $\sum_{j=-\infty}^{\infty}\Phi_j(t) \equiv 1$ for any $t > 0$. We write $m(\xi) = \sum_{j=0}^{\infty} m_j(\xi)$, where, by changing variables,

$$m_j(\xi) \cong 2^{j\alpha}\int_{|y|\leq 1} \frac{\Omega(y')}{|y|^{n+\alpha}} e^{i(2^{j\beta}|y|^{-\beta}-2^{-j}<y,\xi>)}\Phi(y)dy.$$

We define three sets E_1, E_2, E_3 as

$$E_1 = \{j \geq 0 : 2^{j(\beta+1)} \geq C_1|\xi|\},$$

$$E_2 = \{j \geq 0 : 2^{j(\beta+1)} \leq C_2|\xi|\},$$

$$E_3 = \{j \geq 0 : C_2|\xi| < 2^{j(\beta+1)} < C_1|\xi|\},$$

where C_1 is a large number and C_2 is a small number. Both two numbers may depend on n and β, but independent of ξ and j. Thus

$$m(\xi) = \sum_{j\in E_1} m_j(\xi) + \sum_{j\in E_2} m_j(\xi) + \sum_{j\in E_3} m_j(\xi).$$

If $j \in E_1$, we use the polar coordinates to write

$$m_j(\xi) \cong 2^{j\alpha}\int_{S^{n-1}} \Omega(y')\int_0^1 \frac{\Phi(t)}{t^{1+\alpha}} e^{i(2^{j\beta}t^{-\beta}-2^{-j}t|\xi|<y',\xi'>)}dtd\sigma(y').$$

Note that when $t \in [1/2, 2]$ and $C_1 > 100 \cdot 2^\beta/\beta$, we have

$$\left|\frac{d}{dt}\left(2^{j\beta}t^{-\beta} - 2^{-j}t|\xi| < \xi', y' >\right)\right| \gtrsim \left|\beta 2^{j\beta}t^{-\beta-1}\right| - 2^{-j}|\xi|$$
$$\gtrsim 2^{j\beta}.$$

Thus, an easy integration by parts yields $|m_j(\xi)| \lesssim 2^{-j\beta N + j\alpha}\|\Omega\|_{C^{2n}}$ for any $N > 0$.

If $j \in E_2$, without loss of generality, we assume $|\xi_1| \geq |\xi|/n$. It is easy to see that, if $y \in \mathrm{supp}(\Phi)$, we have

$$\left|\frac{\partial}{\partial y_1}\left(2^{j\beta}|y|^{-\beta} - 2^{-j} < \xi, y >\right)\right| \gtrsim 2^{-j}|\xi_1| - \beta|y|^{-\beta-2}2^{j\beta}|y_1|$$
$$\gtrsim 2^{-j}|\xi|.$$

Again, we use integration by parts to obtain, if $j \in E_2$,

$$|m_j(\xi)| \lesssim 2^{j\alpha + Nj}|\xi|^{-N}\|\Omega\|_{C^{N+1}}$$

for any $N > 0$.

We also note that the set E_3 contains only finite number of j and the total number of E_3 is independent of j and ξ. Thus, by Lemma 4.1, if $j \in E_3$, we have

$$|m_j(\xi)| \lesssim 2^{-j\beta n/2}2^{j\alpha}\|\Omega\|_{C^{2n}} \cong |\xi|^{-(\beta n - 2\alpha)/2(\beta+1)}\|\Omega|_{C^{2n}}.$$

Finally, we choose $N = n/2$. Thus

$$\left|\sum_{j\in E_1} m_j(\xi)\right| \lesssim \sum_{2^{-j}\leq|\xi|^{-1/(\beta+1)}} 2^{-j(N\beta-\alpha)}\|\Omega\|_{C^{2n}} \lesssim |\xi|^{-(\beta n-2\alpha)/2(\beta+1)}\|\Omega\|_{C^{2n}},$$

$$\left|\sum_{j\in E_2} m_j(\xi)\right| \lesssim |\xi|^{-N} \sum_{2^{j}\leq|\xi|^{1/(\beta+1)}} 2^{j(\alpha+N)}\|\Omega\|_{C^{2n}} \lesssim |\xi|^{-(\beta n-2\alpha)/2(\beta+1)}\|\Omega\|_{C^{2n}}.$$

It is also easy to check that $|m(\xi)| \lesssim \|\Omega\|_{C^{2n}}$. The lemma is proved. □

Our theorem will be followed from Lemma 4.2 and the Sobolev imbedding theorem. In [8], we know that

$$\|T_{\alpha,\beta}(f,1)\|_{L^{p_0}(\mathbf{R}^n)} \lesssim \|f\|_{L^{p_0}(\mathbf{R}^n)}$$

for $p_0 = n\beta/(n\beta - \alpha)$ and $p_0 = n\beta/\alpha$, and

$$\|T_{0,\beta}(f,1)\|_{L^1(\mathbf{R}^n)} \lesssim \|f\|_{H^1(\mathbf{R}^n)}.$$

It is not difficult to see Fefferman and Stein's results can be extended as

$$\|T_{0,\beta}(f,\Omega)\|_{L^1(\mathbf{R}^n)} \precsim \|\Omega\|_{C^{2n}}\|f\|_{H^1(\mathbf{R}^n)},$$

and

$$\|T_{\alpha,\beta}(f,\Omega)\|_{L^{p_0}(\mathbf{R}^n)} \precsim \|\Omega\|_{C^{2n}}\|f\|_{L^{p_0}(\mathbf{R}^n)}$$

for $p_0 = n\beta/(n\beta-\alpha)$ and $p_0 = n\beta/\alpha$.

Using interpolation, we obtain, for any $n\beta/(n\beta-2\alpha) \leq p \leq 2$,

$$\|T_{\alpha,\beta}(f,\Omega)\|_{L^p(\mathbf{R}^n)} \precsim C\|\Omega\|_{C^{2n}}\|f\|_{L^p_{-\theta}(\mathbf{R}^n)}$$

for any $0 \leq \theta \leq (n\beta - \alpha p')/p'(\beta+1)$. We obtain the theorem for $p \leq 2$ by the Sobolev imbedding theorem.

Similarly we can show the case $p \geq 2$. □

References

[1] J. Chen, D. Fan, and Y. Ying, Certain operators with singular kernels, Canadian J. Math. **55** (2003), 504–532.

[2] L. Chen, On a singular integral, Studia Math. **TLXXXV** (1987), 61–72.

[3] A. P. Calderón and A. Zygmund, On existence of certain singular integrals, Acta. Math. **88** (1952), 85–139.

[4] A. P. Calderón and A. Zygmund, On singular integrals, Amer. J. Math. **18** (1956), 289–309.

[5] Y. Domar, On the spectral synthesis problem for $(n-1)$-dimensional subset of $\mathbb{R}^n$, $n \geq 2$, Ark Math. **9** (1971), 23–37.

[6] J. Duoandikoetxea and J. L. Rubio de Francia, Maximal and singular integral operators via Fourier transform estimates, Invent. Math. **84** (1986), 541–561.

[7] R. Fefferman, A note on singular integrals, Proc. Amer. Math. Soc. **74** (1979), 266–270.

[8] C. Fefferman and E. Stein, H^p spaces of several variables, Acta Math. **129** (1972), 137–193.

[9] D. Fan and Y. Pan, Singular integral operators with rough kernels supported by subvarieties, Amer. J. Math. **119** (1997), 799–839.

[10] D. Fan and Y. Pan, L^2 boundedness of a singular integal operator, Publications Matematiques **41** (1997), 317–333.

[11] D. Fan and Y. Pan, A singular integral operator with rough kernel, Proc. Amer. Math Soc. **125** (1997), 3695–3703.

[12] L. Grafakos and A. Stefanov, Convolution Calderón-Zygmund singular operators with rough kernels, Analysis of divergence (Oronon, ME, 1997), 119–143, Appl. Number. Anal., Birkhäuser Boston, Boston, MA, 1999.

[13] I. Hirschman Jr., On multiplier transformations, Duke Math. Journal **26** (1959), 221–242.

[14] W. Littman, Fourier transforms of surface-carried measures and differentiability of surface average, Bull. Amer. Math. Soc. **69** (1963), 766–770.

[15] H. Le, Hypersingular integrgal operators along surfaces, Integr. Equ. Oper. Theory **44** (2002), 451–465.

[16] H. Le, Singular integral operators along surfaces of revolution, J. Math. Anal. Appl. **274** (2002), 608–625.

[17] J. Namazi, A Singular integral, Ph. D. Thesis, Indiana University, Bloomington, 1984.

[18] F. Ricci and G. Weiss, A characterization of $H^1(\Sigma_{n-1})$, Proc. Sympos. Pure Math. (S. Wainger and G. Weiss eds.) **35**, Amer Math. Soc., Providence, RI, 289–294.

[19] E. Stein, Harmonic analysis: real-variable methods, orthogonality, and oscillatory integrals. With the assistance of Timothy S. Murphy. Princeton Mathematical Series **43**. Monographs in Harmonic Analysis, III. Princeton University Press, Princeton, NJ, 1993.

[20] E. Stein and S. Wainger, Problems in harmonic analysis related to curvature, Bull. Amer. Math. Soc. **84** (1978), 1239–1295.

[21] S. Wainger, Special Trigonometric Series in K Dimension, Memoris Amer. Math. Soc **59**, Amer Math. Soc., Providence, RI, 1965.

[22] R. L. Wheeden, On hypersingular integrals and Lebesgue spaces of differentiable functions (II), Trans. Amer Math Soc., (1969), 37–53.

Some Topics in Harmonic Analysis and Applications
ALM 34, pp. 93–112

On the Dimension Dependence of Some Weighted Inequalities

Alberto Criado* and Fernando Soria†

Abstract

In the context of radial weights we study the dimension dependence of some weighted inequalities for maximal operators. We study the growth of the A_1-constants for radial weights and show the equivalence between the uniform boundedness of these constants, a dimension-free weak L^1 estimate for the maximal operator on annuli and the condition on the weight to be decreasing and essentially constant over dyadic annuli. Each one of these conditions is shown to provide dimension-free weighted weak type L^1 estimates for the centred maximal Hardy-Littlewood operator acting on radial functions. Finally we show that the universal maximal operator is of restricted weak type on weighted $L^n(\mathbb{R}^n)$ with constants uniformly bounded in dimension whenever we consider an A_1 weight.

2000 Mathematics Subject Classification: 42B25.

Keywords and Phrases: radial weights, dimension-free estimates.

1 Introduction

In this paper we will study the dimension dependence of the bounds for some maximal operators when acting over weighted spaces. First, we consider the centered Hardy-Littlewood maximal operator over Euclidean balls. For a locally integrable g on $\mathbb{R}^n$ it is defined as

$$Mg(x) = \sup_{R>0} \fint_{B_R(x)} |g(y)|\, dy,$$

where $B_R(x)$ is the Euclidean ball of radius R centered at x. It is well known since the time of Hardy and Littlewood that for each n, this operator is bounded

*Departamento de Matemáticas, Universidad Autónoma de Madrid, 28049 Madrid, Spain.
†Departamento de Matemáticas and Instituto de Ciencias Matemáticas CSIC-UAM-UC3M-UCM, Universidad Autónoma de Madrid, 28049 Madrid, Spain.

on $L^p(\mathbb{R}^n)$ for $p > 1$ and weakly bounded on $L^1(\mathbb{R}^n)$. Much later, Stein raised the questions whether the operator norm in these inequalities could be bounded independently of the dimension and whether this uniformity in dimension could be related to an infinite dimensional phenomenon. As far as we are concerned, very little is known about the second question. As for the first one, Stein himself showed in [30] (details in [31]) that for all $p > 1$ one has $\|M\|_{L^p(\mathbb{R}^n)\to L^p(\mathbb{R}^n)} \le C_p$, with C_p independent of n. In joint work with Strömberg [31] he also proved that $\|M\|_{L^1(\mathbb{R}^n)\to L^{1,\infty}(\mathbb{R}^n)} = O(n)$ as $n \to \infty$. Although this does not solve the still open problem of deciding whether these weak L^1 operator norms grow to infinity with the dimension or not, it is still the best known result.

This problem of uniform bounds in dimension has also been studied for maximal functions where the averages are taken over balls given by arbitrary norms in $\mathbb{R}^n$. In all cases one has $\|M\|_{L^p(\mathbb{R}^n)\to L^p(\mathbb{R}^n)} \le C$ with C depending only on p for $p > 3/2$ (see [5, 6, 7, 10]). For the balls given by the ℓ^q metrics in $\mathbb{R}^n$, with $1 \le q \le \infty$, for all $p > 1$ one has $\|M\|_{L^p(\mathbb{R}^n)\to L^p(\mathbb{R}^n)} \le C_{p,q}$ with $C_{p,q}$ independent of n (see [26] for the case $1 \le q < \infty$ and [9] for $q = \infty$). As for the weak L^1 inequalities in this case, Stein and Strömberg proved that, in general, $\|M\|_{L^1(\mathbb{R}^n)\to L^{1,\infty}(\mathbb{R}^n)} = O(n \log n)$ as $n \to \infty$. It is still unknown if these operator norms remain bounded in all dimensions, except when those averages are taken over the balls of the ℓ^∞ metric, that is, the case of cubes with sides parallel to the coordinate axes. In this case Aldaz showed in [2] that these weak L^1 operator norms grow to infinity with the dimension (see also [4]).

A variation of the problem arises when the maximal operator is defined using measures different from the Lebesgue one. For a Radon measure μ on $\mathbb{R}^n$ we define the associated maximal operator

$$M_\mu g(x) = \sup_{\substack{R>0 \\ \mu(B_R(x))>0}} \frac{1}{\mu(B_R(x))} \int_{B_R(x)} |g(y)|\, d\mu(y).$$

If μ has a radial density w, then μ and M_μ can be defined in all dimensions. We may ask then if the operator norm of M_μ in $L^p(\mu)$ is uniformly bounded in dimension. If μ is finite this is not true in general (see [1, 13, 3]). For instance, when μ is the Gaussian measure, the $L^p(\mu)$ operator norms of M_μ grow exponentially to infinity with the dimension for all $p < \infty$ (see [14]). The situation is very different when μ satisfies a doubling condition. One says that μ is uniformly strong n-microdoubling if there exist $K > 0$ and $N > 0$ so that for all $n \ge N$, $x \in \mathbb{R}^n$, $R > 0$ and $y \in B_R(x)$ one has

$$\mu(B_{(1+1/n)R}(x)) \le K\mu(B_R(x)) \quad \text{and} \quad \mu(B_R(y)) \le K\mu(B_R(x)).$$

Roughly speaking this property guarantees that small dilations and translation do not alter essentially the measure of a ball. If μ is such a measure, then one recovers the Stein and Strömberg bound $\|M_\mu\|_{L^1(\mathbb{R}^n,d\mu)\to L^{1,\infty}(\mathbb{R}^n,d\mu)} = O(n \log n)$ (see [27]). Moreover one has the uniform bound $\|M_\mu\|_{L^p(\mathbb{R}^n,d\mu)\to L^p(\mathbb{R}^n,d\mu)} \le C_{p,\mu}$ for all $n \ge N$ and $p > 1$ (see [15]).

Still another problem is the one that considers weighted inequalities for the maximal operator. A weight w is an a.e. nonnegative and locally integrable function over $\mathbb{R}^n$. A weight is often regarded as the density of a measure over $\mathbb{R}^n$ that is absolutely continuous with respect to the Lebesgue measure. Following the usual notation we will also denote this measure by w, i.e. for a measurable E we will write $w(E) = \int_E w$ and we will say that a function $f \in L^p(\mathbb{R}^n, w)$ if $\int_{\mathbb{R}^n} |f|^p w < \infty$. For $p \geq 1$ we say that a weight w is in the class $A_p(\mathbb{R}^n)$ if M sends $L^p(\mathbb{R}^n, w)$ into $L^{p,\infty}(\mathbb{R}^n, w)$ boundedly[1]. It is well known that for $p > 1$ this is equivalent to $M : L^p(\mathbb{R}^n, w) \to L^p(\mathbb{R}^n, w)$ boundedly. These bounds have been studied extensively. For more information see [20] or [22].

In this work we will consider radial weights, so that we can define them in all dimensions. For such a weight w we will write $w(x) = w_0(|x|)$ with $w_0 : [0, \infty) \to [0, \infty]$. Duoandikoetxea and Vega announced in [19] the following result.

Theorem 1.1. *Let w_0 be a nonnegative function on $[0, \infty)$, so that $w = w_0(|\cdot|) \in A_p(\mathbb{R}^N)$ with $p < 1$. Then for all $n \geq N$ one has $w \in A_p(\mathbb{R}^n)$ and, moreover,*

$$\|Mf\|_{L^p(\mathbb{R}^n,w)} \leq C\|f\|_{L^p(\mathbb{R}^n,w)},$$

with a constant C that might depend on p and w_0 but not on n.

This solves completely the problem of finding weighted L^p bounds that are uniform in dimension for $p > 1$. In this paper we present some partial results in the case $p = 1$. In the next section we present some result relating the growth of the A_1 constant of a weight with the uniformity of the weak L^1 weighted bounds for the maximal operator over radial functions. The proofs of these results are contained in Section 3. Finally, Section 4 is devoted to show some uniform bounds for the universal maximal operator and the Kakeya maximal operator over radial functions.

2 The maximal operator over radial functions

First, let us describe briefly the concepts and notation that we are going to deal with. We recall that $w \in A_1(\mathbb{R}^n)$ if and only if for some $C > 0$ one has

$$\frac{w(B_R(x))}{|B_R(x)|} \leq C \operatorname*{ess\,inf}_{y \in B_R(x)} w(y), \quad \text{a.e. } x \in \mathbb{R}^n, \ \forall R > 0.$$

Equivalently, $w \in A_1(\mathbb{R}^n)$ if

$$Mw(x) \leq C\, w(x), \quad \text{a.e. } x \in \mathbb{R}^n.$$

The smallest values of C for which the previous inequalities hold will be denoted by $[w]_{A_1(\mathbb{R}^n)}$ and $[w]^*_{A_1(\mathbb{R}^n)}$ respectively. Both are usually called the A_1 constant of the weight.

[1] In what follows, the expression $T : X \to Y$ will denote that the (sublinear) operator T is bounded between the spaces X and Y.

For radial weights there is still another characterization of the A_1 class. We use that if w is radial, then Mw is pointwise comparable with $\mathcal{A}w$, where $\mathcal{A}$ is the maximal operator over centered rings given by

$$\mathcal{A}u(x) := \sup_{0\le a\le |x|\le b} \fint_{a\le|y|\le b} |u(y)|\,dy.$$

More precisely one has:

Lemma 2.1. *Let $u(x) = u_0(|x|)$ be a radial function in $L^1_{\text{loc}}(\mathbb{R}^n)$. Then there exists a constant K_n only depending on the dimension such that for all $x \in \mathbb{R}^n$ one has*

$$\frac{1}{K_n}\,\mathcal{A}u(x) \le Mu(x) \le 2\mathcal{A}u(x). \tag{2.1}$$

When acting on radial functions $\mathcal{A}$ can be written as a one-dimensional maximal operator. Given a weight v over $[0,\infty)$ the associated uncentered maximal operator is defined as

$$\tilde{M}_v g(x) = \sup_{0\le a\le x\le b} \frac{1}{v([a,b])}\int_a^b |g(y)|v(y)\,dy,$$

for $g \in L^1_{\text{loc}}([0,\infty), v)$. If $u(x) = u_0(|x|)$ is a radial function over $\mathbb{R}^n$, writing $v_n(t) = t^{n-1}$, note that $\mathcal{A}u(x) = \tilde{M}_{v_n}u_0(|x|)$.

As a consequence $w(x) = w_0(|x|)$ is in $A_1(\mathbb{R}^n)$ if and only if there exists a constant $C>0$ so that for a.e. x one has

$$\mathcal{A}w(x) = \tilde{M}_{v_n}w_0(|x|) \le Cw_0(|x|) = Cw(x).$$

The smallest of such constants will be denoted by $(w)_{A_1(\mathbb{R}^n)}$. With the usual arguments in weight theory one can see that the previous condition is equivalent to the existence of a constant $C>0$ so that for each interval $I \subset [0,\infty)$ one has

$$\frac{w_0v_n(I)}{v_n(I)} \le C \operatorname*{ess\,inf}_{x\in I} w_0(x).$$

The smallest of such constants is again $(w)_{A_1(\mathbb{R}^n)}$.

Inequality (2.1) has already appeared in the literature. In [25] Menárguez and the second author used the inequality $Mu(x) \le 2\tilde{M}_{v_n}u_0(|x|)$ to prove (2.4) below. Inequality (2.1) in the form $c_n\mathcal{A}u(x) \le Mu(x) \le C_n\mathcal{A}u(x)$, with the constants depending on the dimension, was used by Duoandikoetxea, Moyua, Oruetxebarria and Seijo in [18]. From it they deduced that for any $p \ge 1$ the radial weights w so that $M : L^p_{\text{rad}}(w) \to L^p_{\text{rad}}(w)$ boundedly are exactly the radial A_p weights.

First, we point out that if $w \in A_1(\mathbb{R}^m)$ for some $m \in \mathbb{N}$, then $w \in A_1(\mathbb{R}^n)$ for all $n \ge m$ and $[w]^*_{A_1(\mathbb{R}^n)}$ and $(w)_{A_1(\mathbb{R}^n)}$ grow at most linearly with n.

Lemma 2.2. *Let w_0 be a non-negative function over $[0,\infty)$ and set $w(x) = w_0(|x|)$ and $d\mu(x) = w(x)\,dx$. If for some $m \in \mathbb{N}$ one has $w \in A_1(\mathbb{R}^m)$, then $w \in A_1(\mathbb{R}^n)$ for all $n \geq m$ and $[w]^*_{A_1(\mathbb{R}^n)} \leq C_m \frac{n}{m}[w]^*_{A_1(\mathbb{R}^m)}$. The same holds for $(w)_{A_1(\mathbb{R}^n)}$.*

These bounds are almost optimal, let us see this with an example. Consider $w(x) = (1-|x|)^{-\alpha}$ with $0 < \alpha < 1$. It is easy to see that $w \in A_1(\mathbb{R}^n)$ for all $n \in \mathbb{N}$. Note that in $\mathbb{R}^n$ one has

$$\begin{aligned}\frac{w(B_1(0))}{|B_1(0)|} &= \frac{1}{\omega_{n-1}/n}\int_{B_1}(1-|x|)^{-\alpha}\,dx = n\int_0^1 (1-t)^{-\alpha}t^{n-1}\,dt \\ &= n\,\frac{\Gamma(1-\alpha)\Gamma(n)}{\Gamma(n+1-\alpha)},\end{aligned} \tag{2.2}$$

where we have used the equality

$$\frac{\Gamma(x)\Gamma(y)}{\Gamma(x+y)} = \int_0^1 (1-t)^{x-1}t^{y-1}\,dt.$$

Since Γ is logarithmically convex, writing $n+1-\alpha = \alpha n + (1-\alpha)(n+1)$ one has $\Gamma(n+1-\alpha) \leq \Gamma(n)^\alpha\Gamma(n+1)^{1-\alpha}$, which together with (2.2) says that

$$\frac{w(B_1(0))}{|B_1(0)|} \geq \Gamma(1-\alpha)\,n^\alpha. \tag{2.3}$$

But note that $\inf_{y\in B_1(0)} w(y) = 1$. This gives $[w]^*_{A_1(\mathbb{R}^n)} \geq \Gamma(1-\alpha)\,n^\alpha$. Since α can be taken arbitrarily close to 1, this shows that the upper bound for $[w]^*_{A_1(\mathbb{R}^n)}$ in Lemma 2.2 is near to be optimal. The same calculation works for $(w)_{A_1(\mathbb{R}^n)}$.

There are radial weights for which the A_1 constants remain uniformly bounded in dimension. Indeed, we have the following characterization.

Proposition 2.3. *Let $w(x) = w_0(|x|)$ be a radial weight. The following statements are equivalent:*

(a) *There exists $N > 0$ so that $[w]_{A_1(\mathbb{R}^n)}$, $[w]^*_{A_1(\mathbb{R}^n)}$ or $(w)_{A_1(\mathbb{R}^n)}$ are uniformly bounded for all $n > N$.*

(b) *There exists a constant $C > 0$ so that for all $n > N$, all $f \in L^1_{\mathrm{rad}}(w)$ and $\lambda > 0$ we have*

$$w(\{x \in \mathbb{R}^n : \mathcal{A}f(x) > \lambda\}) \leq \frac{C}{\lambda}\int_{\mathbb{R}^n}|f(x)|w(x)\,dx.$$

(c) *w_0 is essentially constant over dyadic intervals and decreasing up to a constant. This means that there exist positive constants β and η so that*

$$\begin{aligned}&\operatorname*{ess\,sup}_{r\in[R,2R]} w_0(r) \leq \beta \operatorname*{ess\,inf}_{r\in[R,2R]} w_0(r), \quad \forall R > 0,\\ &w_0(s) \leq \eta\, w_0(t), \quad \forall s \geq t \geq 0.\end{aligned}$$

Power weights with negative powers, that is $w_0(t) = t^{-\alpha}$ with $\alpha \geq 0$, are examples of weights with these properties. It is easy to check that they satisfy condition (c) with $\beta = 2^\alpha$ and $\eta = 1$.

Moreover, for the weights satisfying the properties in Proposition 2.3, the maximal operator M acting over radial functions is weakly bounded in weighted L^1 with constants uniformly bounded in dimension. This was already known for $w \equiv 1$. Menárguez and the second author proved in [25] that for all radial f over $\mathbb{R}^n$ and $\lambda > 0$ one has

$$|\{x \in \mathbb{R}^n : Mf(x) > \lambda\}| \leq \frac{4}{\lambda}\|f\|_{L^1(\mathbb{R}^n,dx)}. \tag{2.4}$$

We point out the following extension, that is a corollary of part (b) of Proposition 2.3 and (2.1).

Theorem 2.4. *Let $w(x) = w_0(|x|)$ be a radial weight. Assume that there exist $N > 0$ and $C > 0$ so that $(w)_{A_1(\mathbb{R}^n)} < C$ for all $n > N$. Then for all radial f over $\mathbb{R}^n$ with $n \geq N$ and $\lambda > 0$ one has*

$$w(\{x \in \mathbb{R}^n : Mf(x) > \lambda\}) \leq \frac{4C}{\lambda}\int |f(x)|w(x)\,dx. \tag{2.5}$$

Note that if w is a radial weight so that (2.5) holds for radial functions over $\mathbb{R}^k$, then $w \in A_1(\mathbb{R}^k)$. If w is decreasing up to a constant, the argument leading to (3.3) in the proof of Proposition 2.3 below, shows that w is essentially constant over dyadic intervals. Hence, by Proposition 2.3 we are in the hypothesis of Theorem 2.4 and we have (2.5) in $\mathbb{R}^n$ for all $n \geq k$ with a constant independent of n.

We finish remarking that for the weights w characterized in Proposition 2.3 Mw and $\mathcal{A}w$ are comparable with constants independent of the dimension because one always has

$$\mathcal{A}w(x) \leq (w)_{A_1(\mathbb{R}^n)}w(x) \leq (w)_{A_1(\mathbb{R}^n)}Mw(x).$$

It is easy to find examples of radially increasing functions, for instance $w_0 = t^\alpha$ with $\alpha > 0$, so that $\mathcal{A}w(x) = Mw(x) = \infty$ for all x. Therefore $\mathcal{A}w \leq CMw$ with C independent of the dimension does not imply any of the conditions (a), (b), (c).

3 Proofs of the main results

We begin proving Proposition 2.3. We mention that the equivalence (a) $\Leftrightarrow$ (c) already appeared in [15] for $[w]_{A_1}$. Here we will use similar arguments. Among them, the method of differentiation through dimensions also presented in [15] that is contained in the following Lemma.

Lemma 3.1. *Take $w_0 \in L^1_{\mathrm{loc}}([0,\infty), t^{N-1}\,dt)$ for some $N \geq 1$. Then, for almost every $T > 0$ and for all $s \geq 0$ and $R > 0$ so that $s^2 + R^2 = T^2$, if we take points*

$z^n \in \mathbb{R}^n$ with $|z^n| = s$ and we denote $B(z^n, R) = \{y \in \mathbb{R}^n : |z^n - y| < R\}$, the following holds

$$\lim_{n\to\infty} \fint_{B(z^n,R)} w_0(|x|)\,dx = w_0(T).$$

We now proceed to prove Proposition 2.3.

Proof of Proposition 2.3. It is easy to see that

$$[w]^*_{A_1} \le [w]_{A_1}, \tag{3.1}$$

$$[w]^*_{A_1} \le 2(w)_{A_1}. \tag{3.2}$$

Let us prove (a) $\Rightarrow$ (c). Assume that for $n \ge N$ one has $[w]^*_{A_1(\mathbb{R}^n)} \le C_*$ with C_* independent of n. First, we prove that w_0 is decreasing up to a constant. Assume that $t \ge s \ge 0$ and for each $n \ge N$ take $x_n, y_n \in \mathbb{R}^n$ so that $x_n = \alpha y_n$, for some $\alpha \in (0,1)$, and $|x_n| = s$, $|y_n| = t$. Consider the ball $B_R(x_n)$ with $R^2 = t^2 - s^2$. By hypothesis we have, in the almost everywhere sense,

$$\frac{w(B_R(x_n))}{|B_R(x_n)|} \le C_*\ w(x_n) = C_*\ w_0(s).$$

In view of Lemma 3.1 we take limits as $n \to \infty$ and obtain

$$w_0(t) = w_0(\sqrt{s^2+R^2}) \le C_*\ w_0(s),$$

for almost every $s \le t$.

In order to ensure that w_0 is essentially constant over dyadic intervals, we only need to prove that for all $R > 0$, $w_0(R) \le Cw_0(2R)$ with C independent of R. Take $n \ge N$ and consider $x \in \mathbb{R}^n$ with $|x| = 1$ and the balls $B = B_{R/2}((R/2)x)$, $B^* = B_{R/2}((3R/2)x)$ and $B^{**} = B_{R/2}((5R/2)x)$. Since $w \in A_1(\mathbb{R}^n)$, it is doubling and for some constant K we have $w(B) \le Kw(B^*) \le K^2 w(B^{**})$. By the decreasing property of w_0 that we have just proved, for all $y \in B$ we have $w_0(R) \le C_* w(y)$ and for all $y \in B^{**}$ we have $w(y) \le C_* w_0(2R)$. This yields

$$w_0(R) \le C_*\ \frac{w(B)}{|B|} \le C_* K^2\ \frac{w(B^{**})}{|B^{**}|} \le (C_*K)^2\ w_0(2R). \tag{3.3}$$

We now proceed with (c) $\Rightarrow$ (a). In view of (3.1) and (3.2) it is enough to see that $(w)_{A_1}$ and $[w]_{A_1}$ are uniformly bounded. To see the first, consider an interval $[a,b] \subset [0,\infty)$. If $b \le 2a$, we have

$$\frac{w_0 v_n([a,b])}{v_n([a,b])} \le \operatorname*{ess\,sup}_{a\le r\le b} w_0(r) \le \operatorname*{ess\,sup}_{a\le r\le 2a} w_0(r) \le \beta \operatorname*{ess\,inf}_{a\le r\le 2a} w_0(r) \le \beta \operatorname*{ess\,inf}_{a\le r\le b} w_0(r).$$

If $b > 2a$ we use the fact that under the hypothesis that w_0 is essentially constant over dyadic intervals for any $R > 0$ we have

$$\begin{aligned} w_0 v_n\ ([0,R]) &= \sum_{k=1}^{\infty} \int_{2^{-k}R}^{2^{-k+1}R} w_0(t) t^{n-1}\,dt\ \le\ \sum_{k=1}^{\infty} \beta^k w_0(R)\ \frac{2^{(-k+1)n} - 2^{-kn}}{n} \\ &\le \frac{2^n R^n}{n}\ w_0(R) \sum_{k=1}^{\infty} \left(\frac{\beta}{2^n}\right)^k\ \le\ 2\beta\ w_0(R)\ v_n([0,R]), \end{aligned}$$

the last inequality provided we take $n > N = \log_2 \beta + 1$. Using this

$$\frac{w_0 v_n([a,b])}{v_n([a,b])} \leq \frac{w_0 v_n([0,b])}{v_n([a,b])} \leq 2\beta w_0(b) \, \frac{v_n([0,b])}{v_n([a,b])},$$

and this is all we need. Note that by the decreasing property of w_0 we have $w_0(b) \leq \operatorname{ess\,inf}_{a \leq r \leq b} w_0(r)$ and

$$\frac{v_n([0,b])}{v_n([a,b])} = \frac{b^n}{b^n - a^n} \leq \frac{b^n}{b^n - (b/2)^n} \leq 2.$$

One proves the uniform boundedness of $[w]_{A_1}$ in a similar way (see [15] for the details).

To see the equivalence (a) $\Leftrightarrow$ (b) observe that (b) is equivalent to

$$w_0 v_n\Big(\{r \geq 0 : \tilde{M}_{v_n} f_0(r) > \lambda\}\Big) \leq \frac{C}{\lambda} \int_0^\infty |f_0(r)| w_0(r) v_n(r)\, dr, \qquad (3.4)$$

with C independent of f_0, n and λ. For the implication (a) $\Rightarrow$ (b) assume that $(w)_{A_1}$ is uniformly bounded. Consider a radial function $f(x) = f_0(|x|)$ over $\mathbb{R}^n$ and $\lambda > 0$. The set $E_\lambda = \{r \geq 0 : \tilde{M}_{v_n} f_0(r) > \lambda\}$ is the union of all the intervals $I \subset [0,\infty)$ verifying

$$\frac{1}{v_n(I)} \int_I |f_0(t)| v_n(t)\, dt > \lambda.$$

We use Young's selection principle (see Lemma 4.2.1 in [21]) to obtain a subset $\mathcal{I}$ of such intervals so that $E_\lambda \subset \bigcup_{I \in \mathcal{I}}$ and $\sum_{I \in \mathcal{I}} \chi_I \leq 2$.

For each $I \in A$ by our assumption we have

$$\begin{aligned} w_0 v_n(I) &\leq (w)_{A_1(\mathbb{R}^n)}\, v_n(I)\, \operatorname*{ess\,inf}_I w_0 \leq \frac{(w)_{A_1(\mathbb{R}^n)}}{\lambda} \int_I |f_0(t)|\, v_n(t)\, dt\, \operatorname*{ess\,inf}_I w_0 \\ &\leq \frac{(w)_{A_1(\mathbb{R}^n)}}{\lambda} \int_I |f_0(t)|\, w_0(t)\, v_n(t)\, dt \end{aligned}$$

and hence,

$$w_0 v_n(E_\lambda) \leq \sum_{I \in \mathcal{I}} w_0 v_n(I) \leq \frac{2(w)_{A_1(\mathbb{R}^n)}}{\lambda} \int_0^\infty |f_0(t)|\, w_0(t)\, v_n(t)\, dt.$$

We finish with the implication (b) $\Rightarrow$ (a). Consider intervals $J \subset I \subset [0,\infty)$, take $f_0 = \chi_J$ and $\lambda = w_0 v_n(J)/v_n(I)$. Then $I \subset E_\lambda$ and by (3.4) we have

$$w_0 v_n(I) \leq w_0 v_n(E_\lambda) \leq \frac{C}{\lambda}\, w_0 v_n(J),$$

or equivalently

$$\frac{w_0 v_n(I)}{v_n(I)} \leq C \frac{w_0 v_n(J)}{v_n(J)}.$$

Since J is arbitrary, this implies that

$$\frac{w_0 v_n(I)}{v_n(I)} \leq C \operatorname*{ess\,inf}_{t \in I} w_0(t).$$

□

Our next goal is to show Lemma 2.2 and Theorem 1.1. The technical parts of the proofs are summarized in the following lemmas. The first one, due to Stein (see [31]), provides a method of rotations that reduces the dimension when estimating the mean values over balls.

Lemma 3.2. *Let $k < n$ be natural numbers. For each $x = (x^1, \dots, x^n) \in \mathbb{R}^n$ we call $x_1 = (x^1, \dots, x^k) \in \mathbb{R}^k$ and $x_2 = (x^{k+1}, \dots, x^n) \in \mathbb{R}^{n-k}$. By abuse of the language we will write $x = (x_1, x_2)$. For any positive and measurable function f on $\mathbb{R}^n$ one has*

$$\frac{\int_{|x|<R} f(x)\, dx}{\int_{|x|<R} dx} = \frac{\int_{SO(\mathbb{R}^n)} \int_{|x_1|<R} f(\tau(x_1, 0))|x_1|^{n-k}\, dx_1\, d\tau}{\int_{|x_1|<R} |x_1|^{n-k}\, dx_1},$$

where $SO(\mathbb{R}^n) = \{\tau \in \mathcal{M}_{n\times n}(\mathbb{R}) : \tau\tau^t = \tau^t\tau = I\}$, i.e. the special orthonormal group in $\mathbb{R}^n$ equipped with its Haar measure.

Roughly speaking, it asserts that an integral mean over a ball in $\mathbb{R}^n$ can be transformed into an integral mean over a ball in $\mathbb{R}^k$ combined with all possible rotations in $\mathbb{R}^n$. In [31] Lemma 3.2 was used to obtain the following pointwise controls of the maximal function.

Lemma 3.3. *Let $k < n$ be natural numbers. With the notation of the previous Lemma 3.2, if g is a function over $\mathbb{R}^n$ we write $g_{x_2}(x_1) := g(x_1, x_2) = g(x)$. Denoting by M_m the maximal operator in $\mathbb{R}^m$, we have the following bounds:*

(a) $M_n f(x) \leq \dfrac{n}{k} \displaystyle\int_{SO(n)} M_k \left[(f \circ \tau)_{\tau^{-1}(x)_2}\right] (\tau^{-1}(x)_1)\, d\tau,$

(b) $M_n f(x) \leq \displaystyle\int_{SO(n)} \mathcal{M}_k \left[(f \circ \tau)_{\tau^{-1}(x)_2}\right] (\tau^{-1}(x)_1)\, d\tau,$

where $\mathcal{M}_k$ stands for the k-dimensional spherical maximal operator.

We will also employ a technical result for radial weights.

Lemma 3.4. *Let $w(x) = w_0(|x|)$ be a radial weight over $\mathbb{R}^k$ so that for some $p \geq 1$ we have $w \in A_p(\mathbb{R}^k)$. For each $\rho > 0$ consider the weights $w_\rho(x) = w_0(\sqrt{\rho^2 + |x|^2})$. Then one has $w_\rho \in A_p(\mathbb{R}^k)$, and moreover there exists a constant $C_k > 0$ only depending on k so that for all $\rho \geq 0$ one has $[w_\rho]_{A_p(\mathbb{R}^k)} \leq C_k\, [w]_{A_p(\mathbb{R}^k)}$.*

As a consequence, there exists a constant $\tilde{C}_k > 0$ *only depending on* k *and on* w *so that for all* $\rho \geq 0$ *and* $f \in L^p(w_\rho)$ *one has*

$$\|M_k f\|_{L^p(w_\rho)} \leq \tilde{C}_k \|f\|_{L^p(w_\rho)}.$$

Now we are in conditions to prove Lemma 2.2.

Proof of Lemma 2.2. Assume that $w(x_1) = w_0(|x_1|)$ is an $A_1(\mathbb{R}^k)$ weight. In view of Lemma 3.3 part (a) and Lemma 3.4 if $n \geq k$ and $x \in \mathbb{R}^n$ we have

$$\begin{aligned} M_n w(x) &\leq \frac{n}{k} \int_{SO(\mathbb{R}^n)} M_k[w_{|\tau^{-1}(x)_2|}](\tau^{-1}(x)_1)\, d\tau \\ &\leq C_k [w]^*_{A_1(\mathbb{R}^k)} \frac{n}{k} \int_{SO(\mathbb{R}^n)} w_{|\tau^{-1}(x)_2|}(\tau^{-1}(x)_1)\, d\tau \\ &= C_k [w]^*_{A_1(\mathbb{R}^k)} \frac{n}{k}\, w(x). \end{aligned}$$

The bound for $(w)_{A_1}$ is immediate once we observe that for any $[a,b] \subset [0,\infty)$ one has

$$\begin{aligned} \frac{w_0 v_n([a,b])}{v_n([a,b])} &= \frac{n}{b^n - a^n} \int_a^b w_0(t) t^{n-1}\, dt \\ &\leq b^{n-k} \frac{b^k - a^k}{b^n - a^n} \frac{n}{k} \frac{k}{b^k - a^k} \int_a^b w_0(t) t^{k-1}\, dt \\ &\leq \frac{n}{k} \frac{w_0 v_k([a,b])}{v_k([a,b])}. \end{aligned}$$

The second inequality uses that $b^{n-k}(b^k - a^k) \leq b^n - a^n$, whenever $0 \leq a \leq b$. From the previous calculation one deduces that $\tilde{M}_{v_n} w_0 \leq n/k\, \tilde{M}_{v_k} w_0$. □

In order to prove Theorem 1.1 we will use Lemma 3.3, part (b), where the maximal spherical operator appears. We recall that if $n \geq 2$, for a suitable smooth function f the maximal spherical operator is defined as

$$\mathcal{M}_n f(x) = \sup_{r>0} \frac{1}{\omega_{n-1}} \int_{\mathbb{S}^{n-1}} |f(x+ry)|\, d\sigma_{n-1}(y).$$

This operator is known to be bounded on $L^p(\mathbb{R}^n)$ if and only if $p > n/(n-1)$. Stein proved this in [29] in the case that $n \geq 3$, and Bourgain in [8] for $n = 2$. This allows to define the maximal spherical operator over functions in $L^p(\mathbb{R}^n)$ with $p > n/(n-1)$.

We say that a weight w is in the class $W_p(\mathbb{R}^n)$ if $\mathcal{M}_n$ is bounded on $L^p(w)$. If w is radial we have the following relation with the A_p classes.

Lemma 3.5. *Consider the radial weight* $w(x) = w_0(|x|)$. *If for certain* k *one has* $w \in A_p(\mathbb{R}^k)$ *then there exists* $m \geq k$ *so that* $w \in W_p(\mathbb{R}^m)$. *Moreover* $\|\mathcal{M}_m\|_{L^p(\mathbb{R}^m,w)\to L^p(\mathbb{R}^m,w)}$ *is controlled by* $\|M_k\|_{L^p(\mathbb{R}^k,w)\to L^p(\mathbb{R}^k,w)}$.

Both, Lemmas 3.4 and 3.5 are proved below. Now we are ready to prove Theorem 1.1.

Proof of Theorem 1.1. Since $w \in A_p(\mathbb{R}^k)$, by Lemma 3.4, for all $a \geq 0$ one has $w_a \in A_p(\mathbb{R}^k)$. Furthermore by Lemma 3.5 there exist $m \geq k$ so that for all $a \geq 0$ one has $w_a \in W_p(\mathbb{R}^m)$. Moreover $\|\mathcal{M}_m\|_{L^p(\mathbb{R}^m,w_a)\to L^p(\mathbb{R}^m,w_a)}$ is controlled by $\|M_k\|_{L^p(\mathbb{R}^k,w)\to L^p(\mathbb{R}^k,w)}$. This means that there exists $C_m > 0$ so that for all $a \geq 0$ and $g \in L^p(w_a)$ one has

$$\|\mathcal{M}_m g\|_{L^p(\mathbb{R}^m,w_a)} \leq C_m \|f\|_{L^p(\mathbb{R}^m,w_a)}.$$

Given $f \in L^p(\mathbb{R}^n, w)$ with $n > m$, by Lemma 3.3, part (b) and Minkowski inequality we have

$$\|M_n f\|_{L^p(\mathbb{R}^n,w)} \leq \int_{SO(n)} \left(\int_{\mathbb{R}^n} \left|\mathcal{M}_m[(f\circ\tau)_{\tau^{-1}(x)_2}](\tau^{-1}(x)_1)\right|^p w(x)\, dx \right)^{1/p} d\tau.$$

Applying obvious changes of integration variables and Lemma 3.4, the previous is bounded by

$$\begin{aligned}
&\int_{SO(n)} \left(\int_{\mathbb{R}^n} |\mathcal{M}_m[(f\circ\tau)_{y_2}](y_1)|^p\, w(y)\, dy \right)^{1/p} d\tau \\
&= \int_{SO(n)} \left(\int_{\mathbb{R}^{n-k}} \int_{\mathbb{R}^k} |\mathcal{M}_m[(f\circ\tau)_{y_2}](y_1)|^p\, w_{|y_2|}(y_1)\, dy_1\, dy_2 \right)^{1/p} d\tau \\
&\leq \int_{SO(n)} \left(\int_{\mathbb{R}^{n-k}} C_m \int_{\mathbb{R}^k} |(f\circ\tau)_{y_2}(y_1)|^p\, w_{|y_2|}(y_1)\, dy_1\, dy_2 \right)^{1/p} d\tau \\
&= C_m \|f\|_{L^p(w)}.
\end{aligned}$$

□

We finish justifying Lemmas 3.4 and 3.5.

Proof of Lemma 3.4. If $w \in A_p(\mathbb{R}^k)$ there exist $u,\, v \in A_1(\mathbb{R}^k)$ so that $w = uv^{1-p}$ and $[w]_{A_p(\mathbb{R}^k)} \leq [u]_{A_1(\mathbb{R}^k)}[v]^{p-1}_{A_1(\mathbb{R}^k)}$. Moreover if w is radial, u and v can be chosen to be radial by their construction (see [20]). Therefore it is enough to prove the result in the case $p = 1$.

Assuming $w \in A_1(\mathbb{R}^k)$, we are going to show that there exist a constant $C > 0$, so that for all $x \in \mathbb{R}^k$ and $\rho \geq 0$ one has $Mw_\rho(x) \leq Cw_\rho(x)$. As we observed after Lemma 2.1, if $u(x) = u_0(x)$ is a radial and locally integrable function we have $Mu(x) \leq Cu(x)$ a.e. if and only if $\tilde{M}_{v_k} u_0(|x|) \leq C' u_0(|x|)$ a.e. By hypothesis this last condition is true for w_0. Now we check it for $w_0(\sqrt{\rho^2 + (\,\cdot\,)^2})$. We take

$0 \leq a \leq |x| \leq b$. With the change of variables $s^2 = \rho^2 + t^2$ we obtain

$$
\begin{aligned}
&\frac{k}{b^k - a^k} \int_a^b w_0(\sqrt{\rho^2 + t^2})\, t^{k-1}\, dt \\
&= \frac{k}{b^k - a^k} \int_{\sqrt{\rho^2+a^2}}^{\sqrt{\rho^2+b^2}} w_0(s)\, (s^2 - \rho^2)^{k/2-1} s\, ds \\
&\leq \frac{k^2/2}{(\rho^2 + b^2)^{k/2} - (\rho^2 + a^2)^{k/2}} \int_{\sqrt{\rho^2+a^2}}^{\sqrt{\rho^2+b^2}} w_0(s)\, s^{k-1}\, ds \\
&\leq \frac{k}{2}\, \tilde{M}_{v_k} w_0(\sqrt{\rho^2 + |x|^2}) \leq \frac{k}{2}\, (w)_{A_1(\mathbb{R}^k)}\, w_0(\sqrt{\rho^2 + |x|^2}).
\end{aligned}
$$

The only step that is not immediate is the first inequality. Clearly it would follow from

$$
\left[(\rho^2 + b^2)^{k/2} - (\rho^2 + a^2)^{k/2}\right] \left(\frac{s^2 - \rho^2}{s^2}\right)^{k/2-1} \leq \frac{k}{2}(b^k - a^k).
$$

Observe that the left hand side of this inequality is increasing in s for $s \geq \rho$. Thus, it is enough to check the case $s^2 = \rho^2 + b^2$, that is

$$
\left[(\rho^2 + b^2)^{k/2} - (\rho^2 + a^2)^{k/2}\right] \left(\frac{b^2}{\rho^2 + b^2}\right)^{k/2-1} \leq \frac{k}{2}(b^k - a^k). \tag{3.5}
$$

Applying the Mean Value Theorem to the function $\phi(t) = (\rho^2 + t)^{n/2}$ yields

$$
(\rho^2 + b^2)^{k/2} - (\rho^2 + a^2)^{k/2} \leq \frac{k}{2}(\rho^2 + b^2)^{k/2-1}(b^2 - a^2).
$$

This, together with the observation that $b^{k-2}(b^2 - a^2) \leq b^k - a^k$, proves (3.5). □

Proof of Lemma 3.5. We assume $w \in A_p(\mathbb{R}^k)$. By the reverse Hölder property there exists $s < 1$ so that $w^{1/s}$ is also an $A_p(\mathbb{R}^k)$ weight. Observe that $w^{1/s} \in A_p(\mathbb{R}^m)$ for all $m \geq k$. This is an easy consequence of the factorization and Lemma 2.2.

Let us use the notation $\mathcal{M}_m f(x) = \sup_{t>0} S_t f(x)$, where

$$
S_t f(x) = \frac{1}{\omega_{m-1}} \int_{\mathbb{S}^{m-1}} |f(x + ty)|\, d\sigma_{m-1}(y).
$$

Following Rubio de Francia in [28] we perform a dyadic decomposition of the spherical maximal function. Let $\psi : [0, \infty) \to [0, \infty)$ be a smooth function supported in $[1/2, 2]$, so that for all $r > 0$,

$$
\sum_{j=-\infty}^{\infty} \psi(2^{-j} r) = 1.
$$

Let $\phi(r) = \sum_{j=-\infty}^{-1} \psi(2^{-j}r)$. We can define the operators S_t^j and B_t by $(S_t^j f)^\wedge(\xi) = (S_t f)^\wedge(\xi)\psi(2^{-j}|\xi|)$ and $(B_t f)^\wedge(\xi) = (S_t f)^\wedge \phi(2^{-j}|\xi|)$. It is obvious that

$$\mathcal{M}_m f(x) = \sup_{t>0} \left| B_t f(x) + \sum_{j=0}^{\infty} S_t^j f(x) \right| \leq M_m f(x) + \sum_{j=0}^{\infty} \sup_{t>0} |S_t^j f(x)|. \tag{3.6}$$

It is known that if $p < m/(m-1)$ then

$$\left\| \sup_{t>0} \left| S_t^j f \right| \right\|_{L^p(\mathbb{R}^m, dx)} \leq 2^{j(1-m/p')} \|f\|_{L^p(\mathbb{R}^m, dx)},$$

and that $S_t^j f(x) \leq C 2^j M_m f(x)$ (see [28] and [19]). Then since $w^{1/s} \in A_p(\mathbb{R}^m)$, one has

$$\left\| \sup_{t>0} \left| S_t^j f \right| \right\|_{L^p(w^{1/s})} \leq C 2^j \|f\|_{L^p(w^{1/s})}.$$

By interpolation with change of measure (see [32]), one has

$$\left\| \sup_{t>0} \left| S_t^j f \right| \right\|_{L^p(\mathbb{R}^m, w)} \leq C 2^{sj+(1-s)j(1-m/p')} \|f\|_{L^p(\mathbb{R}^m, w)}.$$

If $m > p'/(1-s)$ the exponent in this bound is negative and then we can sum in j to obtain

$$\begin{aligned} \|\mathcal{M}_m f\|_{L^p(\mathbb{R}^m, w)} &\leq C \|M_m f\|_{L^p(\mathbb{R}^m, w)} + \sum_{j=0}^{\infty} \left\| \sup_{t>0} \left| S_t^j f \right| \right\|_{L^p(\mathbb{R}^m, w)} \\ &\leq C \|f\|_{L^p(\mathbb{R}^m, w)} + C \sum_{j=0}^{\infty} 2^{[s+(1-s)(1-m/p')]j} \|f\|_{L^p(\mathbb{R}^m, w)} \\ &\leq C \|f\|_{L^p(\mathbb{R}^m, w)}. \end{aligned}$$

Here C may depend on m, p and w but is independent of f. C is indeed controlled by the operator norm of M_m in $L^p(\mathbb{R}^m, w)$. □

4 Kakeya maximal operator

Fixed $N > 0$, we denote by $\mathcal{R}_N$ the family of all parallelepipeds in $\mathbb{R}^n$ with edge lengths $h \times h \times \cdots \times h \times Nh$, where $h > 0$ is arbitrary. The Kakeya maximal operator of eccentricity N is defined as

$$\mathcal{K}_N f(x) = \sup_{x \in R \in \mathcal{R}_N} \frac{1}{R} \int_R |f(y)|\, dy.$$

It is easy to prove that $\mathcal{K}_N f(x) \leq N^{(n-1)} M f(x)$ where Mf is here the usual maximal function over all rotated cubes. One just has to replace $R \in \mathcal{R}_N$ by the smallest cube that contains it. Then $\mathcal{K}_N$ is weakly bounded on $L^1(\mathbb{R}^n)$ with a constant growing with N at most at the rate N^{n-1}. By interpolation with the L^∞ case the operator norm on $L^p(\mathbb{R}^n)$ grows at most like $N^{(n-1)/p}$ for $1 < p < \infty$. However, it is conjectured that for $p = n$ it grows no faster than $C_\varepsilon N^\varepsilon$ for each $\varepsilon > 0$. Córdoba proved in [12] that the conjecture is true in the case $n = 2$. In higher dimensions Carbery, Hernández and the second author showed in [11] that the conjecture holds when restricting the action of $\mathcal{K}_N$ to radial functions. Alternative proofs and extensions are due to Duoandikoetxea, Naibo and Oruetxebarria [17] and Duoandikoetxea, Moyua and Oruetxebarria [16].

In this last three papers the result is obtained as a corollary of boundedness results for the universal maximal operator. This is defined as

$$\mathcal{K}f(x) = \sup_{u\in\mathbb{S}^{n-1}} \sup_{a\le 0\le b} \frac{1}{b-a}\int_a^b |f(x+su)|\,ds.$$

$\mathcal{K}$ is related to the Kakeya maximal operator in the sense that it can be regarded as its extremal case, where the eccentricity N is infinity and rectangles become segments. Moreover $\mathcal{K}$ majorizes all the $\mathcal{K}_N$ but turns out to be unbounded on every L^p, except for $p = \infty$ (see [23]). In spite of this, [11] established that $\mathcal{K} : L^{n,1}_{\text{rad}}(\mathbb{R}^n) \to L^{n,\infty}_{\text{rad}}(\mathbb{R}^n)$.

The basic idea to give an alternative proof of this last bound in [17] is that for $f = \chi_A$, the characteristic function of a radial set, we have $\mathcal{K}f(x) \le C_n\,(\mathcal{A}f(x))^{1/n}$. This was further refined in [16] to obtain that for a f radial

$$\mathcal{K}f(x) \le C_q(\tilde{M}_{v_2} f_0^q(|x|)^{1/q}$$

for any $q > 2$, and for $q \ge 2$ if f is the characteristic function of a radial set. These last constants C_q are independent of the dimension, although the weighted inequalities obtained from them were not. Here we prove

Lemma 4.1. *Let E be a radial subset of $\mathbb{R}^n$ and $f = \chi_E$. Then for all $k \ge 2$ one has the pointwise inequality*

$$\mathcal{K}f(x) \le 2(\tilde{M}_{v_k} f_0(|x|))^{1/k}.$$

The constant 2 in this inequality is sharp.

As a consequence, via Proposition 2.3, we obtain

Theorem 4.2. *Let f be a radial function over $\mathbb{R}^n$, with $n \ge 2$, and let w be a radial weight in $A_1(\mathbb{R}^n)$, then*

$$\|\mathcal{K}f\|^*_{L^{n,\infty}(\mathbb{R}^n,w)} \le \frac{2n}{n-1}\,[2(w)_{A_1(\mathbb{R}^n)}]^{1/n}\,\|f\|^*_{L^{n,1}(\mathbb{R}^n,w)}.$$

Observe that in view of Lemma 2.2 the previous implies a bound that is uniform in dimension. As a consequence one also has such a bound for the Kakeya maximal operator $\mathcal{K}_N$. We remark that the only original results claimed in this section are the sharp bound in Lemma 4.1 and the uniformity in the bound of Theorem 4.2.

Assuming Lemma 4.1 for the moment, we provide a proof of the above theorem.

Proof of Theorem 4.2. By density, we just need to prove the result for a simple function of the form

$$f(x) = \sum_{j=1}^{J} c_j \chi_{E_j}(x),$$

where $E_1 \supset \cdots \supset E_J$ are radial sets and $c_1, \ldots, c_J$ are positive reals. If E is a radial set, by Lemma 4.1 for $k = n$ and following the argument in the proof of Proposition 2.3 one has

$$\begin{aligned} w(\{x \in \mathbb{R}^n : \mathcal{K}\chi_E(x) > \lambda\}) &\leq w(\{x \in \mathbb{R}^n : 2\mathcal{A}\chi_E(x)^{1/n} > \lambda\}) \\ &\leq \left(\frac{2}{\lambda}\right)^n 2(w)_{A_1(\mathbb{R}^n)}\, w(E). \end{aligned}$$

Hence $\|\mathcal{K}\chi_E\|^*_{L^{n,\infty}(\mathbb{R}^n,w)} \leq 2\,[2(w)_{A_1(\mathbb{R}^n)} w(E)]^{1/n}$. For a general f we use the standard procedure:

$$\begin{aligned} \|Kf\|^*_{L^{n,\infty}(\mathbb{R}^n,w)} \leq \|Kf\|_{L^{n,\infty}(\mathbb{R}^n,w)} &\leq \sum_{j=1}^{J} c_j \|K\chi_{E_j}\|_{L^{n,\infty}(\mathbb{R}^n,w)} \\ &\leq \frac{n}{n-1} \sum_{j=1}^{J} c_j \|K\chi_{E_j}\|^*_{L^{n,\infty}(\mathbb{R}^n,w)} \\ &\leq \frac{2n}{n-1} [2(w)_{A_1(\mathbb{R}^n)}]^{1/n} \sum_{j=1}^{J} c_j w(E_j)^{1/n} \\ &= \frac{2n}{n-1} [2(w)_{A_1(\mathbb{R}^n)}]^{1/n} \|f\|^*_{L^{n,1}(\mathbb{R}^n,w)}. \end{aligned}$$

□

Proof of Lemma 4.1. We need some notation. Given $w, z \in \mathbb{R}^n$ we denote by $S_{w,z}$ the segment whose extremal points are w, z. We may assume that $|w| \leq |z|$ and will call y to the point in $S_{w,z}$ which is closest to the origin. Consider a radial set $A \subset \mathbb{R}^n$, we define its radial projection over $S_{z,w}$ as $A_0 = \{|y| \leq t \leq |z| : \exists x \in A \text{ with } |x| = t\}$. Denoting by $|E|_d$ the d-dimensional Hausdorff measure of a set E it is enough to prove that one has

$$\frac{|S_{w,z} \cap A|_1}{|z-w|} \leq 2 \left(\frac{k}{|z|^k - |y|^k} \int_{|y|}^{|z|} \chi_{A_0}(s) s^{k-1}\, ds \right)^{1/k}.$$

We proceed in several steps, first we show how to reduce to the case $y = w$. Define z' as the point aligned with w and z so that $|z' - y| = |z - y|$. Note that

$$\frac{|S_{w,z} \cap A|_1}{|z-w|} \leq \frac{|S_{z',z} \cap A|_1}{|z-y|} = 2\, \frac{|S_{y,z} \cap A|_1}{|z-y|}.$$

For the second step we assume $y = w$. We call $L := |z - y|$ and $\ell := |S_{y,z} \cap A|_1$. Consider the point $u \in S_{y,z}$ so that $\ell := |u - y|$. Defining $A^* = \{x \in \mathbb{R}^n : |y| \leq$

$|x| \leq |u|\}$, we are done if we show that we have

$$\frac{|S_{y,z} \cap A|_1}{L} = \frac{|S_{y,z} \cap A^\star|_1}{L} \leq \left(\frac{k}{|z|^k - |y|^k} \int_{|y|}^{|z|} \chi_{A_0^\star}(s) s^{k-1}\, ds \right)^{1/k}$$

$$\leq \left(\frac{k}{|z|^k - |y|^k} \int_{|y|}^{|z|} \chi_{A_0}(s) s^{k-1}\, ds \right)^{1/k}. \qquad (4.1)$$

The equality in (4.1) is a trivial consequences of the definition of $A^\star$. Now we prove the first inequality in (4.1) in the case $k = 2$. Denoting by γ the angle determined by $S_{0,y}$ and $S_{y,z}$ at y, the inequality can be rewritten as

$$\frac{\ell}{L} \leq \left(\frac{|u|^2 - |y|^2}{|z|^2 - |y|^2} \right)^{1/2} = \left(\frac{\ell^2 - 2\ell |y| \cos\gamma}{L^2 - 2L|y| \cos\gamma} \right)^{1/2},$$

where the last expression comes from the cosine law. This is equivalent to

$$\frac{\ell}{L} \leq \frac{\ell - 2|y| \cos\gamma}{L - 2|y| \cos\gamma},$$

which is obviously true since $\ell \leq L$ and $\cos\gamma \leq 0$. For $k \geq 2$ it is enough to show then that

$$\left(\frac{|z|^k - |y|^k}{|u|^k - |y|^k} \right)^{1/k} \geq \left(\frac{|z|^2 - |y|^2}{|u|^2 - |y|^2} \right)^{1/2}.$$

Dividing by $|y|$ and renaming $\alpha = (|z|/|y|)^2$ and $\beta = (|u|/|y|)^2$ the previous inequality becomes

$$\left(\frac{\alpha^{k/2} - 1}{\beta^{k/2} - 1} \right)^{1/k} \geq \left(\frac{\alpha - 1}{\beta - 1} \right)^{1/2},$$

or equivalently

$$(\alpha - 1)^{k/2} (\alpha^{k/2} - 1) \geq (\beta - 1)^{k/2} (\beta^{k/2} - 1),$$

which is true for $\alpha > \beta \geq 1$ since $s \mapsto (s-1)^{k/2}(s^{k/2} - 1)$ is clearly an increasing function for $s \geq 1$.

As for the second inequality in (4.1), let us define $T = \{|v - y| : v \in A \cap S_{y,z}\}$. Note that $|T| = \ell$ and that $T = \{s \geq 0 : s - 2|y| \cos\gamma \in A_0\}$. Therefore, by the change of variables $t = (s^2 + |y|^2 - 2s|y|\cos\gamma)^{1/2}$ one has

$$\int_{A_0} t^{k-1}\, dt = \int_T (s^2 + |y|^2 - 2s|y| \cos\gamma)^{(k-2)/2} (s - 2|y| \cos\gamma)\, ds$$

$$\geq \int_0^\ell (s^2 + |y|^2 - 2s|y| \cos\gamma)^{(k-2)/2} (s - 2|y| \cos\gamma)\, ds$$

$$= \int_{|y|}^{|u|} t^{k-1}\, dt = \int_{A_0^\star} t^{k-1}\, dt.$$

To get the above inequality we have used that the integrated function is increasing with s. □

Remark. The constant 2 in this Lemma is optimal. To see this assume that $C > 0$ is a constant such that for all $z \in \mathbb{R}^n$ and all radial set $A \subset \mathbb{R}^n$ one has

$$\mathcal{K}\chi_A(z) \le C\tilde{M}_{v_2}\chi_{A_0}(|z|)^{1/2}. \tag{4.2}$$

Consider the segment $S_{w,z}$ and define y as the point in $S_{w,z}$ that is closest to the origin. Assume that $|z| > |w|$ take $A = \{x \in \mathbb{R}^n : |y| \le |x| \le |w|\}$. By orthogonality

$$\tilde{M}_{v_2}\chi_{A_0}(|z|) = \sup_{|y|\le t\le |w|} \frac{|w|^2 - t^2}{|z|^2 - t^2} = \frac{|w|^2 - |y|^2}{|z|^2 - |y|^2} = \frac{|w - y|^2}{|z - y|^2} = \frac{(\ell/2)^2}{(L - \ell/2)^2}.$$

Calling as before $L = |z - w|$ and $\ell = 2|w - y|$ we have $\mathcal{K}\chi_A(z) \ge \ell/L$. Then inequality (4.2) implies

$$C \ge 2\,\frac{L - \ell/2}{L}.$$

Since, choosing w appropriately, ℓ can be taken as small as wanted, necessarily we must have $C \ge 2$.

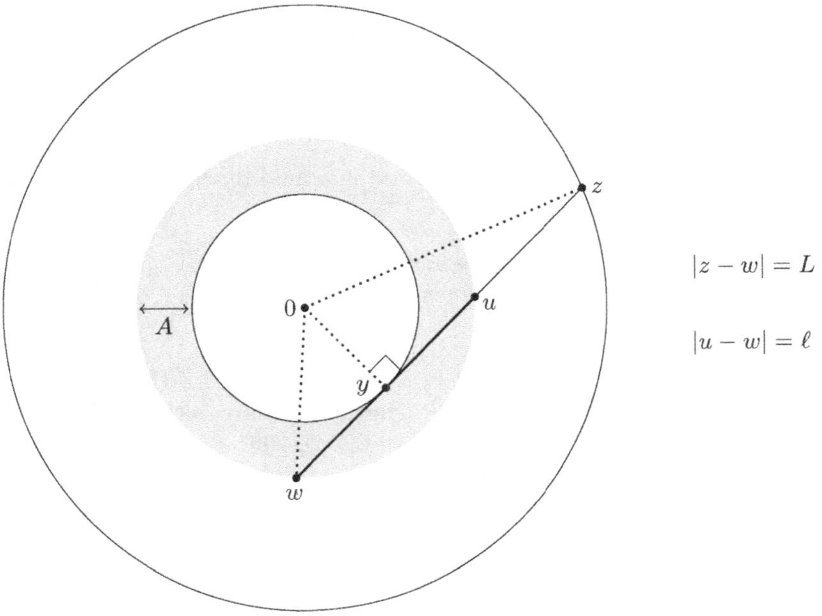

Acknowledgements

Both authors were partially supported by DGU grant MTM2010-16518.

References

[1] J.M. Aldaz, Dimension dependency of the weak type (1,1) bounds for maximal functions associated to finite radial measures, Bull. Lond. Math. Soc. **39** (2007), 203–208.

[2] J.M. Aldaz, The weak type (1,1) bounds for the maximal function associated to cubes grow to infinity with the dimension, Ann. of Math. (2) **173** (2011), no. 2, 1013–1023.

[3] J.M. Aldaz and J. Pérez Lázaro, Dimension dependency of L^p bounds for maximal functions associated to radial measures, Positivity **15** (2011), 199–213.

[4] G. Aubrun, Maximal inequality for high-dimensional cubes, Confluentes Math. **1** (2009), no. 2, 169–179.

[5] J. Bourgain, On high-dimensional maximal function associated to convex bodies, Amer. J. Math. **108** (1986), no. 6, 1467–1476.

[6] J. Bourgain, On the L^p-bounds for maximal functions associated to convex bodies in R^n, Israel J. Math. **54** (1986), no. 3, 257–265.

[7] J. Bourgain, On dimension free maximal inequalities for convex symmetric bodies in $\mathbb{R}^n$, Geometrical aspects of functional analysis (1985/86), 168–176, Lecture Notes in Math. **1267**, Springer, Berlin, 1987.

[8] J. Bourgain, Averages in the plane over convex curves and maximal operators, J. Analyse Math. **47** (1986), 69–85.

[9] J. Bourgain, On the Hardy-Littlewood maximal function for the cube, Israel J. Math. **203** (2014), 275–293.

[10] A. Carbery, An almost-orthogonality principle with applications to maximal functions associated to convex bodies, Bull. Amer. Math. Soc. (N.S.) **14** (1986), no. 2, 269–273.

[11] A. Carbery, E. Hernández, and F. Soria, Estimates for the Kakeya maximal operator on radial functions in $\mathbb{R}^n$, Harmonic analysis (Sendai, 1990), 41–50, ICM-90 Satell. Conf. Proc., Springer, Tokyo, 1991.

[12] A. Córdoba, The Kakeya maximal function and the spherical summation multipliers, Amer. J. Math. **99** (1977), no. 1, 1–22.

[13] A. Criado, On the lack of dimension free estimates in L^p for maximal functions associated to radial measures, Proc. Roy. Soc. Edinburgh Sect. A **140** (2010), no. 3, 541–552.

[14] A. Criado and P. Sjögren, Bounds for maximal functions associated with rotational invariant measures in high dimensions, J. Geom. Anal. **24** (2014), no. 2, 595–612.

[15] A. Criado and F. Soria, Localization and dimension free estimates for maximal functions. J. Funct. Anal. **256** (2013), no. 10, 2553–2583.

[16] J. Duoandikoetxea, A. Moyua, and O. Oruetxebarria, The spherical maximal operator on radial functions, J. Math. Anal. Appl. **387** (2012), no. 2, 655–666.

[17] J. Duoandikoetxea, V. Naibo, and O. Oruetxebarria, k-plane transforms and related operators on radial functions, Michigan Math. J. **49** (2001), no. 2, 265–276.

[18] J. Duoandikoetxea, A. Moyua, O. Oruetxebarria, and E. Seijo, Radial A_p weights with applications to the disc multiplier and the Bochner-Riesz operators, Indiana Univ. Math. J. **57** (2008), no. 3, 1261–1281.

[19] J. Duoandikoetxea and L. Vega, Spherical means and weighted inequalities, J. London Math. Soc. (2) **53** (1996), no. 2, 343–353.

[20] J. García-Cuerva and J.L. Rubio de Francia, Weighted norm inequalities and related topics, North-Holland Mathematics Studies, **116**. Notas de Matemática [Mathematical Notes] **104**. North-Holland Publishing Co., Amsterdam, 1985.

[21] A.M. Garsia, Topics in almost everywhere convergence, Lectures in Advanced Mathematics **4**, Markham Publishing Co., Chicago, Ill. 1970.

[22] L. Grafakos, Modern Fourier analysis, Second edition. Graduate Texts in Mathematics **250**, Springer, New York, 2009.

[23] M. de Guzmán, Differentiation of integrals in $\mathbb{R}^n$, Lecture Notes in Math. **481**, Springer-Verlag, Berlin-New York, 1975.

[24] M.T. Menárguez and F. Soria, Weak type (1,1) inequalities for maximal convolution operators, Rend. Circ. Mat. Palermo (2) **41** (1992), no. 3, 342–352.

[25] M.T. Menárguez and F. Soria, On the maximal operator associated to a convex body in $\mathbb{R}^n$, Collect. Math. **3** (1992), 243–251.

[26] D. Müller, A geometric bound for maximal functions associated to convex bodies, Pacific J. Math. **142** (1990), no. 2, 297–312.

[27] A. Naor and T. Tao, Random martingales and localization of maximal inequalities, J. Funct. Anal. **259** (2010), no. 3, 731–779.

[28] J.L. Rubio de Francia, Maximal functions and Fourier transforms, Duke Math. J. **53** (1986), no. 2, 395–404.

[29] E.M. Stein, Maximal functions. I. Spherical means, Proc. Nat. Acad. Sci. U.S.A. **73** (1976), no. 7, 2174–2175.

[30] E.M. Stein, The development of square functions in the work of A. Zygmund, Bull. Amer. Math. Soc. (N.S.) **7** (1982), no. 2, 359–376.

[31] E.M. Stein and J.O. Strömberg, Behavior of maximal functions in R^n for large n, Ark. Mat. **21** (1983), no. 2, 250–269.

[32] E.M. Stein and G. Weiss, Introduction to Fourier analysis on Euclidean spaces, Princ. Math. Ser. **32**, Princeton University Press, Princeton, N.J., 1971.

Some Topics in Harmonic Analysis and Applications
ALM 34, pp. 113–144

L^p Estimates for Multi-parameter and Multilinear Fourier Multipliers and Pseudo-differential Operators

Wei Dai*, Guozhen Lu† and Lu Zhang ‡

Abstract

In this paper, we will describe some recent works on L^p estimates for multi-parameter and multi-linear paraproducts, Coifman-Meyer Fourier multipliers and pseudo-differential operators and L^p estimates for bilinear and multi-parameter Hilbert transform type Fourier multipliers. We will also discuss the L^p estimates for bilinear operators given by non-smooth symbols with one-dimensional singularity set in the range $1/2 < p \leq 2/3$ and bilinear and bi-parameter Fourier multipliers when the singularity set in one parameter is of dimension one and of dimension zero in the other parameters. Sketches of proofs are given here and detailed proofs can be found in [14, 15, 16, 44, 45].

2000 Mathematics Subject Classification: Primary: 42B20; Secondary: 42B15.

Keywords and Phrases: multi-parameter, Fourier multipliers, pseudo-differential operators, Bilinear Hilbert transforms; L^p estimates; non-smooth symbols; flag symbols, paraproducts.

1 Introduction

In this section, we give a brief review of some well-known results on L^p estimates for classical and multi-linear Fourier multipliers and pseudo-differential operators.

*School of Mathematical Sciences, Beijing Normal University, Beijing 100875, China, and School of Mathematics and Systems Science, Beihang University (BUAA), Beijing 100191, China, *E-mail*: weidai@buaa.edu.cn

†Department of Mathematics, Wayne State University, Detroit, MI 48202, USA, *E-mail*: gzlu@wayne.edu

‡Department of Mathematics, Wayne State University, Detroit, MI 48202, USA, *E-mail*: eu4347@wayne.edu

1.1 Classical Fourier multipliers and pseudo-differential operators

Let the spatial dimension be $d \geq 1$ and $m \in C^{[\frac{d}{2}]+1}(\mathbb{R}^d \setminus \{0\})$ be a bounded function and obey the homogeneous symbol estimates of order 0:[1]

$$|\partial^\alpha m(\xi)| \lesssim_{\alpha,d} |\xi|^{-|\alpha|} \tag{1.1}$$

for all $\xi \neq 0$ and $0 \leq |\alpha| \leq [\frac{d}{2}] + 1$. The following result is well known (for the proof, refer to [57, 61, 63]).

Theorem 1.1 (Mikhlin multiplier theorem). *The linear multiplier operator T_m defined by*

$$T_m f(x) := \int_{\mathbb{R}^d_\xi} m(\xi)\hat{f}(\xi)e^{2\pi i x\cdot\xi}d\xi \tag{1.2}$$

is bounded on $L^p(\mathbb{R}^d)$ and $\|T_m\|_{L^p(\mathbb{R}^d)\to L^p(\mathbb{R}^d)} \lesssim_{p,d} 1$ for every $1 < p < \infty$.

In general, if we consider more general symbols m with limited smoothness assumptions (1.4) in terms of the Sobolev regularity, then the boundedness in L^p spaces still hold for the corresponding multipliers T_m. More precisely,

Theorem 1.2 (Hörmander multiplier theorem, [28]). *Let $s > \frac{d}{2}$ and $\Psi \in \mathcal{S}(\mathbb{R}^d)$ be such that*

$$\operatorname{supp}\Psi \subset \left\{\xi \in \mathbb{R}^d : \frac{1}{2} \leq |\xi| \leq 2\right\}, \quad \sum_{j\in\mathbb{Z}} \Psi(2^{-j}\xi) = 1 \tag{1.3}$$

for all $\xi \in \mathbb{R}^d \setminus \{0\}$. If a symbol $m \in L^\infty(\mathbb{R}^d)$ satisfies

$$\sup_{j\in\mathbb{Z}} \|m(2^j\cdot)\Psi\|_{H^s(\mathbb{R}^d)} < \infty, \tag{1.4}$$

then the multiplier T_m associated with symbol m is bounded on $L^p(\mathbb{R}^d)$ for all $1 < p < \infty$.

Clearly, Hörmander's theorem is stronger than Mikhlin's and the number $\frac{d}{2}$ cannot be improved further.

Definition 1.3 (Standard pseudo-differential operators, see [57, 63, 64]). A C^∞ smooth symbol $a : \mathbb{R}^d \times \mathbb{R}^d \to \mathbb{C}$ is a (standard) symbol of order k for some $k \in \mathbb{R}$ if we have the estimates

$$|\partial_x^\alpha \partial_\xi^\beta a(x,\xi)| \lesssim_{\alpha,\beta,d,k} \langle\xi\rangle^{k-|\beta|} \tag{1.5}$$

[1]Throughout this paper, $A \lesssim B$ means that there exists a universal constant $C > 0$ such that $A \leq CB$. If necessary, we use explicitly $A \lesssim_{\star,\cdots,\star} B$ to indicate that there exists a positive constant $C_{\star,\cdots,\star}$ depending only on the quantities appearing in the subscript continuously such that $A \leq C_{\star,\cdots,\star}B$.

for sufficiently many multi-indices α, β and all $x, \xi \in \mathbb{R}^d$, where $\langle\xi\rangle^2 = 1 + |\xi|^2$ (in particular, $a = O(\langle\xi\rangle^k)$ has at most polynomial growth). In this case, we write $a \in S^k := S^k_{1,0}$ and refer to the linear operator $a(X, D)$ defined by

$$a(X, D)f(x) := \int_{\mathbb{R}^d} a(x, \xi)\hat{f}(\xi)e^{2\pi i x\cdot\xi}d\xi \tag{1.6}$$

as a (standard) pseudo-differential operator (or ΨDO for short) of order k.

The following theorem concerns the boundedness of pseudo-differential operators with bounded symbols.

Theorem 1.4 (Calderón-Vaillancourt theorem, [13]). *Suppose the symbol $a \in C^\infty(\mathbb{R}^d \times \mathbb{R}^d)$ is such that*

$$\sup_{x,\xi\in\mathbb{R}^d} (|\partial_x^\alpha a(x, \xi)| + |\partial_\xi^\alpha a(x, \xi)|) \leq B \tag{1.7}$$

for all multi-indices $|\alpha| \leq 2d + 1$, then the linear operator $a(X, D)$ is bounded on $L^2(\mathbb{R}^d)$. In particular, every (standard) pseudo-differential operator $a(X, D)$ of order 0 is a Calderón-Zygmund operator, therefore, it is bounded on $L^p(\mathbb{R}^d)$ for every $1 < p < \infty$ with

$$\|a(X, D)\|_{L^p(\mathbb{R}^d)\to L^p(\mathbb{R}^d)} \lesssim_{p,d} 1. \tag{1.8}$$

1.2 Multilinear Fourier multipliers and pseudo-differential operators

If we consider the estimates for bilinear operators of two functions f and g, for instance, the Leibnitz rule of the product $f \cdot g$, that is, for every $\alpha > 0$ and $1 < p_i, q_i \leq \infty$ satisfying $\frac{1}{r} = \frac{1}{p_i} + \frac{1}{q_i}$ $(i = 1, 2)$ with $\frac{1}{1+\alpha} < r < \infty$, one can prove the following Kato-Ponce type estimates (for proofs, see [35, 36, 54, 57]; and for applications, see [8, 35, 36, 57]):

$$\|D^\alpha(fg)\|_{L^r} \lesssim \|D^\alpha f\|_{L^{p_1}} \cdot \|g\|_{L^{q_1}} + \|f\|_{L^{p_2}} \cdot \|D^\alpha g\|_{L^{q_2}}, \tag{1.9}$$

where the homogeneous differential operator D is defined by $\widehat{D^\alpha u}(\xi) := (2\pi|\xi|)^\alpha \hat{u}(\xi)$ as usual. The proof of Kato-Ponce type estimates can be reduced to proving the following bilinear Coifman-Meyer multiplier theorem.

Theorem 1.5 (Coifman and Meyer, [11, 12]). *For $d \geq 1$, let $m \in C^L(\mathbb{R}^{2d} \setminus \{0\})$ satisfy the Hörmander-Mihlin conditions*

$$|\partial_\xi^\alpha \partial_\eta^\beta m(\xi, \eta)| \lesssim (|\xi| + |\eta|)^{-(|\alpha|+|\beta|)} \tag{1.10}$$

for all multi-indices α and β satisfying $|\alpha| + |\beta| \leq L$, where $L \in \mathbb{N}$ is sufficiently large. If the bilinear operator T_m is defined by

$$T_m(f, g)(x) := \int_{\mathbb{R}^{2d}} m(\xi, \eta)\hat{f}(\xi)\hat{g}(\eta)e^{2\pi i x\cdot(\xi+\eta)}d\xi d\eta, \tag{1.11}$$

then T_m maps $L^p(\mathbb{R}^d) \times L^q(\mathbb{R}^d)$ into $L^r(\mathbb{R}^d)$ boundedly, as long as $1 < p, q, r \leq \infty$ and $\frac{1}{r} = \frac{1}{p} + \frac{1}{q} > 0$.

In general, for $n \geq 1$ and $d \geq 1$, let m be a bounded function in $\mathbb{R}^{nd}$, smooth away from the origin and satisfying Hörmander-Mikhlin conditions:

$$|\partial^\alpha m(\xi)| \lesssim \frac{1}{|\xi|^{|\alpha|}} \tag{1.12}$$

for sufficiently many multi-indices α. Denote by T_m the n-linear operator defined by

$$T_m(f_1, \ldots, f_n)(x) := \int_{\mathbb{R}^{nd}} m(\xi)\hat{f}_1(\xi_1) \cdots \hat{f}_n(\xi_n) e^{2\pi i x \cdot (\xi_1 + \cdots + \xi_n)} d\xi, \tag{1.13}$$

where $\xi = (\xi_1, \ldots, \xi_n) \in \mathbb{R}^{nd}$ and $f_1, \ldots, f_n$ are Schwartz functions on $\mathbb{R}^d$. The above bilinear Coifman-Meyer theorem (Theorem 1.5) has been extended to multilinear Calderón-Zygmund operators and to the range $0 < r \leq 1$ (see [9, 26, 27, 37]), therefore, we know that the operator T_m given by (1.13) extends to a bounded n-linear operator from $L^{p_1}(\mathbb{R}^d) \times \cdots \times L^{p_n}(\mathbb{R}^d)$ into $L^p(\mathbb{R}^d)$, provided that $1 < p_1, \ldots, p_n \leq \infty$ and $\frac{1}{p} = \frac{1}{p_1} + \cdots + \frac{1}{p_n} > 0$. When $n = 2$, as a consequence of the bilinear T(1) theorem (see [1, 9, 26]), there is also a pseudo-differential variant of the Coifman-Meyer multiplier theorem (Theorem 1.5) for symbol $a \in BS^0_{1,0}(\mathbb{R}^{3d})$, that is, a satisfies the differential inequalities

$$|\partial_x^\gamma \partial_\xi^\alpha \partial_\eta^\beta a(x, \xi, \eta)| \lesssim_{d,\alpha,\beta,\gamma} (1 + |\xi| + |\eta|)^{-|\alpha|-|\beta|} \tag{1.14}$$

for sufficiently many multi-indices α, β, γ. Namely, let T_a be the corresponding bilinear pseudo-differential operators defined by replacing m with a in (1.11), then T_a is bounded from $L^p(\mathbb{R}^d) \times L^q(\mathbb{R}^d)$ into $L^r(\mathbb{R}^d)$, provided that $1 < p, q \leq \infty$ and $\frac{1}{r} = \frac{1}{p} + \frac{1}{q} > 0$ (see [3], and see [4, 51, 57] for $d = 1$ case).

In many situations where m has only limited smoothness, we cannot apply the Coifman-Meyer theorem since the order L of the smoothness of the multiplier needs to be sufficiently large. Finding the best possible number of L thus becomes an interesting problem. By reducing the bilinear Fourier multiplier operators to linear Calderón-Zygmund operators, Coifman and Meyer obtained the L^r estimates under the assumption $L = 2d + 1$. In [27], the authors also proved that the condition (1.10) with $L = 2d + 1$ assures the boundedness of T_m by using the bilinear T(1) theorem. However, this number seems to be too large in view of the linear case (that is, half of the number of variables in m plus one). Recently, Tomita [67] proved the following Hörmander type theorem for bilinear multipliers with limited smoothness in terms of the Sobolev regularity.

Theorem 1.6. *Let $s > \frac{nd}{2}$, $1 < p, p_1, \ldots, p_n < \infty$ and $\frac{1}{p} = \frac{1}{p_1} + \cdots + \frac{1}{p_n}$. If a symbol $m \in L^\infty(\mathbb{R}^{nd})$ satisfies*

$$\sup_{k \in \mathbb{Z}} \|m(2^k \cdot, \ldots, 2^k \cdot)\Psi\|_{H^s(\mathbb{R}^{nd})} < \infty, \tag{1.15}$$

where Ψ is defined as in (1.3) with spatial dimension equal to nd, then the n-linear operator T_m defined by (1.13) is bounded from $L^{p_1}(\mathbb{R}^d) \times \cdots \times L^{p_n}(\mathbb{R}^d)$ into $L^p(\mathbb{R}^d)$.

For further improvement on these Hörmander type theorems in the cases $0 < p \leq 1$ or the cases where p_i $(i = 1, \ldots, n)$ can be smaller than or equal to 1, we refer to [23, 24, 49].

For a large amount of literature involving estimates for multi-linear Calderón-Zygmund operators and multi-linear pseudo-differential operators, refer to e.g. [2, 9, 12, 21, 26, 27, 37, 46, 48, 57, 58] and the references therein.

For the trilinear Coifman-Meyer type theorem, Muscalu [52] proved the following theorem where the multiplier involved is a product of two symbols and has *flag singularities*, that is, for m_1, m_2 satisfying

$$|\partial_\xi^\alpha \partial_\eta^\beta m_1(\xi,\eta)| \lesssim \frac{1}{(|\xi| + |\eta|)^{\alpha+\beta}},$$
$$|\partial_\eta^\beta \partial_\zeta^\gamma m_2(\eta,\zeta)| \lesssim \frac{1}{(|\eta| + |\zeta|)^{\beta+\gamma}} \tag{1.16}$$

for every $\xi, \eta, \zeta \in \mathbb{R}^n$ and sufficiently many indices α, β and γ, we define

$$T_{m_1,m_2}(f_1, f_2, f_3)(x) := \int_{\mathbb{R}^{3n}} m_1(\xi,\eta) m_2(\eta,\zeta) \hat{f}_1(\xi) \hat{f}_2(\eta) \hat{f}_3(\zeta) e^{2\pi i(\xi+\eta+\zeta)\cdot x} d\xi d\eta d\zeta, \tag{1.17}$$

where f_1, f_2, f_3 are Schwartz functions. Then we have

Theorem 1.7 ([52]). *The operator defined in* (1.17) *maps* $L^{p_1} \times L^{p_2} \times L^{p_3} \to L^r$ *for* $1 < p_1, p_2, p_3 < \infty$ *with* $1/p_1 + 1/p_2 + 1/p_3 = 1/r$ *and* $0 < r < \infty$. *In addition,* T_{m_1,m_2} *also maps* $L^\infty \times L^p \times L^q \to L^s$, $L^p \times L^\infty \times L^q \to L^s$, $L^\infty \times L^t \times L^\infty \to L^t$ *for every* $1 < p, q, t < \infty$ *and* $1/p + 1/q = 1/s$.

Moreover, for the above theorem, the estimates like $L^\infty \times L^\infty \times L^t \to L^t$ or $L^\infty \times L^\infty \times L^\infty \to L^\infty$ are false, and these can be checked if we set f_2 to be identically 1.

In [45], the second and third authors of the current article consider a pseudo-differential operator corresponding to the above theorem, that is, let $a(x, \xi, \eta)$, $b(x, \eta, \zeta) \in BS^0_{1,0}$ be symbols satisfying the conditions

$$|\partial_x^l \partial_\xi^\alpha \partial_\eta^\beta a(x,\xi,\eta)| \lesssim \frac{1}{(1 + |\xi| + |\eta|)^{\alpha+\beta}},$$
$$|\partial_x^l \partial_\eta^\beta \partial_\zeta^\gamma b(x,\eta,\zeta)| \lesssim \frac{1}{(1 + |\eta| + |\zeta|)^{\beta+\gamma}} \tag{1.18}$$

for every $x, \xi, \eta, \zeta \in \mathbb{R}^n$ and sufficiently many indices α, β and γ, define the operator

$$T_{ab}(f, g, h)(x) := \int_{\mathbb{R}^{3n}} a(x,\xi,\eta) b(x,\eta,\zeta) \hat{f}(\xi) \hat{g}(\eta) \hat{h}(\zeta) e^{2\pi i x(\xi+\eta+\zeta)} d\xi d\eta d\zeta. \tag{1.19}$$

It's easy to see that the symbol $a(x, \xi, \eta) \cdot b(x, \eta, \zeta)$ satisfies a less restrictive condition than the condition (1.14) for the symbol a. The main result of [45] is the following

Theorem 1.8. *The operator T_{ab} defined as (1.19) is bounded from $L^{p_1}\times L^{p_2}\times L^{p_3}$ to L^r for $1<p_1,p_2,p_3<\infty$ with $1/p_1+1/p_2+1/p_3=1/r$ and $0<r<\infty$. In addition, T_{ab} also maps $L^\infty\times L^p\times L^q\to L^s$, $L^p\times L^\infty\times L^q\to L^s$, $L^\infty\times L^t\times L^\infty\to L^t$ for every $1<p,q,t<\infty$ and $1/p+1/q=1/s$.*

The proof of Theorem 1.8 is to reduce the trilinear pseudo-differential operator with the symbol of flag singularity to a localized version and takes advantage of the *flag paraproducts* from Muscalu's work [52] on the L^p estimates for the Fourier multipliers with symbols of flag singularity. Namely, we need to prove an equivalent localized version of Theorem 1.8 (see [57], and also [14] for the multi-parameter setting). Moreover, the key to prove the localized result is that, conditions (1.18) allow us to only consider the dyadic intervals with lengths at most 1 in the *flag paraproducts*.

More precisely, we can show that Theorem 1.8 can be reduced to an estimate for a localized operator

$$T_{ab}^{0,0}(f,g,h)(x)=\left(\int_{\mathbb{R}^n} a_0(\xi,\eta)b_0(\eta,\zeta)\hat{f}(\xi)\hat{g}(\eta)\hat{h}(\zeta)e^{2\pi ix(\xi+\eta+\zeta)}d\xi d\eta d\zeta\right)\varphi_0(x),$$

where $\varphi_0(x)$ is a Schwartz function supported near the origin and a_0,b_0 satisfy a stronger decay condition than the classical Hörmander-Mikhlin condition.

To this end, we will decompose the operator $T_{ab}^{0,0}$ to some operators of different forms. Among these operators, some of them could be reduced to the classical pseudo-differential operators, and the others could be written as *flag paraproducts*, which are used in the proof of Theorem 1.7, in the forms of

$$(T_1(f,g,h)\cdot\varphi_0)(x)=\sum_{I\in\mathcal{I}}\frac{1}{|I|^{\frac{1}{2}}}\langle f,\phi_I^1\rangle\langle B_I^1(g,h),\phi_I^2\rangle\phi_I^3\varphi_0,$$

$$\text{where}\qquad B_I^1(g,h)=\sum_{\substack{J\in\mathcal{J}\\ |\omega_J^3|\leq|\omega_I^2|}}\frac{1}{|J|^{\frac{1}{2}}}\langle g,\phi_J^1\rangle\langle h,\phi_J^2\rangle\phi_J^3,$$

but with dyadic intervals having lengths at most 1.

Then by taking advantage of the *flag paraproducts* mentioned above, we will be able to prove the desired estimate for the localized version of our theorem. For more details, see [45].

2 L^p estimates for multi-parameter and multi-linear paraproducts, multipliers and pseudo-differential operators

When we come into the situations that a differential operator (with different behaviors on different spatial variables x_i, $i=1,\ldots,d$) acts on a product of several functions, for instance, we consider the bi-parameter Leibnitz rule of $\mathcal{D}_1^\alpha\mathcal{D}_2^\beta(fg)$, where $\widehat{\mathcal{D}_1^\alpha f}(\xi_1,\xi_2) := |\xi_1|^\alpha \hat{f}(\xi_1,\xi_2)$ and $\widehat{\mathcal{D}_2^\beta f}(\xi_1,\xi_2) := |\xi_2|^\beta \hat{f}(\xi_1,\xi_2)$ for $\alpha,\ \beta > 0$, that is, to prove

$$\begin{aligned}\|\mathcal{D}_1^\alpha \mathcal{D}_2^\beta (fg)\|_{L^r} \lesssim & \|\mathcal{D}_1^\alpha \mathcal{D}_2^\beta f\|_{L^{p_1}} \|g\|_{L^{q_1}} + \|f\|_{L^{p_2}} \|\mathcal{D}_1^\alpha \mathcal{D}_2^\beta g\|_{L^{q_2}} \\ & + \|\mathcal{D}_1^\alpha f\|_{L^{p_3}} \|\mathcal{D}_2^\beta g\|_{L^{q_3}} + \|\mathcal{D}_2^\beta f\|_{L^{p_4}} \|\mathcal{D}_1^\alpha g\|_{L^{q_4}}\end{aligned} \tag{2.1}$$

for every $1/p_i + 1/q_i = 1/r$, $1 < p_i,\ q_i \leq \infty$ $(1 \leq i \leq 4)$ and $\max\{1/(1+\alpha),\ 1/(1+\beta)\} < r < \infty$. We realize that the necessity to investigate the bilinear and bi-parameter operators $T_m^{(2)}$ defined by

$$T_m^{(2)}(f,g)(x) := \int_{\mathbb{R}^4} m(\xi,\eta) \hat{f}(\xi) \hat{g}(\eta) e^{2\pi i x\cdot(\xi+\eta)} d\xi d\eta, \tag{2.2}$$

where the symbol m is smooth away from the planes $(\xi_1,\eta_1) = (0,0)$ and $(\xi_2,\eta_2) = (0,0)$ in $\mathbb{R}^2 \times \mathbb{R}^2$ and satisfying the less restrictive Marcinkiewicz conditions

$$|\partial_{\xi_1}^{\alpha_1} \partial_{\xi_2}^{\beta_1} \partial_{\eta_1}^{\alpha_2} \partial_{\eta_2}^{\beta_2} m(\xi,\eta)| \lesssim \frac{1}{|(\xi_1,\eta_1)|^{\alpha_1+\alpha_2}} \cdot \frac{1}{|(\xi_2,\eta_2)|^{\beta_1+\beta_2}} \tag{2.3}$$

for sufficiently many multi-indices $\alpha = (\alpha_1,\alpha_2)$, $\beta = (\beta_1,\beta_2)$.

It becomes more complicated and difficult to establish the L^p estimates for $T_m^{(2)}$ than in the one-parameter multi-linear situations or L^p estimates for linear multi-parameter singular integrals (see e.g., [20] and [31]). In [54], the authors reduced the proof of the L^p estimates for bilinear and bi-parameter operators $T_m^{(2)}$ to proving L^p estimates for bilinear and bi-parameter discretized paraproducts. To state the results in [54], we first review the following definitions and properties about discretized paraproducts.

An interval I on the real line $\mathbb{R}$ is called dyadic if it is of the form $I = 2^{-k}[n,\ n+1]$ for some $k,\ n \in \mathbb{Z}$, a rectangle R on the plane $\mathbb{R}^2$ is called dyadic if there exist some dyadic intervals $I,\ J$ such that $R = I \times J$. Following [57], we first give the following definitions.

Definition 2.1. For $J \subseteq \mathbb{R}$ an arbitrary interval, we say that a smooth function Φ_J is a bump adapted to J, if and only if the following inequalities hold:

$$|\Phi_J^{(l)}(x)| \lesssim_{l,\alpha} \frac{1}{|J|^l} \cdot \frac{1}{(1+\frac{\mathrm{dist}(x,J)}{|J|})^\alpha} \tag{2.4}$$

for every integer $\alpha \in \mathbb{N}$ and for sufficiently many derivatives $l \in \mathbb{N}$. If Φ_J is a bump adapted to J, we say that $|J|^{-\frac{1}{2}}\Phi_J$ is an L^2-normalized bump adapted to J.

Definition 2.2. A family of L^2-normalized adapted bump functions $(\varphi_I)_I$ is said to be nonlacunary if and only if for every I one has

$$\operatorname{supp} \widehat{\varphi_I} \subseteq [-4|I|^{-1},\ 4|I|^{-1}].$$

A family of L^2-normalized adapted bump functions $(\varphi_I)_I$ is said to be lacunary if and only if for any I one has

$$\operatorname{supp} \widehat{\varphi_I} \subseteq \left[-4|I|^{-1},\ -\frac{1}{4}|I|^{-1}\right] \cup \left[\frac{1}{4}|I|^{-1},\ 4|I|^{-1}\right].$$

Definition 2.3. Let $\mathcal{I}$ be a finite set of dyadic intervals. A bilinear expression of the type

$$\Pi_{\mathcal{I}}(f,g)=\sum_{I\in\mathcal{I}}c_I\frac{1}{|I|^{\frac{1}{2}}}\langle f,\varphi_I^1\rangle\langle g,\varphi_I^2\rangle\varphi_I^3 \tag{2.5}$$

is called a bilinear discretized paraproduct if and only if $(c_I)_I$ is a bounded sequence of complex numbers and at least two of the families of L^2-normalized bump functions $(\varphi_I^j)_I$ for $j=1,2,3$ are lacunary in the sense of Definition 2.2.

Then the following is well known, for the proof, see e.g. [41, 57, 54, 56].

Theorem 2.4. *Any bilinear discretized paraproduct $\Pi_{\mathcal{I}}$ has a bounded mapping $L^p\times L^q$ to L^r as long as $1<p,\,q\le\infty$ and $\frac{1}{r}=\frac{1}{p}+\frac{1}{q}>0$. Moreover, the implicit constants in the bounds depend only on p,q,r and are independent of the cardinality of $\mathcal{I}$, provided that the sequence $(c_I)_I$ in (2.5) is bounded by a universal constant.*

Consider two discretized classical paraproducts given by

$$\Pi_{1,\mathcal{I}}(f_1,g_1)=\sum_{I\in\mathcal{I}}c_I\frac{1}{|I|^{\frac{1}{2}}}\langle f_1,\varphi_I^1\rangle\langle g_1,\varphi_I^2\rangle\varphi_I^3$$

and

$$\Pi_{2,\mathcal{J}}(f_2,g_2)=\sum_{J\in\mathcal{J}}c_J\frac{1}{|J|^{\frac{1}{2}}}\langle f_2,\varphi_J^1\rangle\langle g_2,\varphi_J^2\rangle\varphi_J^3,$$

and define the bi-parameter discretized paraproduct $\overrightarrow{\Pi}_{\mathcal{R}}$ by $\overrightarrow{\Pi}_{\mathcal{R}}=\Pi_{1,\mathcal{I}}\otimes\Pi_{2,\mathcal{J}}$ or, more precisely, by

$$\overrightarrow{\Pi}_{\mathcal{R}}(f,g)=\sum_{R\in\mathcal{R}}c_R\frac{1}{|R|^{\frac{1}{2}}}\langle f,\varphi_R^1\rangle\langle g,\varphi_R^2\rangle\varphi_R^3, \tag{2.6}$$

where the numbers c_R are all bounded, the sum is over dyadic rectangles of the form $R=I\times J$ and φ_R^j is defined by $\varphi_R^j:=\varphi_I^j\otimes\varphi_J^j$ for $j=1,2,3$. In [54], by using the duality lemma of $L^{p,\infty}$ presented in [58], the Journé's covering lemma in [32], the $L^{1,\infty}$ sizes and energies technique developed in [59], the "stopping-time decompositions" arguments and multi-linear interpolation (see e.g. [25, 59]), Muscalu, Pipher, Tao and Thiele proved the following L^p estimates for bi-parameter discretized paraproduct $\overrightarrow{\Pi}_{\mathcal{R}}$ (for the proof, refer to [41, 57, 54, 56]).

Theorem 2.5 ([54]). *Any discrete bi-parameter paraproduct (2.6) is bounded from $L^p(\mathbb{R}^2)\times L^q(\mathbb{R}^2)\to L^r(\mathbb{R}^2)$ provided that $1<p,\,q\le\infty$ and $\frac{1}{r}=\frac{1}{p}+\frac{1}{q}>0$.*

As a consequence, the authors obtained in [54] the following L^p estimates for $T_m^{(2)}$ (see also [57], and for subsequent endpoint estimates see [41]).

Theorem 2.6 ([54]). *The bilinear operator $T_m^{(2)}$ defined by (2.2) maps $L^p(\mathbb{R}^2)\times L^q(\mathbb{R}^2)\to L^r(\mathbb{R}^2)$ boundedly, as long as $1<p,\,q\le\infty$ and $\frac{1}{r}=\frac{1}{p}+\frac{1}{q}>0$.*

In general, any collection of n generic vectors $\xi_1 = (\xi_1^i)_{i=1}^d, \ldots, \xi_n = (\xi_n^i)_{i=1}^d$ in $\mathbb{R}^d$ generates naturally the following collection of d vectors in $\mathbb{R}^n$:

$$\bar{\xi}_1 = (\xi_j^1)_{j=1}^n, \quad \bar{\xi}_2 = (\xi_j^2)_{j=1}^n, \quad \ldots, \quad \bar{\xi}_d = (\xi_j^d)_{j=1}^n. \tag{2.7}$$

Let $m = m(\xi) = m(\bar{\xi})$ be a bounded symbol in $L^\infty(\mathbb{R}^{dn})$ that is smooth away from the subspaces $\{\bar{\xi}_1 = 0\} \cup \cdots \cup \{\bar{\xi}_d = 0\}$ and satisfying

$$|\partial_{\bar{\xi}_1}^{\alpha_1} \cdots \partial_{\bar{\xi}_d}^{\alpha_d} m(\bar{\xi})| \lesssim \prod_{i=1}^d |\bar{\xi}_i|^{-|\alpha_i|} \tag{2.8}$$

for sufficiently many multi-indices $\alpha_1, \ldots, \alpha_d$. Denote by $T_m^{(d)}$ the n-linear multiplier operator defined by

$$T_m^{(d)}(f_1, \ldots, f_n)(x) := \int_{\mathbb{R}^{dn}} m(\xi) \hat{f}_1(\xi_1) \cdots \hat{f}_n(\xi_n) e^{2\pi i x \cdot (\xi_1 + \cdots + \xi_n)} d\xi. \tag{2.9}$$

In [56], by proving a generic decomposition lemma, the authors were able to extend Theorem 2.5 to the general n-linear and d-parameter settings (for the proof, see [41, 56]), that is, there is an analogue of the Hölder-type L^p estimates stated in Theorem 2.5 for the discretized d-parameter paraproducts of the form

$$\overrightarrow{\Pi}_{\mathcal{R}}(f_1, \ldots, f_n) = \sum_{R \in \mathcal{R}} c_R \frac{1}{|R|^{\frac{n-1}{2}}} \langle f_1, \varphi_R^1 \rangle \cdots \langle f_n, \varphi_R^n \rangle \varphi_R^{n+1}, \tag{2.10}$$

where the sum runs over the dyadic parallelepipeds $R = I_1 \times \cdots \times I_d \subseteq \mathbb{R}^d$. As a consequence, Muscalu, Pipher, Tao and Thiele [56] generalized Theorem 2.6 to the n-linear and d-parameter setting for any $n \geq 1$, $d \geq 2$, their result is stated in the following theorem (see also [57]).

Theorem 2.7 ([56]). *For any $n \geq 1$, $d \geq 2$, the n-linear, d-parameter multiplier operator $T_m^{(d)}$ given by (2.9) maps $L^{p_1}(\mathbb{R}^d) \times \cdots \times L^{p_n}(\mathbb{R}^d) \to L^p(\mathbb{R}^d)$ boundedly, provided that $1 < p_1, \ldots, p_n \leq \infty$ and $\frac{1}{p} = \frac{1}{p_1} + \cdots + \frac{1}{p_n} > 0$.*

Recently, in the work [10] by Chen and the second author of the current article, they offer a different (and perhaps more elementary) proof than those in [54, 56] to establish a Hörmander type theorem of L^p estimates (and weighted estimates as well) for multi-linear and multi-parameter Fourier multiplier operators with limited smoothness in multi-parameter Sobolev spaces. The method in [10] is mainly using the Littlewood-Paley theory and the boundedness of the L^p norm of a function by its L^p norm of the Littlewood-Paley square function, which is actually its multi-parameter Hardy space H^p norm of the function when $0 < p \leq 1$. The advantage of this method is that it also enables us to identify the minimal smoothness required for the symbol to guarantee the L^p estimates. Their main results in the bilinear and bi-parameter setting of [10] states that:

Theorem 2.8. *Let $s_1, s_2 > d$, $\max\{1, \frac{2d}{\min\{s_1,s_2\}}\} < p, q < \infty$ and $\frac{1}{p} + \frac{1}{q} = \frac{1}{r} > 0$. If $m \in L^\infty(\mathbb{R}^{4d})$ satisfies*

$$\sup_{j,k\in\mathbb{Z}} \|m(2^j\xi_1, 2^k\xi_2, 2^j\eta_1, 2^k\eta_2)\Psi_1(\xi_1,\eta_1)\Psi_2(\xi_2,\eta_2)\|_{H^{s_1,s_2}(\mathbb{R}^{4d})} < \infty,$$

where Ψ_1, Ψ_2 are defined as in (1.3) with spatial dimension equal to 2d, then the bilinear and bi-parameter multiplier operator $T_m^{(2)}$ is bounded from $L^p(\mathbb{R}^{2d}) \times L^q(\mathbb{R}^{2d})$ to $L^r(\mathbb{R}^{2d})$.

Weighted results are also given in [10]. For more details, we refer the reader to [10].

In [14], the first two authors of the current article proved the pseudo-differential variant of the L^p estimates for multi-linear and multi-parameter Coifman-Meyer multiplier operators obtained in [54, 56] (see Theorem 2.6, Theorem 2.7).

Suppose that $a(x,\xi,\eta)$ is a smooth symbol satisfying

$$|\partial_{x_1}^{\gamma_1}\partial_{x_2}^{\gamma_2}\partial_{\xi_1}^{\alpha_1}\partial_{\xi_2}^{\beta_1}\partial_{\eta_1}^{\alpha_2}\partial_{\eta_2}^{\beta_2}a(x,\xi,\eta)| \lesssim \frac{1}{(1+|(\xi_1,\eta_1)|)^{\alpha_1+\alpha_2}} \cdot \frac{1}{(1+|(\xi_2,\eta_2)|)^{\beta_1+\beta_2}} \tag{2.11}$$

for sufficiently many multi-indices $\alpha = (\alpha_1,\alpha_2)$, $\beta = (\beta_1,\beta_2)$, $\gamma = (\gamma_1,\gamma_2)$, and denote by $T_a^{(2)}$ the bilinear operator given by

$$T_a^{(2)}(f,g)(x) := \int_{\mathbb{R}^4} a(x,\xi,\eta)\hat{f}(\xi)\hat{g}(\eta)e^{2\pi i x\cdot(\xi+\eta)}d\xi d\eta. \tag{2.12}$$

In [14], we proved that the same L^p estimates as $T_m^{(2)}$ in Theorem 2.6 hold true for the operator $T_a^{(2)}$ and thus answered an open question raised by Muscalu and Schlag in their book in 2013 [57].

Theorem 2.9 ([14]). *The bilinear and bi-parameter pseudo-differential operator $T_a^{(2)}$ defined by (2.12) maps $L^p(\mathbb{R}^2) \times L^q(\mathbb{R}^2) \to L^r(\mathbb{R}^2)$ boundedly, as long as $1 < p$, $q \le \infty$ and $\frac{1}{r} = \frac{1}{p} + \frac{1}{q} > 0$.*

In general, let symbol $a = a(x,\xi) = a(x,\bar{\xi}) \in C^\infty(\mathbb{R}^{d(n+1)})$ and satisfy the differential inequalities

$$|\partial_x^\gamma \partial_{\bar{\xi}_1}^{\alpha_1} \cdots \partial_{\bar{\xi}_d}^{\alpha_d} a(x,\bar{\xi})| \lesssim \prod_{i=1}^{d}(1+|\bar{\xi}_i|)^{-|\alpha_i|} \tag{2.13}$$

for sufficiently many multi-indices $\alpha_1, \ldots, \alpha_d$ and γ. Denote by $T_a^{(d)}$ the n-linear pseudo-differential operator defined by

$$T_a^{(d)}(f_1,\ldots,f_n)(x) := \int_{\mathbb{R}^{dn}} a(x,\xi)\hat{f}_1(\xi_1)\cdots\hat{f}_n(\xi_n)e^{2\pi i x\cdot(\xi_1+\cdots+\xi_n)}d\xi, \tag{2.14}$$

then we can naturally generalize Theorem 2.9 to the n-linear and d-parameter setting for any $n \ge 1$, $d \ge 2$ and obtain the pseudo-differential variant of Theorem 2.7. Our generalized theorem in [14] is the following.

Theorem 2.10 ([14]). *For any $n \geq 1$, $d \geq 2$, the n-linear, d-parameter pseudo-differential operator $T_a^{(d)}$ maps $L^{p_1}(\mathbb{R}^d) \times \cdots \times L^{p_n}(\mathbb{R}^d) \to L^p(\mathbb{R}^d)$ boundedly, provided that $1 < p_1, \ldots, p_n \leq \infty$ and $\frac{1}{p} = \frac{1}{p_1} + \cdots + \frac{1}{p_n} > 0$.*

We now give an overview of our proof strategy of main theorems in [14] and indicate its additional difficulty and complexity compared with the cases of L^p estimates for one-parameter and multi-linear pseudo-differential operators of Muscalu [51, 57] and the L^p estimates for multi-linear and multi-parameter multiplier theorem of Muscalu, Pipher, Tao and Thiele [54, 56]. For simplicity, we will only discuss the proof strategy of Theorem 2.9 (the bilinear and bi-parameter case, $n = d = 2$) in this article. However, it will be clear from the proof (see [14]) that we can extend the argument to the general n-linear, d-parameter setting (Theorem 2.10) straightforwardly.

By using the idea presented by Muscalu in [51, 57] to prove the L^p estimates for one-parameter ($d = 1$) and bilinear pseudo-differential operators $T_a = T_a^{(1)}$, we will first show that the proof of Theorem 2.9 can be essentially reduced to proving a localized variant of the bilinear and bi-parameter Coifman-Meyer theorem (Theorem 2.6), that is, some kind of localized L^p estimates of the localized bilinear and bi-parameter operator $T_a^{(2),(0,0,\vec{0})}$ given by

$$\begin{aligned} T_a^{(2),(0,0,\vec{0})}(f,g)(x) &= T_a^{(2),\vec{0}}(f,\, g)(x) \cdot \varphi_0' \otimes \varphi_0''(x) \\ &:= \int_{\mathbb{R}^4} m_{\vec{0}}(\xi,\eta) \hat{f}(\xi)\hat{g}(\eta) e^{2\pi i x\cdot(\xi+\eta)} d\xi d\eta \cdot \varphi_0' \otimes \varphi_0''(x), \end{aligned} \tag{2.15}$$

where $\varphi_0'(x_1)$ and $\varphi_0''(x_2)$ are smooth cut-off functions and satisfy $\operatorname{supp}\varphi_0' \subseteq [-1,1] =: I_0$, $\operatorname{supp}\varphi_0'' \subseteq [-1,1] =: J_0$.

Then, we discretize the operator $T_a^{(2),(0,0,\vec{0})}$ into localized bilinear and bi-parameter paraproducts. Since the symbol $m_{\vec{0}}(\xi,\eta)$ of the operator $T_a^{(2),(0,0,\vec{0})}$ doesn't have singularity near the planes $(\xi_1,\eta_1) = (0,0)$, $(\xi_2,\eta_2) = (0,0)$ in $\mathbb{R}^2 \times \mathbb{R}^2$ and satisfies differential estimates

$$|\partial_{\xi_1}^{\alpha_1}\partial_{\xi_2}^{\beta_1}\partial_{\eta_1}^{\alpha_2}\partial_{\eta_2}^{\beta_2} m_{\vec{0}}(\xi,\eta)| \lesssim \frac{1}{(1+|(\xi_1,\eta_1)|)^{|\alpha|}} \cdot \frac{1}{(1+|(\xi_2,\eta_2)|)^{|\beta|}} \tag{2.16}$$

for sufficiently many multi-indices $\alpha = (\alpha_1,\alpha_2)$ and $\beta = (\beta_1,\beta_2)$, which is stronger than the Marcinkiewicz condition (2.3), thus in the present case, it will be clear from the discretization procedure that one can use the L^1-normalized bump functions $\{\psi_k\}_{k\geq -1}$ (which are adapted to intervals of sizes $2^{-k} \lesssim 1$ and of heights 2^k) to carry out inhomogeneous Littlewood-Paley dyadic decomposition with respect to x_1 and x_2 variables respectively. That is the reason why we can restrict further that the summation in (2.18) runs over dyadic intervals having the property that $|I|$, $|J| \lesssim 1$. By making use of the inhomogeneous Littlewood-Paley dyadic decomposition and Bony's paraproducts decomposition, we can discretize the bilinear and bi-parameter operator $T_a^{(2),\vec{0}}$ into a sum of sixteen bilinear and bi-parameter paraproduct operators as follows:

$$\begin{aligned} &T_a^{(2),\vec{0}}(f,g) \\ &= (\Pi^1_{lh}\otimes\Pi^2_{lh})(f,g) + (\Pi^1_{lh}\otimes\Pi^2_{hl})(f,g) + (\Pi^1_{lh}\otimes\Pi^2_{hh})(f,g) + (\Pi^1_{lh}\otimes\Pi^2_{ll})(f,g) \\ &\quad +(\Pi^1_{hl}\otimes\Pi^2_{lh})(f,g) + (\Pi^1_{hl}\otimes\Pi^2_{hl})(f,g) + (\Pi^1_{hl}\otimes\Pi^2_{hh})(f,g) + (\Pi^1_{hl}\otimes\Pi^2_{ll})(f,g) \\ &\quad +(\Pi^1_{hh}\otimes\Pi^2_{lh})(f,g) + (\Pi^1_{hh}\otimes\Pi^2_{hl})(f,g) + (\Pi^1_{hh}\otimes\Pi^2_{hh})(f,g) + (\Pi^1_{hh}\otimes\Pi^2_{ll})(f,g) \\ &\quad +(\Pi^1_{ll}\otimes\Pi^2_{lh})(f,g) + (\Pi^1_{ll}\otimes\Pi^2_{hl})(f,g) + (\Pi^1_{ll}\otimes\Pi^2_{hh})(f,g) + (\Pi^1_{ll}\otimes\Pi^2_{ll})(f,g), \end{aligned} \tag{2.17}$$

where Π^i denotes one of the "low-high", "high-low", "high-high" and "low-low" paraproducts with respect to x_i variable for $i = 1, 2$, for instance, one of these operators can be expressed as

$$\begin{aligned} &T^{(2),\vec{0}}_{a,(lh,hl)}(f,g)(x) := (\Pi^1_{lh}\otimes\Pi^2_{hl})(f,g)(x) \qquad (2.18) \\ &:= \sum_{k,l\geq 1}\int_{\mathbb{R}^4} m_{\vec{0}}(\xi,\eta)(f*(\widetilde{\varphi_k}\otimes\psi_l))^\wedge(\xi)(g*(\psi_k\otimes\widetilde{\varphi_l}))^\wedge(\eta)e^{2\pi i x\cdot(\xi+\eta)}d\xi d\eta, \end{aligned}$$

where $\widetilde{\varphi_k}$ are nonlacunary L^1-normalized bump functions while ψ_k are lacunary L^1-normalized bump functions and satisfy $\operatorname{supp}\widehat{\widetilde{\varphi_k}} \subseteq [-\frac{2}{3}\cdot 2^k, \frac{2}{3}\cdot 2^k]$, $\operatorname{supp}\hat{\psi}_k \subseteq [-\frac{8}{3}\cdot 2^k, -\frac{3}{4}\cdot 2^k]\cup[\frac{3}{4}\cdot 2^k, \frac{8}{3}\cdot 2^k]$. By splitting the integral region $\mathbb{R}^2$ into the union of unit squares, one can apply the discretization and the L^2-normalization procedure to the corresponding tri-linear forms. As a consequence, we can reduce the proof of localized L^p estimates of $T_a^{(2),(0,0,\vec{0})}$ to proving the localized L^p estimates for discrete and localized bilinear and bi-parameter paraproduct operators of the form

$$\overrightarrow{\Pi}^{(2),(0,0,\vec{0})}_{a,\mathcal{R}}(f,g)(x) = \Big\{\sum_{\substack{R=I\times J\in\mathcal{R},\\ |I|,|J|\leq 1}} c_R\frac{1}{|R|^{\frac{1}{2}}}\langle f,\varphi^1_R\rangle\langle g,\varphi^2_R\rangle\varphi^3_R(x)\Big\}\cdot\varphi_0'\otimes\varphi_0''(x). \tag{2.19}$$

It's actually an inhomogeneous variant of the discretization procedure presented in [54] to prove the bi-parameter Coifman-Meyer theorem (Theorem 2.6).

Now, in order to prove Theorem 2.9, we only have the task of proving localized L^p estimates for the localized bilinear and bi-parameter paraproduct operator $\overrightarrow{\Pi}^{(2),(0,0,\vec{0})}_{a,\mathcal{R}}$. One can observe that the supports of functions f, g and the dyadic rectangles $R = I\times J$ may get close to or far away from the integral region $R_{00} = I_0\times J_0$ in two different directions x_1 and x_2 due to the bi-parameter setting, and that the dyadic rectangles R with comparable area may overlap considerably, thus the situations will be more complicated than the one-parameter case ($d = 1$) considered in [4, 51, 57]. Since rapid decay factors can be derived from φ^3_R when R is sufficiently far away from R_{00} in x_1 or x_2 directions (for example, $R\subseteq(5R_{00})^c$), we will split the localized bilinear and bi-parameter paraproduct operator $\overrightarrow{\Pi}^{(2),(0,0,\vec{0})}_{a,\mathcal{R}}$ into a summation of a "main term", "hybrid terms" and an "error term" (see [14]).

Compared with the one-parameter case, there are mainly two key ingredients in our estimates of $\overrightarrow{\Pi}^{(2),(0,0,\vec{0})}_{a,\mathcal{R}}$, one is the estimates of the "main term" and "hybrid terms", the other is the estimates of the discrete bilinear operators $\overrightarrow{\Pi}^{(2),(0,0,\vec{0})}_{a,\mathcal{R}}$ corresponding to bilinear operators involved in decomposition (2.17) which contain at least one of Π^1_{ll} or Π^2_{ll} in the tensor products, such as $\Pi^1_{lh}\otimes\Pi^2_{ll}$, $\Pi^1_{hl}\otimes\Pi^2_{ll}$, $\Pi^1_{hh}\otimes\Pi^2_{ll}$, $\Pi^1_{ll}\otimes\Pi^2_{ll}$, $\Pi^1_{ll}\otimes\Pi^2_{hh}$, $\Pi^1_{ll}\otimes\Pi^2_{hl}$ and $\Pi^1_{ll}\otimes\Pi^2_{lh}$, here we will only consider the case $\Pi^1_{ll}\otimes\Pi^2_{ll}$ without loss of generality. For the estimates of the "main term" and "hybrid terms", if the supports of f, g are close to R_{00} in both x_1 and x_2 directions ($\operatorname{supp} f$, $\operatorname{supp} g\subseteq 15R_{00}$, say), we can apply the Coifman-Meyer theorem (Theorem 2.6) or Theorem 2.5 directly; if for $i=1,2$, at least one of the supports of f, g or dyadic rectangle R is far away from R_{00} in x_i direction, we will obtain enough decay factors from $\langle f,\varphi^1_R\rangle\cdot\langle g,\varphi^2_R\rangle$ or φ^3_R; otherwise, if the supports of f, g and dyadic rectangle R are all close to R_{00} in x_1 (or x_2) direction while at least one of the supports of f, g are far away from R_{00} in x_2 (or x_1) direction, we can apply the one-parameter paraproducts estimates (Theorem 2.4) with respect to x_1 (or x_2) variable directly and obtain sufficient decay factors in x_2 (or x_1) direction so as to reach our conclusions. As to the estimates of the discrete bilinear operator $\overrightarrow{\Pi}^{(2),(0,0,\vec{0})}_{a,\mathcal{R}}$ corresponding to $\Pi^1_{ll}\otimes\Pi^2_{ll}$, one easily observe that at least two of the families of L^2-normalized bump functions $(\varphi^i_I)_{I\in\mathcal{I}}$ for $i=1,2,3$ and two of $(\varphi^j_J)_{J\in\mathcal{J}}$ for $j=1,2,3$ are nonlacunary respectively, which means that, when the supports of f, g and dyadic rectangle R are all close to R_{00} in one direction (i.e. $I\subseteq 2^3I_0$ or $J\subseteq 2^3J_0$) but at least one of the supports of f, g are far away from R_{00} in the other direction, we won't be able to apply the one-parameter paraproducts estimates (Theorem 2.4) with respect to x_1 or x_2 variable any more; however, we can take advantage of the additional properties that $|I|\sim|J|\sim 1$ for every dyadic intervals $I\in\mathcal{I}$ and $J\in\mathcal{J}$ to obtain the convergence of both $\sum_{I\subseteq 2^3I_0}$ and $\sum_{J\subseteq 2^3J_0}$; the other parts of the estimates for the discrete bilinear operator $\overrightarrow{\Pi}^{(2),(0,0,\vec{0})}_{a,\mathcal{R}}$ corresponding to $\Pi^1_{ll}\otimes\Pi^2_{ll}$ are similar to the estimates of the standard discrete paraproduct operator corresponding to $\Pi^1_{lh}\otimes\Pi^2_{hl}$.

We like to end this section with some recent work of the second and third authors on Calderón-Vaillancourt type theorem for bi-parameter and bilinear pseudo-differential operators with minimal smoothness. For similar result in one-parameter case, see [50]. Let $m\in\mathbb{R}$ and $0\le\rho,\delta\le 1$. We consider the following type of bi-parameter and bilinear pseudo-differential operators defined for $f,g\in\mathcal{S}(\mathbb{R}^{2n})$:

$$T_\sigma(f,g)=\iint_{\mathbb{R}^{2n}\times\mathbb{R}^{2n}}\sigma(x,\xi,\eta)\cdot\widehat{f}(\xi)\cdot\widehat{g}(\eta)\cdot e^{ix(\xi+\eta)}d\xi d\eta,$$

where $x=(x_1,x_2),\xi=(\xi_1,\xi_2),\eta=(\eta_1,\eta_2)\in\mathbb{R}^n\times\mathbb{R}^n$ and σ satisfies

$$\begin{aligned}|\partial^{\alpha_1}_{x_1}\partial^{\alpha_2}_{x_2}\partial^{\beta_1}_{\xi_1}\partial^{\gamma_1}_{\eta_1}\partial^{\beta_2}_{\xi_2}\partial^{\gamma_2}_{\eta_2}\sigma(x,\xi,\eta)|&\le C_{\alpha,\beta,\gamma}(1+|\xi_1|+|\eta_1|)^{\frac{m}{2}+\delta|\alpha_1|-\rho(|\beta_1|+|\gamma_1|)}\\&\cdot(1+|\xi_2|+|\eta_2|)^{\frac{m}{2}+\delta|\alpha_2|-\rho(|\beta_2|+|\gamma_2|)}\end{aligned}\tag{2.20}$$

for all multi-indices $\alpha=(\alpha_1,\alpha_2),\beta=(\beta_1,\beta_2),\gamma=(\gamma_1,\gamma_2)$, and some positive constants $C_{\alpha,\beta,\gamma}$ depending on α,β,γ.

We denote the class of such symbols by $BBS^m_{\rho,\delta}$. We also denote by $Op(BBS^m_{\rho,\delta})$ the class of all operators T_σ with $\sigma \in BBS^m_{\rho,\delta}$.

Recently, the second and third authors proved the following theorem in [44].

Theorem 2.11. *Let $m \in \mathbb{R}$, $1 \le p, q, r \le \infty$, and $\frac{1}{p} + \frac{1}{q} = \frac{1}{r}$.*
(a) *All the operators of class $Op(BBS^m_{0,0})$ are bounded in $L^p \times L^q \to L^r$ if*

$$m < m(p,q) = -2n \max\left\{\frac{1}{2}, \frac{1}{p}, \frac{1}{q}, 1 - \frac{1}{r}\right\}.$$

(b) *If the operators of class $Op(BBS^m_{0,0})$ are bounded in $L^p \times L^q \to L^r$, then we must have*

$$m \le m(p,q) = -2n \max\left\{\frac{1}{2}, \frac{1}{p}, \frac{1}{q}, 1 - \frac{1}{r}\right\}$$

The proof of Theorem 2.11 mainly consists of two parts: the boundedness of $L^\infty \times L^\infty \to L^\infty$ when $m < -2n$, and the boundedness of $L^2 \times L^2 \to L^1$ when $m < -n$, and then our theorem follows from the interpolation argument.
The boundedness on $L^\infty \times L^\infty \to L^\infty$ follows from the following theorem.

Theorem 2.12. *When $m < -2n(1-\rho)$, for $\sigma \in BBS^m_{\rho,\delta}$ where $0 \le \delta, \rho \le 1, \delta < 1$, we then have that $T_\sigma : L^\infty \times L^\infty \to L^\infty$.*

As a consequence of Theorem 2.12, and the duality argument, we have

Corollary 2.13. *For T_σ with $m < 2n(\rho - 1)$ and $\sigma \in BBS^m_{\rho,\delta}$ and $0 \le \delta \le \rho \le 1$ we have:*

$$T_\sigma : L^\infty \times L^\infty \to L^\infty, \quad L^1 \times L^\infty \to L^1, \quad L^\infty \times L^1 \to L^1.$$

To prove Theorem 2.12, one needs to establish the following local estimates.

Lemma 2.14. *Let $m \in \mathbb{R}, 0 \le \delta, \rho \le 1, \sigma \in BBS^m_{\rho,\delta}$.*
(a) *If $0 < R_1, R_2 \le 1$ and $\mathrm{supp}(\sigma) \subseteq \{(x,\xi,\eta) : |\xi_i| + |\eta_i| \le R_i, i = 1,2\}$, then*

$$\|T_\sigma(f,g)\|_{L^\infty} \lesssim (R_1R_2)^{2n} \|f\|_{L^\infty} \|g\|_{L^\infty}, \quad f, g \in L^\infty.$$

(b) *If $R_1, R_2 \ge 1$ and $\mathrm{supp}(\sigma) \subseteq \{(x,\xi,\eta) : R_i \le |\xi_i| + |\eta_i| \le 4R_i, i = 1,2\}$, then*

$$\|T_\sigma(f,g)\|_{L^\infty} \lesssim (R_1R_2)^{(1-\rho)2n+m} \|f\|_{L^\infty} \|g\|_{L^\infty}, \quad f, g \in L^\infty.$$

(c) *If $0 < R_1 \le 1, R_2 \ge 1$ and $\mathrm{supp}(\sigma) \subseteq \{(x,\xi,\eta) : |\xi_1| + |\eta_1| \le R_1, R_2 \le |\xi_2| + |\eta_2| \le 4R_2\}$, then*

$$\|T_\sigma(f,g)\|_{L^\infty} \lesssim (R_1)^{2n} (R_2)^{(1-\rho)2n+m} \|f\|_{L^\infty} \|g\|_{L^\infty}, \quad f, g \in L^\infty.$$

The second part of proof of Theorem 2.11 is the following boundedness of $L^2 \times L^2 \to L^1$ for $m < -n$.

Theorem 2.15. *The operators of class $Op(BBS^m_{0,0})$ are bounded in $L^2 \times L^2 \to L^1$ for $m < -n$.*

To prove Theorem 2.15, we take a function $\Psi(x,y) \in \mathcal{S}(\mathbb{R}^{2n})$ satisfying $\Psi = 1$ on $\{|x|+|y| \le 20\}$ and supported on $\{|x|+|y| \le 30\}$. Then we have

$$\begin{aligned}
&\sigma(x,\xi,\eta)\\
&= \sigma(x,\xi,\eta)\Psi(\xi_1,\eta_1)\Psi(\xi_2,\eta_2) + \sigma(x,\xi,\eta)(1-\Psi(\xi_1,\eta_1))(1-\Psi(\xi_2,\eta_2))\\
&\quad +\sigma(x,\xi,\eta)\Psi(\xi_2,\eta_2)(1-\Psi(\xi_1,\eta_1)) + \sigma(x,\xi,\eta)\Psi(\xi_1,\eta_1)(1-\Psi(\xi_2,\eta_2))\\
&= \mathrm{I} + \mathrm{II} + \mathrm{III} + \mathrm{IV}.
\end{aligned}$$

We take the example of estimating II. We decompose σ as follows:

$$\begin{aligned}
&\sigma(x,\xi,\eta)\\
&= \sigma(x,\xi,\eta)\Phi_{(1,1)}(\xi_1,\eta_1)\Phi_{(1,1)}(\xi_2,\eta_2) + \sigma(x,\xi,\eta)\Phi_{(1,1)}(\xi_1,\eta_1)\Phi_{(0,1)}(\xi_2,\eta_2)\\
&\quad +\sigma(x,\xi,\eta)\Phi_{(1,1)}(\xi_1,\eta_1)\Phi_{(1,0)}(\xi_2,\eta_2)\\
&= \sigma(x,\xi,\eta)\Phi_{(0,1)}(\xi_1,\eta_1)\Phi_{(1,1)}(\xi_2,\eta_2) + \sigma(x,\xi,\eta)\Phi_{(0,1)}(\xi_1,\eta_1)\Phi_{(0,1)}(\xi_2,\eta_2)\\
&\quad +\sigma(x,\xi,\eta)\Phi_{(0,1)}(\xi_1,\eta_1)\Phi_{(1,0)}(\xi_2,\eta_2)\\
&= \sigma(x,\xi,\eta)\Phi_{(1,0)}(\xi_1,\eta_1)\Phi_{(1,1)}(\xi_2,\eta_2) + \sigma(x,\xi,\eta)\Phi_{(1,0)}(\xi_1,\eta_1)\Phi_{(0,1)}(\xi_2,\eta_2)\\
&\quad +\sigma(x,\xi,\eta)\Phi_{(1,0)}(\xi_1,\eta_1)\Phi_{(1,0)}(\xi_2,\eta_2)\\
&= \sigma_{11}(x,\xi,\eta) + \sigma_{12}(x,\xi,\eta) + \sigma_{13}(x,\xi,\eta)\\
&\quad +\sigma_{21}(x,\xi,\eta) + \sigma_{22}(x,\xi,\eta) + \sigma_{23}(x,\xi,\eta)\\
&\quad +\sigma_{31}(x,\xi,\eta) + \sigma_{32}(x,\xi,\eta) + \sigma_{33}(x,\xi,\eta),
\end{aligned}$$

such that

$$\begin{aligned}
&\operatorname{supp}\sigma_{11} \subseteq \{(x,\xi,\eta) : |\xi_i|+|\eta_i| \ge 20, |\xi_i|/8 \le |\eta_i| \le 8|\xi_i|, i=1,2\},\\
&\operatorname{supp}\sigma_{12} \subseteq \{(x,\xi,\eta) : |\xi_i|+|\eta_i| \ge 20, |\xi_1|/8 \le |\eta_1| \le 8|\xi_1|, |\xi_2| \le |\eta_2|/2, i=1,2\},\\
&\operatorname{supp}\sigma_{13} \subseteq \{(x,\xi,\eta) : |\xi_i|+|\eta_i| \ge 20, |\xi_1|/8 \le |\eta_1| \le 8|\xi_1|, |\eta_2| \le |\xi_2|/2, i=1,2\},\\
&\operatorname{supp}\sigma_{21} \subseteq \{(x,\xi,\eta) : |\xi_i|+|\eta_i| \ge 20, |\xi_1| \le |\eta_1|/2, |\xi_2|/8 \le |\eta_2| \le 8|\xi_2|, i=1,2\},\\
&\operatorname{supp}\sigma_{22} \subseteq \{(x,\xi,\eta) : |\xi_i|+|\eta_i| \ge 20, |\xi_i| \le |\eta_i|/2, i=1,2\},\\
&\operatorname{supp}\sigma_{23} \subseteq \{(x,\xi,\eta) : |\xi_i|+|\eta_i| \ge 20, |\xi_1| \le |\eta_1|/2, |\eta_2| \le |\xi_2|/2, i=1,2\},\\
&\operatorname{supp}\sigma_{31} \subseteq \{(x,\xi,\eta) : |\xi_i|+|\eta_i| \ge 20, |\eta_1| \le |\xi_1|/2, |\xi_2|/8 \le |\eta_2| \le 8|\xi_2|, i=1,2\},\\
&\operatorname{supp}\sigma_{32} \subseteq \{(x,\xi,\eta) : |\xi_i|+|\eta_i| \ge 20, |\eta_1| \le |\xi_1|/2, |\xi_2| \le |\eta_2|/2, i=1,2\},\\
&\operatorname{supp}\sigma_{33} \subseteq \{(x,\xi,\eta) : |\xi_i|+|\eta_i| \ge 20, |\eta_i| \le |\xi_i|/2, i=1,2\},
\end{aligned}$$

and we can conclude that

$$|\partial_\xi^\alpha \partial_\eta^\beta \sigma_{jk}(x,\xi,\eta)| \le (1+|\xi_1|+|\eta_1|)^{\frac{-n}{2}}(1+|\xi_2|+|\eta_2|)^{\frac{-n}{2}} \tag{2.21}$$

for $j,k=1,2,3$ all multi-indices α,β.

Then the operator T_σ with a symbol $\sigma=\sigma_{ij}$ are bounded from $L^2 \times L^2$ to L^1 for $m < -n$. For more details, see [44].

As a corollary of Theorem 2.15, by the duality argument, we have

Corollary 2.16. *For T_σ with $\sigma(x,\xi,\eta) \in BBS^m_{0,0}$ and $m < -n$ we have:*

$$T_\sigma : L^2 \times L^2 \to L^1, \quad L^2 \times L^\infty \to L^2, \quad L^\infty \times L^2 \to L^2.$$

Combining Theorems 2.12 and 2.15, we can conclude Theorem 2.11. For more details, see [44].

3 L^p estimates for bilinear and multi-parameter Hilbert transforms

The bilinear Hilbert transform is defined by

$$BHT(f_1,f_2)(x) := p.\,v. \int_{\mathbb{R}} f_1(x-t)f_2(x+t)\frac{dt}{t}, \tag{3.1}$$

or equivalently, it can also be written as the bilinear multiplier operator

$$BHT: \quad (f_1,f_2) \mapsto \int_{\mathbb{R}^2} \operatorname{sgn}(\xi_1-\xi_2)\hat{f}_1(\xi_1)\hat{f}_2(\xi_2)e^{2\pi i x(\xi_1+\xi_2)}d\xi_1 d\xi_2, \tag{3.2}$$

where f_1 and f_2 are Schwartz functions on $\mathbb{R}$.

The boundedness problem of the bilinear Hilbert transform was originally raised by Calderón in connection with the Cauchy integral along Lipschitz curves. He conjectured that the bilinear Hilbert transform was bounded from $L^2 \times L^2$ to L^1. Inspired by the two classic proofs of almost everywhere convergence of Fourier series by Carleson [7] and Fefferman [19], and using delicate orthogonality estimates and combinatorial selection of trees, and deep time-frequency analysis, Lacey and Thiele proved in [42, 43] the following celebrated L^p estimates for the bilinear Hilbert transform.

Theorem 3.1 ([42, 43]). *The bilinear operator BHT maps $L^p(\mathbb{R}) \times L^q(\mathbb{R})$ into $L^r(\mathbb{R})$ boundedly for any $1 < p,\, q \leq \infty$ with $\frac{1}{p}+\frac{1}{q}=\frac{1}{r}$ and $\frac{2}{3} < r < \infty$.*

In general, we consider a symbol $m(\xi_1,\xi_2)$ that is smooth away from the singularity line $\Gamma := \{(\xi_1,\xi_2) \in \mathbb{R}^2 : \xi_1 = \xi_2\}$ and satisfies

$$|\partial^\alpha m(\xi)| \lesssim \frac{1}{\operatorname{dist}(\xi,\Gamma)^{|\alpha|}} \tag{3.3}$$

for every $\xi = (\xi_1,\xi_2) \in \mathbb{R}^2 \setminus \Gamma$ and sufficiently many multi-indices α. Gilbert and Nahmod [21] proved that the L^p estimates as BHT are valid for the generalized bilinear multiplier operators T_m associated with symbol m in the same range of $p > \frac{2}{3}$ as that in [42, 43]. There has been much work related to the bilinear operators of BHT type. Bernicot [4] proved a pseudo-differential variant of the multiplier estimates in [21]. Uniform estimates were obtained by Thiele [65], Grafakos and Li [22] and Li [39]. A two-dimensional bilinear Hilbert transform was studied by Demeter and Thiele [18]. A maximal variant of Theorem 3.1 was proved by Lacey [38] and generalized by Demeter, Tao and Thiele [17]. In Muscalu, Tao and Thiele [59] and Jung [33], the authors investigated various trilinear variants of the bilinear Hilbert transform. For more related results involving estimates for multi-linear singular multiplier operators, we refer to the works, e.g., [1, 9, 11, 12, 16, 20, 26, 31, 37, 57, 58, 55, 66] and the references therein.

In multi-parameter cases, there is also a large amount of literature devoted to studying the estimates of multi-parameter and multi-linear operators (see [10, 14, 15, 29, 34, 47, 41, 57, 54, 56, 62] and the references therein). In the bilinear and bi-parameter cases, let Γ_i $(i = 1, 2)$ be subspaces in $\mathbb{R}^2$, we consider operators T_m defined by

$$T_m(f_1, f_2)(x) := \int_{\mathbb{R}^4} m(\xi, \eta)\hat{f}_1(\xi_1, \eta_1)\hat{f}_2(\xi_2, \eta_2)e^{2\pi ix\cdot((\xi_1,\eta_1)+(\xi_2,\eta_2))}d\xi d\eta, \tag{3.4}$$

where the symbol m satisfies

$$|\partial_\xi^\alpha \partial_\eta^\beta m(\xi, \eta)| \lesssim \frac{1}{\operatorname{dist}(\xi, \Gamma_1)^{|\alpha|}} \cdot \frac{1}{\operatorname{dist}(\eta, \Gamma_2)^{|\beta|}} \tag{3.5}$$

for sufficiently many multi-indices $\alpha = (\alpha_1, \alpha_2)$ and $\beta = (\beta_1, \beta_2)$. If $\dim \Gamma_1 = \dim \Gamma_2 = 0$, as we mentioned before, Muscalu, Pipher, Tao and Thiele proved in [54, 56] that Hölder type L^p estimates are available for T_m; Chen and the second author established in [10] such estimates for multipliers with limited smoothness on the symbols. A multi-linear and multi-parameter pseudo-differential operator analogue has also been studied and L^p estimates have been obtained by the first two authors in [14]. Furthermore, symbolic calculus has been carried out and boundedness of multi-parameter and multi-linear pseudo-differential operators in the Hörmander classes have been established by Hong and the second author [29]. However, if $\dim \Gamma_1 = \dim \Gamma_2 = 1$, let T_m be the double bilinear Hilbert transform on polydisks $BHT \otimes BHT$ defined by

$$BHT \otimes BHT(f_1, f_2)(x, y) := p.v. \int_{\mathbb{R}^2} f_1(x - s, y - t)f_2(x + s, y + t)\frac{ds}{s}\frac{dt}{t}, \tag{3.6}$$

the authors in [54] also proved that the operator $BHT \otimes BHT$ does not satisfy any L^p estimates of Hölder type by constructing a counterexample. In fact, consider bounded functions $f_1(x, y) = f_2(x, y) = e^{ixy}$, one has formally

$$\begin{aligned} BHT \otimes BHT(f_1, f_2)(x, y) &= (f_1 \cdot f_2)(x, y) \int_{\mathbb{R}^2} \frac{e^{2ist}}{st} dsdt \\ &= i\pi(f_1 \cdot f_2)(x, y) \int_{\mathbb{R}} \frac{\operatorname{sgn}(s)}{s} ds, \end{aligned}$$

then localize functions f_1, f_2 and let

$$f_1^N(x, y) = f_2^N(x, y) = e^{ixy}\chi_{[-N,N]}(x)\chi_{[-N,N]}(y),$$

one can verify the pointwise estimate

$$\begin{aligned} |BHT \otimes BHT(f_1^N, f_2^N)(x, y)| &\geq \left| \int_{-\frac{N}{10}}^{\frac{N}{10}} \int_{-\frac{N}{10}}^{\frac{N}{10}} \frac{e^{2ist}}{st} dsdt \right| + O(1) \\ &\geq C \log N + O(1) \end{aligned} \tag{3.7}$$

for every $x, y \in [-\frac{N}{100}, \frac{N}{100}]$ and sufficiently large $N \in \mathbb{Z}^+$, which indicates that no Hölder type L^p estimates are available for the bilinear operator $BHT \otimes BHT$. When $\dim \Gamma_1 = 0$ and $\dim \Gamma_2 = 1$, Muscalu, Pipher, Tao and Thiele raised the following problem in Question 8.2 in [54].

Question 3.2 ([54]). *Let* $\dim \Gamma_1 = 0$ *and* $\dim \Gamma_2 = 1$ *with* Γ_2 *non-degenerate in the sense of* [58]. *If* m *is a multiplier satisfying* (3.5), *does the corresponding operator* T_m *defined by* (3.4) *satisfy any* L^p *estimates?*

In [62], Silva answered this question partially and proved that T_m defined by (3.4), (3.5) with $\dim \Gamma_1 = 0$ and $\dim \Gamma_2 = 1$ maps $L^p \times L^q \to L^r$ boundedly when $\frac{1}{p} + \frac{1}{q} = \frac{1}{r}$ with $p, q > 1$, $\frac{1}{p} + \frac{2}{q} < 2$ and $\frac{1}{q} + \frac{2}{p} < 2$. One should observe that the admissible range for these tuples (p, q, r) is a proper subset of the region $p, q > 1$ and $\frac{3}{4} < r < \infty$, which is also properly contained in the admissible range of BHT (see Theorem 3.1).

Naturally, we may wonder whether the bi-parameter bilinear operator T_m given by (3.4), (3.5) (with appropriate decay assumptions on the symbol m and singularity sets Γ_1, Γ_2 satisfying $\dim \Gamma_1 = 0 \,\text{or}\, 1$, $\dim \Gamma_2 = 1$) satisfies the same L^p estimates as BHT.

To study this problem, we must find the implicit decay assumptions on symbol m to preclude the existence of those kinds of counterexamples constructed in the above (3.7) for $BHT \otimes BHT$. To this end, let us consider first the bilinear operator $T_m \otimes BHT$ of tensor-product type, which is defined by

$$T_m \otimes BHT(f_1, f_2)(x, y) := p.v. \int_{\mathbb{R}^2} f_1(x-s, y-t) f_2(x+s, y+t) \frac{K(s)}{t} dsdt, \quad (3.8)$$

where the symbol $m(\xi_1^1, \xi_2^1) = m(\zeta) := \hat{K}(\zeta)$ with $\zeta := \xi_1^1 - \xi_2^1$ has one dimensional non-degenerate singularity set Γ_1. Let $f_1(x, y) = f_2(x, y) = e^{ixy}$, one can easily derive that

$$T_m \otimes BHT(f_1, f_2)(x, y) = (f_1 \cdot f_2)(x, y) \int_{\mathbb{R}^2} K(s) \frac{e^{2ist}}{t} dsdt. \quad (3.9)$$

From (3.9) and the above counterexample constructed in (3.7) for operator $BHT \otimes BHT$, we observe that one sufficient condition for precluding the existence of these kinds of counterexamples is $K \in L^1$, or equivalently, $m = \hat{K} \in \mathcal{F}(L^1)$. From the Riemann-Lebesgue theorem, we know that a necessary condition for $m \in \mathcal{F}(L^1)$ is $m(\zeta) \to 0$ as $|\zeta| \to \infty$. Moreover, if $K \in L^1(\mathbb{R})$ is odd, one can even derive that

$$\left| \int_{\mathbb{R}} \frac{m(\zeta)}{\zeta} d\zeta \right| \lesssim \|K\|_{L^1}$$

(this indicates that there are many uniformly continuous functions with logarithmic decay rate do not belong to $\mathcal{F}(L^1)$). Therefore, in order to guarantee that the same L^p estimates as the bilinear Hilbert transform are available for bilinear operators $T_m \otimes BHT$ and $BHT \otimes BHT$, we need some appropriate decay assumptions on the symbol.

In [15], the first two authors of the current article proved the same L^p estimates as BHT for modified bilinear operators $T_m \otimes BHT$ with arbitrary non-smooth symbols which decay faster than the logarithmic rate.

For $d \geq 2$, any two generic vectors $\xi_1 = (\xi_1^i)_{i=1}^d$, $\xi_2 = (\xi_2^i)_{i=1}^d$ in $\mathbb{R}^d$ generates naturally the following collection of d vectors in $\mathbb{R}^2$:

$$\bar{\xi}_1 = (\xi_1^1, \xi_2^1), \quad \bar{\xi}_2 = (\xi_1^2, \xi_2^2), \quad \ldots, \quad \bar{\xi}_d = (\xi_1^d, \xi_2^d). \tag{3.10}$$

Let $m = m(\xi) = m(\bar{\xi})$ be a bounded symbol in $L^\infty(\mathbb{R}^{2d})$ that is smooth away from the subspaces $\Gamma_1 \cup \cdots \cup \Gamma_{d-1} \cup \Gamma_d$ and satisfying

$$\operatorname{dist}(\bar{\xi}_d, \Gamma_d)^{|\alpha_d|} \cdot \int_{\mathbb{R}^{2(d-1)}} \frac{|\partial_{\bar{\xi}_1}^{\alpha_1} \cdots \partial_{\bar{\xi}_d}^{\alpha_d} m(\bar{\xi})|}{\prod_{i=1}^{d-1} \operatorname{dist}(\bar{\xi}_i, \Gamma_i)^{2-|\alpha_i|}} d\bar{\xi}_1 \cdots d\bar{\xi}_{d-1} \leq B < +\infty \tag{3.11}$$

for sufficiently many multi-indices $\alpha_1, \ldots, \alpha_d$, where $\dim \Gamma_i = 0$ for $i = 1, \ldots, d-1$ and $\Gamma_d := \{(\xi_1^d, \xi_2^d) \in \mathbb{R}^2 : \xi_1^d = \xi_2^d\}$. Denote by $T_m^{(d)}$ the bilinear multiplier operator defined by

$$T_m^{(d)}(f_1, f_2)(x) := \int_{\mathbb{R}^{2d}} m(\xi) \hat{f}_1(\xi_1) \hat{f}_2(\xi_2) e^{2\pi i x \cdot (\xi_1 + \xi_2)} d\xi. \tag{3.12}$$

The result in [15] for bilinear operators $T_m^{(d)}$ satisfying (3.11) and (3.12) is the following Theorem 3.3.

Theorem 3.3 ([15]). *For any $d \geq 2$, the bilinear and d-parameter multiplier operator $T_m^{(d)}$ maps $L^{p_1}(\mathbb{R}^d) \times L^{p_2}(\mathbb{R}^d) \to L^p(\mathbb{R}^d)$ boundedly for any $1 < p_1, p_2 \leq \infty$ with $\frac{1}{p} = \frac{1}{p_1} + \frac{1}{p_2}$ and $\frac{2}{3} < p < \infty$. The implicit constants in the bounds depend only on p_1, p_2, p, d and B.*

Remark 3.4. One can observe that for arbitrarily small $\varepsilon > 0$, let $m^\varepsilon = m^\varepsilon(\xi) = m^\varepsilon(\bar{\xi})$ be a bounded symbol in $L^\infty(\mathbb{R}^{2d})$ that is smooth away from the subspaces $\Gamma_1 \cup \cdots \cup \Gamma_{d-1} \cup \Gamma_d$ defined as in Theorem 3.3 and satisfying differential estimates

$$|\partial_{\bar{\xi}_1}^{\alpha_1} \cdots \partial_{\bar{\xi}_d}^{\alpha_d} m^\varepsilon(\bar{\xi})| \lesssim \prod_{i=1}^{d-1} \left(\frac{1}{\operatorname{dist}(\bar{\xi}_i, \Gamma_i)^{|\alpha_i|}} \cdot \langle \log_2 \operatorname{dist}(\bar{\xi}_i, \Gamma_i) \rangle^{-(1+\varepsilon)} \right) \cdot \frac{1}{\operatorname{dist}(\bar{\xi}_d, \Gamma_d)^{|\alpha_d|}} \tag{3.13}$$

for sufficiently many multi-indices $\alpha_1, \ldots, \alpha_d$, then m^ε satisfies conditions (3.11).

As shown in [54], the bilinear and bi-parameter Hilbert transform does not satisfy any L^p estimates. This is the case when the singularity sets Γ_1 and Γ_2 satisfy $\dim \Gamma_1 = \dim \Gamma_2 = 1$. Thus, it is natural to ask if the L^p estimates will break down for any bilinear and bi-parameter Fourier multiplier operator with $\dim \Gamma_1 = \dim \Gamma_2 = 1$. In other words, will a non-smooth symbol with the same dimensional singularity sets but with a slightly better decay than that for the bilinear and bi-parameter Hilbert transform assure the L^p estimates? Our next two theorems in [15] address this issue.

For $d=2$ and arbitrarily small $\varepsilon>0$, let $\widetilde{m}^\varepsilon=\widetilde{m}^\varepsilon(\xi)=\widetilde{m}^\varepsilon(\bar{\xi})$ be a bounded symbol in $L^\infty(\mathbb{R}^4)$ that is smooth away from the subspaces $\Gamma_1\cup\Gamma_2$ and satisfying

$$|\partial_{\bar{\xi}_1}^{\alpha_1}\partial_{\bar{\xi}_2}^{\alpha_2}\widetilde{m}^\varepsilon(\bar{\xi})|\lesssim\prod_{i=1}^{2}\frac{1}{\operatorname{dist}(\bar{\xi}_i,\Gamma_i)^{|\alpha_i|}}\cdot\langle\log_2\operatorname{dist}(\bar{\xi}_1,\Gamma_1)\rangle^{-(1+\varepsilon)} \tag{3.14}$$

for sufficiently many multi-indices α_1,α_2, where $\langle x\rangle:=\sqrt{1+x^2}$ and $\Gamma_i:=\{(\xi_1^i,\xi_2^i)\in\mathbb{R}^2:\xi_1^i=\xi_2^i\}$ for $i=1,2$. Denote by $T^{(2)}_{\widetilde{m}^\varepsilon}$ the bilinear multiplier operator defined by

$$T^{(2)}_{\widetilde{m}^\varepsilon}(f_1,f_2)(x):=\int_{\mathbb{R}^4}\widetilde{m}^\varepsilon(\xi)\hat{f}_1(\xi_1)\hat{f}_2(\xi_2)e^{2\pi ix\cdot(\xi_1+\xi_2)}d\xi. \tag{3.15}$$

Our result in [15] for bilinear operators $T^{(2)}_{\widetilde{m}^\varepsilon}$ satisfying (3.14) and (3.15) is the following Theorem 3.5.

Theorem 3.5 ([15]). *For $d=2$ and any $\varepsilon>0$, the bilinear and bi-parameter multiplier operator $T^{(2)}_{\widetilde{m}^\varepsilon}$ maps $L^{p_1}(\mathbb{R}^2)\times L^{p_2}(\mathbb{R}^2)\to L^p(\mathbb{R}^2)$ boundedly for any $1<p_1,\, p_2\leq\infty$ with $\frac{1}{p}=\frac{1}{p_1}+\frac{1}{p_2}$ and $\frac{2}{3}<p<\infty$. The implicit constants in the bounds depend only on p_1, p_2, p, ε and tend to infinity as $\varepsilon\to 0$.*

Our result in [15] for modified bilinear and bi-parameter Hilbert transform of tensor-product type with a slightly better decay than that of $BHT\otimes BHT$ is the following Theorem 3.6.

Theorem 3.6 ([15]). *For any $\varepsilon>0$, let the bilinear and bi-parameter operator $BHT^\varepsilon\otimes BHT$ be defined by*

$$BHT^\varepsilon\otimes BHT(f_1,f_2)(x_1,x_2)=p.v.\int_{\mathbb{R}^2}f_1(x-s)f_2(x+s)\frac{\Psi^\varepsilon(s_1)}{s_2}ds_1ds_2$$

with the function Ψ^ε satisfying

$$|\partial_{\xi_1}^{\alpha_1}\widehat{\Psi^\varepsilon}(\xi_1^1-\xi_2^1)|\lesssim|\xi_1^1-\xi_2^1|^{-|\alpha_1|}\cdot\langle\log_2|\xi_1^1-\xi_2^1|\rangle^{-(1+\varepsilon)} \tag{3.16}$$

for sufficiently many multi-indices α_1, then it satisfies the same L^p estimates as $T^{(2)}_{\widetilde{m}^\varepsilon}$.

We will give an outline of our proof strategy of main theorems (Theorems 3.3, 3.5 and 3.6) in [15] here. For simplicity, we will only consider the bi-parameter case $d=2$ and $\Gamma_i=\{(0,0)\}$ $(i=1,\dots,d-1)$ in the proof of Theorem 3.3. It will be clear from the proof (see [15]) that we can extend the argument to the general d-parameter and $\dim\Gamma_i=0$ $(i=1,\dots,d-1)$ cases straightforwardly. We will only discuss the proof of Theorem 3.5 and omit the proof of Theorem 3.6, since one can observe from the discretization procedure that the bilinear and bi-parameter operator $BHT^\varepsilon\otimes BHT$ can be reduced to the same bilinear model operators $\widetilde{\Pi}^\varepsilon_{\mathbb{P}}$ as $T^{(2)}_{\widetilde{m}^\varepsilon}$.

As we can see from the study of multi-parameter and multi-linear Coifman-Meyer multiplier operators (see e.g. [54, 56, 58, 59]), a standard approach to obtain L^p estimates of bilinear operators $T_m^{(d)}$ and $T_{\widetilde{m}^\varepsilon}^{(2)}$ is to reduce them into discrete sums $\Pi_{\vec{\mathbb{P}}}$ and $\widetilde{\Pi}_{\vec{\mathbb{P}}}^\varepsilon$ of inner products with wave packets (see [66]), where

$$\Pi_{\vec{\mathbb{P}}}(f_1,f_2)(x) := \sum_{\vec{P}\in\vec{\mathbb{P}}} \frac{C_{Q_{\vec{P}}}}{|I_{\vec{P}}|^{\frac{1}{2}}} \langle f_1, \Phi^1_{\vec{P}_1}\rangle\langle f_2, \Phi^2_{\vec{P}_2}\rangle \Phi^3_{\vec{P}_3}(x) \tag{3.17}$$

and

$$\widetilde{\Pi}^\varepsilon_{\vec{\mathbb{P}}}(f_1,f_2)(x) := \sum_{\vec{P}\in\vec{\mathbb{P}}} \frac{\widetilde{C}^\varepsilon_{Q_{\vec{P}}}}{|I_{\vec{P}}|^{\frac{1}{2}}} \langle f_1, \Phi^1_{\vec{P}_1}\rangle\langle f_2, \Phi^2_{\vec{P}_2}\rangle \Phi^3_{\vec{P}_3}(x), \tag{3.18}$$

while $\Phi^j_{\vec{P}_j} := \Phi^{j,(0,0)}_{\vec{P}_j}$ for $j=1,2,3$ respectively, $\vec{\mathbb{P}} = \widetilde{\mathbb{P}'}\times\mathbb{P}''$ in (3.17) with arbitrary finite collection $\widetilde{\mathbb{P}'}$ of tri-tiles and arbitrary finite sparse collection $\mathbb{P}''$ of rank 1, the finite set $\vec{\mathbb{P}} = \mathbb{P}'\times\mathbb{P}''$ in (3.18) is an arbitrary sparse collection (of bi-parameter tri-tiles) of rank 1 (for more details, refer to [15]).

By using multi-linear interpolation (see [25, 30, 57, 58]) and the symmetry of operators $\Pi_{\vec{\mathbb{P}}}$ and $\widetilde{\Pi}^\varepsilon_{\vec{\mathbb{P}}}$, we can reduce further the proof of L^p estimates for bilinear operators $T_m^{(d)}$ and $T_{\widetilde{m}^\varepsilon}^{(2)}$ to proving the following restricted weak type estimates for the model operators $\Pi_{\vec{\mathbb{P}}}$ and $\widetilde{\Pi}^\varepsilon_{\vec{\mathbb{P}}}$.

Proposition 3.7 ([15]). *Let p_1 and p_2 be such that p_1 is strictly larger than 1 and arbitrarily close to 1 and p_2 is strictly smaller than 2 and arbitrarily close to 2 and such that for $\frac{1}{p} := \frac{1}{p_1} + \frac{1}{p_2}$, one has $\frac{2}{3} < p < 1$. Then both the model operators $\Pi_{\vec{\mathbb{P}}}$ and $\widetilde{\Pi}^\varepsilon_{\vec{\mathbb{P}}}$ defined in (3.17) and (3.18) are of the restricted weak type (p_1, p_2, p). Moreover, the implicit constants in the bounds depend only on p_1, p_2, p, ε or B and are independent of the particular choice of the finite collection $\vec{\mathbb{P}}$.*

Indeed, first we should observe that if p_1, p_2, p are as in Theorems 3.3, 3.5 and 3.6, then the 3-tuple $(\frac{1}{p_1}, \frac{1}{p_2}, \frac{1}{p'})$ lies in the interior of the convex hull of the following six extremal points: $\beta^1 := (-\frac{1}{2}, \frac{1}{2}, 1)$, $\beta^2 := (-\frac{1}{2}, 1, \frac{1}{2})$, $\beta^3 := (\frac{1}{2}, -\frac{1}{2}, 1)$, $\beta^4 := (1, -\frac{1}{2}, \frac{1}{2})$, $\beta^5 := (\frac{1}{2}, 1, -\frac{1}{2})$ and $\beta^6 := (1, \frac{1}{2}, -\frac{1}{2})$. Then, if we assume that Proposition 3.7 has been proved, from the symmetry of operators $\Pi_{\vec{\mathbb{P}}}$ and $\widetilde{\Pi}^\varepsilon_{\vec{\mathbb{P}}}$ and their adjoints, we deduce that both the tri-linear forms Λ associated to bilinear operators $\Pi_{\vec{\mathbb{P}}}$ and $\widetilde{\Pi}^\varepsilon_{\vec{\mathbb{P}}}$ are of *uniformly restricted weak type* β for 3-tuples $\beta = (\beta_1, \beta_2, \beta_3)$ arbitrarily close to the six extremal points $\beta^1, \dots, \beta^6$ inside the convex hull of them and satisfying if β_j is close to $\frac{1}{2}$ for some $j = 1, 2, 3$, then β_j is strictly larger than $\frac{1}{2}$. By using multi-linear interpolation lemmas 9.4 and 9.6 in [57] or lemma 3.8 in [58], we first obtain restricted weak type estimates of Λ for good tuples inside the smaller convex hull of the three coordinate points $(1,0,0)$, $(0,1,0)$ and $(0,0,1)$. After that, we use the interpolation lemma 9.5 in [57] or lemma 3.10 in [58] to obtain restricted weak type estimates of Λ for bad tuples and finally conclude that restricted weak type estimates of Λ hold for all tuples β inside the convex hull of the six extremal points $\beta^1, \dots, \beta^6$.

It only remains to convert these restricted weak type estimates into strong type estimates. To do this, one just has to apply (exactly as in [58]) the multi-linear

Marcinkiewicz interpolation theorem in [30] in the case of good tuples and the interpolation lemma 3.11 in [58] in the case of bad tuples. This concludes the proof of our main results in [15], Theorems 3.3, 3.5 and 3.6. Therefore, we only have the task of proving Proposition 3.7.

It's well known that a standard approach to prove L^p estimates for one-parameter n-linear operators with singular symbols (e.g., Coifman-Meyer multiplier, BHT and one-parameter paraproducts) is the generic estimates of the corresponding $(n+1)$-linear forms consisting of estimates for different *sizes* and *energies* (see [33, 57, 58, 59]), which relied on the one dimensional BMO theory, or more precisely, the John-Nirenberg type inequalities to get good control over the relevant *sizes*. Unfortunately, there is no routine generalization of such approach to multi-parameter settings, for instance, we don't have analogues of the John-Nirenberg inequalities for *dyadic rectangular* BMO spaces in two-parameter case (see [57]). To overcome these difficulties, in [54] Muscalu, Pipher, Tao and Thiele developed a completely new approach to prove L^p estimates for bi-parameter paraproducts, their essential ideas is to apply the *stopping-time decompositions* based on hybrid square and maximal operators MM, MS, SM and SS, the one dimensional BMO theory and Journé's lemma, and hence could not be extended to solve the general d-parameter ($d \geq 3$) cases. As to the general d-parameter ($d \geq 3$) cases, by proving a generic decomposition in [56], the authors simplified the arguments introduced by them in [54] and this simplification works equally well in all d-parameter settings.

However, in [15], in order to prove our main Theorems 3.3 and 3.5 in bi-parameter settings, we have at least two different difficulties from [54, 56]. First, observe that if one restricts the sum of *tri-tiles* $P'' \in \mathbb{P}''$ in the definitions of discrete model operators (see [15]) to a *tree* then one essentially gets a tensor product of two discrete paraproducts on x_1 and x_2 variable respectively, which can be estimated by the MM, MS, SM and SS functions, but due to the *extra degree of freedom* in frequency in x_2 direction, there are infinitely many such tensor products of paraproducts in the summation, so it's difficult for us to carry out the *stopping-time decompositions* by using the hybrid square and maximal operators as in [54, 56]. Second, in the proof of Theorem 3.5, note that there are infinitely many *tri-tiles* $P' \in \mathbb{P}'$ with the property that $I_{P'} = I_0$ for a certain fixed dyadic interval I_0 of the same length as $I_{P'}$, so we can't get estimate $\sum_{P'} |I_{P'}| \lesssim |\widetilde{I}|$ for all dyadic intervals $I_{P'} \subseteq \widetilde{I}$ with comparable lengths, and hence we can't apply the Journé's lemma as in [54] either. By making use of the L^2 *sizes* and L^2 *energies* estimates of the tri-linear forms, the *almost orthogonality* of *wave packets* associated with different *tiles* of distinct *trees* and the decay assumptions on the symbols, we are able to overcome these difficulties in the proof of Theorems 3.3 and 3.5 in bi-parameter settings.

Nevertheless, in the proof of Theorem 3.5 in general d-parameter settings ($d \geq 3$), one easily observe that the generic decomposition will destroy the *perfect orthogonality* of *wave packets* associated with distinct *tiles* which have disjoint frequency intervals in both x_1 and x_2 directions, thus we can't apply the generic decomposition to extend the results of Theorem 3.5 to higher parameters $d \geq 3$ as in [56]. For the proof of Theorem 3.3, we are able to apply the generic decomposition lemma to the $d-1$ variables $x_1, \dots, x_{d-1}$. Although one can't obtain that

$\operatorname{supp} \Phi^{3,\ell}_{\widetilde{P'_3}} \otimes \Phi^3_{P''_3}$ is entirely contained in the exceptional set U as in [56], but one can observe that the support set is contained in U in all the $x_1, \ldots, x_{d-1}$ variables except the last x_d. Therefore, we only need to consider the distance from the support set to the set E'_3 in x_d direction and obtain enough decay factors for summation, the extension of the proof to the general d-parameter ($d \geq 3$) cases is straightforward.

4 L^p estimates for bilinear operators given by non-smooth symbols with one-dimensional singularity set in the range $1/2 < p \leq 2/3$

One can observe that the L^p estimates for the bilinear operators of BHT type derived in these previous works are available only for $p > \frac{2}{3}$. In [38], by constructing a counterexample, Lacey proved that the discrete model operators associated with the bilinear maximal functions cannot be a uniformly bounded bilinear map from $L^p \times L^q$ into L^r for $\frac{1}{2} < r < \frac{2}{3}$. For the endpoint case $r = \frac{2}{3}$, Bilyk and Grafakos [5] proved some distributional estimates for BHT of log type, then Plinio and Thiele [60] improved the distributional estimates by replacing the single logarithmic term with a double logarithmic term.

Since Lacey and Thiele established the L^p estimates for $\frac{2}{3} < p < \infty$ in [42, 43], whether the bilinear operators of BHT type satisfy L^p estimates all the way down to $\frac{1}{2}$ has remained an open problem. Though we do not have a counterexample yet for the L^p estimates for the bilinear Hilbert transform in the range of $\frac{1}{2} < p < \frac{2}{3}$, we will provide a counterexample for a modified version. We denote by $\mathcal{F}L^p(\mathbb{R})$ the space consisting of all functions f whose Fourier transform $\hat{f}$ satisfies $\hat{f} \in L^p(\mathbb{R})$. The Hausdorff-Young inequality tells us that $\|\hat{f}\|_{L^{p'}(\mathbb{R})} \lesssim_p \|f\|_{L^p(\mathbb{R})}$ for $1 \leq p \leq 2$. Then by Theorem 3.1 due to Lacey and Thiele, it implies that the bilinear Hilbert transform maps $\mathcal{F}L^{p'_1} \times L^{p_2} \to L^p$ for $p_1 \geq 2$ and maps $L^{p_1} \times \mathcal{F}L^{p'_2} \to L^p$ for $p_2 \geq 2$ with $\frac{1}{p_1} + \frac{1}{p_2} = \frac{1}{p}$. Thus it is interesting to ask whether the bilinear operators of BHT type map $\mathcal{F}L^{p'_1} \times L^{p_2} \to L^p$ for $p_1 < 2$ or $L^{p_1} \times \mathcal{F}L^{p'_2} \to L^p$ for $p_2 < 2$ boundedly with $\frac{1}{p_1} + \frac{1}{p_2} = \frac{1}{p}$. In [16], we address this issue. The first main result proved by the first two authors of the current article in [16] is a negative answer to the boundedness of $\mathcal{F}L^{p'_1} \times L^{p_2} \to L^p$ for $p_1 < 2$ and $L^{p_1} \times \mathcal{F}L^{p'_2} \to L^p$ for $p_2 < 2$.

Theorem 4.1 ([16]). *For $p \in [\frac{1}{2}, \infty)$ and $1 \leq p_1, p_2 \leq \infty$ satisfying $\frac{1}{p_1} + \frac{1}{p_2} = \frac{1}{p}$, the necessary condition for the bilinear Hilbert transform to map $\mathcal{F}L^{p'_1}(\mathbb{R}) \times L^{p_2}(\mathbb{R}) \to L^p(\mathbb{R})$ boundedly is $p_1 \geq 2$ and to map $L^{p_1}(\mathbb{R}) \times \mathcal{F}L^{p'_2}(\mathbb{R}) \to L^p(\mathbb{R})$ boundedly is $p_2 \geq 2$, respectively. In particular, the bilinear Hilbert transform maps neither $\mathcal{F}L^{p'_1} \times L^{p_2} \to L^p$ nor $L^{p_1} \times \mathcal{F}L^{p'_2} \to L^p$ boundedly for any $p \in [1/2, 2/3)$ and $p_1, p_2 \geq 1$ satisfying $\frac{1}{p_1} + \frac{1}{p_2} = \frac{1}{p}$.*

Indeed, without loss of generality, we first assume on the contrary that the conclusions in Theorem 4.1 are not true, that is, there exists some $r_0 \in [\frac{1}{2}, 2)$ and

$1 \le p_0 \le \infty$, $1 \le q_0 < 2$ such that $\frac{1}{r_0} = \frac{1}{p_0} + \frac{1}{q_0}$ and the bilinear Hilbert transform maps $L^{p_0}(\mathbb{R}) \times \mathcal{F}L^{q_0'}(\mathbb{R})$ into $L^{r_0}(\mathbb{R})$ boundedly. Let N be an arbitrarily fixed large positive integer, we constructed in [16] two functions $\mathcal{F}$ and $\mathcal{G}_N$ by

$$\mathcal{F}(x) := \Phi^1\big(x - \frac{1}{2}\big), \quad \mathcal{G}_N(x) := \sum_{k=1}^{N} \mathcal{G}_{N,k}(x) := \sum_{k=1}^{N} \Phi^2\big(x - \frac{1}{2}\big) e^{4\pi i(k+2\varepsilon_k N)x}, \tag{4.1}$$

where $\Phi^1(x)$ and $\Phi^2(x)$ are two L^1-normalized even Schwartz functions adapted to the interval $[-\frac{1}{2}, \frac{1}{2}]$, such that $\hat{\Phi}^j \ge 0$ and $\operatorname{supp} \hat{\Phi}^j \subseteq [-\frac{1}{2}, \frac{1}{2}]$ for $j = 1, 2$, $\{\varepsilon_k\}_{k=1}^N$ are independent and identically distributed random variables with $P(\varepsilon_k = \pm 1) = \frac{1}{2}$ for every $k = 1, \ldots, N$, i.e., $\{\varepsilon_k\}_{k=1}^N$ is a sequence with Rademacher distribution. By using the Khinchine's inequality, we get the lower bound estimates:

$$\mathbb{E}\Bigg[\Bigg\| \sum_{k=1}^{N} BHT\Big(\mathcal{F}, \mathcal{G}_{N,k}\Big) \Bigg\|_{L^{r_0}}^{r_0}\Bigg] \gtrsim N^{\frac{r_0}{2}}, \tag{4.2}$$

while the upper bound yields:

$$\mathbb{E}\Bigg[\Bigg\| \sum_{k=1}^{N} BHT\Big(\mathcal{F}, \mathcal{G}_{N,k}\Big) \Bigg\|_{L^{r_0}}^{r_0}\Bigg] \lesssim \|\mathcal{F}\|_{L^{p_0}}^{r_0} \cdot \|\widehat{\mathcal{G}_N}\|_{L^{q_0'}}^{r_0} \lesssim N^{(1-\frac{1}{q_0})r_0}, \tag{4.3}$$

for arbitrary positive integer N large enough. Combining the estimates (4.2) and (4.3), we must have

$$N^{\frac{r_0}{2}} \lesssim N^{(1-\frac{1}{q_0})r_0} \tag{4.4}$$

for arbitrary positive integer N large enough, which implies that $\frac{r_0}{2} \le (1 - \frac{1}{q_0})r_0$ holds true, that is, we must have $q_0 \ge 2$, which contradicts the assumption $1 \le q_0 < 2$.

Although there are no uniform L^p estimates that are known for the bilinear Hilbert transform in the range $p \in (1/2, 2/3)$, by decomposing the bilinear multiplier operator T_m into a summation of infinitely many bilinear paraproducts without modulation invariance, we can prove that there exists a class of symbols m (with one-dimensional singularity sets) which also satisfy the symbol estimates of BHT type operators (see (3.3)) investigated in [21] and are arbitrarily close to the symbols of BHT type operators (see (3.2) and (3.3)), such that the corresponding bilinear multiplier operators T_m associated with symbols m satisfy L^p estimates all the way down to $\frac{1}{2}$. Our next result in [16] is the following theorem.

Theorem 4.2 ([16]). *For arbitrarily given $\delta > 0$, let $m(\xi_1, \xi_2)$ be a symbol that is smooth away from the singularity line $\Gamma := \{(\xi_1, \xi_2) \in \mathbb{R}^2 : \xi_1 = \xi_2\}$ and satisfies*

$$|\partial^\alpha m(\xi)| \lesssim \frac{1}{\operatorname{dist}(\xi, \Gamma)^{|\alpha|}} \cdot \exp\Bigg\{ - \delta\big(1 - \frac{|\alpha|}{3}\big) \frac{|(\xi_1, \xi_2)|}{\operatorname{dist}(\xi, \Gamma)} \Bigg\}, \quad 0 \le |\alpha| \le 2 \tag{4.5}$$

and

$$|\partial^\alpha m(\xi)| \lesssim \frac{1}{\operatorname{dist}(\xi, \Gamma)^{|\alpha|}}, \quad |\alpha| \ge 3 \tag{4.6}$$

for every $\xi = (\xi_1, \xi_2) \in \mathbb{R}^2 \setminus \Gamma$ *and sufficiently many multi-indices* α, *then the bilinear multiplier operator* T_m *defined by*

$$T_m(f_1, f_2)(x) := \int_{\mathbb{R}^2} m(\xi_1, \xi_2)\hat{f}_1(\xi_1)\hat{f}_2(\xi_2)e^{2\pi i x(\xi_1+\xi_2)}d\xi_1 d\xi_2 \tag{4.7}$$

maps $L^{p_1} \times L^{p_2} \to L^p$ *boundedly for any* $\frac{1}{p} = \frac{1}{p_1} + \frac{1}{p_2}$ *with* $1 < p_1, p_2 \leq \infty$ *and* $\frac{1}{2} < p < \infty$. *The implicit constants in the bounds depend only on* p_1, p_2, p *when* $p > \frac{2}{3}$, *also depend on* δ *and tend to infinity as* $\delta \to 0$ *when* $\frac{1}{2} < p \leq \frac{2}{3}$.

We would like to mention several ingredients in the proof of Theorem 4.2. Since a standard approach to obtain L^p estimates of the bilinear operators T_m is to reduce it into discrete sums of inner products with wave packets (see [66]), we first need to decompose the symbol $m(\xi)$ in a natural way.

By writing the characteristic function $1_{\xi_1 \neq \xi_2}$ of the region $\{\xi \in \mathbb{R}^2 : \xi_1 \neq \xi_2\}$ into finite sum of smoothed versions of characteristic functions of the cones $\{\xi_2 > |\xi_1|\}$, $\{\xi_2 < -|\xi_1|\}$, $\{\xi_1 > |\xi_2|\}$ and $\{\xi_1 < -|\xi_2|\}$, we can decompose the bilinear operator T_m into a finite sum of four parts. Since all the operators obtained in this decomposition can be treated in the same way, without loss of generality, we will discuss in detail only one of them, for instance, the bilinear operator $T_{m_{\mathcal{A}}}$ given by smoothly truncating the symbol m on the cone $\mathcal{A} := \{\xi = (\xi_1, \xi_2) \in \mathbb{R}^2 : \xi_2 > |\xi_1|\}$.

For this purpose, we consider the collection $\mathcal{Q}_n$ of all shifted dyadic squares (for definitions, refer to [59, 15, 16]) $Q = Q_1 \times Q_2$ satisfying the property that

$$Q \cap \Gamma = \emptyset \quad \text{and} \quad \operatorname{diam}(Q) \simeq 10^{-4}\left(\frac{1}{\sqrt{n+1}} - \frac{1}{\sqrt{n+2}}\right) \cdot \operatorname{dist}(Q, (0,0)) \tag{4.8}$$

for arbitrary positive integer $n = 1, 2, \ldots$, where the singularity line $\Gamma = \{\xi = (\xi_1, \xi_2) \in \mathbb{R}^2 : \xi_1 = \xi_2\}$. Correspondingly, we decompose the cone $\mathcal{A}$ into a sequence of cones $\mathcal{C}_1, \mathcal{C}_2, \ldots, \mathcal{C}_n, \ldots$ defined by

$$\mathcal{C}_1 := \left\{\xi = (\xi_1, \xi_2) \in \mathbb{R}^2 : \frac{\sqrt{3}}{2} \leq \frac{\operatorname{dist}(\xi, \Gamma)}{|(\xi_1, \xi_2)|} \leq 1,\ \xi_2 > |\xi_1|\right\}, \tag{4.9}$$

$$\mathcal{C}_2 := \left\{\xi = (\xi_1, \xi_2) \in \mathbb{R}^2 : \frac{1}{2} \leq \frac{\operatorname{dist}(\xi, \Gamma)}{|(\xi_1, \xi_2)|} \leq \frac{\sqrt{3}}{2},\ \xi_2 > |\xi_1|\right\} \tag{4.10}$$

and

$$\mathcal{C}_n := \left\{\xi = (\xi_1, \xi_2) \in \mathbb{R}^2 : \frac{1}{\sqrt{n+2}} \leq \frac{\operatorname{dist}(\xi, \Gamma)}{|(\xi_1, \xi_2)|} \leq \frac{1}{\sqrt{n+1}},\ \xi_2 > |\xi_1|\right\} \tag{4.11}$$

for every $n \geq 3$. The borderlines $\{\mathcal{L}_n\}_{n=1}^{\infty}$ of these cones are defined by

$$\mathcal{L}_1 := \left\{\xi = (\xi_1, \xi_2) \in \mathbb{R}^2 : \xi_2 = \left(\tan\frac{7}{12}\pi\right)\xi_1 > |\xi_1|\right\}, \tag{4.12}$$

$$\mathcal{L}_2 := \left\{\xi = (\xi_1, \xi_2) \in \mathbb{R}^2 : \xi_2 = \left(\tan \frac{5}{12}\pi\right)\xi_1 > |\xi_1|\right\} \tag{4.13}$$

and

$$\mathcal{L}_n := \left\{\xi = (\xi_1, \xi_2) \in \mathbb{R}^2 : \xi_2 = \left(\tan\left(\frac{\pi}{4} + \arcsin\frac{1}{\sqrt{n+2}}\right)\right)\xi_1 > |\xi_1|\right\} \tag{4.14}$$

for every $n \geq 3$.

Now let us define the following disjoint collections of shifted dyadic squares

$$\mathbb{Q}_n := \{Q = Q_1 \times Q_2 : Q \in \mathcal{Q}_n,\ \ Q \cap \mathcal{C}_n \neq \emptyset \text{ and } Q \cap \mathcal{L}_n = \emptyset\} \tag{4.15}$$

for $n = 1, 2, \ldots$. Since the set of squares $\{\frac{7}{10}Q : Q \in \bigcup_{n=1}^{\infty} \mathbb{Q}_n\}$ also forms a finitely overlapping cover of the cone $\mathcal{A} := \bigcup_{n=1}^{\infty} \mathcal{C}_n = \{\xi = (\xi_1, \xi_2) : |\xi_1| < \xi_2\}$, by a standard partition of unity, we can write the smoothed characteristic function $\tilde{\chi}_{\{|\xi_1|<\xi_2\}}$ of the cone $\mathcal{A}$ as

$$\tilde{\chi}_{\{|\xi_1|<\xi_2\}} = \sum_{n=1}^{\infty} \sum_{Q \in \mathbb{Q}_n} \phi_Q(\xi_1, \xi_2), \tag{4.16}$$

where each ϕ_Q is a smooth bump function adapted to Q and supported in $\frac{8}{10}Q$. One observes that all the collections $\mathbb{Q}_n$ $(n = 1, 2, \ldots)$ can be decomposed further into at most $10^8 n^{\frac{3}{2}}$ (modulo some fixed constant C_0 that is independent of $n = 1, 2, \ldots$) disjoint sub-collections $\mathbb{Q}_n^k$ $(1 \leq k \leq C_0 10^8 n^{\frac{3}{2}})$ which contains only one unique shifted dyadic square Q with the fixed scale $\ell(Q) = 2^{l_0}$ for some arbitrarily given integer $l_0 \in \mathbb{Z}$, that is,

$$\mathbb{Q}_n = \bigcup_{k=1}^{\sim 10^8 n^{\frac{3}{2}}} \mathbb{Q}_n^k \tag{4.17}$$

for every $n = 1, 2, \ldots$.

In order to prove L^p estimates for T_m (Theorem 4.2), it's enough for us to investigate the bilinear operator $T_{m_{\mathcal{A}}}$ given by

$$T_{m_{\mathcal{A}}}(f_1, f_2)(x) := \int_{\mathbb{R}^2} m(\xi_1, \xi_2)\tilde{\chi}_{\{|\xi_1|<\xi_2\}}(\xi_1, \xi_2)\hat{f}_1(\xi_1)\hat{f}_2(\xi_2)e^{2\pi i x(\xi_1+\xi_2)}d\xi_1 d\xi_2. \tag{4.18}$$

By using (4.17) and (4.18), we can decompose $T_{m_{\mathcal{A}}}$ into a summation of bilinear multiplier operators:

$$T_{m_{\mathcal{A}}} = \sum_{n=1}^{\infty} \sum_{k=1}^{\sim 10^8 n^{\frac{3}{2}}} T_{m_{\mathcal{A}},n}^k, \tag{4.19}$$

where the bilinear operators $T_{m_{\mathcal{A}},n}^k$ are given by

$$T_{m_{\mathcal{A}},n}^k(f_1, f_2)(x) := \sum_{Q \in \mathbb{Q}_n^k} \int_{\mathbb{R}^2} m(\xi_1, \xi_2)\phi_Q(\xi_1, \xi_2)\hat{f}_1(\xi_1)\hat{f}_2(\xi_2)e^{2\pi i x(\xi_1+\xi_2)}d\xi_1 d\xi_2 \tag{4.20}$$

for every $n = 1, 2, \ldots$ and $1 \leq k \lesssim 10^8 n^{\frac{3}{2}}$. Therefore, the proof of Theorem 4.2 can be reduced to proving L^p estimates for each single bilinear multipliers $T^k_{m_{\mathcal{A}},n}$ with the constants in the bounds being independent of k and having enough decay which is acceptable for summation with respect to n, that is, the following proposition.

Proposition 4.3 ([16]). *For every $\delta > 0$, $n = 1, 2, \ldots$ and $1 \leq k \lesssim 10^8 n^{\frac{3}{2}}$, let $T^k_{m_{\mathcal{A}},n}$ be the bilinear multiplier operator defined by (4.20) with the symbol m satisfies the differential estimates (4.5) and (4.6), then we have*

$$\|T^k_{m_{\mathcal{A}},n}(f_1, f_2)\|_{L^p(\mathbb{R})} \lesssim_{n,p,p_1,p_2,\delta} \|f_1\|_{L^{p_1}(\mathbb{R})} \cdot \|f_2\|_{L^{p_2}(\mathbb{R})}, \tag{4.21}$$

provided that $\frac{1}{p} = \frac{1}{p_1} + \frac{1}{p_2}$ with $1 < p_1, p_2 \leq \infty$ and $\frac{1}{2} < p < \infty$. Moreover, the implicit constants in the bounds satisfy

$$C_{n,p,p_1,p_2,\delta} \lesssim_{p,p_1,p_2} e^{-\delta\sqrt{n}} + n^{-3}. \tag{4.22}$$

Then, by a standard discretization and normalization procedure, we can reduce further the proof of Proposition 4.3 to proving the following L^p estimates for an arbitrary single model operator $\Pi^{n,k}_{\mathbb{P}}$ (for the definition of the collections $\mathbb{P}^k_n$ of tri-tiles, we refer the reader to [16]).

Proposition 4.4 ([16]). *For every $n = 1, 2, \ldots$ and $1 \leq k \lesssim 10^8 n^{\frac{3}{2}}$, if the finite collection $\mathbb{P} \subseteq \mathbb{P}^k_n$ is chosen arbitrarily, then the discrete model operator $\Pi^{n,k}_{\mathbb{P}}$ defined by*

$$\Pi^{n,k}_{\mathbb{P}}(f_1, f_2)(x) := \sum_{P \in \mathbb{P} \subseteq \mathbb{P}^k_n} \frac{1}{|I_P|^{\frac{1}{2}}} \langle f_1, \Phi^1_{P_1}\rangle \langle f_2, \Phi^2_{P_2}\rangle \Phi^3_{P_3}(x) \tag{4.23}$$

maps $L^{p_1}(\mathbb{R}) \times L^{p_2}(\mathbb{R}) \to L^p(\mathbb{R})$ boundedly for any $1 < p_1, p_2 \leq \infty$ and $\frac{1}{2} < p < \infty$ satisfying $\frac{1}{p} = \frac{1}{p_1} + \frac{1}{p_2}$. Moreover, the implicit constants in the bounds depend only on p_1, p_2, p and are independent of n, k and the particular choice of finite sub-collection $\mathbb{P}$.

By the multi-linear interpolations (see [25, 30, 57, 58]) and the symmetry of operators $\Pi^{n,k}_{\mathbb{P}}$, in order to prove L^p estimates in the general cases $1 < p_1, p_2 \leq \infty$ and $\frac{1}{2} < p < \infty$, we only need to prove that the bilinear model operators $\Pi^{n,k}_{\mathbb{P}}$ satisfy the $L^{\frac{1}{2}+\varepsilon,\infty}$ estimates for $\varepsilon > 0$ arbitrarily small. To do this, by using the duality lemma for $L^{r,\infty}$ (see [57, 58]) and scaling invariance, we fix p_1, p_2 two numbers larger than 1 and arbitrarily close to 1, let functions f_1, f_2 such that $\|f_1\|_{L^{p_1}} = \|f_2\|_{L^{p_2}} = 1$ and a measurable set $E \subseteq \mathbb{R}$ satisfying $|E| = 1$, our task is to find a dominative subset $E' \subseteq E$ with comparable measure $|E'| \simeq 1$ such that the corresponding trilinear forms $\Lambda^k_{\mathbb{P},n}$ satisfy the estimates:

$$|\Lambda^k_{\mathbb{P},n}(f_1, f_2, f_3)| = \left| \sum_{P \in \mathbb{P} \subseteq \mathbb{P}^k_n} \frac{1}{|I_P|^{\frac{1}{2}}} \langle f_1, \Phi^1_{P_1}\rangle \langle f_2, \Phi^2_{P_2}\rangle \langle f_3, \Phi^3_{P_3}\rangle \right| \lesssim 1, \tag{4.24}$$

where $f_3 := \chi_{E'}$.

Since one can essentially regard the operators $\Pi_{\mathbb{P}}^{n,k}$ given by (4.23) as one-parameter paraproducts, we can use the *stopping-time decomposition* arguments based on square operators S and maximal operators M developed in [54, 55, 56] to prove (4.24), which ends the proof of Proposition 4.4, and also concludes the proof of Theorem 4.2.

Acknowledgements

The research of the first author was partly supported by a grant of NNSF of China (No. 11371056) and the China Postdoctoral Science Foundation (No. 2013M540057). The research of the second and third authors was partly supported by a US NSF grant DMS-1301595.

References

[1] A. Bényi, C. Demeter, A. Nahmod, C. Thiele, R. Torres, and P. Villarroya, Modulation invariant bilinear T(1) theorem, J. Anal. Math. **109** (2009), 279–352.

[2] A. Bényi, D. Maldonado, V. Naibo, and R. H. Torres, On the Hörmander classes of bilinear pseudodifferential operators, Integr. Equ. Oper. Theory **67** (2010), 341–364.

[3] A. Bényi and R. H. Torres, Symbolic calculus and the transposes of bilinear pseudodifferential operators, Comm. Partial Diff. Eq. **28** (2003), 1161–1181.

[4] F. Bernicot, Local estimates and global continuities in Lebesgue spaces for bilinear operators, Anal. PDE **1** (2008), 1–27.

[5] D. Bilyk and L. Grafakos, Distributional estimates for the bilinear Hilbert transform, J. Geom. Anal. **16** (2006), no. 4, 563–584.

[6] J. M. Bony, Calcul symbolique et propagation des singularités pour les équations aux dérivées partielles non linéaires, Ann. Sci. École Norm. Sup.(4) **14** (1981), no. 2, 209–246.

[7] L. Carleson, On convergence and growth of partial sums of Fourier series, Acta Math. **116** (1966), 135–157.

[8] D. Cao, W. Dai, and W. Yang, Continuous dependence of Cauchy problem for nonlinear Schrödinger equation in H^s, Journal of Differential Equations **255** (2013), 2018–2064.

[9] M. Christ and J. Journé, Polynomial growth estimates for multilinear singular integrals, Acta Math. **159** (1987), 51–80.

[10] J. Chen and G. Lu, Hörmander type theorems for multi-linear and multi-parameter Fourier multiplier operators with limited smoothness, Nonlinear Analysis **101** (2014), 98–112.

[11] R. Coifman and Y. Meyer, Au delá des opérateurs pseudo-différentieles, Astérisque **57** (1978).

[12] R. Coifman and Y. Meyer, Wavelets, Calderón Zygmund Operators and Multilinear Operators, translated from the 1990 and 1991 French originals by D. Salinger, Cambridge Studies in Advanced Mathematics **48**, Cambridge University Press, Cambridge, 1997.

[13] A. P. Calderón and R. Vaillancourt, A class of bounded pseudo-differential operators, Proc. Nat. Acad. Sci. USA **69** (1972), 1185–1187.

[14] W. Dai and G. Lu, L^p estimates for multi-linear and multi-parameter pseudo-differential operators, to appear in Bull. Soc. Math. Fr., arXiv:1308.4062.

[15] W. Dai and G. Lu, L^p estimates for bilinear and multi-parameter Hilbert transforms, Analysis & PDE **8** (2015), no. 3, 675–712.

[16] W. Dai and G. Lu, L^p estimates for the bilinear Hilbert transform for $1/2 < p \leq 2/3$: a counterexample and generalizations to non-smooth symbols, preprint, arXiv:1409.3875.

[17] C. Demeter, T. Tao, and C. Thiele, Maximal multi-linear operators, Trans. Amer. Math. Soc. **360** (2008), no. 9, 4989–5042.

[18] C. Demeter and C. Thiele, On the two dimensional bilinear Hilbert transform, Amer. J. Math. **132** (2010), no. 1, 201–256.

[19] C. Fefferman, Pointwise convergence of Fourier series, Ann. Math. **98** (1973), 551–571.

[20] R. Fefferman and E. M. Stein, Singular integrals on product spaces, Adv. Math. **45** (1982), no. 2, 117–143.

[21] J. Gilbert and A. Nahmod, Bilinear operators with non-smooth symbols I, J. Fourier Anal. Appl. **7** (2001), no. 5, 435–467.

[22] L. Grafakos and X. Li, Uniform estiamtes for the bilinear Hilbert transform I, Ann. Math. **159** (2004), 889–933.

[23] L. Grafakos, A. Miyachi, and N. Tomita, On multilinear Fourier multipliers of limited smoothness, Canad. J. Math. **65** (2013), no. 2, 299–330.

[24] L. Grafakos and Z. Si, The Hörmander multiplier theorem for multilinear operators, J. Reine Angew. Math. **668** (2012), 133–147.

[25] L. Grafakos and T. Tao, Multilinear interpolation between adjoint operators, J. Funct. Anal. **199** (2003), 379–385.

[26] L. Grafakos and R. H. Torres, Multilinear Calderón Zygmund theory, Adv. Math. **165** (2002), 124–164.

[27] L. Grafakos and R. H. Torres, Discret decomposition for bilinear operators and almost diagonal coditions, Trans. Amer. Math. Soc. **354** (2002), 1153–1176.

[28] L. Hörmander, Estimates for translation invariant operators in L^p spaces, Acta Math. **104** (1960), 93–140.

[29] Q. Hong and G. Lu, Symbolic calculus and boundedness of multi-parameter and multi-linear pseudo-differential operators, Advanced Nonlinear Studies **14** (2014), no. 4, 1055–1082.

[30] S. Janson, On interpolation of multilinear operators, Lecture Notes in Mathematics **1302**, Springer, 290–302.

[31] J. Journé, Calderón-Zygmund operators on product spaces, Rev. Mat. Iberoamericana **1** (1985), no. 3, 55–91.

[32] J. Journé, A covering lemma for product spaces, Ann. Inst. Fourier **96** (1986), 593–598.

[33] J. Jung, Iterated trilinear Fourier integrals with arbitrary symbols, preprint, arXiv:1311.1574.

[34] R. Kesler, Mixed estimates for degenerate multilinear oscillatory integrals and their tensor product generalizations, preprint, arXiv:1311.2322.

[35] T. Kato and G. Ponce, Commutator estimates and the Euler and Navier-Stokes equations, Comm. Pure. Appl. Math. **41** (1988), 891–907.

[36] C. Kenig, G. Ponce, and L. Vega, On the (generalized) KdV equations, Duke Math. J. **59** (1989), 585–610.

[37] C. Kenig and E. M. Stein, Multilinear estimates and fractional integration, Math. Res. Lett. **6** (1999), 1–15.

[38] M. T. Lacey, The bilinear maximal functions map into L^p for $2/3 < p \leq 1$, Ann. Math. **155** (2000), 35–57.

[39] X. Li, Uniform estimates for the bilinear Hilbert transform II, Revista Mat. Iberoamericana **22** (2006), 1069–1126.

[40] X. Li, Bilinear Hilbert transforms along curves, I: the monomial case, Anal. PDE **6** (2013), no. 1, 197–220.

[41] M. T. Lacey and J. Metcalfe, Paraproducts in one and several parameters, Forum Math. **19** (2007), no. 2, 325–351.

[42] M. T. Lacey and C. Thiele, L^p estimates for the bilinear Hilbert transform, Ann. Math. **146** (1997), 693–724.

[43] M. T. Lacey and C. Thiele, On Calderón's conjecture, Ann. Math. **150** (1999), 475–496.

[44] G. Lu and L. Zhang, Bi-parameter and bilinear Calderón-Vaillancourt theorem with minimal smoothness, preprint, 2014.

[45] G. Lu and L. Zhang, L^p estimates for a trilinear pseudo-differential operator with flag symbols, preprint, 2014.

[46] G. Lu and P. Zhang, Multilinear Caldern-Zygmund operators with kernels of Dini's type and applications, Nonlinear Anal. **107** (2014), 92–117.

[47] P. Luthy, On a bi-parameter maximal multilinear operator, J. Funct. Anal. **268** (2015), no. 5, 1105–1152.

[48] D. Maldonado and V. Naibo, Weighted norm inequalities for paraproducts and bilinear pseudo-differential operators with mild regularity, J. Fourier Anal. Appl. **15** (2009), no. 2, 218–261.

[49] A. Miyachi and N. Tomita, Minimal smoothness conditions for bilinear Fourier multipliers, Rev. Mat. Iberoam. **29** (2013), no. 2, 495–530.

[50] A. Miyachi and N. Tomita, Calderón-Vaillancourt type theorem for bilinear pseudo-differential operators, Indiana Univ. Math. J. **62** (2013), no. 4, 1165–1201.

[51] C. Muscalu, Unpublished notes I, II, IAS Princeton, 2003.

[52] C. Muscalu, Paraproducts with flag singularities I. A case study, Revista Mat. Iberoamericana **23** (2007), 705–742.

[53] C. Muscalu, Flag paraproducts, Contemp. Math. **505** (2010), 131–151.

[54] C. Muscalu, J. Pipher, T. Tao, and C. Thiele, Bi-parameter paraproducts, Acta Math. **193** (2004), no. 2, 269–296.

[55] C. Muscalu, J. Pipher, T. Tao, and C. Thiele, A short proof of the Coifman-Meyer multilinear theorem, 2004.

[56] C. Muscalu, J. Pipher, T. Tao, and C. Thiele, Multi-parameter paraproducts, Revista Mat. Iberoamericana **22** (2006), 963–976.

[57] C. Muscalu and W. Schlag, Classical and Multilinear Harmonic Analysis, I, II, Cambridge Studies in Advanced Mathematics **138**, Cambridge University Press, Cambridge, 2013.

[58] C. Muscalu, T. Tao, and C. Thiele, Multilinear multipliers given by singular symbols, J. Amer. Math. Soc. **15** (2002), 469–496.

[59] C. Muscalu, T. Tao, and C. Thiele, L^p estimates for the biest II. The Fourier case, Math. Ann. **329** (2004), 427–461.

[60] F. D. Plinio and C. Thiele, Endpoint bounds for the bilinear Hilbert transform, preprint, arXiv:1403.5978.

[61] E. M. Stein, Harmonic Analysis: Real-variable Methods, Orthogonality, and Oscillatory Integrals, Princeton Mathematical Series **43**. Monographs in Harmonic Analysis, III. Princeton University Press, Princeton, NJ, 1993.

[62] P. Silva, Vector valued inequalities for families of bilinear Hilbert transforms and applications to bi-parameter problems, preprint, arXiv:1203.3251.

[63] T. Tao, Lecture notes on harmonic analysis, UCLA, Department of Mathematics, Los Angeles, 2002.

[64] M. Taylor, Tools for PDEs: Pseudodifferential Operators, Paradifferential Operators and Layer Potentials, Mathematical Surveys Monographs **10**, American Mathematical Society, Providence, RI, 1998.

[65] C. Thiele, A uniform estimate, Ann. Math. **156** (2002), 519–563.

[66] C. Thiele, Wave Packet Analysis, CBMS Conference Series in Mathematics **105**, American Mathematical Society, Providence, RI, 2006.

[67] N. Tomita, A Hörmander type multiplier theorem for multilinear operators, J. Funct. Anal. **259** (2010), 2028–2044.

Some Topics in Harmonic Analysis and Applications
ALM 34, pp. 145–162

Existence and Uniqueness Theory for the Fractional Schrödinger Equation on the Torus

Seckin Demirbas*, M. Burak Erdoğan† and Nikolaos Tzirakis‡

Abstract

We study the Cauchy problem for the 1-d periodic fractional Schrödinger equation with cubic nonlinearity. In particular we prove local well-posedness in Sobolev spaces, for solutions evolving from rough initial data. In addition we show the existence of global-in-time infinite energy solutions. Our tools include a new Strichartz estimate on the torus along with ideas that Bourgain developed in studying the periodic cubic NLS.

2000 Mathematics Subject Classification: 35Q55 (35B10, 35B65).

Keywords and Phrases: fractional Schrödinger equation, smoothing, local well-posedness, global well-posedness, Strichartz estimate.

1 Introduction

In this paper we study a fractional semilinear Schrödinger type equation with periodic boundary conditions,

$$\begin{cases} iu_t + (-\Delta)^{\alpha} u = \pm |u|^2 u, & x \in \mathbb{T}, \quad t \in \mathbb{R}, \\ u(x,0) = u_0(x) \in H^s(\mathbb{T}), \end{cases} \tag{1}$$

where $\alpha \in (1/2, 1)$. The equation is called defocusing when the sign in front of the nonlinearity is a minus and focusing when the sign is a plus.

*The author is supported by the NSF grant DMS-1201872. Department of Mathematics, University of Illinois, Urbana, IL 61801, USA. *E-mail*: demirba2@illinois.edu

†The author is supported by the NSF grant DMS-1201872. Department of Mathematics, University of Illinois, Urbana, IL 61801, USA. *E-mail*: berdogan@math.uiuc.edu

‡The author is supported by Research Board fund RB-14054. Department of Mathematics, University of Illinois, Urbana, IL 61801, USA. *E-mail*: tzirakis@math.uiuc.edu

Posed on the real line the equation has appeared at a formal level in many recent articles, see [12] and the references therein. For example it is a basic model equation in the theory of fractional quantum mechanics introduced by Laskin [13]. A rigorous derivation of the equation can be found in [12] starting from a family of models describing charge transport in bio polymers like the DNA. The starting point is a discrete nonlinear Schrödinger equation with general lattice interactions. Equation (1) with $\alpha \in (\frac{1}{2}, 1)$ appears as the continuum limit of the long-range interactions between quantum particles on the lattice. Whereas, allowing only the short-range interactions (e.g. neighboring particle interactions) the authors obtain the standard Schrödinger equation ($\alpha = 1$) which is completely integrable, see [1].

In this work we study the periodic problem mainly for two reasons. First due to the lack of strong dispersion the mathematical theory for the fractional Schrödinger equations are less developed than the cubic nonlinear Schrödinger equation (NLS). Secondly when we consider periodic boundary conditions the analysis becomes harder, for any dispersion relation, since the dispersive character of the equation can only be exploited after employing averaging arguments and a careful analysis of the resonant set of frequencies [8].

The local and global well-posedness for the periodic NLS was established by Bourgain in [2]. He used number theoretic arguments to obtain periodic Strichartz estimates along with a new scale of spaces adapted to the dispersive relation of the linear group. More precisely he proved the existence and uniqueness of local-in-time strong $L^2(\mathbb{T})$ solutions. Since it is known that smooth solutions of the NLS satisfy mass conservation

$$M(u)(t) = \int_{\mathbb{T}} |u(t,x)|^2 = M(u)(0),$$

Bourgain's result showed the existence of global-in-time strong $L^2(\mathbb{T})$ solutions in the focusing and defocusing case. The L^2 theorem of Bourgain is sharp since as it was shown in [4], the solution operator is not uniformly continuous on $H^s(\mathbb{T})$ for $s < 0$.

The local well-posedness for the fractional NLS on the real line was recently studied in [6]. The authors showed that the equation is locally well-posed in $H^s(\mathbb{R})$, for $s \geq \frac{1-\alpha}{2}$. They also proved that the solution operator fails to be uniformly continuous in time for $s < \frac{1-\alpha}{2}$. Since the periodic case is less dispersive, we expect the range $s \geq \frac{1-\alpha}{2}$ to be the optimal range for the local theory also in the periodic case.

In this paper we obtain the following results for the fractional NLS. We first establish a Strichartz estimate that reads as follows

$$\|e^{it(-\Delta)^\alpha} f\|_{L^4_{t\in\mathbb{T}} L^4_{x\in\mathbb{T}}} \lesssim \|f\|_{H^s(\mathbb{T})},$$

for $s > \frac{1-\alpha}{4}$. To use this estimate and prove local well-posedness of the equation one has to overcome the derivative loss on the right hand side of the inequality. In principle this can be done by the method in [5] and [7] which gives local well-posedness in the $H^s(\mathbb{T})$ level, for $s > \frac{1-\alpha}{2}$. However, since the proof in [5] and [7] is quite involved, we choose to establish the local theory by obtaining trilinear

$X^{s,b}$ estimates directly. Then a standard iteration finishes the proof without any further analysis. We remark that for classical solutions in $H^s(\mathbb{T})$, $s > \frac{1}{2}$, local theory in the space $C([0,T];H^s(\mathbb{T}))$ is known. The proof is the same both on the real line and on the torus and it is based on the Banach algebra property of the Sobolev spaces for $s > \frac{1}{2}$. Moreover, the length of the local interval of existence is lower bounded by $\frac{1}{\|u_0\|^2_{H^s(\mathbb{T})}}$. To lower the regularity of the local existence theory and to prove the smoothing estimate of Section 5 we have to reprove the local theory in the $X^{s,b}$ spaces. In this case the solution is controlled on the larger $X^{s,b}$ norm, since $X^{s,b}_T \in C([0,T];H^s(\mathbb{T}))$ for any $b > \frac{1}{2}$, and thus the length of the interval of existence is smaller. In our case for $s > \frac{1}{2}$ it is lower bounded by $\frac{1}{\|u_0\|^{4+}_{H^s(\mathbb{T})}}$.

We note that in addition to the conservation of mass, smooth solutions of (1) satisfy energy conservation

$$E(u)(t) = \int_{\mathbb{T}} \left||\nabla|^\alpha u(t,x)\right|^2 \mp \frac{1}{2}\int_{\mathbb{T}} \left|u(t,x)\right|^4 = E(u)(0).$$

Note that local theory in H^α level along with the conservation of mass and energy imply the existence of global-in-time energy solutions. Since the equation is mass and energy sub-critical [14], one also obtains global solutions also in the focusing case. This follows from the Gagliardo-Nirenberg inequality

$$\|u\|^4_{L^4} \lesssim \||\nabla|^\alpha u\|^{\frac{1}{\alpha}}_{L^2}\|u\|^{4-\frac{1}{\alpha}}_{L^2}$$

which controls the potential energy via the kinetic energy $\||\nabla|^\alpha u\|_{L^2}$. One can then control the Sobolev norm of the solution for all times even in the focusing case since $\frac{1}{\alpha} < 2$. We omit the standard details.

In the second part of the paper we use the high-low frequency decomposition of Bourgain [3], to prove global solutions below the energy level. Bourgain's method consists of estimating separately the evolution of the low frequencies and of the high frequencies of the initial data. The low frequency part is smooth and thus, by conservation of energy, globally defined. The difference equation which is high frequency has small norm. By using smoothing estimates this decomposition can be iterated as long as the norm of the nonlinear part is controlled by the initial energy of the smooth part. As a byproduct of the method one obtains that the nonlinear part of the solution is actually smoother than the linear propagator and stays always in the energy space. Moreover, the global solutions satisfy polynomial-in-time bounds. We summarize the results in the following two theorems:

Theorem 1. *For any $\alpha \in (\frac{1}{2},1)$, and any $b > \frac{1}{2}$ sufficiently close to $\frac{1}{2}$, the equation (1) is locally well-posed in the space $X^{s,b}_T \subset C([0,T];H^s(\mathbb{T}))$ for any $s > \frac{1-\alpha}{2}$, where $T = T(\|u_0\|_{H^s(\mathbb{T})})$. Moreover, for $s > \frac{1}{2}$ the local existence time $T \gtrsim \|u_0\|^{-4-}_{H^s(\mathbb{T})}$.*

Theorem 2. *For any $\alpha \in (\frac{1}{2},1)$, the equation (1) is globally well-posed in $H^s(\mathbb{T})$ for any $s > \frac{10\alpha+1}{12}$. Moreover,*

$$u(t) - e^{it(-\Delta)^\alpha \pm iPt}u_0 \in H^\alpha(\mathbb{T})$$

for all times, where $P = \frac{1}{\pi}\|u_0\|_2^2$.

Remark. We will prove Theorem 2 only for the defocusing case. As we mentioned in our introductory remarks since the problem is mass sub-critical for $\alpha > \frac{1}{2}$, we can also control the H^α norm of the solution by Gagliardo-Nirenberg inequality in the focusing case. Once we have the control of the norm in terms of the initial energy, the proof of the theorem follows along the same lines. In particular we obtain the same global well-posedness results with the same global-in-time bounds for the focusing problem.

The paper is organized as follows: in Section 2 we introduce our notation and define the spaces that the iteration will take place. In addition we state two elementary lemmas that we use in proving the Strichartz estimates and the multilinear estimates. Section 3 contains the proof of the Strichartz estimate. It is obtained by a careful analysis of the resonant terms and non resonant interacting terms. Section 4 contains the local well-posedness theory for the model equation. We prove multilinear estimates in the $X^{s,b}$ spaces defined in Section 2. In Section 5 we prove the main smoothing estimate of this paper. The reader should notice that the estimate is sharp within the tools used and for $\alpha = 1$ it coincides with the smoothing estimate for the NLS that was recently obtained in [10]. Finally in Section 6 we use the established local theory and the smoothing estimate to prove global well-posedness for infinite energy solutions. As a final remark we note that our global-in-time results are not optimal.

2 Notation and preliminaries

First of all recall that for $s \geq 0$, $H^s(\mathbb{T})$ is defined as a subspace of L^2 via the norm

$$\|f\|_{H^s(\mathbb{T})} := \sqrt{\sum_{k\in\mathbb{Z}} \langle k\rangle^{2s} |\widehat{f}(k)|^2},$$

where

$$\langle k\rangle := (1+k^2)^{1/2} \text{ and } \widehat{f}(k) = \frac{1}{2\pi}\int_0^{2\pi} f(x)e^{-ikx}dx$$

are the Fourier coefficients of f. Plancherel's theorem takes the form

$$\sum_{k\in\mathbb{Z}} |\widehat{f}(k)|^2 = \frac{1}{2\pi}\int_0^{2\pi} |f(x)|^2 dx.$$

We denote the linear propagator of the Schrödinger equation as $e^{it(-\Delta)^\alpha}$, where it is defined on the Fourier side as $(\widehat{e^{it(-\Delta)^\alpha} f})(n) = e^{itn^{2\alpha}}\widehat{f}(n)$. Similarly, $|\nabla|^\alpha$ is defined as $(\widehat{|\nabla|^\alpha f})(n) = n^\alpha \widehat{f}(n)$. We also use $(\cdot)^+$ to denote $(\cdot)^\epsilon$ for all $\epsilon > 0$ with implicit constants depending on ϵ.

The Bourgain spaces, $X^{s,b}$, will be defined as the closure of compactly supported smooth functions under the norm

$$\|u\|_{X^{s,b}} \doteq \|e^{-it(-\Delta)^\alpha} u\|_{H_t^b(\mathbb{R})H_x^s(\mathbb{T})} = \|\langle\tau - |n|^{2\alpha}\rangle^b \langle n\rangle^s \widehat{u}(n,\tau)\|_{L_\tau^2 l_n^2},$$

and the restricted norm will be given as

$$\|u\|_{X_T^{s,b}} \doteq \inf(\|v\|_{X^{s,b}}, \text{ for } v = u \text{ on } [0,T]).$$

In this paper, by local and global well-posedness we mean the following

Definition 3. We say the equation (1) is locally well-posed in H^s, if there exist a time $T_{LWP} = T_{LWP}(\|u_0\|_{H^s})$ such that the solution exists and is unique in $X^{s,b}_{T_{LWP}} \subset C([0,T_{LWP}), H^s)$ and depends continuously on the initial data. We say that the equation is globally well-posed when T_{LWP} can be taken arbitrarily large.

We close this section by presenting two elementary lemmas that will be used repeatedly. For the proof of the first lemma see the Appendix of [9].

Lemma 1. *If $\beta \geq \gamma \geq 0$ and $\beta + \gamma > 1$, then*

$$\sum_n \frac{1}{\langle n - k_1\rangle^\beta \langle n - k_2\rangle^\gamma} \lesssim \langle k_1 - k_2\rangle^{-\gamma} \phi_\beta(k_1 - k_2),$$

and

$$\int_{\mathbb{R}} \frac{1}{\langle \tau - k_1\rangle^\beta \langle \tau - k_2\rangle^\gamma} d\tau \lesssim \langle k_1 - k_2\rangle^{-\gamma} \phi_\beta(k_1 - k_2),$$

where

$$\phi_\beta(k) := \sum_{|n|\leq|k|} \frac{1}{\langle n\rangle^\beta} \sim \begin{cases} 1, & \beta > 1, \\ \log(1 + \langle k\rangle), & \beta = 1, \\ \langle k\rangle^{1-\beta}, & \beta < 1. \end{cases}$$

Lemma 2. *Fix $\alpha \in (1/2, 1)$. For $n, j, k \in \mathbb{Z}$, we have*

$$\begin{aligned} g(j,k,n) &:= |(n+k)^{2\alpha} - (n+j+k)^{2\alpha} + (n+j)^{2\alpha} - n^{2\alpha}| \\ &\gtrsim \frac{|k||j|}{(|k|+|j|+|n|)^{2-2\alpha}}, \end{aligned}$$

where the implicit constant depends on α.

Proof. Let $f_c(x) = (x+c)^{2\alpha} - (x-c)^{2\alpha}$. We have

$$g(j,k,n) = \left| f_{\frac{j}{2}}\left(n + \frac{j}{2}\right) - f_{\frac{j}{2}}\left(n + k + \frac{j}{2}\right)\right|.$$

We claim that

$$f_c'(x) \gtrsim \frac{|c|}{\max(|c|,|x|)^{2-2\alpha}}.$$

Using the claim, we have by the mean value theorem (for $j, k \neq 0$)

$$\begin{aligned} g(j,k,n) &= \left| f_{\frac{j}{2}}\left(n + \frac{j}{2}\right) - f_{\frac{j}{2}}\left(n + k + \frac{j}{2}\right)\right| \\ &\gtrsim |k||j| \min_{\gamma \in (n+\frac{j}{2}, n+k+\frac{j}{2})} \frac{1}{\max(\frac{|j|}{2}, |\gamma|)^{2-2\alpha}} \\ &\gtrsim \frac{|k||j|}{(|k|+|j|+|n|)^{2-2\alpha}}. \end{aligned}$$

It remains to prove the claim. Since f_c is odd, and $j \neq 0$, it suffices to consider $x \geq 0$ and $c \gtrsim 1$. We have

$$f_c'(x) = 2\alpha\big[(x+c)^{2(\alpha-1)}|x+c| - (x-c)^{2(\alpha-1)}|x-c|\big].$$

We consider three cases:

Case 1: $0 \leq x \leq c \Rightarrow f_c'(x) = 2\alpha\big[(x+c)^{2\alpha-1} + (x-c)^{2\alpha-1}\big]$. Thus

$$f_c'(x) \gtrsim c^{2\alpha-1}.$$

Case 2: $c \leq x \lesssim c \Rightarrow f_c'(x) = 2\alpha\big[(x+c)^{2\alpha-1} - (x-c)^{2\alpha-1}\big]$. Then we get

$$f_c'(x) \gtrsim c^{2\alpha-1}\Big(\Big(\frac{x}{c}+1\Big)^{2\alpha-1} - \Big(\frac{x}{c}-1\Big)^{2\alpha-1}\Big) \gtrsim c^{2\alpha-1}.$$

Case 3: $x \gg c \Rightarrow f_c'(x) = 2\alpha\big[(x+c)^{2\alpha-1} - (x-c)^{2\alpha-1}\big]$. Then we have

$$f_c'(x) = 2\alpha x^{2\alpha-1}\Big(\Big(1+\frac{c}{x}\Big)^{2\alpha-1} - \Big(1-\frac{c}{x}\Big)^{2\alpha-1}\Big) \approx x^{2\alpha-1}\frac{c}{x} = x^{2\alpha-2}c.$$

Hence, in all cases we have

$$f_c'(x) \gtrsim \frac{|c|}{\max(|c|,|x|)^{2-2\alpha}}.$$

□

3 Strichartz estimates

In this section we prove

Theorem 4. $\|e^{it(-\Delta)^\alpha} f\|_{L^4_{t\in\mathbb{T}} L^4_{x\in\mathbb{T}}} \lesssim \|f\|_{H^s}$ *for* $s > \frac{1-\alpha}{4}$.

Proof. Notice that in this proof we can always take $s < \frac{1}{4}$. Calling $g = \langle\nabla\rangle^s f$, and denoting $\widehat{g}(k)$ by g_k, we write

$$\begin{aligned}
&\|e^{it(-\Delta)^\alpha} f\|^4_{L^4_t L^4_x} \\
&= \int_0^{2\pi}\int_0^{2\pi} \sum_{k_1,k_2,k_3,k_4} \frac{e^{it(k_1^{2\alpha}-k_2^{2\alpha}+k_3^{2\alpha}-k_4^{2\alpha})} e^{ix(k_1-k_2+k_3-k_4)} g_{k_1}\overline{g_{k_2}} g_{k_3}\overline{g_{k_4}}}{\langle k_1\rangle^s\langle k_2\rangle^s\langle k_3\rangle^s\langle k_4\rangle^s} dxdt \\
&= \int_0^{2\pi} \sum_{k_1-k_2+k_3-k_4=0} \frac{e^{it(k_1^{2\alpha}-k_2^{2\alpha}+k_3^{2\alpha}-k_4^{2\alpha})} g_{k_1}\overline{g_{k_2}} g_{k_3}\overline{g_{k_4}}}{\langle k_1\rangle^s\langle k_2\rangle^s\langle k_3\rangle^s\langle k_4\rangle^s} dt \\
&\lesssim \sum_{k_1-k_2+k_3-k_4=0} \frac{|g_{k_1}||g_{k_2}||g_{k_3}||g_{k_4}|}{\langle k_1\rangle^s\langle k_2\rangle^s\langle k_3\rangle^s\langle k_4\rangle^s} \frac{1}{\max(1, |k_1^{2\alpha}-k_2^{2\alpha}+k_3^{2\alpha}-k_4^{2\alpha}|)}.
\end{aligned}$$

Renaming the variables as $k_1 = n+j$, $k_2 = n+k+j$, $k_3 = n+k$, and $k_4 = n$, and using Lemma 2, we get

$$\begin{aligned}
\|e^{it(-\Delta)^\alpha} f\|^4_{L^4_t L^4_x} &\lesssim \sum_{n,k,j} \frac{|g_n||g_{n+j}||g_{n+k}||g_{n+k+j}|}{\langle n\rangle^s\langle n+k\rangle^s\langle n+j\rangle^s\langle n+k+j\rangle^s} \frac{1}{\max\big(1, \frac{|kj|}{(|k|+|j|+|n|)^{2-2\alpha}}\big)} \\
&:= \mathrm{I} + \mathrm{II},
\end{aligned}$$

where I contains the terms with $|kj| \ll (|k|+|j|+|n|)^{2-2\alpha}$ and II contains the remaining terms.

First note that the summation set in I does not contain any terms with both $n=0$ and $|kj| \neq 0$ since $\alpha \in (1/2,1)$. Also noting that if $kj \neq 0$, then

$$|kj| \ll (|k|+|j|+|n|)^{2-2\alpha} \lesssim |k|^{2-2\alpha}+|j|^{2-2\alpha}+|n|^{2-2\alpha} \lesssim |kj|+|n|^{2-2\alpha},$$

since $\alpha \in (1/2,1)$. We can thus write

$$I \lesssim \sum_{\substack{n,k,j \\ 0<|kj|\lesssim |n|^{2-2\alpha}}} \frac{|g_n||g_{n+j}||g_{n+k}||g_{n+k+j}|}{\langle n\rangle^s \langle n+k\rangle^s \langle n+j\rangle^s \langle n+k+j\rangle^s} + \sum_{j,n} |g_n|^2|g_{n+j}|^2 + \sum_{k,n} |g_n|^2|g_{n+k}|^2.$$

The last two sums are equal to $\|g\|_{L^2}^4$. We estimate the first sum by Cauchy-Schwarz inequality to get

$$\begin{aligned}
&\lesssim \Big(\sum_{n,k,j} |g_{n+j}|^2|g_{n+k}|^2|g_{n+k+j}|^2\Big)^{1/2} \\
&\quad \cdot \Big(\sum_{\substack{n,k,j \\ 0<|kj|\lesssim |n|^{2-2\alpha}}} \frac{|g_n|^2}{\langle n\rangle^{2s} \langle n+k\rangle^{2s} \langle n+j\rangle^{2s} \langle n+k+j\rangle^{2s}}\Big)^{1/2} \\
&\lesssim \|g\|_{L^2}^4 \sup_n \Big(\sum_{\substack{k,j \\ 0<|kj|\lesssim |n|^{2-2\alpha}}} \frac{1}{\langle n\rangle^{2s} \langle n+k\rangle^{2s} \langle n+j\rangle^{2s} \langle n+k+j\rangle^{2s}}\Big)^{1/2}.
\end{aligned}$$

The condition on the sum implies, except for finitely many n's, that $|k| \ll |n|$ and $|j| \ll |n|$. Therefore

$$\begin{aligned}
&\sum_{\substack{k,j \\ 0<|kj|\lesssim |n|^{2-2\alpha}}} \frac{1}{\langle n\rangle^{2s} \langle n+k\rangle^{2s} \langle n+j\rangle^{2s} \langle n+k+j\rangle^{2s}} \\
&\lesssim \frac{1}{\langle n\rangle^{8s}} \sum_{0<|kj|\lesssim |n|^{2-2\alpha}} 1 \lesssim \langle n\rangle^{2-2\alpha-8s} \log\langle n\rangle \lesssim 1
\end{aligned}$$

provided that $s > \frac{1-\alpha}{4}$.

For the second sum we have

$$\mathrm{II} \lesssim \sum_{\substack{n,k,j \\ |kj|\gtrsim |n|^{2-2\alpha}}} \frac{|g_n||g_{n+j}||g_{n+k}||g_{n+k+j}|(|n|+|k|+|j|)^{2-2\alpha}}{\langle n\rangle^s \langle n+k\rangle^s \langle n+j\rangle^s \langle n+k+j\rangle^s |kj|}.$$

Using the symmetry in k and j, we have

$$\mathrm{II} \lesssim \sum_{\substack{n,k,j \\ |kj|\gtrsim |n|^{2-2\alpha},\ |k|\geq |j|}} \frac{|g_n||g_{n+j}||g_{n+k}||g_{n+k+j}|(|n|+|k|)^{2-2\alpha}}{\langle n\rangle^s \langle n+k\rangle^s \langle n+j\rangle^s \langle n+k+j\rangle^s |kj|}.$$

To estimate the sum we consider three frequency regions, $|k| \approx |n|$, $|k| \ll |n|$, and $|k| \gg |n|$.

Region 1: $|k| \approx |n|$. In this region, using Cauchy-Schwarz inequality as above, it suffices to show that the sum

$$\sum_{\substack{|k|\geq|j|\\|k|\approx|n|}} \frac{(|n|+|k|)^{4-4\alpha}}{\langle n\rangle^{2s}\langle n+k\rangle^{2s}\langle n+j\rangle^{2s}\langle n+k+j\rangle^{2s}k^2j^2}$$

is bounded in n. We bound this by

$$\sum_{\substack{|k|\geq|j|\\|k|\approx|n|}} \frac{|n|^{2-4\alpha-2s}}{\langle n+k\rangle^{2s}\langle n+j\rangle^{2s}\langle n+k+j\rangle^{2s}j^2}.$$

Using the inequality

$$\langle m+j\rangle\langle j\rangle \gtrsim \langle m\rangle,$$

and recalling that $s < \frac{1}{4}$, we obtain

$$\lesssim \sum_{\substack{|k|\geq|j|\\|k|\approx|n|}} \frac{|n|^{2-4\alpha-4s}}{\langle n+k\rangle^{4s}j^{2-4s}} \lesssim \langle n\rangle^{2-4\alpha-4s+1-4s}.$$

Here we first summed in j and then in k. The sum is bounded in n provided that $s > \frac{3-4\alpha}{8}$.

Region 2: $|k| \ll |n|$. As in Region 1, it suffices to show that the sum

$$\sum_{\substack{|j|\leq|k|\ll|n|\\|kj|\gtrsim|n|^{2-2\alpha}}} \frac{|n|^{4-4\alpha}}{\langle n\rangle^{2s}\langle n+k\rangle^{2s}\langle n+j\rangle^{2s}\langle n+k+j\rangle^{2s}k^2j^2} \approx \sum_{\substack{|j|\leq|k|\ll|n|\\|kj|\gtrsim|n|^{2-2\alpha}}} \frac{|n|^{4-4\alpha-8s}}{k^2j^2}$$

is bounded in n. To this end, notice that

$$\sum_{\substack{|j|\leq|k|\ll|n|\\|kj|\gtrsim|n|^{2-2\alpha}}} \frac{|n|^{4-4\alpha-8s}}{k^2j^2} \lesssim \sum_{|j|\leq|k|\ll|n|} \frac{|n|^{4-4\alpha-8s}}{|j||k|\langle n\rangle^{2-2\alpha}} \lesssim \sup_n |n|^{2-2\alpha-8s}\log(|n|)^2,$$

which is finite provided that $s > \frac{1-\alpha}{4}$.

Region 3: $|k| \gg |n|$. In this region we bound the sum by Cauchy-Schwarz inequality as follows:

$$\sum_{\substack{|j|\le|k|,\,|n|\ll|k|\\|kj|\gtrsim|n|^{2-2\alpha}}}\frac{|g_n||g_{n+j}||g_{n+k+j}||g_{n+k}||k|^{2-2\alpha}}{\langle n\rangle^s\langle n+k\rangle^s\langle n+j\rangle^s\langle n+k+j\rangle^s|kj|}$$

$$\lesssim\Big(\sum_{n,k,j}|g_n|^2|g_{n+j}|^2|g_{n+k+j}|^2\Big)^{1/2}$$

$$\cdot\Big(\sum_{|j|\le|k|,\,|n|\ll|k|}\frac{|g_{n+k}|^2|k|^{4-4\alpha}}{\langle n\rangle^{2s}\langle n+k\rangle^{2s}\langle n+j\rangle^{2s}\langle n+k+j\rangle^{2s}k^2j^2}\Big)^{1/2}$$

$$\lesssim\|g\|_{L^2}^3\Big(\sum_{|j|\le|k|,\,|n|\ll|k|}\frac{|g_{n+k}|^2|k|^{2-4\alpha-2s}}{\langle n\rangle^{2s}\langle n+j\rangle^{2s}\langle n+k+j\rangle^{2s}j^2}\Big)^{1/2}.$$

Estimating the j sum in parenthesis as in Region 1, we have

$$\lesssim\sum_{|n|\ll|k|}\frac{|g_{n+k}|^2|k|^{2-4\alpha-2s}}{\langle n\rangle^{4s}\langle n+k\rangle^{2s}}\lesssim\sum_{|n|\ll|k|}\frac{|g_{n+k}|^2|k|^{2-4\alpha-4s}}{\langle n\rangle^{4s}}$$

$$\lesssim\sum_{n,k}|g_{n+k}|^2|k|^{1-2\alpha-4s}\langle n\rangle^{1-2\alpha-4s}.$$

We estimate this by Cauchy-Schwarz

$$\Big(\sum_{n,k}|g_{n+k}|^2|k|^{2-4\alpha-8s}\Big)^{\frac12}\Big(\sum_{n,k}|g_{n+k}|^2\langle n\rangle^{2-4\alpha-8s}\Big)^{\frac12}\lesssim\|g\|_{L^2}^2,$$

provided that $2-4\alpha-8s<-1$, i.e. $s>\frac38-\frac{\alpha}{2}$. In the last inequality we summed in n and k separately.

Thus, for $s>\max(\frac{1-\alpha}{4},\frac{3-4\alpha}{8})=\frac{1-\alpha}{4}$, for $\alpha>\frac12$, we obtain the Strichartz estimates. □

4 Local well-posedness via the $X^{s,b}$ method

We will prove Theorem 1 for the defocusing equation by obtaining multilinear estimates in $X^{s,b}$ spaces. With the change of variable $u(x,t)\to u(x,t)e^{iPt}$ in the equation (1), where $P=\frac{1}{\pi}\|u_0\|_2^2$, we obtain the equation

$$iu_t+(-\Delta)^\alpha u+|u|^2u-Pu=0,\quad t\in\mathbb{R},\ x\in\mathbb{T},$$

with initial data in $u_0\in H^s(\mathbb{T})$, $s>0$.

Note the following identity which follows from Plancherel's theorem:

$$\begin{aligned}\widehat{|u|^2u}(k)&=\sum_{k_1,k_2}\widehat{u}(k_1)\overline{\widehat{u}(k_2)}\widehat{u}(k-k_1+k_2)\\&=\frac{1}{\pi}\|u\|_2^2\widehat{u}(k)-|\widehat{u}(k)|^2\widehat{u}(k)+\sum_{k_1\ne k,k_2\ne k_1}\widehat{u}(k_1)\overline{\widehat{u}(k_2)}\widehat{u}(k-k_1+k_2)\\&:=P\widehat{u}(k)+\widehat{\rho(u)}(k)+\widehat{R(u)}(k).\end{aligned}\qquad(2)$$

Using this in the Duhamel's formula, we have

$$u(t)=e^{it(-\Delta)^\alpha}u_0(x)-i\int_0^t e^{i(-\Delta)^\alpha(t-t')}(\rho(u)+R(u))dt'.$$

By standard iteration techniques, it suffices to obtain an estimate of the form:

$$\Big\|\int_0^t e^{i(-\Delta)^\alpha(t-t')}(\rho(u)+R(u))dt'\Big\|_{X_T^{s,b}}\lesssim T^\delta\|u\|^3_{X_T^{s,b}},$$

for $s>\frac{1-\alpha}{2}$ and for some $b>\frac12$, $\delta>0$.

To prove this estimate and obtain a lower bound for the local existence time we need the following lemma:

Lemma 3 ([11]). *For b,b' such that $0\le b+b'<1$, $0\le b'<1/2$, then we have*

$$\Big\|\int_0^t e^{i(-\Delta)^\alpha(t-\tau)}f(\tau)d\tau\Big\|_{X_T^{s,b}}\lesssim T^{1-b-b'}\|f\|_{X_T^{s,-b'}},$$

for $T\in[0,1]$.

Proposition 1. *Let $\alpha\in(\frac12,1)$ and $s>\frac{1-\alpha}{2}$, then for $b>1/2$ we have*

$$\big\|\rho(u)+R(u)\big\|_{X^{s,-b'}}\lesssim\|u\|^3_{X^{s,b}},$$

provided that $b'<\frac12$ is sufficiently close to $\frac12$. Moreover, for $s>\frac12$ we can take $b'=0$.

As we remarked in the introduction, in the case that $s>\frac12$, the condition $b'=0$ implies the existence of the local solution in $[0,\delta]$ as long as $\delta^{\frac12-}\|u_0\|^2_{H^s(\mathbb{T})}\sim1$. This bound although sub-optimal, it is necessary for the proof of the global well-posedness below the energy space that we establish in Section 6.

Proof. We present the proof for $R(u)$. The proof for $\rho(u)$ is easier and in what follows it corresponds to the terms given by $j=k=0$.

First note that

$$\big\|R(u)\big\|_{X^{s,-b'}}=\Big\|\int_{\tau_1-\tau_2+\tau_3=\tau}\sum_{\substack{k_1-k_2+k_3=n\\k_1\neq n,k_2}}\frac{\widehat{u}(\tau_1,k_1)\overline{\widehat{u}(\tau_2,k_2)}\widehat{u}(\tau_3,k_3)\langle n\rangle^s}{\langle\tau-n^{2\alpha}\rangle^{b'}}\Big\|_{L^2_\tau l^2_n},$$

By a duality argument and denoting $|\widehat{u}(\tau,n)|\langle n\rangle^s\langle\tau-n^{2\alpha}\rangle^b=v(\tau,n)$, we get

$$\begin{aligned}&\big\|R(u)\big\|_{X^{s,-b'}}\\&\le\sup_{\|g\|_{L^2_\tau l^2_n}=1}\int_{\tau_1-\tau_2+\tau_3-\tau=0}\sum_{\substack{k_1-k_2+k_3-n=0\\k_1\neq n,k_2}}\frac{\langle n\rangle^s v(\tau_1,k_1)v(\tau_2,k_2)v(\tau_3,k_3)g(\tau,n)}{\langle k_1\rangle^s\langle k_2\rangle^s\langle k_3\rangle^s\langle\tau-n^{2\alpha}\rangle^{b'}}\\&\quad\times\frac{1}{\langle\tau_1-k_1^{2\alpha}\rangle^b\langle\tau_2-k_2^{2\alpha}\rangle^b\langle\tau_3-k_3^{2\alpha}\rangle^b},\end{aligned}$$

and thus, by Cauchy-Schwarz and then integrating in τ variables as in [9], we have

$$\begin{aligned}
&\big\|R(u)\big\|^2_{X^{s,-b'}} \\
&\leq \|v\|^6_{L^2_\tau l^2_n} \sup_{\tau,n} \int_{\tau_1-\tau_2+\tau_3=\tau} \sum_{\substack{k_1-k_2+k_3=n\\ k_1\neq n,k_2}} \frac{\langle n\rangle^{2s}}{\langle k_1\rangle^{2s}\langle k_2\rangle^{2s}\langle k_3\rangle^{2s}\langle \tau-n^{2\alpha}\rangle^{2b'}} \\
&\times \frac{1}{\langle \tau_1-k_1^{2\alpha}\rangle^{2b}\langle \tau_2-k_2^{2\alpha}\rangle^{2b}\langle \tau_3-k_3^{2\alpha}\rangle^{2b}}. \\
&\lesssim \|u\|^6_{X^{s,b}} \sup_n \sum_{\substack{k_1-k_2+k_3=n\\ k_1\neq n,k_2}} \frac{\langle n\rangle^{2s}}{\langle k_1\rangle^{2s}\langle k_2\rangle^{2s}\langle k_3\rangle^{2s}\langle k_1^{2\alpha}-k_2^{2\alpha}+k_3^{2\alpha}-n^{2\alpha}\rangle^{2b'}}.
\end{aligned}$$

Hence, we need to show that

$$M_n = \sum_{\substack{k_1-k_2+k_3=n\\ k_1\neq n,k_2}} \frac{\langle n\rangle^{2s}}{\langle k_1\rangle^{2s}\langle k_2\rangle^{2s}\langle k_3\rangle^{2s}\langle k_1^{2\alpha}-k_2^{2\alpha}+k_3^{2\alpha}-n^{2\alpha}\rangle^{2b'}}$$

is bounded in n. Renaming the variables as $k_1=n+j$, $k_2=n+k+j$, $k_3=n+k$, and using Lemma 2, we get

$$\begin{aligned}
M_n &\lesssim \sum_{j,k\neq 0} \frac{\langle n\rangle^{2s}}{\langle n+j\rangle^{2s}\langle n+k+j\rangle^{2s}\langle n+k\rangle^{2s}\max\big(1,\frac{|kj|^{2b'}}{(|k|+|j|+|n|)^{4(1-\alpha)b'}}\big)} \\
&:= \mathrm{I}+\mathrm{II},
\end{aligned}$$

where I contains the terms with $|kj|\ll(|k|+|j|+|n|)^{2-2\alpha}$ and II contains the remaining terms. Here we note that M_n is bounded in n for $b'=0$ in the case $s>\frac{1}{2}$. From now on we consider the range $\frac{1-\alpha}{2}<s\leq\frac{1}{2}$, and take $b'=\frac{1}{2}-$. To estimate I, as in the proof of Theorem 4, we write

$$\mathrm{I}\lesssim \sum_{0<|kj|\lesssim|n|^{2-2\alpha}} \frac{\langle n\rangle^{2s}}{\langle n+k\rangle^{2s}\langle n+k+j\rangle^{2s}\langle n+j\rangle^{2s}} \lesssim \langle n\rangle^{2-2\alpha-4s}\log\langle n\rangle,$$

which is bounded provided that $s>\frac{1-\alpha}{2}$. Similarly,

$$\begin{aligned}
\mathrm{II} &\lesssim \sum_{|kj|\gtrsim|n|^{2-2\alpha}} \frac{\langle n\rangle^{2s}(|k|+|j|+|n|)^{2(1-\alpha)}}{\langle n+k\rangle^{2s}\langle n+k+j\rangle^{2s}\langle n+j\rangle^{2s}|kj|^{1-}} \\
&\lesssim \sum_{\substack{|kj|\gtrsim|n|^{2-2\alpha}\\ |k|\geq|j|}} \frac{\langle n\rangle^{2s}(|k|+|n|)^{2(1-\alpha)}}{\langle n+k\rangle^{2s}\langle n+k+j\rangle^{2s}\langle n+j\rangle^{2s}|kj|^{1-}}.
\end{aligned}$$

Second line follows from the kj symmetry of the sum. To estimate the sum we consider three regions:

Region 1: $|k|\gg|n|$. The sum is

$$\lesssim \sum_{\substack{|k|\geq|j|\\ |k|\gg|n|}} \frac{\langle n\rangle^{2s}|k|^{2(1-\alpha)-2s-1+}}{\langle n+j\rangle^{2s}\langle n+k+j\rangle^{2s}|j|^{1-}}.$$

Note that for $\frac{1}{2} \geq s > \frac{1-\alpha}{2}$, we can bound it by

$$\lesssim \sum_{\substack{|k|\geq|j| \\ |k|\gg|n|}} \frac{\langle n\rangle^{2s}|k|^{2(1-\alpha)-4s+}}{\langle n+j\rangle^{2s}|k|^{1-2s+}\langle n+k+j\rangle^{2s}|j|^{1-}}$$

$$\lesssim \sum_{\substack{|k|\geq|j| \\ |k|\gg|n|}} \frac{\langle n\rangle^{2(1-\alpha)-2s+}}{\langle n+j\rangle^{2s}|k|^{1-2s+}\langle n+k+j\rangle^{2s}|j|^{1-}}$$

$$\lesssim \sum_{j} \frac{\langle n\rangle^{2(1-\alpha)-2s+}}{\langle n+j\rangle^{2s}|j|^{1-}} \lesssim \langle n\rangle^{2(1-\alpha)-4s+},$$

which is bounded in n. In the k and j sums we used Lemma 1.

Region 2: $|k| \approx |n|$. In this region we have the bound

$$\lesssim \sum_{\substack{|k|\geq|j| \\ |k|\approx|n|}} \frac{\langle n\rangle^{2s+1-2\alpha+}}{\langle n+j\rangle^{2s}\langle n+k\rangle^{2s}\langle n+k+j\rangle^{2s}|j|^{1-}}$$

$$\lesssim \sum_{j} \frac{\langle n\rangle^{2s+1-2\alpha+}A}{\langle n+j\rangle^{2s}|j|^{1-}},$$

where $A = |j|^{1-4s}$ if $4s > 1$, $A = |n|^{1-4s}$ if $4s < 1$ and $A = \log(|n|)$ if $4s = 1$. Then, by considering these cases separately and using Lemma 1 in the j sums, one obtains boundedness in n for $s > \frac{1-\alpha}{2}$ and $\alpha > \frac{1}{2}$.

Region 3: $|k| \ll |n|$. We have the bound

$$\lesssim \sum_{|j|\leq|k|\ll|n|} \frac{\langle n\rangle^{-4s+2-2\alpha}}{|kj|^{1-}} \lesssim \langle n\rangle^{-4s+2-2\alpha+},$$

which is bounded in n. □

5 A smoothing estimate

We first note that

$$\|\rho(u)\|_{H^{s+c}} = \sqrt{\sum_{k} |\widehat{u}(k)|^{6}\langle k\rangle^{2s+2c}} \lesssim \|u\|_{H^s}^{3} \tag{3}$$

for $0 \leq c \leq 2s$, which implies that the contribution of $\rho(u)$ to the Duhamel formula is smoother than u. One can also obtain the same level of smoothing in $X^{s,b}$ spaces: For $c \leq 2s$,

$$\|\rho(u)\|_{X^{s+c,-\frac{1}{2}+}} \lesssim \|u\|_{X^{s,\frac{1}{2}+}}^{3}.$$

To prove the same for the non resonant terms $R(u)$ we have the following proposition:

Proposition 2. *For* $s > \frac{1-\alpha}{2}$ *and* $c < \min(\alpha - \frac{1}{2}, 2s + \alpha - 1)$ *, we have*

$$\|R(u)\|_{X^{s+c,-\frac{1}{2}+}} \lesssim \|u\|^3_{X^{s,\frac{1}{2}+}}.$$

Proof. Repeating the steps in the proof of Proposition 1, it suffices to prove that

$$M(n) = \sum_{kj \neq 0} \frac{\langle n \rangle^{2s+2c}}{\langle n+j \rangle^{2s} \langle n+k \rangle^{2s} \langle n+j+k \rangle^{2s} \langle \frac{|kj|}{(|n|+|k|+|j|)^{2-2\alpha}} \rangle^{1-}}$$

is bounded in n.

For the terms with $0 < |kj| \lesssim |n|^{2-2\alpha}$, since $|k|, |j| \ll |n|$, we have the bound

$$\lesssim \sum_{0<|kj|\lesssim |n|^{2-2\alpha}} \langle n \rangle^{-4s+2c} \lesssim \langle n \rangle^{-4s+2c+2-2\alpha} \log(n),$$

which is bounded provided that $c < 2s + \alpha - 1$.

For the remaining terms, we have to consider the cases $s > 1/2$ and $s \leq 1/2$ separately. Again by symmetry in j and k, it is enough to consider $|k| \geq |j|$.

Case 1: $s > 1/2$. As before, we will consider three regions:

Region 1.1: $|k| \gg |n|$. Then we have

$$\begin{aligned} &\lesssim \sum_{\substack{|k|\geq|j|>0 \\ |k|\gg|n|}} \frac{\langle n \rangle^{2s+2c} |k|^{1-2\alpha-2s+}}{\langle n+j \rangle^{2s} \langle n+k+j \rangle^{2s} |j|^{1-}} \\ &\lesssim \sum_{\substack{j \\ |k|\gg|n|}} \frac{\langle n \rangle^{2c+1-2\alpha+}}{\langle n+j \rangle^{2s} \langle n+k+j \rangle^{2s} |j|^{1-}} \\ &\lesssim \sum_{j} \frac{\langle n \rangle^{2c+1-2\alpha+}}{\langle n+j \rangle^{2s} \langle j \rangle^{1-}} \lesssim \langle n \rangle^{2c-2\alpha+}, \end{aligned}$$

which is bounded for $c < \alpha$. In the forth inequality we used Lemma 1.

Region 1.2: $|k| \approx |n|$. In this region we have

$$\begin{aligned} &\lesssim \sum_{\substack{|k|\geq|j|>0 \\ |k|\approx|n|}} \frac{\langle n \rangle^{2c+2s+1-2\alpha+}}{\langle n+k \rangle^{2s} \langle n+j \rangle^{2s} \langle n+j+k \rangle^{2s} |j|^{1-}} \\ &\lesssim \sum_{\substack{|k|\geq|j|>0 \\ |k|\approx|n|}} \frac{\langle n \rangle^{2c+1-2\alpha+}}{\langle n+k \rangle^{2s} |j|^{1-}} \lesssim \langle n \rangle^{2c+1-2\alpha+} \end{aligned}$$

for $c < \alpha - \frac{1}{2}$.

Region 1.3: $|k| \ll |n|$. We have

$$\lesssim \sum_{\substack{|k|\geq|j|>0 \\ |k|\ll|n|}} \frac{\langle n \rangle^{-4s+2c+2-2\alpha+}}{|kj|^{1-}} \lesssim \langle n \rangle^{2c-4s+2-2\alpha+},$$

which is bounded for $c < 2s + \alpha - 1$. This finishes the case $s > 1/2$.

Case 2: $\frac{1-\alpha}{2} < s \leq 1/2$.

Region 2.1: $|k| \gg |n|$. As in the proof of Proposition 1, we have

$$\lesssim \sum_{\substack{|k|\geq|j|>0 \\ |k|\gg|n|}} \frac{\langle n\rangle^{2s+2c-4s+2-2\alpha+}}{\langle n+j\rangle^{2s}\langle n+k+j\rangle^{2s}|k|^{1-2s+}|j|^{1-}} \lesssim \langle n\rangle^{2c-4s+2-2\alpha+},$$

which is bounded for $c < 2s+\alpha-1$.

Region 2.2: $|k| \approx |n|$. In this region we have

$$\lesssim \sum_{\substack{|k|\geq|j|>0 \\ |k|\approx|n|}} \frac{\langle n\rangle^{2s+2c+1-2\alpha+}}{\langle n+j\rangle^{2s}\langle n+k\rangle^{2s}\langle n+k+j\rangle^{2s}|j|^{1-}} \lesssim \sum_j \frac{\langle n\rangle^{2s+2c+1-2\alpha+}A}{\langle n+j\rangle^{2s}|j|^{1-}},$$

where $A = \langle j\rangle^{1-4s}$ for $\frac{1}{4} \leq s \leq \frac{1}{2}$, and $A = \langle n\rangle^{1-4s}$ for $0 < s < \frac{1}{4}$. Hence,

$$\lesssim \langle n\rangle^{2c+1-2\alpha+} \qquad \text{for } s \geq \frac{1}{4},$$
$$\lesssim \langle n\rangle^{2c-4s+2-2\alpha+} \qquad \text{for } 0 < s < \frac{1}{4},$$

which is bounded for $c < 2s+\alpha-1$ when $s \in (0, \frac{1}{4})$ and $c < \alpha - \frac{1}{2}$ when $s \geq \frac{1}{4}$.

Region 2.3: $|k| \ll |n|$. We have

$$\lesssim \sum_{\substack{|k|\geq|j|>0 \\ |k|\ll|n|}} \frac{\langle n\rangle^{2c+2-2\alpha-4s+}}{|kj|^{1-}} \lesssim \langle n\rangle^{2c+2-2\alpha-4s+},$$

which is bounded for $c < 2s+\alpha-1$.

Hence, for all s, collecting the results we get the proposition. □

This implies that (see [10] for more details):

Theorem 5. *For $\alpha \in (\frac{1}{2}, 1)$, $s > \frac{1-\alpha}{2}$ and $c < \min(2s+\alpha-1, \alpha-\frac{1}{2})$ we have*

$$\|u(t) - e^{it(-\Delta)^\alpha - iPt}u_0\|_{H^{s+c}} \lesssim \|u_0\|^3_{H^s}$$

for $t < T$, where T is the local existence time.

We finish this section by noting that if we define the multilinear versions of ρ and R via

$$\widehat{\rho(u,v,w)}(k) = \widehat{u}(k)\overline{\widehat{v}(k)}\widehat{w}(k), \quad \widehat{R(u,v,w)}(k) = \sum_{k_1\neq k, k_2\neq k_1} \widehat{u}(k_1)\overline{\widehat{v}(k_2)}\widehat{w}(k-k_1+k_2),$$

then the assertions of Proposition 1 and Proposition 2 remain valid.

6 Global well-posedness via high-low frequency decomposition

From the local theory along with energy and mass conservation, the existence of global solutions in H^α follows easily. In this case, one can control the H^α norm and apply the local theory with a uniform in time step to reach any time. In this section we use Bourgain's high-low frequency decomposition together with the smoothing estimate from the previous section to obtain global well-posedness for initial data with infinite energy.

Proof of Theorem 2. Fix $s \in (\frac{1}{2}, \alpha)$. With the change of variable

$$u(x,t) \to u(x,t)e^{iPt}$$

in equation (1), where $P = \frac{1}{\pi}\|u_0\|_2^2$, we obtain the equation

$$iu_t + (-\Delta)^\alpha u + |u|^2 u - Pu = 0, \quad t \in \mathbb{R}, \ x \in \mathbb{T},$$

with initial data in $u_0 \in H^s(\mathbb{T})$. In what follows, the implicit constants will depend on $\|u_0\|_{H^s}$. We fix N large and decompose the equation into two equations, $u = v + w$:

$$\begin{cases} iv_t + (-\Delta)^\alpha v + |v|^2 v - Pv = 0, \\ v(x,0) = P_N u_0(x) \doteq \Phi_0, \end{cases} \tag{4}$$

and

$$\begin{cases} iw_t + (-\Delta)^\alpha w + |v+w|^2(v+w) - Pw - |v|^2 v = 0, \\ w(x,0) = u_0(x) - \Phi_0 \doteq \Psi_0, \end{cases} \tag{5}$$

where P_N is the projection onto the frequencies $|n| \leq N$.

First note that $\|\Phi_0\|_{H^\alpha} \lesssim N^{\alpha-s}$. Moreover, by the local existence theory we presented in H^α and H^s levels, noting that $\alpha > s > \frac{1}{2}$, we have for $\delta \approx N^{-4(\alpha-s)}$,

$$\|v\|_{X_\delta^{\alpha,b}} \lesssim \|\Phi_0\|_{H^\alpha} \lesssim N^{\alpha-s}, \qquad \|v\|_{X_\delta^{s,b}} \lesssim \|\Phi_0\|_{H^s} \lesssim 1.$$

Since equation (4) has the same Hamiltonian, we have

$$H(v(t)) = H(\Phi_0) \lesssim N^{2\alpha-2s}$$

by the Gagliardo-Nirenberg inequality.

Now pick an $s_0 < s$ to be determined later. Note that $\|\Psi_0\|_{H^{s_0}} \lesssim N^{s_0-s}$. The local existence for w equation follows similarly by the multilinear estimates from the previous sections with the same δ as above (since the norm of w is small). We thus have

$$\|w\|_{X_\delta^{s_0,b}} \lesssim \|\Psi_0\|_{H^{s_0}} \lesssim N^{s_0-s}, \qquad \|w\|_{X_\delta^{s,b}} \lesssim \|\Psi_0\|_{H^s} \lesssim 1.$$

Now using the decomposition (2) for the nonlinearity $\mathcal{N} := |v+w|^2(v+w) - Pw - |v|^2 v$ in (5) we have (with $u = v + w$)

$$\begin{aligned}\mathcal{N} &= Pu - Pw - \frac{1}{\pi}\|v\|_{L^2}^2 v + \rho(u) - \rho(v) + R(u) - R(v) \\ &= \frac{1}{\pi}(\|u_0\|_2^2 - \|\Phi_0\|_{L^2}^2)v + \rho(u) - \rho(v) + R(u) - R(v).\end{aligned}$$

Using the multilinear smoothing estimate and the multilinearity of ρ and R, we have

$$\|\mathcal{N}\|_{X_\delta^{\alpha,-\frac{1}{2}+}} \lesssim \big|\|u_0\|_2^2 - \|\Phi_0\|_{L^2}^2\big| \|v\|_{X_\delta^{\alpha,-\frac{1}{2}+}} + \|w\|_{X_\delta^{s_0,b}}^3 + \|w\|_{X_\delta^{s_0,b}} \|v\|_{X_\delta^{s_0,b}}^2,$$

for $\alpha - s_0 < \min(2s_0 + \alpha - 1, \alpha - \frac{1}{2})$, in particular for $s_0 > \frac{1}{2}$.

Ignoring the support condition of Φ_0 and Ψ_0, we have

$$\big|\|u_0\|_2^2 - \|\Phi_0\|_{L^2}^2\big| \lesssim \|\Psi_0\|_{L^2} + \|\Psi_0\|_{L^2}^2 \lesssim N^{-s}.$$

Therefore, we obtain

$$\begin{aligned}\|\mathcal{N}\|_{X_\delta^{\alpha,-\frac{1}{2}+}} &\lesssim N^{-s}\delta^{1-}\|v\|_{X_\delta^{\alpha,b}} + \|w\|_{X_\delta^{s_0,b}}^3 + \|w\|_{X_\delta^{s_0,b}}\|v\|_{X_\delta^{\alpha,b}}^2 \\ &\lesssim N^{-s}\delta^{1-}N^{\alpha-s} + N^{3(s_0-s)} + N^{s_0-s}N^{2(\alpha-s)} \lesssim N^{2\alpha+s_0-3s}.\end{aligned}$$

Taking $t_1 = \delta$, we write

$$u(t_1) = w(t_1) + v(t_1) = e^{it_1(-\Delta)^\alpha + iPt_1}\Psi_0 + w_1(t_1) + v(t_1).$$

By the bound on $\mathcal{N}$ and Duhamel's formula, we have

$$\|w_1(t_1)\|_{H^\alpha} \lesssim N^{2\alpha+s_0-3s}.$$

We repeat this process by decomposing $u(t_1) = \Phi_1 + \Psi_1$, where

$$\Psi_1 = e^{it_1(-\Delta)^\alpha + iPt_1}\Psi_0, \qquad \Phi_1 = w_1(t_1) + v(t_1).$$

Since $e^{it_1(-\Delta)^\alpha + iPt_1}$ is unitary, Ψ_1 satisfies all the properties of Ψ_0. To control the H^α norm of Φ_1, we note

$$\begin{aligned}H(\Phi_1) &= H(\Phi_1) - H(v(t_1)) + H(v(t_1)) \\ &= H(w_1(t_1) + v(t_1)) - H(v(t_1)) + H(\Phi_0),\end{aligned}$$

where the second equality follows from the conservation of the Hamiltonian.

Note that

$$\begin{aligned}\big|H(f+g) - H(f)\big| &\lesssim \big|\||\nabla|^\alpha(f+g)\|_2^2 - \||\nabla|^\alpha f\|_2^2\big| + \int \big||f+g|^4 - |f|^4\big| \\ &\lesssim \|g\|_{H^\alpha}^2 + \|g\|_{H^\alpha}\|f\|_{H^\alpha} + \int |g|\big(|f|^3 + |g|^3\big) \\ &\lesssim \|g\|_{H^\alpha}^2 + \|g\|_{H^\alpha}\|f\|_{H^\alpha} + \|g\|_{H^{\frac{1}{4}+}}^4 + \|g\|_{H^{\frac{1}{4}+}}\|f\|_{H^{\frac{1}{4}+}}^3 \\ &\lesssim \|g\|_{H^\alpha}^2 + \|g\|_{H^\alpha}\|f\|_{H^\alpha} + \|g\|_{H^\alpha}^4 + \|g\|_{H^\alpha}\|f\|_{H^\alpha}^3.\end{aligned}$$

Using this for $f = v(t_1)$ and $g = w_1(t_1)$, we obtain

$$H(w_1(t_1) + v(t_1)) - H(v(t_1)) \lesssim N^{2\alpha+s_0-3s}N^{3(\alpha-s)} = N^{5\alpha+s_0-6s}.$$

To reach time T we have to iterate this process $\frac{T}{\delta}$ times. To bound the Hamiltonian at time T by a constant multiple of the initial value, we need

$$N^{5\alpha+s_0-6s}\frac{T}{\delta} = TN^{9\alpha+s_0-10s}$$

to be $\lesssim N^{2\alpha-2s}$. This holds for $s > \frac{7\alpha}{8} + \frac{1}{16}$ by taking $s_0 = \frac{1}{2}+$ and N sufficiently large.

The calculation above can be improved by interpolating between H^α and L^2 to bound the $H^{\frac{1}{4}+}$ norms. For example, by Duhamel's formula and Minkowski inequality, we have

$$\|w_1(t_1)\|_{L^2} \lesssim \int_0^{t_1} \|\mathcal{N}\|_{L^2} dt.$$

The worst term in $\mathcal{N}$ is of the form $|v^2 w|$ which can be bounded as follows

$$\delta^{\frac{1}{2}}\|v\|^2_{L_t^4 L_x^4}\|w\|_{L_t^\infty L_x^\infty} \lesssim \delta^{\frac{1}{2}}\|v\|^2_{L_t^4 L_x^4}\|w\|_{X_\delta^{s_0,b}} \lesssim \delta H(v)^{\frac{1}{2}}\|w\|_{X_\delta^{s_0,b}} \lesssim \delta N^{\alpha+s_0-2s}.$$

After, $\frac{T}{\delta}$ steps, the L^2 norm remains $\lesssim N^{\alpha+s_0-2s} \lesssim 1$, for $s > \frac{\alpha}{2} + \frac{1}{4}$. Therefore the L^2 norm of the low frequency part also remains $\lesssim 1$.

Using this in the bound for the Hamiltonian, we get

$$\begin{aligned}
&H(w_1(t_1) + v(t_1)) - H(v(t_1)) \\
&\lesssim N^{2\alpha+s_0-3s}N^{\alpha-s} + N^{(1-\frac{1}{4\alpha})(\alpha+s_0-2s)}N^{\frac{2\alpha+s_0-3s}{4\alpha}}N^{\frac{3(\alpha-s)}{4\alpha}}N^{+} \\
&\lesssim N^{3\alpha+\frac{1}{2}-4s+} + N^{\alpha+\frac{3}{2}-2s-\frac{s}{\alpha}+} \lesssim N^{3\alpha+\frac{1}{2}-4s+}.
\end{aligned}$$

After $\frac{T}{\delta}$ steps we get the bound $TN^{7\alpha+\frac{1}{2}-8s+}$. This term is less than similar the initial energy of the high frequency part which is of order $N^{2\alpha-2s}$ for $s > \frac{5\alpha}{6} + \frac{1}{12}$. We can then iterate our result to reach any time T by sending N to infinity. □

References

[1] M. Ablowitz and Y. Ma, The periodic cubic Schrödinger equation, Stud. Appl. Math. **65** (1981), 113–158.

[2] J. Bourgain, Fourier transform restriction phenomena for certain lattice subsets and applications to nonlinear evolution equations, Geometric and Functional Analysis **3** (1993), 107–156.

[3] J. Bourgain, Refinements of Strichartz' inequality and applications to 2-D NLS with critical nonlinearity, Int. Math. Res. Not. **5** (1998), 253–283.

[4] N. Burq, P. Gerard, and N. Tzvetkov, An instability property of the nonlinear Schrödinger equation on S^d, Math. Res. Let. **9** (2002), 323–335.

[5] F. Catoire and W.-M. Wang, Bounds on Sobolev norms for the nonlinear Schrödinger equation on general tori, Communications in Pure and Applied Analysis **9** (2010), no. 2, 483–491.

[6] Y. Cho, G. Hwang, S. Kwon, and S. Lee, Well-posedness and ill-posedness for the cubic fractional Schrödinger equations, preprint, arXiv:1311.0082.

[7] S. Demirbas, Local well-posedness for 2-D Schrödinger equation on irrational tori and bounds on Sobolev norms, preprint, arXiv:1307.0051.

[8] M. B. Erdogan and N. Tzirakis, Global smoothing for the periodic KdV evolution, Int. Math. Res. Not. (2012), rns189, 26pp, DOI: 10.1093/imrn/rns189.

[9] M. B. Erdogan and N. Tzirakis, Smoothing and global attractors for the Zakharov system on the torus, Anal. PDE **6** (2013), no. 3, 723–750.

[10] M. B. Erdogan and N. Tzirakis, Talbot effect for the cubic nonlinear Schrödinger equation on the torus, Math. Res. Lett. **20** (2013), no. 6, 1081–1090.

[11] J. Ginibre, Le problème de Cauchy pour des EDP semi-linéaires périodiques en variables d'espace, Inventiones Mathematicae **37** (1995), 163–187.

[12] K. Kirkpatric, E. Lenzmann, and G. Staffilani, On the continuum limit for discrete NLS with long-range lattice interactions, Comm. Math. Phys. **317** (2013), no. 3, 563–591.

[13] N, Laskin, Fractional quantum mechanics and Lévy path integrals, Phys. Lett. A **268** (2000), 298–305.

[14] T. Tao, Nonlinear Dispersive Equations: Local and Global Analysis, Amererican Mathematical Society, CBMS **106**, 2006.

Some Topics in Harmonic Analysis and Applications
ALM 34, pp. 163–175

Compactness of Maximal Commutators of Bilinear Calderón-Zygmund Singular Integral Operators *

Yong Ding[†], Ting Mei[‡] and Qingying Xue[§]

Abstract

Let T be a bilinear Calderón-Zygmund singular integral operator and T_* be its corresponding truncated maximal operator. The commutators in the i-th entry and the iterated commutators of T_* are defined respectively by

$$T_{*,b,1}(f,g)(x) = \sup_{\delta>0}\left|\iint_{|x-y|+|x-z|>\delta} K(x,y,z)(b(y)-b(x))f(y)g(z)dydz\right|,$$

$$T_{*,b,2}(f,g)(x) = \sup_{\delta>0}\left|\iint_{|x-y|+|x-z|>\delta} K(x,y,z)(b(z)-b(x))f(y)g(z)dydz\right|,$$

$$\begin{aligned} T_{*,(b_1,b_2)}(f,g)(x) = \sup_{\delta>0}\bigg|\iint_{|x-y|+|x-z|>\delta} & K(x,y,z) \\ & \times (b_1(y)-b_1(x))(b_2(z)-b_2(x))f(y)g(z)dydz\bigg|. \end{aligned}$$

In this paper, the compactness of the commutators $T_{*,b,1}$, $T_{*,b,2}$ and $T_{*,(b_1,b_2)}$ on $L^r(\mathbb{R}^n)$ is established.

2010 Mathematics Subject Classification: 42B25, 42B99.

Keywords and Phrases: bilinear maximal Calderón-Zygmund singular integral, iterated commutators, CMO space, compactness.

*The first author is supported by NSFC (No.11371057) and SRFDP (No.20130003110003), the third author is supported by NSFC (No. 11471041), the Fundamental Research Funds for the Central Universities (No. 2014KJJCA10) and NCET-13-0065.

[†]School of Mathematical Sciences, Beijing Normal University, Laboratory of Mathematics and Complex Systems Ministry of Education, Beijing, 100875, China; *E-mail:* `dingy@bnu.edu.cn`

[‡]School of Mathematical Sciences, Beijing Normal University, Laboratory of Mathematics and Complex Systems Ministry of Education, Beijing, 100875, China; *E-mail:* `meiting@mail.bnu.edu.cn`

[§]School of Mathematical Sciences, Beijing Normal University, Laboratory of Mathematics and Complex Systems Ministry of Education, Beijing, 100875, China; *E-mail:* `qyxue@bnu.edu.cn`

1 Introduction and main results

Let T_Ω be the well-known Calderón-Zygmund singular integral operator defined by

$$T_\Omega f(x) = p.v. \int_{\mathbb{R}^n} \frac{\Omega(x-y)}{|x-y|^n} f(y)\, dy.$$

In 1976, Coifman, Rochberg and Weiss [11] defined the following commutator of T_Ω with a local integrable function b,

$$\begin{aligned}[b, T_\Omega]f(x) &= b(x)T_\Omega(f)(x) - T_\Omega(bf)(x)\\ &= p.v. \int_{\mathbb{R}^n} (b(x)-b(y)) \frac{\Omega(x-y)}{|x-y|^n} f(y) dy. \end{aligned} \tag{1.1}$$

The authors of [11] proved that $[b, T_\Omega]$ is bounded on L^p for $1 < p < \infty$ when $b \in BMO$ and Ω satisfies:

(i) $\Omega(\lambda x) = \Omega(x)$, for any $\lambda > 0$ and $x \neq 0$,

(ii) $\displaystyle\int_{S^{n-1}} \Omega(x')\, d\sigma(x') = 0$,

(iii) $|\Omega(x') - \Omega(y')| \leq |x' - y'|$, for any x', $y' \in S^{n-1}$.

In 1978, Uchiyama gave the L^p-compactness characterization of $[b, T_\Omega]$.

Theorem A ([20]). Let Ω satisfy (i)–(iii). If $b \in \cup_{q>1} L^q_{loc}(\mathbb{R}^n)$, then $[b, T_\Omega]$ is a compact operator on $L^p(\mathbb{R}^n)$ for $1 < p < \infty$ if and only if $b \in CMO(\mathbb{R}^n)$, here and in the sequel, $CMO(\mathbb{R}^n)$ denotes the closure of $C_c^\infty(\mathbb{R}^n)$ in $BMO(\mathbb{R}^n)$.

In 1993, Beatrous and Li [3] gave the boundedness and the compactness characterizations for $[b, T_\Omega]$ in L^p space and some spaces of homogeneous type. In 2001, Krantz and Li applied the compactness characterization of the commutator $[b, T_\Omega]$ to study Hankel type operators on Bergman spaces (see [15, 16]). As for the compactness of commutators for the other type operators, such as the Riesz potential, singular integral with variable kernel, parabolic singular integral, Littlewood-Paley operators, one may see [19] and the recent works [5–10].

Let T be a bilinear Calderón-Zygmund operator and assume that the kernel K satisfies the usual conditions in such a theory, that is, $K \in 2\text{-}CZK(A, \gamma)$ (see [13] for its definition). Let T_* be the corresponding bilinear maximal singular integral operator defined by

$$T_*(f, g)(x) = \sup_{\delta > 0} \left| \iint_{|x-y|+|x-z|>\delta} K(x, y, z) f(y) g(z)\, dydz \right|. \tag{1.2}$$

In 2002, Grafakos and Torres [14] obtained the following L^r-estimate of T_*:

$$\|T_*(f, g)\|_r \leq C \|f\|_p \|g\|_q, \tag{1.3}$$

for $1 < p, q < \infty$, $1/2 < r < \infty$ with $\frac{1}{r} = \frac{1}{p} + \frac{1}{q}$.

Let b, b_1, $b_2 \in BMO(\mathbb{R}^n)$. We are interested in the following three maximal commutators of bilinear Calderón-Zygmund operators:

$$T_{*,b,1}(f, g)(x) = \sup_{\delta>0} |[T_\delta, b]_1(f, g)(x)| = \sup_{\delta>0} |(T_\delta(bf, g) - bT_\delta(f, g))(x)|$$

$$:= \sup_{\delta>0} \left| \iint_{|x-y|+|x-z|>\delta} K(x,y,z)(b(y)-b(x))f(y)g(z)\, dydz \right|, \tag{1.4}$$

$$\begin{aligned} T_{*,b,2}(f,g)(x) &= \sup_{\delta>0} |[T_\delta, b]_2(f,g)(x)| = \sup_{\delta>0} |(T_\delta(f,bg) - bT_\delta(f,g))(x)| \\ &:= \sup_{\delta>0} \left| \iint_{|x-y|+|x-z|>\delta} K(x,y,z)(b(z)-b(x))f(y)g(z)\, dydz \right|, \end{aligned} \tag{1.5}$$

$$\begin{aligned} T_{*,(b_1,b_2)}(f,g)(x) &= \sup_{\delta>0} |([T_\delta, b_1]_1(f, b_2 g) - b_2[T_\delta, b_1]_1(f,g))(x)| \\ &:= \sup_{\delta>0} \Big| \iint_{|x-y|+|x-z|>\delta} K(x,y,z) \\ &\quad \times (b_1(y)-b_1(x))(b_2(z)-b_2(x))f(y)g(z)\, dydz \Big|. \end{aligned} \tag{1.6}$$

By the results in [18], the operator defined by (1.6) maps $L^p(\mathbb{R}^n) \times L^q(\mathbb{R}^n) \to L^r(\mathbb{R}^n)$ with $\frac{1}{r} = \frac{1}{p} + \frac{1}{q}$ for all $\frac{1}{2} < r < \infty$, $1 < p, q < \infty$, with the following estimate:

$$\|T_{*,(b_1,b_2)}(f,g)\|_{L^r(\mathbb{R}^n)} \le C \prod_{j=1}^{2} \|b_j\|_{BMO} \|f\|_{L^p(\mathbb{R}^n)} \|g\|_{L^q(\mathbb{R}^n)}. \tag{1.7}$$

Remark 1.1. We can't find any results for the L^p-boundedness of the maximal commutators defined by (1.4) and (1.5). However, we may prove the following results by the same idea in [18].

$$\|T_{*,b,i}(f,g)\|_{L^r(\mathbb{R}^n)} \le C\|b\|_{BMO} \|f\|_{L^p(\mathbb{R}^n)} \|g\|_{L^q(\mathbb{R}^n)}, \tag{1.8}$$

with $\frac{1}{r} = \frac{1}{p} + \frac{1}{q}$ for all $\frac{1}{2} < r < \infty$, $1 < p, q < \infty$.

Compare with the classical compact results in Theorem A, for the compactness of bilinear operators, recently, Bényi et al. [1] studied the compactness for commutators of bilinear Calderón-Zygmund singular integral operators. Bényi and Torres [2] also considered compactness properties of commutators of bilinear fractional integrals. We first give the definition of the compactness for a bilinear or bi-sublinear operator.

Definition 1.2. Let $B_{r,X} = \{x \in X : \|x\| \le r\}$ be the closed ball of radius r centered at the origin in the normed space X. A bilinear or bi-sublinear operator $T : X \times Y \to Z$ is called compact if $T(B_{1,X} \times B_{1,Y})$ is precompact in Z.

It is natural to ask whether the compact results still hold for the maximal commutators $T_{*,b,i}$, $T_{*,(b_1,b_2)}$ of the bilinear singular integral operators or not.

The main purpose of the present paper is to show that the maximal commutators $T_{*,b,1}$, $T_{*,b,2}$ and $T_{*,(b_1,b_2)}$ of bilinear Calderón-Zygmund operators are all the compact operators on L^p if the b, b_1, $b_2 \in CMO(\mathbb{R}^n)$.

Theorem 1.1. *Let $1 \leq r < \infty$, $1 < p, q < \infty$ with $\frac{1}{r} = \frac{1}{p} + \frac{1}{q}$. For any $f \in L^p(\mathbb{R}^n)$, $g \in L^q(\mathbb{R}^n)$, let $T_{*,b,1}(f,g)$, $T_{*,b,2}(f,g)$ be defined in* (1.4) and (1.5). If $b \in CMO(\mathbb{R}^n)$, then $T_{*,b,1}$ and $T_{*,b,2}$ are compact operators from $L^p(\mathbb{R}^n) \times L^q(\mathbb{R}^n)$ to $L^r(\mathbb{R}^n)$.

Theorem 1.2. *Let $1 \leq r < \infty$, $1 < p, q < \infty$ with $\frac{1}{r} = \frac{1}{p} + \frac{1}{q}$. For any $f \in L^p(\mathbb{R}^n)$, $g \in L^q(\mathbb{R}^n)$, let $T_{*,(b_1,b_2)}(f,g)$ be defined in* (1.6). If b_1, $b_2 \in CMO(\mathbb{R}^n)$, then $T_{*,(b_1,b_2)}$ is a compact operator from $L^p(\mathbb{R}^n) \times L^q(\mathbb{R}^n)$ to $L^r(\mathbb{R}^n)$.

Remark 1.3. Theorems 1.1 and 1.2 also hold for m-linear maximal Calderón-Zygmund singular integral operators (including linear case $m = 1$). The essential ideas in the proof are similar, of course, with more complicated and delicate division in the main steps of the proof in Section 2,3. For simplicity, we omit the proof.

Remark 1.4. The compact results are new even for the linear case which can be proved by using similar ideas and steps as in the proof of Theorem 1.1.

2 The proof of Theorem 1.1

Before proving Theorem 1.1, we first give the following lemmas.

Lemma 2.1 (Frechet-Kolmogorov, [21]). *A subset G of $L^p(\mathbb{R}^n)(1 \leq p < \infty)$ is strongly pre-compact if and only if G satisfies the following conditions:*

(a) $\sup_{f \in G} \|f\|_p < \infty$;

(b) $\lim_{\alpha \to \infty} \|f\chi_{E_\alpha}\|_p = 0$, *uniformly for $f \in G$, where $E_\alpha = \{x \in \mathbb{R}^n : |x| > \alpha\}$;*

(c) $\lim_{|h| \to 0} \|f(\cdot + h) - f(\cdot)\|_p = 0$, *uniformly for $f \in G$.*

Lemma 2.2. *For any $\delta > 0$, $0 < \epsilon < \frac{1}{2}$, we have the following inequalities*

$$\iint_{\frac{\delta}{1+2\epsilon} \leq |y|+|z| \leq \delta} \frac{dydz}{(|y| + |z|)^{2n}} \leq C[1 - (1 + 2\epsilon)^{-n}], \tag{2.1}$$

$$\iint_{\delta \leq |y|+|z| \leq \frac{\delta}{1-2\epsilon}} \frac{dydz}{(|y| + |z|)^{2n}} \leq C[(1 - 2\epsilon)^{-n} - 1], \tag{2.2}$$

where the constant C is independent of δ and ϵ.

Proof. We first give the estimate for (2.1). Simple computation and spherical coordinates transformations give that

$$\begin{aligned} &\iint_{\frac{\delta}{1+2\epsilon} \leq |y|+|z| \leq \delta} \frac{dydz}{(|y| + |z|)^{2n}} \\ \leq &\int_{|z| \leq \frac{\delta}{1+2\epsilon}} \int_{\frac{\delta}{1+2\epsilon} - |z| \leq |y| \leq \delta - |z|} \frac{dydz}{(|y| + |z|)^{2n}} + \int_{\frac{\delta}{1+2\epsilon} \leq |z| \leq \delta} \int_{|y| \leq \delta - |z|} \frac{dydz}{(|y| + |z|)^{2n}} \end{aligned}$$

$$\leq C\int_{|z|\leq\frac{\delta}{1+2\epsilon}}\int_{\frac{\delta}{1+2\epsilon}-|z|}^{\delta-|z|}\frac{r^{n-1}}{(r+|z|)^{2n}}drdz+C\int_{\frac{\delta}{1+2\epsilon}\leq|z|\leq\delta}\int_0^{\delta-|z|}\frac{r^{n-1}}{(r+|z|)^{2n}}drdz$$
$$\leq C\int_{|z|\leq\frac{\delta}{1+2\epsilon}}[\delta^{-n}(1+2\epsilon)^n-\delta^{-n}]dz+C\int_{\frac{\delta}{1+2\epsilon}\leq|z|\leq\delta}(|z|^{-n}-\delta^{-n})dz$$
$$\leq C[1-(1+2\epsilon)^{-n}]+C\delta^{-n}[\delta^n-\delta^n(1+2\epsilon)^{-n}]\leq C[1-(1+2\epsilon)^{-n}].$$

Analogously, we also can obtain (2.2).

$$\iint_{\delta\leq|y|+|z|\leq\frac{\delta}{1-2\epsilon}}\frac{dydz}{(|y|+|z|)^{2n}}$$
$$\leq\int_{|z|\leq\delta}\int_{\delta-|z|\leq|y|\leq\frac{\delta}{1-2\epsilon}-|z|}\frac{dydz}{(|y|+|z|)^{2n}}+\int_{\delta\leq|z|\leq\frac{\delta}{1-2\epsilon}}\int_{|y|\leq\frac{\delta}{1-2\epsilon}-|z|}\frac{dydz}{(|y|+|z|)^{2n}}$$
$$\leq C\int_{|z|\leq\delta}[\delta^{-n}-\delta^{-n}(1-2\epsilon)^n]dz+C\int_{\delta\leq|z|\leq\frac{\delta}{1-2\epsilon}}(|z|^{-n}-\delta^{-n}(1-2\epsilon)^n)dz$$
$$\leq C[1-(1-2\epsilon)^n]+C\delta^{-n}[\delta^n(1-2\epsilon)^{-n}-\delta^n]\leq C[(1-2\epsilon)^{-n}-1].$$

We therefore complete the proof of the lemma. □

Proof of Theorem 1.1. We only prove $i=1$. Without loss of generality, let B_1, B_2 be unit balls in $L^p(\mathbb{R}^n)$ and $L^q(\mathbb{R}^n)$, respectively. We need to show the set $\{T_{*,b,1}(f,g): f\in B_1,\, g\in B_2\}$ is strongly pre-compact in $L^r(\mathbb{R}^n)$ with $b\in CMO(\mathbb{R}^n)$.

We first show that if the set $\{T_{*,b,1}(f,g): f\in B_1,\, g\in B_2\}$ is strongly pre-compact in $L^r(\mathbb{R}^n)$ for $b\in C_c^\infty(\mathbb{R}^n)$, then the set $\{T_{*,b,1}(f,g): f\in B_1,\, g\in B_2\}$ is also strongly pre-compact in $L^r(\mathbb{R}^n)$ for $b\in CMO$. In fact, suppose that $b\in CMO$, then for any $\epsilon>0$, there exists $b^\epsilon\in C_c^\infty(\mathbb{R}^n)$ such that $\|b-b^\epsilon\|_{BMO}<\epsilon$. It is easy to see that

$$|T_{*,b,1}(f,g)(x)-T_{*,b^\epsilon,1}(f,g)(x)|$$
$$\leq\sup_{\delta>0}\left|\iint_{|x-y|+|x-z|>\delta}\Big[(b(y)-b(x))-(b^\epsilon(y)-b^\epsilon(x))\Big]K(x,y,z)f(y)g(z)\,dydz\right|$$
$$\leq T_{*,b-b^\epsilon,1}(f,g)(x).$$

Then combine with the above inequality and (1.8), we have

$$\begin{aligned}\|T_{*,b,1}(f,g)-T_{*,b^\epsilon,1}(f,g)\|_r&\leq\|T_{*,b-b^\epsilon,1}(f,g)\|_r\\&\leq C\|b-b^\epsilon\|_{BMO}\|f\|_p\|g\|_q\leq C\epsilon.\end{aligned}\tag{2.3}$$

Denote $F_1:=\{T_{*,b^\epsilon,1}(f,g): f\in B_1,\, g\in B_2\}$, then (a), (b) and (c) in Lemma 2.1 hold for F_1. We need to show that (a), (b) and (c) also hold for the set $\widetilde{F}_1:=\{T_{*,b,1}(f,g): f\in B_1,\, g\in B_2\}$. (2.3) gives that

$$\sup_{f\in B_1,g\in B_2}\|T_{*,b,1}(f,g)\|_r\leq\sup_{f\in B_1,g\in B_2}\|T_{*,b^\epsilon,1}(f,g)\|_r+C\epsilon<\infty.\tag{2.4}$$

On the other hand,

$$\begin{aligned}\lim_{\alpha\to\infty}\|T_{*,b,1}(f,g)\chi_{E_\alpha}\|_r \le& \lim_{\alpha\to\infty}\|T_{*,b^\epsilon,1}(f,g)\chi_{E_\alpha}\|_r + \|T_{*,b-b^\epsilon,1}(f,g)\|_r\\ \le& C\epsilon\to 0 \quad (\epsilon\to 0).\end{aligned}\tag{2.5}$$

$$\begin{aligned}&\lim_{|h|\to 0}\|T_{*,b,1}(f,g)(\cdot+h)-T_{*,b,1}(f,g)(\cdot)\|_r\\ \le&\lim_{|h|\to 0}\|T_{*,b^\epsilon,1}(f,g)(\cdot+h)-T_{*,b^\epsilon,1}(f,g)(\cdot)\|_r+2\|T_{*,b-b^\epsilon,1}(f,g)\|_r\\ \le& C\epsilon\to 0 \quad (\epsilon\to 0).\end{aligned}\tag{2.6}$$

It is obvious to see that the above limits hold uniformly in $\widetilde{F}_1$. Therefore, we know $\widetilde{F}_1$ is strongly pre-compact in $L^r(\mathbb{R}^n)$ for $b\in CMO$. Thus, to prove Theorem 1.1, it suffices to verify that the set $F_1 := \{T_{*,b,1}(f,g) : f\in B_1,\ g\in B_2\}$ is strongly pre-compact in $L^r(\mathbb{R}^n)$ for $b\in C_c^\infty(\mathbb{R}^n)$. By Lemma 2.1, we need only to prove (a), (b) and (c) hold uniformly in F_1.

For (a), by (1.8), we easily obtain

$$\sup_{f\in B_1, g\in B_2}\|T_{*,b,1}(f,g)\|_r \le C\|b\|_{BMO}\sup_{f\in B_1, g\in B_2}\|f\|_p\|g\|_q < C<\infty. \tag{2.7}$$

Notice $b\in C_c^\infty(\mathbb{R}^n)$, without loss of generality, we can assume that $\operatorname{supp} b\subset\{x\in\mathbb{R}^n : |x|\le\beta\}$ with $\beta>1$. For any $\epsilon>0$, $0<s<\frac{n}{q}$, we take $\alpha>2\beta$ such that $\alpha^{-n-s+n/r}<\epsilon$, then we have

$$\|T_{*,b,1}(f,g)\chi_{E_\alpha}\|_r < C\epsilon. \tag{2.8}$$

In fact, combine with the support set of b and notice that $K\in 2\text{-}CZK(A,\gamma)$, for $|x|>\alpha$, Hölder's inequality gives

$$\begin{aligned}&T_{*,b,1}(f,g)(x)\\ \le& C\int_{|y|\le\beta}\frac{|b(y)|}{|x-y|^{n+s}}|f(y)|\int_{\mathbb{R}^n}\frac{|g(z)|}{(|x-y|+|x-z|)^{n-s}}\,dzdy\\ \le& C\|b\|_\infty\int_{|y|\le\beta}\frac{|f(y)|}{|x-y|^{n+s}}\,dy\Big(\int_{\mathbb{R}^n}\frac{1}{(1+|x-z|)^{(n-s)q'}}\,dz\Big)^{1/q'}\|g\|_q\\ \le& C\beta^{\frac{n}{p'}}|x|^{-n-s}\|f\|_p\|g\|_q\le C|x|^{-n-s}.\end{aligned}$$

Thus, we have

$$\Big(\int_{\mathbb{R}^n}|T_{*,b,1}(f,g)(x)|^r\chi_{E_\alpha}(x)\,dx\Big)^{1/r}\le C\Big(\int_{|x|>\alpha}|x|^{-(n+s)r}\,dx\Big)^{\frac{1}{r}}\le C\epsilon.$$

That is, (2.8) holds uniformly for $f\in B_1, g\in B_2$.

It remains to prove that (c) holds also for $T_{*,b,1}(f,g)$ uniformly with $f\in B_1, g\in B_2$. That is, we need to verify that for any $0<\epsilon<\frac{1}{4}$, if $|h|$ is sufficiently small and dependent only on ϵ, then

$$\|T_{*,b,1}(f,g)(\cdot+h)-T_{*,b,1}(f,g)(\cdot)\|_r < C\epsilon \tag{2.9}$$

holds uniformly for $f \in B_1, g \in B_2$.

In fact, for any $h \in \mathbb{R}^n$, we denote $\widetilde{K}_\delta(x,y,z) = K(x,y,z)\chi_{|x-y|+|x-z|>\delta}$, then

$$\begin{aligned}
&|T_{*,b,1}(f,g)(x+h) - T_{*,b,1}(f,g)(x)| \\
&\leq \sup_{\delta>0}\left| \int_{\mathbb{R}^n}\int_{\mathbb{R}^n} \widetilde{K}_\delta(x+h,y,z)(b(y)-b(x+h))f(y)g(z)dydz \right. \qquad (2.10)\\
&\left. - \int_{\mathbb{R}^n}\int_{\mathbb{R}^n} \widetilde{K}_\delta(x,y,z)(b(y)-b(x))f(y)g(z)dydz \right|.
\end{aligned}$$

We can control the right hand side of the above inequality by the sum of the following four terms:

$$\begin{aligned}
J_1 &:= \sup_{\delta>0}\left| \iint_{|x-y|+|x-z|>\epsilon^{-1}|h|} \widetilde{K}_\delta(x,y,z)(b(x)-b(x+h))f(y)g(z)dydz \right|,\\
J_2 &:= \sup_{\delta>0}\left| \iint_{|x-y|+|x-z|>\epsilon^{-1}|h|} (\widetilde{K}_\delta(x+h,y,z)-\widetilde{K}_\delta(x,y,z)) \right.\\
&\quad \left. \times (b(y)-b(x+h))f(y)g(z)dydz \right|,\\
J_3 &:= \sup_{\delta>0}\left| \iint_{|x-y|+|x-z|\leq\epsilon^{-1}|h|} \widetilde{K}_\delta(x,y,z)(b(y)-b(x))f(y)g(z)dydz \right|,\\
J_4 &:= \sup_{\delta>0}\left| \iint_{|x-y|+|x-z|\leq\epsilon^{-1}|h|} \widetilde{K}_\delta(x+h,y,z)(b(y)-b(x+h))f(y)g(z)dydz \right|.
\end{aligned}$$

We will give the estimates for J_1, J_2, J_3, J_4, respectively in the following.

Estimate for J_1. It is easy to see that

$$\begin{aligned}
J_1 \leq& |b(x+h)-b(x)| \sup_{\delta>0}\left| \iint_{|x-y|+|x-z|>\epsilon^{-1}|h|} \widetilde{K}_\delta(x,y,z)f(y)g(z)\, dydz \right|\\
\leq& |h|\|\nabla b\|_\infty \sup_{\delta>0}\left| \iint_{\substack{|x-y|+|x-z|>\epsilon^{-1}|h|\\ |x-y|+|x-z|>\delta}} K(x,y,z)f(y)g(z)\, dydz \right|\\
\leq& C|h|T_*(f,g)(x).
\end{aligned}$$

Applying (1.3), we obtain

$$\|J_1\|_r \leq C|h|. \qquad (2.11)$$

Estimate for J_2. Notice that

$$\begin{aligned}
J_2 \leq& \sup_{\delta>0}\left| \iint_{|x-y|+|x-z|>\epsilon^{-1}|h|} (K(x+h,y,z)-K(x,y,z))\chi_{|x+h-y|+|x+h-z|>\delta} \right.\\
&\left. \times (b(y)-b(x+h))f(y)g(z)dydz \right|\\
&+ \sup_{\delta>0}\left| \iint_{|x-y|+|x-z|>\epsilon^{-1}|h|} K(x,y,z)(\chi_{|x+h-y|+|x+h-z|>\delta} - \chi_{|x-y|+|x-z|>\delta}) \right.
\end{aligned}$$

$$\times (b(y) - b(x+h))f(y)g(z)dydz\Big|$$

$$:= J_{21} + J_{22}.$$

Observe that if $|x-y|+|x-z| > \epsilon^{-1}|h|$ and $0 < \epsilon < 1/4$, then $|h| \le \frac{1}{2}\max\{|x-y|, |x-z|, |y-z|\}$. Since $K \in 2\text{-}CZK(A,\gamma)$, we have

$$J_{21} \le C\|b\|_\infty \iint_{|x-y|+|x-z|>\epsilon^{-1}|h|} \frac{|h|^\gamma}{(|x-y|+|x-z|)^{2n+\gamma}} |f(y)g(z)|\, dydz.$$

Minkowski's inequality and Hölder's inequality give that

$$\|J_{21}\|_r \le C \iint_{|y|+|z|>\epsilon^{-1}|h|} \frac{|h|^\gamma}{(|y|+|z|)^{2n+\gamma}} \left(\int_{\mathbb{R}^n} |f(x-y)g(x-z)|^r dx\right)^{\frac{1}{r}} dydz$$
$$\le C\epsilon^\gamma.$$

For J_{22}, it is easy to see that

$$J_{22} \le \sup_{\delta>0} \Bigg| \iint_{\substack{|x-y|+|x-z|>\epsilon^{-1}|h| \\ |x+h-y|+|x+h-z|>\delta \\ |x-y|+|x-z|\le\delta}} K(x,y,z)(b(y)-b(x+h))f(y)g(z)dydz \Bigg|$$
$$+ \sup_{\delta>0} \Bigg| \iint_{\substack{|x-y|+|x-z|>\epsilon^{-1}|h| \\ |x+h-y|+|x+h-z|\le\delta \\ |x-y|+|x-z|>\delta}} K(x,y,z)(b(y)-b(x+h))f(y)g(z)dydz \Bigg|$$
$$:= J_{221} + J_{222}.$$

For J_{221}, as $|x-y|+|x-z| > \epsilon^{-1}|h|$, $|x+h-y|+|x+h-z| > \delta$ and $0 < \epsilon < 1/4$, then $|x-y|+|x-z| \ge \frac{1}{2\epsilon+1}(|x+h-y|+|x+h-z|) \ge \frac{\delta}{2\epsilon+1}$. Then for any $1 < r_0 < \min\{p,q\}$, Hölder's inequality and (2.1) give that

$$J_{221} \le C\|b\|_\infty \sup_{\delta>0} \iint_{\frac{\delta}{2\epsilon+1}\le|x-y|+|x-z|\le\delta} \frac{|f(y)g(z)|}{(|x-y|+|x-z|)^{2n}} dydz$$
$$\le C \sup_{\delta>0} \left(\iint_{\frac{\delta}{2\epsilon+1}\le|y|+|z|\le\delta} \frac{|f(x-y)g(x-z)|^{r_0}}{(|y|+|z|)^{2n}} dydz\right)^{\frac{1}{r_0}}$$
$$\times \sup_{\delta>0} \left(\iint_{\frac{\delta}{2\epsilon+1}\le|y|+|z|\le\delta} \frac{dydz}{(|y|+|z|)^{2n}}\right)^{\frac{1}{r_0'}}$$
$$\le C \sup_{\delta>0} \left((2\epsilon+1)^{2n}\delta^{-2n} \iint_{|y|+|z|\le\delta} |f(x-y)g(x-z)|^{r_0}\, dydz\right)^{\frac{1}{r_0}}$$
$$\times [1-(2\epsilon+1)^{-n}]^{\frac{1}{r_0'}}$$
$$\le C\epsilon^{\frac{1}{r_0'}} M(|f|^{r_0})(x)^{\frac{1}{r_0}} M(|g|^{r_0})(x)^{\frac{1}{r_0}}.$$

Notice that $\frac{1}{r} = \frac{1}{p} + \frac{1}{q}$, and $\frac{p}{r_0}, \frac{q}{r_0} > 1$. Thus, Hölder's inequality and the L^p-boundedness of Hardy-Littlewood maximal operator M give

$$\|J_{221}\|_r \le C\epsilon^{\frac{1}{r_0'}} \|M(|f|^{r_0})^{\frac{1}{r_0}} M(|g|^{r_0})^{\frac{1}{r_0}}\|_r \le C\epsilon^{\frac{1}{r_0'}} \|M(|f|^{r_0})^{\frac{1}{r_0}}\|_p \|M(|g|^{r_0})^{\frac{1}{r_0}}\|_q$$

$$\leq C\epsilon^{\frac{1}{r_0'}}\||f|^{r_0}\|_{p/r_0}^{\frac{1}{r_0}}\||g|^{r_0}\|_{q/r_0}^{\frac{1}{r_0}} \leq C\epsilon.$$

The estimate for J_{222} is completely similar. It is easy to see that as $|x-y|+|x-z| > \epsilon^{-1}|h|$, $|x+h-y|+|x+h-z| \leq \delta$ and $0<\epsilon<1/4$, then $|x-y|+|x-z| \leq \frac{1}{1-2\epsilon}(|x+h-y|+|x+h-z|) \leq \frac{\delta}{1-2\epsilon}$. Therefore, analogous to J_{221}, using Hölder's inequality and (2.2), then for any $1<r_0<\min\{p,q\}$,

$$\begin{aligned}
J_{222} \leq& C\|b\|_\infty \sup_{\delta>0}\iint_{\delta\leq|x-y|+|x-z|\leq\frac{\delta}{1-2\epsilon}} \frac{|f(y)g(z)|}{(|x-y|+|x-z|)^{2n}}\,dydz\\
\leq& CM(|f|^{r_0})(x)^{\frac{1}{r_0}}M(|g|^{r_0})(x)^{\frac{1}{r_0}}[(1-2\epsilon)^{-n}-1]^{\frac{1}{r_0'}}\\
\leq& C\epsilon^{\frac{1}{r_0'}}M(|f|^{r_0})(x)^{\frac{1}{r_0}}M(|g|^{r_0})(x)^{\frac{1}{r_0}}.
\end{aligned}$$

Thus, we also can obtain

$$\|J_{222}\|_r \leq C\epsilon.$$

Combine with the estimates for J_{21}, J_{221} and J_{222}, then

$$\|J_2\|_r \leq C\epsilon. \tag{2.12}$$

Estimate for J_3**.** Note that $|b(x)-b(y)| \leq \|\nabla b\|_\infty|x-y|$ and $K \in 2\text{-}CZK(A,\gamma)$, we have

$$J_3 \leq C\|\nabla b\|_\infty \iint_{|x-y|+|x-z|\leq\epsilon^{-1}|h|} \frac{|f(y)g(z)|}{(|x-y|+|x-z|)^{2n-1}}\,dydz.$$

Therefore, Minkowski's inequality and Hölder's inequality give that

$$\begin{aligned}
\|J_3\|_r \leq& C\iint_{|y|+|z|\leq\epsilon^{-1}|h|}\left(\int_{\mathbb{R}^n}|f(x-y)g(x-z)|^r\,dx\right)^{\frac{1}{r}}\frac{dydz}{(|y|+|z|)^{2n-1}}\\
\leq& C\epsilon^{-1}|h|.
\end{aligned} \tag{2.13}$$

Estimate for J_4**.** With the same way, we have

$$\|J_4\|_r \leq C(2+\epsilon^{-1})|h|. \tag{2.14}$$

Note that the constants C in (2.11)–(2.14) are independent of h and ϵ. Taking $|h|$ to be sufficiently small, we obtain (2.9). Therefore, (c) holds for $T_{*,b,1}(f,g)$ uniformly for $f\in B_1, g\in B_2$. We complete the proof.

3 The proof of Theorem 1.2

First note that $T_{*,(b_1,b_2)}$ is bounded from $L^p(\mathbb{R}^n)\times L^q(\mathbb{R}^n)$ to $L^r(\mathbb{R}^n)$ by (1.7). Let B_1, B_2 be unit balls in $L^p(\mathbb{R}^n)$ and $L^q(\mathbb{R}^n)$, respectively. We need to prove that the set $G=\{T_{*,(b_1,b_2)}(f,g): f\in B_1, g\in B_2\}$ is strongly pre-compact in $L^r(\mathbb{R}^n)$. Notice that

$$|T_{*,(b_1,b_2)}(f,g)(x)-T_{*,(b_1^\epsilon,b_2^\epsilon)}(f,g)(x)|$$

$$\leq \sup_{\delta>0}\Big|\iint_{|x-y|+|x-z|>\delta}\Big[(b_1(y)-b_1(x))(b_2(z)-b_2(x))$$
$$-(b_1^\epsilon(y)-b_1^\epsilon(x))(b_2^\epsilon(z)-b_2^\epsilon(x))\Big]K(x,y,z)f(y)g(z)\,dydz\Big|$$
$$\leq T_{*,(b_1-b_1^\epsilon,b_2)}(f,g)(x)+T_{*,(b_1^\epsilon,b_2-b_2^\epsilon)}(f,g)(x).$$

Therefore, if $b_j^\epsilon \in C_c^\infty(\mathbb{R}^n)$, such that $\|b_j-b_j^\epsilon\|_{BMO}<\epsilon$ $(j=1,2)$, then (1.7) gives that

$$\|T_{*,(b_1,b_2)}(f,g)-T_{*,(b_1^\epsilon,b_2^\epsilon)}(f,g)\|_r$$
$$\leq \|T_{*,(b_1-b_1^\epsilon,b_2)}(f,g)\|_r+\|T_{*,(b_1^\epsilon,b_2-b_2^\epsilon)}(f,g)\|_r$$
$$\leq C(\|b_2\|_{BMO}\|b_1-b_1^\epsilon\|_{BMO}+\|b_1^\epsilon\|_{BMO}\|b_2-b_2^\epsilon\|_{BMO})\|f\|_p\|g\|_q\leq C\epsilon. \tag{3.1}$$

Thus, we only need to prove that G is strongly pre-compact in $L^r(\mathbb{R}^n)$ for $b_j\in C_c^\infty(\mathbb{R}^n)$. By Lemma 2.1, we only need to show that (a), (b) and (c) in Lemma 2.1 hold for $T_{*,(b_1,b_2)}(f,g)$ uniformly in G with $b_1,b_2\in C_c^\infty(\mathbb{R}^n)$.

By (1.7), we easily obtain that (a) holds for $T_{*,(b_1,b_2)}(f,g)$ uniformly with $f\in B_1, g\in B_2$. Notice $b_j\in C_c^\infty(\mathbb{R}^n)$, without loss of generality, we can also assume that $\operatorname{supp} b_j\subset\{x\in\mathbb{R}^n : |x|\leq\beta\}$ with $\beta>1$. For any $\epsilon>0$, we take $\alpha>2\beta$ such that $(\alpha-\beta)^{-2n+n/r}<\epsilon$, then we have

$$\Big(\int_{\mathbb{R}^n}|T_{*,(b_1,b_2)}(f,g)(x)|^r\chi_{E_\alpha}(x)\,dx\Big)^{\frac{1}{r}}$$
$$\leq C\Big(\int_{E_\alpha}\Big(\int_{B(0,\beta)}\int_{B(0,\beta)}\frac{|b_1(y)||b_2(z)|}{(|x-y|+|x-z|)^{2n}}|f(y)||g(z)|\,dydz\Big)^r dx\Big)^{\frac{1}{r}}$$
$$\leq C\|b_1\|_\infty\|b_2\|_\infty\int_{B(0,\beta)}\int_{B(0,\beta)}\Big(\int_{|x|>\alpha-\beta}\frac{dx}{|x|^{2nr}}\Big)^{1/r}|f(y)||g(z)|\,dydz$$
$$\leq C\|b_1\|_\infty\|b_2\|_\infty\beta^{n(2-1/r)}\epsilon, \tag{3.2}$$

which shows that (b) holds for $T_{*,(b_1,b_2)}(f,g)$ uniformly with $f\in B_1, g\in B_2$. It remains to prove that for any $0<\epsilon<1/4$, if $|h|$ is sufficiently small and dependent only on ϵ, then

$$\|T_{*,(b_1,b_2)}(f,g)(\cdot+h)-T_{*,(b_1,b_2)}(f,g)(\cdot)\|_r<C\epsilon \tag{3.3}$$

holds uniformly for $f\in B_1, g\in B_2$.

In fact, for any $h\in\mathbb{R}^n$, we denote $\vec{b}(x,y,z)=(b_1(y)-b_1(x))(b_2(z)-b_2(x))$ and $\vec{b}_h(x,y,z)=\vec{b}(x+h,y,z)-\vec{b}(x,y,z)$, using the notation in the proof of Theorem 1.1, we have

$$|T_{*,(b_1,b_2)}(f,g)(x+h)-T_{*,(b_1,b_2)}(f,g)(x)|$$
$$\leq\sup_{\delta>0}\Big|\int_{\mathbb{R}^n}\int_{\mathbb{R}^n}\widetilde{K}_\delta(x+h,y,z)\vec{b}(x+h,y,z)f(y)g(z)dydz$$
$$-\int_{\mathbb{R}^n}\int_{\mathbb{R}^n}\widetilde{K}_\delta(x,y,z)\vec{b}(x,y,z)f(y)g(z)dydz\Big|.$$

Similar to the decomposition for (2.10), we can control the right hand side of the above inequality by $L_1 + L_2 + L_3 + L_4$, where

$$L_1 := \sup_{\delta>0} \left| \iint_{|x-y|+|x-z|>\epsilon^{-1}|h|} \widetilde{K}_\delta(x,y,z)\vec{b}_h(x,y,z)f(y)g(z)dydz \right|,$$

$$L_2 := \sup_{\delta>0} \left| \iint_{|x-y|+|x-z|>\epsilon^{-1}|h|} (\widetilde{K}_\delta(x+h,y,z) - \widetilde{K}_\delta(x,y,z))\vec{b}(x+h,y,z)f(y)g(z)dydz \right|,$$

$$L_3 := \sup_{\delta>0} \left| \iint_{|x-y|+|x-z|\le\epsilon^{-1}|h|} \widetilde{K}_\delta(x,y,z)\vec{b}(x,y,z)f(y)g(z)dydz \right|,$$

$$L_4 := \sup_{\delta>0} \left| \iint_{|x-y|+|x-z|\le\epsilon^{-1}|h|} \widetilde{K}_\delta(x+h,y,z)\vec{b}(x+h,y,z)f(y)g(z)dydz \right|.$$

In the following, we will give the estimates for L_1, L_2, L_3, L_4, respectively.
Estimate for L_1. Observe

$$\begin{aligned} \vec{b}_h(x,y,z) = &(b_1(x) - b_1(x+h))(b_2(x) - b_2(x+h)) \\ &+ (b_1(x) - b_1(x+h))(b_2(z) - b_2(x)) \\ &+ (b_1(y) - b_1(x))(b_2(x) - b_2(x+h)). \end{aligned} \tag{3.4}$$

Note that $|b_j(x) - b_j(x+h)| \le |h|\|\nabla b_j\|_\infty$, then we have

$$\begin{aligned} &L_1 \\ \le&|h|^2\|\nabla b_1\|_\infty\|\nabla b_2\|_\infty \sup_{\delta>0} \left| \iint_{|x-y|+|x-z|>\epsilon^{-1}|h|} \widetilde{K}_\delta(x,y,z)f(y)g(z)\,dydz \right| \\ &+ |h|\|\nabla b_1\|_\infty \sup_{\delta>0} \left| \iint_{|x-y|+|x-z|>\epsilon^{-1}|h|} \widetilde{K}_\delta(x,y,z)(b_2(z) - b_2(x))f(y)g(z)\,dydz \right| \\ &+ |h|\|\nabla b_2\|_\infty \sup_{\delta>0} \left| \iint_{|x-y|+|x-z|>\epsilon^{-1}|h|} \widetilde{K}_\delta(x,y,z)(b_1(y) - b_1(x))f(y)g(z)\,dydz \right| \\ \le&C|h|^2T_*(f,g)(x) + C|h|T_{*,b_2,2}(f,g)(x) + C|h|T_{*,b_1,1}(f,g)(x). \end{aligned}$$

Then (1.3) and (1.8) give that

$$\|L_1\|_r \le C(|h|^2 + |h|\|b_1\|_{BMO} + |h|\|b_2\|_{BMO})\|f\|_p\|g\|_q \le C(|h| + |h|^2). \tag{3.5}$$

Estimate for L_2. The estimate for L_2 is similar to J_2. We have

$$\begin{aligned} L_2 \le&\sup_{\delta>0} \left| \iint_{|x-y|+|x-z|>\epsilon^{-1}|h|} (K(x+h,y,z) - K(x,y,z))\chi_{|x+h-y|+|x+h-z|>\delta} \right. \\ &\left. \times \vec{b}(x+h,y,z)f(y)g(z)dydz \right| \\ &+ \sup_{\delta>0} \left| \iint_{\substack{|x-y|+|x-z|>\epsilon^{-1}|h| \\ |x+h-y|+|x+h-z|>\delta \\ |x-y|+|x-z|\le\delta}} K(x,y,z)\vec{b}(x+h,y,z)f(y)g(z)dydz \right| \end{aligned}$$

$$+\sup_{\delta>0}\left|\iint_{\substack{|x-y|+|x-z|>\epsilon^{-1}|h| \\ |x+h-y|+|x+h-z|\le\delta \\ |x-y|+|x-z|>\delta}} K(x,y,z)\vec{b}(x+h,y,z)f(y)g(z)dydz\right|$$
$$:=L_{21}+L_{22}+L_{23}.$$

The estimates for L_{21}, L_{22}, L_{23} are completely analogous to J_{21}, J_{221}, J_{222}. Then,

$$\|L_2\|_r \le C\epsilon. \tag{3.6}$$

Estimate for L_3. Note that $|b_j(x)-b_j(y)| \le \|\nabla b_j\|_\infty |x-y|$ $(j=1,2)$ and $K \in 2$-$CZK(A,\gamma)$, we have

$$L_3 \le C\|\nabla b_1\|_\infty \|\nabla b_2\|_\infty \iint_{|x-y|+|x-z|\le\epsilon^{-1}|h|} \frac{|f(y)g(z)}{(|x-y|+|x-z|)^{2n-2}}\,dydz.$$

Therefore, by Minkowski's inequality and Hölder's inequality, we have

$$\begin{aligned}&\|L_3\|_r\\ \le& C\|\nabla b_1\|_\infty \|\nabla b_2\|_\infty \iint_{|y|+|z|\le\epsilon^{-1}|h|}\left(\int_{\mathbb{R}^n}|f(x-y)g(x-z)|^r dx\right)^{\frac{1}{r}}\frac{dydz}{(|y|+|z|)^{2n-2}}\\ \le& C(\epsilon^{-1}|h|)^2\|f\|_p\|g\|_q \le C(\epsilon^{-1}|h|)^2.\end{aligned} \tag{3.7}$$

Estimate for L_4. With the same way, we have

$$\|L_4\|_r \le C(2+\epsilon^{-1})^2|h|^2. \tag{3.8}$$

Note that the constants C in (3.5)–(3.8) are independent of h and ϵ. Taking $|h|<(2+\epsilon^{-1})^{-1}\epsilon$, we obtain (3.3). Therefore, (c) holds for $T_{*,(b_1,b_2)}(f,g)$ uniformly for $f\in B_1, g\in B_2$. Hence, we complete the proof of Theorem 1.2.

References

[1] Á. Bényi, W. Damián, K. Moen, and R. Torres, Compactness properties of commutators of bilinear fractional integrals, Math. Z. **280** (2015), 569–582.

[2] Á. Bényi and R. Torres, Compact bilinear operators and commutators, Proc. Amer. Math. Soc. **141** (2013), 3609–3621.

[3] F. Beatrous and S.-Y. Li, On the boundedness and compactness of operators of Hankel type, J. Funct. Anal. **111** (1993), 350–379.

[4] X. Chen and Q. Xue, Weighted estimates for a class of multilinear fractional type operators, J. Math. Anal. Appl. **362** (2010), 355–373.

[5] Y. Chen and Y. Ding, Compactness of commutators of singular integrals with variable kernels (in chinese), Chinese Ann. Math. **30(A)** (2009), 201–212.

[6] Y. Chen and Y. Ding, Compactness of commutators for Littlewood-Paley operators, Kodai Math. Jour. **32** (2009), 256–323.

[7] Y. Chen and Y. Ding, Compactness of the commutators for parabolic singular integrals, Science in China (A) **53** (2010), 2633–2648.

[8] Y. Chen, Y. Ding, and X. Wang, Compactness of commutators of Riesz potential on Morrey spaces, Potential Anal. **30** (2009), 301–313.

[9] Y. Chen, Y. Ding, and X. Wang, Compactness for commutators of Marcinkiewicz integral in Morrey spaces, Taiwanese Jour. Math. **15** (2011), 633–658.

[10] Y. Chen, Y. Ding, and X. Wang, Compactness of commutators for singular integrals on Morrey spaces, Canadian Jour. Math. **64** (2012), 257–281.

[11] R. Coifman, R. Rochberg, and G. Weiss, Fractorization theorems for Hardy spaces in several variables, Ann. of Math. (2) **103** (1976), 611–635.

[12] C. Fefferman and E. M. Stein, H^p spaces of several variables, Acta Math. **129** (1972), 137–193.

[13] L. Grafakos and R. Torres, Multilinear Calderón-Zygmund theory, Adv. Math. **165** (2002), 124–164.

[14] L. Grafakos and R. Torres, Maximal operator and weighted norm inequalities for multilinear singular integrals, Indiana. Univ. Math. J. **51** (2002), 1261–1276.

[15] S. Krantzl and S.-Y. Li, Boundedness and compactness of integral operators on spaces of homogeneous type and applications, I. J. Math. Anal. Appl. **258** (2001), 629–641.

[16] S. Krantzl and S.-Y. Li, Boundedness and compactness of integral operators on spaces of homogeneous type and applications, II. J. Math. Anal. Appl. **258** (2001), 642–657.

[17] A. Lerner, S. Ombrosi, C. Pérez, R. Torres, and R. González, New maximal functions and multiple weights for the multilinear Calderón-Zygmund theory, Adv. Math. **220** (2009), 1222–1264.

[18] Q. Xue, Weighted estimates for the iterated commutators of multilinear maximal and fractional type operators, Studia Math. **217** (2013), 97–122.

[19] S. Wang, The compactness of the commutator of fractional integral operator (in Chinese), Chinese Ann. Math. **8(A)** (1987), 475–482.

[20] A. Uchiyama, On the compactness of operators of Hankel type, Tohoku Math. Journ. **30** (1978), 163–171.

[21] K. Yosida, Functional Analysis, Springer-Verlag, Berlin, 1995.

Some Topics in Harmonic Analysis
and Applications
ALM 34, pp. 177–202

Weak Hardy Spaces

Loukas Grafakos*† and Danqing He ‡

Abstract

We provide a careful treatment of the weak Hardy spaces $H^{p,\infty}(\mathbf{R}^n)$ for all indices $0 < p < \infty$. The study of these spaces presents differences from the study of the Hardy-Lorentz spaces $H^{p,q}(\mathbf{R}^n)$ for $q < \infty$, due to the lack of a good dense subspace of them. We obtain several properties of weak Hardy spaces and we discuss a new square function characterization for them, obtained by He [16].

2010 Mathematics Subject Classification: Primary 42B20; Secondary 46E35.

Keywords and Phrases: Besov spaces, Calderón-Zygmund operators.

1 Introduction

The impact of the theory of Hardy spaces in the last forty years has been significant. The higher dimensional Euclidean theory of Hardy spaces was developed by Fefferman and Stein [10] who proved a variety of characterizations for them. A deep atomic decomposition characterization of these spaces was given by Coifman [3] in dimension one and by Latter [18] in higher dimensions. The treatments of Hardy spaces in Lu [19], García-Cuerva and Rubio de Francia [11], Grafakos [13], Stein [24], and Triebel [25] cover the main aspects of their classical theory in the Euclidean setting. Among hundreds of references on this topic, the works of Coifman and Weiss [4], Macías and Segovia [22], Duong and Yan [7], Han, Müller and Yang [15], Hu, Yang and Zhou [17] contain powerful extensions of the theory of Hardy spaces to the setting of spaces of homogeneous type. A new type of Hardy space, called Herz-type Hardy space was introduced by Lu and Yang [21] to measure the localization fine-tuned on cubical shells centered at the origin.

*Grafakos' research partially supported by the NSF under grant DMS 0900946.

†Department of Mathematics University of Missouri Columbia, MO 65211, USA; *E-mail:* grafakosl@missouri.edu

‡Department of Mathematics University of Missouri Columbia, MO 65211, USA; *E-mail:* dhd27@mail.missouri.edu

In this work we provide a careful treatment of the weak Hardy space H^p (henceforth $H^{p,\infty}$) on $\mathbf{R}^n$ for $0 < p < \infty$. This is defined as the space of all bounded tempered distributions whose Poisson maximal function lies in weak L^p (or $L^{p,\infty}$). The study of these spaces presents certain crucial differences from the study of the Hardy-Lorentz spaces $H^{p,q}$ when $q < \infty$, due to the fact that they lack a good dense subspace of smooth functions such as the Schwartz class; this is explained in Theorem 6 and it was also observed by Fefferman and Soria [9]. As a consequence, certain results concerning these spaces cannot be proved by restricting attention to the Schwartz class. In this article we bypass this difficulty to obtain several maximal characterizations of weak Hardy spaces working directly from the definition and we prove an interpolation result for these spaces working with general bounded distributions.

It is well known that function spaces can be characterized in terms of Littlewood-Paley square function expressions. Such characterizations are given in Ding, Lu and Xue [6], Lu and Yang [20], Peetre [23], and Triebel [25], but we discuss here a new square function characterization of $H^{p,\infty}$ in terms of the Littlewood-Paley operators, a result that was recently obtained by He [16].

Fefferman, Riviere and Sagher [8], Fefferman and Soria [9], Alvarez [2], Abu-Shammala and Torchinsky [1] have obtained a variety of results concerning the weak H^p spaces. Fefferman, Riviere and Sagher [8] have studied interpolation between the Hardy-Lorentz spaces $H^{p,q}$. Fefferman and Soria [9] carefully investigated the space $H^{1,\infty}$, and they provided its atomic decomposition. Alvarez [2] provided the atomic decomposition of the spaces $H^{p,\infty}$ and she also studied the action of singular integrals on them. In [8], Fefferman, Riviere and Sagher obtained the spaces $H^{p,q}$ as an intermediate interpolation space of Hardy spaces, but their proof was only given for Schwartz functions which are not dense in $H^{p,\infty}$, and thus their proof contains an incomplete deduction in the case $q = \infty$. He [16] overcomes the technical issues arising from the lack of density of Schwartz functions via a Calderón-Zygmund type decomposition for general weak H^p distributions. Some results in the literature of weak Hardy spaces are based on the interpolation result in [8] and possibly on the assumption that locally integrable functions are dense in this space. Although the latter is unknown as of this writing, the former is possible and is explained here.

Our exposition builds the theory of weak Hardy spaces, starting from the classical definition of the Poisson maximal function. We discuss various maximal characterizations of these spaces and we state an interpolation theorem for $H^{p,\infty}$ from initial strong H^{p_0} and H^{p_1} estimates with $p_0 < p < p_1$. Using this interpolation result, the second author [16] has obtained a new square function characterization for the spaces $H^{p,\infty}$, which is presented here without proof. This characterization is based on a singular integral estimate for vector-valued weak H^p spaces. For this reason, we develop the theory of weak Hardy spaces in the vector-valued setting.

2 Relevant background

To introduce the vector-valued weak Hardy spaces we need a sequence of definitions given in this section. We denote by ℓ^2 the space $\ell^2(\mathbf{Z})$ of all square-integrable sequences and by $\ell^2(L)$ the finite-dimensional space of all sequences of length $L \in \mathbf{Z}^+$ with the ℓ^2 norm. We say that a sequence of distributions $\{f_j\}_j$ lies in $\mathcal{S}'(\mathbf{R}^n, \ell^2)$ if there are constants $C, M > 0$ such that for every $\varphi \in \mathcal{S}(\mathbf{R}^n)$ we have

$$\big\|\{\langle f_j, \varphi\rangle\}_j\big\|_{\ell^2} = \Big(\sum_j |\langle f_j, \varphi\rangle|^2\Big)^{1/2} \le C \sum_{|\alpha|,|\beta|\le M} \sup_{y\in\mathbf{R}^n} |y^\beta \partial^\alpha \varphi(y)|.$$

And this sequence of distributions $\vec{f} = \{f_j\}_j$ is called bounded if for any $\varphi \in \mathcal{S}(\mathbf{R}^n)$ we have

$$\big\|\{\varphi * f_j\}_j\big\|_{\ell^2} = \Big(\sum_j |\varphi * f_j|^2\Big)^{1/2} \in L^\infty(\mathbf{R}^n). \tag{1}$$

Let $a, b > 0$ and let Φ be a Schwartz function on $\mathbf{R}^n$.

Definition 1. For a sequence $\vec{f} = \{f_j\}_{j\in\mathbf{Z}}$ of tempered distributions on $\mathbf{R}^n$ we define the smooth maximal function of $\vec{f}$ with respect to Φ as

$$\boldsymbol{M}(\vec{f}; \Phi)(x) = \sup_{t>0} \big\|\{(\Phi_t * f_j)(x)\}_j\big\|_{\ell^2}.$$

We define the nontangential maximal function with aperture a of $\vec{f}$ with respect to Φ as

$$\boldsymbol{M}_a^*(\vec{f}; \Phi)(x) = \sup_{t>0} \sup_{\substack{y\in\mathbf{R}^n \\ |y-x|\le at}} \big\|\{(\Phi_t * f_j)(y)\}_j\big\|_{\ell^2}.$$

We also define the auxiliary maximal function

$$\boldsymbol{M}_b^{**}(\vec{f}; \Phi)(x) = \sup_{t>0} \sup_{y\in\mathbf{R}^n} \frac{\big\|\{(\Phi_t * f_j)(x-y)\}_j\big\|_{\ell^2}}{(1+t^{-1}|y|)^b}.$$

For a fixed positive integer N we define the grand maximal function of $\vec{f}$ as

$$\mathcal{M}_N(\vec{f}) = \sup_{\varphi\in\mathcal{F}_N} \boldsymbol{M}_1^*(\vec{f}; \varphi), \tag{2}$$

where

$$\mathcal{F}_N = \Big\{\varphi \in \mathcal{S}(\mathbf{R}^n) : \ \mathfrak{N}_N(\varphi) \le 1\Big\}, \tag{3}$$

and

$$\mathfrak{N}_N(\varphi) = \int_{\mathbf{R}^n} (1+|x|)^N \sum_{|\alpha|\le N+1} |\partial^\alpha \varphi(x)|\, dx$$

is the "norm" of φ. More generally, we define the "norm" of φ adapted to the pair $(x_0, R) \in \mathbf{R}^n \times \mathbf{R}^+$ by setting

$$\mathfrak{N}_N(\varphi; x_0, R) = \int_{\mathbf{R}^n} \Big(1 + \Big|\frac{x-x_0}{R}\Big|\Big)^N \sum_{|\alpha|\le N+1} R^{|\alpha|} |\partial^\alpha \varphi(x)|\, dx.$$

Note that $\mathfrak{N}_N(\varphi; 0, 1) = \mathfrak{N}_N(\varphi)$.

If the function Φ is the Poisson kernel, then the maximal functions $M(\vec{f};\Phi)$, $M_a^*(\vec{f};\Phi)$ and $M_b^{**}(\vec{f};\Phi)$ are well defined for sequences of bounded tempered distributions $\vec{f}=\{f_j\}_j$ in view of (1).

We note that the following simple inequalities

$$M(\vec{f};\Phi)\le M_a^*(\vec{f};\Phi)\le (1+a)^b M_b^{**}(\vec{f};\Phi) \tag{4}$$

are valid. We now define the vector-valued Hardy space $H^{p,\infty}(\mathbf{R}^n,\ell^2)$.

Definition 2. Let $\vec{f}=\{f_j\}_j$ be a sequence of bounded tempered distributions on $\mathbf{R}^n$ and let $0<p<\infty$. We say that $\vec{f}$ lies in the vector-valued weak Hardy space $H^{p,\infty}(\mathbf{R}^n,\ell^2)$ vector-valued Hardy space if the Poisson maximal function

$$M(\vec{f};P)(x)=\sup_{t>0}\big\|\{(P_t*f_j)(x)\}_j\big\|_{\ell^2}$$

lies in $L^{p,\infty}(\mathbf{R}^n)$. If this is the case, we set

$$\big\|\vec{f}\big\|_{H^{p,\infty}(\mathbf{R}^n,\ell^2)}=\big\|M(\vec{f};P)\big\|_{L^{p,\infty}(\mathbf{R}^n)}=\Big\|\sup_{\varepsilon>0}\Big(\sum_j|P_\varepsilon*f_j|^2\Big)^{\frac{1}{2}}\Big\|_{L^{p,\infty}(\mathbf{R}^n)}.$$

The next theorem provides a characterization of $H^{p,\infty}$ in terms of different maximal functions.

Theorem 1. *Let $0<p<\infty$. Then the following statements are valid:*

(a) *There exists a Schwartz function Φ with integral 1 and a constant C_1 such that*

$$\big\|M(\vec{f};\Phi)\big\|_{L^{p,\infty}(\mathbf{R}^n,\ell^2)}\le C_1\big\|\vec{f}\big\|_{H^{p,\infty}(\mathbf{R}^n,\ell^2)} \tag{5}$$

for every sequence $\vec{f}=\{f_j\}_j$ of tempered distributions.

(b) *For every $a>0$ and Φ in $\mathcal{S}(\mathbf{R}^n)$ there exists a constant $C_2(n,p,a,\Phi)$ such that*

$$\big\|M_a^*(\vec{f};\Phi)\big\|_{L^{p,\infty}(\mathbf{R}^n,\ell^2)}\le C_2(n,p,a,\Phi)\big\|M(\vec{f};\Phi)\big\|_{L^{p,\infty}(\mathbf{R}^n,\ell^2)} \tag{6}$$

for every sequence $\vec{f}=\{f_j\}_j$ of tempered distributions.

(c) *For every $a>0$, $b>n/p$, and Φ in $\mathcal{S}(\mathbf{R}^n)$ there exists a constant $C_3(n,p,a,b,\Phi)$ such that*

$$\big\|M_b^{**}(\vec{f};\Phi)\big\|_{L^{p,\infty}(\mathbf{R}^n,\ell^2)}\le C_3(n,p,a,b,\Phi)\big\|M_a^*(\vec{f};\Phi)\big\|_{L^{p,\infty}(\mathbf{R}^n,\ell^2)} \tag{7}$$

for every sequence $\vec{f}=\{f_j\}_j$ of tempered distributions.

(d) *For every $b>0$ and Φ in $\mathcal{S}(\mathbf{R}^n)$ with $\int_{\mathbf{R}^n}\Phi(x)\,dx\neq 0$ there exists a constant $C_4(b,\Phi)$ such that if $N=[b]+1$ we have*

$$\big\|\mathcal{M}_N(\vec{f})\big\|_{L^{p,\infty}(\mathbf{R}^n,\ell^2)}\le C_4(b,\Phi)\big\|M_b^{**}(\vec{f};\Phi)\big\|_{L^{p,\infty}(\mathbf{R}^n,\ell^2)} \tag{8}$$

for every sequence $\vec{f}=\{f_j\}_j$ of tempered distributions.

(e) *For every positive integer N there exists a constant $C_5(n,N)$ such that every sequence $\vec{f} = \{f_j\}_j$ of tempered distributions that satisfies $\big\|\mathcal{M}_N(\vec{f})\big\|_{L^{p,\infty}(\mathbf{R}^n,\ell^2)} < \infty$ is bounded and satisfies*

$$\|\vec{f}\|_{H^{p,\infty}(\mathbf{R}^n,\ell^2)} \le C_5(n,N)\big\|\mathcal{M}_N(\vec{f})\big\|_{L^{p,\infty}(\mathbf{R}^n,\ell^2)}, \tag{9}$$

that is, it lies in the Hardy space $H^{p,\infty}(\mathbf{R}^n,\ell^2)$.

We conclude that for bounded distributions $\vec{f} = \{f_j\}$ the following equivalence of quasi-norms holds

$$\big\|\mathcal{M}_N(\vec{f})\big\|_{L^{p,\infty}} \approx \big\|M_b^{**}(\vec{f};\Phi)\big\|_{L^{p,\infty}} \approx \big\|M_a^{*}(\vec{f};\Phi)\big\|_{L^{p,\infty}} \approx \big\|M(\vec{f};\Phi)\big\|_{L^{p,\infty}}$$

with constants that depend only on Φ, a, n, p, and all the preceding quasi-norms are also equivalent with $\|\vec{f}\|_{H^{p,\infty}(\mathbf{R}^n,\ell^2)}$.

There is an alternative characterization of the weak H^p quasi-norm via the weak L^p quasi-norm of the associated square function. As usual, we denote by

$$\Delta_j(f) = \Delta_j^{\Psi}(f) = \Psi_{2^{-j}} * f$$

the Littlewood-Paley operator of f, where $\Psi_t(x) = t^{-n}\Psi(x/t)$.

Theorem 2 ([16]). *Let Ψ be a radial Schwartz function on $\mathbf{R}^n$ whose Fourier transform is nonnegative, supported in $\frac{1}{2} + \frac{1}{10} \le |\xi| \le 2 - \frac{1}{10}$, and satisfies* (40). *Let Δ_j be the Littlewood-Paley operators associated with Ψ and let $0 < p < \infty$. Then there exists a constant $C = C_{n,p,\Psi}$ such that for all $f \in H^p(\mathbf{R}^n)$ we have*

$$\Big\|\Big(\sum_{j\in\mathbf{Z}} |\Delta_j(f)|^2\Big)^{\frac{1}{2}}\Big\|_{L^{p,\infty}} \le C\big\|f\big\|_{H^{p,\infty}}. \tag{10}$$

Conversely, suppose that a tempered distribution f satisfies

$$\Big\|\Big(\sum_{j\in\mathbf{Z}} |\Delta_j(f)|^2\Big)^{\frac{1}{2}}\Big\|_{L^{p,\infty}} < \infty. \tag{11}$$

Then there exists a unique polynomial $Q(x)$ such that $f - Q$ lies in $H^{p,\infty}$ and satisfies the estimate

$$\frac{1}{C}\big\|f - Q\big\|_{H^{p,\infty}} \le \Big\|\Big(\sum_{j\in\mathbf{Z}} |\Delta_j(f)|^2\Big)^{\frac{1}{2}}\Big\|_{L^{p,\infty}}. \tag{12}$$

The following version of the classical Fefferman-Stein vector-valued inequality [10] is useful. This result for upper Boyd indices less than infinity is contained in [5] (page 85). A self-contained proof of the following result is contained in [16].

Proposition 1. *If $1 < p, q < \infty$, then for all sequences of functions $\{f_j\}_j$ in $L^{p,\infty}(\ell^q)$ we have*

$$\big\|\|\{M(f_j)\}\|_{\ell^q}\big\|_{L^{p,\infty}} \le C_{p,q}\,\big\|\|\{f_j\}\|_{\ell^q}\big\|_{L^{p,\infty}},$$

where M is the Hardy-Littlewood maximal function.

3 The proof of Theorem 1

The proof of this theorem is based on the following lemma whose proof can be found in [13].

Lemma 1. *Let $m \in \mathbf{Z}^+$ and let Φ in $\mathcal{S}(\mathbf{R}^n)$ satisfy $\int_{\mathbf{R}^n} \Phi(x)\,dx = 1$. Then there exists a constant $C_0(\Phi, m)$ such that for any Ψ in $\mathcal{S}(\mathbf{R}^n)$, there are Schwartz functions $\Theta^{(s)}$, $0 \le s \le 1$, with the properties*

$$\Psi(x) = \int_0^1 (\Theta^{(s)} * \Phi_s)(x)\,ds \tag{13}$$

and

$$\int_{\mathbf{R}^n} (1+|x|)^m |\Theta^{(s)}(x)|\,dx \le C_0(\Phi, m)\, s^m\, \mathfrak{N}_m(\Psi). \tag{14}$$

We now prove Theorem 1.

Proof. (a) Consider the function $\psi(s)$ defined on the interval $[1,\infty)$ as follows:

$$\psi(s) = \frac{e}{\pi}\frac{1}{s} e^{-\frac{\sqrt{2}}{2}(s-1)^{\frac{1}{4}}} \sin\left(\frac{\sqrt{2}}{2}(s-1)^{\frac{1}{4}}\right). \tag{15}$$

Clearly $\psi(s)$ decays faster than any negative power of s and satisfies

$$\int_1^\infty s^k\,\psi(s)\,ds = \begin{cases} 1 & \text{if } k=0, \\ 0 & \text{if } k = 1,2,3,\dots. \end{cases} \tag{16}$$

We now define the function

$$\Phi(x) = \int_1^\infty \psi(s) P_s(x)\,ds\,, \tag{17}$$

where P_s is the Poisson kernel. Note that the double integral

$$\int_{\mathbf{R}^n}\int_1^\infty \frac{s}{(s^2+|x|^2)^{\frac{n+1}{2}}}\, s^{-N}\,ds\,dx$$

converges and so it follows from (16) and (17) that $\int_{\mathbf{R}^n} \Phi(x)\,dx = 1$. Furthermore, we have that

$$\widehat{\Phi}(\xi) = \int_1^\infty \psi(s)\widehat{P_s}(\xi)\,ds = \int_1^\infty \psi(s) e^{-2\pi s|\xi|}\,ds$$

using that $\widehat{P_s}(\xi) = e^{-2\pi s|\xi|}$. This function is rapidly decreasing as $|\xi| \to \infty$ and the same is true for all the derivatives

$$\partial^\alpha \widehat{\Phi}(\xi) = \int_1^\infty \psi(s)\partial_\xi^\alpha\big(e^{-2\pi s|\xi|}\big)\,ds\,. \tag{18}$$

Moreover, the function $\widehat{\Phi}$ is clearly smooth on $\mathbf{R}^n \setminus \{0\}$ and we will show that it also smooth at the origin. Notice that for all multiindices α we have

$$\partial_\xi^\alpha (e^{-2\pi s|\xi|}) = s^{|\alpha|} p_\alpha(\xi) |\xi|^{-m_\alpha} e^{-2\pi s|\xi|}$$

for some $m_\alpha \in \mathbf{Z}^+$ and some polynomial $p_\alpha(\xi)$. By Taylor's theorem, for some function $v(s,|\xi|)$ with $0 \le v(s,|\xi|) \le 2\pi s|\xi|$, we have

$$e^{-2\pi s|\xi|} = \sum_{k=0}^{L} (-2\pi)^k \frac{|\xi|^k}{k!} s^k + \frac{(-2\pi s|\xi|)^{L+1}}{(L+1)!} e^{-v(s,|\xi|)} .$$

Choosing $L > m_\alpha$ gives

$$\partial_\xi^\alpha (e^{-2\pi s|\xi|}) = \sum_{k=0}^{L} (-2\pi)^k \frac{|\xi|^k}{k!} s^{k+|\alpha|} \frac{p_\alpha(\xi)}{|\xi|^{m_\alpha}} + s^{|\alpha|} \frac{p_\alpha(\xi)}{|\xi|^{m_\alpha}} \frac{(-2\pi s|\xi|)^{L+1}}{(L+1)!} e^{-v(s,|\xi|)},$$

which inserted in (18) and in view of (16), yields that when $|\alpha| > 0$, the derivative $\partial^\alpha \widehat{\Phi}(\xi)$ tends to zero as $\xi \to 0$ and when $\alpha = 0$, $\widehat{\Phi}(\xi) \to 1$ as $\xi \to 0$. We conclude that $\widehat{\Phi}$ is continuously differentiable, and hence smooth at the origin, hence it lies in the Schwartz class, and thus so does Φ.

Finally, we have the estimate

$$\begin{aligned} M(\vec{f};\Phi)(x) &= \sup_{t>0} \Big(\sum_j |(\Phi_t * f_j)(x)|^2 \Big)^{1/2} \\ &= \sup_{t>0} \Big(\sum_j \Big| \int_1^\infty \psi(s)(P_{ts} * f_j)(x)\, ds \Big|^2 \Big)^{1/2} \\ &\le \Big(\int_1^\infty |\psi(s)|\, ds \Big)^{1/2} \sup_{t>0} \Big(\sum_j \int_1^\infty |(P_{ts} * f_j)(x)|^2\, |\psi(s)|\, ds \Big)^{1/2} \\ &\le \Big(\int_1^\infty |\psi(s)|\, ds \Big)^{1/2} \Big(\int_1^\infty \sup_{t>0} \sum_j |(P_{ts} * f_j)(x)|^2\, |\psi(s)|\, ds \Big)^{1/2} \\ &\le \int_1^\infty |\psi(s)|\, ds\, M(\vec{f};P)(x) , \end{aligned}$$

and the required conclusion follows since $\int_1^\infty |\psi(s)|\, ds \le C_1$. We have actually obtained the pointwise estimate $M(\vec{f};\Phi) \le C_1\, M(\vec{f};P)$ which clearly implies (5).

(b) We present the proof only in the case when $a = 1$ since the case of general $a > 0$ is similar. We derive (6) as a consequence of the estimate

$$\big\| M_1^*(\vec{f};\Phi) \big\|_{L^{p,\infty}}^p \le C_2''(n,p,\Phi)^p \big\| M(\vec{f};\Phi) \big\|_{L^{p,\infty}}^p + \frac{1}{2} \big\| M_1^*(\vec{f};\Phi) \big\|_{L^{p,\infty}}^p , \tag{19}$$

which requires a priori knowledge of the fact that $\|M_1^*(\vec{f};\Phi)\|_{L^{p,\infty}} < \infty$. This presents a significant hurdle that needs to be overcome by an approximation. For this reason we introduce a family of maximal functions $M_1^*(\vec{f};\Phi)^{\varepsilon,N}$ for $0 \le \varepsilon, N < \infty$ such that $\|M_1^*(\vec{f};\Phi)^{\varepsilon,N}\|_{L^p} < \infty$ and such that $M_1^*(\vec{f};\Phi)^{\varepsilon,N} \uparrow M_1^*(\vec{f};\Phi)$ as $\varepsilon \downarrow 0$ and we prove (19) with $M_1^*(\vec{f};\Phi)^{\varepsilon,N}$ in place of $M_1^*(\vec{f};\Phi)^{\varepsilon,N}$, i.e., we prove

$$\| M_1^*(\vec{f};\Phi)^{\varepsilon,N}\|_{L^{p,\infty}}^p \le C_2'(n,p,\Phi,N)^p \|M(\vec{f};\Phi)\|_{L^{p,\infty}}^p + \frac{1}{2}\|M_1^*(\vec{f};\Phi)^{\varepsilon,N}\|_{L^{p,\infty}}^p , \tag{20}$$

where there is an additional dependence on N in the constant $C_2'(n,p,\Phi,N)$, but there is no dependence on ε. The $M_1^*(\vec{f};\Phi)^{\varepsilon,N}$ are defined as follows: for a bounded distribution $\vec{f}$ in $\mathcal{S}'(\mathbf{R}^n,\ell^2)$ such that $M(\vec{f};\Phi) \in L^p$ we define

$$M_1^*(\vec{f};\Phi)^{\varepsilon,N}(x) = \sup_{0<t<\frac{1}{\varepsilon}} \sup_{|y-x|<t} \Big(\sum_j \big|(\Phi_t * f_j)(y)\big|^2\Big)^{1/2} \Big(\frac{t}{t+\varepsilon}\Big)^N \frac{1}{(1+\varepsilon|y|)^N} .$$

We first show that $M_1^*(\vec{f};\Phi)^{\varepsilon,N}$ lies in $L^p(\mathbf{R}^n) \cap L^\infty(\mathbf{R}^n)$ if N is large enough depending on $\vec{f}$. Indeed, using that $(\Phi_t * f_j)(x) = \langle f_j, \Phi_t(x-\cdot)\rangle$ and the fact that $\vec{f}$ is in $\mathcal{S}'(\mathbf{R}^n,\ell^2)$, we obtain constants $C = C_{\vec{f}}$ and $m = m_{\vec{f}}$, $m > n$ such that:

$$\begin{aligned}
&\Big(\sum_j \big|(\Phi_t * f_j)(y)\big|^2\Big)^{\frac{1}{2}} \\
&\le C \sum_{|\gamma|\le m, |\beta|\le m} \sup_{w\in\mathbf{R}^n} |w^\gamma (\partial^\beta \Phi_t)(y-w)| \\
&\le C \sum_{|\beta|\le m} \sup_{z\in\mathbf{R}^n} (1+|y|^m+|z|^m)|(\partial^\beta\Phi_t)(z)| \\
&\le C\,(1+|y|^m) \sum_{|\beta|\le m} \sup_{z\in\mathbf{R}^n} (1+|z|^m)|(\partial^\beta\Phi_t)(z)| \\
&\le C\,\frac{(1+|y|^m)}{\min(t^n,t^{n+m})} \sum_{|\beta|\le m} \sup_{z\in\mathbf{R}^n} (1+|z|^m)|(\partial^\beta\Phi)(z/t)| \\
&\le C\,\frac{(1+|y|)^m}{\min(t^n,t^{n+m})}(1+t^m) \sum_{|\beta|\le m} \sup_{z\in\mathbf{R}^n} (1+|z/t|^m)|(\partial^\beta\Phi)(z/t)| \\
&\le C_{\vec{f},\Phi}(1+\varepsilon|y|)^m \varepsilon^{-m}(1+t^m)(t^{-n}+t^{-n-m}) .
\end{aligned}$$

Multiplying by $(\frac{t}{t+\varepsilon})^N(1+\varepsilon|y|)^{-N}$ for some $0<t<\frac{1}{\varepsilon}$ and $|y-x|<t$ yields

$$\begin{aligned}
&\Big(\sum_j \big|(\Phi_t * f_j)(y)\big|^2\Big)^{1/2} \Big(\frac{t}{t+\varepsilon}\Big)^N \frac{1}{(1+\varepsilon|y|)^N} \\
&\le C_{\vec{f},\Phi} \frac{\varepsilon^{-m}(1+\varepsilon^{-m})(\varepsilon^{n-N}+\varepsilon^{n+m-N})}{(1+\varepsilon|y|)^{N-m}} ,
\end{aligned}$$

and using that $1+\varepsilon|y| \ge \frac{1}{2}(1+\varepsilon|x|)$, we obtain for some $C''(\vec{f},\Phi,\varepsilon,n,m,N) < \infty$,

$$M_1^*(\vec{f};\Phi)^{\varepsilon,N}(x) \le \frac{C''(\vec{f},\Phi,\varepsilon,n,m,N)}{(1+\varepsilon|x|)^{N-m}} .$$

Taking $N > m + n/p$, we have that $M_1^*(\vec{f};\Phi)^{\varepsilon,N}$ lies in $L^{p,\infty}(\mathbf{R}^n)$. This choice of N depends on m and hence on the sequence of distributions $\vec{f} = \{f_j\}_j$.

We now introduce functions

$$U(\vec{f};\Phi)^{\varepsilon,N}(x) = \sup_{0<t<\frac{1}{\varepsilon}} \sup_{|y-x|<t} t\Big(\sum_j \big|\nabla(\Phi_t * f_j)(y)\big|^2\Big)^{1/2}\Big(\frac{t}{t+\varepsilon}\Big)^N \frac{1}{(1+\varepsilon|y|)^N}$$

and

$$\begin{aligned} &V(\vec{f};\Phi)^{\varepsilon,N}(x) \\ &= \sup_{0<t<\frac{1}{\varepsilon}} \sup_{y\in\mathbf{R}^n} \Big(\sum_j \big|(\Phi_t * f_j)(y)\big|^2\Big)^{\frac{1}{2}}\Big(\frac{t}{t+\varepsilon}\Big)^N \frac{1}{(1+\varepsilon|y|)^N}\Big(\frac{t}{t+|x-y|}\Big)^{[\frac{2n}{p}]+1}. \end{aligned}$$

We need the norm estimate

$$\big\|V(\vec{f};\Phi)^{\varepsilon,N}\big\|_{L^{p,\infty}} \le C(n)^{\frac{2}{p}}\big\|M_1^*(\vec{f};\Phi)^{\varepsilon,N}\big\|_{L^{p,\infty}} \tag{21}$$

and the pointwise estimate

$$U(\vec{f};\Phi)^{\varepsilon,N} \le A(n,p,\Phi,N)\, V(\vec{f};\Phi)^{\varepsilon,N}\,, \tag{22}$$

where

$$A(\Phi,N,n,p) = 2^{[\frac{2n}{p}]+1}\sum_{j=1}^n C_0(\partial_j\Phi, N+[\tfrac{2n}{p}]+1)\,\mathfrak{N}_{N+[\frac{2n}{p}]+1}(\partial_j\Phi)\,.$$

To prove (21) we observe that when $z \in B(y,t) \subseteq B(x,|x-y|+t)$ we have

$$\Big(\sum_j \big|(\Phi_t * f_j)(y)\big|^2\Big)^{\frac{1}{2}}\Big(\frac{t}{t+\varepsilon}\Big)^N \frac{1}{(1+\varepsilon|y|)^N} \le M_1^*(f;\Phi)^{\varepsilon,N}(z)\,,$$

from which it follows that for any $y \in \mathbf{R}^n$,

$$\begin{aligned} &\Big(\sum_j \big|(\Phi_t * f_j)(y)\big|^2\Big)^{\frac{1}{2}}\Big(\frac{t}{t+\varepsilon}\Big)^N \frac{1}{(1+\varepsilon|y|)^N} \\ \le&\Big(\frac{1}{|B(y,t)|}\int_{B(y,t)}\big[M_1^*(\vec{f};\Phi)^{\varepsilon,N}(z)\big]^{\frac{p}{2}}\,dz\Big)^{\frac{2}{p}} \\ \le&\Big(\frac{|x-y|+t}{t}\Big)^{\frac{2n}{p}}\Big(\frac{1}{|B(x,|x-y|+t)|}\int_{B(x,|x-y|+t)}\big[M_1^*(\vec{f};\Phi)^{\varepsilon,N}(z)\big]^{\frac{p}{2}}\,dz\Big)^{\frac{2}{p}} \\ \le&\Big(\frac{|x-y|+t}{t}\Big)^{[\frac{2n}{p}]+1} M\Big(\big[M_1^*(\vec{f};\Phi)^{\varepsilon,N}\big]^{\frac{p}{2}}\Big)^{\frac{2}{p}}(x)\,. \end{aligned}$$

We now use the boundedness of the Hardy-Littlewood maximal operator M from $L^{2,\infty}$ to $L^{2,\infty}$ to obtain (21) as follows:

$$\big\|V(\vec{f};\Phi)^{\varepsilon,N}\big\|_{L^{p,\infty}} = \big\|M\big((M_1^*(\vec{f};\Phi)^{\varepsilon,N})^{\frac{p}{2}}\big)^{\frac{2}{p}}\big\|_{L^{2,\infty}}$$

$$
\begin{aligned}
&= \big\|M\big((M_1^*(\vec f;\Phi)^{\varepsilon,N})^{\frac{p}{2}}\big)\big\|_{L^{2,\infty}}^{\frac{2}{p}}\\
&\le C(n)^{\frac{2}{p}}\big\|(M_1^*(\vec f;\Phi)^{\varepsilon,N})^{\frac{p}{2}}\big\|_{L^{2,\infty}}^{\frac{2}{p}}\\
&= C(n)^{\frac{2}{p}}\big\|M_1^*(\vec f;\Phi)^{\varepsilon,N}\big\|_{L^{p,\infty}}.
\end{aligned}
$$

In proving (22), we may assume that Φ has integral 1; otherwise we can multiply Φ by a suitable constant to arrange for this to happen. We note that for each $x\in\mathbf{R}^n$ we have

$$
t\big\|\nabla(\Phi_t*\vec f)(x)\big\|_{\ell^2} = \big\|(\nabla\Phi)_t*\vec f(x)\big\|_{\ell^2} \le \sqrt{n}\sum_{j=1}^n \big\|(\partial_j\Phi)_t*\vec f(x)\big\|_{\ell^2},
$$

and it suffices to work with each partial derivative $\partial_j\Phi$. Using Lemma 1 we write

$$
\partial_j\Phi = \int_0^1 \Theta^{(s)}*\Phi_s\,ds
$$

for suitable Schwartz functions $\Theta^{(s)}$. Fix $x\in\mathbf{R}^n$, $t>0$, and y with $|y-x|<t<1/\varepsilon$. Then we have

$$
\begin{aligned}
&\big\|\big((\partial_j\Phi)_t*\vec f\big)(y)\big\|_{\ell^2}\Big(\frac{t}{t+\varepsilon}\Big)^N\frac{1}{(1+\varepsilon|y|)^N}\\
=&\Big(\frac{t}{t+\varepsilon}\Big)^N\frac{1}{(1+\varepsilon|y|)^N}\Big\|\int_0^1\big((\Theta^{(s)})_t*\Phi_{st}*\vec f\big)(y)\,ds\Big\|_{\ell^2}\\
\le&\Big(\frac{t}{t+\varepsilon}\Big)^N\int_0^1\int_{\mathbf{R}^n}t^{-n}\big|\Theta^{(s)}(t^{-1}z)\big|\,\frac{\big\|\big(\Phi_{st}*\vec f\big)(y-z)\big\|_{\ell^2}}{(1+\varepsilon|y|)^N}\,dz\,ds\,. \qquad (23)
\end{aligned}
$$

Inserting the factor 1 written as

$$
\Big(\frac{ts}{ts+|x-(y-z)|}\Big)^{[\frac{2n}{p}]+1}\Big(\frac{ts}{ts+\varepsilon}\Big)^N\Big(\frac{ts+|x-(y-z)|}{ts}\Big)^{[\frac{2n}{p}]+1}\Big(\frac{ts+\varepsilon}{ts}\Big)^N
$$

in the preceding z-integral and using that

$$
\frac{1}{(1+\varepsilon|y|)^N}\le\frac{(1+\varepsilon|z|)^N}{(1+\varepsilon|y-z|)^N}
$$

and the fact that $|x-y|<t<1/\varepsilon$, we obtain the estimate

$$
\begin{aligned}
&\Big(\frac{t}{t+\varepsilon}\Big)^N\int_0^1\int_{\mathbf{R}^n}t^{-n}\big|\Theta^{(s)}(t^{-1}z)\big|\,\frac{\big\|\big(\Phi_{st}*\vec f\big)(y-z)\big\|_{\ell^2}}{(1+\varepsilon|y|)^N}\,dz\,ds\\
\le&V(\vec f;\Phi)^{\varepsilon,N}(x)\int_0^1\int_{\mathbf{R}^n}(1+\varepsilon|z|)^N\Big(\frac{ts+|x-(y-z)|}{ts}\Big)^{[\frac{2n}{p}]+1}t^{-n}\big|\Theta^{(s)}(t^{-1}z)\big|\,dz\,\frac{ds}{s^N}\\
\le&V(\vec f;\Phi)^{\varepsilon,N}(x)\int_0^1\int_{\mathbf{R}^n}s^{-[\frac{2n}{p}]-1-N}(1+\varepsilon t|z|)^N(s+1+|z|)^{[\frac{2n}{p}]+1}\big|\Theta^{(s)}(z)\big|\,dz\,ds
\end{aligned}
$$

$$\leq 2^{[\frac{2n}{p}]+1} C_0 \left(\partial_j \Phi, N + \left[\frac{2n}{p}\right] + 1\right) \mathfrak{N}_{N+[\frac{2n}{p}]+1}(\partial_j \Phi)\, V(\vec{f};\Phi)^{\varepsilon,N}(x)$$

in view of conclusion (14) of Lemma 1. Combining this estimate with (23), we deduce (22). Estimates (21) and (22) together yield

$$\big\|U(\vec{f};\Phi)^{\varepsilon,N}\big\|_{L^{p,\infty}} \leq C(n)^{\frac{2}{p}} A(n,p,\Phi,N)\big\|M_1^*(\vec{f};\Phi)^{\varepsilon,N}\big\|_{L^{p,\infty}}. \tag{24}$$

We now set

$$E_\varepsilon = \{x \in \mathbf{R}^n : \; U(\vec{f};\Phi)^{\varepsilon,N}(x) \leq K\, M_1^*(\vec{f};\Phi)^{\varepsilon,N}(x)\}$$

for some constant K to be determined shortly. With $A = A(n,p,\Phi,N)$ we have

$$\begin{aligned}\big\|M_1^*(\vec{f};\Phi)^{\varepsilon,N}\big\|^p_{L^{p,\infty}((E_\varepsilon)^c)} &\leq \frac{1}{K^p}\big\|U(\vec{f};\Phi)^{\varepsilon,N}\big\|^p_{L^{p,\infty}((E_\varepsilon)^c)} \\ &\leq \frac{1}{K^p}\big\|U(\vec{f};\Phi)^{\varepsilon,N}\big\|^p_{L^{p,\infty}} \\ &\leq \frac{C(n)^2 A^p}{K^p}\big\|M_1^*(\vec{f};\Phi)^{\varepsilon,N}\big\|^p_{L^{p,\infty}} \\ &\leq \frac{1}{2}\big\|M_1^*(\vec{f};\Phi)^{\varepsilon,N}\big\|^p_{L^{p,\infty}},\end{aligned} \tag{25}$$

provided we choose K such that $K^p = 2\,C(n)^p A(n,p,\Phi,N)^p$. Obviously K is a function of n, p, Φ, N and in particular depends on N.

It remains to estimate the weak $L^{p,\infty}$ quasi-norm of $M_1^*(f;\Phi)^{\varepsilon,N}$ over the set E_ε. We claim that the following pointwise estimate is valid:

$$M_1^*(\vec{f};\Phi)^{\varepsilon,N}(x) \leq 4\,C'(n,N,K)^{\frac{1}{q}} \Big[M\big(M(\vec{f};\Phi)^q\big)(x)\Big]^{\frac{1}{q}} \tag{26}$$

for any $x \in E_\varepsilon$ and $0 < q < \infty$ and some constant $C'(n,N,K)$, where M is the Hardy-Littlewood maximal operator. For the proof of (26) we cite [13] in the scalar case, but we indicate below why the proof also holds in the vector-valued setting.

To prove (26) we fix $x \in E_\varepsilon$ and we also fix y such that $|y - x| < t$.

By the definition of $M_1^*(f;\Phi)^{\varepsilon,N}(x)$ there exists a point $(y_0,t) \in \mathbf{R}^{n+1}_+$ such that $|x - y_0| < t < \frac{1}{\varepsilon}$ and

$$\big\|(\Phi_t * \vec{f})(y_0)\big\|_{\ell^2}\left(\frac{t}{t+\varepsilon}\right)^N \frac{1}{(1+\varepsilon|y_0|)^N} \geq \frac{1}{2} M_1^*(\vec{f};\Phi)^{\varepsilon,N}(x). \tag{27}$$

By the definitions of E_ε and $U(\vec{f};\Phi)^{\varepsilon,N}$, for any $x \in E_\varepsilon$ we have

$$t\big\|\nabla(\Phi_t * \vec{f})(\xi)\big\|_{\ell^2}\left(\frac{t}{t+\varepsilon}\right)^N \frac{1}{(1+\varepsilon|\xi|)^N} \leq K\, M_1^*(\vec{f};\Phi)^{\varepsilon,N}(x) \tag{28}$$

for all ξ satisfying $|\xi - x| < t < \frac{1}{\varepsilon}$. It follows from (27) and (28) that

$$t\big\|\nabla(\Phi_t * \vec{f})(\xi)\big\| \leq 2K\big\|(\Phi_t * \vec{f})(y_0)\big\|_{\ell^2}\left(\frac{1+\varepsilon|\xi|}{1+\varepsilon|y_0|}\right)^N \tag{29}$$

for all ξ satisfying $|\xi - x| < t < \frac{1}{\varepsilon}$. We let z be such that $|z-x| < t$ and $|z-y_0| < t$. Applying the mean value theorem and using (29), we obtain, for some ξ between y_0 and z,

$$\begin{aligned}\big\|(\Phi_t * \vec{f})(z) - (\Phi_t * \vec{f})(y_0)\big\|_{\ell^2} &= \big\|\nabla(\Phi_t * \vec{f})(\xi)\big\|_{\ell^2} |z - y_0| \\ &\leq \frac{2K}{t}\big\|(\Phi_t * \vec{f})(\xi)\big\|_{\ell^2}\left(\frac{1+\varepsilon|\xi|}{1+\varepsilon|y_0|}\right)^N |z-y_0| \\ &\leq \frac{2^{N+1}K}{t}\big\|(\Phi_t * \vec{f})(y_0)\big\|_{\ell^2} |z-y_0| \\ &\leq \frac{1}{2}\big\|(\Phi_t * \vec{f})(y_0)\big\|_{\ell^2},\end{aligned}$$

provided z also satisfies $|z-y_0| < 2^{-N-2}K^{-1}t$ in addition to $|z-x| < t$. Therefore, for z satisfying $|z-y_0| < 2^{-N-2}K^{-1}t$ and $|z-x| < t$ we have

$$\big\|(\Phi_t * \vec{f})(z)\big\|_{\ell^2} \geq \frac{1}{2}\big\|(\Phi_t * \vec{f})(y_0)\big\|_{\ell^2} \geq \frac{1}{4}M_1^*(\vec{f};\Phi)^{\varepsilon,N}(x),$$

where the last inequality uses (27). Thus we have

$$\begin{aligned}M\big(M(\vec{f};\Phi)^q\big)(x) &\geq \frac{1}{|B(x,t)|}\int_{B(x,t)}\big[M(\vec{f};\Phi)(w)\big]^q\,dw \\ &\geq \frac{1}{|B(x,t)|}\int_{B(x,t)\cap B(y_0,2^{-N-2}K^{-1}t)}\big[M(\vec{f};\Phi)(w)\big]^q\,dw \\ &\geq \frac{1}{|B(x,t)|}\int_{B(x,t)\cap B(y_0,2^{-N-2}K^{-1}t)}\frac{1}{4^q}\big[M_1^*(\vec{f};\Phi)^{\varepsilon,N}(x)\big]^q\,dw \\ &\geq \frac{|B(x,t)\cap B(y_0,2^{-N-2}K^{-1}t)|}{|B(x,t)|}\frac{1}{4^q}\big[M_1^*(\vec{f};\Phi)^{\varepsilon,N}(x)\big]^q \\ &\geq C'(n,N,K)^{-1}4^{-q}\big[M_1^*(\vec{f};\Phi)^{\varepsilon,N}(x)\big]^q,\end{aligned}$$

where we used the simple geometric fact that if $|x-y_0| \leq t$ and $\delta > 0$, then

$$\frac{|B(x,t)\cap B(y_0,\delta t)|}{|B(x,t)|} \geq c_{n,\delta} > 0,$$

the minimum of this constant being obtained when $|x-y_0| = t$. This proves (26).

Taking $q = p/2$ and applying the boundedness of the Hardy-Littlewood maximal operator on $L^{2,\infty}$ yields

$$\big\|M_1^*(\vec{f};\Phi)^{\varepsilon,N}\big\|_{L^{p,\infty}(E_\varepsilon)} \leq C_2'(n,p,\Phi,N)\big\|M(\vec{f};\Phi)\big\|_{L^{p,\infty}}. \tag{30}$$

Combining this estimate with (25), we finally prove (20).

Recalling the fact (obtained earlier) that $\|M_1^*(\vec{f};\Phi)^{\varepsilon,N}\|_{L^{p,\infty}} < \infty$, we deduce from (20) that

$$\big\|M_1^*(\vec{f};\Phi)^{\varepsilon,N}\big\|_{L^{p,\infty}} \leq 2^{\frac{1}{p}}C_2'(n,p,\Phi,N)\big\|M(\vec{f};\Phi)\big\|_{L^{p,\infty}}. \tag{31}$$

The preceding constant depends on $\vec{f}$ but is independent of ε. Notice that

$$M_1^*(\vec{f};\Phi)^{\varepsilon,N}(x) \geq \frac{2^{-N}}{(1+\varepsilon|x|)^N} \sup_{0<t<1/\varepsilon} \left(\frac{t}{t+\varepsilon}\right)^N \sup_{|y-x|<t} \left\|(\Phi_t * \vec{f})(y)\right\|_{\ell^2}$$

and that the previous expression on the right increases to

$$2^{-N} M_1^*(\vec{f};\Phi)(x)$$

as $\varepsilon \downarrow 0$. Since the constant in (31) does not depend on ε, an application of the Lebesgue monotone convergence theorem yields

$$\|M_1^*(\vec{f};\Phi)\|_{L^{p,\infty}} \leq 2^{N+\frac{1}{p}} C_2'(n,p,\Phi,N) \|M(\vec{f};\Phi)\|_{L^{p,\infty}}. \tag{32}$$

The problem with this estimate is that the finite constant $2^N C_2'(n,p,\Phi,N)$ depends on N and thus on $\vec{f}$. However, we have managed to show that under the assumption $\|M(\vec{f};\Phi)\|_{L^{p,\infty}} < \infty$, one must necessarily have $\|M_1^*(\vec{f};\Phi)\|_{L^{p,\infty}} < \infty$.

Keeping this significant observation in mind, we repeat the preceding argument from the point where the functions $U(\vec{f};\phi)^{\varepsilon,N}$ and $V(\vec{f};\phi)^{\varepsilon,N}$ are introduced, setting $\varepsilon = N = 0$. Then we arrive to (19) with a constant $C_2''(n,p,\Phi) = C_2'(n,p,\Phi,0)$ which is independent of N and thus of $\vec{f}$. We conclude the validity of (6) with $C_2(n,p,1,\Phi) = 2^{1/p} C_2''(n,p,\Phi)$ when $a = 1$. An analogous constant is obtained for different values of $a > 0$.

(c) Let $B(x,R)$ denote a ball centered at x with radius R. Recall that

$$M_b^{**}(\vec{f};\Phi)(x) = \sup_{t>0} \sup_{y\in\mathbf{R}^n} \frac{\|(\Phi_t * \vec{f})(x-y)\|_{\ell^2}}{\left(\frac{|y|}{t}+1\right)^b}.$$

It follows from the definition

$$M_a^*(\vec{f};\Phi)(z) = \sup_{t>0} \sup_{|w-z|<at} \|(\Phi_t * \vec{f})(w)\|_{\ell^2}$$

that

$$\|(\Phi_t * \vec{f})(x-y)\|_{\ell^2} \leq M_a^*(\vec{f};\Phi)(z) \qquad \text{if } z \in B(x-y, at).$$

But the ball $B(x-y,at)$ is contained in the ball $B(x,|y|+at)$; hence it follows that

$$\begin{aligned}
\|(\Phi_t * \vec{f})(x-y)\|_{\ell^2}^{\frac{n}{b}} &\leq \frac{1}{|B(x-y,at)|} \int_{B(x-y,at)} M_a^*(\vec{f};\Phi)(z)^{\frac{n}{b}}\, dz \\
&\leq \frac{1}{|B(x-y,at)|} \int_{B(x,|y|+at)} M_a^*(\vec{f};\Phi)(z)^{\frac{n}{b}}\, dz \\
&\leq \left(\frac{|y|+at}{at}\right)^n M\big(M_a^*(\vec{f};\Phi)^{\frac{n}{b}}\big)(x) \\
&\leq \max(1,a^{-n}) \left(\frac{|y|}{t}+1\right)^n M\big(M_a^*(\vec{f};\Phi)^{\frac{n}{b}}\big)(x),
\end{aligned}$$

from which we conclude that for all $x \in \mathbf{R}^n$ we have

$$M_b^{**}(\vec{f};\Phi)(x) \le \max(1,a^{-b})\left\{M\big(M_a^*(\vec{f};\Phi)^{\frac{n}{b}}\big)(x)\right\}^{\frac{b}{n}}.$$

We now take $L^{p,\infty}$ norms on both sides of this inequality and using the fact that $pb/n > 1$ and the boundedness of the Hardy-Littlewood maximal operator M from $L^{pb/n,\infty}$ to itself, we obtain the required conclusion (7).

(d) In proving (d) we may replace b by the integer $b_0 = [b]+1$. Let Φ be a Schwartz function with integral equal to 1. Applying Lemma 1 with $m = b_0$, we write any function φ in $\mathcal{F}_N$ as

$$\varphi(y) = \int_0^1 (\Theta^{(s)} * \Phi_s)(y)\,ds$$

for some choice of Schwartz functions $\Theta^{(s)}$. Then we have

$$\varphi_t(y) = \int_0^1 ((\Theta^{(s)})_t * \Phi_{ts})(y)\,ds$$

for all $t > 0$. Fix $x \in \mathbf{R}^n$. Then for y in $B(x,t)$ we have

$$\begin{aligned}
&\|(\varphi_t * \vec{f})(y)\|_{\ell^2}\\
&\le \int_0^1\int_{\mathbf{R}^n} |(\Theta^{(s)})_t(z)|\,\|(\Phi_{ts} * \vec{f})(y-z)\|_{\ell^2}\,dz\,ds\\
&\le \int_0^1\int_{\mathbf{R}^n} |(\Theta^{(s)})_t(z)|\,M_{b_0}^{**}(\vec{f};\Phi)(x)\left(\frac{|x-(y-z)|}{st}+1\right)^{b_0} dz\,ds\\
&\le \int_0^1 s^{-b_0}\int_{\mathbf{R}^n} |(\Theta^{(s)})_t(z)|\,M_{b_0}^{**}(\vec{f};\Phi)(x)\left(\frac{|x-y|}{t}+\frac{|z|}{t}+1\right)^{b_0} dz\,ds\\
&\le 2^{b_0} M_{b_0}^{**}(\vec{f};\Phi)(x)\int_0^1 s^{-b_0}\int_{\mathbf{R}^n} |\Theta^{(s)}(w)|\,\big(|w|+1\big)^{b_0}\,dw\,ds\\
&\le 2^{b_0} M_{b_0}^{**}(\vec{f};\Phi)(x)\int_0^1 s^{-b_0}C_0(\Phi,b_0)\,s^{b_0}\,\mathfrak{N}_{b_0}(\varphi)\,ds\,,
\end{aligned}$$

where we applied conclusion (14) of Lemma 1. Setting $N = b_0 = [b]+1$, we obtain for y in $B(x,t)$ and $\varphi \in \mathcal{F}_N$,

$$\|(\varphi_t * \vec{f})(y)\|_{\ell^2} \le 2^{b_0}C_0(\Phi,b_0)\,M_{b_0}^{**}(\vec{f};\Phi)(x)\,.$$

Taking the supremum over all y in $B(x,t)$, over all $t > 0$, and over all φ in $\mathcal{F}_N$, we obtain the pointwise estimate

$$\mathcal{M}_N(\vec{f})(x) \le 2^{b_0}C_0(\Phi,b_0)\,M_{b_0}^{**}(\vec{f};\Phi)(x)\,, \qquad x \in \mathbf{R}^n,$$

where $N = b_0$. This clearly yields (8) if we set $C_4 = 2^{b_0}C_0(\Phi,b_0)$.

(e) We fix an $\vec{f} \in \mathcal{S}'(\mathbf{R}^n)$ that satisfies $\|\mathcal{M}_N(\vec{f})\|_{L^{p,\infty}} < \infty$ for some fixed positive integer N. To show that $\vec{f}$ is a bounded distribution, we fix a Schwartz function φ and we observe that for some positive constant $c = c_\varphi$, we have that $c\varphi$ is an element of $\mathcal{F}_N$ and thus $M_1^*(\vec{f}; c\varphi) \le \mathcal{M}_N(\vec{f})$. Then

$$c\,\|(\varphi * \vec{f})(x)\|_{\ell^2} \le \sup_{|z-y|\le 1} \|(c\varphi * \vec{f})(z)\|_{\ell^2} \le M_1^*(\vec{f}; c\varphi)(y) \le M_N(\vec{f})(y)$$

for $|y-x| \le 1$. So let $\lambda = c\,\|(\varphi * \vec{f})(x)\|_{\ell^2}$ and then the inequality

$$v_n^{\frac{1}{p}} \frac{\lambda}{2} \le \frac{\lambda}{2} |\{M_N(\vec{f}) > \frac{\lambda}{2}\}|^{\frac{1}{p}} \le \|M_N(\vec{f})\|_{L^{p,\infty}} < \infty \tag{33}$$

shows that λ is finite and can be controlled by $2\|M_N(\vec{f})\|_{L^{p,\infty}} v_n^{-\frac{1}{p}}$. Here $v_n = |B(0,1)|$ is the volume of the unit ball in $\mathbf{R}^n$. This implies that $\|\varphi * \vec{f}\|_{\ell^2}$ is a bounded function. We conclude that $\vec{f}$ is a bounded distribution. We now show that $\vec{f}$ is an element of $H^{p,\infty}$. We fix a smooth function with compact support θ such that

$$\theta(x) = \begin{cases} 1 & \text{if} \quad |x| < 1, \\ 0 & \text{if} \quad |x| > 2. \end{cases}$$

We observe that the identity

$$\begin{aligned} P(x) &= P(x)\theta(x) + \sum_{k=1}^{\infty} \big(\theta(2^{-k}x)P(x) - \theta(2^{-(k-1)}x)P(x)\big) \\ &= P(x)\theta(x) + \frac{\Gamma(\frac{n+1}{2})}{\pi^{\frac{n+1}{2}}} \sum_{k=1}^{\infty} 2^{-k} \left(\frac{\theta(\,\cdot\,) - \theta(2(\,\cdot\,))}{(2^{-2k} + |\cdot|^2)^{\frac{n+1}{2}}} \right)_{2^k} (x) \end{aligned}$$

is valid for all $x \in \mathbf{R}^n$. Setting

$$\Phi^{(k)}(x) = \big(\theta(x) - \theta(2x)\big) \frac{1}{(2^{-2k} + |x|^2)^{\frac{n+1}{2}}},$$

we note that for some fixed constant $c_0 = c_0(n, N)$, the functions $c_0\,\theta\,P$ and $c_0\Phi^{(k)}$ lie in $\mathcal{F}_N$ uniformly in $k = 1, 2, 3, \dots$.

Lemma 2. *Let f be a bounded distribution on $\mathbf{R}^n$. Then we have*

$$(P * f)(x) = ((\theta P) * f)(x) + \frac{\Gamma(\frac{n+1}{2})}{\pi^{\frac{n+1}{2}}} \sum_{k=1}^{\infty} 2^{-k} (\Phi^{(k)}_{2^{-k}} * f)(x)$$

for all $x \in \mathbf{R}^n$, where the series converges in $\mathcal{S}'(\mathbf{R}^n)$.

Combining this observation with the identity for $P(x)$ obtained earlier, we conclude that

$$\sup_{t>0} \|P_t * \vec{f}\|_{\ell^2} \le \sup_{t>0} \|(\theta P)_t * \vec{f}\|_{\ell^2} + \frac{1}{c_0} \frac{\Gamma(\frac{n+1}{2})}{\pi^{\frac{n+1}{2}}} \sup_{t>0} \sum_{k=1}^{\infty} 2^{-k} \|(c_0\Phi^{(k)})_{2^k t} * \vec{f}\|_{\ell^2}$$

$$\leq C_5(n,N)\mathcal{M}_N(\vec{f})\,,$$

which proves the required conclusion (9).

We observe that the last estimate also yields the stronger estimate

$$M_1^*(\vec{f};P)(x) = \sup_{t>0} \sup_{\substack{y\in\mathbf{R}^n\\ |y-x|\leq t}} |(P_t * \vec{f})(y)| \leq C_5(n,N)\mathcal{M}_N(\vec{f})(x)\,. \tag{34}$$

It follows that the quasinorm $\|M_1^*(\vec{f};P)\|_{L^{p,\infty}}$ is at most a constant multiple of $\|\mathcal{M}_N(\vec{f})\|_{L^{p,\infty}}$ and thus it is also equivalent to $\|\vec{f}\|_{H^{p,\infty}}$.

This concludes the proof of Theorem 1. □

It remains to prove Lemma 2.

Proof of Lemma 2. We begin with the identity

$$\begin{aligned} P(x) &= P(x)\theta(x) + \sum_{k=1}^{\infty} \big(\theta(2^{-k}x)P(x) - \theta(2^{-(k-1)}x)P(x)\big) \\ &= P(x)\theta(x) + \frac{\Gamma(\frac{n+1}{2})}{\pi^{\frac{n+1}{2}}} \sum_{k=1}^{\infty} 2^{-k} \bigg(\frac{\theta(\cdot) - \theta(2(\cdot))}{(2^{-2k} + |\cdot|^2)^{\frac{n+1}{2}}} \bigg)_{2^k}(x) \end{aligned}$$

which is valid for all $x \in \mathbf{R}^n$.

Fix a function $\phi \in \mathcal{S}(\mathbf{R}^n)$ whose Fourier transform is equal to 1 in a neighborhood of zero. Then $f = \phi * f + (\delta_0 - \phi) * f$ and we also have $P_t * f = P_t * \phi * f + P_t * (\delta_0 - \phi) * f$. Given a function ψ in $\mathcal{S}(\mathbf{R}^n)$ we need to show that

$$\big\langle (\theta P)_t * \phi * f, \psi \big\rangle + \frac{\Gamma(\frac{n+1}{2})}{\pi^{\frac{n+1}{2}}} \sum_{k=1}^{N} 2^{-k} \big\langle (\Phi^{(k)})_{2^k t} * \phi * f, \psi \big\rangle \to \langle P_t * \phi * f, \psi \rangle$$

and

$$\big\langle (\theta P)_t * (\delta_0 - \phi) * f, \psi \big\rangle + \frac{\Gamma(\frac{n+1}{2})}{\pi^{\frac{n+1}{2}}} \sum_{k=1}^{N} 2^{-k} \big\langle (\Phi^{(k)})_{2^k t} * (\delta_0 - \phi) * f, \psi \big\rangle \to \langle P_t * (\delta_0 - \phi) * f, \psi \rangle$$

as $N \to \infty$.

The first of these claims is equivalent to

$$\big\langle \phi * f, \psi * (\theta P)_t \big\rangle + \frac{\Gamma(\frac{n+1}{2})}{\pi^{\frac{n+1}{2}}} \sum_{k=1}^{N} 2^{-k} \big\langle \phi * f, \psi * (\Phi^{(k)})_{2^k t} \big\rangle \to \langle \phi * f, \psi * P_t \rangle$$

as $N \to \infty$. Here $\phi * f \in L^\infty$ and the actions $\langle \cdot, \cdot \rangle$ are convergent integrals in all three cases. This claim will be a consequence of the Lebesgue dominated convergence theorem since:

$$\psi * (\theta P)_t + \frac{\Gamma(\frac{n+1}{2})}{\pi^{\frac{n+1}{2}}} \sum_{k=1}^{N} 2^{-k} \psi * (\Phi^{(k)})_{2^k t} \to \psi * P_t$$

pointwise (which is also a consequence of the Lebesgue dominated convergence theorem) as $N \to \infty$ and hence the same is true after multiplying by the bounded function $\phi * f$ and also

$$|\phi * f| \left| \psi * (\theta P)_t + \frac{\Gamma(\frac{n+1}{2})}{\pi^{\frac{n+1}{2}}} \sum_{k=1}^{N} 2^{-k} \psi * (\Phi^{(k)})_{2^k t} \right| \le |\phi * f| \ (|\psi| * P_t) \in L^1(\mathbf{R}^n).$$

We now turn to the corresponding assertion where ϕ is replaced by $\delta - \phi$. Using the Fourier transform, this assertion is equivalent to

$$\langle \widehat{f}, \widehat{(\theta P)_t}(1-\widehat{\phi})\widehat{\psi}\rangle + \frac{\Gamma(\frac{n+1}{2})}{\pi^{\frac{n+1}{2}}} \sum_{k=1}^{N} 2^{-k} \langle \widehat{f}, \widehat{(\Phi^{(k)})_{2^k t}}(1-\widehat{\phi})\widehat{\psi}\rangle \to \langle \widehat{f}, \widehat{P_t}(1-\widehat{\phi})\widehat{\psi}\rangle$$

Since $\widehat{f} \in \mathcal{S}'(\mathbf{R}^n)$, this assertion will be a consequence of the fact that

$$\widehat{(\theta P)_t}(1-\widehat{\phi})\widehat{\psi} + \frac{\Gamma(\frac{n+1}{2})}{\pi^{\frac{n+1}{2}}} \sum_{k=1}^{N} 2^{-k} \widehat{(\Phi^{(k)})_{2^k t}}(1-\widehat{\phi})\widehat{\psi} \to \widehat{P_t}(1-\widehat{\phi})\widehat{\psi}$$

in $\mathcal{S}(\mathbf{R}^n)$. It will therefore be sufficient to show that for all multiindices α and β we have

$$\sup_{\xi \in \mathbf{R}^n} \left| \partial_\xi^\alpha \left[\left\{ \widehat{(\theta P)_t}(\xi) + \frac{\Gamma(\frac{n+1}{2})}{\pi^{\frac{n+1}{2}}} \sum_{k=1}^{N} 2^{-k} \widehat{(\Phi^{(k)})_{2^k t}}(\xi) - \widehat{P_t}(\xi) \right\} (1-\widehat{\phi}(\xi))\widehat{\psi}(\xi)\xi^\beta \right] \right| \to 0.$$

The term inside the curly brackets is equal to the Fourier transform of the function $(\theta(2^{-N}x) - 1)P_t(x)$, which is

$$\int_{\mathbf{R}^n} (e^{-2\pi t|\xi-\zeta|} - e^{-2\pi t|\xi|}) 2^{Nn} \widehat{\theta}(2^N \zeta)\, d\zeta$$

since $\widehat{\theta}$ has integral 1. Note that $|\xi| \ge c > 0$ since $1 - \widehat{\phi}(\xi) = 0$ in a neighborhood of zero. Let $\varepsilon_N > 0$. First consider the integral

$$\int_{|\xi-\zeta|>\varepsilon_N} (e^{-2\pi t|\xi-\zeta|} - e^{-2\pi t|\xi|}) 2^{Nn} \widehat{\theta}(2^N \zeta)\, d\zeta\, .$$

Both exponentials are differentiable in this range and differentiation gives

$$\begin{aligned} &\partial_\xi^\gamma \int_{|\xi-\zeta|>\varepsilon_N} (e^{-2\pi t|\xi-\zeta|} - e^{-2\pi t|\xi|}) 2^{Nn} \widehat{\theta}(2^N \zeta)\, d\zeta \\ &= \int_{|\xi-\zeta|>\varepsilon_N} \left[Q_\gamma\Big(\frac{\xi-\zeta}{|\xi-\zeta|}, \frac{1}{|\xi-\zeta|}\Big) e^{-2\pi t|\xi-\zeta|} - Q_\gamma\Big(\frac{\xi}{|\xi|}, \frac{1}{|\xi|}\Big) e^{-2\pi t|\xi|} \right] 2^{Nn} \widehat{\theta}(2^N \zeta)\, d\zeta\, , \end{aligned}$$

where Q_γ is a polynomial of the following $n+1$ variables

$$Q_\gamma\Big(\frac{\xi}{|\xi|}, \frac{1}{|\xi|}\Big) = Q_\gamma\Big(\frac{\xi_1}{|\xi|}, \dots, \frac{\xi_n}{|\xi|}, \frac{1}{|\xi|}\Big)$$

that depends on γ. Note that

$$\left| Q_\gamma\Big(\frac{\xi-\zeta}{|\xi-\zeta|},\frac{1}{|\xi-\zeta|}\Big)e^{-2\pi t|\xi-\zeta|} - Q_\gamma\Big(\frac{\xi}{|\xi|},\frac{1}{|\xi|}\Big)e^{-2\pi t|\xi|}\right| \le C\,|\zeta|\,.$$

Thus the integral is bounded by

$$\int_{|\xi-\zeta|>\varepsilon_N} C\,2^{Nn}|\zeta|\widehat{\theta}(2^N\zeta)\,d\zeta = C'2^{-N}\,,$$

which tends to zero as $N\to\infty$. Now consider the integral

$$\partial_\xi^\gamma \int_{|\xi-\zeta|\le\varepsilon_N} (e^{-2\pi t|\xi-\zeta|} - e^{-2\pi t|\xi|})2^{Nn}\widehat{\theta}(2^N\zeta)\,d\zeta$$

$$= \partial_\xi^\gamma \int_{|\zeta|\le\varepsilon_N} (e^{-2\pi t|\zeta|} - e^{-2\pi t|\xi|})2^{Nn}\widehat{\theta}(2^N(\xi-\zeta))\,d\zeta$$

in which $|\xi|\ge c>0$. We obtain another expression which is bounded by

$$C\,\varepsilon_N^n\,2^{N|\gamma|} \le C\,\varepsilon_N^n\,2^{N|\alpha|}\,,$$

which tends to zero if we pick $\varepsilon_N = 2^{-N|\alpha|/n}/N$. Finally

$$\sup_{\xi\in\mathbf{R}^n}\left|\partial_\xi^{\alpha-\gamma}\Big\{(1-\widehat{\phi}(\xi))\widehat{\psi}(\xi)\xi^\beta\Big\}\right| < \infty$$

and so the claimed conclusion follows by applying Leibniz's rule to the expression on which ∂_ξ^α is acting.

Fix a smooth radial nonnegative compactly supported function θ on $\mathbf{R}^n$ such that $\theta=1$ on the unit ball and vanishing outside the ball of radius 2. Set $\Phi^{(k)}(x) = \big(\theta(x)-\theta(2x)\big)(2^{-2k}+|x|^2)^{-\frac{n+1}{2}}$ for $k\ge 1$. Prove that for all bounded tempered distributions f and for all $t>0$ we have

$$P_t * f = (\theta P)_t * f + \frac{\Gamma(\frac{n+1}{2})}{\pi^{\frac{n+1}{2}}}\sum_{k=1}^\infty 2^{-k}(\Phi^{(k)})_{2^k t} * f\,,$$

where the series converges in $\mathcal{S}'(\mathbf{R}^n)$. Here

$$P(x) = \Gamma(\frac{n+1}{2})/\pi^{\frac{n+1}{2}}(1+|x|^2)^{\frac{n+1}{2}}$$

is the Poisson kernel. □

4 Properties of $H^{p,\infty}$

The spaces $H^{p,\infty}$ have several properties analogous to those of the classical Hardy spaces H^p. Here we provide a list of these properties and we provide proofs for some of them. The missing proofs can be found in [16].

Theorem 3. *Let $1 < p < \infty$. Then we have $L^{p,\infty} = H^{p,\infty}$ and $\|f\|_{L^{p,\infty}} \approx \|f\|_{H^{p,\infty}}$.*

Theorem 4. (a) *For any $0 < p < \infty$, every $\vec{f} = \{f_j\}_j$ in $H^{p,\infty}(\mathbf{R}^n, \ell^2)$, and any $\varphi \in \mathcal{S}(\mathbf{R}^n)$ we have*

$$\Big(\sum_j |\langle f_j, \varphi\rangle|^2\Big)^{1/2} \le \mathfrak{N}_N(\varphi) \inf_{|z|\le 1} \mathcal{M}_N(\vec{f})(z), \tag{35}$$

where $N = [\frac{n}{p}] + 1$, and consequently there is a constant $C_{n,p}$ such that

$$\Big(\sum_j |\langle f_j, \varphi\rangle|^2\Big)^{1/2} \le \mathfrak{N}_N(\varphi)\, C_{n,p} \,\|\vec{f}\,\|_{H^{p,\infty}}. \tag{36}$$

(b) *Let $0 < p \le 1$, $N = [n/p] + 1$, and $p < r \le \infty$. Then there is a constant $C(p, n, r)$ such that for any $\vec{f} \in H^{p,\infty}$ and $\varphi \in \mathcal{S}(\mathbf{R}^n)$ we have*

$$\Big\|\Big(\sum_j |f_j * \varphi|^2\Big)^{1/2}\Big\|_{L^r} \le C(p, n, r)\mathfrak{N}_N(\varphi)\|\vec{f}\,\|_{H^{p,\infty}}. \tag{37}$$

(c) *For any $x_0 \in \mathbf{R}^n$, for all $R > 0$, and any $\psi \in \mathcal{S}(\mathbf{R}^n)$ we have*

$$\Big(\sum_j |\langle f_j, \psi\rangle|^2\Big)^{1/2} \le \mathfrak{N}_N(\psi; x_0, R) \inf_{|z-x_0|\le R} \mathcal{M}_N(\vec{f})(z). \tag{38}$$

Proposition 2. *Let $0 < p < \infty$. The following triangle inequality holds for all f, g in $H^{p,\infty}$:*

$$\|f + g\|^p_{H^{p,\infty}} \le 2^p(\|f\|^p_{H^{p,\infty}} + \|g\|^p_{H^{p,\infty}}).$$

Moreover, we have

$$\|\{f_j\}\|_{H^{p,\infty}}(\mathbf{R}^n, \ell^2) \approx \sup_{0<|E|<\infty} |E|^{-\frac{1}{r}+\frac{1}{p}} \Big(\int_E \sup_{t>0} \|\{(\varphi_t * f_j)(x)\}_j\|^r_{\ell^2}\, dx\Big)^{\frac{1}{r}}$$

for $0 < r < p$.

Proposition 3. *For $0 < p < \infty$, $H^{p,\infty}(\mathbf{R}^n, \ell^2(L))$ are complete quasi-normed spaces.*

Theorem 5. *L^r is not dense in $L^{p,\infty}$, whenever $0 < r \le \infty$ and $0 < p < \infty$.*

The distribution $\delta_1 - \delta_{-1}$ is used to prove the following result.

Theorem 6. *The space of Schwartz functions $\mathcal{S}$ is not dense in $H^{1,\infty}$.*

Interpolation is a powerful tool that yields results in the theory of weak Hardy spaces. The following lemma will be useful in the proof of Theorem 9.

Lemma 3 ([16]). *Let $0 < p_1 < p < p_2 < \infty$. Given $\vec{F} = \{f_k\}_{k=1}^L \in H^{p,\infty}(\mathbf{R}^n, \ell^2(L))$ and $\alpha > 0$, then there exists $\vec{G} = \{g^k\}_{k=1}^L$ and $\vec{B} = \{b^k\}_{k=1}^L$ such that $\vec{F} = \vec{G} + \vec{B}$ and*

$$\|\vec{B}\|_{H^{p_1}(\mathbf{R}^n,\ell^2(L))}^{p_1} \le C\alpha^{p_1-p}\|F\|_{H^{p,\infty}(\mathbf{R}^n,\ell^2(L))}^p$$

and

$$\|\vec{G}\|_{H^{p_2}(\mathbf{R}^n,\ell^2(L))}^{p_2} \le C\alpha^{p_2-p}\|\vec{F}\|_{H^{p,\infty}(\mathbf{R}^n,\ell^2(L))}^p$$

where $C = C(p_1, p_2, p, n)$, in particular is independent of L.

Next, we will discuss the cancellation of Schwartz functions in $H^{p,\infty}$ for $p \le 1$. We denote by $[a]$ the integer part of a real number a. We have the following result:

Theorem 7. *If $f \in H^p(\mathbf{R}^n)$, then $\int_{\mathbf{R}^n} x^\gamma f(x)\,dx = 0$ for $|\gamma| \le [\frac{n}{p} - n]$ if these integrals converge absolutely. If $f \in H^{p,\infty}(\mathbf{R}^n)$, then $\int_{\mathbf{R}^n} x^\gamma f(x)\,dx = 0$ for $|\gamma| \le -[n - \frac{n}{p}] - 1$ if these integrals converge absolutely.*

Proof. For $f \in H^p$, we have $|\eta_t * f(x)| \le \frac{C}{t^{\frac{n}{p}}}\|f\|_{H^p}$ and $\|\eta_t * f\|_{L^p} \le C\|f\|_{H^p}$, where $\eta \in \mathcal{S}$ with $1 \ge \widehat{\eta} \ge 0$, $\widehat{\eta}(\xi) = 1$ for $|\xi| \le 1$ and 0 for $|\xi| \ge 2$. Therefore $\|\eta_t * f\|_{L^1} \le Ct^{n-\frac{n}{p}}\|f\|_{H^p}$. For each fixed $|\xi|$ we can take $t = \frac{1}{2|\xi|}$, which would imply that $|\widehat{f}(\xi)| \le C|\xi|^{\frac{n}{p}-n}\|f\|_{H^p}$. All Schwartz functions ψ with $\int_{\mathbf{R}^n} x^\gamma \psi = 0$ for all γ are dense in H^p. For such a ψ we have $\lim_{|\xi|\to 0} \widehat{\psi}(\xi)/|\xi|^{\frac{n}{p}-n} = 0$. By the density we mentioned we know that for all f in H^p we must have $\lim_{|\xi|\to 0} \widehat{f}(\xi)/|\xi|^{\frac{n}{p}-n} = 0$. And this limit gives us that $\partial^\gamma \widehat{f}(0) = 0$ for $|\gamma| \le [\frac{n}{p} - n]$, i.e. $\int_{\mathbf{R}^n} x^\gamma f(x)dx = 0$, once we notice that $\partial^\gamma \widehat{f}(\xi)$ are well-defined and continuous.

Next we will prove the corresponding result for $H^{p,\infty}$. By lemma 3 we know that if $f \in H^{p,\infty}$, then $f = h + g$ with $h \in H^{p_1}$ and $g \in H^{p_2}$, where $p_1 < p < p_2$. Moreover h and g have the same integrability as f since they're truncation of f. We have $\lim_{|\xi|\to 0} \widehat{h}(\xi)/|\xi|^{\frac{n}{p_1}-n} = 0$ and $\lim_{|\xi|\to 0} \widehat{g}(\xi)/|\xi|^{\frac{n}{p_2}-n} = 0$, which implies $\lim_{|\xi|\to 0} \widehat{f}(\xi)/|\xi|^{\frac{n}{p_2}-n} = 0$ for any $p_2 > p$. This result will give us the same cancellation for f as that in H^p if $\frac{n}{p} - n$ is not an integer but one degree less if it's an integer, namely $\int_{\mathbf{R}^n} x^\gamma f(x)\,dx = 0$ for all $|\gamma| \le -[n - \frac{n}{p}] - 1$. □

A natural question is that if the number $-[n - \frac{n}{p}] - 1$ is sharp. The fact that $L^1 \subset H^{1,\infty}$, which is the result of weak type $(1,1)$ boundedness of Hardy-Littlewood maximal function, suggests that this number is sharp. Indeed this is the case and we give a precise proof below.

We can get the converse of the previous theorem, i.e. every Schwartz function satisfying the cancellation of the theorem must be in the corresponding space. We need the following lemma which can be found in Appendix K.2 of [12].

Lemma 4. *Let $x_\mu, x_\nu \in \mathbf{R}^n$, $M, N > 0$, and L a nonnegative integer. Suppose that φ_μ and φ_ν are functions on $\mathbf{R}^n$ that satisfy*

$$|\partial_x^\alpha \varphi_\mu(x)| \le \frac{A_\alpha 2^{\mu n} 2^{\mu L}}{(1 + 2^\mu|x - x_\mu|)^M} \qquad \text{for all } |\alpha| = L,$$

$$|\varphi_\nu(x)| \le \frac{B2^{\nu n}}{(1+2^\nu|x-x_\nu|)^N},$$

for some A_α and B positive, and $\int_{\mathbf{R}^n}\varphi_\nu(x)x^\beta dx = 0$ for all $|\beta| \le L-1$, where the last condition is supposed to be vacuous when $L=0$. Suppose that $N > M+L+n$ and $\nu \ge \mu$. Then we have

$$\left|\int_{\mathbf{R}^n}\varphi_\mu\varphi_\nu\,dx\right| \le C\frac{2^{\mu n}2^{-(\nu-\mu)L}}{(1+2^\mu|x_\mu - x_\nu|)^M}.$$

Theorem 8. *Any $f \in \mathcal{S}$ with $\int_{\mathbf{R}^n} x^\gamma f(x)dx = 0$ for $|\gamma| \le [\frac{n}{p}-n]$ lies in $H^p(\mathbf{R}^n)$. Any $f \in \mathcal{S}$ with $\int_{\mathbf{R}^n} x^\gamma f(x)dx = 0$ for all $|\gamma| \le -[n-\frac{n}{p}]-1$ lies in $H^{p,\infty}(\mathbf{R}^n)$.*

Proof. We want to estimate

$$f^+(x) = \sup_{t>0}|(\psi_t * f)(x)| = \sup_{t>0}\left|\int_{\mathbf{R}^n} f(y)\psi_t(x-y)dy\right|.$$

Take $\psi_t = \varphi_\mu$, $f = \varphi_\nu$, $L = [\frac{n}{p}-n]+1$ for the first case and $-[n-\frac{n}{p}]$ for the second, $\nu = 0$. The condition $\mu \le \nu$ forces $t^{-1} = 2^\mu \le 1$. So

$$\begin{aligned}\sup_{t\ge1}\left|\int_{\mathbf{R}^n} f(y)\psi_t(x-y)dy\right| &\le C\sup_{t\ge1}\frac{t^{-n}t^{-L}}{(1+t^{-1}|x|)^M}\\ &= C\sup_{t\ge1}\frac{1}{(t+|x|)^{n+L}}\frac{1}{(1+t^{-1}|x|)^{M-n-L}}\\ &\le C\frac{1}{(1+|x|)^{n+L}}.\end{aligned}$$

If we take $f = \varphi_\mu$, $\psi_t = \varphi_\nu$, $L_0 = 0$, $\mu = 0$, then $t^{-1} = 2^\nu \ge 1$. We have

$$\sup_{0<t\le1}\left|\int_{\mathbf{R}^n} f(y)\psi_t(x-y)dy\right| \le C\sup_{0<t\le1}\frac{2^{\nu L_0}}{(1+|x|)^M} \le \frac{C}{(1+|x|)^{n+L}}.$$

Thus $f^+(x)$ is controlled by $C(1+|x|)^{-[\frac{n}{p}]-1}$, which is in L^p, while the second kind of cancellation implies that $f^+(x)$ is in $H^{p,\infty}$. □

Next we will get a corollary of these two theorems to characterize the class of Schwartz functions

$$\mathcal{S}_k = \left\{f \in \mathcal{S}(\mathbf{R}^n) : \int_{\mathbf{R}^n} x^\gamma f(x)dx = 0,\ |\gamma| \le k\right\}$$

in terms of the decay of the corresponding smooth maximal functions of their elements.

Corollary 1. (i) *For a Schwartz function f the following equivalence is valid:*

$$f \in \mathcal{S}_k \Leftrightarrow f^+(x) \le \frac{C}{(1+|x|)^{n+k+1}};$$

(ii) *For a Schwartz function f if $f^+(x) \le \frac{C}{(1+|x|)^{n+k+\epsilon}}$ for some $\epsilon > 0$, then $f^+(x) \le \frac{C}{(1+|x|)^{n+k+1}}$.*

Proof. (i) The forward direction comes from the calculation of Theorem 8. The backwards direction is a result of Theorem 7 since such a function f lies in some H^p with $k=[\frac{n}{p}-n]$.

(ii) It's easy to see that this f is in H^p for $p=\frac{n}{n+k}$, therefore $f\in\mathcal{S}_k$ by Theorem 7, and the conclusion follows by (i). □

A result similar to that $L^p\cap\mathcal{S}=L^q\cap\mathcal{S}$ for $p,q\ge 1$ will be revealed by next corollary.

Corollary 2. (i) *$H^p\cap\mathcal{S}=H^q\cap\mathcal{S}$ for $p,q\in(\frac{n}{n+k+1},\frac{n}{n+k}]$, $k\in\mathbf{N}$.*

(ii) *$H^{p,\infty}\cap\mathcal{S}=H^{q,\infty}\cap\mathcal{S}$ for $p,q\in[\frac{n}{n+k+1},\frac{n}{n+k})$, $k\in\mathbf{N}$.*

(iii) *$H^{p,\infty}\cap\mathcal{S}=H^p\cap\mathcal{S}$ for $p\neq\frac{n}{n+k}$.*

(iv) *The statement in* (i) *fails for $p=\frac{n}{n+k+1}$ and* (ii) *fails for $p=\frac{n}{n+k}$.*

(v) *For all $f\in\mathcal{S}_k:=H^p\cap\mathcal{S}$, $p\in(\frac{n}{n+k+1},\frac{n}{n+k}]$ and all $x\in\mathbf{R}^n$, the best estimate for $f^+=\sup_{t>0}|(f*\varphi_t)|$ is $f^+(x)\le\frac{C}{(1+|x|)^{n+k+1}}$.*

Proof. (i) Suppose that $f\in H^p\cap\mathcal{S}$. Then $\int_{\mathbf{R}^n}x^\gamma f(x)dx=0$ for all $|\gamma|\le[\frac{n}{p}-n]$. This implies that $f^+(x)\le C(1+|x|)^{-([\frac{n}{p}-n]+1+n)}$ and in turn this implies that $f\in H^q$ for $[\frac{n}{q}-n]\le[\frac{n}{p}-n]$. This implies the required conclusion since $[\frac{n}{p}]=[\frac{n}{q}]$ for $p,q\in(\frac{n}{n+k+1},\frac{n}{n+k}]$, whenever $k\ge 0$.

(ii) The proof is similar to (i) but notice that a given $f\in H^{p,\infty}$ satisfies the cancellation condition that $\int_{\mathbf{R}^n}x^\gamma f(x)\,dx=0$ for $|\gamma|\le-[n-\frac{n}{p}]-1$.

(iii) This is a consequence of the fact that $[r]=-[-r]-1$ for r is not an integer, Theorem 7 and Theorem 8.

(iv) Let's consider only the H^p case since it's equivalent to the $H^{p,\infty}$ case. We can consider $g\in\mathcal{S}_k$ such that $g\notin\mathcal{S}_{k+1}$ (e.g. $g(x)=x^k e^{-|x|^2}$). This g is in H^p for $p\in(\frac{n}{n+k+1},\frac{n}{n+k}]$ but not $p=\frac{n}{n+k+1}$, otherwise $g\in\mathcal{S}_{k+1}$ by Theorem 7.

(v) If we had $f^+(x)\le\frac{C}{(1+|x|)^{n+k+1+\epsilon}}$ for some $\epsilon>0$ and all $x\in\mathbf{R}^n$, then it follows from Corollary 1 that f would belong to $H^{\frac{n}{n+k+1}}$, which is not true by (iv). □

The existence of a function g in (iv) shows that we cannot improve the cancellation of Theorem 7 for H^p and $H^{p,\infty}$.

5 Square function characterization of $H^{p,\infty}$

In this section we outline the proof of a new characterization of weak Hardy spaces in terms of Littlewood–Paley square functions; details can be found in [16]. We begin by stating an interpolation theorem and a consequence of it first.

Theorem 9. *Let J and L be positive integers and let $0<p_1<p<p_2<\infty$, moreover $p_1\le 1$.*

(a) *Let T be a sublinear operator defined on $H^{p_1}(\mathbf{R}^n,\ell^2(L))+H^{p_2}(\mathbf{R}^n,\ell^2(L))$. Assume that T maps $H^{p_1}(\mathbf{R}^n,\ell^2(L))$ to $H^{p_1}(\mathbf{R}^n,\ell^2(J))$ with constant A_1 and*

$H^{p_2}(\mathbf{R}^n,\ell^2(L))$ to $H^{p_2}(\mathbf{R}^n,\ell^2(J))$ with constant A_2. Then there exists a constant $c_{p_1,p_2,p,n}$ independent of J and L such that

$$\|T(\vec{F})\|_{H^{p,\infty}(\mathbf{R}^n,\ell^2(J))} \le c_{p_1,p_2,p,n}\, A_1^{\frac{\frac{1}{p}-\frac{1}{p_2}}{\frac{1}{p_1}-\frac{1}{p_2}}} A_2^{\frac{\frac{1}{p_1}-\frac{1}{p}}{\frac{1}{p_1}-\frac{1}{p_2}}} \|\vec{F}\|_{H^{p,\infty}(\mathbf{R}^n,\ell^2(L))}$$

for $\vec{F}\in H^{p,\infty}(\mathbf{R}^n,\ell^2(L))$.

(b) *Suppose that T is a sublinear operator defined on $H^{p_1}(\mathbf{R}^n,\ell^2(L))+H^{p_2}(\mathbf{R}^n,\ell^2(L))$. Assume that T maps $H^{p_1}(\mathbf{R}^n,\ell^2(L))$ to $L^{p_1}(\mathbf{R}^n,\ell^2(J))$ with constant A_1 and also maps $H^{p_2}(\mathbf{R}^n,\ell^2(L))$ to $L^{p_2}(\mathbf{R}^n,\ell^2(J))$ with constant A_2. Then there exists a constant C independent of J and L such that*

$$\|T(\vec{F})\|_{L^{p,\infty}(\mathbf{R}^n,\ell^2(J))} \le c_{p_1,p_2,p,n}\, A_1^{\frac{\frac{1}{p}-\frac{1}{p_2}}{\frac{1}{p_1}-\frac{1}{p_2}}} A_2^{\frac{\frac{1}{p_1}-\frac{1}{p}}{\frac{1}{p_1}-\frac{1}{p_2}}} \|\vec{F}\|_{H^{p,\infty}(\mathbf{R}^n,\ell^2(L))}$$

for all distributions $\vec{F}\in H^{p,\infty}(\mathbf{R}^n,\ell^2(L))$.

Corollary 3. *Let $0<p<\infty$ and suppose that $\{K_j(x)\}_{j=1}^L$ is a family of kernels defined on $\mathbf{R}^n\setminus\{0\}$ satisfying*

$$\sum_{j=1}^{L} |\partial^\alpha K_j(x)| \le A|x|^{-n-|\alpha|} < \infty$$

for all $|\alpha|\le \max\{[n/p]+2, n+1\}$ and

$$\sup_{\xi\in\mathbf{R}^n} \sum_{j=1}^{L} |\widehat{K_j}(\xi)| \le B < \infty\,.$$

Then for some $0<p$ there exists a constant $C_{n,p}$ independent of L such that

$$\Big\|\sum_{j=1}^{L} K_j * f_j\Big\|_{H^{p,\infty}(\mathbf{R}^n)} \le C_{n,p}(A+B)\|\{f_j\}_{j=1}^L\|_{H^{p,\infty}(\mathbf{R}^n,\ell^2(L))}. \tag{39}$$

We fix a radial function $\Psi\in\mathcal{S}(\mathbf{R}^n)$ such that $\widehat{\Psi}$ is nonnegative, supported in the annulus $\frac{1}{2}+\frac{1}{10}\le|\xi|\le 2-\frac{1}{10}$, and satisfies

$$\sum_{j\in\mathbf{Z}} \widehat{\Psi}(2^{-j}\xi) = 1 \tag{40}$$

for all $\xi\neq 0$. We define the related Littlewood-Paley operators Δ_j by

$$\Delta_j(f) = \Delta_j^{\Psi}(f) = \Psi_{2^{-j}} * f\,. \tag{41}$$

We also define the function Φ by $\widehat{\Phi}(\xi) = \sum_{j\le 0} \widehat{\Psi}(2^{-j}\xi)$ for $\xi\neq 0$ and $\widehat{\Phi}(0)=1$. We now provide a sketch of the proof of Theorem 2.

Proof. Choose $f \in H^{p,\infty}$ and denote $f_M = \sum_{|j|\le M} \Delta_j(f) = \Phi_{2^{-M}} * f - \Phi_{2^M} * f$ and $S(f) = (\sum_{|j|\le M} |\Delta_j(f)|^2)^{\frac{1}{2}}$. Then by Theorem 9 it follows that S maps $H^{p,\infty}$ to $L^{p,\infty}$ bounded for $p \in (p_1, p_2)$, so

$$\|S(f)\|_{L^{p,\infty}} \le C\|f\|_{H^{p,\infty}}.$$

Applying Fatou's lemma for $L^{p,\infty}$ spaces we have

$$\|(\sum_{j\in\mathbf{Z}} |\Delta_j(f)|^2)^{\frac{1}{2}}\|_{L^{p,\infty}} \le \liminf_{M\to\infty} \|(\sum_{|j|\le M} |\Delta_j(f)|^2)^{\frac{1}{2}}\|_{L^{p,\infty}} \le C\|f\|_{H^{p,\infty}}.$$

Now let's assume we have a distribution $f \in \mathcal{S}'$ such that

$$\|(\sum_{j\in\mathbf{Z}} |\Delta_j(f)|^2)^{\frac{1}{2}}\|_{L^{p,\infty}} < \infty.$$

By lemma 6.5.3 of [13] and Proposition 1, we can show that $\{\Delta_j * f\}_{j\in\mathbf{Z}} \in H^{p,\infty}(\mathbf{R}^n, l^2)$ with that

$$\|\sup_{t>0}(\sum_{j\in\mathbf{Z}} |\varphi_t * \Delta_j(f)|^2)^{\frac{1}{2}}\|_{L^{p,\infty}} \le C_p'\|(\sum_{j\in\mathbf{Z}}(|\Delta_j(f)|)^2)^{\frac{1}{2}}|_{L^{p,\infty}}.$$

Let $\widehat{\eta}(\xi) = \widehat{\Psi}(\xi/2) + \widehat{\Psi}(\xi) + \widehat{\Psi}(2\xi)$, then by Corollary 3

$$\|\sum_{|j|\le M} \Delta_j(f)\|_{H^{p,\infty}} \le C\|(\sum_{j\in\mathbf{Z}}(|\Delta_j(f)|)^2)^{\frac{1}{2}}\|_{L^{p,\infty}}.$$

So $\{\sum_{|j|\le M} \Delta_j(f)\}_M$ is a bounded sequence in $H^{p,\infty}$ uniformly in M. Then we use the following lemma contained in [16].

Lemma 5. *If $\{f_j\}$ is bounded by B in $H^{p,\infty}$ (or H^p), then there exists a subsequence $\{f_{j_k}\}$ such that $f_{j_k} \to f$ in $\mathcal{S}'$ for some f in $H^{p,\infty}$ (or H^p) with $\|f\|_{H^{p,\infty}} \le B$ (or $\|f\|_{H^p} \le B$).*

By the lemma $\sum_{|j|\le M_k} \Delta_j(f) \to g$ in $\mathcal{S}'$ with

$$\|g\|_{H^{p,\infty}} \le C\|(\sum_{j\in\mathbf{Z}} |\Delta_j(f)|^2)^{\frac{1}{2}}\|_{L^{p,\infty}}.$$

Moreover there is a unique polynomial Q such that $g = f - Q$. □

Different functions Ψ provide spaces with equivalent norms. This is easily seen by the above characterization.

Corollary 4. *The definition of space*

$$F^{p,\infty} := \{f \in \mathcal{S}' : \|(\sum_{j\in\mathbf{Z}} |\Delta_j(f)|^2)^{\frac{1}{2}}\|_{L^{p,\infty}} < \infty\}$$

is independent of the choice of Ψ.

This is a consequence of Theorem 2 and can also be a consequence of Proposition 1. The argument also applies to

$$F_p^{\alpha,q} = \{f \in \mathcal{S}' : \ \|(\sum_{j\in\mathbf{Z}} |2^{j\alpha}\Delta_j(f)|^q)^{\frac{1}{q}}\|_{L^p} < \infty\},$$

so we are allowed to define more general spaces generalizing $H^{p,\infty}$,

$$F_{p,\infty}^{\alpha,q} = \{f \in \mathcal{S}' : \ \|(\sum_{j\in\mathbf{Z}} |2^{j\alpha}\Delta_j(f)|^q)^{\frac{1}{q}}\|_{L^{p,\infty}} < \infty\}.$$

The square function characterization of weak H^p has useful applications; for instance it was used in [14] to obtain weak type endpoint estimates for multilinear paraproducts.

References

[1] W. Abu-Shammala and A. Torchinsky, The Hardy-Lorentz spaces $H^{p,q}(R^n)$, Studia Mathematica **182** (2007), 283–294.

[2] J. Alvarez, H^p and weak H^p continuity of Calderón-Zygmund type operators, Fourier Analysis: Analytic and Geometric Aspects **157** (1994), 17–34.

[3] R. R. Coifman, A real variable characterization of H^p, Studia Math. **51** (1974), 269–274.

[4] R. R. Coifman and G. Weiss, Extensions of Hardy spaces and their use in analysis, Bull. Amer. Math. Soc. (N.S.) **83** (1977), 569–645.

[5] D. Cruz-Uribe, J.-M. Martell, and C. Pérez, Weights, Extrapolation and the Theory of Rubio de Francia, Operator Theory, Advances and Applications **215**, Birkhäuser, Springer, Basel AG 2011.

[6] Y. Ding, S. Lu, and Q. Xue, Parametrized Littlewood-Paley operators on Hardy and weak Hardy spaces, Math. Nachr. **280** (2007), 351–363.

[7] X. T. Duong and L. Yan, Hardy spaces of spaces of homogeneous type, Proc. Amer. Math. Soc. **131** (2003), 3181–3189.

[8] C. Fefferman, N. M. Riviere, and Y. Sagher, Interpolation between H^p spaces: the real method, Trans. Amer. Math. Soc. **191** (1974), 75–81.

[9] R. Fefferman and F. Soria, The space Weak H^1, Studia Math. **85** (1987), 1–16.

[10] C. Fefferman and E. M. Stein, Some maximal inequalities, Amer. J. Math. **93** (1971), 107–115.

[11] J. García-Cuerva and J.-L. Rubio de Francia, Weighted Norm Inequalities and Related Topics, North-Holland Math. Studies **116**, North-Holland, Amsterdam, 1985.

[12] L. Grafakos, Classical Fourier Analysis, Second Edition, Graduate Texts in Math. **249**, Springer, New York, 2008.

[13] L. Grafakos, Modern Fourier Analysis, Second Edition, Graduate Texts in Math. **250**, Springer, New York, 2008.

[14] L. Grafakos, D. He, N. Kalton, and M. Mastyło, Multilinear paraproducts revisited, Revista Matematica Iber, to appear.

[15] Y. Han, D. Müller, and D. Yang, Littlewood-Paley characterizations for Hardy spaces on spaces of homogeneous type, Math. Nachr. **279** (2006), 1505–1537.

[16] D. He, Square function characterization of weak Hardy spaces, J. Fourier Anal. Appl., to appear.

[17] G. Hu, D. Yang, and Y. Zhou, Boundedness of singular integrals in Hardy spaces on spaces of homogeneous type, Taiwanese J. Math. **13** (2009), 91–135.

[18] R. H. Latter, A decomposition of $H^p(\mathbf{R}^n)$ in terms of atoms, Studia Math. **62** (1977), 92–101.

[19] S. Lu, Four Lectures on Real H^p Spaces, World Scientific, Singapore, 1995.

[20] S. Lu and D. Yang, The Littlewood-Paley function and φ-transform characterizations of a new Hardy space HK_2 associated with the Herz space, Studia Math. **101** (1992), 285–298.

[21] S. Lu and D. Yang, Multiplier theorems for Herz type Hardy spaces, Proc. Amer. Math. Soc. **126** (1998), 3337–3346.

[22] R. A. Macías and C. Segovia, A decomposition into atoms of distributions on spaces of homogeneous type, Adv. in Math. **33** (1979), 271–309.

[23] J. Peetre, New Thoughts on Besov Spaces, Duke University Math. Series **1**, Durham, 1976.

[24] E. M. Stein, Harmonic Analysis: Real-Variable Methods, Orthogonality, and Oscillatory Integrals, Princeton University Press, 1993.

[25] H. Triebel, Theory of Function Spaces, Birkhäuser-Verlag, Basel, Boston, Stuttgart, 1983; and Akad. Verlagsges. Geest & Portig, Leipzig, 1983.

Some Topics in Harmonic Analysis
and Applications
ALM 34, pp. 203–229

A Local Tb Theorem with Vector-valued Testing Functions

Ana Grau de la Herrán* and Steve Hofmann†

Abstract

A local Tb theorem is an L^2 boundedness criterion by which the question of the global behavior of an operator is reduced to its local behavior, acting on a family of test functions b_Q indexed by the dyadic cubes. Typically, these test functions are either scalar valued (in the case of a scalar valued operator T), or matrix valued (in the case of vector valued T). In the present paper, we present a variant of the local Tb theorem for square functions with a vector valued kernel, in which the testing functions b_Q are also vector valued (rather than matrix valued). We then discuss the applicability of this result to the question of boundedness of layer potentials associated to certain complex elliptic operators in divergence form.

2000 Mathematics Subject Classification: 42B20.

Keywords and Phrases: local Tb Theorems, square functions, layer potentials.

1 Introduction, history, preliminaries

The Tb theorem, like its predecessor, the $T1$ Theorem, is an L^2 boundedness criterion, originally proved by McIntosh and Meyer [21], and by David, Journé and Semmes [10] in the context of singular integrals, but later extended by Semmes [25] to the setting of "square functions". The latter arise in many applications in complex function theory and in PDE, and may be viewed as singular integrals taking values in a Hilbert space.

The essential idea of Tb and $T1$ type theorems, is that they reduce the question of L^2 boundedness to verifying the behavior of an operator on a single test function b (or the constant function 1). In the present paper, we present a variant of the

*Department of Mathematics and Statistics, University of Helsinki, Finland.
†Department of Mathematics, University of Missouri, Columbia, USA.

"local" version of such a result. So-called "local" Tb theorems are related to previous work of Christ [7], who proved the first local Tb theorem in the singular integral setting. The term "local" in this context, refers to the fact that, instead of one globally defined testing function b, one is allowed to test the operator locally, say on each dyadic cube, with a local testing function that is adapted to that cube. The advantage here, in applications, is the additional flexibility that one gains: it may be easier to verify "good" behavior of the operator locally, when the testing functions are allowed to vary. The point is that sometimes particular properties of the operator may be exploited to verify the appropriate testing criterion.

Typically, these test functions are either scalar valued (in the case of a scalar valued operator T), or matrix valued (in the case of vector valued T). We remark that, for example, a specialized local Tb theorem for square functions, with matrix valued testing functions, lies at the heart of the solution of the Kato square root problem [19, 17, 5]. In the present paper, we present a variant of the local Tb theorem for square functions built with a vector valued "Littlewood-Paley" operator, in which the testing functions b_Q are also vector valued (rather than matrix valued). To illustrate the use of this theorem, we apply it to obtain L^2 boundedness of layer potentials associated to certain complex divergence form elliptic equations and systems, a result which has previously been established by Rosén [24] by another method (albeit based also on local Tb theory). We present here only a restricted version of Rosén's result, in which we assume that interior estimates of De Giorgi/Nash/Moser ("DG/N/M") type hold for solutions of $Lu := -\operatorname{div} A\nabla u = 0$. Rosén's theorem does not require the DG/N/M hypothesis. We expect that the general case of Rosén's theorem can be treated by our methods, and we plan to do this in a future paper, but for now we content ourselves with the restricted version, our main purpose here being to illustrate the utility of the local Tb theorem with vector valued testing functions. For the sake of those readers who may be interested in the PDE application, we note that DG/N/M estimates are available for complex L^∞ perturbations of real operators (or of diagonal systems), and, for the class of operators under consideration (namely, those with "radially independent" coefficients), hold always when the ambient dimension is at most 3.

1.1 Summary of results

We present here a brief synopsis of the results that we prove in this paper, and provide some explanatory context. The precise statements of our theorems will appear in the sequel, as noted. In these results, $\{\Theta_t\}$ will be a family of operators mapping $L^2(\mathbb{R}^n, \mathbb{C}^m)$, $m \geq 1$, uniformly into $L^2(\mathbb{R}^n, \mathbb{C})$, for which we seek to prove the square function bound

$$\iint_{\mathbb{R}^{n+1}_+} |\Theta_t f(x)|^2 \frac{dtdx}{t} \leq C \, \|f\|^2_{L^2(\mathbb{R}^n)}. \tag{1.1}$$

Theorem 2.13. We prove a local Tb theorem for square functions, in which the testing functions are vector-valued, rather than matrix-valued. This theorem is stated and proved in Section 2.

Theorem 3.23. We apply Theorem 2.13 to establish L^2 bounds for layer potentials associated to a divergence form (complex) elliptic operator L in $\mathbb{R}^{n+1}_+$, with t-independent coefficients, assuming that null solutions of L satisfy local Hölder continuity estimates of De Giorgi/Nash type. Thus, we recover a result obtained in [24], by a direct proof which bypasses the functional calculus formalism elaborated in [4]. As noted above, this result is a restricted version of that of [24], as the De Giorgi/Nash hypothesis is not assumed in the latter paper. This theorem is stated and proved in Section 3.

Remark. As mentioned above, the novelty of Theorem 2.13 is that the testing functions are *vector*-valued rather than matrix-valued. This point of view turns out to be essential to our approach to the application to layer potentials.

In an appendix, we also prove

Theorem 4.5. This is an auxiliary result, which generalizes the "$T1$ Theorem for square functions" of [8]. The proof is a routine adaptation of that of [8].

1.2 Notation and some definitions

- We shall use the letters c, C to denote positive constants, not necessarily the same at each occurrence, which depend only on dimension and the constants appearing in the hypotheses of the theorems. We shall also write $A \lesssim B$ and $A \approx B$ to mean, respectively, that $A \leq CB$ and $0 < c \leq A/B \leq C$, where the constants c and C are as above, unless explicitly noted. Moreover if we want to specify any particular dependency of the constant we will denote it by subscript or by $C(\cdot)$, e.g., C_n or $C(n)$ is a constant that depends on dimension n.

- We denote points in $\mathbb{R}^{n+1}$ by $(x,t) \in \mathbb{R}^n \times \mathbb{R} = \mathbb{R}^{n+1}$ (we use the notational convention that $x_{n+1} = t$), or sometimes, for convenience, by capital letter X.

- We set $\mathbb{R}^{n+1}_+ := \mathbb{R}^n \times (0, +\infty)$ and $\partial\mathbb{R}^{n+1}_+ := \mathbb{R}^n \times \{0\}$.

- For a Borel set $A \subset \mathbb{R}^{n+1}$, we let $\mathbb{1}_A$ denote the usual indicator function of A, i.e. $\mathbb{1}_A(x) = 1$ if $x \in A$, and $\mathbb{1}_A(x) = 0$ if $x \notin A$.

- The letter Q will be used to denote a cube in $\mathbb{R}^n$, we let $\ell(Q)$ denote its side length, and we shall write Q_r to denote that the cube has side length $\ell(Q) = r$.

- We let $\mathbb{D}$ denote the collection of all closed dyadic cubes in $\mathbb{R}^n$, and let $\mathbb{D}_k$ denote the grid of dyadic cubes of side length 2^{-k}.

- For a Borel set $A \subset \mathbb{R}^n$, a Borel measure μ defined on $\mathbb{R}^n$, and a Borel measurable function f, we set $\fint_A f\,d\mu = \frac{1}{\mu(A)} \int_A f\,d\mu$.

- Let $q \in [1,\infty]$, we denote by $q' \in [1,\infty]$, the number such that we have $\frac{1}{q} + \frac{1}{q'} = 1$, where as usual we define $1/\infty := 0$.

- For a Banach space X, we let $\mathcal{B}(X)$ denote the space of bounded linear operators on X.

Next, we define some auxiliary operators.

Definition 1.2. We say that a family of convolution operators $\{Q_s\}_{s>0}$ is a **CLP family** ("Calderón-Littlewood-Paley" family) if for some $\sigma > 0$, and some $\psi \in L^1(\mathbb{R}^n)$, with $|\psi(x)| \lesssim (1+|x|)^{-n-\sigma}$, we have

$$Q_s f = s^{-n}\psi(\cdot/s) * f \,, \text{ and } \widehat{\psi}(\xi) \leq C \min\left(|\xi|^\sigma, |\xi|^{-\sigma}\right),$$

$$\sup_{s>0}\left(\|Q_s f\|_{L^2(\mathbb{R}^n)} + \|s\nabla Q_s f\|_{L^2(\mathbb{R}^n)}\right) \leq C\|f\|_{L^2(\mathbb{R}^n)},$$

$$\int_{\mathbb{R}^n}\int_0^\infty |Q_s f(x)|^2 \frac{dsdx}{s} \leq C\|f\|^2_{L^2(\mathbb{R}^n)},$$

$$\int_0^\infty Q_s^2 \frac{ds}{s} = I\,,$$

where convergence to the identity in the last formula is in the strong operator topology on $\mathcal{B}(L^2)$.

Definition 1.3. We define the Hardy-Littlewood maximal operator $\mathcal{M}$, acting on $f \in L^1_{loc}(\mathbb{R}^n)$, by

$$\mathcal{M}(f)(x) := \sup_{r>0} \frac{1}{|B_r(x)|}\int_{B_r(x)} f(y)\,dy\,,$$

where $B_r(x)$ is the ball centered at x and radius r.

Definition 1.4. We say that P_t is a nice approximate identity, if P_t is an operator of convolution type, with a smooth, compactly supported kernel Φ. That means that for a function $f : \mathbb{R}^n \to \mathbb{C}$,

$$P_t f = \Phi_t * f\,, \text{ with } \Phi_t = t^{-n}\Phi\left(\frac{x}{t}\right), \int \Phi(x)dx = 1\,, \quad \Phi_t \in C_0^\infty(\mathbb{R}^n).$$

In particular then, $P_t 1 = 1$, and

$$(P_t f)(x) \leq C\mathcal{M}f(x)\,.$$

Definition 1.5. For $0 < s \leq 1$, the homogeneous Sobolev space $\dot{L}^2_s$ is the completion of C_0^∞ with respect to the norm $\|f\|_{\dot{L}^2_s} := \|(-\Delta)^{s/2})f\|_{L^2}$, where Δ is the usual Laplacian.

1.3 Some useful results

We collect here some known results that we shall invoke in the sequel.

Lemma 1.6 ([1, Lemma 3.11]). *Suppose that Θ_t is an operator satisfying*

$$\|\Theta_t(f\mathbb{1}_{2^{k+1}Q\setminus 2^kQ})\|_{L^2(Q)} \leq C2^{-k(n+2)/2}\,\|f\|_{L^2(2^{k+1}Q\setminus 2^kQ)}\,, \tag{1.7}$$

whenever $t \approx \ell(Q)$, and that $\|\Theta_t\|_{2->2} \leq C$. Let $b \in L^\infty(\mathbb{R}^n)$, and let $\mathcal{A}_t$ denote a self-adjoint "averaging operator" whose kernel $\varphi_t(x)$ satisfies $0 \leq \varphi_t(x) \leq Ct^{-n}\mathbb{1}_{\{|x|<Ct\}}$, $\int \varphi_t(x)dx = 1$. Then $\Theta_t b$ is well-defined as an element of $L^2_{loc}(\mathbb{R}^n)$, and

$$\sup_{t>0}\|(\Theta_t b)\mathcal{A}_t f\|_{L^2(\mathbb{R}^n)} \leq C\|b\|_{L^\infty(\mathbb{R}^n)}\|f\|_{L^2(\mathbb{R}^n)}. \tag{1.8}$$

Lemma 1.9 ([1, Lemma 3.5]). *Suppose that* $\{R_t\}_{t\in\mathbb{R}}$ *is a family of operators satisfying*

$$\|R_t(f\mathbb{1}_{2^{k+1}Q\setminus 2^kQ})\|_{L^2(Q)} \le C2^{-k(n+2+\beta)/2}\,\|f\|_{L^2(2^{k+1}Q\setminus 2^kQ)}\,, \tag{1.10}$$

for all $t \approx \ell(Q)$, *and for some* $\beta > 0$. *Suppose also that* $\sup_t \|R_t\|_{2\to 2} \le C$, *and that* $R_t 1 = 0$ *for all* $t > 0$. *Then for* $h \in \dot{L}^2_1(\mathbb{R}^n)$,

$$\int_{\mathbb{R}^n} |R_t h|^2 \le Ct^2 \int_{\mathbb{R}^n} |\nabla_x h|^2. \tag{1.11}$$

If, in addition, $\|R_t\nabla_x\|_{2\to 2} \le C/t$, *then also*

$$\iint_{\mathbb{R}^{n+1}_+} |R_t f(x)|^2 \frac{dxdt}{t} \le C\,\|f\|_2^2. \tag{1.12}$$

Remark 1.13 ([1, Lemma 3.11 and Lemma 3.5]). are stated for operators satisfying the off-diagonal decay estimates (1.7) and (1.10) respectively, for all $t \lesssim \ell(Q)$, but the proof in [1] actually requires only that off-diagonal decay bounds hold with $t \approx \ell(Q)$. Also [1, Lemma 3.5] is stated under the hypothesis that (1.10) holds with $\beta \ge 2$, but the proof given there actually requires only that $\beta > 0$, as we assume here.

2 A local Tb theorem with vector-valued testing functions

In this section, we consider a version of the local Tb theorem for vector-valued Θ_t, in which the testing functions themselves are vector-valued rather than matrix-valued. In the following section, we shall then apply this version to establish L^2 bounds for layer potentials associated to a class of divergence form elliptic operators in the half-space $\mathbb{R}^{n+1}_+$.

We shall impose the following basic hypothesis.

Hypothesis 2.1. We consider a family of operators $\{\Theta_t\}_{t>0}$, with

$$\Theta_t := (\Theta_t^0, \Theta_t') := (\Theta_t^0, \Theta_t^1, \Theta_t^2, ..., \Theta_t^m)\,, \quad m \ge 1\,,$$

where each Θ_t^j, $0 \le j \le m$, maps $L^2(\mathbb{R}^n, \mathbb{C})$ into itself, and where for $g = (g^0, g') := (g^0, g^1, ..., g^m) \in L^2(\mathbb{R}^n, \mathbb{C}^{m+1})$, we set

$$\Theta_t g = \sum_{j=0}^m \Theta_t^j g^j\,, \qquad \Theta_t' g' = \sum_{j=1}^m \Theta_t^j g^j.$$

We suppose that Θ_t satisfies the following properties:

(a) (Uniform L^2 bounds and off-diagonal decay in L^2).

$$\sup_{t>0} ||\Theta_t g||_{L^2(\mathbb{R}^n)} \le C||g||_{L^2(\mathbb{R}^n, \mathbb{C}^{m+1})}\,, \tag{2.2}$$

$$||\Theta_t g_j||_{L^2(Q)} \leq C2^{-j(n+2+\beta)/2}||g_j||_{L^2(2^{j+1}Q\setminus 2^jQ)}\,, \quad \ell(Q) \leq t \leq 2\ell(Q)\,, \tag{2.3}$$

for some $\beta > 0$, where $g_j := g\mathbb{1}_{2^{j+1}Q\setminus 2^jQ}$;

(b) (Quasi-orthogonality on a subspace of L^2). There exists a subspace H of $L^2(\mathbb{R}^n, \mathbb{C}^m)$, such that for some (hence every) CLP family $\{Q_s\}$ (cf. Definition 1.2), there is a $\beta > 0$ for which we have

$$||\Theta_t Q_s h||_{L^2(\mathbb{R}^n)} \leq C\left(\frac{s}{t}\right)^\beta ||h||_{L^2(\mathbb{R}^n)}, \quad \forall h = (h^0, h') \in L^2(\mathbb{R}^n, \mathbb{C}) \times H\,,\ \forall s \leq t\,. \tag{2.4}$$

We further define the action of Θ_t (respectively, Θ'_t) on matrix-valued $B = (B_{ij})_{0\leq i,j\leq m} : \mathbb{R}^n \to \mathbb{M}^{m+1}$ (resp., $B' = (B_{ij})_{1\leq i,j\leq m} : \mathbb{R}^n \to \mathbb{M}^m$) in the obvious way:

$$\Theta_t B(x) = \left(\sum_{i=0}^m \Theta_t^i(x,y)\, B_{ij}\right)_{0\leq j\leq m}\,, \qquad \Theta'_t B'(x) = \left(\sum_{i=1}^m \Theta_t^i(x,y)\, B_{ij}\right)_{1\leq j\leq m}\,. \tag{2.5}$$

Remark 2.6. Let us note that it is only in the quasi-orthogonality condition (b), that we restrict to a sub-space $H \subset L^2(\mathbb{R}^n, \mathbb{C}^m)$; in condition (a), we suppose that the operator Θ_t is allowed to act on general C^{m+1}-valued functions in L^2.

For Θ_t as in Hypothesis 2.1 above, for convenience of notation, we set

$$\zeta(x,t) := \Theta_t 1(x)\,, \quad \zeta'(x,t) := \Theta'_t 1(x)\,, \quad \zeta^0(x,t) := \Theta_t^0 1(x)\,, \tag{2.7}$$

where in the previous line, "1" denotes, respectively, the $(m+1)\times(m+1)$ identity matrix. the $m \times m$ identity matrix, and the scalar constant 1. For $\delta \in [0,1)$, we also set

$$(\Theta_t 1)_\delta = \zeta_\delta(x,t) := \zeta(x,t) 1_{\{\delta<t<1/\delta\}}\,, \tag{2.8}$$

and similarly for $(\Theta'_t 1)_\delta = \zeta'_\delta$ and $\left(\Theta_t^0 1\right)_\delta = \zeta_\delta^0$.

Definition 2.9. Given Υ (either scalar or vector valued), defined on $\mathbb{R}^{n+1}_+$, we set

$$\|\Upsilon\|_C := \sup_Q \frac{1}{|Q|}\int_0^{\ell(Q)}\int_Q |\Upsilon(x,t)|^2 \frac{dxdt}{t}\,.$$

Definition 2.10. Given a dyadic cube Q, and a positive constant C_0, we shall say that a Borel measure μ_Q on $\mathbb{R}^n$ is "adapted to Q with constant C_0", if

$$d\mu_Q \;=\; \phi_Q\, dx\,,$$

where $\phi_Q : \mathbb{R}^n \to [0,1]$ is a Lipschitz function satisfying

$$\|\nabla\phi_Q\|_{L^\infty(\mathbb{R}^n)} \leq C_0\, \ell(Q)^{-1}\,, \tag{2.11}$$

and also

$$\frac{1}{C_0} \leq \phi_Q\,, \quad \text{on } Q\,. \tag{2.12}$$

We observe that ϕ_Q, being Lipschitz, is not supported in Q, by (2.12).
The main result of this paper is the following

Theorem 2.13. *Let Θ_t and H be as in Hypothesis 2.1, define ζ, ζ_δ, etc. as in (2.7)–(2.8), and suppose that there exist positive constants σ, η, C_0 and C_1, a system $\{b_Q\} = \{(b_Q^0, b_Q')\} \subset L^2(\mathbb{R}^n, \mathbb{C}) \times H \subset L^2(\mathbb{R}^n, \mathbb{C}^{m+1})$, indexed by dyadic cubes $Q \subset \mathbb{R}^n$, and a system of measures $\{\mu_Q\}$, also indexed by dyadic cubes, with each μ_Q "adapted to Q with constant C_0" as in Definition 2.10, such that for every dyadic cube Q, we have*

$$\int_{\mathbb{R}^n} |b_Q(x)|^2 dx \leq C_0 |Q|, \tag{2.14}$$

$$\int_0^{\ell(Q)} \int_Q |\Theta_t b_Q(x)|^2 \frac{dxdt}{t} \leq C_0 |Q|, \tag{2.15}$$

$$\sigma \leq \mathcal{R}e \left(\fint_Q b_Q^0 \, d\mu_Q \right), \tag{2.16}$$

$$\left| \fint_Q b_Q' \, d\mu_Q \right| \leq \eta\sigma, \tag{2.17}$$

with $\eta \leq 1/(2C_1 + 4)$, where, for all sufficiently small $\delta > 0$ (hence also for $\delta = 0$),

$$\| (\Theta_t' 1)_\delta \|_{\mathcal{C}} \leq C_1 \left(1 + \| (\Theta_t^0 1)_\delta \|_{\mathcal{C}}\right). \tag{2.18}$$

Then for all $f^0 \in L^2(\mathbb{R}^n, \mathbb{C})$,

$$\iint_{\mathbb{R}^{n+1}_+} |\Theta_t^0 f^0(x)|^2 \frac{dxdt}{t} \leq C \|f^0\|_2^2. \tag{2.19}$$

Remark 2.20. Given the hypotheses of Theorem 2.13, we also have that for all $h' \in H$,

$$\iint_{\mathbb{R}^{n+1}_+} |\Theta_t' h'(x)|^2 \frac{dxdt}{t} \leq C \|h'\|_2^2. \tag{2.21}$$

Indeed, by a well known argument of [12], (2.19) implies that $\|\Theta_t^0 1\|_{\mathcal{C}} \leq C$, whence (2.21) follows immediately by (2.18) and a generalization (Theorem 4.5 below) of the "$T1$ Theorem for square functions" of [8].

Remark 2.22. It may be that in some applications, one could simply take $\phi_Q \equiv 1$, i.e., $d\mu_Q = dx$, but in our application in Section 3, it is useful to have the extra flexibility inherent in (2.16)–(2.17).

Remark 2.23. As a practical matter, in applications, one expects to construct b_Q' depending on η, where the parameter η in (2.17) is at one's disposal and can therefore be made sufficiently small. This is precisely what we shall do in applying Theorem 2.13 to the study of variable coefficient layer potentials, in Section 3. We emphasize that η is required to be small depending only upon C_1, but not on C_0; thus, in applications, there is no harm if C_0 becomes larger as η becomes smaller, as indeed, will be the case in our application in Section 3.

Remark 2.24. We observe that condition (2.18), in which the Carleson measure norm of $\Theta_t^0 1$ controls that of $\Theta_t' 1$, reflects a fairly common occurrence in PDE: namely, control of a vector valued function, in some norm, by one of its components. Indeed, in our application to layer potentials, the verification of (2.18) depends on exploiting a PDE relationship between Θ_t^0 and Θ_t', and using the solution of the Kato problem. Similar considerations are exploited in [14] and (without the Kato problem) in [26].

Proof. As is always the case with local Tb theorems for square functions, the proof is based on the arguments used to solve the Kato square root problem [19, 17, 5]. By the case $m = 1$, $H = L^2(\mathbb{R}^n, \mathbb{C})$ of Theorem 4.5 (the generalized $T1$ Theorem of [8]), it is enough to show that $\|\zeta^0\|_{\mathcal{C}} \leq C$ (we are using the notation of (2.7)). We proceed under the *a priori* qualitative assumption that $\|\zeta^0\|_{\mathcal{C}}$ is finite, but with no particular quantitative bound. We may do this by working with ζ_δ in place of ζ, as long as we obtain bounds that are independent of δ. To simplify the notation, we shall simply write ζ, not ζ_δ. We define

$$F_1 := \{(x,t) \in \mathbb{R}^{n+1}_+ : |\zeta^0(x,t)| \leq \sqrt{\eta}\,|\zeta'(x,t)|\},$$

$$F_2 := \{(x,t) \in \mathbb{R}^{n+1}_+ : |\zeta^0(x,t)| > \sqrt{\eta}\,|\zeta'(x,t)|\},$$

so that

$$\|\zeta^0\|_{\mathcal{C}} \leq \|\zeta^0 1_{F_1}\|_{\mathcal{C}} + \|\zeta^0 1_{F_2}\|_{\mathcal{C}}. \tag{2.25}$$

By Definition 2.9, and (2.18), we have

$$\|\zeta^0 1_{F_1}\|_{\mathcal{C}} \leq \eta\,\|\zeta'\|_{\mathcal{C}} \leq C_1\eta(1 + \|\zeta^0\|_{\mathcal{C}}).$$

Since $\eta \leq 1/(2C_1)$, we may hide the term $C_1\eta\,\|\zeta^0\|_{\mathcal{C}}$ on the left hand side of (2.25). Thus, it suffices to show that $\|\zeta^0 1_{F_2}\|_{\mathcal{C}} \leq C$. To this end, we first note that by (2.16) and (2.17),

$$\begin{aligned}\sigma|\zeta^0| &\leq \left|\zeta^0 \fint_Q b_Q^0\,d\mu_Q\right| \leq \left|\zeta \cdot \fint_Q b_Q\,d\mu_Q\right| + \left|\zeta' \cdot \fint_Q b_Q'\,d\mu_Q\right| \\ &\leq \left|\zeta \cdot \fint_Q b_Q\,d\mu_Q\right| + \eta\sigma|\zeta'|,\end{aligned}$$

for every dyadic cube Q. In F_2, we then have (again for every Q)

$$\sigma|\zeta^0| \leq \left|\zeta \cdot \fint_Q b_Q\,d\mu_Q\right| + \sqrt{\eta}\sigma|\zeta^0|,$$

and also

$$|\zeta| \leq |\zeta^0| + |\zeta'| \leq (1 + \eta^{-1/2})|\zeta^0| \leq 2\eta^{-1/2}|\zeta^0|.$$

Combining the latter two estimates, we obtain that for all $(x,t) \in F_2$, and for every Q,

$$\frac{\sqrt{\eta}}{2}(1 - \sqrt{\eta})\sigma|\zeta(x,t)| \leq (1 - \sqrt{\eta})\sigma|\zeta^0(x,t)| \leq \left|\zeta(x,t) \cdot \fint_Q b_Q\,d\mu_Q\right|. \tag{2.26}$$

We now observe that, as above, in order to estimate $\|\zeta^0 1_{F_2}\|_{\mathcal{C}}$, it suffices to prove that, for $\epsilon > 0$ chosen small enough,

$$\|\zeta^0 1_{F_2} \mathbb{1}_{\Gamma^\epsilon}(\zeta)\|_{\mathcal{C}} \leq C\,, \tag{2.27}$$

where Γ^ϵ is an arbitrary cone of aperture ϵ, i.e.,

$$\Gamma^\epsilon = \Gamma^\epsilon(\nu) := \{z \in \mathbb{C}^{m+1} : |(z/|z|) - \nu| < \epsilon\}\,,$$

for some unit vector $\nu \in \mathbb{C}^{m+1}$. Indeed, given (2.27), we may then sum over an appropriate collection of such cones, covering $\mathbb{C}^{m+1}$, to obtain the bound $\|\zeta^0 1_{F_2}\|_{\mathcal{C}} \leq C(\epsilon, m)$. We therefore fix such a cone $\Gamma^\epsilon = \Gamma^\epsilon(\nu)$, and observe that, for $(x,t) \in F_2$, and $\zeta(x,t) \in \Gamma^\epsilon$, since $\eta \leq 1/4$, by (2.26), we have for every dyadic cube Q,

$$\begin{aligned}\frac{\sqrt{\eta}}{4}\sigma &\leq \left|\frac{\zeta(x,t)}{|\zeta(x,t)|}\cdot \fint_Q b_Q\, d\mu_Q\right| \leq \left|\left(\frac{\zeta(x,t)}{|\zeta(x,t)|} - \nu\right)\cdot \fint_Q b_Q\, d\mu_Q\right| + \left|\nu \cdot \fint_Q b_Q\, d\mu_Q\right| \\ &\leq C_0\epsilon + \left|\nu\cdot \fint_Q b_Q\, d\mu_Q\right|\,,\end{aligned}$$

where in the last step we have used Schwarz's inequality, the fact that

$$1/C_0 \leq d\mu_Q/dx = \phi_Q \leq 1$$

on Q, and (2.14). Since ϵ is at our disposal, we may choose it small enough, say $\epsilon \leq \sigma\sqrt{\eta}/(8C_0)$, and then hide the small term, to obtain

$$\frac{\sqrt{\eta}}{8}\sigma := \theta \leq \left|\nu\cdot \fint_Q b_Q\, d\mu_Q\right|. \tag{2.28}$$

We observe that the latter bound does not depend on (x,t), but on the other hand, was deduced from the existence of some $(x,t) \in F_2$, for which $\zeta(x,t) \in \Gamma^\epsilon$. Thus, (2.28) holds for any Q such that

$$\iint_{R_Q} |\zeta^0(x,t)|^2 1_{F_2}(x,t) \mathbb{1}_{\Gamma^\epsilon}(\zeta(x,t))\frac{dx dt}{t} \neq 0\,,$$

where as usual, $R_Q := Q \times (0, \ell(Q))$ denotes the standard Carleson box above Q. Consequently, in proving (2.27), we may suppose henceforth, without loss of generality, that (2.28) holds in any dyadic cube Q under consideration.

We therefore fix a dyadic cube Q such that (2.28) holds, and we use a stopping time procedure to extract a family $\{Q_j\}$ of non-overlapping dyadic sub-cubes of Q, which are maximal with respect to the property that at least one of the following conditions holds:

$$\fint_{Q_j} |b_Q|\, d\mu_Q > \frac{\theta}{4\epsilon} \qquad \text{(type I)}, \tag{2.29}$$

$$\left|\nu\cdot\fint_{Q_j} b_Q\, d\mu_Q\right| \leq \frac{\theta}{2} \qquad \text{(type II)}. \tag{2.30}$$

If some Q_j happens to satisfy both the type I and type II conditions, then we arbitrarily assign it to be of type II; for simplicity of notation, we write $Q_j \in I$, or $Q_j \in II$, to mean that the cube is of type I, or type II, respectively. We now claim that for this family of cubes,

$$\sum_j |Q_j| \le (1-\beta)|Q|\,, \tag{2.31}$$

for some uniform $\beta > 0$, and that

$$|\zeta(x,t)|^2 \mathbb{1}_{\Gamma^\epsilon}(\zeta(x,t)) \le C_\theta\, |\zeta(x,t)\cdot A_t^{\mu_Q} b_Q(x)|^2\,, \qquad \forall (x,t) \in E_Q^* := R_Q \backslash \big(\bigcup_j R_{Q_j}\big)\,, \tag{2.32}$$

where $A_t^{\mu_Q}$ is the dyadic averaging operator adapted to the measure μ_Q, i.e.,

$$A_t^{\mu_Q} f(x) := \fint_{Q(x,t)} f\, d\mu_Q\,,$$

where $Q(x,t)$ denotes the smallest dyadic cube, of side length at least t, that contains x.

Let us verify (2.32) first. Observe first that for any dyadic sub-cube $Q' \subset Q$, which is not contained in any Q_j, we have that the opposite inequalities to (2.29) and (2.30) hold, for the average of b_Q over Q', by maximality of the cubes in the family $\{Q_j\}$; i.e., by definition of $A_t^{\mu_Q}$, we have

$$\frac{\theta}{2} \le \big|\nu \cdot A_t^{\mu_Q}(b_Q)(x)\big| \text{ and } |A_t^{\mu_Q}(b_Q)(x)| \le \frac{\theta}{4\epsilon}\,, \qquad \forall (x,t) \in E_Q^*\,.$$

Thus, if $z \in \Gamma^\epsilon$, and $(x,t) \in E_Q^*$, we have

$$\begin{aligned}\frac{\theta}{2} \le \big|\nu \cdot A_t^{\mu_Q}(b_Q)(x)\big| &\le \big|(z/|z|)\cdot A_t^{\mu_Q}(b_Q)(x)\big| + \big|(\nu - (z/|z|))\cdot A_t^{\mu_Q}(b_Q)(x)\big| \\ &\le \big|(z/|z|)\cdot A_t^{\mu_Q}(b_Q)(x)\big| + \frac{\theta}{4}\,.\end{aligned}$$

We now obtain (2.32), by setting $z = \zeta(x,t)$.

Next, we establish (2.31). Set $E := Q \setminus (\cup_j Q_j)$, and $B_1 := \cup_{Q_j \in I} Q_j$. We note that by definition of the type I cubes, $B_1 \subset \{M(b_Q) > \theta/(4\epsilon)\}$, where M denotes the Hardy-Littlewood maximal operator, whence it follows by (2.14) that,

$$|B_1| \le C\left(\frac{\epsilon}{\theta}\right)^2 \int_{\mathbb{R}^n} |b_Q|^2 \le C C_0 \left(\frac{\epsilon}{\theta}\right)^2 |Q|\,. \tag{2.33}$$

Let us note that by Definition 2.10,

$$\frac{1}{C_0}|Q| \le \mu_Q(Q) \le |Q|\,. \tag{2.34}$$

By (2.28), and then Definition 2.10, (2.30), (2.33), and (2.14), we have

$$\begin{aligned}\theta\,\mu_Q(Q) &\leq \left|\nu\cdot\int_Q b_Q\,d\mu_Q\right| \\ &\leq \left|\nu\cdot\int_E b_Q\,d\mu_Q\right| + \int_{B_1}|b_Q|\,d\mu_Q + \sum_{Q_j\in II}\left|\nu\cdot\int_{Q_j} b_Q\,d\mu_Q\right| \\ &\leq |E|^{1/2}\,\|b_Q\|_2 + |B_1|^{1/2}\|b_Q\|_2 + \frac{\theta}{2}\sum_j \mu_Q(Q_j) \\ &\leq C\,|E|^{1/2}\,|Q|^{1/2} + C_\theta\epsilon\,|Q| + \frac{\theta}{2}\mu(Q)\,.\end{aligned}$$

Choosing ϵ small enough, and using (2.34), we have $|Q| \leq C_\theta\,|E|$, which is equivalent to (2.31).

With (2.31) and (2.32) in hand, we turn to the proof of (2.27). We note that by (2.31), and a standard John-Nirenberg lemma for Carleson measures (see, e.g., [13, Lemma 1.37]),

$$\begin{aligned}\|\zeta^0 1_{F_2}\mathbb{1}_{\Gamma^\epsilon}(\zeta)\|_{\mathcal{C}} &\lesssim \sup_Q \frac{1}{|Q|}\iint_{E_Q^*}|\zeta^0(x,t)|^2 1_{F_2}(x,t)\mathbb{1}_{\Gamma^\epsilon}(\zeta(x,t))\frac{dxdt}{t} \\ &\lesssim \sup_Q \frac{1}{|Q|}\iint_{R_Q}|\zeta(x,t)\cdot A_t^{\mu_Q}b_Q(x)|^2\frac{dxdt}{t}\,,\end{aligned}$$

where in the last step we have used the trivial bound $|\zeta^0| \leq |\zeta|$, and (2.32).

Thus it remains only to show that the last term is bounded. To this end, recalling that $\zeta = \Theta_t 1$, we fix Q, and use the familiar trick of Coifman-Meyer [9] to write

$$\begin{aligned}\zeta\cdot A_t^{\mu_Q}b_Q &= ((\Theta_t 1)A_t^{\mu_Q} - (\Theta_t 1)A_t^{\mu_Q}P_t)b_Q + ((\Theta_t 1)A_t^{\mu_Q}P_t - \Theta_t)b_Q + \Theta_t b_Q \\ &:= R_t^{(1)}b_Q + R_t^{(2)}b_Q + \Theta_t b_Q\,, \qquad (2.35)\end{aligned}$$

where as above P_t denotes a nice approximate identity operator (cf. Definition 1.4). The contribution of $\Theta_t b_Q$ may be handled immediately by hypothesis (2.15). The contributions of the two remainder terms $R_t^{(i)}b_Q$, $i = 1, 2$, may be handled as follows. By (2.14), and a standard orthogonality argument, it is enough to show that for some $\beta_0 > 0$, and for all $t \in (0, \ell(Q))$, we have

$$\int_Q |R_t^{(i)}Q_s h(y)|^2 dy \lesssim \min\left(\frac{s}{t},\frac{t}{s}\right)^{\beta_0}\int_{\mathbb{R}^n}|h(x)|^2 dx\,. \qquad (2.36)$$

We establish the latter fact as follows. We first note that by (2.2) and (2.3), we may invoke [1, Lemma 3.11] (cf. Lemma 1.6 above), to deduce that $\Theta_t 1$ is well defined, and satisfies

$$\sup_{t>0}\|(\Theta_t 1)A_t^{\mu_Q}\|_{L^2(\mathbb{R}^n)\to L^2(\mathbb{R}^n)} \leq C\,, \qquad (2.37)$$

and therefore also that each $R_t^{(i)}$ satisfies (2.2) and (2.3). Thus, since in addition, we have that $R_t^{(i)}1 = 0$ for $i = 1, 2$, we may invoke [1, Lemma 3.5] (cf. Lemma 1.9 above), to obtain that

$$\|R_t^{(i)} Q_s h\|_2 \lesssim t\, \|\nabla Q_s h\|_2\,, \qquad \forall h \in L^2\,,$$

which yields (2.36), in the case $t \leq s$, with $\beta_0 = 2$, by properties of Q_s (cf. Definition 1.2).

Consider now the case $s \leq t$ of (2.36). Since $A_t^{\mu Q}$ is a projection operator, we have

$$R_t^{(1)} = (\Theta_t 1) A_t^{\mu Q} (A_t^{\mu Q} - P_t)\,,$$

so that the desired bound for this term, for all $h \in L^2$, follows from (2.37), and the readily verified fact that there is some $\beta > 0$ such that

$$\|A_t^{\mu Q} Q_s\|_{L^2 \to L^2} + \|P_t Q_s\|_{L^2 \to L^2} \lesssim \left(\frac{s}{t}\right)^{\beta}\,, \qquad s \leq t \leq \ell(Q)\,.$$

Similarly, the case $s \leq t$ of (2.36) follows for $R_t^{(2)}$ from the latter fact, and the fact that (2.4) holds for all $h \in L^2(\mathbb{R}^n, \mathbb{C}) \times H$. □

3 Application of Theorem 2.13 to the theory of layer potentials

We consider layer potentials associated to divergence form complex coefficient equations $Lu = 0$, where

$$L = -\operatorname{div} A\nabla := -\sum_{i,j=1}^{n+1} \frac{\partial}{\partial x_i}\left(A_{i,j}\frac{\partial}{\partial x_j}\right)$$

is defined in $\mathbb{R}^{n+1} = \{(x,t) \in \mathbb{R}^n \times \mathbb{R}\}, n \geq 2$ (we recall that we use the notational conventions that $x_{n+1} = t$, and that capital letters may be used to denote points in $\mathbb{R}^{n+1}$, e.g., $X = (x,t)$). Here, $A = A(x)$ is an $(n+1) \times (n+1)$ matrix of complex-valued L^∞ coefficients, defined on $\mathbb{R}^n$ (i.e., independent of the t variable) and satisfying the uniform ellipticity condition

$$\lambda|\xi|^2 \leq \Re e\,\langle A(x)\xi, \xi\rangle := \Re e \sum_{i,j=1}^{n+1} A_{ij}(x)\xi_j\bar{\xi_i}, \quad \|A\|_{L^\infty(\mathbb{R}^n)} \leq \Lambda, \tag{3.1}$$

for some $\lambda > 0$, $\Lambda < \infty$, and for all $\xi \in \mathbb{C}^{n+1}$, $x \in \mathbb{R}^n$.

In the present work, we further suppose that weak solutions of the equations $Lu = 0$ (and also $L^*u = 0$, where L^* denotes the Hermitian adjoint of L) satisfy De Giorgi/Nash ("DG/N") estimates, i.e., that there is a constant C and an exponent $\alpha > 0$ such that for any ball $B = B(X,R) \subset \mathbb{R}^{n+1}$, of radius R, for which $Lu = 0$

in the concentric double $2B := B(X, 2R)$, we have the local Hölder continuity estimate

$$|u(Y) - u(Z)| \leq C\left(\frac{|Y-Z|}{R}\right)^{\alpha}\left(\fint_{2B}|u|^2\right)^{1/2}, \tag{3.2}$$

whenever $Y, Z \in B$. Observe that any u satisfying (3.2) also satisfies Moser's "local boundedness" estimate [22]

$$\sup_{Y\in B}|u(Y)| \leq C\left(\fint_{2B}|u|^2\right)^{1/2}. \tag{3.3}$$

Estimates (3.2) and (3.3) always hold for real coefficients [11, 23], and are stable under small complex, L^∞ perturbations [2] (see also [6, 16]); moreover they hold always for t-independent complex operators in ambient dimension $n+1=3$ [1, Section 11]. In the presence of the DG/N estimates (for L and L^*), by [15], the operators L and L^*, respectively, have fundamental solutions $E(X, Y)$, and $E^*(X, Y) = \overline{E(Y, X)}$, satisfying the following properties: first, that

$$L_{x,t}\, E(x,t,y,s) = \delta_{(y,s)}, \;\; L^*_{y,s}\, E^*(y,s,x,t) := L^*_{y,s}\, \overline{E(x,t,y,s)} = \delta_{(x,t)}, \tag{3.4}$$

where δ_X denotes the Dirac mass at the point X; second, by the t-independence of our coefficients, that

$$E(x,t,y,s) = E(x,t-s,y,0); \tag{3.5}$$

and finally, for $j \geq 0$, that there exists a constant C_j depending only on j, dimension, ellipticity and (3.2) and (3.3), such that for for all $(x,t) \neq (y,s) \in \mathbb{R}^{n+1}$, we have

$$\left|(\partial_t)^j E(x,t,y,s)\right| \leq C_j\left(|t-s|+|x-y|\right)^{1-n-j}, \tag{3.6}$$

$$\left|\left(\Delta_h(\partial_t)^j E(\cdot,t,y,0)\right)(x)\right| + \left|\left(\Delta_h(\partial_t)^j E(x,t,\cdot,0)\right)(y)\right| \leq C_j\frac{|h|^\alpha}{(|t|+|x-y|)^{n+j+\alpha-1}}, \tag{3.7}$$

whenever $2|h| \leq |x-y|$ or $|h| < 20|t|$, for some $\alpha > 0$, where $(\Delta_h f)(x) := f(x+h) - f(x)$.

We define the single layer potential operator, associated to L, by

$$\mathcal{S}_t f(x) := \int_{\mathbb{R}^n} E(x,t,y,0)\, f(y)\, dy, \;\; t \in \mathbb{R}; \tag{3.8}$$

the single layer potential associated to L^* is defined analogously, with E^* in place of E.

We shall encounter operators whose kernels involve derivatives applied to the second set of variables in the fundamental solution $E(x,t,y,s)$, and we denote this by appropriate parenthetic grouping; e.g., for $f : \mathbb{R}^n \to \mathbb{C}^{n+1}$, we set

$$(\mathcal{S}_t\nabla) f(x) := \int_{\mathbb{R}^n} \left(\nabla_{y,s} E(x,t,y,s)\right)\Big|_{s=0} \cdot f(y)\, dy. \tag{3.9}$$

We similarly denote the individual components of the vector-valued operator $(\mathcal{S}_t\nabla)$, thus for $f:\mathbb{R}^n\to\mathbb{C}^n$,

$$(\mathcal{S}_t\nabla y)f = -\mathcal{S}_t\,(\mathrm{div}_y\, f)\,, \tag{3.10}$$

and, by translation invariance in t (3.5),

$$(\mathcal{S}_t\partial_s) = -\partial_t\mathcal{S}_t\,. \tag{3.11}$$

With these notational conventions in hand, we record for future reference the following estimates taken from [1]:

Lemma 3.12 ([1, Lemma 2.8, Lemma 2.9 and Lemma 2.10]). *Suppose that L, L^* are t-independent divergence form complex elliptic operators as above, whose null solutions satisfy the DGN bounds (3.2). Then for some C depending only on n, (3.1) and (3.2), for every fixed $x\in\mathbb{R}^n$ and $t\neq 0$, we have*

$$\int_{\mathbb{R}^n} |\nabla_{y,s}E(y,s,x,t)|_{s=0}|^2\,dy \le C|t|^{-n}. \tag{3.13}$$

Moreover, for $k\geq 1$ and $j\geq 0$, there is a constant C_j depending on n, (3.1) and (3.2), such that

$$\int_{2^{k+1}Q\setminus 2^kQ} \left|\left(2^k\ell(Q)\right)^j(\partial_t)^j\,\nabla_{y,s}E(x,t,y,s)\big|_{s=0}\right|^2 dy \le C_j\left(2^k\ell(Q)\right)^{-n}\,,\quad \forall t\in\mathbb{R}\,, \tag{3.14}$$

and

$$\int_{2Q}\left|t^j\,(\partial_t)^j\,\nabla_{y,s}E(x,t,y,s)\big|_{s=0}\right|^2 dy \le C_j\,(\ell(Q))^{-n}\,,\quad \frac{1}{2}\ell(Q)\le t\le 2\ell(Q)\,. \tag{3.15}$$

In addition, for $f:\mathbb{R}^n\to\mathbb{C}^{n+1}$, and for every cube Q, for all integers $k\geq 1$, and $j\geq 0$, and for all $t\in\mathbb{R}$,

$$\|(\partial_t)^j(\mathcal{S}_t\nabla)\,(f1_{2^{k+1}Q\setminus 2^kQ})\|^2_{L^2(Q)} \le C_j\,2^{-nk}(2^k\ell(Q))^{-2j}\|f\|^2_{L^2(2^{k+1}Q\setminus 2^kQ)}\,, \tag{3.16}$$

Moreover, for each $j\geq 1$,

$$\|t^j(\partial_t)^j\,(\mathcal{S}_t\nabla)\,f\|_{L^2(\mathbb{R}^n)}\le C_j\,\|f\|_2. \tag{3.17}$$

We note that by (3.11), for scalar valued f we obtain from (3.17) that, for $j\geq 1$,

$$\|t^j(\partial_t)^{j+1}\mathcal{S}_tf\|_{L^2(\mathbb{R}^n)}\le C_j\,\|f\|_2\,. \tag{3.18}$$

We further note for future reference that (3.16) can be reformulated as

$$\|t^j(\partial_t)^j(\mathcal{S}_t\nabla)\,(f1_{2^{k+1}Q\setminus 2^kQ})\|^2_{L^2(Q)}\le C_j\,2^{-nk}\left(\frac{t}{2^k\ell(Q)}\right)^{2j}\|f\|^2_{L^2(2^{k+1}Q\setminus 2^kQ)}\,, \tag{3.19}$$

so in particular,

$$\|t^2(\partial_t)^2(\mathcal{S}_t\nabla)\,(f1_{2^{k+1}Q\setminus 2^kQ})\|^2_{L^2(Q)} \lesssim 2^{-(n+4)k}\,\|f\|^2_{L^2(2^{k+1}Q\setminus 2^kQ)}\,, \quad \text{if } t\approx \ell(Q)\,, \tag{3.20}$$

which by (3.11) yields also that

$$\|t^2(\partial_t)^3\mathcal{S}_t\,(f1_{2^{k+1}Q\setminus 2^kQ})\|^2_{L^2(Q)} \lesssim 2^{-(n+4)k}\,\|f\|^2_{L^2(2^{k+1}Q\setminus 2^kQ)}\,, \quad \text{if } t\approx \ell(Q)\,. \tag{3.21}$$

Finally, we note, again for future reference, that by applying (3.14)–(3.15), with $t\approx \ell(Q)$, we obtain by Cauchy-Schwarz that

$$\int_{\mathbb{R}^n} \left| t^j\,(\partial_t)^j\,\nabla_{y,s}E(x,t,y,s)\big|_{s=0}\right| dy \le C_j\,, \qquad j\ge 1\,. \tag{3.22}$$

The main result of this section is the following, which is originally due to Rosén [24][1].

Theorem 3.23. *Let L be a t-independent complex elliptic operator as above, for which solutions of $Lu=0$ and $L^*u=0$ satisfy the DG/N bounds* (3.2) *(hence also* (3.3)*). Then the single layer potential satisfies the following square function bound:*

$$\iint_{\mathbb{R}^{n+1}_+} \left|t\,(\partial_t)^2\mathcal{S}_tf(x)\right|^2 \frac{dxdt}{t} \le C\,\|f\|^2_{L^2(\mathbb{R}^n)}\,, \tag{3.24}$$

with C depending only upon n,λ,Λ, and the constants in (3.2) *and* (3.3)*. Analogous bounds hold in the lower half space $\mathbb{R}^{n+1}_-$, and for the single layer potential associated to L^*.*

This is the fundamental result concerning the layer potentials. Indeed, let N_* denotes the non-tangential maximal operator, i.e.,

$$N_*(u)(x) := \sup_{(y,t):\,|x-y|<t} |u(y,t)|\,.$$

By [1, Lemma 5.2], we have

$$\sup_{t\neq 0}\|\nabla\mathcal{S}_tf\|^2_{L^2(\mathbb{R}^n)} \lesssim \|N_*\,(\partial_t\mathcal{S}_tf)\,\|^2_{L^2(\mathbb{R}^n)} + \int_{-\infty}^{\infty}\int_{\mathbb{R}^n} \left|t\partial_t^2\mathcal{S}_tf(x)\right|^2\frac{dxdt}{|t|} + \|f\|^2_{L^2(\mathbb{R}^n)}\,,$$

and also

$$\int_{-\infty}^{\infty}\int_{\mathbb{R}^n} |t\nabla\partial_t\mathcal{S}_tf(x)|^2\,\frac{dxdt}{|t|} \lesssim \int_{-\infty}^{\infty}\int_{\mathbb{R}^n} \left|t\partial_t^2\mathcal{S}_tf(x)\right|^2\frac{dxdt}{|t|} + \|f\|^2_{L^2(\mathbb{R}^n)}\,.$$

Moreover, by [3], we have

$$\|N_*\,(\partial_t\mathcal{S}_tf)\,\|^2_{L^2(\mathbb{R}^n)} \lesssim \int_{-\infty}^{\infty}\int_{\mathbb{R}^n} |t\nabla\partial_t\mathcal{S}_tf(x)|^2\,\frac{dxdt}{|t|}\,.$$

[1] as noted above, even without the DG/N bounds.

Combining these last three estimates yields, in particular, the bound

$$\sup_{t\neq 0} \|\nabla \mathcal{S}_t f\|_{L^2(\mathbb{R}^n)} \lesssim \|f\|_{L^2(\mathbb{R}^n)} .$$

Bounds for the associated double layer potential (see [1, (1.5)] for a precise definition) follow by duality, while boundedness of an appropriate non-tangential maximal function of $\nabla \mathcal{S}_t f$ follows by [1, Lemma 4.8]. The analogous L^p and H^p bounds will appear in [20].

Let us proceed now to give the proof of Theorem 3.23.

Proof. We first make the following claim:

$$\iint_{\mathbb{R}^{n+1}_+} \left|t\,(\partial_t)^2 \mathcal{S}_t f(x)\right|^2 \frac{dxdt}{t} \lesssim \iint_{\mathbb{R}^{n+1}_+} \left|t^2\,(\partial_t)^3 \mathcal{S}_t f(x)\right|^2 \frac{dxdt}{t} + \|f\|^2_{L^2(\mathbb{R}^n)}. \tag{3.25}$$

Indeed, the left hand side of (3.25) equals the limit, as $\delta \to 0$, of

$$\int_\delta^{1/\delta}\int_{\mathbb{R}^n} \left|(\partial_t)^2 \mathcal{S}_t f(x)\right|^2 t\,dxdt \lesssim \int_\delta^{1/\delta}\int_{\mathbb{R}^n} \left|(\partial_t)^2 \mathcal{S}_t f(x)\right| \left|(\partial_t)^3 \mathcal{S}_t f(x)\right| t^2\,dxdt + \|f\|_2^2 ,$$

where we have integrated by parts in t, and then used (3.18), with $j = 1$, to control the boundary terms. Applying "Cauchy's inequality with ε's", hiding a small term on the left hand side of the last inequality, and then letting $\delta \to 0$, we obtain (3.25). Thus, in lieu of (3.24), it is now enough to prove

$$\iint_{\mathbb{R}^{n+1}_+} \left|t^2\,(\partial_t)^3 \mathcal{S}_t f(x)\right|^2 \frac{dxdt}{t} \leq C\,\|f\|^2_{L^2(\mathbb{R}^n)} . \tag{3.26}$$

We therefore turn to the proof of (3.26). We shall utilize the following notation: given a vector $\vec{V} := (V_1, ..., V_n, V_{n+1}) \subset \mathbb{R}^{n+1}$, we denote its "horizontal component" by

$$V_{\|} := (V_1, ..., V_n) .$$

Similarly, the horizontal component of the $(n+1)$-dimensional gradient operator is

$$\nabla_{\|} := (\partial_{x_1}, ..., \partial_{x_n}) . \tag{3.27}$$

For convenience, we write

$$\left(\nabla_{y,s} u\right)(y,0) := \left(\nabla_{y,s} u(y,s)\right)\Big|_{s=0} .$$

We now define

$$\Theta_t^0 := t^2\,(\partial_t)^3 \mathcal{S}_t ,$$

and for $f : \mathbb{R}^n \to \mathbb{C}^n$, we set

$$\Theta_t' f(x) := t^2\,(\partial_t)^2 \int_{\mathbb{R}^n} \left(\overline{A^*}(y)\,(\nabla_{y,s} E)\,(x,t,y,0)\right)_{\|} \cdot f(y)\,dy . \tag{3.28}$$

It is then enough to show that the hypotheses of Theorem 2.13 are verified for this choice of $\Theta_t = (\Theta_t^0, \Theta_t')$, with $m = n$, where the "0" term corresponds to the $t = x_{n+1}$ direction. We first observe that (2.2), and (2.3) (with $\beta = 2$), follow directly from (3.17)–(3.18) (with $j = 2$), and (3.20)–(3.21), respectively. We shall establish the quasi-orthogonality estimate (2.4), with $m = n$, and with $H := \{\nabla f : f \in \dot{L}_1^2(\mathbb{R}^n)\}$. We treat separately the contributions of $\Theta_t^0 Q_s h^0$, and of $\Theta_t' Q_s h'$, where $h^0 \in L^2(\mathbb{R}^n, \mathbb{C})$, and $h' \in H$. For the former, we use a standard "Calderón-Zygmund" argument exploiting (3.7) (with $j = 3$), and the cancellation of the convolution kernel of Q_s; we leave the routine details to the reader. Consider now $\Theta_t' Q_s h'$, where $h' = \nabla f$, with $f \in \dot{L}_1^2(\mathbb{R}^n)$. Recall that, modulo constants, we may identify $f \in \dot{L}_1^2(\mathbb{R}^n)$, with $I_1 g$, where $g \in L^2(\mathbb{R}^n)$, and $\|g\|_2 \approx \|\nabla f\|_2$, and $I_1 = (-\Delta)^{-1/2}$ is the standard fractional integral operator of order 1. Moreover, for $(x,t) \in \mathbb{R}^{n+1}_+$ fixed, and for $y_{n+1} := s < t$, using that $E(x,t,y,s) = \overline{E^*(y,s,x,t)}$, we have (in the weak sense)

$$\begin{aligned} -\operatorname{div}_y \left(\overline{A^*}(y) \nabla_{y,s} E(x,t,y,s) \right)_{\|} &= -\sum_{i=1}^{n} \sum_{j=1}^{n+1} \overline{\partial_{y_i} A^*_{i,j}(y)\, \partial_{y_j} E^*(y,s,x,t)} \\ &= \sum_{j=1}^{n+1} \overline{\partial_s A^*_{n+1,j}(y)\, \partial_{y_j} E^*(y,s,x,t)} \\ &= -\sum_{j=1}^{n+1} \overline{A^*_{n+1,j}(y)\, \partial_t\, \partial_{y_j} E^*(y,s,x,t)}\,, \end{aligned} \tag{3.29}$$

where we have used that $E^*(\cdot,x,t)$ is an adjoint solution away from the pole at (x,t), and, in the last step, t-independence of the coefficients (cf. (3.5).) Thus, for $h' = \nabla f \in H$, again using that $\overline{E^*(y,s,x,t)} = E(x,t,y,s)$, we have, by (3.29) and the definition of Θ_t',

$$\Theta_t' Q_s h' = \Theta_t' \nabla_{\|} Q_s f = -t^2 (\partial_t)^3 \left(\mathcal{S}_t \nabla\right) \left(\vec{\alpha} Q_s I_1 g\right),$$

where $\vec{\alpha} := (A_{1,n+1}, ..., A_{n+1,n+1})$. By (3.17) (with $j = 3$), since $\|s^{-1} Q_s I_1\|_{L^2 \to L^2} \le C$, we then obtain (2.4), with $\beta = 1$. We have therefore established that Θ_t^0 and Θ_t' satisfy all the conditions of Definition 2.1.

It remains to construct a system $\{b_Q\} \subset L^2(\mathbb{R}^n, \mathbb{C}) \times H$, and a family of Borel measures $\{\mu_Q\}$ as in Definition 2.10, satisfying the hypotheses (2.14)–(2.17) of Theorem 2.13, and also to verify hypothesis (2.18). In fact, the latter estimate is known: it has been proved in [18, Section 3], with constant C_1 depending only on dimension, ellipticity, and the DGN constants. However, for the reader's convenience, we shall supply a proof here, at the end of the present section. For now, we turn to the heart of the matter, namely, the construction of the system $\{b_Q\}$, and the family $\{\mu_Q\}$, verifying (2.14)–(2.17).

Given a cube $Q \subset \mathbb{R}^n$, we let x_Q denote its center, and we let

$$X_Q^{\pm} := \left(x_Q, \pm \tau \,\ell(Q)\right),$$

denote the upper and lower "Corkscrew points" relative to Q, where $\tau \in (0, 1/8)$

is a small number at our disposal, to be chosen. Set

$$F_Q(y,s) := E(y,s,X_Q^+) - E(y,s,X_Q^-)\,,$$

and define

$$b_Q^0(y) := |Q|\left(\partial_{\nu_A^-} F_Q\right)(y,0) := |Q|\, e_{n+1}\cdot\, A(y)\,(\nabla_{y,s}F_Q)\,(y,0)\,,$$

where $e_{n+1} := (0,...,0,1)$ is the inner unit normal to the half space $\mathbb{R}^{n+1}_+$ (thus, $\partial_{\nu_A^-}$ is the outer co-normal derivative, relative to A, on the boundary of the *lower* half-space $\mathbb{R}^{n+1}_-$), and

$$b_Q'(y) := |Q|\,\nabla_{\|} F_Q(y,0)\,.$$

We then have, by (3.13), with $t = \pm\tau\,\ell(Q)$,

$$\int_{\mathbb{R}^n} |b_Q|^2 \le C\tau^{-n}|Q|\,, \tag{3.30}$$

which is (2.14) with $C_0 \approx \tau^{-n}$.

Next, by (3.29) and the definition of Θ_t', we have that

$$\begin{aligned}\Theta_t' b_Q'(x) &= t^2|Q|\,(\partial_t)^3\int_{\mathbb{R}^n}(-e_{n+1})\cdot\overline{A^*(y)\,(\nabla_{y,s}E^*)\,(y,0,x,t)}\,F_Q(y,0)\,dy\\ &:= -\,t^2|Q|\,(\partial_t)^3\int_{\mathbb{R}^n}\overline{\left(\partial_{\nu_{A^*}^-}E^*\right)(y,0,x,t)}\,F_Q(y,0)\,dy\,,\end{aligned} \tag{3.31}$$

so that $\partial_{\nu_{A^*}^-}$ is the outer co-normal derivative, relative to A^*, for the lower half-space. Let $\langle\cdot,\cdot\rangle_{\mathbb{R}^{n+1}_-}$ denote the distributional pairing between continuous functions and measures in the lower half-space $\mathbb{R}^{n+1}_-$. Combining (3.31) with the definitions of b_Q^0 and Θ_t^0, and recalling that $\overline{E^*(y,s,x,t)} = E(x,t,y,s)$, we have for each fixed $(x,t)\in\mathbb{R}^{n+1}_+$,

$$\begin{aligned}&\Theta_t b_Q(x)\\ &= \Theta_t^0 b_Q^0(x) + \Theta_t' b_Q'(x)\\ &= t^2|Q|\,(\partial_t)^3\int_{\mathbb{R}^n}\left(\overline{E^*(y,0,x,t)}\left(\partial_{\nu_A^-}F_Q\right)(y,0) - \overline{\left(\partial_{\nu_{A^*}^-}E^*\right)(y,0,x,t)}\,F_Q(y,0)\right)dy\\ &= -\,t^2|Q|\,(\partial_t)^3\Big(\left\langle\overline{E^*(\cdot,\cdot,x,t)},LF_Q\right\rangle_{\mathbb{R}^{n+1}_-} - \iint_{\mathbb{R}^{n+1}_-}\overline{L^*E^*(y,s,x,t)}\,F_Q(y,s)\,dyds\Big)\\ &= t^2|Q|\,(\partial_t)^3\overline{E^*(X_Q^-,x,t)} = t^2|Q|\,(\partial_t)^3E(x,t,X_Q^-)\,,\end{aligned}$$

where we have used the definition of F_Q, (3.4), and the fact that, in the lower half-space, $L^*E^*(\cdot,\cdot,x,t) = 0 = LE(\cdot,\cdot,X_Q^+)$. Therefore, by (3.6), with $j=3$, we have that for $(x,t)\in R_Q := Q\times(0,\ell(Q))$,

$$|\Theta_t b_Q(x)| \lesssim \tau^{-n-2}\left(\frac{t}{\ell(Q)}\right)^2.$$

Consequently,

$$\iint_{R_Q} |\Theta_t b_Q(x)|^2 \frac{dxdt}{t} \lesssim \tau^{-2n-4}\,|Q| \int_0^{\ell(Q)} \left(\frac{t}{\ell(Q)}\right)^4 \frac{dt}{t} \approx \tau^{-2n-4}\,|Q|\,,$$

which yields (2.15) with $C_0 \approx \tau^{-2n-4}$.

We now turn to the proofs of (2.16)–(2.17). We begin by defining the measure μ_Q. Let ω be a small, positive constant, to be chosen, and let $\phi_Q : \mathbb{R}^n \to [0,1]$ be a smooth bump function, supported in $(1+\omega)Q$, with $\phi_Q \equiv 1$ on $(1/2)Q$, where for any positive constant κ, we let κQ denote the cube of side length $\kappa\ell(Q)$, concentric with Q. Clearly, we may choose ϕ_Q so that $\|\nabla\phi_Q\|_{L^\infty(\mathbb{R}^n)} \leq 2\ell(Q)^{-1}$, with $\phi_Q \gtrsim \omega$ on Q; thus (2.11) holds with $C_0 = 2$, and (2.12) holds with $C_0 \approx 1/\omega$. In accordance with Definition 2.10 we then set $d\mu_Q := \phi_Q dx$. Let $\Phi_Q : \mathbb{R}^{n+1} \to [0,1]$ be a smooth extension of ϕ_Q, i.e., $\Phi_Q(x,0) = \phi_Q(x)$, with Φ_Q supported in $I_{(1+\omega)Q}$, and $\Phi_Q \equiv 1$ in $I_{(1/2)Q}$, where in general, for any cube $Q \subset \mathbb{R}^n$, we let $I_Q := Q \times (-\ell(Q), \ell(Q))$ denote the "two-sided Carleson box". We choose the extension Φ_Q so that

$$\|\nabla\Phi_Q\|_{L^\infty(\mathbb{R}^{n+1})} \leq 2\,\ell(Q)^{-1}\,.$$

We note that, by (3.30),

$$\int_{\mathbb{R}^n\setminus Q} |b_Q|\,\phi_Q \leq \big((1+\omega)Q\setminus Q\big)^{1/2}\|b_Q\|^2_{L^2(\mathbb{R}^n)} \leq C\,\omega^{1/2}\tau^{-n/2}\,|Q|\,. \qquad (3.32)$$

Observe also that

$$\begin{aligned}\int_{\mathbb{R}^n} b_Q^0\,d\mu_Q &= \int_{\mathbb{R}^n} b_Q^0\,\phi_Q = |Q|\int_{\mathbb{R}^n}\left(\partial_{\nu_A^-}F_Q\right)(y,0)\,\phi_Q(y)\,dy\\ &= |Q|\left(-\,\langle\Phi_Q, LF_Q\rangle_{\mathbb{R}^{n+1}_-} + \iint_{\mathbb{R}^{n+1}_-} A\nabla F_Q\cdot\nabla\Phi_Q\right) := |Q|\,(\mathrm{I}+\mathrm{II})\,.\end{aligned}$$

By definition of F_Q and Φ_Q,

$$\mathrm{I} = \Phi_Q(X_Q^-) = 1\,.$$

Also, by the construction of Φ_Q, we have that

$$\begin{aligned}|\mathrm{II}| &\lesssim \frac{1}{\ell(Q)}\iint_{I_{(1+\omega)Q}\setminus I_{(1/2)Q}} |\nabla F_Q|\\ &\lesssim \ell(Q)^{(n-1)/2}\left(\iint_{I_{(1+\omega)Q}\setminus I_{(1/2)Q}} |\nabla F_Q|^2\right)^{1/2}\\ &\lesssim \ell(Q)^{(n-3)/2}\left(\iint_{I_{2Q}\setminus I_{(1/4)Q}} |F_Q|^2\right)^{1/2}\\ &= \ell(Q)^{(n-3)/2}\left(\iint_{I_{2Q}\setminus I_{(1/4)Q}}\left|\int_{-\tau\ell(Q)}^{\tau\ell(Q)} \partial_t E(y,s,x_Q,t)\,dt\right|^2 dyds\right)^{1/2} \lesssim \tau\,,\end{aligned}$$

where the implicit constants depend only on dimension, ellipticity, and the DGN constants, and where in the last three steps we have used Caccioppoli's inequality, the definition of F_Q, and (3.6) with $j = 1$. Combining our estimates for terms I and II, we obtain

$$\mathcal{R}e \int_{\mathbb{R}^n} b_Q^0 \, d\mu_Q \geq |Q|\big(1 - C\tau\big).$$

In conjunction with (3.32), the latter bound implies

$$\mathcal{R}e \int_Q b_Q^0 \, d\mu_Q \geq |Q|\big(1 - C\tau - C\omega^{1/2}\tau^{-n/2}\big) \geq \frac{1}{2}|Q|,$$

if we set $\omega := \tau^{n+2}$, and choose τ sufficiently small. Thus, we obtain (2.16) with $\sigma = 1/2$.

Finally, we verify (2.17). By definition of b'_Q, μ_Q, ϕ_Q, and F_Q, we have

$$\left|\int_{\mathbb{R}^n} b'_Q d\mu_Q\right| = |Q| \left|\int_{\mathbb{R}^n} \nabla_y F_Q(y,0)\, \phi_Q(y)\, dy\right| = |Q| \left|\int_{\mathbb{R}^n} F_Q(y,0)\, \nabla_y \phi_Q(y)\, dy\right|$$
$$\leq C\,\ell(Q)^{n-1} \int_{(1+\omega)Q\setminus(1/2)Q} \left|\int_{-\tau\ell(Q)}^{\tau\ell(Q)} \partial_t E(y,0,x_Q,t)\, dt\right| dy \leq C\tau\,|Q|,$$

where in the last step we have used (3.6) with $j = 1$. Combining the latter estimate with (3.32), we have

$$\left|\int_Q b'_Q d\mu_Q\right| \leq C\big(\tau + \omega^{1/2}\tau^{-n/2}\big)\,|Q| \leq C\tau\,|Q|,$$

by our choice of $\omega = \tau^{n+2}$. Since the constant τ is at our disposal (cf. Remark 2.23), we obtain (2.17), with $\eta \approx \tau$. This concludes the proof of our application, Theorem 3.23, modulo estimate (2.18). We conclude this section with a proof of the latter, following [18], where the estimate originally appeared (at least implicitly).

3.1 Verification of hypothesis (2.18)

To lighten the exposition, we treat the case $\delta = 0$; the case $\delta > 0$ may be handled by the same argument. We recall that

$$\Theta_t^0 := t^2\,(\partial_t)^3 \mathcal{S}_t,$$

so by (3.9)–(3.11), and by definition of Θ'_t (cf. (3.28)),

$$\Theta'_t 1 = -\Theta_t^0(\vec{b}) + t^2\,(\partial_t)^2\,\big(\mathcal{S}_t \nabla_{\|}\big)\,(A_{\|}), \tag{3.33}$$

where $\nabla_{\|}$ is the "horizontal gradient" defined in (3.27), and where

$$\vec{b} := (A_{n+1,j})_{1\leq j\leq n},$$

and $A_{\|}$ is the $n \times n$ sub matrix of A, with entries $A_{i.j}, 1 \leq i,j \leq n$.

Our goal is to control the "Carleson norm" of $\Theta_t' 1$ in terms of that of $\Theta_t^0 1$, where in the former case, "1" is interpreted as the $n \times n$ identity matrix, and in the latter case it is just the real number 1. To this end, we treat the two terms on the right hand side of (3.33) separately. The first of these equals minus

$$\Theta_t^0(\vec{b}) = \left(\Theta_t^0 - \Theta_t^0 1 P_t\right)(\vec{b}) + (\Theta_t^0 1) P_t(\vec{b}) := R_t(\vec{b}) + (\Theta_t^0 1) P_t(\vec{b}),$$

where P_t is a nice approximate identity as in Definition 1.4. By (3.6), with $s = 0$ and $j = 3$, we have that the kernel of Θ_t^0, and hence also that of R_t, is bounded by

$$Ct^2 \Big(t + |x - y|\Big)^{-n-2},$$

and in addition, by (3.7), satisfies an appropriate local Hölder continuity condition in the y variable. Moreover, $R_t 1 = 0$, so by standard arguments, we have

$$\sup_Q \frac{1}{|Q|} \int_0^{\ell(Q)} \int_Q |R_t(\vec{b})(x)|^2 \frac{dxdt}{t} \leq C \|\vec{b}\|_\infty^2 \leq C\Lambda^2,$$

where Λ is the bound in (3.1). Moreover, since P_t is bounded on L^∞, uniformly in t, we have

$$\sup_Q \frac{1}{|Q|} \int_0^{\ell(Q)} \int_Q \left|\Theta_t^0 1(x) P_t(\vec{b})(x)\right|^2 \frac{dxdt}{t} \leq \Lambda^2 \, \|\Theta_t^0 1\|_{\mathcal{C}}.$$

Thus, the desired bound (2.18) holds for the contribution of $\Theta_t^0(\vec{b})$.

Next we consider the second term on the right hand side of (3.33). To this end, we invoke a key fact in the proof of the Kato conjecture. By [5], there exists, for each Q, a mapping $F_Q = \mathbb{R}^n \to \mathbb{C}^n$ such that

$$\begin{aligned}
&\text{(i)} \quad \int_{\mathbb{R}^n} |\nabla_\| F_Q|^2 \leq C|Q|, \\
&\text{(ii)} \quad \int_{\mathbb{R}^n} |L_\| F_Q|^2 \leq C \frac{|Q|}{\ell(Q)^2}, \\
&\text{(iii)} \quad \sup_Q \int_0^{\ell(Q)} \fint_Q |\Phi(x,t)|^2 \frac{dxdt}{t} \\
&\qquad \leq C \sup_Q \int_0^{\ell(Q)} \fint_Q |\Phi(x,t)\, A_t \nabla_\| F_Q(x)|^2 \frac{dxdt}{t},
\end{aligned} \tag{3.34}$$

for every mapping $\Phi : \mathbb{R}_+^{n+1} \to \mathbb{C}^n$, where $A_t = A_t^Q$ denotes the dyadic averaging operator, i.e. if $Q(x,t)$ is the minimal dyadic cube (with respect to the grid induced by Q) containing x, with side length at least t, then

$$A_t g(x) := \fint_{Q(x,t)} g(y) dy.$$

Here $\nabla_{\|} F_Q$ is the Jacobian matrix $(D_i(F_Q)_j)_{1\leq i,j\leq n}$, and the product

$$\Phi A_t \nabla_{\|} F_Q = \sum_{i=1}^{n} \Phi_i A_t D_i F_Q$$

is a vector. Given the existence of a family of mappings F_Q with these properties, we see that by (iii), applied with $\Phi(\cdot,t) = t^2(\partial_t)^2(\mathcal{S}_t\nabla_{\|})(A_{\|}) := T_t A_{\|}$, it is enough to show that

$$\int_0^{\ell(Q)}\int_Q |\left(T_t A_{\|}\right)(x)\ \left(A_t \nabla_{\|} F_Q\right)(x)|^2 \frac{dxdt}{t} \lesssim \left(1 + \|\Theta_t^0 1\|_{\mathcal{C}}\right)|Q|.$$

But as in [6], we may exploit the idea of [9] to write

$$\begin{aligned}(T_t A_{\|})A_t\nabla_{\|}F_Q &= \left\{(T_t A_{\|})A_t - T_t A_{\|}\right\}\nabla_{\|}F_Q + T_t A_{\|}\nabla_{\|}F_Q\\ &= (T_t A_{\|})(A_t - P_t)\nabla_{\|}F_Q + \left\{(T_t A_{\|})P_t - T_t A_{\|}\right\}\nabla_{\|}F_Q + T_t A_{\|}\nabla_{\|}F_Q\\ &:= R_t^{(1)}\nabla_{\|}F_Q + R_t^{(2)}\nabla_{\|}F_Q + T_t A_{\|}\nabla_{\|}F_Q,\end{aligned}$$

where as above, P_t is a nice approximate identify. The last of these terms is the "main term". We treat it as follows. By definition of T_t, we have that

$$T_t A_{\|}\nabla_{\|}F_Q = t^2(\partial_t)^2\mathcal{S}_t L_{\|}F_Q\,.$$

Thus, by the case $j=1$ of (3.18), and (3.34)(ii), we then obtain

$$\begin{aligned}\int_0^{\ell(Q)}\int_Q \left|\left(T_t A_{\|}\nabla_{\|}F_Q\right)(x)\right|^2\frac{dxdt}{t} &\leq \int_0^{\ell(Q)}\int_{\mathbb{R}^n}\left|t\,(\partial_t)^2\mathcal{S}_t L_{\|}F_Q(x)\right|^2 t\,dxdt\\ &\lesssim \frac{|Q|}{\ell(Q)^2}\int_0^{\ell(Q)} tdt \approx |Q|\,,\end{aligned}$$

as desired.

Turning now to the remainder terms, we first observe that by the case $j=2$ of (3.22), we have that $T_t A_{\|}\in L^\infty$, uniformly in t, so the contribution of the term $R_t^{(1)}$ may be handled by a standard orthogonality argument, given (3.34)(i).

Next, let us consider the operator $R_t^{(2)}$, which we rewrite as

$$R_t^{(2)} = \left[\left(T_t A_{\|}\right)P_t\ -\ T_t A_{\|}P_t\right]\ +\ T_t A_{\|}(P_t - I) := R_t' + R_t''\,.$$

Note that $R_t' 1 = 0$. Thus, by the case $j=2$ of (3.16)–(3.17), along with our previous observation that $t^2(\partial_t)^2(\mathcal{S}_t\nabla_{\|})(A_{\|}) = T_t A_{\|}$ belongs to L^∞, the operator R_t' satisfies (all of) the hypotheses of Lemma 1.9, and therefore,

$$\int_0^{\ell(Q)}\int_Q |R_t'\,\nabla_{\|}F_Q(x)|^2\frac{dxdt}{t}\ \leq\ \int_0^{\infty}\int_{\mathbb{R}^n}|R_t'\,\nabla_{\|}F_Q(x)|^2\frac{dxdt}{t}\lesssim\|\nabla_{\|}F_Q\|_2^2\lesssim|Q|\,,$$

as desired, where in the last step we have used (3.34)(i).

Finally, we consider R_t''. Since the kernel of $\mathcal{S}_t$ is $E(x,t,y,0)$, where $\overline{E(x,t,y,s)} = E^*(y,s,x,t)$ is an adjoint solution in (y,s) away from the pole at (x,t), and is jointly translation invariant in the arguments (t,s), we have that

$$R_t''(\nabla_\| F_Q) = t^2 \Big(((\partial_t)^3 \mathcal{S}_t \nabla)\,(\vec{a}\,(P_t - I)F_Q)\Big) - t^2\,\partial_t^3 \mathcal{S}_t \left(\vec{b}\nabla_\|(P_t - I)F_Q\right)$$
$$:= I_1 + I_2\,, \tag{3.35}$$

where $\vec{a} := (A_{1,n+1},...,A_{n+1,n+1})$, and as above, $\vec{b} := (A_{n+1,1},...,A_{n+1,n})$. By the case $j=3$ of (3.17), $t^3(\partial_t)^3(\mathcal{S}_t\nabla)$ is bounded on $L^2(\mathbb{R}^n)$, uniformly in t. Moreover,

$$\iint_{\mathbb{R}^{n+1}_+} |t^{-1}(I - P_t)F_Q(x)|^2 \frac{dxdt}{t} \lesssim \|\nabla_\| F_Q\|_2^2 \lesssim |Q|\,,$$

by a standard argument using Plancherel's theorem, and then (3.34)(i). Thus, the desired bound holds for the contribution of I_1. To handle I_2, we further decompose it as

$$I_2 = t^2\,\partial_t^3 \mathcal{S}_t\left(\vec{b}\nabla_\| F_Q\right) - t^2\left\{\partial_t^3 \mathcal{S}_t \vec{b} P_t - \left(\partial_t^3 \mathcal{S}_t \vec{b}\right) P_t\right\}\nabla_\| F_Q$$
$$-t^2\left(\partial_t^3 \mathcal{S}_t \vec{b}\right) P_t(\nabla_\| F_Q)$$
$$:= I_2' + I_2'' + I_2'''\,.$$

The term I_2'' is of the form $R_t(\nabla_\| F_Q)$, with R_t satisfying the hypotheses of Lemma 1.9, and so can be handled like R_t' above. By definition $I_2' = \Theta_t^0(\vec{b}\nabla_\| F_Q)$, so by the $T1$ theorem for square functions of [8] (cf. Theorem 4.5 below),

$$\int_0^{\ell(Q)}\int_Q |I_2'|^2 \frac{dxdt}{t} \le \int_0^\infty \int_{\mathbb{R}^n} \left|\Theta_t^0\left(\vec{b}\nabla_\| F_Q\right)(x)\right|^2 \frac{dxdt}{t}$$
$$\lesssim \|\vec{b}\|_\infty^2 \|\nabla_\| F_Q\|_2^2\left(1 + \|\Theta_t^0 1\|_{\mathcal{C}}\right) \lesssim |Q|\left(1 + \|\Theta_t^0 1\|_{\mathcal{C}}\right)\,,$$

where in the last step we have used (3.34)(i), and the fact that $\|\vec{b}\|_\infty \le \Lambda$. Similarly, by the same result of [8], combined with a well-known argument of [12], we obtain that

$$\|t^2\,\partial_t^3 \mathcal{S}_t(\vec{b})\|_{\mathcal{C}} \lesssim 1 + \|t^2\,\partial_t^3 \mathcal{S}_t 1\|_{\mathcal{C}} = 1 + \|\Theta_t^0 1\|_{\mathcal{C}}\,.$$

The desired bound for the contribution of I_2''' now follows by Carleson's lemma. □

4 Appendix: a generalized Christ-Journé $T1$ Theorem for square functions

In this Appendix, we shall impose the following hypothesis.

Hypothesis 4.1. We consider a family of operators $\{\Theta_t\}_{t>0}$, taking values in $\mathbb{C}^m$, $m \ge 1$, so that $\Theta_t := (\Theta_t^1, \Theta_t^2, ..., \Theta_t^m)$, and for $f = (f^1, f^2, ..., f^m) \in L^2(\mathbb{R}^n, \mathbb{C}^m)$, we set

$$\Theta_t f := \sum_{j=1}^m \Theta_t^j f^j\,.$$

We also define the action of Θ_t on an $m \times m$ matrix valued function $b = (b_{j,k})$, in the obvious way, i.e., $\Theta_t b = ((\Theta_t b)_1, (\Theta_t b)_2, ..., (\Theta_t b)_m)$ is a $\mathbb{C}^m$ valued function, with

$$(\Theta_t b)_k := \sum_{j=1}^{m} \Theta_t^j b_{j,k} .$$

We suppose that Θ_t satisfies the following properties:

(a) (Uniform L^2 bounds and off-diagonal decay in L^2).

$$\sup_{t>0} ||\Theta_t f||_{L^2(\mathbb{R}^n)} \leq C||f||_{L^2(\mathbb{R}^n)} , \tag{4.2}$$

$$||\Theta_t f_j||_{L^2(Q)} \leq C 2^{-j(n+2+\beta)/2} ||f_j||_{L^2(2^{j+1}Q \setminus 2^j Q)} , \quad \ell(Q) \leq t \leq 2\ell(Q) , \tag{4.3}$$

for some $\beta > 0$, where $f_j := f \mathbb{1}_{2^{j+1}Q \setminus 2^j Q}$.

(b) (Quasi-orthogonality on a subspace of L^2). There exists a subspace H of $L^2(\mathbb{R}^n, \mathbb{C}^m)$, such that for some (hence every) CLP family $\{Q_s\}$ (cf. Definition 1.2), there is a $\beta > 0$ for which we have

$$||\Theta_t Q_s h||_{L^2(\mathbb{R}^n)} \leq C \left(\frac{s}{t}\right)^{\beta} ||h||_{L^2(\mathbb{R}^n)}, \quad \forall h \in H , \ \forall s \leq t . \tag{4.4}$$

We have the following generalization of the Christ-Journé $T1$ theorem for square functions [8].

Theorem 4.5 ($T1$ Theorem). *Suppose that Θ_t satisfies Hypothesis* 4.1, *and suppose further that we have the Carleson measure estimate*

$$\sup_Q \frac{1}{|Q|} \int_Q \int_0^{\ell(Q)} |\Theta_t 1(x)|^2 \frac{dxdt}{t} \leq C. \tag{4.6}$$

Then we have the square function estimate

$$\iint_{\mathbb{R}^{n+1}_+} |\Theta_t f(x)|^2 \frac{dxdt}{t} \leq C||f||^2_{L^2(\mathbb{R}^n)} , \quad \forall f \in H. \tag{4.7}$$

We view "1" in this context as the $m \times m$ identity matrix.

The proof will follow that of the $T1$ theorem of [8]; indeed, the lack of pointwise kernel bounds, and the fact that our square function acts only on elements of H, do not present serious obstacles.

Proof of Theorem 4.5. We consider

$$\iint_{\mathbb{R}^{n+1}_+} |\Theta_t h(x)|^2 \frac{dxdt}{t} ,$$

with $h \in H$, where H is a subspace of $L^2(\mathbb{R}^n, \mathbb{C}^m)$. Let us set

$$d\mu(x,t) := |\Theta_t 1(x)|^2 \frac{dxdt}{t} ,$$

so that, by hypothesis, we have the Carleson measure estimate

$$\|\mu\|_{\mathcal{C}} := \sup_Q \frac{1}{|Q|} \iint_{R_Q} d\mu(x,t) < \infty .$$

We follow the familiar idea of Coifman-Meyer to write

$$\Theta_t h = \Big(\Theta_t - (\Theta_t 1)A_t P_t\Big)h + (\Theta_t 1)A_t P_t h := R_t h + (\Theta_t 1)P_t h ,$$

where as usual, A_t is the dyadic averaging operator, and P_t is a nice approximate identity. By Carleson's embedding lemma, and the well-known non-tangential estimate for $A_t P_t h$,

$$\iint_{\mathbb{R}^{n+1}_+} |(\Theta_t 1)A_t P_t h(x)|^2 \frac{dxdt}{t} \lesssim \|\mu\|_{\mathcal{C}} \|h\|_2^2 ,$$

as desired (we remark that for this term, the estimate holds for all $h \in L^2(\mathbb{R}^n, \mathbb{C}^m)$). Moreover,

$$\iint_{\mathbb{R}^{n+1}_+} |R_t h(x)|^2 \frac{dxdt}{t} \lesssim \|h\|_2^2 , \qquad \forall h \in H ,$$

by the same argument used to treat the remainder terms in the proof of Theorem 2.13. Indeed, the present R_t is the same (up to a minus sign) as $R_t^{(2)}$ considered in (2.35), except with $\phi_Q \equiv 1$, i.e., with $d\mu_Q \equiv dx$. □

References

[1] M. A. Alfonseca, P. Auscher, A. Axelsson, S. Hofmann, and S. Kim, Analyticity of layer potentials and L^2 solvability of boundary value problems for divergence form elliptic equations with complex L^∞ coefficients, Advances in Math. **226** (2011), 4533–4606.

[2] P. Auscher, Regularity theorems and heat kernel for elliptic operators. J. London Math. Soc. (2) **54** (1996), no. 2, 284–296.

[3] P. Auscher and A. Axelsson, Weighted maximal regularity estimates and solvability of non-smooth elliptic systems I, Invent. Math. **184** (2011), 47–115.

[4] P. Auscher, A. Axelsson, and A. McIntosh, Solvability of elliptic systems with square integrable boundary data, Ark. Mat. **48** (2010), 253–287.

[5] P. Auscher, S. Hofmann, M. Lacey, A. McIntosh, and P. Tchamitchian, The solution of the Kato square root problem for second order elliptic operators on $\mathbb{R}^n$, Annals of Math. **156** (2002), 633–654.

[6] P. Auscher and P. Tchamitchian, Square root problem for divergence operators and related topics, Astérisque **249** (1998), Société Mathématique de France.

[7] M. Christ, A $T(b)$ theorem with remarks on analytic capacity and the Cauchy integral, Colloquium Mathematicum **LX/LXI** (1990), 601–628.

[8] M. Christ and J.-L. Journé, Polynomial growth estimates for multilinear singular integral operators. Acta Math. **159** (1987), no. 1–2, 51–80.

[9] R. Coifman and Y. Meyer. Non-linear harmonic analysis and PDE, in E. M. Stein, editor, Beijing Lectures in Harmonic Analysis, volume 112 of Annals of Math. Studies, Princeton University Press, 1986.

[10] G. David, J.-L. Journé, and S. Semmes, Opérateurs de Calderón-Zygmund, fonctions para-accrétives et interpolation, Rev. Mat. Iberoamericana **1** (1985), 1–56.

[11] E. De Giorgi, Sulla differenziabilità e l'analiticità delle estremali degli integrali multipli regolari, Mem. Accad. Sci. Torino. Cl. Sci. Fis. Mat. Nat. (3) **3** (1957), 25–43.

[12] C. Fefferman and E. M. Stein, H^p spaces of several variables. Acta Math. **129** (1972), no. 3–4, 137–193.

[13] S. Hofmann, Local $T(b)$ theorems and application in PDE, Harmonic analysis and partial differential equations 29–52, Contemp. Math **505**, Amer. Math. Soc., Providence, RI, 2010.

[14] S. Hofmann, C. Kenig, S. Mayboroda, and J. Pipher, The regularity problem for second order elliptic operators with complex-valued bounded measurable coefficients, to appear in Math. Ann.

[15] S. Hofmann and S. Kim, The Green function estimates for strongly elliptic systems of second order, Manuscripta Math. **124** (2007), no. 2, 139–172.

[16] S. Hofmann and S. Kim, Gaussian estimates for fundamental solutions to certain parabolic systems, Publ. Mat. **48** (2004), no. 2, 481–496.

[17] S. Hofmann, M. Lacey, and A. McIntosh, The solution of the Kato problem for divergence form elliptic operators with Gaussian heat kernel bounds, Annals of Math. **156** (2002), 623–631.

[18] S. Hofmann, S. Mayboroda, and M. Mourgoglou, L^p and endpoint solvability results for divergence form elliptic equations with complex L^∞ coefficients, to appear in Advances in Math.

[19] S. Hofmann and A. McIntosh, The solution of the Kato problem in two dimensions, Proceedings of the Conference on Harmonic Analysis and PDE held in El Escorial, Spain in July 2000, Publ. Mat. Vol. extra, (2002), 143–160.

[20] S. Hofmann, M. Mitrea, and A. Morris, The method of layer potentials in L^p and endpoint spaces for elliptic operators with L^∞ coefficients, to appear in Proc. London Math. Soc.

[21] A. McIntosh and Y. Meyer, Algèbres dopérateurs définis par des intégrales singulières, C. R. Acad. Sci. Paris 301 Série **1** (1985), 395–397.

[22] J. Moser, On Harnack's theorem for elliptic differential equations, Comm. Pure Appl. Math. **14** (1961), 577–591.

[23] J. Nash, Continuity of solutions of parabolic and elliptic equations, Amer. J. Math. **80** (1958), 931–954.

[24] A. Rosén, Layer potentials beyond singular integral operators, Publicacions Matemàtiques **57** (2013), no. 2, 429–454.

[25] S. Semmes, Square function estimates and the $T(b)$ Theorem, Proc. Amer. Math. Soc. **110** (1990), no. 3, 721–726.

[26] G. Verchota, Layer potentials and regularity for the Dirichlet problem for Laplace's equation in Lipschitz domains, J. Funct. Anal. **59** (1984), no. 3, 572–611.

Some Topics in Harmonic Analysis and Applications
ALM 34, pp. 231-264

Non-homogeneous Local $T1$ Theorem: Dual Exponents

Michael T. Lacey* and Antti V. Vähäkangas†

Abstract

We provide an alternative proof of a (local) $T1$ theorem for dual exponents in the non-homogeneous setting of upper doubling measures. This previously known theorem provides necessary and sufficient conditions for the L^p-boundedness of Calderón-Zygmund operators in the described setting, and the novelty lies in the method of proof.

2000 Mathematics Subject Classification: Primary: 42B20; Secondary: 42B25, 60G46.

Keywords and Phrases: local Tb, non-homogeneous, martingale transforms, quasi-orthogonality, surgery.

1 Introduction

1.1 Background and motivation

The subject of local Tb theorems in the classical setting of $\mathbf{R}^n$ with Lebesgue measure is rather well understood by now. We refer, in particular, to [17] and to [1–3, 8, 22, 23]. These theorems extend the David-Journé $T1$ Theorem [7], and the Tb theorem of Christ [6] by giving flexible conditions under which an operator T with a Calderón-Zygmund kernel extends to a bounded linear operator on L^2. By "local" we understand that the Tb conditions involve a family of test functions b_Q, one for each cube Q, which should satisfy a non-degeneracy condition on its "own" Q. Furthermore, both b_Q and Tb_Q are subject to normalized integrability

*School of Mathematics, Georgia Institute of Technology, Atlanta GA 30332, USA. *E-mail:* lacey@math.gatech.edu

†Department of Mathematics and Statistics, P.O. Box 68, FI-00014 University of Helsinki, Finland. *E-mail:* antti.vahakangas@jyu.fi

conditions on Q (with suitable exponents). Symmetric assumptions are imposed for the adjoint T^*.

In the non-homogeneous setting less is know. In the relevant literature [16, 19, 26] one usually encounters stronger $L^\infty(\mathbf{R}^n)$ (sometimes BMO) conditions on Tb_Q's, as well as on test functions b_Q. In the search after relaxation of these conditions one faces complications that arise from the feature that the underlying measure μ need not be doubling.

We provide an alternative proof of a local $T1$ theorem—which is, in fact, a $T1$ theorem in its local formulation—in the non-homogeneous setting of upper doubling measures [13, 15]. The local testing functions are indicators of cubes: $b_Q = \mathbf{1}_Q$, and integrability conditions on $\mathbf{1}_Q T \mathbf{1}_Q$ and $\mathbf{1}_Q T^* \mathbf{1}_Q$ are those of dual exponents $1 < p_1 < \infty$ and $p_2 = p_1/(p_1-1)$. This result is already known and available in the literature, see Remark 6, and the motivation stems from the fact that our novel proof possibly lends itself to other situations. In particular, a non-homogeneous local Tb theorem, say, for dual exponents, has not yet been established, and it seems plausible that the new techniques in the present paper can be used to attack this open and difficult problem.

More precisely, our proof relies upon a so called corona decomposition, adapted to the maximal averages of given two functions f_1 and f_2. The advantage of this approach is that one has powerful quasi orthogonality inequalities, useful throughout the proof. A direct argument can be used to control a difficult "inside" term, thereby we avoid the typical use of paraproducts and Carleson measures. This argument can be viewed as an extension of its "homogeneous" counterparts that are developed in [22, 23].

1.2 A local $T1$ theorem

Let μ be a compactly supported Borel measure on $\mathbf{R}^n$. We assume the upper doubling conditions of Hytönen [13]: there is a dominating function $\lambda : \mathbf{R}^n \times \mathbf{R}_+ \to \mathbf{R}_+$, and a constant $C_\lambda > 0$, such that for all $x \in \mathbf{R}^n$ and $r > 0$:

$$\mu(B(x,r)) \le \lambda(x,r) \le C_\lambda \lambda(x, r/2)\,. \tag{1}$$

Moreover, we assume that $r \mapsto \lambda(x,r)$ is non-decreasing for all $x \in \mathbf{R}^n$. The number $d = \log_2 C_\lambda$ can be thought of as the dimension of μ.

We assume that a linear operator T is bounded on $L^2(d\mu)$, and it is adapted to λ in the following sense. There is a kernel $K : \mathbf{R}^n \times \mathbf{R}^n \to \mathbf{R}$ such that for all compactly supported $f \in L^2(\mathbf{R}^n)$,

$$Tf(x) = \int_{\mathbf{R}^n} K(x,y) f(y)\, d\mu(y)\,, \qquad x \notin \operatorname{supp}(f).$$

We assume that these kernel estimates hold for some $\eta \in (0,1)$:

$$|K(x,y)| \le \min\left\{ \frac{1}{\lambda(x,|x-y|)}, \frac{1}{\lambda(y,|x-y|)} \right\}, \qquad x \ne y\,, \tag{2}$$

$$|K(x,y) - K(x',y)| \le \frac{|x-x'|^\eta}{|x-y|^\eta \lambda(x,|x-y|)}, \qquad |x-y| \ge 2|x-x'|\,, \tag{3}$$

and

$$|K(x,y) - K(x,y')| \le \frac{|y-y'|^\eta}{|x-y|^\eta \lambda(y, |x-y|)}, \qquad |x-y| \ge 2|y-y'|.$$

The operator T is said to be a *Calderón-Zygmund operator*. We are interested in *quantitative* estimates for the operator norm of T on $L^p(\mu)$ for $1 < p < \infty$, and the following hypothesis, together with kernel assumptions, provides the essential quantitative information.

- *Local Testing Condition Hypothesis.* For given two exponents $p_1, p_2 \in (1,\infty)$, there is a constant $\mathbf{T}_{\mathrm{loc}}$ as follows. For all cubes Q in $\mathbf{R}^n$,

$$\int_Q |T\mathbf{1}_Q|^{p_1}\, d\mu(x) \le \mathbf{T}_{\mathrm{loc}}^{p_1} \mu(Q), \qquad \int_Q |T^*\mathbf{1}_Q|^{p_2}\, d\mu(x) \le \mathbf{T}_{\mathrm{loc}}^{p_2} \mu(Q). \tag{4}$$

We provide a novel proof of the following previously known theorem.

Theorem 5. *Let T be a Calderón-Zygmund operator. Fix $1 < p_1, p_2 < \infty$, $1/p_1 + 1/p_2 \le 1$. Assume the following two conditions* (1)–(2)*:*

(1) *T is (a priori) bounded on $L^{p_1}(d\mu)$;*

(2) *T satisfies a Local Testing Condition Hypothesis with exponents p_1 and p_2.*

Under these assumptions, we have a quantitative norm estimate

$$\mathbf{T} := \|T\|_{L^{p_1}(d\mu) \to L^{p_1}(d\mu)} \lesssim 1 + \mathbf{T}_{\mathrm{loc}},$$

where the implied constant depends on n, p_1, p_2, η, μ.

In the sequel, unless otherwise specified, we assume that p_1 and p_2 are in duality: $p_2 = \frac{p_1}{p_1 - 1}$.

Remark 6. Theorem 5 is known and available in the literature. Indeed, under the assumptions of this theorem, it is straightforward to verify that T satisfies a "weak boundedness property" and "testing conditions", namely for all cubes Q in $\mathbf{R}^n$, and an appropriate $\sigma \ge 1$,

$$\left| \int_Q T\mathbf{1}_Q \, d\mu \right| \le \mathbf{T}_{\mathrm{loc}} \mu(Q), \qquad T\mathbf{1} \in \mathrm{BMO}_\sigma^{p_1}(\mu), \qquad T^*\mathbf{1} \in \mathrm{BMO}_\sigma^{p_2}(\mu); \tag{7}$$

we refer to Remark 19 for further details. It remains to apply a non-homogeneous $T1$ theorem, see [25] or [11, Tb theorem 2] for $\lambda(x,r) = r^d$ dominating the measure, and [24, Theorem 2.1] for the general case. Moreover, by using the last theorem, it is even possible to relax the integrability conditions in (4) to exponents $p_1 = 1 = p_2$. Let us also remark that the case of $p_1 = 2 = p_2$ has been addressed in [28] with a function $\lambda(x,r) = \max\{\delta(x)^d, r^d\}$ dominating the measure, where $\delta(x) = \mathrm{dist}(x, \mathbf{R}^n \setminus H)$ for an an open set H in $\mathbf{R}^n$.

Remark 8. The p-independence property of Calderón-Zygmund operators, i.e., if their L^2 boundedness is equivalent to their L^p boundedness, has been addressed, for instance, in [9, 14]. It is an interesting question, if our proof can be adapted to obtain a quantitative p-independence result for Calderón-Zygmund operators, under an appropriate set of local testing hypotheses.

1.3 Structure of the paper

We use the non-homogeneous techniques of [25], in particular, good and bad cubes are applied in a partially novel manner. Martingale techniques, including L^p estimates for martingale transforms and Stein's inequality, are fundamental. These techniques are also applied in a related paper [19], from which we borrow also some other ideas, e.g., treatments of "separated" and "nearby" terms. Our main technical contribution is treatment of the most difficult "inside" term by a strong definition of goodness and a corona decomposition, avoiding (a) explicit construction of paraproduct operators; and (b) Carleson embedding theorems.

The heart of the matter is estimation of a form $|\langle Tf_1, f_2\rangle|$, where f_j's are perturbed functions, supported on large dyadic cubes $Q_{j,0} \in \mathcal{D}_j$. Here $\mathcal{D}_j$ is a random dyadic system. The perturbation is simply a projection to good cubes, and results in that the usual martingale differences $\Delta_Q f_j$ vanish if $Q \subset Q_{j,0}$ is a bad cube. After a probabilistic absorption argument, the focus will be on a triangular form

$$\Big| \sum_{P,Q \text{ good}} \mathbf{1}_{\ell Q \leq \ell P} \cdot \langle T\Delta_P f_1, \Delta_Q f_2\rangle \Big| , \tag{9}$$

where always $P \subset Q_{1,0}$ and $Q \subset Q_{2,0}$. This form is further split into "inside", "separated", and "nearby" terms. The analysis of the inside term, in which Q is deeply inside P, is taken up in Sections 5 and 6—the argument is transparent, and our strong definition of *goodness of cubes* has a key role. The construction of paraproducts is avoided, and even Carleson embedding theorems are not needed; in this we follow [20, 21]. We apply a corona decomposition, and the associated stopping tree is constructed in Section 4, where we also record the basic "quasi-orthogonality" properties. The separated term, in which Q is always far away from P, is analysed in Section 7, and the (usual) goodness is crucial. Throughout Sections 8–10, we treat the nearby term, where cubes are close to each other both in position and size. The usual surgery is performed.

Acknowledgements. The authors would like to thank Tuomas Hytönen and Henri Martikainen for indicating the connection of Theorem 5 to the $T1$ theorems that are available in the literature. Research supported in part by grant NSF-DMS 0968499, and a grant from the Simons Foundation (#229596 to Michael Lacey), and the Australian Research Council through grant ARC-DP120100399. ML and AVV benefited from the research program Operator Related Function Theory and Time-Frequency Analysis at the Centre for Advanced Study at the Norwegian Academy of Science and Letters in Oslo during 2012–2013. AVV was supported by the Finnish Academy of Science and Letters, Vilho, Yrjö and Kalle Väisälä Foundation.

2 Preliminaries

2.1 Notation

The implied constants are allowed to depend upon parameters $r, n, p_1, p_2, \eta, \mu$. The distances are measured in supremum norm, $|x| = \|x\|_\infty$ for $x \in \mathbf{R}^n$. We denote $L^p = L^p(d\mu)$ if $1 \le p \le \infty$. For a cube Q and $f \in L^1_{\text{loc}}$, write $\langle f \rangle_Q := \mu(Q)^{-1} \int_Q f \, d\mu$ with the convention $\langle f \rangle_Q = 0$ if $\mu(Q) = 0$. The side length of a cube Q is written as ℓQ, and the midpoint as x_Q. The "long distance" between cubes Q and P is $D(Q,P) = \ell Q + \text{dist}(Q,P) + \ell P$.

A "dyadic cube" is any cube in either random grid $\mathcal{D}_j$ with $j \in \{1,2\}$, Section 2.2. By $\mathcal{D}_{j,k}$ we denote those dyadic cubes $Q \in \mathcal{D}_j$ for which $\ell Q = 2^k$, $k \in \mathbf{Z}$. The dyadic children of $Q \in \mathcal{D}_j$ are $\{Q_1, \ldots, Q_{2^n}\} = \text{ch}(Q)$, its dyadic parent is $\pi_j Q = \pi_j^1 Q$, and $\pi_j^t Q = \pi_j(\pi_j^{t-1} Q)$ for $t \in \{2,3,\ldots\}$. For $\mathcal{S}_j \subset \mathcal{D}_j$ the family $\text{ch}_{\mathcal{S}_j}(S) = \text{ch}^1_{\mathcal{S}_j}(S)$ consists of the $\mathcal{S}_j$-children of $S \in \mathcal{S}_j$: the maximal cubes in $\mathcal{S}_j$ that are strictly contained in S. We also denote $\text{ch}^0_{\mathcal{S}_j}(S) = \{S\}$ and, for $t > 1$, write $S' \in \text{ch}^t_{\mathcal{S}_j}(S)$ if $S' \in \text{ch}_{\mathcal{S}_j}(S'')$ for some $S'' \in \text{ch}^{t-1}_{\mathcal{S}_j}(S)$. For *any* cube Q which is contained in a cube in $\mathcal{S}_j$, we take $\pi_{\mathcal{S}_j} Q = \pi^0_{\mathcal{S}_j} Q$ to be the $\mathcal{S}_j$-parent of Q: the minimal $\mathcal{S}_j$-cube containing Q (if Q is not contained in a cube in $\mathcal{S}_j$, we set $\pi_{\mathcal{S}_j} Q = \mathbf{R}^n$). For $t \ge 1$ and any cube Q, contained in at least $t+1$ cubes in $\mathcal{S}_j$, we let $\pi^t_{\mathcal{S}_j} Q$ to be $\pi^{t-1}_{\mathcal{S}_j} S'$, where $\pi_{\mathcal{S}_j} Q \in \text{ch}_{\mathcal{S}_j}(S')$.

2.2 Random grids

We use the foundational tool of random grids, initiated by Nazarov-Treil-Volberg [26], which has in turn been used repeatedly. We refer, e.g., to [10, 15, 21, 27]. Throughout the paper, we shall use two random dyadic grids (systems) $\mathcal{D}_j$, $j \in \{1,2\}$. A third random grid $\mathcal{D}_3$ appears at the very end. These are constructed as follows; we refer to [12] for further details.

The random grids $\mathcal{D}_j$ are parametrized by sequences $\omega_j \in (\{0,1\}^n)^{\mathbf{Z}}$, $j \in \{1,2,3\}$, where we tacitly assume three independent copies of $(\{0,1\}^n)^{\mathbf{Z}}$. More precisely, for a cube $\widehat{Q} \in \widehat{\mathcal{D}}$ in the standard dyadic grid, the *position* of an ω_j-translated cube is

$$Q = \widehat{Q} \dot{+} \omega_j := \widehat{Q} + \sum_{k \,:\, 2^{-k} < \ell \widehat{Q}} 2^{-k} \omega_{j,k} \,,$$

which is a function of $\omega_j \in (\{0,1\}^n)^{\mathbf{Z}}$. A dyadic grid (system)

$$\mathcal{D}_j = \mathcal{D}(\omega_j) = \{\widehat{Q} \dot{+} \omega_j \,:\, \widehat{Q} \in \widehat{\mathcal{D}}\}$$

is the family of these ω_j-translated cubes. The natural uniform probability measure $\mathbf{P}_{\omega_j}$ is placed upon the respective copy of $(\{0,1\}^n)^{\mathbf{Z}}$. Each component $\omega_{j,k}$, $k \in \mathbf{Z}$, has an equal probability 2^{-n} of taking any of the 2^n values, and all components are independent of each other. The expectation with respect to $\mathbf{P}_{\omega_j}$ is denoted by $\mathbf{E}_{\omega_j}$. We will usually simply write P or S for a cube in $\mathcal{D}_1$, and Q or R for a cube in $\mathcal{D}_2$, instead of the heavier notation $\widehat{Q} \dot{+} \omega_j$ with $\widehat{Q} \in \widehat{\mathcal{D}}$.

Choose, once and for all, a constant $\gamma \in (0,1)$ such that

$$d\gamma/(1-\gamma) \le \eta/4, \quad \gamma \le \frac{\eta}{2(d+\eta)}\,, \qquad d = \log_2 C_\lambda \,. \tag{10}$$

Here η is the constant appearing in the kernel condition (3). We also denote

$$\theta(j) = \left\lceil \frac{\gamma j + r}{1-\gamma} \right\rceil \quad \text{for } j = 0, 1, 2, \ldots . \tag{11}$$

Throughout $r \in \mathbf{N}$ should be thought of as a large integer, whose exact value is assigned later.

2.3 Goodness of cubes

We impose a strong definition of goodness: by doing so, we ensure that good cubes $Q \in \mathcal{D}_1 \cup \mathcal{D}_2$ from either system are always far away from the boundaries of much larger cubes in *either* one of these two systems.

A cube $Q \in \mathcal{D}_j$ is *k-bad* for $j, k \in \{1, 2\}$ if there is a cube $P \in \mathcal{D}_k$ such that $\ell P \geq 2^r \ell Q$ and $\operatorname{dist}(Q, \partial P) \leq (\ell Q)^\gamma (\ell P)^{1-\gamma}$. Otherwise, Q is *k-good.* The following properties are known [12].

1. For $\widehat{Q} \in \widehat{\mathcal{D}}$, position and k-goodness of $Q = \widehat{Q} \dot{+} \omega_j$ are independent random variables.

2. The probability $\pi_{j,k,\text{good}} := \mathbf{P}_{\omega_k}(\widehat{Q} \dot{+} \omega_j \text{ is } k\text{-good})$ is independent of $\widehat{Q} \in \widehat{\mathcal{D}}$.

3. $\pi_{j,k,\text{bad}} := 1 - \pi_{j,k,\text{good}} \lesssim 2^{-\gamma r}$, with implied constant independent of r.

A cube $Q \in \mathcal{D}_j$ with $j \in \{1, 2\}$ is *bad* if it is k-bad for some $k \in \{1, 2\}$. Otherwise, we say that Q is *good.* To state this condition otherwise, if $Q \in \mathcal{D}_j$ is good, we have inequality

$$(\ell Q)^\gamma (\ell P)^{1-\gamma} < \operatorname{dist}(Q, \partial P) ,$$

if $P \in \mathcal{D}_1 \cup \mathcal{D}_2$ and $2^r \ell Q \leq \ell P$. Define bad and good projections by $I = P_{j,\text{bad}} + P_{j,\text{good}}$, where

$$P_{j,\text{bad}} \phi := \sum_{Q \in \mathcal{D}_j \,:\, Q \text{ is bad}} \Delta_Q \phi , \qquad \phi \in L^q \quad (1 < q < \infty) .$$

Here $\Delta_Q \phi = \sum_{Q' \in \text{ch}(Q)} \{\langle \phi \rangle_{Q'} - \langle \phi \rangle_Q\} \mathbf{1}_{Q'}$ is the martingale difference with respect to μ. The following proposition is a straightforward modification of [23, Proposition 2.4].

Proposition 12. *For every $j \in \{1, 2\}$ and $1 < q < \infty$ there is a constant $c_q > 0$ so that*

$$\mathbf{E}_{\omega_1} \mathbf{E}_{\omega_2} \| P_{j,\text{bad}} \phi \|_q^q \lesssim 2^{-\gamma r / c_q} \| \phi \|_q^q , \tag{13}$$

where $\phi \in L^q$ is any function, independent of both random grids $\mathcal{D}_k$ with $k \in \{1, 2\}$. Moreover, the implied constant is independent of r.

Proof. We apply Marcinkiewicz interpolation theorem to the linear operator

$$P_{j,\text{bad}} : L^q(d\mu) \to L^q(\mathbf{P}_{\omega_1} \otimes \mathbf{P}_{\omega_2} \otimes d\mu) .$$

The projection to bad cubes is a martingale transform: by inequality (14), the following inequality with no decay holds,

$$\mathbf{E}_{\omega_1}\mathbf{E}_{\omega_2}\|P_{j,\mathrm{bad}}\phi\|_p^p \le \sup_{\omega_1,\omega_2} \|P_{j,\mathrm{bad}}\phi\|_p^p \lesssim \|\phi\|_p^p\,, \quad 1<p<\infty\,.$$

Thus, it suffices to verify the claimed decay for $q=2$. To this end, we have by orthogonality of martingale differences,

$$\begin{aligned}
\mathbf{E}_{\omega_1}\mathbf{E}_{\omega_2}\|P_{j,\mathrm{bad}}\phi\|_2^2 &= \mathbf{E}_{\omega_1}\mathbf{E}_{\omega_2}\sum_{\widehat{Q}\in\widehat{\mathcal{D}}} \mathbf{1}_{\widehat{Q}\dot{+}\omega_j \text{ is bad}}\|\Delta_{\widehat{Q}\dot{+}\omega_j}\phi\|_2^2 \\
&\le \sum_{k=1}^{2} \mathbf{E}_{\omega_1}\mathbf{E}_{\omega_2}\sum_{\widehat{Q}\in\widehat{\mathcal{D}}} \mathbf{1}_{\widehat{Q}\dot{+}\omega_j \text{ is } k\text{-bad}}\|\Delta_{\widehat{Q}\dot{+}\omega_j}\phi\|_2^2 \\
&\le \sum_{k=1}^{2} \pi_{j,k,\mathrm{bad}}\mathbf{E}_{\omega_1}\mathbf{E}_{\omega_2}\sum_{\widehat{Q}\in\widehat{\mathcal{D}}} \|\Delta_{\widehat{Q}\dot{+}\omega_j}\phi\|_2^2 \\
&\le (\pi_{j,1,\mathrm{bad}}+\pi_{j,2,\mathrm{bad}})\|\phi\|_2^2\,.
\end{aligned}$$

In the third step, we used Fubini's theorem, linearity of expectation, and the fact that $\|\Delta_{\widehat{Q}\dot{+}\omega_j}\phi\|_2^2$ and k-badness of $\widehat{Q}\dot{+}\omega_j$ are independent random variables. □

2.4 Square function inequalities

The martingale transform inequality is this, see e.g. [5]. For all functions $f\in L^p$, and constants satisfying $\sup_{Q\in\mathcal{D}_j}|\varepsilon_Q|\le 1$,

$$\Big\|\sum_{Q\in\mathcal{D}_j}\varepsilon_Q\Delta_Q f\Big\|_p \lesssim \|f\|_p, \quad 1<p<\infty\,, \quad j\in\{1,2\}\,. \tag{14}$$

A consequence of Khintchine's inequality and inequality (14) is the following

$$\Big\|\Big(\sum_{k\in\mathbf{Z}}|\Delta_{j,k}f|^2\Big)^{1/2}\Big\|_p \lesssim \|f\|_p\,, \tag{15}$$

where $f\in L^p$ with $1<p<\infty$, and $\Delta_{j,k}f=\sum_{Q\in\mathcal{D}_{j,k}}\Delta_Q f$ for $k\in\mathbf{Z}$ and $j\in\{1,2\}$.

We will use the following *Stein's inequality*, see e.g. [4]. For $1<p<\infty$ and $j\in\{1,2\}$,

$$\Big\|\Big(\sum_{k\in\mathbf{Z}}|\mathbf{E}_{j,k}f_k|^2\Big)^{1/2}\Big\|_p \lesssim \Big\|\Big(\sum_{k\in\mathbf{Z}}|f_k|^2\Big)^{1/2}\Big\|_p\,, \tag{16}$$

where $(f_k)_{k\in\mathbf{Z}}$ is *any* sequence in $L^p(d\mu)$, $\mathbf{E}_{j,k}f=\sum_{Q\in\mathcal{D}_{j,k}}\mathbf{E}_Q f$, and $\mathbf{E}_Q f=\langle f\rangle_Q\mathbf{1}_Q$. We don't rely on Fefferman-Stein inequalities for the vector-valued maximal function. Stein's inequality is their replacement in the present, non-homogeneous, setting.

2.5 Off-diagonal estimates

Here we collect useful off-diagonal estimates.

Lemma 17. *Let $Q \subset P \subset R$ be cubes in $\mathbf{R}^n$ such that $\ell Q \leq \operatorname{dist}(Q, R \setminus P)$. Then,*

$$|T\mathbf{1}_{R\setminus P}(x) - T\mathbf{1}_{R\setminus P}(x_Q)| \lesssim \left(\frac{\ell Q}{\operatorname{dist}(Q, R \setminus P)}\right)^{\eta}, \qquad x \in Q\,. \tag{18}$$

Proof. The kernel condition (3) applies,

$$LHS(18) \leq \int_{R\setminus P} |K(x,y) - K(x_Q, y)|\, d\mu(y) \leq \int_{R\setminus P} \frac{|x - x_Q|^{\eta}}{|x-y|^{\eta}\lambda(x,\, x-y|)}\, d\mu(y)\,.$$

Let us denote $\delta := \operatorname{dist}(Q, R \setminus P)$ and $A_j = \{y \in \mathbf{R}^n : 2^j\delta \leq |x-y| < 2^{j+1}\delta\}$ for $j \geq 0$. Observe that $R \setminus P \subset \cup_{j=0}^{\infty} A_j$. Since $A_j \subset B(x, 2|x-y|)$ for each $y \in A_j$, we can bound the last integral by $C_\lambda (\ell Q)^\eta \sum_{j=0}^{\infty} (2^j\delta)^{-\eta} \lesssim (\ell Q/\delta)^\eta$ as required. □

Remark 19. Let us verify that the a priori boundedness of T on L^{p_1}, and Local Testing Condition Hypothesis, together imply the assumptions of a $T1$ theorem; namely, conditions (7) with $\sigma = 3$. The first condition therein is, indeed, a trivial consequence of inequality (4). Hence, it suffices to verify that $b := T\mathbf{1}$ satisfies $b \in \mathrm{BMO}^{p_1}_\sigma(\mu)$, i.e.,

$$\|b\|_{\mathrm{BMO}^{p_1}_\sigma(\mu)} := \sup_Q \left\{ \frac{1}{\mu(\sigma Q)} \int_Q |b(x) - \langle b\rangle_Q|^{p_1}\, d\mu(x) \right\}^{1/p_1} \lesssim 1 + \mathbf{T}_{\mathrm{loc}}\,, \tag{20}$$

where the supremum is taken over all cubes Q in $\mathbf{R}^n$. Indeed, a completely analogous argument then shows that $T^*\mathbf{1} \in \mathrm{BMO}^{p_2}_\sigma(\mu)$.

In order to verify inequality (20), let us fix a cube Q in which the supremum above is (almost) attained. Let us then fix a large cube R in $\mathbf{R}^n$, containing both $3Q$ and the compact support of the measure μ. In particular, $T\mathbf{1} = T\mathbf{1}_R \in L^{p_1}$, and we can estimate

$$\begin{aligned}
\|b\|^{p_1}_{\mathrm{BMO}^{p_1}_\sigma(\mu)} &\lesssim \frac{1}{\mu(3Q)} \int_Q |T\mathbf{1}_R(x) - T\mathbf{1}_{R\setminus 3Q}(x_Q)|^{p_1}\, d\mu(x) \\
&\lesssim \frac{1}{\mu(3Q)} \int_Q |T\mathbf{1}_{3Q}|^{p_1}\, d\mu + \frac{1}{\mu(3Q)} \int_Q |T\mathbf{1}_{R\setminus 3Q} - T_{R\setminus 3Q}(x_Q)|^{p_1}\, d\mu\,.
\end{aligned}$$

By inequality (4), the first term in the last line is dominated by $\mathbf{T}^{p_1}_{\mathrm{loc}}$. And by Lemma 17, the last term is seen to be bounded by $\lesssim 1$.

In the following two lemmata, we write $D(Q,P)/\ell P \sim 2^u$ if $2^u < D(Q,P)/\ell P \leq 2^{u+1}$.

Lemma 21. *Suppose $P \in \mathcal{D}_{1,k}$ and $Q \in \mathcal{D}_{2,k-m}$ is a good cube such that $D(Q,P)/\ell P \sim 2^u$, where $k \in \mathbf{Z}$ and $u, m \in \mathrm{N}_0$. Then, we have $Q \subset \pi_1^{u+\theta(u+m)} P$.*

Proof. Denote $t = u+\theta(u+m) \geq r$. By goodness, either $Q \subset \pi^t P$ or $Q \subset \mathbf{R}^n \setminus \pi^t P$. In the former case, we are done. In the latter case, we obtain a contradiction. Indeed, by goodness,

$$(\ell Q)^\gamma (\ell \pi^t P)^{1-\gamma} < \operatorname{dist}(Q, \partial \pi^t P) = \operatorname{dist}(Q, \pi^t P) \leq D(Q,P) \leq 2^{u+1} \ell P .$$

Substituting $\ell Q = 2^{k-m}$ and $\ell P = 2^k$ yields $u + \theta(u+m) = t < u + \theta(u+m)$ after elementary manipulations. This is a contradiction. □

Lemma 22. *Suppose that P and Q are as in Lemma 21. Assume also that $\ell Q \leq$ $\operatorname{dist}(Q,P)$. Then, by denoting $S := \pi_1^{u+\theta(u+m)} P$,*

$$|K(x,y) - K(x_Q, y)| \lesssim \frac{2^{-\eta(u+m)/4}}{\mu(S)}, \qquad (x,y) \in Q \times P . \tag{23}$$

Proof. By inequalities (3) and $\ell Q \leq \operatorname{dist}(Q,P)$, we obtain $LHS(23) \leq \alpha \cdot \beta$ where $\alpha = C_\lambda^\kappa (\ell Q)^\eta / \operatorname{dist}(Q,P)^\eta$ and $\beta = 1/\mu(B(x, 2^\kappa |x-y|))$ with κ specified in the two case studies.

Case $\ell P < \operatorname{dist}(Q,P)$: choose $\kappa = 2 + \theta(u+m)$. Observe the inequality $2^{u+k} <$ $4\operatorname{dist}(Q,P)$. Combined with Lemma 21 this implies a relation $S \subset B(x, 2^\kappa |x-y|)$ and, in particular, that $\beta \leq \mu(S)^{-1}$. The inequality $2^{u+k} < 4\operatorname{dist}(Q,P)$, followed by $C_\lambda^\kappa = 2^{d\kappa}$ and (10), shows that $\alpha \lesssim 2^{d\theta(u+m) - \eta(u+m)} \lesssim 2^{-\eta(u+m)/4}$.

Case $\ell P \geq \operatorname{dist}(Q,P)$: choose $\kappa \in \mathbf{N}$ in such a way that

$$2^{\kappa-1} < \frac{c \ell S}{(\ell Q)^\gamma (\ell P)^{1-\gamma}} \leq 2^\kappa , \qquad c = 2^{r(1-\gamma)} .$$

A useful consequence of goodness is this

$$\operatorname{dist}(Q,P) > (\ell Q)^\gamma (\ell P)^{1-\gamma} / c . \tag{24}$$

Lemma 21 and inequality (24) yield $S \subset B(x, 2^\kappa |x-y|)$, hence $\beta \leq \mu(S)^{-1}$. Inequality (24) also allows us to estimate

$$\alpha \lesssim 2^{d\kappa} \left(\frac{\ell Q}{\ell P} \right)^{\eta(1-\gamma)} \lesssim 2^{d(m+u+\theta(u+m)) - m(d+\eta)(1-\gamma)} \lesssim 2^{-\eta(u+m)/4} .$$

In the last step, we used the fact that $u \leq 1$ and both of the inequalities (10). □

3 Perturbations and a basic decomposition

Let us denote $\mathbf{T} := \|T\|_{L^{p_1} \to L^{p_1}}$. We fix functions $\widetilde{f}_j \in L^{p_j}(d\mu)$, $j = 1,2$, supported in $\operatorname{supp}(\mu)$, and satisfying $\mathbf{T} \leq 2|\langle T\widetilde{f}_1, \widetilde{f}_2 \rangle|$ and $\|\widetilde{f}_1\|_{p_1} = 1 = \|\widetilde{f}_2\|_{p_2}$.

For almost every pair $\{\mathcal{D}_j : j \in \{1,2\}\}$ we will define certain perturbations $f_j = f_j(\widetilde{f}_j, \mathcal{D}_1, \mathcal{D}_2)$ of functions $\widetilde{f}_j$. The role of these perturbations is indicated by following proposition.

Proposition 25. *Under assumptions of Theorem* 5, *the following statement holds for a fixed* $t > p_1 \vee p_2$. *For every sufficiently large* $r \in \mathbf{N}$ *and every* $\epsilon, \upsilon \in (0,1)$,

$$\mathbf{E}_{\omega_1}\mathbf{E}_{\omega_2}\|\widetilde{f}_j - f_j\|_{p_j}^{p_j} \leq c2^{-\gamma r/c}\,, \qquad j = 1,2\,, \tag{26}$$

$$\mathbf{E}_{\omega_1}\mathbf{E}_{\omega_2}\big|\langle Tf_1, f_2\rangle\big| \leq C(r,\upsilon,\epsilon)(1+\mathbf{T}_{\mathrm{loc}}) + \big(C(r,\upsilon)\epsilon^{1/t} + C(r)\upsilon^{1/t}\big)\mathbf{T}\,. \tag{27}$$

Aside from the parameters indicated, constants c, $C(r,\upsilon,\epsilon)$, $C(r,\upsilon)$, *and* $C(r)$ *are also allowed to depend upon* n, p_1, p_2, η, μ.

Proposition 25 and an absorption argument provide a proof of Theorem 5. Hence, we are left with proving this proposition. During this section, we select functions f_j by using projections to good cubes, and then begin with the analysis of the resulting form $\big|\langle Tf_1, f_2\rangle\big|$.

3.1 Perturbations of $\widetilde{f}_j$

For $j \in \{1,2\}$ we denote by $Q_{j,0}$ a cube in $\mathcal{D}_j = \mathcal{D}(\omega_j)$, containing the support $\mathrm{supp}(\mu)$ of the measure μ. Such a cube exists almost surely with respect to ω_j [19, Lemma 2.8]. In the sequel, we will restrict ourselves to such sequences ω_j. Let $\mathcal{G}_j$ be the family of all good cubes in $\mathcal{D}_j$ that are contained in $Q_{j,0}$, and denote $\mathcal{G}_{j,k} = \mathcal{D}_{j,k} \cap \mathcal{G}_j$ for $k \in \mathbf{Z}$.

We define approximates of the functions $\widetilde{f}_j$ to be the following perturbations,

$$f_j := \langle \widetilde{f}_j\rangle_{Q_{j,0}}\mathbf{1}_{Q_{j,0}} + \sum_{Q\in\mathcal{G}_j}\Delta_Q\widetilde{f}_j\,, \qquad j = 1,2\,.$$

Recall the fact that the support of μ is contained in $Q_{j,0}$. Therefore $\Delta_Q\widetilde{f}_j = 0$ almost everywhere w.r.t. μ if $Q \in \mathcal{D}_j$ is not contained in $Q_{j,0}$. Hence, in the view of Proposition 12, we have

$$\mathbf{E}_{\omega_1}\mathbf{E}_{\omega_2}\|\widetilde{f}_j - f_j\|_{p_j}^{p_j} = \mathbf{E}_{\omega_1}\mathbf{E}_{\omega_2}\|P_{\mathrm{j,bad}}\widetilde{f}_j\|_{p_j}^{p_j} \leq c2^{-\gamma r/c}\,. \tag{28}$$

This is inequality (26).

3.2 Decomposition of the bilinear form

During the course of the remaining sections, we prove inequality (27), which then completes the proof of Theorem 5.

By using the facts that $f_j = f_j\mathbf{1}_{Q_{j,0}}$ and $\Delta_R f_j = 0$ if $R \subset Q_{j,0}$ is a bad cube, we easily find that an expansion of the bilinear form is

$$\langle Tf_1, f_2\rangle = \langle T\mathbf{E}_{Q_{1,0}}f_1, f_2\rangle + \langle T\sum_{P\in\mathcal{G}_1}\Delta_P f_1, \mathbf{E}_{Q_{2,0}}f_2\rangle + \sum_{(P,Q)\in\mathcal{G}_1\times\mathcal{G}_2}\langle T\Delta_P f_1, \Delta_Q f_2\rangle. \tag{29}$$

Using the assumptions and inequality (14), it is straightforward to verify that

$$|\langle T\mathbf{E}_{Q_{1,0}}f_1, f_2\rangle| + |\langle T\sum_{P\in\mathcal{G}_1}\Delta_P f_1, \mathbf{E}_{Q_{2,0}}f_2\rangle| \lesssim \mathbf{T}_{\mathrm{loc}}\|f_1\|_{p_1}\|f_2\|_{p_2} \lesssim \mathbf{T}_{\mathrm{loc}}\,.$$

The last term in the right hand side of (29) remains. This main term is further split into dual triangular sums, one of which is the sum over $(P,Q) \in \mathcal{G}_1 \times \mathcal{G}_2$ such that $\ell P \geq \ell Q$. This sum will be our main point of interest, and we only remark that the dual triangular sum, associated with cubes $\ell P < \ell Q$, is estimated in a similar manner.

The family $\{(P,Q) \in \mathcal{G}_1 \times \mathcal{G}_2 : \ell P \geq \ell Q\}$ is partitioned into three subfamilies:

$$\begin{aligned}
\mathcal{P}_{\text{inside}} &:= \{(P,Q) \in \mathcal{G}_1 \times \mathcal{G}_2 \ : \ Q \subset P \text{ and } 2^r \ell Q < \ell P\}\,; \\
\mathcal{P}_{\text{separated}} &:= \{(P,Q) \in \mathcal{G}_1 \times \mathcal{G}_2 \ : \ \ell Q \leq \ell P \text{ and } \ell Q \leq \operatorname{dist}(Q,P)\}\,; \\
\mathcal{P}_{\text{nearby}} &:= \{(P,Q) \in \mathcal{G}_1 \times \mathcal{G}_2 \ : \ 2^{-r}\ell P \leq \ell Q \leq \ell P \text{ and } \operatorname{dist}(Q,P) < \ell Q\}\,.
\end{aligned}$$

The fact that this is a partition relies on the goodness of Q. We refer to [19, Section 13] for further details. The sums over these collections of cubes are handled separately. Let us denote

$$\mathbf{B}_\star(f_1,f_2) = \sum_{(P,Q)\in\mathcal{P}_\star} \langle T\Delta_P f_1, \Delta_Q f_2\rangle\,, \qquad \star \in \{\text{inside}, \text{separated}, \text{nearby}\}\,.$$

The analysis of the (most difficult) inside term is performed within Sections 5 and 6. It relies on a corona decomposition, and the associated stopping tree is first constructed in Section 4. The separated term is analysed in a standard manner in Section 7. Finally, throughout Sections 8–10, we treat the nearby terms via surgery.

4 A stopping tree construction

A stopping tree construction is used in the analysis of the inside-term.

For $j \in \{1,2\}$, let us define a stopping tree $\mathcal{S}_j \subset \mathcal{D}_j$ and a function $\sigma_j : \mathcal{S}_j \mapsto \mathbf{R}_+$ as follows. Take the maximal good $\mathcal{D}_j$-cubes $Q \subset Q_{j,0}$ in $\mathcal{S}_j$, and define $\sigma_j(Q) := \langle |f_j| \rangle_Q$ for these maximal cubes. At inductive stage, if $S \in \mathcal{S}_j$ is a minimal cube, we consider the maximal $\mathcal{D}_j$-cubes $Q \subsetneq S$ subject to both of the conditions (1)–(2):

(1) $\langle |f_j| \rangle_Q > 4\sigma_j(S)$;

(2) Either Q or $\pi_j Q$ is a good cube.

We add these cubes Q to the stopping tree $\mathcal{S}_j$, and define $\sigma_j(Q) := \langle |f_j| \rangle_Q$ for each of them.

Remark 30. Condition (2), imposed in the construction of stopping trees, will be useful to us in many occasions. A minor side effect is that we can rely on inequality $\langle |f_j| \rangle_Q \leq 4\sigma_1(\pi_{\mathcal{S}_j} Q)$ for a $\mathcal{D}_j$-dyadic cube $Q \subsetneq Q_{j,0}$ only if either Q or $\pi_j Q$ is good. But this is, in fact, all we need.

Remark 31. By construction $\mathcal{S}_j$ is a "sparse family of cubes", i.e.,

$$\sum_{S' \in \operatorname{ch}_{\mathcal{S}_j}(S)} \mu(S') \leq 4^{-1}\mu(S)\,, \qquad S \in \mathcal{S}_j\,, \quad j \in \{1,2\}\,. \tag{32}$$

In particular, family $\mathcal{S}_j$ satisfies a "Carleson condition": $\sum_{S' \in \mathcal{S}_j : S' \subset S} \mu(S') \lesssim \mu(S)$ if $S \in \mathcal{S}_j$.

4.1 Quasi-orthogonality

The following is a key inequality,

$$\sum_{S\in\mathcal{S}_j} \sigma_j(S)^{p_j}\mu(S) \lesssim \|f_j\|_{p_j}^{p_j} \lesssim 1\,, \qquad j\in\{1,2\}\,. \tag{33}$$

Proof of (33). We apply a dyadic maximal function:

$$M_{j,\mu}f_j(x) = \sup_{x\in Q\in\mathcal{D}_j} \langle |f_j|\rangle_Q.$$

For $S\in\mathcal{S}_j$, we let E_S be the set S minus all the $\mathcal{S}_j$-children of S. By inequality (32), $\mu(E_S)\geq \frac{3}{4}\mu(S)$, and the sets E_S are pairwise disjoint by definition. Hence,

$$\begin{aligned}\sum_{S\in\mathcal{S}_j} \sigma_j(S)^{p_j}\mu(S) &\leq \frac{4}{3}\sum_{S\in\mathcal{S}_j} \langle |f_j|\rangle_S^{p_j}\mu(E_S) \\ &\lesssim \sum_{S\in\mathcal{S}_j}\int_{E_S} (M_{j,\mu}f_j)^{p_j}\,d\mu \leq \int_{\mathbf{R}^n} (M_{j,\mu}f_j)^{p_j}\,d\mu\,.\end{aligned}$$

Thus, the first inequality in (33) follows from the fact that $M_{j,\mu}$ is bounded on L^{p_j}. The second inequality is a consequence of the martingale transform inequality (14). □

4.2 Martingale projections

For $S\in\mathcal{S}_j$ and $\phi\in L^1_{\text{loc}}$, we define $P_{j,S}\phi = \sum_{Q\in\mathcal{D}_j\,:\,\pi_{\mathcal{S}_j}Q=S}\Delta_Q\phi$. By orthogonality of martingale differences and inequality (14), for all $S\in\mathcal{S}_j$ and all sequences of constants satisfying $\sup_{Q\in\mathcal{G}_j}|\varepsilon_Q|\leq 1$,

$$\Bigg\|\sum_{Q\in\mathcal{G}_j\,:\,\pi_{\mathcal{S}_j}Q=S} \varepsilon_Q\Delta_Q f_j\Bigg\|_{p_j}^{p_j} \lesssim \|P_{j,S}f_j\|_{p_j}^{p_j}\,. \tag{34}$$

Of fundamental importance is the following inequality, which *does not hold* for general families of orthogonal martingale projections, in the case of $1<p_j<2$,

$$\sum_{S\in\mathcal{S}_j} \|P_{j,S}f_j\|_{p_j}^{p_j} \lesssim 1\,. \tag{35}$$

Proof of (35). Let us write

$$\sum_{S\in\mathcal{S}_j} \|P_{j,S}f_j\|_{p_j}^{p_j} = \sum_{S\in\mathcal{S}_j} \|\mathbf{1}_{S\setminus E_S}P_{j,S}f_j\|_{p_j}^{p_j} + \sum_{S\in\mathcal{S}_j} \|\mathbf{1}_{E_S}P_{j,S}f_j\|_{p_j}^{p_j}\,, \tag{36}$$

where E_S denotes the set $S\setminus\bigcup_{S'\in\operatorname{ch}_{\mathcal{S}_j}(S)} S'$. We estimate the two terms separately. First,

$$|\mathbf{1}_{S\setminus E_S}P_{j,S}f_j| = \Bigg|\sum_{S'\in\operatorname{ch}_{\mathcal{S}_j}(S)} \mathbf{1}_{S'}\{\langle f_j\rangle_{S'} - \langle f_j\rangle_S\}\Bigg| \lesssim \sum_{S'\in\operatorname{ch}_{\mathcal{S}_j}(S)} \mathbf{1}_{S'}\sigma_j(S')\,.$$

Since the family $\mathrm{ch}_{\mathcal{S}_j}(S)$ is disjoint, the upper bound for the first term in $RHS(36)$ follows from inequality (33).

By (33) it remains to show that $|\mathbf{1}_{E_S} P_{j,S} f_j| \lesssim \mathbf{1}_{E_S}\sigma_j(S)$ almost everywhere. We restrict ourselves to points in which $\lim_{k\to-\infty} \mathbf{E}_{j,k} f_j(x) = f_j(x)$, hence

$$|\mathbf{1}_{E_S}(x) P_{j,S} f_j(x)| = |\mathbf{1}_{E_S}(x)\{f_j(x) - \langle f_j\rangle_S\}|.$$

Observe that $|\langle f_j\rangle_S| \le \sigma_j(S)$. Now, there are three cases (1)–(3) for $x \in E_S$ as above:

(1) If there are no good $\mathcal{D}_j$-cubes inside S containing x, we have $P_{j,S} f_j(x) = 0$ by definitions.

(2) There is a minimal good $\mathcal{D}_j$-cube $Q \subset S$ containing x, in which case we let $Q_x \in \mathrm{ch}(Q)$ be the child containing x. If $R \subset Q_x$ is a $\mathcal{D}_j$-cube containing x, we easily find that $\langle f_j\rangle_R = \langle f_j\rangle_{Q_x}$. Thus, by martingale convergence,

$$|f_j(x)| = \lim_{\ell R\to 0} |\langle f_j\rangle_R| = |\langle f_j\rangle_{Q_x}| \le 4\sigma_j(\pi_{\mathcal{S}_j} Q_x) = 4\sigma_j(S)\,.$$

In the penultimate step above, we used Remark 30 and the fact that $\pi_j Q_x = Q$ is good. And, in the last step, we used the fact that $x \in E_S$.

(3) There are arbitrarily small good $\mathcal{D}_j$-cubes $Q \subset S$ containing x. Hence,

$$|f_j(x)| = \lim_{\ell Q\to 0} |\langle f_j\rangle_Q| \le \sup\{\langle |f_j|\rangle_Q \,:\, x \in Q \subset S\} \le 4\sigma_j(S)\,.$$

The limit and supremum above are both restricted to good $\mathcal{D}_j$-cubes satisfying $x \in Q \subset S$. □

4.3 Family $\mathcal{L}_2(S)$ and its layers

This construction is needed as we study the case of $p_j \neq 2$, and in particular it will allow us to more freely use the inequality (35).

For $S \in \mathcal{S}_1$ let us define $\mathcal{L}_2(S) \subset \mathcal{S}_2$ to be the family of cubes of the form $R = \pi_{\mathcal{S}_2} Q$, where $Q \in \mathcal{G}_2$ satisfies $(P, Q) \in \mathcal{P}_{\text{inside}}$ for some cube $P \in \mathcal{G}_1$ with $\pi_{\mathcal{S}_1} P_Q = S$. Here P_Q stands for the child of P containing Q; it exists by goodness of Q.

Lemma 37 records the observation that there are at most $2(r+1)$ layers in $\mathcal{L}_2(S)$ which contain cubes R such that $R \not\subset S$. To be more precise, let $\mathcal{L}_2^k(S)$ denote the layer $k \ge 0$ cubes in $\mathcal{L}_2(S)$, i.e., the cubes R in this family for which $\pi^k_{\mathcal{L}_2(S)} R$ is a maximal cube in $\mathcal{L}_2(S)$.

Lemma 37. *Suppose that $S \in \mathcal{S}_1$ and $R \in \mathcal{L}_2^k(S)$ with $k \ge 2(r+1)$. Then $R \subset S$.*

Proof. We first claim that, if $R \in \mathcal{L}_2^k(S)$ with $k \ge 1$, then

$$2^{k-1}\ell R \le 2^r \ell S\,. \tag{38}$$

The lemma is a consequence of inequality (38). Indeed, if $k \ge 2(r+1)$, we then have $2^{r+1}\ell R \le \ell S$ and $S \cap R \neq \emptyset$. It remains to recall that either R or πR is a good (by construction).

Let us then prove inequality (38). Clearly, it suffices to verify the case of $k = 1$. Suppose that $R \subsetneq R_0$ is a cube in the first layer, and $R_0 \in \mathcal{L}_2^0(S)$ is maximal. Then, by definition, there are cubes $Q, Q' \in \mathcal{G}_2$ such that $Q \cup Q' \subset S$, $Q \subset R$ and $Q' \cap (R_0 \setminus R) \neq \emptyset$. From these facts it easily follows that $S \cap R \neq \emptyset$ and $\mathrm{dist}(S, \partial R) = 0$. Since either S or $\pi_1 S$ is a good cube, $\ell R \leq 2^r \ell S$. □

4.4 Further inequalities

The reader may omit this technical section for the time being. The following important inequality parallels (35); recall definition of $\mathcal{L}_2(S)$ in Section 4.3: for all sequences of constants satisfying $\sup_{\mathcal{G}_2 \times \mathcal{S}_1} |\varepsilon_{Q,S}| \leq 1$,

$$\sum_{S \in \mathcal{S}_1} \sum_{S' \in \mathrm{ch}^t_{\mathcal{S}_1}(S)} \sum_{\substack{R \in \mathcal{L}_2(S) \\ R \not\subset S}} \Big\| \sum_{\substack{Q \in \mathcal{G}_2 : \pi_{\mathcal{S}_2} Q = R \\ \pi_{\mathcal{S}_1} Q = S'}} \varepsilon_{Q,S} \Delta_Q f_2 \Big\|_{p_2}^{p_2} \lesssim 1, \qquad \text{if } t \geq 0. \tag{39}$$

Proof of inequality (39). Let us fix $t \geq 0$. First, by martingale transform inequality (14) and orthogonality of martingale differences, we can assume that $\varepsilon_{Q,S} = 1$ for all Q and S. By Lemma 41, we obtain an upper bound

$$\sum_{R \in \mathcal{S}_2} \sigma_2(R)^{p_2} \sum_{S \in \mathcal{S}_1; R \not\subset S} \mu(S \cap R) + \sum_{R \in \mathcal{S}_2} \sum_{R' \in \mathrm{ch}_{\mathcal{S}_2}(R)} \sum_{S \in \mathcal{S}_1} \sum_{\substack{S' \in \mathrm{ch}^t_{\mathcal{S}_1}(S) \\ \pi_{\mathcal{S}_1}(\pi_2 R') = S'}} \sigma_2(R')^{p_2} \mu(R'). \tag{40}$$

By inequality (33), the second term is bounded by $\lesssim 1$; Indeed, for a fixed R' there is at most one pair of cubes S, S' such that $\pi_{\mathcal{S}_1}(\pi_2 R') = S'$.

Concerning the first term in (40), we observe that $\sum_{S \in \mathcal{S}_1; R \not\subset S} \mu(S \cap R) \lesssim \mu(R)$ and then apply inequality (33). Mentioned observation is reached by splitting the series in two parts, depending if $S \subset R$ or not; The series with $S \subset R$ is estimated by the Carleson condition, Remark 31. The second series, in which $S \not\subset R$, is estimated by using the fact that $2^{-r} \ell S \leq \ell R \leq 2^r \ell S$ if $S \cap R \neq \emptyset$, $S \not\subset R$, and $R \not\subset S$. Indeed, there are at most $c(n, r)$ such cubes $S \in \mathcal{S}_1$ for a fixed $R \in \mathcal{S}_2$. □

Lemma 41. *Let us fix $S \in \mathcal{S}_1$, $S' \in \mathrm{ch}^t_{\mathcal{S}_1}(S)$, and $R \in \mathcal{L}_2(S)$ such that $R \not\subset S$. Then*

$$\Big\| \sum_{\substack{Q \in \mathcal{G}_2 : \pi_{\mathcal{S}_2} Q = R \\ \pi_{\mathcal{S}_1} Q = S'}} \Delta_Q f_2 \Big\|_{p_2}^{p_2} \lesssim \mu(S' \cap R) \sigma_2(R)^{p_2} \tag{42}$$

$$+ \sum_{R' \in \mathrm{ch}_{\mathcal{S}_2}(R)} \mathbf{1}_{\pi_{\mathcal{S}_1}(\pi_2 R') = S'} \mu(R') \sigma_2(R')^{p_2}.$$

Proof. Let E_R be the set R take away all the $\mathcal{S}_2$-children of R. Then, $LHS(42)$ is bounded by

$$\mathbf{A} + \mathbf{B} = \Big\| \mathbf{1}_{R \setminus E_R} \sum_{\substack{Q \in \mathcal{G}_2 : \pi_{\mathcal{S}_2} Q = R \\ \pi_{\mathcal{S}_1} Q = S'}} \Delta_Q f_2 \Big\|_{p_2}^{p_2} + \Big\| \mathbf{1}_{E_R \cap S'} \sum_{\substack{Q \in \mathcal{G}_2 : \pi_{\mathcal{S}_2} Q = R \\ \pi_{\mathcal{S}_1} Q = S'}} \Delta_Q f_2 \Big\|_{p_2}^{p_2}.$$

In order to estimate $\mathbf{A}$, let us consider a child $R' \in \mathrm{ch}_{\mathcal{S}_2}(R)$ for which there are cubes $Q^+ = Q^+(R')$ and $Q^- = Q^-(R')$: these are the maximal and minimal cubes, respectively, subject to conditions $Q \in \mathcal{G}_2$, $\pi_{\mathcal{S}_2} Q = R$, $\pi_{\mathcal{S}_1} Q = S'$, and $R' \subsetneq Q$. Then

$$\Big| \mathbf{1}_{R'} \sum_{\substack{Q \in \mathcal{G}_2 : \pi_{\mathcal{S}_2} Q = R \\ \pi_{\mathcal{S}_1} Q = S'}} \Delta_Q f_2 \Big| = \mathbf{1}_{R' \cap S'} \cdot |\{\langle f_2 \rangle_{Q^-_{R'}} - \langle f_2 \rangle_{Q^+}\}|$$

$$\lesssim \begin{cases} \mathbf{1}_{\pi_{S_1}(\pi_2 R') = S'} \mathbf{1}_{R'} \sigma_2(R') & \text{if } Q^-_{R'} = R'\,; \\ \mathbf{1}_{S' \cap R'} \sigma_2(R) & \text{otherwise}\,. \end{cases} \tag{43}$$

We used the facts that $\Delta_Q f_2 = 0$ if $Q \subset Q_{2,0}$ is bad, and that $\pi_{\mathcal{S}_2} Q^-_{R'} = R$ if $Q^-_{R'} \neq R'$. Writing $R \setminus E_R = \sum_{R' \in \mathrm{ch}_{\mathcal{S}_2}(R)} \mathbf{1}_{R'}$ and using disjointness of these children yields

$$\mathbf{A} \lesssim \mu(S' \cap R) \sigma_2(R)^{p_2} + \sum_{R' \in \mathrm{ch}_{\mathcal{S}2}(R)} \mathbf{1}_{\pi_{S_1}(\pi_2 R') = S'} \mu(R') \sigma_2(R')^{p_2}\,. \tag{44}$$

We turn to term $\mathbf{B}$; we will implicitly use the fact that $\Delta_Q f_2 = 0$ if $Q \subset Q_{2,0}$ is a bad $\mathcal{D}_2$-cube. Let us fix a point $x \in E_R \cap S'$ such that $\lim_{k \to -\infty} \mathbf{E}_{2,k} f_2(x) = f_2(x)$, and there is a maximal cube $Q^+ \in \mathcal{G}_2$, subject to conditions $x \in Q \in \mathcal{G}_2$, $\pi_{\mathcal{S}_2} Q = R$, $\pi_{\mathcal{S}_1} Q = S'$. Then,

$$\Big| \sum_{\substack{Q \in \mathcal{G}_2 : \pi_{\mathcal{S}_2} Q = R \\ \pi_{S_1} Q = S'}} \Delta_Q f_2(x) \Big| = \Big| \sum_{\substack{Q \in \mathcal{G}_2 : \pi_{S_1} Q = S' \\ Q \subset Q^+}} \Delta_Q f_2(x) \Big|\,. \tag{45}$$

We aim to verify that $RHS(45) \lesssim \sigma_2(R)$. This allows us to conclude that $\mathbf{B} \lesssim \mu(S' \cap R) \sigma_2(R)^{p_2}$. There are two cases. First, $\pi_{\mathcal{S}_1} Q = S'$ for all cubes $x \in Q \in \mathcal{G}_2$ with $Q \subset Q^+$; In this case, we proceed as in the proof of (35) in order to see that $RHS(45) = |f_2(x) - \langle f_2 \rangle_{Q^+}| \lesssim \sigma_2(R)$. Second, there is a minimal cube Q^- subject to conditions $\pi_{\mathcal{S}_1} Q = S'$, $x \in Q \in \mathcal{G}_2$, $Q \subset Q^+$. In this case, we find that $RHS(45) = |\langle f_2 \rangle_{Q^-_x} - \langle f_2 \rangle_{Q^+}| \lesssim \sigma_2(R)$, where Q^-_x denotes the child of Q^-, containing x. In the last step, we used the fact that $x \in E_R$ so that $\pi_{\mathcal{S}_2}(Q^-_x) = R$. □

5 The inside-paraproduct term

First we decompose the inside term $\mathbf{B}_{\text{inside}}(f_1, f_2)$, associated with the indexing cubes $\mathcal{P}_{\text{inside}}$. There will be three terms labelled as: "paraproduct", "stopping", and "error". The "paraproduct" term is treated in this section. The other ones are treated in Section 6.

The conditions for $(P, Q) \in \mathcal{P}_{\text{inside}}$ are: $P \in \mathcal{G}_1$, $Q \in \mathcal{G}_2$, $Q \subset P$, and $2^r \ell Q < \ell P$. These are abbreviated to $Q \Subset P$. The child of P containing Q is denoted by P_Q; it exists by goodness of Q. For $(P, Q) \in \mathcal{P}_{\text{inside}}$ we write

$$\Delta_P f_1 = \langle \Delta_P f_1 \rangle_{P_Q} \mathbf{1}_{\pi_{S_1} P_Q} - \langle \Delta_P f_1 \rangle_{P_Q} \mathbf{1}_{\pi_{S_1} P_Q \setminus P_Q} + \Delta_P f_1 \cdot \mathbf{1}_{P \setminus P_Q}\,.$$

This equation is valid pointwise μ-almost everywhere (everywhere if $\mu(P_Q) \neq 0$), and it yields the following expansion, respectively,

$$\begin{aligned}\mathbf{B}_{\text{inside}}(f_1, f_2) &= \sum_{(P,Q)\in\mathcal{P}_{\text{inside}}} \langle T\Delta_P f_1, \Delta_Q f_2\rangle \\ &= \mathbf{B}^{\text{para}}(f_1,f_2) - \mathbf{B}^{\text{stop}}(f_1,f_2) + \mathbf{B}^{\text{error}}(f_1,f_2)\,, \end{aligned} \tag{46}$$

Hence, e.g., $\mathbf{B}^{\text{para}}(f_1,f_2) = \sum_{(P,Q)\in\mathcal{P}_{\text{inside}}} \langle \Delta_P f_1\rangle_{P_Q} \langle T\mathbf{1}_{\pi_{\mathcal{S}_1}P_Q}, \Delta_Q f_2\rangle$. The main result in this section is the following estimate for the paraproduct term.

Proposition 47. *We have inequality* $|\mathbf{B}^{\text{para}}(f_1,f_2)| \lesssim 1 + \mathbf{T}_{\text{loc}}$.

The remainder of this section is dedicated to the proof of this proposition, and the main focus will be on auxiliary inequalities (52) and (57). Let us first examine how these inequalities are used to prove Proposition 47. First, for $S \in \mathcal{S}_1$, recall definition of $\mathcal{L}_2(S)$ given in Section 4.3. We define

$$\begin{aligned}&\mathbf{B}^{\text{para}}_{S,\not\subset}(f_1,f_2) + \mathbf{B}^{\text{para}}_{S,\subset}(f_1,f_2) \\ &= \Bigg\{ \sum_{\substack{R\in\mathcal{L}_2(S)\\ R\not\subset S}} + \sum_{\substack{R\in\mathcal{L}_2(S)\\ R\subset S}} \Bigg\} \sum_{Q\in\mathcal{G}_2 : \pi_{\mathcal{S}_2}Q = R} \ \sum_{\substack{P\in\mathcal{G}_1 : \pi_{\mathcal{S}_1}P_Q = S\\ Q\Subset P}} \langle \Delta_P f_1\rangle_{P_Q} \langle T\mathbf{1}_S, \Delta_Q f_2\rangle\,.\end{aligned}$$

Then, by the auxiliary inequalities mentioned above,

$$|\mathbf{B}^{\text{para}}(f_1,f_2)| \le \Bigg| \sum_{S\in\mathcal{S}_1} \mathbf{B}^{\text{para}}_{S,\not\subset}(f_1,f_2)\Bigg| + \Bigg| \sum_{S\in\mathcal{S}_1} \mathbf{B}^{\text{para}}_{S,\subset}(f_1,f_2)\Bigg| \lesssim 1 + \mathbf{T}_{\text{loc}}\,.$$

This concludes the proof of Proposition 47, assuming the auxiliary inequalities.

5.1 A telescoping identity

For fixed $Q \in \mathcal{G}_2$ and $S \in \mathcal{S}_1$, let us define a constant $\varepsilon_{Q,S}$ by

$$\varepsilon_{Q,S}\sigma_1(S) = \sum_{\substack{P\in\mathcal{G}_1 : \pi_{\mathcal{S}_1}P_Q = S\\ Q\Subset P}} \langle \Delta_P f_1\rangle_{P_Q}\,. \tag{48}$$

It is important to use the condition $\pi_{\mathcal{S}_1}P_Q = S$ instead of $\pi_{\mathcal{S}_1}P = S$. Otherwise, the following important lemma might fail, as the measure μ need not be doubling.

Lemma 49. *For* $Q \in \mathcal{G}_2$ *and* $S \in \mathcal{S}_1$, *we have* $|\varepsilon_{Q,S}| \lesssim 1$.

Proof. Recall our convention that $\langle \Delta_P f_1\rangle_{P_Q} = 0$ if $\mu(P_Q) = 0$. Consider the minimal and maximal dyadic cubes: P^- and P^+, subject to conditions $P \in \mathcal{G}_1$, $\pi_{\mathcal{S}_1}P_Q = S$, $Q \Subset P$, and $\mu(P_Q) \neq 0$. If such cubes do not exist, we are done. Otherwise, we claim that

$$\varepsilon_{Q,S}\sigma_1(S) = \langle f_1\rangle_{P^-_Q} - \langle f_1\rangle_{P^+}\,. \tag{50}$$

By using equation (50) and the construction of the stopping tree, we find that $|\varepsilon_{Q,S}| \leq 8$.

It remains to prove equation (50). Suppose that $P \in \mathcal{G}_1$ is such that $\pi_{\mathcal{S}_1} P_Q = S$, $Q \Subset P$, and $\mu(P_Q) \neq 0$. Then $P^- \subset P \subset P^+$. By this observation,

$$\sum_{\substack{P \in \mathcal{G}_1 : \pi_{\mathcal{S}_1} P_Q = S \\ Q \Subset P : \mu(P_Q) \neq 0}} \Delta_P f_1 \cdot \mathbf{1}_{P_Q^-} = \sum_{\substack{P \in \mathcal{G}_1 \\ P^- \subset P \subset P^+}} \mathbf{1}_{\pi_{\mathcal{S}_1} P_Q = S} \mathbf{1}_{Q \Subset P} \mathbf{1}_{\mu(P_Q) \neq 0} \cdot \Delta_P f_1 \cdot \mathbf{1}_{P_Q^-} . \quad (51)$$

Observe that $\mathbf{1}_{\pi_{\mathcal{S}_1} P_Q = S} \mathbf{1}_{Q \Subset P} \mathbf{1}_{\mu(P_Q) \neq 0} = 1$ inside the summation. Also, $\Delta_P f_1 = 0$ if P is a bad cube with $P^- \subset P \subset P^+$. Thus, by adding the zero contribution from the bad cubes in a formal manner, we obtain a telescoping identity: $LHS(51) = \{\langle f_1 \rangle_{P_Q^-} - \langle f_1 \rangle_{P^+}\} \mathbf{1}_{P_Q^-}$. The equation (50) follows from this: first, we restrict ourselves to cubes P with $\mu(P_Q) \neq 0$ in the series defining $\varepsilon_{Q,S}$. Then, we replace the P_Q averages by P_Q^- averages inside the summation; observe that $P_Q^- \subset P_Q$ and $\mu(P_Q^-) \neq 0$. Finally, we exchange the order of summation and the brackets, and apply the obtained telescoping identity. □

5.2 Summation involving cubes $R \not\subset S$

Our aim in this section is to prove an inequality

$$\Big| \sum_{S \in \mathcal{S}_1} \mathbf{B}^{\mathrm{para}}_{S,\not\subset}(f_1, f_2) \Big| \lesssim 1 + \mathbf{T}_{\mathrm{loc}} . \quad (52)$$

Let us express the series defining $\mathbf{B}^{\mathrm{para}}_{S,\not\subset}(f_1, f_2)$ in a convenient manner. For this purpose, observe that $Q \subset P_Q \subset S$ for any cube Q in the series defining $\mathbf{B}^{\mathrm{para}}_{S,\not\subset}(f_1, f_2)$. In particular, $\pi_{\mathcal{S}_1} Q \subset S$. Thus, by organising the Q-summation in terms of their $\mathcal{S}_1$-parents and defining $\varepsilon_{Q,S}$ as the solution to equation (48), we find that

$$\begin{aligned} &\mathbf{B}^{\mathrm{para}}_{S,\not\subset}(f_1, f_2) \\ &= \sigma_1(S) \sum_{t \geq 0} \sum_{\substack{R \in \mathcal{L}_2(S) \\ R \not\subset S}} \sum_{S' \in \mathrm{ch}^t_{\mathcal{S}_1}(S)} \sum_{\substack{Q \in \mathcal{G}_2 : \pi_{\mathcal{S}_2} Q = R \\ \pi_{\mathcal{S}_1} Q = S'}} \langle \mathbf{1}_R \mathbf{1}_{S'} \{T\mathbf{1}_S - \tau_{t,S'}\}, \varepsilon_{Q,S} \Delta_Q f_2 \rangle . \end{aligned} \quad (53)$$

By using the fact that $\Delta_Q f_2$ has mean zero, we have also subtracted off the constants

$$\tau_{t,S'} = \begin{cases} 0 & \text{if } t \in \{0, \ldots, 2r+1\}\,; \\ T\mathbf{1}_{S \setminus \pi^{\lfloor t/2 \rfloor}_{\mathcal{S}_1} S'}(x_{S'}) & \text{otherwise}. \end{cases}$$

For convenience, let us denote

$$\mathbf{A}_{t,S} = \Big\{ \sum_{\substack{R \in \mathcal{L}_2(S) \\ R \not\subset S}} \sum_{S' \in \mathrm{ch}^t_{\mathcal{S}_1}(S)} \| \mathbf{1}_R \mathbf{1}_{S'} \{T\mathbf{1}_S - \tau_{t,S'}\} \|^{p_1}_{p_1} \Big\}^{1/p_1} ,$$

and

$$\mathbf{B}_{t,S} = \Bigg\{ \sum_{\substack{R\in\mathcal{L}_2(S)\\ R\not\subset S}} \sum_{S'\in \mathrm{ch}^t_{\mathcal{S}_1}(S)} \Bigg\| \sum_{\substack{Q\in\mathcal{G}_2:\pi_{\mathcal{S}_2}Q=R\\ \pi_{\mathcal{S}_1}Q=S'}} \varepsilon_{Q,S}\Delta_Q f_2 \Bigg\|_{p_2}^{p_2} \Bigg\}^{1/p_2}.$$

The useful inequality $\sup_{\mathcal{G}_2\times\mathcal{S}_1}|\varepsilon_{Q,S}| \lesssim 1$ is a consequence of Lemma 49. By equation (53) and Hölder's inequality, combined with Lemma 54,

$$\begin{aligned}
&\Big|\sum_{S\in\mathcal{S}_1} \mathbf{B}^{\mathrm{para}}_{S,\not\subset}(f_1,f_2)\Big| \\
&\lesssim \sum_{t\geq 0}\sum_{S\in\mathcal{S}_1} \sigma_1(S)\mathbf{A}_{t,S}\mathbf{B}_{t,S} \\
&\lesssim (1+\mathbf{T}_{\mathrm{loc}})\sum_{t\geq 0} 2^{-t/p_1}\Big\{\sum_{S\in\mathcal{S}_1}\sigma_1(S)^{p_1}\mu(S)\Big\}^{1/p_1}\Big\{\sum_{S\in\mathcal{S}_1}\mathbf{B}^{p_2}_{t,S}\Big\}^{1/p_2}.
\end{aligned}$$

Inequality (52) is obtained by applying inequalities (33) and (39), and summing the geometric series afterwards.

Lemma 54. *For every $t\geq 0$ and $S\in\mathcal{S}_1$, we have $\mathbf{A}_{t,S}\lesssim (1+\mathbf{T}_{\mathrm{loc}})2^{-t/p_1}\mu(S)^{1/p_1}$.*

Proof. By Lemma 37 and the fact that layers $\mathcal{L}^k_2(S)$, $k\geq 0$, are comprised of disjoint cubes, we can bound $\mathbf{A}^{p_1}_{t,S}$ by

$$\sum_{k=0}^{2r+1}\sum_{S'\in\mathrm{ch}^t_{\mathcal{S}_1}(S)}\sum_{\substack{R\in\mathcal{L}^k_2(S)\\ R\not\subset S}} \|\mathbf{1}_R\mathbf{1}_{S'}\{T\mathbf{1}_S-\tau_{t,S'}\}\|^{p_1}_{p_1} \lesssim \sum_{S'\in\mathrm{ch}^t_{\mathcal{S}_1}(S)} \|\mathbf{1}_{S'}\{T\mathbf{1}_S-\tau_{t,S'}\}\|^{p_1}_{p_1}. \tag{55}$$

Let us first focus on the case of $t\in\{0,\ldots,2r+1\}$. By inequality (55) and the facts that cubes in $\mathrm{ch}^t_{\mathcal{S}_1}(S)$ are disjoint and they are contained in S,

$$\mathbf{A}^{p_1}_{t,S}\lesssim \|\mathbf{1}_S T\mathbf{1}_S\|^{p_1}_{p_1}\leq \mathbf{T}^{p_1}_{\mathrm{loc}}\mu(S)\lesssim (1+\mathbf{T}_{\mathrm{loc}})^{p_1}2^{-t}\mu(S).$$

Let us then focus on the case of $t\geq 2r+2$; we begin by writing

$$RHS(55) = \sum_{S''\in\mathrm{ch}^{\lceil t/2\rceil}_{\mathcal{S}_1}(S)}\sum_{\substack{S'\in\mathrm{ch}^t_{\mathcal{S}_1}(S)\\ \pi^{\lfloor t/2\rfloor}_{\mathcal{S}_1}S'=S''}} \|\mathbf{1}_{S'}\{T\mathbf{1}_S - T\mathbf{1}_{S\setminus S''}(x_{S'})\}\|^{p_1}_{p_1}.$$

To conclude the proof of lemma, it suffices to first verify that for all $S''\in\mathrm{ch}^{\lceil t/2\rceil}_{\mathcal{S}_1}(S)$,

$$\sum_{\substack{S'\in\mathrm{ch}^t_{\mathcal{S}_1}(S)\\ \pi^{\lfloor t/2\rfloor}_{\mathcal{S}_1}S'=S''}} \|\mathbf{1}_{S'}\{T\mathbf{1}_S - T\mathbf{1}_{S\setminus S''}(x_{S'})\}\|^{p_1}_{p_1}\lesssim (1+\mathbf{T}_{\mathrm{loc}})^{p_1}\mu(S''), \tag{56}$$

and then inductively apply the sparseness property of $\mathcal{S}_1$, we refer to Remark 31.

In order to prove the remaining inequality (56), we estimate $LHS(56)$ by $2^{p_1-1}(\alpha+\beta)$, where

$$\alpha+\beta = \sum_{\substack{S'\in \mathrm{ch}^t_{\mathcal{S}_1}(S)\\ \pi^{\lfloor t/2\rfloor}_{\mathcal{S}_1}S'=S''}} \|\mathbf{1}_{S'}T\mathbf{1}_{S''}\|^{p_1}_{p_1} + \sum_{\substack{S'\in \mathrm{ch}^t_{\mathcal{S}_1}(S)\\ \pi^{\lfloor t/2\rfloor}_{\mathcal{S}_1}S'=S''}} \|\mathbf{1}_{S'}\{T\mathbf{1}_{S\setminus S''}-T\mathbf{1}_{S\setminus S''}(x_{S'})\}\|^{p_1}_{p_1}\,.$$

Observe that the cubes S' are contained in S'', and they are disjoint. The Local Testing Condition implies the inequality $\alpha \le \mathbf{T}^{p_1}_{\mathrm{loc}}\mu(S'')$. In order to analyse term β, we fix $S'\in \mathrm{ch}^t_{\mathcal{S}_1}(S)$ such that $\pi^{\lfloor t/2\rfloor}_{\mathcal{S}_1}S'=S''$. Since $\lfloor t/2\rfloor \ge r+1$, we have $2^r\ell S' < \ell S''$. By construction of the stopping cubes, either S' or $\pi_1 S'$ is good. In both of these cases, by goodness[1], we have $\ell S' \le \mathrm{dist}(S',\partial S'')$. Hence, by the off-diagonal estimate (18), we have $|T\mathbf{1}_{S\setminus S''}(x)-T\mathbf{1}_{S\setminus S''}(x_{S'})|\lesssim 1$ if $x\in S'$. This inequality allows us to conclude that $\beta\lesssim \mu(S'')$. □

5.3 Summation involving cubes $R\subset S$

Here we show the inequality

$$\Big|\sum_{S\in\mathcal{S}_1}\mathbf{B}^{\mathrm{para}}_{S,\subset}(f_1,f_2)\Big| \lesssim 1+\mathbf{T}_{\mathrm{loc}}\,. \tag{57}$$

Let us fix $S\in\mathcal{S}_1$, and express the series defining $\mathbf{B}^{\mathrm{para}}_{S,\subset}(f_1,f_2)$ in a convenient manner. For a cube $S'\in\mathcal{S}_1$, we denote by $\mathcal{R}(S')$ the family of maximal cubes in $\{R'\in\mathcal{S}_2 : \pi_{\mathcal{S}_1}R'=S'\}$; this can be an empty family. By defining constants $\varepsilon_{Q,S}$ as solutions to (48), we can write $\mathbf{B}^{\mathrm{para}}_{S,\subset}(f_1,f_2)$ as

$$\sigma_1(S)\sum_{t,k\ge 0}\sum_{S'\in \mathrm{ch}^t_{\mathcal{S}_1}(S)}\sum_{R\in\mathcal{R}(S')}\sum_{\substack{R'\in \mathrm{ch}^k_{\mathcal{S}_2}(R)\\ \pi_{\mathcal{S}_1}R'=S'}}\sum_{Q\in\mathcal{G}_2:\pi_{\mathcal{S}_2}Q=R'}\langle \mathbf{1}_{R'}\{T\mathbf{1}_S-\tau_{t,k,S',R'}\}, \varepsilon_{Q,S}\Delta_Q f_2\rangle\,, \tag{58}$$

where we have denoted

$$\tau_{t,k,S',R'} = \begin{cases} 0 & \text{if } t,k\in\{0,\dots,2r+1\}\,;\\ T\mathbf{1}_{S\setminus \pi^{\lfloor k/2\rfloor}_{\mathcal{S}_2}R'}(x_{R'}) & \text{if } k\ge 2(r+1)\,;\\ T\mathbf{1}_{S\setminus \pi^{\lfloor t/2\rfloor}_{\mathcal{S}_1}S'}(x_{S'}) & \text{otherwise}\,. \end{cases}$$

It will be convenient to denote for all $t\ge 0$,

$$\mathbf{A}_{t,S} = \Bigg\{\sum_{k\ge 0}\sum_{S'\in \mathrm{ch}^t_{\mathcal{S}_1}(S)}\sum_{R\in\mathcal{R}(S')}\sum_{\substack{R'\in \mathrm{ch}^k_{\mathcal{S}_2}(R)\\ \pi_{\mathcal{S}_1}R'=S'}}\|\mathbf{1}_{R'}\{T\mathbf{1}_S-\tau_{t,k,S',R'}\}\|^{p_1}_{p_1}\Bigg\}^{1/p_1}\,.$$

[1]This application is the principal motivation for our definition of goodness; recall that good cubes are neither 1-bad nor 2-bad. The same application arises also later, Lemma 59.

The useful inequality $\sup_{\mathcal{G}_2\times\mathcal{S}_1}|\varepsilon_{Q,S}|\lesssim 1$ is a consequence of Lemma 49. Hence, by Lemma 59 and Hölder's inequality, combined with inequality (34),

$$
\begin{aligned}
&|\mathbf{B}^{\mathrm{para}}_{S,\mathcal{C}}(f_1,f_2)| \\
&\lesssim \sum_{t\geq 0}\sigma_1(S)\mathbf{A}_{t,S}\Bigg\{\sum_{S'\in\mathrm{ch}^t_{\mathcal{S}_1}(S)}\ \sum_{\substack{R'\in\mathcal{S}_2\\ \pi_{\mathcal{S}_1}R'=S'}}\Bigg\|\sum_{Q\in\mathcal{G}_2:\pi_{\mathcal{S}_2}Q=R'}\varepsilon_{Q,S}\Delta_Q f_2\Bigg\|_{p_2}^{p_2}\Bigg\}^{1/p_2} \\
&\lesssim (1+\mathbf{T}_{\mathrm{loc}})\sum_{t\geq 0}2^{-t/p_1}\sigma_1(S)\mu(S)^{1/p_1}\Bigg\{\sum_{S'\in\mathrm{ch}^t_{\mathcal{S}_1}(S)}\ \sum_{\substack{R'\in\mathcal{S}_2\\ \pi_{\mathcal{S}_1}R'=S'}}\|P_{2,R'}f_2\|_{p_2}^{p_2}\Bigg\}^{1/p_2}.
\end{aligned}
$$

The very last upper bound is summable in $S\in\mathcal{S}_1$. Indeed, after changing the order of S and t summations, an application of inequalities (33) and (35) leaves us a geometric series in t. The proof of inequality (57) is complete.

Lemma 59. *For each $S\in\mathcal{S}_1$ and $t\geq 0$, we have $\mathbf{A}_{t,S}\lesssim(1+\mathbf{T}_{\mathrm{loc}})2^{-t/p_1}\mu(S)^{1/p_1}$.*

Proof. Let us make a case study, and first assume that $t\in\{0,\dots,2r+1\}$. We split $\mathbf{A}^{p_1}_{t,S}$ in two subseries, subject to $k\in\{0,\dots,2r+1\}$ and $k\geq 2(r+1)$. For a fixed $k\in\{0,\dots,2r+1\}$, we rely on disjointness properties of layers and maximal cubes in order to see that

$$
\sum_{S'\in\mathrm{ch}^t_{\mathcal{S}_1}(S)}\ \sum_{R\in\mathcal{R}(S')}\ \sum_{\substack{R'\in\mathrm{ch}^k_{\mathcal{S}_2}(R)\\ \pi_{\mathcal{S}_1}R'=S'}}\|\mathbf{1}_{R'}\{T\mathbf{1}_S-\tau_{t,k,S',R'}\}\|_{p_1}^{p_1}\leq\|\mathbf{1}_S T\mathbf{1}_S\|_{p_1}^{p_1}\leq\mathbf{T}^{p_1}_{\mathrm{loc}}\mu(S).
$$

Applying these inequalities with finite number of indices $k\in\{0,\dots,2r+1\}$ shows the required inequality for the first subseries. The second subseries is bounded by

$$
\begin{aligned}
&\sum_{k\geq 2r+2}\ \sum_{S'\in\mathrm{ch}^t_{\mathcal{S}_1}(S)}\ \sum_{R\in\mathcal{R}(S')}\ \sum_{R''\in\mathrm{ch}^{\lceil k/2\rceil}_{\mathcal{S}_2}(R)}\ \sum_{\substack{R'\in\mathrm{ch}^k_{\mathcal{S}_2}(R)\\ \pi^{\lfloor k/2\rfloor}_{\mathcal{S}_2}R'=R''}}\|\mathbf{1}_{R'}\{T\mathbf{1}_S-T\mathbf{1}_{S\setminus R''}(x_{R'})\}\|_{p_1}^{p_1} \\
&\lesssim(1+\mathbf{T}_{\mathrm{loc}})^{p_1}\sum_{k\geq 2r+2}2^{-k}\sum_{S'\in\mathrm{ch}^t_{\mathcal{S}_1}(S)}\mu(S')\lesssim(1+\mathbf{T}_{\mathrm{loc}})^{p_1}\mu(S). \qquad (60)
\end{aligned}
$$

In the first step above, we applied a simple modification of inequality (56) and sparsness property of $\mathcal{S}_2$, we refer to Remark 31.

Let us then focus on the case of $t\geq 2(r+1)$. Again, we split the series $\mathbf{A}^{p_1}_{t,S}$ in two subseries as before. For the first subseries, associated with indices $k\in\{0,\dots,2r+1\}$, we use inequality

$$
\begin{aligned}
&\sum_{S'\in\mathrm{ch}^t_{\mathcal{S}_1}(S)}\ \sum_{R\in\mathcal{R}(S')}\ \sum_{\substack{R'\in\mathrm{ch}^k_{\mathcal{S}_2}(R)\\ \pi_{\mathcal{S}_1}R'=S'}}\|\mathbf{1}_{R'}\{T\mathbf{1}_S-\tau_{t,k,S',R'}\}\|_{p_1}^{p_1} \\
&\leq\sum_{S''\in\mathrm{ch}^{\lceil t/2\rceil}_{\mathcal{S}_1}(S)}\ \sum_{\substack{S'\in\mathrm{ch}^t_{\mathcal{S}_1}(S)\\ \pi^{\lfloor t/2\rfloor}_{\mathcal{S}_1}S'=S''}}\|\mathbf{1}_{S'}\{T\mathbf{1}_S-T\mathbf{1}_{S\setminus S''}(x_{S'})\}\|_{p_1}^{p_1},
\end{aligned}
$$

and then proceed as in the proof of Lemma 54. Finally, the second subseries is bounded by $LHS(60)$ which, in turn, is controlled by $\lesssim (1+\mathbf{T}_{\mathrm{loc}})^{p_1} 2^{-t}\mu(S)$, Remark 31. □

6 The inside-stopping/error term

In the present section, we concentrate on the two terms, labelled as "stopping" and "error", that were introduced in the beginning of Section 5. We aim to prove the following proposition.

Proposition 61. *We have* $\left|\mathbf{B}^{\mathrm{stop}}(f_1,f_2)\right| + \left|\mathbf{B}^{\mathrm{error}}(f_1,f_2)\right| \lesssim 1$.

6.1 The stopping term

The stopping term $\mathbf{B}^{\mathrm{stop}}(f_1,f_2)$ is written as $\sum_{t=r+1}^{\infty} \mathbf{B}_t^{\mathrm{stop}}(f_1,f_2)$,

$$\begin{aligned}
&|\mathbf{B}_t^{\mathrm{stop}}(f_1,f_2)| \\
&= \Bigg| \sum_{P\in\mathcal{G}_1} \sum_{\substack{Q\in\mathcal{G}_2\\ 2^t\ell Q=\ell P}} \mathbf{1}_{Q\subset P} \langle \Delta_P f_1\rangle_{P_Q} \langle T\mathbf{1}_{\pi_{S_1}P_Q\setminus P_Q}, \Delta_Q f_2\rangle \Bigg| \\
&\lesssim 2^{-t\eta(1-\gamma)} \int_{\mathbf{R}^n} \sum_{P\in\mathcal{G}_1} \sum_{\substack{Q\in\mathcal{G}_2\\ 2^t\ell Q=\ell P}} \mathbf{1}_Q(x)|\Delta_P f_1(x)| \cdot \mathbf{1}_{Q\subset P}|\Delta_Q f_2(x)|\, d\mu(x)\,.
\end{aligned}$$

In the last step, we used the off-diagonal estimate (18) and the fact that $\Delta_Q f_2$ has mean zero. Applying Cauchy-Schwarz and Hölder's inequality, and then observing inequalities

$$\sum_{\substack{P\in\mathcal{G}_1\\ 2^t\ell Q=\ell P}} \mathbf{1}_{Q\subset P} \le 1 \quad (Q\in\mathcal{G}_2)\,, \qquad \sum_{\substack{Q\in\mathcal{G}_2\\ 2^t\ell Q=\ell P}} \mathbf{1}_Q \le \mathbf{1}_{\mathbf{R}^n} \quad (P\in\mathcal{G}_1)\,, \qquad (62)$$

we obtain, for a fixed $t\ge r+1$,

$$\begin{aligned}
|\mathbf{B}_t^{\mathrm{stop}}(f_1,f_2)| &\lesssim 2^{-t\eta(1-\gamma)} \Bigg\| \Big(\sum_{P\in\mathcal{G}_1} |\Delta_P f_1|^2 \Big)^{1/2} \Bigg\|_{p_1} \Bigg\| \Big(\sum_{Q\in\mathcal{G}_2} |\Delta_Q f_2|^2 \Big)^{1/2} \Bigg\|_{p_2} \\
&\lesssim 2^{-t\eta(1-\gamma)}\,.
\end{aligned}$$

In the penultimate step, we used inequality (15). The last bound is summable in t, and this concludes analysis of the stopping term.

6.2 The error term

We write $\mathbf{B}^{\mathrm{error}}(f_1,f_2) = \sum_{t=r+1}^{\infty} \mathbf{B}_t^{\mathrm{error}}(f_1,f_2)$,

$$\mathbf{B}_t^{\mathrm{error}}(f_1,f_2) = \sum_{j=1}^{2^n} \sum_{Q\in\mathcal{G}_2} \sum_{\substack{P\in\mathcal{G}_1\\ 2^t\ell Q=\ell P}} \mathbf{1}_{Q\subset P} \mathbf{1}_{P_j\neq P_Q} \langle \Delta_P f_1\rangle_{P_j} \langle T\mathbf{1}_{P_j}, \Delta_Q f_2\rangle\,.$$

Let us denote $T_{P_j,Q} = \mathbf{1}_{P_j \neq P_Q}\{T\mathbf{1}_{P_j} - T\mathbf{1}_{P_j}(x_Q)\}$. By the fact that $\Delta_Q f_2$ has mean zero, we can bound $|\mathbf{B}_t^{\text{error}}(f_1,f_2)|$ with $t \geq r+1$ by

$$\sum_{j=1}^{2^n} \left| \int_{\mathbf{R}^n} \sum_{Q\in\mathcal{G}_2} \Delta_Q f_2(x)\cdot \mathbf{1}_Q(x) \sum_{\substack{P\in\mathcal{G}_1\\ 2^t\ell Q=\ell P}} \mathbf{1}_{Q\subset P}\mathbf{1}_{P_j\neq P_Q}\langle \Delta_P f_1\rangle_{P_j} T_{P_j,Q}(x)\,d\mu(x)\right|$$

$$\leq \mathbf{A}_t \cdot \left\| \Big(\sum_{Q\in\mathcal{G}_2} |\Delta_Q f_2|^2\Big)^{1/2}\right\|_{p_2} \lesssim \mathbf{A}_t\,,$$

where we have denoted

$$\mathbf{A}_t = \sum_{j=1}^{2^n} \left\| \Big(\sum_{k\in\mathbf{Z}} \Big| \sum_{P\in\mathcal{G}_{1,k+t}} \langle \Delta_P f_1\rangle_{P_j} \sum_{Q\in\mathcal{G}_{2,k}} \mathbf{1}_{Q\subset P}\mathbf{1}_{P_j\neq P_Q}\mathbf{1}_Q T_{P_j,Q}\Big|^2\Big)^{1/2}\right\|_{p_1}.$$

By an off-diagonal estimate for $T_{P_j,Q}$, i.e. Lemma 22 applied to cubes P_j and Q,

$$\Big| \sum_{Q\in\mathcal{G}_{2,k}} \mathbf{1}_{Q\subset P}\mathbf{1}_{P_j\neq P_Q}\mathbf{1}_Q(x) T_{P_j,Q}(x)\Big| \lesssim 2^{-t\eta/4}\mathbf{1}_P(x)\mu(P_j)\mu(P)^{-1}, \qquad x\in\mathbf{R}^n\,. \tag{63}$$

Thus, by inequalities (16) and (15),

$$\begin{aligned}\mathbf{A}_t &\lesssim 2^{-t\eta/4} \left\| \Big(\sum_{k\in\mathbf{Z}} \Big| \sum_{P\in\mathcal{G}_{1,k+t}} \langle |\Delta_P f_1|\rangle_P \mathbf{1}_P\Big|^2\Big)^{1/2}\right\|_{p_1} \\ &\leq 2^{-t\eta/4} \left\| \Big(\sum_{k\in\mathbf{Z}} (\mathbf{E}_{1,k+t}|\Delta_{k+t} f_1|)^2\Big)^{1/2}\right\|_{p_1} \\ &\lesssim 2^{-t\eta/4}\|f_1\|_{p_1} \lesssim 2^{-t\eta/4}\,.\end{aligned}$$

The last bound is summable in $t\geq r+1$. The proof of Proposition 61 is complete.

7 The separated term

Here we treat the separated term, we refer to Section 3.2.

Proposition 64. *We have inequality* $|\mathbf{B}_{\text{separated}}(f_1,f_2)| \lesssim 1$.

For the proof, we need preparations. Recall that $D(Q,P) = \ell Q + \operatorname{dist}(Q,P) + \ell P$, and write $D(Q,P)/\ell P \sim 2^u$ if $2^u < D(Q,P)/\ell P \leq 2^{u+1}$. The separated term is a sum over $u,m\in\mathrm{N}_0$ and $j\in\{1,2,\dots,2^n\}$ of terms

$$\mathbf{B}^{u,m,j}(f_1,f_2) = \sum_{k\in\mathbf{Z}} \sum_{Q\in\mathcal{G}_{2,k-m}} \sum_{\substack{P\in\mathcal{G}_{1,k}\\ D(Q,P)/\ell P\sim 2^u}} \mathbf{1}_{\ell Q\leq \operatorname{dist}(Q,P)} \langle \Delta_P f_1\rangle_{P_j} \langle T\mathbf{1}_{P_j}, \Delta_Q f_2\rangle\,.$$

For Q and P_j as in the summation above, let us write $T_{P_j,Q} = \mathbf{1}_{\ell Q \le \operatorname{dist}(Q,P)} \{T\mathbf{1}_{P_j} - T\mathbf{1}_{P_j}(x_Q)\}$. Since $\Delta_Q f_2$ has mean zero, we can write $|\mathbf{B}^{u,m,j}(f_1,f_2)|$ as

$$\left| \int_{\mathbf{R}^n} \sum_{k\in\mathbf{Z}} \sum_{Q\in\mathcal{G}_{2,k-m}} \Delta_Q f_2(x)\cdot \mathbf{1}_Q(x) \sum_{\substack{P\in\mathcal{G}_{1,k}\\ D(Q,P)/\ell P\sim 2^u}} \langle \Delta_P f_1\rangle_{P_j} T_{P_j,Q}(x)\, d\mu(x)\right|$$

$$\le \mathbf{A}_{u,m,j}\cdot \left\| \left(\sum_{k\in\mathbf{Z}} \sum_{Q\in\mathcal{G}_{2,k-m}} |\Delta_Q f_2|^2 \right)^{1/2} \right\|_{p_2} \lesssim \mathbf{A}_{u,m,j}\,,$$

where we have denoted

$$\mathbf{A}_{u,m,j} = \left\| \left(\sum_{k\in\mathbf{Z}} \sum_{Q\in\mathcal{G}_{2,k-m}} \left| \mathbf{1}_Q \sum_{\substack{P\in\mathcal{G}_{1,k}\\ D(Q,P)/\ell P\sim 2^u}} \langle \Delta_P f_1\rangle_{P_j} T_{P_j,Q} \right|^2 \right)^{1/2} \right\|_{p_1}.$$

In order to finish the proof of Proposition 64, we invoke the following lemma.

Lemma 65. *For $u,m\in\mathbf{N}_0$ and $j\in\{1,2,\ldots,2^n\}$, we have $\mathbf{A}_{u,m,j}\lesssim 2^{-\eta(m+u)/4}$.*

Proof. For each $S\in\mathcal{D}_{1,k+u+\theta(u+m)}$, $k\in\mathbf{Z}$, we consider the kernel

$$K_S(x,y) = \sum_{\substack{P\in\mathcal{G}_{1,k}\\ P\subset S}} \sum_{\substack{Q\in\mathcal{G}_{2,k-m}\\ D(Q,P)/\ell P\sim 2^u}} \mathbf{1}_Q(x)\cdot \widetilde{T}_{P_j,Q}(x)\cdot \mathbf{1}_{P_j}(y)\,, \tag{66}$$

where $\widetilde{T}_{P_j,Q} = \mathbf{1}_{\ell Q\le \operatorname{dist}(Q,P)} \widetilde{T}_{P_j,Q}$ is defined by

$$\frac{T_{P_j,Q}(x)}{\mu(P_j)} = 2^{-\eta(m+u)/4}\cdot \frac{\widetilde{T}_{P_j,Q}(x)}{\mu(S)}\,.$$

By Lemma 21 and Lemma 22,

$$|K_S(x,y)| \lesssim \sum_{\substack{P\in\mathcal{G}_{1,k}\\ P\subset S}} \sum_{\substack{Q\in\mathcal{G}_{2,k-m}\\ D(Q,P)/\ell P\sim 2^u}} \mathbf{1}_Q(x)\cdot \mathbf{1}_{P_j}(y) \le \mathbf{1}_S(x)\cdot \mathbf{1}_S(y)\,. \tag{67}$$

We can now finish the proof as follows. Inequality (67) allows us to write

$$2^{\eta p_1(m+u)/4}\mathbf{A}^{p_1}_{u,m,j}$$

as

$$\int_{\mathbf{R}^n} \left(\sum_{k\in\mathbf{Z}} \left| \sum_{S\in\mathcal{D}_{1,k+u+\theta(u+m)}} \frac{1}{\mu(S)} \int_{\mathbf{R}^n} K_S(x,y)\Delta_k f_1(y)\, d\mu(y) \right|^2 \right)^{p_1/2} d\mu(x)$$

$$\lesssim \int_{\mathbf{R}^n} \left(\sum_{k\in\mathbf{Z}} \Bigg| \underbrace{\sum_{S\in\mathcal{D}_{1,k+u+\theta(u+m)}} \langle |\Delta_k f_1|\rangle_S \mathbf{1}_S(x)}_{=\mathbf{E}_{1,k+u+\theta(u+m)}|\Delta_k f_1|(x)} \Bigg|^2 \right)^{p_1/2} d\mu(x)\,.$$

Appealing to inequalities (16) and (15) shows that $\mathbf{A}_{u,m,j}\lesssim 2^{-\eta(m+u)/4}$. $\square$

8 Preparations for the nearby term

The surgery argument for the nearby term follows [19] but there are also essential differences. Let us abbreviate $(P,Q) \in \mathcal{P}_{\text{nearby}}$ as $P \sim Q$. Hence, the conditions for $P \sim Q$ are

$$(P,Q) \in \mathcal{G}_1 \times \mathcal{G}_2\,, \qquad 2^{-r}\ell P \le \ell Q \le \ell P\,, \quad \operatorname{dist}(Q,P) < \ell Q = \ell Q \wedge \ell P. \tag{68}$$

In particular $\ell Q \le \ell P \le D(Q,P) \le (2^r+2)\ell Q$, i.e., these quantities are comparable if $P \sim Q$. During the course of the remaining sections, we will prove the following proposition.

Proposition 69. *For a fixed $t > p_1 \vee p_2$, we have*

$$\mathbf{E}_{\omega_1}\mathbf{E}_{\omega_2}|\mathbf{B}_{\text{nearby}}(f_1,f_2)| \le C(r,\upsilon,\epsilon)(1+\mathbf{T}_{\text{loc}}) + \big(C(r,\upsilon)\epsilon^{1/t} + C(r)\upsilon^{1/t}\big)\mathbf{T}\,.$$

Aside from the indicated absorption parameters, the constants on the right hand side can depend upon the parameters n, p_1, p_2, η, μ.

Remark 70. We shall track dependence of various inequalities on absorption parameters: r, υ, ϵ. There is no need to do this quantitatively, and thus we agree upon the following convenient notation: $C(r)$, $C(r,\upsilon)$, and $C(r,\upsilon,\epsilon)$ denote positive numbers that are allowed to depend on the indicated absorption parameters, but also on parameters n, p_1, p_2, η, μ. Moreover, the value of these numbers is allowed to vary from one occurrence to another.

For a given $P \in \mathcal{G}_1$ there are at most $C(r)$ cubes $Q \in \mathcal{G}_2$ satisfying (68). Hence, without loss of generality, it suffices consider a finite number of subseries of the general form

$$\mathbf{E}_{\omega_1}\mathbf{E}_{\omega_2}\Big|\sum_{P\in\mathcal{G}_1}\langle T\Delta_P f_1, \Delta_Q f_2\rangle\Big|\,, \tag{71}$$

where $Q = Q(P) \in \mathcal{G}_2 \cup \{\emptyset\}$ inside the summation satisfies $P \sim Q$ or $Q = \emptyset$.[1] At the same time, we can also assume that for any $Q \in \mathcal{G}_2$ there is at most one $P \in \mathcal{G}_1$ such that $Q = P(Q)$. We fix one series like this, and the convention that Q is implicitly a function of P will be maintained.

8.1 First reductions

We immediately find that (71) is dominated by

$$\sum_{i,j=1}^{2^n}\mathbf{E}_{\omega_1}\mathbf{E}_{\omega_2}\Big|\sum_{P\in\mathcal{G}_1}\langle\Delta_P f_1\rangle_{P_j}\langle T\mathbf{1}_{P_j},\mathbf{1}_{Q_i}\rangle\langle\Delta_Q f_2\rangle_{Q_i}\Big|\,. \tag{72}$$

Fix $i,j \in \{1,\dots,n\}$. For a cube R in $\mathbf{R}^n$, define an "υ-boundary region": $\delta_R^\upsilon = (1+\upsilon)R \setminus (1-\upsilon)R$. If $P \in \mathcal{D}_1$ and $Q = Q(P) \neq \emptyset$, we write

$$\begin{aligned} &Q_{i,\partial} = Q_i \cap \delta^\upsilon_{P_j}; \quad Q_{i,\text{sep}} = (Q_i \setminus Q_{i,\partial}) \setminus (Q_i \cap P_j); \quad \Delta_{Q_i} = (Q_i \cap P_j) \setminus (Q_{i,\partial});\\ &P_{j,\partial} = P_j \cap \delta^\upsilon_{Q_i}; \quad P_{j,\text{sep}} = (P_j \setminus P_{j,\partial}) \setminus (Q_i \cap P_j); \quad \Delta_{P_j} = (Q_i \cap P_j) \setminus P_{j,\partial}. \end{aligned} \tag{73}$$

[1]We agree that $\Delta_\emptyset f_2 = 0$.

For an illustration of these sets, we refer to Figure 1.

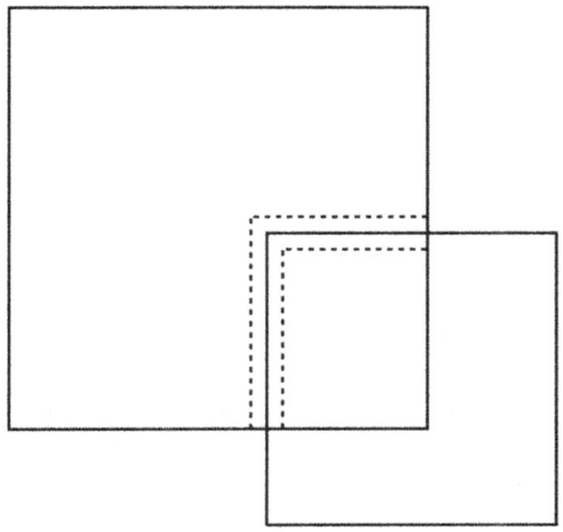

Figure 1: The larger cube is P_j, and the smaller cube is Q_i. The dashed line segments separate sets $P_{j,\mathrm{sep}}$, $P_{j,\partial}$, and Δ_{P_j} from each other.

We write the matrix coefficient $\langle T\mathbf{1}_{P_j}, \mathbf{1}_{Q_i}\rangle$ in (72) as

$$\begin{aligned} &\langle T\mathbf{1}_{P_{j,\mathrm{sep}}}, \mathbf{1}_{Q_i}\rangle + \langle T\mathbf{1}_{P_{j,\partial}}, \mathbf{1}_{Q_i}\rangle + \langle T\mathbf{1}_{\Delta_{P_j}}, \mathbf{1}_{\Delta_{Q_i}}\rangle \\ &+\langle T\mathbf{1}_{\Delta_{P_j}}, \mathbf{1}_{Q_{i,\partial}}\rangle + \langle T\mathbf{1}_{\Delta_{P_j}}, \mathbf{1}_{Q_{i,\mathrm{sep}}}\rangle, \end{aligned} \tag{74}$$

and these are denoted by $M_1(P)+M_2(P)+M_3(P)+M_4(P)+M_5(P)$, respectively.

8.2 Description of different terms

The heart of the argument lies in estimating terms

$$M_3(P) = \langle \mathbf{1}_{\Delta_{P_j}}, T\mathbf{1}_{\Delta_{Q_i}}\rangle = \alpha_1(P) + \alpha_2(P) + \alpha_3(P),$$

where the last decomposition depends on a third random dyadic system $\mathcal{D}_3$, we refer to (76). Terms $\alpha_2(P)$ and $\alpha_3(P)$, along with $M_2(P)$ and $M_4(P)$, are "υ-boundary" terms. The "separated" terms $M_1(P)$ and $M_5(P)$ are treated by kernel size condition.

The term $\alpha_1(P)$ will further be expanded in (77) as

$$\alpha_1(P) = \beta_1(P) + \beta_2(P) + \beta_3(P),$$

where $\beta_1(P)$ and $\beta_2(P)$ are so called "ϵ-boundary" terms. The local testing conditions and kernel size estimates are exploited in estimating "intersecting" term $\beta_3(P)$.

8.3 Decomposition of $M_3(P)$

Without loss of generality, we can assume that $\Delta_{Q_i} \neq \emptyset$ and $\Delta_{P_j} \neq \emptyset$. Indeed, otherwise we already have $M_3(P) = 0$.

We introduce a third random dyadic system $\mathcal{D}_3 = \mathcal{D}(\omega_3)$ that is independent of both $\mathcal{D}_1$ and $\mathcal{D}_2$. Fix $j(v) \in \mathrm{Z}$ such that $v/64 \leq 2^{j(v)} < v/32$. Then, for every $P \in \mathcal{G}_1$ with $Q = Q(P) \neq \emptyset$, we define a layer

$$\mathcal{L} = \mathcal{L}(P, v) := \mathcal{D}_{3,\log_2(s)}$$

of $\mathcal{D}_3$-cubes with side length

$$s = 2^{j(v)}\ell Q_i = 2^{j(v)} \cdot (\ell Q_i \wedge \ell P_j)\,. \tag{75}$$

That is, $\mathcal{L}$ is a layer of $\mathcal{D}_3$ that depends on parameters P and v.

Let $\Delta^{\mathcal{L}}_{Q_i}, \Delta^{\mathcal{L}}_{P_j} \subset Q_i \cap P_j$ be the following adaptations of Δ_{Q_i} and Δ_{P_j} to $\mathcal{L}$. If necessary, we enlarge the latter sets so that, for every $G \in \mathcal{L}$, either $G \cap \Delta^{\mathcal{L}}_{Q_i} = G \cap \Delta^{\mathcal{L}}_{P_j} = G$ or one of the two intersections $G \cap \Delta^{\mathcal{L}}_{Q_i}$ and $G \cap \Delta^{\mathcal{L}}_{P_j}$ is empty. This is done in such a way that we can write

$$\Delta^{\mathcal{L}}_{Q_i} = \Delta_{Q_i} \cup \Delta^{\partial}_{Q_i}\,, \quad \Delta^{\mathcal{L}}_{P_j} = \Delta_{P_j} \cup \Delta^{\partial}_{P_j}\,,$$

both as disjoint unions, such that $\Delta^{\partial}_{Q_i} \subset Q_{i,\partial} \cap P_j$ and $\Delta^{\partial}_{P_j} \subset P_{j,\partial} \cap Q_i$. For an illustration, we refer to Figure 2.

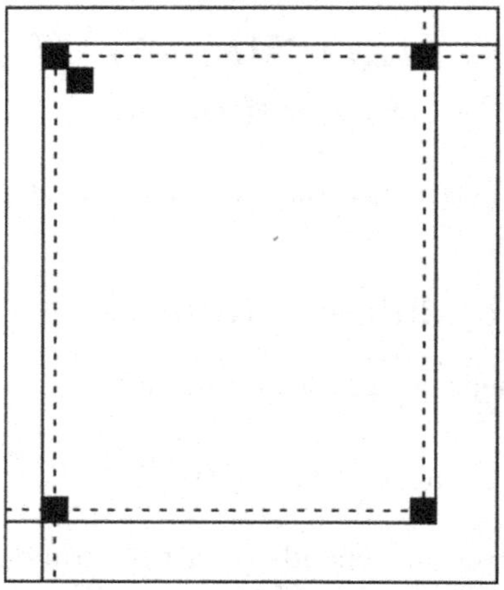

Figure 2: Parallelogram is $Q_i \cap P_j$. Interiors of some black $\mathcal{L}$-cubes intersect dashed line segments, which belong to the boundaries of either Δ_{P_j} or Δ_{Q_i}. The $\mathcal{L}$-adjusted sets $\Delta^{\mathcal{L}}_{Q_i}$ and $\Delta^{\mathcal{L}}_{P_j}$, with solid boundaries, do not intersect the indicated $\mathcal{L}$-cubes.

Now observe that $M_3(P) = \langle T\mathbf{1}_{\Delta_{P_j}}, \mathbf{1}_{\Delta_{Q_i}}\rangle$ can be written as

$$\alpha_1(P) + \alpha_2(P) + \alpha_3(P) = \langle T\mathbf{1}_{\Delta^{\mathcal{L}}_{P_j}}, \mathbf{1}_{\Delta^{\mathcal{L}}_{Q_i}}\rangle - \langle T\mathbf{1}_{\Delta^{\partial}_{P_j}}, \mathbf{1}_{\Delta^{\mathcal{L}}_{Q_i}}\rangle - \langle T\mathbf{1}_{\Delta_{P_j}}, \mathbf{1}_{\Delta^{\partial}_{Q_i}}\rangle\,. \tag{76}$$

We remark that the terms in this decomposition depends on $\mathcal{D}_3$.

In order to define ϵ-boundary terms, we let $P \in \mathcal{G}_1$ and write

$$L_\epsilon = L_\epsilon(P, v) = \bigcup_{G \in \mathcal{L}(P,v)} \delta_G^\epsilon, \quad \delta_G^\epsilon = (1+\epsilon)G \setminus (1-\epsilon)G.$$

We also write $\widetilde{G} = G \setminus L_\epsilon$ if $G \in \mathcal{L} = \mathcal{L}(P, v)$. Define

$$\Delta'_{Q_i} = \Delta^{\mathcal{L}}_{Q_i} \cap L_\epsilon, \quad \widetilde{\Delta}_{Q_i} = \Delta^{\mathcal{L}}_{Q_i} \setminus L_\epsilon, \quad \Delta'_{P_j} = \Delta^{\mathcal{L}}_{P_j} \cap L_\epsilon, \quad \widetilde{\Delta}_{P_j} = \Delta^{\mathcal{L}}_{P_j} \setminus L_\epsilon.$$

Finally, we write $\alpha_1(P) = \langle T\mathbf{1}_{\Delta^{\mathcal{L}}_{P_j}}, \mathbf{1}_{\Delta^{\mathcal{L}}_{Q_i}} \rangle$ as

$$\beta_1(P) + \beta_2(P) + \beta_3(P) = \langle T\mathbf{1}_{\Delta'_{P_j}}, \mathbf{1}_{\Delta^{\mathcal{L}}_{Q_i}} \rangle + \langle T\mathbf{1}_{\widetilde{\Delta}_{P_j}}, \mathbf{1}_{\Delta'_{Q_i}} \rangle + \langle T\mathbf{1}_{\widetilde{\Delta}_{P_j}}, \mathbf{1}_{\widetilde{\Delta}_{Q_i}} \rangle. \tag{77}$$

9 The nearby-non-boundary term

We estimate summations involving the separated terms $M_1(P)$ and $M_5(P)$, and the intersecting term $\beta_3(P)$. All of the estimates will be uniform over all three dyadic grids.

9.1 Separated term

The two indicators appearing in either $M_1(P)$ or $M_5(P)$ are associated with sets separated from each other. This observation will allow us to prove inequality

$$\Big| \sum_{P \in \mathcal{G}_1} \langle \Delta_P f_1 \rangle_{P_j} (M_1(P) + M_5(P)) \langle \Delta_Q f_2 \rangle_{Q_i} \Big| \le C(r, v). \tag{78}$$

Proof of inequality (78). We focus on summation over the terms $M_1(P)$, and the treatment of summation over terms $M_5(P)$ is analogous. We write $T_{P_j,Q_i} = \mathbf{1}_{Q=Q(P)} \langle T\mathbf{1}_{P_{j,\mathrm{sep}}}, \mathbf{1}_{Q_i} \rangle$. Then, by inequalities (68), the term under focus can be written as

$$\Big| \sum_{m=0}^{r} \sum_{u \in \{0,1\}} \sum_{k \in \mathbf{Z}} \sum_{Q \in \mathcal{G}_{2,k-m}} \sum_{\substack{P \in \mathcal{G}_{1,k} \\ D(Q,P)/\ell P \sim 2^u}} \langle \Delta_P f_1 \rangle_{P_j} T_{P_j,Q_i} \langle \Delta_Q f_2 \rangle_{Q_i} \Big|$$

$$\le \sum_{m,u} \mathbf{A}_{m,u,i,j} \cdot \Big\| \Big(\sum_{k \in \mathbf{Z}} \sum_{Q \in \mathcal{G}_{2,k-m}} |\Delta_Q f_2|^2 \Big)^{1/2} \Big\|_{p_2} \lesssim \sum_{m,u} \mathbf{A}_{m,u,i,j},$$

where we have denoted

$$\mathbf{A}_{m,u,i,j} = \Big\| \Big(\sum_{k \in \mathbf{Z}} \sum_{Q \in \mathcal{G}_{2,k-m}} \Big| \mathbf{1}_{Q_i} \sum_{\substack{P \in \mathcal{G}_{1,k} \\ D(Q,P)/\ell P \sim 2^u}} \langle \Delta_P f_1 \rangle_{P_j} \frac{T_{P_j,Q_i}}{\mu(Q_i)} \Big|^2 \Big)^{1/2} \Big\|_{p_1}.$$

The proof of inequality (78) is finished by invoking Lemma 79 below. □

Lemma 79. *For* $m \in \{0,1,\ldots,r\}$ *and* $u \in \{0,1\}$, *we have* $|\mathbf{A}_{m,u,i,j}| \le C(r,\upsilon)$.

Proof. For each $k \in \mathbf{Z}$ and $S \in \mathcal{D}_{1,k+u+\theta(u+m)}$, define a kernel

$$K_S(x,y) = \sum_{\substack{P\in\mathcal{G}_{1,k}\\ P\subset S}} \sum_{\substack{Q\in\mathcal{G}_{2,k-m}\\ D(Q,P)/\ell P\sim 2^u}} \mathbf{1}_{Q_i}(x)\cdot \widetilde{T}_{P_j,Q_i}\cdot \mathbf{1}_{P_j}(y)\,, \qquad x,y\in\mathbf{R}^n\,, \tag{80}$$

where $\widetilde{T}_{P_j,Q_i} = \mathbf{1}_{Q=Q(P)}\widetilde{T}_{P_j,Q_i}$ is defined by

$$\frac{T_{P_j,Q_i}}{\mu(P_j)\mu(Q_i)} = \frac{\widetilde{T}_{P_j,Q_i}}{\mu(S)}\,.$$

Consider cubes P and Q as in the definition of K_S, and let $y \in P_{j,\mathrm{sep}}$ and $x \in Q_i$. By the upper doubling properties of μ, and the facts that $|x-y| \ge \upsilon 2^{-r}\ell P_j$ and $S \subset B(y, 2^{1+u+\theta(u+m)}\ell P)$, we find that $\lambda(y,|x-y|)^{-1} \le C(r,\upsilon)\mu(S)^{-1}$. Hence, by definition,

$$|T_{P_j,Q_i}| \le \int_{Q_i}\int_{P_{j,\mathrm{sep}}} \frac{1}{\lambda(y,|x-y|)}\,d\mu(y)\,d\mu(x) \le C(r,\upsilon)\mu(Q_i)\mu(P_j)\mu(S)^{-1}\,.$$

As a consequence $|\widetilde{T}_{P_j,Q_i}| \le C(r,\upsilon)$ and, by recalling Lemma 21,

$$|K_S(x,y)| \le C(r,\upsilon) \sum_{\substack{P\in\mathcal{G}_{1,k}\\ P\subset S}} \sum_{\substack{P\in\mathcal{G}_{2,k-m}\\ D(Q,P)/\ell P\sim 2^u}} \mathbf{1}_{Q_i}(x)\cdot\mathbf{1}_{P_j}(y) \le C(r,\upsilon)\cdot\mathbf{1}_S(x)\cdot\mathbf{1}_S(y)\,. \tag{81}$$

After these preparations, we finish the proof by proceeding as in Lemma 65. □

9.2 Intersecting term

The following inequality deals with intersecting part, i.e., terms $\beta_3(P)$;

$$\Big|\sum_{P\in\mathcal{G}_1} \langle\Delta_P f_1\rangle_{P_j}\beta_3(P)\langle\Delta_Q f_2\rangle_{Q_i}\Big| \le C(r,\upsilon,\epsilon)(1+\mathbf{T}_{\mathrm{loc}})\,. \tag{82}$$

The proof of this inequality relies on the kernel size estimate and local testing conditions.

Proof of inequality (82). We tacitly restrict all the summations here to cubes $P \in \mathcal{G}_1$ for which $\mu(Q_i\cap P_j) \ne 0$. Indeed, otherwise $\beta_3(P) = 0$. By writing $\mu(Q_i\cap P_j) = \int \mathbf{1}_{Q_i}\mathbf{1}_{P_j}\,d\mu$ and using Cauchy-Schwarz and Hölder's inequality,

$$\begin{aligned}&\Big|\sum_{P\in\mathcal{G}_1} \langle\Delta_P f_1\rangle_{P_j}\beta_3(P)\langle\Delta_Q f_2\rangle_{Q_i}\Big|\\ &\le \Big\|\Big(\sum_{P\in\mathcal{G}_1} |\langle\Delta_P f_1\rangle_{P_j}\mathbf{1}_{P_j}|^2\Big)^{1/2}\Big\|_{p_1}\cdot\Big\|\Big(\sum_{P\in\mathcal{G}_1}\Big|\frac{\beta_3(P)}{\mu(Q_i\cap P_j)}\langle\Delta_Q f_2\rangle_{Q_i}\mathbf{1}_{Q_i}\Big|^2\Big)^{1/2}\Big\|_{p_2}\,.\end{aligned} \tag{83}$$

By inequality (15), the first factor is bounded by $\lesssim 1$. Let us then focus on the second factor; by writing the summation in terms of Q and using Lemma 84, we obtain an upper bound $C(r, \upsilon, \epsilon)(1 + \mathbf{T}_{\text{loc}})$ for the second term. $\square$

Lemma 84. *Let $P \in \mathcal{G}_1$. Then $|\beta_3(P)| \le C(r, \upsilon, \epsilon)(1 + \mathbf{T}_{\text{loc}})\mu(Q_i \cap P_j)$.*

Proof. We can assume that $Q = Q(P) \neq \emptyset$, hence $P \sim Q$. Consider the expansion

$$\begin{aligned}\beta_3(P) &= \langle T\mathbf{1}_{\widetilde{\Delta}_{P_j}}, \mathbf{1}_{\widetilde{\Delta}_{Q_i}}\rangle \\ &= \sum_{\substack{G,H\in\mathcal{L}\\ G\neq H}} \langle T(\mathbf{1}_G\mathbf{1}_{\widetilde{\Delta}_{P_j}}), \mathbf{1}_H\mathbf{1}_{\widetilde{\Delta}_{Q_i}}\rangle + \sum_{G\in\mathcal{L}} \langle T(\mathbf{1}_G\mathbf{1}_{\widetilde{\Delta}_{P_j}}), \mathbf{1}_G\mathbf{1}_{\widetilde{\Delta}_{Q_i}}\rangle.\end{aligned}$$

In both of the series above, the finite number of summands depends on n and υ. Hence, it suffices to obtain estimates for individual summands for fixed $G, H \in \mathcal{L}$. First, if $G \neq H$, then

$$\ell Q_i \le C(r, \upsilon, \epsilon)\,\mathrm{dist}(G \cap \widetilde{\Delta}_{P_j}, H \cap \widetilde{\Delta}_{Q_i}).$$

In particular, $\lambda(x, |x-y|)^{-1} \le C(r, \upsilon, \epsilon)\mu(Q_i)^{-1}$ if $x \in H \cap \widetilde{\Delta}_{Q_i}$ and $y \in G \cap \widetilde{\Delta}_{P_j}$. Hence,

$$\begin{aligned}|\langle T(\mathbf{1}_G\mathbf{1}_{\widetilde{\Delta}_{P_j}}), \mathbf{1}_H\mathbf{1}_{\widetilde{\Delta}_{Q_i}}\rangle| &\le \int_{H\cap\widetilde{\Delta}_{Q_i}} \int_{G\cap\widetilde{\Delta}_{P_j}} \frac{1}{\lambda(x, |x-y|)}\, d\mu(y)\, d\mu(x) \\ &\le C(r, \upsilon, \epsilon)\frac{\mu(Q_i \cap P_j)\mu(Q_i \cap P_j)}{\mu(Q_i)} \\ &\le C(r, \upsilon, \epsilon)\mu(Q_i \cap P_j). \end{aligned} \tag{85}$$

In the last step, we also used the fact that $\widetilde{\Delta}_{P_j} \cup \widetilde{\Delta}_{Q_i} \subset Q_i \cap P_j$.

Then we consider the case of $G = H$. By construction,

$$\langle T(\mathbf{1}_G\mathbf{1}_{\widetilde{\Delta}_{P_j}}), \mathbf{1}_G\mathbf{1}_{\widetilde{\Delta}_{Q_i}}\rangle = \begin{cases} \langle T\mathbf{1}_{\widetilde{G}}, \mathbf{1}_{\widetilde{G}}\rangle & \text{if } G = G \cap \Delta^{\mathcal{L}}_{P_j} = G \cap \Delta^{\mathcal{L}}_{Q_i}\,; \\ 0 & \text{otherwise.}\end{cases} \tag{86}$$

In any case, by local testing conditions $|\langle T(\mathbf{1}_G\mathbf{1}_{\widetilde{\Delta}_{P_j}}), \mathbf{1}_G\mathbf{1}_{\widetilde{\Delta}_{Q_i}}\rangle| \le \mathbf{T}_{\text{loc}}\mu(\widetilde{G}) \le \mathbf{T}_{\text{loc}}\mu(Q_i \cap P_j)$. $\square$

10 The nearby-boundary term

Here we treat the ϵ and υ boundary terms by probabilistic arguments.

10.1 The ϵ-boundary terms

Following inequality controls summation for ϵ-boundary terms. Let $t > p_1 \vee p_2$ be a positive real number. Then

$$\mathbf{E}_{\omega_3}\Big|\sum_{P\in\mathcal{G}_1} \langle \Delta_P f_1\rangle_{P_j}\big(\beta_1(P) + \beta_2(P)\big)\langle \Delta_Q f_2\rangle_{Q_i}\Big| \le C(r, \upsilon)\epsilon^{1/t}\mathbf{T}. \tag{87}$$

The expectations over the dyadic system $\mathcal{D}_3$ are crucial here, and here only.

We let $(\varepsilon_k)_{k\in\mathbf{Z}}$ be a sequence of Rademacher variables, supported on a probability space $(\Omega, \mathbf{P})$. We also associate Rademacher variables to $\mathcal{D}_j$-dyadic cubes with $j \in \{1,2\}$: fix an injection $R \mapsto j(R) : \mathcal{D}_j \to \mathbf{Z}$, and use notation $\varepsilon_R = \varepsilon_{j(R)}$.

We rely on the following improvement of the contraction principle [18, Lemma 3.1].

Proposition 88. *Suppose that* $\{\rho_k : k \in \mathbf{Z}\} \subset L^t(\widetilde{\Omega})$ *for some* σ*-finite measure space* $(\widetilde{\Omega}, \widetilde{\mathbf{P}})$ *and* $t \in (2,\infty)$. *Then, for all complex-valued sequences* $(\xi_k)_{k\in\mathbf{Z}}$,

$$\Big\| \sum_{k=-\infty}^{\infty} \varepsilon_k \rho_k \xi_k \Big\|_{L^t(\widetilde{\Omega};L^2(\Omega))} \lesssim \sup_{k\in\mathbf{Z}} \|\rho_k\|_{L^t(\widetilde{\Omega})} \cdot \Big\| \sum_{k=-\infty}^{\infty} \varepsilon_k \xi_k \Big\|_{L^2(\Omega)} .$$

Proof of inequality (87). Let us focus on the sum involving the terms $\beta_1(P)$; the estimate for the sum involving terms $\beta_2(P)$ is similar. We randomize and use Hölder's inequality

$$\begin{aligned}
&\Big| \sum_{P\in\mathcal{G}_1} \langle \Delta_P f_1\rangle_{P_j} \langle T\mathbf{1}_{\Delta'_{P_j}}, \mathbf{1}_{\Delta^{\mathcal{L}}_{Q_i}}\rangle \langle \Delta_Q f_2\rangle_{Q_i} \Big| \\
&= \Big| \int_\Omega \Big\langle T\Big(\sum_{S\in\mathcal{G}_1} \varepsilon_S \mathbf{1}_{\Delta'_{S_j}} \langle \Delta_S f_1\rangle_{S_j}\Big), \sum_{P\in\mathcal{G}_1} \varepsilon_P \mathbf{1}_{\Delta^{\mathcal{L}}_{Q_i}} \langle \Delta_Q f_2\rangle_{Q_i} \Big\rangle d\mathbf{P}(\epsilon) \Big| \\
&\le \Big\| T\Big(\sum_{S\in\mathcal{G}_1} \varepsilon_S \mathbf{1}_{\Delta'_{S_j}} \langle \Delta_S f_1\rangle_{S_j}\Big) \Big\|_{L^{p_1}(\Omega\times\mathbf{R}^n)} \Big\| \sum_{P\in\mathcal{G}_1} \varepsilon_P \mathbf{1}_{\Delta^{\mathcal{L}}_{Q_i}} \langle \Delta_Q f_2\rangle_{Q_i} \Big\|_{L^{p_2}(\Omega\times\mathbf{R}^n)} .
\end{aligned} \tag{89}$$

Index the very last summation in terms of $\mathcal{D}_2$. This can be done by using our standing assumptions of $P \mapsto Q(P) = Q$. Then, by the contraction principle and inequality $|\mathbf{1}_{\Delta^{\mathcal{L}}_{Q_i}}| \le \mathbf{1}_{Q_i}$, we see that the second factor in the last line of (89) is bounded (up to a constant multiple) by $\|f_2\|_{p_2} \lesssim 1$.

In order to estimate the first factor in the last line of (89) we first extract operator norm $\mathbf{T}$. Then we fix $S \in \mathcal{G}_{1,k}$ with $k \in \mathbf{Z}$. By (68) and (75),

$$\Delta'_{S_j} \subset L_\epsilon(S,v) = \bigcup_{G\in\mathcal{L}(S,v)} \delta^\epsilon_G \subset \bigcup_{m=j(v)+k-r-1}^{j(v)+k-1} \bigcup_{G\in\mathcal{D}_{3,m}} \delta^\varepsilon_G =: \delta^\epsilon(k).$$

Hence, we have $\mathbf{1}_{\Delta'_{S_j}} \le \mathbf{1}_{\delta^\epsilon(k)} \mathbf{1}_{S_j}$. By the contraction principle and assumption $t \ge p_1$, we obtain that

$$\begin{aligned}
&\mathbf{E}_{\omega_3} \Big\| \sum_{S\in\mathcal{G}_1} \varepsilon_S \mathbf{1}_{\Delta'_{S_j}} \langle \Delta_S f_1\rangle_{S_j} \Big\|_{L^{p_1}(\Omega\times\mathbf{R}^n)} \\
&\lesssim \mathbf{E}_{\omega_3} \Big\| \sum_{k\in\mathbf{Z}} \varepsilon_k \mathbf{1}_{\delta^\epsilon(k)} \sum_{S\in\mathcal{D}_{1,k}} \mathbf{1}_{S_j} \langle \Delta_S f_1\rangle_{S_j} \Big\|_{L^{p_1}(\Omega\times\mathbf{R}^n)}
\end{aligned}$$

$$\le \left(\int_{\mathbf{R}^n} \left[\mathbf{E}_{\omega_3} \left\| \sum_{k \in \mathbf{Z}} \varepsilon_k \mathbf{1}_{\delta^\epsilon(k)}(x) \sum_{S \in \mathcal{D}_{1,k}} \mathbf{1}_{S_j}(x) \langle \Delta_S f_1 \rangle_{S_j} \right\|_{L^{p_1}(\mathbf{\Omega})}^t \right]^{p_1/t} d\mu(x) \right)^{1/p_1} . \tag{90}$$

For a fixed $x \in \mathbf{R}^n$, the last integrand evaluated at x is of the form as in Proposition 88 with $\xi_k = \sum_{S \in \mathcal{D}_{1,k}} \mathbf{1}_{S_j}(x) \langle \Delta_S f_1 \rangle_{S_j}$. Moreover, the random variables $\rho_k := \mathbf{1}_{\delta^\epsilon(k)}(x)$ as functions of $\omega_3 \in \Omega_3 = (\{0,1\}^n)^{\mathbf{Z}}$ belong to $L^t(\Omega_3)$, and they satisfy

$$\sup_{k \in \mathbf{Z}} \| \mathbf{1}_{\delta^\epsilon(k)}(x) \|_{L^t(\Omega_3)} = \sup_{k \in \mathbf{Z}} \mathbf{P}_{\omega_3}(\mathbf{1}_{\delta^\epsilon(k)}(x) = 1)^{1/t} \le C(r,\upsilon)\epsilon^{1/t}.$$

Hence, by Proposition 88 and Khintchine's inequality,

$$LHS(90) \le C(r,\upsilon)\epsilon^{1/t} \left\| \sum_{S \in \mathcal{D}_1} \varepsilon_S \mathbf{1}_{S_j} \langle \Delta_S f_1 \rangle_{S_j} \right\|_{L^{p_1}(\Omega \times \mathbf{R}^n)} \le C(r,\upsilon)\epsilon^{1/t} .$$

The proof is complete. □

10.2 The υ-boundary terms

The following inequality controls summation of the υ-boundary terms. Let $t > p_1 \vee p_2$. Then

$$\mathbf{E}_{\omega_1} \mathbf{E}_{\omega_2} \left| \sum_{P \in \mathcal{G}_1} \langle \Delta_P f_1 \rangle_{P_j} (M_2(P) + M_4(P) + \alpha_2(P) + \alpha_3(P)) \langle \Delta_Q f_2 \rangle_{Q_i} \right| \le C(r) \upsilon^{1/t} \mathbf{T} . \tag{91}$$

Before the proof of this inequality, let us remark that although both $\alpha_2(P)$ and $\alpha_3(P)$ depend on the random dyadic system $\mathcal{D}_3$, the inequality is uniform over all such systems.

Proof of inequality (91). First we observe that functions f_j depend on *both* dyadic systems, as they are (essentially) projections to good cubes. This dependency is not allowed in the argument below. Fortunately, this issue can be easily addressed—if $Q(P) \ne \emptyset$ in the series above, we have both $P \in \mathcal{G}_1$ and $Q = Q(P) \in \mathcal{G}_2$. Then, in particular $\Delta_P f_1 = \Delta_P \widetilde{f_1}$ and $\Delta_Q f_2 = \Delta_Q \widetilde{f_2}$. Functions $\widetilde{f_j}$ do not depend on the dyadic systems, and we use them to replace f_j's.

By (74) and (76), $M_2(P) + \alpha_2(P)$ and $M_4(P) + \alpha_3(P)$ are given by

$$\langle T \mathbf{1}_{P_{j,\partial}}, \mathbf{1}_{Q_i} \rangle - \langle T \mathbf{1}_{\Delta^\partial_{P_j}}, \mathbf{1}_{\Delta^\mathcal{L}_{Q_i}} \rangle; \quad \langle T \mathbf{1}_{\Delta_{P_j}}, \mathbf{1}_{Q_{i,\partial}} \rangle - \langle T \mathbf{1}_{\Delta_{P_j}}, \mathbf{1}_{\Delta^\partial_{Q_i}} \rangle ,$$

respectively. Observe that

$$\begin{aligned} &(\mathbf{1}_{P_{j,\partial}} + \mathbf{1}_{\Delta^\partial_{P_j}}) \lesssim \mathbf{1}_{P_{j,\partial}}, \quad (\mathbf{1}_{Q_i} + \mathbf{1}_{\Delta^\mathcal{L}_{Q_i}}) \lesssim \mathbf{1}_{Q_i} , \\ &\mathbf{1}_{\Delta_{P_j}} \lesssim \mathbf{1}_{P_j}, \quad (\mathbf{1}_{Q_{i,\partial}} + \mathbf{1}_{\Delta^\partial_{Q_i}}) \lesssim \mathbf{1}_{Q_{i,\partial}}. \end{aligned} \tag{92}$$

pointwise μ-almost everywhere. By triangle inequality, it suffices to estimate the following sums: one involving terms $m(P) \in \{M_2(P), \alpha_2(P)\}$, and the other involving terms in $\{M_4(P), \alpha_3(P)\}$. We focus on the first sum; the second one is estimated in an analogous manner, using $\mathbf{E}_{\omega_1}$.

By randomizing, using Hölder's inequality, extracting the operator norm of T, and applying the contraction principle with inequalities (92),

$$\begin{aligned} &\mathbf{E}_{\omega_2}\Big|\sum_{P\in\mathcal{G}_1} \langle \Delta_P \widetilde{f}_1\rangle_{P_j} m(P) \langle \Delta_Q \widetilde{f}_2\rangle_{Q_i}\Big| \\ &\lesssim \mathbf{T}\cdot \mathbf{E}_{\omega_2}\Big\{\Big\|\sum_{S\in\mathcal{G}_1} \varepsilon_S \mathbf{1}_{S_{j,\partial}} \langle \Delta_P \widetilde{f}_1\rangle_{S_j}\Big\|_{L^{p_1}(\Omega\times\mathbf{R}^n)} \Big\|\sum_{Q\in\mathcal{D}_2} \varepsilon_Q \mathbf{1}_{Q_i} \langle \Delta_Q \widetilde{f}_2\rangle_{Q_i}\Big\|_{L^{p_2}(\Omega\times\mathbf{R}^n)}\Big\}. \end{aligned} \tag{93}$$

By the contraction principle, we find that the last factor is ω_2-uniformly bounded by $\lesssim \|\widetilde{f}_2\|_{p_2} = 1$. In order to treat the remaining factor, we write

$$\delta^v(k) = \bigcup_{m=k-r-1}^{k-1} \bigcup_{Q\in\mathcal{D}_{2,m}} \delta^v_Q.$$

By (68) and (73), $\mathbf{1}_{S_{j,\partial}} \le \mathbf{1}_{S_j}\mathbf{1}_{\delta^v_{Q_i}} \le \mathbf{1}_{S_j}\mathbf{1}_{\delta^v(k)}$ if $Q = Q(S)$ with $S\in\mathcal{G}_{1,k}$. Fix $x\in\mathbf{R}^n$. The random variables $\rho_k := \mathbf{1}_{\delta^v(k)}(x)$ as functions of $\omega_2 \in (\{0,1\}^n)^{\mathbf{Z}}$ belong to $L^t((\{0,1\}^n)^{\mathbf{Z}})$,

$$\sup_{k\in\mathbf{Z}} \|\mathbf{1}_{\delta^v(k)}(x)\|_{L^t((\{0,1\}^n)^{\mathbf{Z}})} = \sup_{k\in\mathbf{Z}} \mathbf{P}_{\omega_2}(\mathbf{1}_{\delta^v(k)}(x) = 1)^{1/t} \le C(r)v^{1/t}.$$

Hence, proceeding as in connection with (90), we find that

$$\begin{aligned} &\mathbf{E}_{\omega_2}\Big\|\sum_{S\in\mathcal{G}_1} \varepsilon_S \mathbf{1}_{S_{j,\partial}} \langle \Delta_P \widetilde{f}_1\rangle_{S_j}\Big\|_{L^{p_1}(\Omega\times\mathbf{R}^n)} \\ &\le C(r)v^{1/t}\Big\|\sum_{P\in\mathcal{D}_1} \varepsilon_P \mathbf{1}_{P_j} \langle \Delta_P \widetilde{f}_1\rangle_{P_j}\Big\|_{L^{p_1}(\Omega\times\mathbf{R}^n)}. \end{aligned}$$

The last term is bounded by a constant multiple of $C(r)v^{1/t}$. □

References

[1] P. Auscher, S. Hofmann, C. Muscalu, T. Tao, and C. Thiele, Carleson measures, trees, extrapolation, and $T(b)$ theorems, Publ. Mat. **46** (2002), no. 2, 257–325.

[2] P. Auscher and E. Routin, Local Tb theorems and hardy inequalities, J. Geom. Anal. **23** (2013), no. 1, 303–374.

[3] P. Auscher and Q. X. Yang, BCR algorithm and the $T(b)$ theorem, Publ. Mat. **53** (2009), no. 1, 179–196.

[4] J. Bourgain, Vector-valued singular integrals and the H^1-BMO duality, Probability theory and harmonic analysis (Cleveland, Ohio 1983), Monogr. Textbooks Pure Appl. Math. **98**, Dekker, New York, 1986, pp. 1–19.

[5] D. L. Burkholder, Boundary value problems and sharp inequalities for martingale transforms, Ann. Probab. **12** (1984), no. 3, 647–702.

[6] M. Christ, A $T(b)$ theorem with remarks on analytic capacity and the Cauchy integral, Colloq. Math. **60/61** (1990), no. 2, 601–628.

[7] G. David and J.-L. Journé, A boundedness criterion for generalized Calderón-Zygmund operators, Ann. of Math. (2) **120** (1984), no. 2, 371–397.

[8] S. Hofmann, A proof of the local Tb Theorem for standard Calderón-Zygmund operators (2007), available at `http://arxiv.org/abs/0705.0840`.

[9] G. Hu, Y. Meng, and D. Yang, A new characterization of regularized BMO spaces on non-homogeneous spaces and its applications, Ann. Acad. Sci. Fenn. Math. **38** (2013), 3–27.

[10] T. P. Hytönen, The sharp weighted bound for general Calderón-Zygmund operators, Ann. of Math. (2) **175** (2012), no. 3, 1473–1506.

[11] T. P. Hytönen, The vector-valued non homogeneous Tb theorem, Int. Math. Res. Not., IMRN (2014), no. 2, 451–511.

[12] T. Hytönen, Representation of singular integrals by dyadic operators, and the A_2 theorem (2011), available at `http://arxiv.org/pdf/1108.5119.pdf`.

[13] T. Hytönen, A framework for non-homogeneous analysis on metric spaces, and the RBMO space of Tolsa, Publ. Mat. **54** (2010), no. 2, 485–504.

[14] T. Hytönen, S. Liu, D. Yang, and D. Yang, Boundedness of Calderón-Zygmund operators on non-homogeneous metric measure spaces, Canad. J. Math. **64** (2012), no. 4, 892–923.

[15] T. Hytönen and H. Martikainen, Non-homogeneous Tb theorem and random dyadic cubes on metric measure spaces, J. Geom. Anal. **22** (2012), no. 4, 1071–1107.

[16] T. Hytönen and H. Martikainen, On general local Tb theorems, Trans. Amer. Math. Soc. **364** (2012), no. 9, 4819–4846.

[17] T. Hytönen and F. Nazarov, The local Tb theorem with rough test functions (2012), available at `http://arxiv.org/abs/1206.0907`.

[18] T. Hytönen and M. Veraar, R-boundedness of smooth operator-valued functions, Integral Equations Operator Theory **63** (2009), no. 3, 373–402.

[19] T. P. Hytönen and A. V. Vähäkangas, The local non-homogeneous Tb theorem for vector-valued functions, Glasg. Math. J. **57** (2015), no. 1, 17–82.

[20] M. T. Lacey, E. T. Sawyer, C.-Y. Shen, and I. Uriarte-Tuero, The two weight inequality for Hilbert transform, coronas, and energy conditions, available at `http://www.arxiv.org/abs/1108.2319`.

[21] M. T. Lacey, E. T. Sawyer, C.-Y. Shen, and I. Uriarte-Tuero, Two weight inequality for the Hilbert transform: a real variable characterization, I, Duke Math. J. **163** (2014), no. 15, 2795–2820.

[22] M. T. Lacey and A. V. Vähäkangas, The perfect local Tb theorem and twisted martingale transforms, Proc. Amer. Math. Soc., **142** (2014), no. 5, 1689–1700.

[23] M. T. Lacey and A. V. Vähäkangas, On the local Tb theorem: a direct proof under duality assumption, Proc. Edinb. Math. Soc. DOI: 10.1017/S0013091514000340.

[24] H. Martikainen, Vector-valued non-homogeneous Tb theorem on metric measure spaces, Rev. Mat. Iberoam. **28** (2012), no. 4, 961–998.

[25] F. Nazarov, S. Treil, and A. Volberg, The Tb-theorem on non-homogeneous spaces, Acta Math. **190** (2003), no. 2, 151–239.

[26] F. Nazarov, S. Treil, and A. Volberg, Accretive system Tb-theorems on nonhomogeneous spaces, Duke Math. J. **113** (2002), no. 2, 259–312.

[27] S. Petermichl, Dyadic shifts and a logarithmic estimate for Hankel operators with matrix symbol, C. R. Acad. Sci. Paris Sér. I Math. **330** (2000), no. 6, 455–460.

[28] A. Volberg and B. D. Wick, Bergman-type singular integral operators and the characterization of Carleson measures for Besov-Sobolev spaces and the complex ball, Amer. J. Math. **134** (2012), no. 4, 949–992.

Some Topics in Harmonic Analysis and Applications
ALM 34, pp. 265–298

The Dynamics of the NLS with the Combined Terms in Five and Higher Dimensions

Changxing Miao*, Guixiang Xu† and Lifeng Zhao‡

Abstract

In this paper, we continue the study in [42] to show the scattering and blow-up result of the solution for the nonlinear Schrödinger equation with the energy below the threshold m in the energy space $H^1(\mathbb{R}^d)$,

$$iu_t + \Delta u = -|u|^{4/(d-2)}u + |u|^{4/(d-1)}u, \ d \geq 5. \qquad \text{(CNLS)}$$

The threshold is given by the ground state W for the energy-critical NLS: $iu_t+\Delta u = -|u|^{4/(d-2)}u$. Compared with the argument in [42], the new ingredient is that we use the double duhamel formula in [31, 51] to lower the regularity of the critical element in $L^\infty_t H^1_x$ to $L^\infty \dot{H}^{-\epsilon}_x$ for some $\epsilon > 0$ in five and higher dimensions and obtain the compactness of the critical element in L^2_x, which is used to control the spatial center function $x(t)$ of the critical element and furthermore used to defeat the critical element in the reductive argument.

2010 Mathematics Subject Classification: Primary: 35L70, Secondary: 35Q55.

Keywords and Phrases: nonlinear Schrödinger equation, scattering, threshold energy.

*Institute of Applied Physics and Computational Mathematics, P. O. Box 8009, Beijing 100088, China, Beijing Center of Mathematics and Information Sciences, Beijing 100048, China, *E-mail:* miao_changxing@iapcm.ac.cn

†Institute of Applied Physics and Computational Mathematics, P. O. Box 8009, Beijing 100088, China, *E-mail:* xu_guixiang@iapcm.ac.cn

‡University of Science and Technology of China, Hefei 230026, China, *E-mail:* zhaolifengustc@yahoo.cn

1 Introduction

We consider the dynamics of the energy solutions for the nonlinear Schrödinger equation (NLS) with the combined nonlinearities in five and higher dimensions

$$\begin{cases} iu_t + \Delta u = f_1(u) + f_2(u), & (t,x) \in \mathbb{R} \times \mathbb{R}^d, \\ u(0) = u_0(x) \in H^1(\mathbb{R}^d). \end{cases} \tag{1.1}$$

where $u : \mathbb{R} \times \mathbb{R}^d \mapsto \mathbb{C}$, $d \geq 5$ and $f_1(u) = -|u|^{4/(d-2)}u$, $f_2(u) = |u|^{4/(d-1)}u$.

The equation has the following mass and Hamiltonian

$$M(u)(t) = \frac{1}{2}\int_{\mathbb{R}^d} |u(t,x)|^2 \, dx;$$
$$E(u)(t) = \int_{\mathbb{R}^d} \frac{1}{2}|\nabla u(t,x)|^2 \, dx + F_1(u(t)) + F_2(u(t)),$$

where

$$F_1(u(t)) = -\frac{d-2}{2d}\int_{\mathbb{R}^d} |u(t,x)|^{\frac{2d}{d-2}} \, dx, \quad F_2(u(t)) = \frac{d-1}{2d+2}\int_{\mathbb{R}^d} |u(t,x)|^{\frac{2d+2}{d-1}} \, dx.$$

They are conserved for the sufficient smooth solutions of (1.1).

In [50], Tao, Visan and Zhang made the comprehensive study of

$$iu_t + \Delta u = |u|^{\frac{4}{d-2}}u + |u|^{\frac{4}{d-1}}u$$

in the energy space. They made use of the interaction Morawetz estimate established in [8] and the stability theory for the scattering solution. Their result is based on the scattering result of the defocusing, energy-critical NLS in the energy space, which is established by Bourgain [5, 6] for the radial case, I-team [9], Ryckman-Visan [47] and Visan [52] for the nonradial case. Since the classical interaction Morawetz estimate fails for (1.1), Tao, et al., leave the scattering and blow-up dichotomy of (1.1) below the threshold as an open problem in [50]. For other results, please refer to [1, 2, 15, 18, 19, 21, 22, 43, 44, 45, 50, 53, 54].

For the focusing, energy-critical NLS

$$iu_t + \Delta u = -|u|^{\frac{4}{d-2}}u. \tag{1.2}$$

Kenig and Merle first applied the concentration compactness in [4, 27, 28] into the scattering theory of the radial solution of (1.2) in [25] with the energy below that of the ground state of

$$-\Delta W = |W|^{\frac{4}{d-2}}W.$$

Subsequently, Killip and Visan made use of the double Duhamel argument in [29, 51] to removed the radial assumption in [30]. For the applications of the concentration compactness in the scattering and the blow up theory of the NLS, NLW, NLKG and Hartree equations, please see [10, 11, 12, 13, 15, 14, 16, 21, 22, 23, 26, 29, 30, 32, 33, 34, 35, 36, 37, 38, 39, 40, 41].

In [42], we made use of the concentration compactness argument and rigidity argument to show the dichotomy of the radial solution of (1.1) in $H^1(\mathbb{R}^3)$ with energy less than the threshold. In this paper, we continue this study in five and higher dimensions.

Now for $\varphi \in H^1$, we denote the scaling quantity $\varphi^{\lambda}_{d,-2}$ by

$$\varphi^{\lambda}_{d,-2}(x) = e^{d\lambda}\varphi(e^{2\lambda}x).$$

We denote the scaling derivative of E by $K(\varphi)$,

$$\begin{aligned} K(\varphi) = \mathcal{L}E(\varphi) &:= \frac{d}{d\lambda}\Big|_{\lambda=0} E(\varphi^{\lambda}_{d,-2}) \\ &= \int_{\mathbb{R}^d} 2|\nabla\varphi|^2 - 2|\varphi|^{\frac{2d}{d-2}} + \frac{2d}{d+1}|\varphi|^{\frac{2d+2}{d-1}}\, dx, \end{aligned} \tag{1.3}$$

which is connected with the Virial identity, and then plays the important role in the blow-up and scattering of the solution of (1.1).

The threshold is determined by the following constrained minimization of the energy $E(\varphi)$

$$m = \inf\{E(\varphi) : 0 \neq \varphi \in H^1(\mathbb{R}^d),\ K(\varphi) = 0\}. \tag{1.4}$$

Since we consider the $\dot{H}^1$-critical growth with the subcritical perturbation, we need the following modified energy

$$E^c(u) = \int_{\mathbb{R}^d} \left(\frac{1}{2}|\nabla u(t,x)|^2 - \frac{d-2}{2d}|u(t,x)|^{\frac{2d}{d-2}}\right) dx.$$

As the nonlinearity $|u|^{\frac{4}{d-1}}u$ is the defocusing, $\dot{H}^1$-subcritical perturbation, one think that the focusing, $\dot{H}^1$-critical term plays the decisive role of the threshold of the scattering solution of (1.1) in the energy space. Just as the 3d case in [42], the first result is to characterize the threshold energy m as the following

Proposition 1.1 ([42, Proposition 1.1]). *For $d \geq 5$, there is no minimizer for (1.4). But for the threshold energy m, we have*

$$m = E^c(W),$$

where $W \in H^1(\mathbb{R}^d)$ is the ground state of the massless equation

$$-\Delta W = |W|^{\frac{4}{d-2}}W.$$

Main result in this paper is

Theorem 1.2. *For $d \geq 5$. Let $u_0 \in H^1(\mathbb{R}^d)$ with*

$$E(u_0) < m, \tag{1.5}$$

and u be the solution of (1.1) and I be its maximal interval of existence. Then

(a) *If $K(u_0) \geq 0$, then $I = \mathbb{R}$, and u scatters in both time directions as $t \to \pm\infty$ in H^1;*

(b) *If $K(u_0) < 0$ and $xu_0 \in L^2$ or u_0 is radial , then u blows up both forward and backward at finite time in H^1.*

By the above result, we conclude that the focusing, $\dot{H}^1$-critical term make the main contribution to the determination of the threshold of the scattering solution of (1.1). For the case $d=3$, we verify the above result for the radial case in [42]. In this paper, we show the scattering result without the radial assumption in five and higher dimensions. Compared with the argument in [42], the new ingredient in five and higher dimensions is that we can use the double duhamel formula in [31, 51] to lower the regularity of the critical element in $L_t^\infty H_x^1$ to $L^\infty \dot{H}_x^{-\epsilon}$ for some $\epsilon > 0$ and obtain the compactness of the critical element in L_x^2, which is used to control the spatial center function $x(t)$ of the critical element and furthermore used to defeat the critical element in the reductive argument.

At last, from the assumption in Theorem 1.2, we know that the solution starts from the following subsets of the energy space,

$$\mathcal{K}^+ = \Big\{\varphi \in H^1(\mathbb{R}^3) : \varphi \text{ is radial}, \ E(\varphi) < m, \ K(\varphi) \geq 0\Big\},$$
$$\mathcal{K}^- = \Big\{\varphi \in H^1(\mathbb{R}^3) : \varphi \text{ is radial}, \ E(\varphi) < m, \ K(\varphi) < 0\Big\}.$$

By the similar scaling argument to that in [42], we know that $\mathcal{K}^\pm \neq \emptyset$.

This paper is organized as follows. In Section 2, we give the basic well-known results, including the linear and nonlinear estimates, the local well-posedness, the perturbation theory and the monotonicity formula. In Section 3, we show the threshold by the variational method, which also give the proof of Proposition 1.1 and various variational estimates, which will be used in the proof of Theorem 1.2. In Section 4, we give the proof of the blow up in Theorem 1.2 in the radial case. In Section 5, we show the linear and nonlinear profile decompositions of the H^1-bound sequences of solution of (1.1). In Section 6, we make use of the stability theory and compactness argument to show the global wellposedness and scattering result in Theorem 1.2 in a reductive argument.

2 Preliminaries

In this section, we give some notation and some well-known results.

2.1 Littlewood-Paley decomposition and Besov space

Let $\Lambda_0(x) \in \mathcal{S}(\mathbb{R}^d)$ such that its Fourier transform $\widetilde{\Lambda}_0(\xi) = 1$ for $|\xi| \leq 1$ and $\widetilde{\Lambda}_0(\xi) = 0$ for $|\xi| \geq 2$. Then we define $\Lambda_k(x)$ for any $k \in \mathbb{Z}\backslash\{0\}$ and $\Lambda_{(0)}(x)$ by the Fourier transforms:

$$\widetilde{\Lambda}_k(\xi) = \widetilde{\Lambda}_0(2^{-k}\xi) - \widetilde{\Lambda}_0(2^{-k+1}\xi), \quad \widetilde{\Lambda}_{(0)}(\xi) = \widetilde{\Lambda}_0(\xi) - \widetilde{\Lambda}_0(2\xi).$$

Let $s \in \mathbb{R}$, $1 \leq p, q \leq \infty$. The inhomogeneous Besov space $B^s_{p,q}$ is defined by

$$B^s_{p,q} = \left\{f : f \in \mathcal{S}'(\mathbb{R}^d), \Big\|2^{ks}\|\Lambda_k * f\|_{L^p_x}\Big\|_{l^q_{k\geq 0}} < \infty\right\},$$

where $\mathcal{S}'(\mathbb{R}^d)$ denotes the space of tempered distributions. The homogeneous Besov space $\dot{B}^s_{p,q}$ can be defined by

$$\dot{B}^s_{p,q} = \left\{ f : f \in \mathcal{Z}'(\mathbb{R}^d), \left(\sum_{k\in\mathbb{Z}\backslash\{0\}} 2^{qks} \|\Lambda_k * f\|^q_{L^p_x} + \left\|\|\Lambda_{(0)} * f\|_{L^p_x}\right\|^q \right)^{1/q} < \infty \right\}.$$

$\mathcal{Z}'(\mathbb{R}^d)$ denotes the dual space of

$$\mathcal{Z}(\mathbb{R}^d) = \{f \in \mathcal{S}(\mathbb{R}^d); \partial^\alpha \hat{f}(0) = 0; \forall \alpha \in \mathbb{N}^d \text{ multi-index}\}$$

and can be identified by the quotient space of $\mathcal{S}'/\mathcal{P}$ with the polynomials space $\mathcal{P}$.

2.2 Linear estimates and nonlinear estimates

We say that a pair of exponents (q, r) is Schrödinger $\dot{H}^s$-admissible in $d \geq 5$ if

$$\frac{2}{q} + \frac{d}{r} = \frac{d}{2} - s$$

and $2 \leq q, r \leq \infty$. If $I \times \mathbb{R}^d$ is a space-time slab, we define the $\dot{S}^0(I \times \mathbb{R}^d)$ Strichartz norm by

$$\|u\|_{\dot{S}^0(I\times\mathbb{R}^d)} := \sup \|u\|_{L^q_t L^r_x(I\times\mathbb{R}^d)}$$

where the sup is taken over all L^2-admissible pairs (q, r). We define the $\dot{S}^s(I \times \mathbb{R}^d)$ and $S^s(I \times \mathbb{R}^d)$ Strichartz norm to be

$$\|u\|_{\dot{S}^s(I\times\mathbb{R}^d)} := \|D^s u\|_{\dot{S}^0(I\times\mathbb{R}^d)}, \quad \|u\|_{S^s(I\times\mathbb{R}^d)} := \|\langle\nabla\rangle^s u\|_{\dot{S}^0(I\times\mathbb{R}^d)}.$$

We also use $\dot{N}^0(I \times \mathbb{R}^d)$ to denote the dual space of $\dot{S}^0(I \times \mathbb{R}^d)$ and

$$\dot{N}^k(I \times \mathbb{R}^d) := \{u; D^k u \in \dot{N}^0(I \times \mathbb{R}^d)\}.$$

Before we introduce the linear estimate, we first give some exponents, which will be frequently used in the paper. For

$$\begin{aligned} S(I) :=& L^\infty\left(I; L^2\right) \cap L^2\left(I; L^{2^*}\right), \\ W_1(I) :=& L^{2\frac{d+2}{d-2}}\left(I; L^{2\frac{d+2}{d-2}}\right), \quad V_1(I) := L^{2\frac{d+2}{d-2}}\left(I; L^{2\frac{d(d+2)}{d^2+4}}\right), \\ W_2(I) :=& L^{2\frac{d+2}{d-1}}\left(I; L^{2\frac{d+2}{d-1}}\right), \quad V_2(I) := L^{2\frac{d+2}{d-1}}\left(I; L^{2\frac{d(d+2)}{d^2+2}}\right). \end{aligned}$$

By definition and Sobolev's inequality, we have

Lemma 2.1. *For any $\dot{S}^1$ function u on $I \times \mathbb{R}^d$, we have*

$$\|\nabla u\|_{S(I)} + \|\nabla u\|_{V_1(I)} + \|u\|_{W_1(I)} + \|\nabla u\|_{V_2(I)} + \||\nabla|^{1/2} u\|_{W_2(I)} \lesssim \|u\|_{\dot{S}^1}.$$

Now we state the standard Strichartz estimate.

Lemma 2.2 ([7, 24, 49]). *Let I be a compact time interval, $k \in [0,1]$, and let $u : I \times \mathbb{R}^d \to \mathbb{C}$ be an $\dot{S}^k$ solution to the forced Schrödinger equation*

$$iu_t + \Delta u = F$$

for a function F. Then we have

$$\|u\|_{\dot{S}^k(I\times\mathbb{R}^d)} \lesssim \|u(t_0)\|_{\dot{H}^k(\mathbb{R}^d)} + \|F\|_{\dot{N}^k(I\times\mathbb{R}^d)},$$

for any time $t_0 \in I$.

For $d \geq 5$, let

$$s_\alpha := \frac{d}{2} - \frac{2}{\alpha-1} = 1 - \frac{2}{d}, \text{ then } \alpha = \frac{d^2+2d+4}{d^2-2d+4}.$$

Let (γ, ρ) be the $\dot{H}^{s_\alpha}$-admissible pair such that

$$\rho = \frac{\alpha+1+2^*}{2}, \quad \frac{2}{\gamma} = d\left(\frac{1}{2} - \frac{1}{\rho}\right) - s_\alpha.$$

Lemma 2.3 ([17]). *For $d \geq 5$, and any $F \in L_t^2\left(I; \dot{B}^{\frac{2}{d}}_{\frac{2d^2}{d^2+4},2}\right)$, we have*

$$\left\|\int_0^t e^{i(t-s)\Delta}F(s)\,ds\right\|_{L^\gamma\left(I;\dot{B}^{\frac{2}{d}}_{\rho,2}\right)} \lesssim \|F\|_{L_t^2\left(I;\dot{B}^{\frac{2}{d}}_{\frac{2d^2}{d^2+4},2}\right)}.$$

Let (q_i, r_i), $i = 1, \ldots, 6$ be the exponentials such that

$$\begin{aligned}
\frac{1}{2} &= \frac{4/(d-2)}{q_1} + \frac{1}{\gamma} = \frac{1}{q_2} + \frac{1}{q_3} + \frac{1}{\gamma} = \frac{4/(d-1)}{q_4} + \frac{1}{\gamma} = \frac{1}{q_5} + \frac{1}{q_6} + \frac{1}{\gamma},\\
\frac{d^2+4}{2d^2} &= \frac{4/(d-2)}{r_1} + \frac{1}{\rho} = \frac{1}{r_2} + \frac{1}{r_3} + \frac{d^2-2\rho}{d^2\rho} = \frac{4/(d-1)}{r_4} + \frac{1}{\rho}\\
&= \frac{1}{r_5} + \frac{1}{r_6} + \frac{d^2-2\rho}{d^2\rho},
\end{aligned}$$

where

$$\frac{1}{q_2} = \frac{1}{r_2} = \frac{d-2}{2(d+2)} \times \left(\frac{4}{d-2} - \frac{4}{d}\right), \quad \frac{1}{q_5} = \frac{1}{r_5} = \frac{d-1}{2(d+2)} \times \left(\frac{4}{d-1} - \frac{4}{d}\right),$$

then

1. $A_1 = (q_1, r_1)$ is $\dot{H}^1$-admissible pair; $W_1 = \left(\left(\frac{4}{d-2} - \frac{4}{d}\right)q_2, \left(\frac{4}{d-2} - \frac{4}{d}\right)r_2\right)$ is the diagonal $\dot{H}^1$-admissible pair; $B_1 = (\frac{4}{d}q_3, \frac{4}{d}r_3)$ is $\dot{H}^{1/2}$-admissible pair.

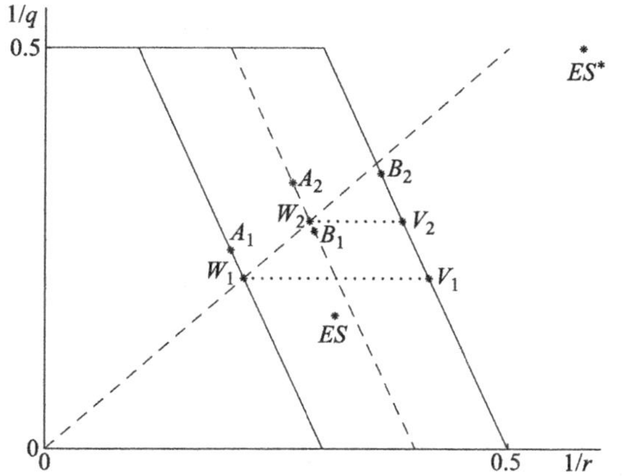

Figure 1: Admissible pairs: A_i, B_i, V_i, W_i, $i = 1, 2$, and ES, ES^*

2. $A_2 = (q_4, r_4)$ is $\dot{H}^{1/2}$-admissible pair; $W_2 = \left(\left(\frac{4}{d-1} - \frac{4}{d}\right)q_5, \left(\frac{4}{d-1} - \frac{4}{d}\right)r_5\right)$ is the diagonal $\dot{H}^{1/2}$-admissible pair; $B_2 = (\frac{4}{d}q_6, \frac{4}{d}r_6)$ is L^2-admissible pair.

3. $ES = (\gamma, \rho)$, $ES^* = \left(2, \frac{2d^2}{d^2+4}\right)$.

Lemma 2.4 ([2, 52]). *For $d \geq 5$, assume that h_1 and h_2 are Hölder continuous functions of order $\frac{4}{d-2}$ and $\frac{4}{d-1}$, respectively. Let I be an interval, then we have*

$$\begin{aligned}
&\big\|h_1(v)w\big\|_{L_t^2\left(I; B^{\frac{2}{d}}_{\frac{2d^2}{d^2+4},2}\right)} \\
&\lesssim \|v\|^{\frac{4}{d-2}}_{L_t^{q_1} L_x^{r_1}} \|w\|_{L^\gamma\left(I; B^{\frac{2}{d}}_{\rho,2}\right)} \\
&\quad + \|v\|^{\frac{4}{d-2} - \frac{4}{d}}_{L_t^{\left(\frac{4}{d-2} - \frac{4}{d}\right)q_2} L_x^{\left(\frac{4}{d-2} - \frac{4}{d}\right)r_2}} \big\||\nabla|^{1/2} v\big\|^{\frac{4}{d}}_{L_t^{\frac{4}{d}q_3} L_x^{\frac{4}{d}r_3}} \|w\|_{L^\gamma\left(I; B^{\frac{2}{d}}_{\rho,2}\right)},
\end{aligned}$$

$$\begin{aligned}
&\big\|h_2(v)w\big\|_{L_t^2\left(I; B^{\frac{2}{d}}_{\frac{2d^2}{d^2+4},2}\right)} \\
&\lesssim \|v\|^{\frac{4}{d-1}}_{L_t^{q_4} L_x^{r_4}} \|w\|_{L^\gamma\left(I; B^{\frac{2}{d}}_{\rho,2}\right)} \\
&\quad + \|v\|^{\frac{4}{d-1} - \frac{4}{d}}_{L_t^{\left(\frac{4}{d-1} - \frac{4}{d}\right)q_5} L_x^{\left(\frac{4}{d-1} - \frac{4}{d}\right)r_5}} \big\||\nabla|^{1/2} v\big\|^{\frac{4}{d}}_{L_t^{\frac{4}{d}q_6} L_x^{\frac{4}{d}r_6}} \|w\|_{L^\gamma\left(I; B^{\frac{2}{d}}_{\rho,2}\right)}.
\end{aligned}$$

2.3 Local wellposedness and perturbation theory

Let us denote $ST(I)$ by $W_1(I) \cap W_2(I)$. By the analogue analysis as those in [2, 42], we have

Theorem 2.5 (Local wellposedness, [2, 42, 50]). *Let $u_0 \in H^1$, then for every $T > 0$, there exists $\eta = \eta(T)$ such that if*

$$\| \langle \nabla \rangle e^{it\Delta} u_0 \|_{V_2([-T,T])} \leq \eta,$$

then (1.1) *admits a unique strong H^1_x-solution u defined on $[-T, T]$. Let $(-T_{\min}, T_{\max})$ be the maximal time interval on which u is well-defined. Then, $u \in S^1(I \times \mathbb{R}^d)$ for every compact time interval $I \subset (-T_{\min}, T_{\max})$ and the following properties hold:*

1. *If $T_{\max} < \infty$, then*

$$\|u\|_{ST((0,T_{\max})\times\mathbb{R}^d)} = \infty.$$

Similarly, if $T_{\min} < \infty$, then

$$\|u\|_{ST((-T_{\min},0)\times\mathbb{R}^d)} = \infty.$$

2. *The solution u depends continuously on the initial data u_0 in the following sense: The functions $T_{\min}$ and $T_{\max}$ are lower semicontinuous from H^1_x to $(0, +\infty]$. Moreover, if $u_0^{(m)} \to u_0$ in H^1_x and $u^{(m)}$ is the maximal solution to* (1.1) *with initial data $u_0^{(m)}$, then $u^{(m)} \to u$ in $ST(I \times \mathbb{R}^d)$ and every compact subinterval $I \subset (-T_{\min}, T_{\max})$.*

Proposition 2.6 ([2, 42]). *Let I be a compact time interval and let w be an approximate solution to* (1.1) *on $I \times \mathbb{R}^d$ in the sense that*

$$i\partial_t w + \Delta w = -|w|^{\frac{4}{d-2}} w + |w|^{\frac{4}{d-1}} w + e$$

for some suitable small function e. Assume that for some constants $L, E_0 > 0$, we have

$$\|w\|_{ST(I)} \leq L, \quad \|w(t_0)\|_{H^1_x(\mathbb{R}^d)} \leq E_0.$$

for some $t_0 \in I$. Let $u(t_0)$ close to $w(t_0)$ in the sense that for some $E' > 0$, we have

$$\|u(t_0) - w(t_0)\|_{H^1_x} \leq E'.$$

Assume also that for some ε, we have

$$\left\| \langle \nabla \rangle e^{i(t-t_0)\Delta} \big(u(t_0) - w(t_0)\big) \right\|_{V_2(I)} \leq \varepsilon, \quad \| \langle \nabla \rangle e \|_{L^{2\frac{d+2}{d+4}}(I)} \leq \varepsilon, \tag{2.1}$$

where $0 < \varepsilon \leq \varepsilon_0 = \varepsilon_0(E_0, E', L)$ is a small constant. Then there exists a solution u to (1.1) *on $I \times \mathbb{R}^d$ with initial data $u(t_0)$ at time $t = t_0$ satisfying*

$$\| \langle \nabla \rangle (u - w) \|_{V_1(I)} \leq C(E_0, E', L)\, \varepsilon, \quad \textit{and} \quad \| \langle \nabla \rangle u \|_{S(I)} \leq C(E_0, E', L).$$

2.4 Monotonicity formula

Lemma 2.7 ([20]). *Let $\phi \in C_0^\infty(\mathbb{R}^3)$, and u be the solution of (1.1). Then we have*

$$\begin{aligned}\partial_t \int_{\mathbb{R}^d} \phi(x)\big|u(t,x)\big|^2 \, dx &= -2\Im \int_{\mathbb{R}^d} \nabla\phi \cdot \nabla \bar{u}\, u \, dx,\\ \partial_t^2 \int_{\mathbb{R}^d} \phi(x)\big|u(t,x)\big|^2 \, dx &= 4\int_{\mathbb{R}^d} \phi_{ij}(x) u_i(t,x)\bar{u}_j(t,x)\, dx - \int_{\mathbb{R}^d} \Delta^2\phi\big|u(t,x)\big|^2 \, dx\\ &\quad - \frac{4}{d}\int_{\mathbb{R}^d} \Delta\phi\big|u(t,x)\big|^{2^*} dx + \frac{4}{d+1}\int_{\mathbb{R}^3} \Delta\phi\big|u(t,x)\big|^{\frac{2d+2}{d-1}} dx.\end{aligned}$$

3 Variational characterization

In this section, we show the threshold energy m (Proposition 1.1) by the variational method, and various estimates for the solutions of (1.1) with the energy below the threshold. The argument is the analogue as the case $d = 3$ in [42].

Let us denote the quadratic and nonlinear parts of K by K^Q and K^N, that is,

$$K(\varphi) = K^Q(\varphi) + K^N(\varphi),$$

where

$$K^Q(\varphi) = 2\int_{\mathbb{R}^d} |\nabla\varphi|^2 \, dx, \text{ and } K^N(\varphi) = \int_{\mathbb{R}^d} \left(-2|\varphi|^{\frac{2d}{d-2}} + \frac{2d}{d+1}|\varphi|^{\frac{2d+2}{d-1}}\right) dx.$$

Lemma 3.1. *For any $\varphi \in H^1(\mathbb{R}^d)$, we have*

$$\lim_{\lambda\to-\infty} K^Q(\varphi_{d,-2}^\lambda) = 0. \tag{3.1}$$

Proof. It is obvious by the definition of K^Q. □

Now we show the positivity of K near 0 in the energy space.

Lemma 3.2. *For any bounded sequence $\varphi_n \in H^1(\mathbb{R}^d)\backslash\{0\}$ with*

$$\lim_{n\to+\infty} K^Q(\varphi_n) = 0,$$

then for large n, we have

$$K(\varphi_n) > 0.$$

Proof. By the fact that $K^Q(\varphi_n) \to 0$, we know that

$$\lim_{n\to+\infty} \big\|\nabla\varphi_n\big\|_{L^2}^2 = 0.$$

Then by the Sobolev and Gagliardo-Nirenberg inequalities, we have for large n

$$\big\|\varphi_n\big\|_{L_x^{2^*}}^{2^*} \lesssim \big\|\nabla\varphi_n\big\|_{L_x^2}^{2^*} = o(\big\|\nabla\varphi_n\big\|_{L^2}^2),$$

$$\left\|\varphi_n\right\|_{L_x^{\frac{2d+2}{d-1}}}^{\frac{2d+2}{d-1}} \lesssim \left\|\varphi_n\right\|_{L^2}^{\frac{2}{d-1}} \left\|\nabla\varphi_n\right\|_{L^2}^{\frac{2d}{d-1}} = o(\left\|\nabla\varphi_n\right\|_{L^2}^2),$$

where we use the boundedness of $\left\|\varphi_n\right\|_{L^2}$. Hence for large n, we have

$$K(\varphi_n) = \int_{\mathbb{R}^d} \left(2|\nabla\varphi_n|^2 - 2|\varphi_n|^{2^*} + \frac{2d}{d+1}|\varphi_n|^{\frac{2d+2}{d-1}}\right) dx \approx \int_{\mathbb{R}^d} |\nabla\varphi_n|^2 \, dx > 0.$$

This concludes the proof. □

By the definition of K, we denote two real numbers

$$\bar{\mu} = \max\left\{4, 0, \frac{4d}{d-1}\right\} = \frac{4d}{d-1}, \quad \underline{\mu} = \min\left\{4, 0, \frac{4d}{d-1}\right\} = 0.$$

Next, we show the behavior of the scaling derivative functional K with respect to the scaling $\varphi_{d,-2}^{\lambda}$.

Lemma 3.3. *For any $\varphi \in H^1$, we have*

$$(\bar{\mu} - \mathcal{L})\, E(\varphi) = \int_{\mathbb{R}^d} \left(\frac{2}{d-1}\left|\nabla\varphi\right|^2 + \frac{2}{d-1}\left|\varphi\right|^{2^*}\right) dx,$$

$$\mathcal{L}\,(\bar{\mu} - \mathcal{L})\, E(\varphi) = \int_{\mathbb{R}^d} \left(\frac{8}{d-1}\left|\nabla\varphi\right|^2 + \frac{8d}{(d-1)(d-2)}\left|\varphi\right|^{2^*}\right) dx,$$

where the scaling derivative $\mathcal{L}$ is defined by (1.3).

Proof. By the definition of $\mathcal{L}$, we have

$$\mathcal{L}\left\|\nabla\varphi\right\|_{L^2}^2 = 4\left\|\nabla\varphi\right\|_{L^2}^2, \quad \mathcal{L}\left\|\varphi\right\|_{L^{2^*}}^{2^*} = \frac{4d}{d-2}\left\|\varphi\right\|_{L^{2^*}}^{2^*}, \quad \mathcal{L}\left\|\varphi\right\|_{L^{\frac{2d+2}{d-1}}}^{\frac{2d+2}{d-1}} = \frac{4d}{d-1}\left\|\varphi\right\|_{L^{\frac{2d+2}{d-1}}}^{\frac{2d+2}{d-1}}$$

which implies that

$$\begin{aligned}
(\bar{\mu} - \mathcal{L})\, E(\varphi) &= \bar{\mu}E(\varphi) - K(\varphi) \\
&= \int_{\mathbb{R}^d} \left(\frac{2}{d-1}\left|\nabla\varphi\right|^2 + \frac{2}{d-1}\left|\varphi\right|^{2^*}\right) dx, \\
\mathcal{L}\,(\bar{\mu} - \mathcal{L})\, E(\varphi) &= \frac{2}{d-1}\mathcal{L}\left\|\nabla\varphi\right\|_{L^2}^2 + \frac{2}{d-1}\mathcal{L}\left\|\varphi\right\|_{L^{2^*}}^{2^*} \\
&= \int_{\mathbb{R}^d} \left(\frac{8}{d-1}\left|\nabla\varphi\right|^2 + \frac{8d}{(d-1)(d-2)}\left|\varphi\right|^{2^*}\right) dx.
\end{aligned}$$

This completes the proof. □

According to the above analysis, we will replace the functional E in (1.4) with a positive functional H, while extending the minimizing region from "the mountain ridge $K(\varphi) = 0$" to "the mountain flank $K(\varphi) \leq 0$". Let

$$H(\varphi) := \left(1 - \frac{\mathcal{L}}{\bar{\mu}}\right) E(\varphi) = \int_{\mathbb{R}^d} \left(\frac{1}{2d}\left|\nabla\varphi\right|^2 + \frac{1}{2d}\left|\varphi\right|^{2^*}\right) dx,$$

then for any $\varphi \in H^1\backslash\{0\}$, we have

$$H(\varphi) > 0, \quad \mathcal{L}H(\varphi) \geq 0.$$

Now by the similar argument to that in [42], we can characterization the minimization problem (1.4) by making use of H.

Lemma 3.4 ([42, Lemma 2.9]). *For the minimization m in (1.4), we have*

$$\begin{aligned} m &= \inf\{H(\varphi) : \varphi \in H^1(\mathbb{R}^d),\ \varphi \neq 0,\ K(\varphi) \leq 0\} \\ &= \inf\{H(\varphi) : \varphi \in H^1(\mathbb{R}^d),\ \varphi \neq 0,\ K(\varphi) < 0\}. \end{aligned} \tag{3.2}$$

Next we will use the ($\dot{H}^1$-invariant) scaling argument to remove the $\dot{H}^1$-subcritica growth term

$$\int_{\mathbb{R}^d} |\varphi|^{\frac{2d+2}{d-1}}\, dx \text{ in } K,$$

that is, to replace the constrained condition $K(\varphi) < 0$ with $K^c(\varphi) < 0$, where

$$K^c(\varphi) := \int_{\mathbb{R}^d} \left(2|\nabla\varphi|^2 - 2|\varphi|^{2^*}\right) dx.$$

In fact, we have

Lemma 3.5 ([42, Lemma 2.10]). *For the minimization m in (1.4), we have*

$$\begin{aligned} m &= \inf\{H(\varphi) : \varphi \in H^1(\mathbb{R}^d),\ \varphi \neq 0,\ K^c(\varphi) < 0\} \\ &= \inf\{H(\varphi) : \varphi \in H^1(\mathbb{R}^d),\ \varphi \neq 0,\ K^c(\varphi) \leq 0\}. \end{aligned}$$

The above result holds for the defocusing perturbation, which implies that $K^c(\varphi) \leq K(\varphi)$. While the argument does not hold for the focusing perturbation from the proof in [42]. Please refer to [1, 2] for the related discussions. After these preparations, we can now make use of the sharp constant of the Sobolev inequality in [3, 48] to compute the minimization m as following.

Lemma 3.6 ([42, Lemma 2.11]). *For the minimization m in (1.4), we have*

$$m = E^c(W).$$

After the computation of the minimization m in (1.4), we now give some useful variational estimates.

Lemma 3.7 ([42, Lemma 5.4]). *Let $k \in \mathbb{N}$ and $\varphi_0, \ldots, \varphi_k \in H^1(\mathbb{R}^d)$. Assume that there exist some δ, $\varepsilon > 0$ with $(3d-1)\varepsilon < 2d\delta$ such that*

$$\sum_{j=0}^{k} E(\varphi_j) - \varepsilon \leq E\left(\sum_{j=0}^{k} \varphi_j\right) < m - \delta, \ \text{and} \ -\varepsilon \leq K\left(\sum_{j=0}^{k} \varphi_j\right) \leq \sum_{j=0}^{k} K(\varphi_j) + \varepsilon.$$

Then $\varphi_j \in \mathcal{K}^+$ for all $j = 0, \ldots, k$.

Lemma 3.8. *For $d \geq 5$ and any $\varphi \in H^1$ with $K(\varphi) \geq 0$, we have*

$$\int_{\mathbb{R}^d} \left(\frac{1}{2d} |\nabla \varphi|^2 + \frac{1}{2d} |\varphi|^{2^*} \right) dx \leq E(\varphi) \leq \int_{\mathbb{R}^d} \left(\frac{1}{2} |\nabla \varphi|^2 + \frac{d-1}{2d+2} |\varphi|^{\frac{2d+2}{d-1}} \right) dx. \tag{3.3}$$

Proof. On one hand, the second inequality in (3.3) is trivial. On the other hand, by the definition of E and K, we have

$$E(\varphi) = \int_{\mathbb{R}^d} \left(\frac{1}{2d} |\nabla \varphi|^2 + \frac{1}{2d} |\varphi|^{2^*} \right) dx + \frac{d-1}{4d} K(\varphi),$$

which implies the first inequality in (3.3). □

At the last of this part, we give the uniform bounds on the scaling derivative functional $K(\varphi)$ with the energy $E(\varphi)$ below the threshold m, which plays an important role for the blow-up and scattering analysis in Section 4 and Section 6.

Lemma 3.9 ([42, Lemma 2.13]). *For any $\varphi \in H^1$ with $E(\varphi) < m$, then there exists a constant $\delta > 0$ such that*

1. *If $K(\varphi) < 0$, then*

$$K(\varphi) \leq -\bar{\mu}\big(m - E(\varphi)\big). \tag{3.4}$$

2. *If $K(\varphi) \geq 0$, then*

$$K(\varphi) \geq \min\left(\bar{\mu}\Big(m - E(\varphi)\Big), \frac{2}{2d-3} \big\|\nabla \varphi\big\|_{L^2}^2 + \frac{2d}{(d+1)(2d-3)} \big\|\varphi\big\|_{L^{\frac{2d+2}{d-1}}}^{\frac{2d+2}{d-1}} \right). \tag{3.5}$$

Proof. By Lemma 3.3, for any $\varphi \in H^1$, we have

$$\mathcal{L}^2 E(\varphi) = \bar{\mu} \mathcal{L} E(\varphi) - \frac{8}{d-1} \big\|\nabla \varphi\big\|_{L^2}^2 - \frac{8d}{(d-1)(d-2)} \big\|\varphi\big\|_{L^{2^*}}^{2^*}.$$

Let $j(\lambda) = E(\varphi_{d,-2}^{\lambda})$, then we have

$$j''(\lambda) = \bar{\mu} j'(\lambda) - \frac{8e^{4\lambda}}{d-1} \big\|\nabla \varphi\big\|_{L^2}^2 - \frac{8de^{\frac{4d}{d-2}\lambda}}{(d-1)(d-2)} \big\|\varphi\big\|_{L^{2^*}}^{2^*}. \tag{3.6}$$

Case 1: if $K(\varphi) < 0$, then by (3.1), Lemma 3.2 and the continuity of K in λ, there exists a negative number $\lambda_0 < 0$ such that $K(\varphi_{d,-2}^{\lambda_0}) = 0$, and

$$K(\varphi_{d,-2}^{\lambda}) < 0, \quad \forall\ \lambda \in (\lambda_0, 0).$$

By (1.4), we obtain $E(\varphi_{d,-2}^{\lambda_0}) \geq m$. Now by integrating (3.6) over $[\lambda_0, 0]$, we have

$$\int_{\lambda_0}^{0} j''(\lambda)\, d\lambda \leq \bar{\mu} \int_{\lambda_0}^{0} j'(\lambda)\, d\lambda,$$

which implies that

$$K(\varphi) = j'(0) - j'(\lambda_0) \leq \bar{\mu}\,(j(0) - j(\lambda_0)) \leq -\bar{\mu}(m - E(\varphi)),$$

which implies (3.4).

Case 2: $K(\varphi) \geq 0$. We divide it into two subcases:

When

$$2\bar{\mu}K(\varphi) \geq \frac{8d}{(d-1)(d-2)}\|\varphi\|_{L^{2^*}}^{2^*}.$$

Since

$$\begin{aligned}\frac{8d}{(d-1)(d-2)}\|\varphi\|_{L^{2^*}}^{2^*} = &-\frac{4d}{(d-1)(d-2)}K(\varphi)\\ &+\int_{\mathbb{R}^d}\Big(\frac{8d}{(d-1)(d-2)}|\nabla\varphi|^2\\ &+\frac{8d^2}{(d+1)(d-1)(d-2)}|\varphi|^{\frac{2d+2}{d-1}}\Big)\,dx,\end{aligned}$$

then we have

$$\begin{aligned}2\bar{\mu}K(\varphi) \geq &-\frac{4d}{(d-1)(d-2)}K(\varphi)\\ &+\int_{\mathbb{R}^d}\Big(\frac{8d}{(d-1)(d-2)}|\nabla\varphi|^2+\frac{8d^2}{(d+1)(d-1)(d-2)}|\varphi|^{\frac{2d+2}{d-1}}\Big)\,dx,\end{aligned}$$

which implies that

$$K(\varphi) \geq \frac{2}{2d-3}\|\nabla\varphi\|_{L^2}^2 + \frac{2d}{(d+1)(2d-3)}\|\varphi\|_{L^{\frac{2d+2}{d-1}}}^{\frac{2d+2}{d-1}}.$$

When

$$2\bar{\mu}K(\varphi) \leq \frac{8d}{(d-1)(d-2)}\|\varphi\|_{L^{2^*}}^{2^*}.$$

By (3.6), we have for $\lambda = 0$,

$$\begin{aligned}0 <&2\bar{\mu}j'(\lambda) < \frac{8de^{\frac{4d}{d-2}\lambda}}{(d-1)(d-2)}\|\varphi\|_{L^{2^*}}^{2^*},\\ j''(\lambda) =&\bar{\mu}j'(\lambda) - \frac{8e^{4\lambda}}{d-1}\|\nabla\varphi\|_{L^2}^2 - \frac{8de^{\frac{4d}{d-2}\lambda}}{(d-1)(d-2)}\|\varphi\|_{L^{2^*}}^{2^*} \leq -\bar{\mu}j'(\lambda).\end{aligned}\tag{3.7}$$

By the continuity of j' and j'' in λ, we know that j' is an accelerated decreasing function as λ increases until $j'(\lambda_0) = 0$ for some finite number $\lambda_0 > 0$ and (3.7) holds on $[0, \lambda_0]$.

By $K(\varphi_{d,-2}^{\lambda_0}) = j'(\lambda_0) = 0$, we know that

$$E(\varphi_{d,-2}^{\lambda_0}) \geq m.$$

Now integrating (3.7) over $[0, \lambda_0]$, we obtain that

$$-K(\varphi) = j'(\lambda_0) - j'(0) \leq -\bar{\mu}\big(j(\lambda_0) - j(0)\big) \leq -\bar{\mu}(m - E(\varphi)).$$

This completes the proof. □

4 Part I: blow up for $\mathcal{K}^-$

In this section, we prove the blow-up result of Theorem 1.2 in the case that u_0 is radial. The case $xu_0 \in L^2$ is trivial. We can also refer to [42] for the similar discussions to the case $d = 3$. The spatial localization argument comes from [46]. Now let ϕ be a smooth, radial function satisfying $\partial_r^2\phi(r) \leq 2$, $\phi(r) = r^2$ for $r \leq 1$, and $\phi(r)$ is constant for $r \geq 3$. For some R, we define

$$V_R(t) := \int_{\mathbb{R}^d} \phi_R(x)|u(t,x)|^2\,dx, \quad \phi_R(x) = R^2\phi\left(\frac{|x|}{R}\right).$$

By Lemma 2.7, $\Delta\phi_R(r) = 2d$ for $r \leq R$, and $\Delta^2\phi_R(r) = 0$ for $r \leq R$, we have

$$\begin{aligned}
\partial_t^2 V_R(t) =& 4\int_{\mathbb{R}^d} \partial_{ij}(\phi_R) u_i(t,x)\bar{u}_j(t,x)\,dx - \int_{\mathbb{R}^d} \Delta^2\phi_R\big|u(t,x)\big|^2\,dx \\
& - \frac{4}{d}\int_{\mathbb{R}^d} \Delta\phi_R\big|u(t,x)\big|^{2^*}\,dx + \frac{4}{d+1}\int_{\mathbb{R}^3} \Delta\phi_R\big|u(t,x)\big|^{\frac{2d+2}{d-1}}\,dx \\
\leq & 4\int_{\mathbb{R}^d} \left(2|\nabla u(t)|^2 - 2|u(t)|^{2^*} + \frac{2d}{d+1}|u(t)|^{\frac{2d+2}{d-1}}\right)\,dx \\
& + \frac{c}{R^2}\int_{R\leq|x|\leq 3R} \big|u(t)\big|^2\,dx + c\int_{R\leq|x|\leq 3R} \left(\big|u(t)\big|^{\frac{2d+2}{d-1}} + \big|u(t)\big|^{2^*}\right)\,dx.
\end{aligned}$$

By the radial Sobolev inequality, we have

$$\big\|f\big\|_{L^\infty(|x|\geq R)} \leq \frac{c}{R^{(d-1)/2}}\big\|f\big\|_{L^2(|x|\geq R)}^{1/2}\big\|\nabla f\big\|_{L^2(|x|\geq R)}^{1/2}.$$

Therefore, by the mass conservation and Young's inequality, we know that for any $\epsilon > 0$ there exist sufficiently large R such that

$$\begin{aligned}
&\partial_t^2 V_R(t) \\
\leq & 4K(u(t)) + \epsilon\big\|\nabla u(t)\big\|_{L^2}^2 + \epsilon^2 \\
= & \frac{16d}{d-2}E(u) - \left(\frac{16}{d-2} - \epsilon\right)\big\|\nabla u(t)\big\|_{L^2}^2 - \frac{8d}{(d+1)(d-2)}\big\|u(t)\big\|_{L^{\frac{2d+2}{d-1}}}^{\frac{2d+2}{d-1}} + \epsilon^2. \quad (4.1)
\end{aligned}$$

By $K(u) < 0$, the mass and energy conservations, Lemma 3.9 and the continuity argument, we know that for any $t \in I$, we have

$$K(u(t)) \leq -\bar{\mu}\,(m - E(u(t))) < 0.$$

By Lemma 3.4, we have

$$m \leq H(u(t)) < \frac{1}{d}\big\|u(t)\big\|_{L^{2^*}}^{2^*},$$

where we have used the fact that $K(u(t)) < 0$ in the second inequality. By the fact $m = \frac{1}{d}\left(C_d^*\right)^{-d}$ and the sharp Sobolev inequality, we have

$$\big\|\nabla u(t)\big\|_{L^2}^{2^*} \geq \left(C_d^*\right)^{-2^*}\big\|u(t)\big\|_{L^{2^*}}^{2^*} > (dm)^{\frac{d}{d-2}},$$

which implies that $\|\nabla u(t)\|_{L^2}^2 > dm$.

In addition, by $E(u_0) < m$ and energy conservation, there exists $\delta_1 > 0$ such that $E(u(t)) \le (1-\delta_1)m$. Thus, if we choose ϵ sufficiently small, we have

$$\partial_t^2 V_R(t) \le \frac{16d}{d-2}(1-\delta_1)m - d(\frac{16}{d-2} - \epsilon)m + \epsilon^2 \le -\frac{8d}{d-2}\delta_1 m,$$

which implies that u must blow up at finite time. □

5 Profile decomposition

In this part, we will use the method in [4, 23, 27, 42] to show the linear and nonlinear profile decompositions of the H^1-bounded sequences of solutions of (1.1) in five and higher dimensions, which will be used to construct the critical element (minimal energy non-scattering solution) and show its properties, especially the compactness and regularity. In order to do it, we cite the similar notation to those in [23, 42]. Now we introduce the complex-valued function $\vec{v}(t,x)$ by

$$\vec{v}(t,x) = \langle\nabla\rangle v(t,x), \quad v(t,x) = \langle\nabla\rangle^{-1}\vec{v}(t,x).$$

Given $(t_n^j, x_n^j, h_n^j) \in \mathbb{R}\times\mathbb{R}^d\times(0,1]$, let τ_n^j, T_n^j denote the scaled time drift, the unitary operator in $L^2(\mathbb{R}^d)$, defined by

$$\tau_n^j = -\frac{t_n^j}{\left(h_n^j\right)^2}, \quad T_n^j\varphi(x) = \frac{1}{(h_n^j)^{d/2}}\varphi\left(\frac{x-x_n^j}{h_n^j}\right).$$

5.1 Linear profile decomposition

By the similar arguments to that in [42], we can establish that

Proposition 5.1 ([42, Proposition 5.1 and Lemma 5.3]). *Let*

$$\vec{v}_n(t,x) = e^{it\Delta}\vec{v}_n(0)$$

be a sequence of the free Schrödinger solutions with bounded L^2 norm. Then up to a subsequence, there exist $K \in \{0,1,2,\ldots,\infty\}$, $\{\varphi^j\}_{j\in[0,K)} \subset L^2(\mathbb{R}^d)$ and $\{t_n^j, x_n^j, h_n^j\}_{n\in\mathbb{N}} \subset \mathbb{R}\times\mathbb{R}^d\times(0,1]$ satisfying

$$\vec{v}_n(t,x) = \sum_{j=0}^{k-1}\vec{v}_n^j(t,x) + \vec{w}_n^k(t,x), \tag{5.1}$$

where $\vec{v}_n^j(t,x) = e^{i(t-t_n^j)\Delta}T_n^j\varphi^j$, and

$$\lim_{k\to K}\overline{\lim_{n\to+\infty}}\left\|\vec{w}_n^k\right\|_{L_t^\infty(\mathbb{R};B_{\infty,\infty}^{-d/2}(\mathbb{R}^d))} = 0, \tag{5.2}$$

and for any $l < j < k \le K$,

$$\lim_{n\to+\infty}\left(\log\left|\frac{h_n^j}{h_n^l}\right| + \left|\frac{t_n^j - t_n^l}{(h_n^l)^2}\right| + \left|\frac{x_n^j - x_n^l}{h_n^l}\right|\right) = \infty, \tag{5.3}$$

$$\lim_{k\to K}\overline{\lim_{n\to+\infty}}\left|M(v_n(0))-\sum_{j=0}^{k-1}M(v_n^j(0))-M(w_n^k(0))\right|=0, \tag{5.4}$$

$$\lim_{k\to K}\overline{\lim_{n\to+\infty}}\left|E(v_n(0))-\sum_{j=0}^{k-1}E(v_n^j(0))-E(w_n^k(0))\right|=0, \tag{5.5}$$

$$\lim_{k\to K}\overline{\lim_{n\to+\infty}}\left|K(v_n(0))-\sum_{j=0}^{k-1}K(v_n^j(0))-K(w_n^k(0))\right|=0. \tag{5.6}$$

Moreover, each sequence $\{h_n^j\}_{n\in\mathbb{N}}$ *is either going to* 0 *or identically* 1 *for all* n.

We call $\vec{v}_n^j$ and $\vec{w}_n^k$ the free concentrating wave and the remainder, respectively. According to the above result and Lemma 3.7, we conclude

Proposition 5.2 ([42, Proposition 5.5]). *Let* $\vec{v}_n(t,x)$ *be a sequence of the free Schrödinger solutions satisfying* $v_n(0)\in\mathcal{K}^+$ *and* $E(v_n(0))<m$. *Let*

$$\vec{v}_n(t,x)=\sum_{j=0}^{k-1}\vec{v}_n^j(t,x)+\vec{w}_n^k(t,x)$$

be the linear profile decomposition given by Proposition 5.1. *Then for large* n *and all* $j<K$, *we have*

$$v_n^j(0)\in\mathcal{K}^+,\quad w_n^K(0)\in\mathcal{K}^+,$$

such that (5.4)–(5.6). *Moreover for all* $j<K$, *we have*

$$0\le\lim_{n\to+\infty}E(v_n^j(0))\le\overline{\lim_{n\to+\infty}}E(v_n^j(0))\le\overline{\lim_{n\to+\infty}}E(v_n(0)),$$

where the last inequality becomes equality only if $K=1$ *and* $w_n^1\to0$ *in* $L_t^\infty\dot{H}_x^1$.

5.2 Nonlinear profile decomposition

After the linear profile decomposition of a sequence of initial data in the last subsection, we now give the nonlinear profile decomposition of a sequence of solutions of (1.1) with the same initial data in the energy space $H^1(\mathbb{R}^d)$. The procedure is the same as the 3d case in [42].

Let $v_n(t,x)$ be a sequence of solutions for the free Schrödinger equation with initial data in $\mathcal{K}^+$, that is, $v_n\in H^1(\mathbb{R}^d)$ and

$$(i\partial_t+\Delta)v_n=0,\quad v_n(0)\in\mathcal{K}^+.$$

Let

$$\vec{v}_n(t,x)=\langle\nabla\rangle v_n(t,x),$$

then by Proposition 5.1, we have a sequence of the free concentrating wave $\vec{v}_n^j(t,x)$ with $\vec{v}_n^j(t_n^j) = T_n^j\varphi^j$, $v_n^j(0) \in \mathcal{K}^+$ for $j = 0, \ldots, K$, such that

$$\vec{v}_n(t,x) = \sum_{j=0}^{k-1} \vec{v}_n^j(t,x) + \vec{w}_n^k(t,x) = \sum_{j=0}^{k-1} e^{i(t-t_n^j)\Delta} T_n^j \varphi^j + \vec{w}_n^k$$
$$= \sum_{j=0}^{k-1} T_n^j e^{i\left(\frac{t-t_n^j}{(h_n^j)^2}\right)\Delta} \varphi^j + \vec{w}_n^k.$$

Now for any concentrating wave $\vec{v}_n^j$, $j = 0, \ldots, K$, we undo the group action, i.e., the scaling and translation transformation T_n^j, to look for the linear profile V^j. Let

$$\vec{v}_n^j(t,x) = T_n^j \vec{V}^j\left(\frac{t-t_n^j}{(h_n^j)^2}\right),$$

then we have

$$(i\partial_t + \Delta)\vec{V}^j = 0, \quad \vec{V}^j(0) = \varphi^j.$$

Now let $u_n^j(t,x)$ be the nonlinear solution of (1.1) with initial data $v_n^j(0)$, that is

$$(i\partial_t + \Delta)\vec{u}_n^j(t,x) = \langle\nabla\rangle f_1(\langle\nabla\rangle^{-1}\vec{u}_n^j) + \langle\nabla\rangle f_2(\langle\nabla\rangle^{-1}\vec{u}_n^j),$$
$$\vec{u}_n^j(0) = \vec{v}_n^j(0) = T_n^j\vec{V}^j(\tau_n^j), \quad u_n^j(0) \in \mathcal{K}^+,$$

where $\tau_n^j = -t_n^j/(h_n^j)^2$. In order to look for the nonlinear profile $\vec{U}_\infty^j$ associated to the free concentrating wave $(\vec{v}_n^j;\, h_n^j, t_n^j, x_n^j)$, we also need undo the group action. We denote

$$\vec{u}_n^j(t,x) = T_n^j\vec{U}_n^j\left(\frac{t-t_n^j}{(h_n^j)^2}\right),$$

then we have

$$(i\partial_t + \Delta)\vec{U}_n^j = \left(\langle\nabla\rangle_n^j\right) f_1\left(\left(\langle\nabla\rangle_n^j\right)^{-1}\vec{U}_n^j\right)$$
$$+ (h_n^j)^{\frac{2}{d-1}} \cdot \left(\langle\nabla\rangle_n^j\right) f_2\left(\left(\langle\nabla\rangle_n^j\right)^{-1}\vec{U}_n^j\right),$$
$$\vec{U}_n^j(\tau_n^j) = \vec{V}^j(\tau_n^j).$$

Up to a subsequence, we may assume that there exist $h_\infty^j \in \{0,1\}$ and $\tau_\infty^j \in [-\infty,\infty]$ for every $j = \{0,\ldots,K\}$, such that

$$h_n^j \to h_\infty^j, \quad \text{and } \tau_n^j \to \tau_\infty^j.$$

As $n \to +\infty$, the limit equation of $\vec{U}_n^j$ is given by

$$\begin{aligned}(i\partial_t + \Delta)\,\vec{U}_\infty^j &= \left(\langle\nabla\rangle_\infty^j\right) f_1\left(\left(\langle\nabla\rangle_\infty^j\right)^{-1}\vec{U}_\infty^j\right)\\ &\quad + \left(h_\infty^j\right)^{\frac{2}{d-1}}\cdot\left(\langle\nabla\rangle_\infty^j\right) f_2\left(\left(\langle\nabla\rangle_\infty^j\right)^{-1}\vec{U}_\infty^j\right),\\ \vec{U}_\infty^j(\tau_\infty^j) &= \vec{V}^j(\tau_\infty^j) \in L^2(\mathbb{R}^d).\end{aligned}$$

Let

$$\widehat{U}_\infty^j := \left(\langle\nabla\rangle_\infty^j\right)^{-1}\vec{U}_\infty^j,$$

then

$$\begin{aligned}(i\partial_t + \Delta)\,\widehat{U}_\infty^j &= f_1\left(\widehat{U}_\infty^j\right) + \left(h_\infty^j\right)^{\frac{2}{d-1}}\cdot f_2\left(\widehat{U}_\infty^j\right),\\ \widehat{U}_\infty^j(\tau_\infty^j) &= \left(\langle\nabla\rangle_\infty^j\right)^{-1}\vec{V}^j(\tau_\infty^j).\end{aligned}$$

The unique existence of a local solution $\vec{U}_\infty^j$ around τ_∞^j is known in all cases, including $h_\infty^j = 0$ and $\tau_\infty^j = \pm\infty$. $\vec{U}_\infty^j$ on the maximal existence interval is called the nonlinear profile associated with the free concentrating wave $\left(\vec{v}_n^j;\ h_n^j, t_n^j, x_n^j\right)$.

The nonlinear concentrating wave $u_{(n)}^j$ associated with $\left(\vec{v}_n^j;\ h_n^j, t_n^j, x_n^j\right)$ is defined by

$$\vec{u}_{(n)}^j(t,x) = T_n^j\vec{U}_\infty^j\left(\frac{t - t_n^j}{(h_n^j)^2}\right),$$

then we have

$$\begin{aligned}(i\partial_t + \Delta)\,\vec{u}_{(n)}^j &= \langle\nabla\rangle_\infty^j f_1\left(\left(\langle\nabla\rangle_\infty^j\right)^{-1}\vec{u}_{(n)}^j\right)\\ &\quad + \left(h_\infty^j\right)^{\frac{2}{d-1}}\cdot\langle\nabla\rangle_\infty^j f_2\left(\left(\langle\nabla\rangle_\infty^j\right)^{-1}\vec{u}_{(n)}^j\right),\\ \vec{u}_{(n)}^j(0) &= T_n^j\vec{U}_\infty^j(\tau_n^j),\end{aligned}$$

which implies that

$$\begin{aligned}\left\|\vec{u}_{(n)}^j(0) - \vec{u}_n^j(0)\right\|_{L^2} &= \left\|T_n^j\vec{U}_\infty^j(\tau_n^j) - T_n^j\vec{V}^j(\tau_n^j)\right\|_{L^2} = \left\|\vec{U}_\infty^j(\tau_n^j) - \vec{V}^j(\tau_n^j)\right\|_{L^2}\\ &\le \left\|\vec{U}_\infty^j(\tau_n^j) - \vec{U}_\infty^j(\tau_\infty^j)\right\|_{L^2} + \left\|\vec{V}^j(\tau_n^j) - \vec{V}^j(\tau_\infty^j)\right\|_{L^2} \to 0.\end{aligned}$$

We denote

$$\vec{u}_{(n)}^j = \langle\nabla\rangle\, u_{(n)}^j.$$

If $h_\infty^j = 1$, we have $h_n^j \equiv 1$, then $u_{(n)}^j \in H^1(\mathbb{R}^d)$ and satisfies

$$(i\partial_t + \Delta)\,u_{(n)}^j = f_1(u_{(n)}^j) + f_2(u_{(n)}^j).$$

If $h_\infty^j = 0$, then $u_{(n)}^j \in H^1(\mathbb{R}^d)$ satisfies

$$(i\partial_t + \Delta)\, u_{(n)}^j = \frac{|\nabla|}{\langle\nabla\rangle} f_1\left(\frac{\langle\nabla\rangle}{|\nabla|} u_{(n)}^j\right).$$

Let u_n be a sequence of (local) solutions of (1.1) with initial data in $\mathcal{K}^+$ at $t = 0$, and let v_n be the sequence of the free solutions with the same initial data. We consider the linear profile decomposition given by Proposition 5.1

$$\vec{v}_n(t,x) = \sum_{j=0}^{k-1} \vec{v}_n^j(t,x) + \vec{w}_n^k(t,x), \quad \vec{v}_n^j(t_n^j) = T_n^j \varphi^j, \quad v_n^j(0) \in \mathcal{K}^+.$$

With each free concentrating wave $\{\vec{v}_n^j\}_{n\in\mathbb{N}}$, we associate the nonlinear concentrating wave $\{\vec{u}_{(n)}^j\}_{n\in\mathbb{N}}$. A nonlinear profile decomposition of u_n is given by

$$\vec{u}_{(n)}^{<k}(t,x) := \sum_{j=0}^{k-1} \vec{u}_{(n)}^j(t,x) = \sum_{j=0}^{k-1} T_n^j \vec{U}_\infty^j\left(\frac{t-t_n^j}{(h_n^j)^2}\right). \tag{5.7}$$

Since the smallness condition (5.2) and the orthogonality condition (5.3) ensure that every nonlinear concentrating wave and the remainder interacts weakly with the others, we will show that $\vec{u}_{(n)}^{<k} + \vec{w}_n^k$ is a good approximation for $\vec{u}_n$ provided that each nonlinear profile has the finite global Strichartz norm.

Now we define the Strichartz norms. First let $ST(I)$ and $ST^*(I)$ be the function spaces on $I \times \mathbb{R}^d$ defined as Section 2.3

$$ST(I) := W_1(I) \cap W_2(I), \quad ST^*(I) := L_t^2\left(I; B^{\frac{2}{d}}_{\frac{2d^2}{d^2+4},2}\right).$$

The Strichartz norm for the nonlinear profile $\widehat{U}_\infty^j$ depends on the scaling h_∞^j.

$$ST_\infty^j(I) := \begin{cases} W_1(I) \cap W_2(I) & \text{for } h_\infty^j = 1, \\ W_1(I) & \text{for } h_\infty^j = 0. \end{cases}$$

By the similar arguments to that in [23, 42], we have

Lemma 5.3 ([42, Lemma 5.6]). *In the nonlinear profile decomposition* (5.7). *Suppose that for each $j < K$, we have*

$$\|\widehat{U}_\infty^j\|_{ST_\infty^j(\mathbb{R})} + \|\vec{U}_\infty^j\|_{L_t^\infty L_x^2(\mathbb{R}^d)} < \infty.$$

Then for any finite interval I, any $j < K$ and any $k \leq K$, we have

$$\varlimsup_{n\to+\infty} \|u_{(n)}^j\|_{ST(I)} \lesssim \|\widehat{U}_\infty^j\|_{ST_\infty^j(\mathbb{R})}, \tag{5.8}$$

$$\varlimsup_{n\to+\infty} \|u_{(n)}^{<k}\|_{ST(I)}^2 \lesssim \varlimsup_{n\to+\infty} \sum_{j<k} \|u_{(n)}^j\|_{ST(I)}^2, \tag{5.9}$$

where the implicit constants do not depend on I, j or k. We also have

$$\lim_{n\to+\infty}\left\| f_1\left(u_{(n)}^{<k}\right) - \sum_{j<k}\frac{\langle\nabla\rangle_\infty^j}{\langle\nabla\rangle} f_1\left(\frac{\langle\nabla\rangle}{\langle\nabla\rangle_\infty^j} u_{(n)}^j\right)\right\|_{ST^*(I)} = 0, \tag{5.10}$$

$$\lim_{n\to+\infty}\left\| f_2\left(u_{(n)}^{<k}\right) - \sum_{j<k}\left(h_\infty^j\right)^{\frac{2}{d-1}}\frac{\langle\nabla\rangle_\infty^j}{\langle\nabla\rangle} f_2\left(\frac{\langle\nabla\rangle}{\langle\nabla\rangle_\infty^j} u_{(n)}^j\right)\right\|_{ST^*(I)} = 0. \tag{5.11}$$

After this preliminaries, we now show that $\vec{u}_{(n)}^{<k} + \vec{w}_n^k$ is a good approximation for $\vec{u}_n$ provided that each nonlinear profile has finite global Strichartz norm.

Proposition 5.4 ([42, Proposition 5.7]). *Let u_n be a sequence of local solutions of* (1.1) *around $t = 0$ in $\mathcal{K}^+$ satisfying*

$$M(u_n) < \infty, \quad \varlimsup_{n\to\infty} E(u_n) < m.$$

Suppose that in the nonlinear profile decomposition (5.7), *every nonlinear profile $\widehat{U}_\infty^j$ has finite global Strichartz and energy norms we have*

$$\|\widehat{U}_\infty^j\|_{ST_\infty^j(\mathbb{R})} + \|\vec{U}_\infty^j\|_{L_t^\infty L_x^2(\mathbb{R}^d)} < \infty.$$

Then u_n is bounded for large n in the Strichartz and the energy norms

$$\varlimsup_{n\to\infty}\|u_n\|_{ST(\mathbb{R})} + \|\vec{u}_n\|_{L_t^\infty L_x^2(\mathbb{R})} < \infty.$$

Proof. We only need to verify the condition of Proposition 2.6. Note that $u_{(n)}^{<k} + w_n^k$ satisfies that

$$\begin{aligned}
&(i\partial_t + \Delta)\left(u_{(n)}^{<k} + w_n^k\right)\\
&= f_1\left(u_{(n)}^{<k} + w_n^k\right) + f_2\left(u_{(n)}^{<k} + w_n^k\right) + f_1\left(u_{(n)}^{<k}\right)\\
&\quad - f_1\left(u_{(n)}^{<k} + w_n^k\right) + f_2\left(u_{(n)}^{<k}\right) - f_2\left(u_{(n)}^{<k} + w_n^k\right)\\
&\quad + \sum_{j<k}\frac{\langle\nabla\rangle_\infty^j}{\langle\nabla\rangle} f_1\left(\frac{\langle\nabla\rangle}{\langle\nabla\rangle_\infty^j} u_{(n)}^j\right) - f_1\left(u_{(n)}^{<k}\right)\\
&\quad + \sum_{j<k}\left(h_\infty^j\right)^{\frac{2}{d-1}}\frac{\langle\nabla\rangle_\infty^j}{\langle\nabla\rangle} f_2\left(\frac{\langle\nabla\rangle}{\langle\nabla\rangle_\infty^j} u_{(n)}^j\right) - f_2\left(u_{(n)}^{<k}\right).
\end{aligned}$$

First, by the construction of $\vec{u}_{(n)}^{<k}$, we know that

$$\left\|\left(\vec{u}_{(n)}^{<k}(0) + \vec{w}_n^k(0)\right) - \vec{u}_n(0)\right\|_{L_x^2} \le \sum_{j<k}\left\|\vec{u}_{(n)}^j(0) - \vec{u}_n^j(0)\right\|_{L_x^2} \to 0,$$

as $n \to +\infty$, which also implies that for large n, we have

$$\left\| \vec{u}^{<k}_{(n)}(0) + \vec{w}^k_n(0) \right\|_{L^2_x} \le E_0.$$

Next, by the linear profile decomposition in Proposition 5.1, we know that

$$\begin{aligned}\left\|u_n(0)\right\|^2_{L^2} &= \|v_n(0)\|^2_{L^2_x} = \sum_{j<k} \left\|v^j_n(0)\right\|^2_{L^2_x} + \left\|w^k_n(0)\right\|^2_{L^2_x} + o_n(1)\\ &\ge \sum_{j<k} \left\|v^j_n(0)\right\|^2_{L^2_x} + o_n(1) = \sum_{j<k} \left\|u^j_{(n)}(0)\right\|^2_{L^2_x} + o_n(1),\\ \|u_n(0)\|^2_{\dot H^1_x} &= \|v_n(0)\|^2_{\dot H^1_x} = \sum_{j<k} \left\|v^j_n(0)\right\|^2_{\dot H^1_x} + \left\|w^k_n(0)\right\|^2_{\dot H^1_x} + o_n(1)\\ &\ge \sum_{j<k} \left\|v^j_n(0)\right\|^2_{\dot H^1_x} + o_n(1) = \sum_{j<k} \left\|u^j_{(n)}(0)\right\|^2_{\dot H^1_x} + o_n(1),\end{aligned}$$

which means except for a finite set $J \subset \mathbb{N}$, the energy of $u^j_{(n)}$ with $j \notin J$ is smaller than the iteration threshold, hence we have

$$\left\|u^j_{(n)}\right\|_{ST(\mathbb{R})} \lesssim \left\|\vec{u}^j_{(n)}(0)\right\|_{L^2_x},$$

thus, for any finite interval I, by Lemma 5.3, we have

$$\begin{aligned}\sup_k \overline{\lim_{n\to+\infty}} \left\|u^{<k}_{(n)}\right\|^2_{ST(I)} &\lesssim \sup_k \overline{\lim_{n\to+\infty}} \sum_{j<k} \left\|u^j_{(n)}\right\|^2_{ST(I)}\\ &= \sup_k \overline{\lim_{n\to+\infty}} \left[\sum_{j<k, j\in J} \left\|u^j_{(n)}\right\|^2_{ST(I)} + \sum_{j<k, j\notin J} \left\|u^j_{(n)}\right\|^2_{ST(I)} \right]\\ &\lesssim \sum_{j<k, j\in J} \left\|\widehat{U}^j_\infty\right\|^2_{ST^j_\infty(I)} + \sup_k \overline{\lim_{n\to+\infty}} \sum_{j<k, j\notin J} \left\|\vec{u}^j_{(n)}(0)\right\|^2_{L^2_x}\\ &< \infty.\end{aligned}$$

This together with the Strichartz estimate for w^k_n implies that

$$\sup_k \overline{\lim_{n\to+\infty}} \left\|u^{<k}_{(n)} + w^k_n\right\|^2_{ST(I)} < \infty.$$

Last we need show the nonlinear perturbation is small in some sense. By Proposition 5.1 and Lemma 5.3, we have [1]

$$\begin{aligned}&\left\| f_1\left(u^{<k}_{(n)}\right) - f_1\left(u^{<k}_{(n)} + w^k_n\right)\right\|_{ST^*(I)} \to 0,\\ &\left\| f_2\left(u^{<k}_{(n)}\right) - f_2\left(u^{<k}_{(n)} + w^k_n\right)\right\|_{ST^*(I)} \to 0,\end{aligned}$$

[1] Since we use the full derivative $\langle \nabla \rangle$, we need use the local smoothing effect about the free solution $w(t,x)$ as that in [2, 30] to verify the weak interaction between the nonlinear concentrating waves $u^j_{(n)}(t,x)$ and the remainder $w(t,x)$.

and

$$\left\| \sum_{j<k} \frac{\langle\nabla\rangle^j_\infty}{\langle\nabla\rangle} f_1\left(\frac{\langle\nabla\rangle}{\langle\nabla\rangle^j_\infty} u^j_{(n)}\right) - f_1\left(u^{<k}_{(n)}\right)\right\|_{ST^*(I)} \to 0,$$

$$\left\| \sum_{j<k} h^j_\infty \frac{\langle\nabla\rangle^j_\infty}{\langle\nabla\rangle} f_2\left(\frac{\langle\nabla\rangle}{\langle\nabla\rangle^j_\infty} u^j_{(n)}\right) - f_2\left(u^{<k}_{(n)}\right)\right\|_{ST^*(I)} \to 0,$$

as $n \to +\infty$. Therefore, by Proposition 2.6, we can obtain the desired result, which concludes the proof. □

6 Part II: GWP and scattering for $\mathcal{K}^+$

We now use the stability analysis of the scattering solution of (1.1) and the compactness analysis of a sequence of the energy solutions of (1.1) to show the scattering result of Theorem 1.2 by contradiction.

For any finite positive number $C < \infty$, let E^*_C be the threshold for the uniform Strichartz norm bound, i.e.,

$$E^*_C := \sup\{A > 0, ST(A) < \infty\}$$

where $ST(A)$ denotes the supremum of $\|u\|_{ST(I)}$ for any strong solution u of (1.1) in $\mathcal{K}^+$ on any interval I satisfying $E(u) \le A$, $M(u) \le C$.

The small solution scattering theory gives us $E^*_C > 0$. We are going to show that $E^*_C \ge m$ by contradiction. From now on, suppose that $E^*_C \ge m$ fails, that is, we assume that

$$E^*_C < m. \tag{6.1}$$

6.1 Existence of a critical element

This part is similar to Section 6.1 in [42]. By the definition of E^*_C and the fact that $E^*_C < m$, there exist a sequence of solutions $\{u_n\}_{n\in\mathbb{N}}$ of (1.1) in $\mathcal{K}^+$, which have the maximal existence interval I_n and satisfy that for some finite number C

$$M(u_n) \le C, \quad E(u_n) \to E^*_C < m, \quad \|u_n\|_{ST(I_n)} \to +\infty, \quad \text{as } n \to +\infty,$$

then we have $\|u_n\|_{H^1} < \infty$ by Lemma 3.8. By the compact argument (profile decomposition) and the stability theory, we can show that

Theorem 6.1. *For $d \ge 5$. Let u_n be a sequence of solutions of (1.1) in $\mathcal{K}^+$ on $I_n \subset \mathbb{R}$ satisfying*

$$M(u_n) \le C, \quad E(u_n) \to E^*_C < m, \quad \|u_n\|_{ST(I_n)} \to +\infty, \quad \textit{as } n \to +\infty.$$

Then there exists a global solution u_c of (1.1) in $\mathcal{K}^+$ satisfying

$$E(u_c) = E_C^* < m, \quad K(u_c) > 0, \quad \|u_c\|_{ST(\mathbb{R})} = \infty.$$

In addition, there are a sequence $(t_n, x_n) \in \mathbb{R} \times \mathbb{R}^d$ and $\varphi \in L^2(\mathbb{R}^d)$ such that, up to a subsequence, we have as $n \to +\infty$,

$$\left\| \frac{|\nabla|}{\langle \nabla \rangle} \left(\vec{u}_n(0,x) - e^{-it_n\Delta}\varphi(x - x_n) \right) \right\|_{L^2} \to 0. \tag{6.2}$$

Proof. By the time translation symmetry of (1.1), we can translate u_n in t such that $0 \in I_n$ for all n. Then by the linear and nonlinear profile decomposition of u_n, we have

$$\begin{aligned} e^{it\Delta}\vec{u}_n(0,x) &= \sum_{j<k} \vec{v}_n^j(t,x) + \vec{w}_n^k(t,x), \quad \vec{v}_n^j(t,x) = e^{i(t-t_n^j)\Delta} T_n^j \varphi^j, \\ \vec{u}_{(n)}^{<k}(t,x) &= \sum_{j<k} \vec{u}_{(n)}^j(t,x), \quad \vec{u}_{(n)}^j(t,x) = T_n^j \vec{U}_\infty^j \left(\frac{x - x_n^j}{(h_n^j)^2} \right), \\ & \left\| \vec{u}_{(n)}^j(0) - \vec{v}_n^j(0) \right\|_{L^2} \to 0. \end{aligned}$$

By Proposition 5.2 and the following observations that

1. Every solution of (1.1) in $\mathcal{K}^+$ with the energy less than E_C^*, the mass less than C has global finite Strichartz norm by the definition of E_C^*.

2. Proposition 5.4 precludes that all the nonlinear profiles $\vec{U}_\infty^j$ have finite global Strichartz norm.

we deduce that there is only one profile and

$$E(u_{(n)}^0(0)) \to E_C^*, \quad u_{(n)}^0(0) \in \mathcal{K}^+, \quad \|\widehat{U}_\infty^0\|_{ST_\infty^0(I)} = \infty, \quad \|w_n^1\|_{L_t^\infty \dot{H}_x^1} \to 0.$$

If $h_n^0 \to 0$, then $\widehat{U}_\infty^0 = |\nabla|^{-1}\vec{U}_\infty^0$ solves the $\dot{H}^1$-critical NLS

$$(i\partial_t + \Delta)\widehat{U}_\infty^0 = f_1(\widehat{U}_\infty^0)$$

and satisfies

$$E^c\left(\widehat{U}_\infty^0(\tau_\infty^0)\right) = E_C^* < m, \;\; K^c\left(\widehat{U}_\infty^0(\tau_\infty^0)\right) \geq 0, \;\; \|\widehat{U}_\infty^0\|_{W_1(I)} = \infty.$$

However, it is in contradiction with Killip-Visan's result in [30]. Hence $h_n^0 \equiv 1$, which implies (6.2).

Now we show that $\widehat{U}_\infty^0 = \langle \nabla \rangle^{-1} \vec{U}_\infty^j$ is a global solution, which is the consequence of the compactness of (6.2). Suppose not, then we can choose a sequence $t_n \in \mathbb{R}$ which approaches the maximal existence time. Since $\widehat{U}_\infty^0(t + t_n)$ satisfies the assumption of this theorem, then applying the above argument to it, we obtain that for some $\psi \in L^2$ and another sequence $(t'_n, x'_n) \in \mathbb{R} \times \mathbb{R}^d$, as $n \to +\infty$

$$\left\| \frac{|\nabla|}{\langle \nabla \rangle} \left(\vec{U}^0_\infty(t_n) - e^{-it'_n\Delta}\psi(x - x'_n) \right) \right\|_{L^2} \to 0. \tag{6.3}$$

Let $\vec{v}(t) := e^{it\Delta}\psi$. For any $\varepsilon > 0$, there exist $\delta > 0$ with $I = [-\delta, \delta]$ such that

$$\left\| \vec{v}(t - t'_n) \right\|_{V_2(I)} \leq \varepsilon,$$

which together with (6.3) implies that for sufficiently large n

$$\left\| e^{it\Delta} \vec{U}^0_\infty(t_n) \right\|_{V_2(I)} \leq \varepsilon.$$

If ε is small enough, this implies that the solution $\widehat{U}^0_\infty$ exists on $[t_n - \delta, t_n + \delta]$ for large n by the small data theory. This contradicts the choice of t_n. Hence $\widehat{U}^0_\infty$ is a global solution and it is just the desired critical element u_c. By Proposition 1.1, we know that $K(u_c) > 0$. □

6.2 Compactness of the critical element

Since (1.1) is symmetric in t, we may assume that

$$\left\| u_c \right\|_{ST(0,+\infty)} = \infty. \tag{6.4}$$

We call it a forward critical element.

Proposition 6.2. *For $d \geq 5$. Let u_c be a forward critical element. Then there exists $x(t) : (0, \infty) \to \mathbb{R}^d$ such that the set*

$$\{u_c(t, x - x(t)); 0 < t < \infty\}$$

is precompact in $\dot{H}^s$ for any $s \in (0, 1]$.

Proof. By the conservation of the mass, it suffices to prove the precompactness of $\{u_c(t_n)\}$ in $\dot{H}^1$ for any positive time $t_1, t_2, \ldots$. Suppose t_n converges, then it is trivial from the continuity in t.

Suppose $t_n \to +\infty$. Applying Theorem 6.1 to the sequence of solutions $\vec{u}_c(t + t_n)$, we get another sequence $(t'_n, x'_n) \in \mathbb{R} \times \mathbb{R}^d$ and $\varphi \in L^2$ such that

$$\frac{|\nabla|}{\langle \nabla \rangle} \left(\vec{u}_c(t_n, x) - e^{-it'_n\Delta}\varphi(x - x'_n) \right) \to 0 \quad \text{in } L^2.$$

1. If $t'_n \to -\infty$, then we have

$$\left\| \langle \nabla \rangle^{-1} e^{it\Delta} \vec{u}_c(t_n) \right\|_{ST(0,+\infty)} = \left\| \langle \nabla \rangle^{-1} e^{it\Delta} \varphi \right\|_{ST(-t'_n,+\infty)} + o_n(1) \to 0.$$

   Hence u_c can solve (1.1) for $t > t_n$ with large n globally by iteration with small Strichartz norms, which contradicts (6.4).

2. If $t'_n \to +\infty$, then we have

$$\| \langle\nabla\rangle^{-1} e^{it\Delta}\vec{u}_c(t_n)\|_{ST(-\infty,0)} = \| \langle\nabla\rangle^{-1} e^{it\Delta}\varphi\|_{ST(-\infty,-t'_n)} + o_n(1) \to 0.$$

Hence u_c can solve (1.1) for $t < t_n$ with large n with diminishing Strichartz norms, which implies $u_c = 0$ by taking the limit, which is a contradiction.

Thus t'_n is bounded, which implies that t'_n is precompact, so is $u_c(t_n, x + x'_n)$ in $\dot{H}^1$. □

As a consequence, the energy of u_c stays within a fixed radius for all positive time, modulo arbitrarily small rest (that is, the spatial scaling function of u_c is 1). More precisely, we define the exterior energy by

$$E_{R,c}(u;t) = \int_{|x-c|\geq R} \left(\left|\nabla u(t,x)\right|^2 + \left|u(t,x)\right|^{2^*} + \left|u(t,x)\right|^{\frac{2d+2}{d-1}}\right) dx$$

for any $R > 0$ and $c \in \mathbb{R}^d$. Then we have

Corollary 6.3. *For $d \geq 5$. Let u_c be a forward critical element. then for any ε, there exist $R_0(\varepsilon) > 0$ and $x(t) : (0, +\infty) \to \mathbb{R}^d$ such that at any time $t > 0$, we have*

$$E_{R_0,x(t)}(u_c; t) \leq \varepsilon E(u_c).$$

Corollary 6.4 ([51]). *For $d \geq 5$. Let u_c be the critical element as shown in Theorem 6.1. Then for all $t \in \mathbb{R}$,*

$$\begin{aligned} u(t) &= \lim_{T\to+\infty} i \int_t^T e^{i(t-s)\Delta}\Big(f_1(u(s)) + f_2(u(s))\Big) ds \\ &= \lim_{T\to-\infty} -i \int_T^t e^{i(t-s)\Delta}\Big(f_1(u(s)) + f_2(u(s))\Big) ds \end{aligned}$$

as weak limits in $\dot{H}^s_x$ for any $s \in (0, 1]$.

6.3 Zero momentum of the critical element

Next we show that the critical element can not move with any positive speed in the sense of energy by it's Galilean invariance.

Proposition 6.5. *For $d \geq 5$. Let u_c be the critical element as shown in Theorem 6.1. then its total momentum, which is a conserved quantity, vanishes:*

$$P(u_c) := 2\Im \int_{\mathbb{R}^d} \overline{u_c(t,x)} \cdot \nabla u_c(t,x)\, dx \equiv 0.$$

Proof. We drop the subscript c for simplicity. Note that the momentum $P(u)$ and the mass $M(u)$ are finite and conserved. Moreover, $M(u) \neq 0$, otherwise u would be identically zero and not a critical element.

Let $\widetilde{u}$ be the Galilean boost of u by ξ_0, which is determined later,

$$\widetilde{u}(t,x) := e^{ix\cdot\xi_0}e^{-it|\xi_0|^2}u(t,x-2\xi_0 t),$$

then we have

$$\big\|\nabla\widetilde{u}(t)\big\|_{L^2}^2 = \big\|\nabla u(t)\big\|_{L^2}^2 + |\xi_0|^2 M(u) + \xi_0\cdot P(u).$$

Equivalently, we have $M(\widetilde{u}) = M(u)$ and

$$E(\widetilde{u}) = E(u) + \frac{1}{2}|\xi_0|^2 M(u) + \frac{1}{2}\xi_0\cdot P(u),$$
$$K(\widetilde{u}) = K(u) + 2|\xi_0|^2 M(u) + 2\xi_0\cdot P(u).$$

If $P(u)\neq 0$, choosing $\xi_0\in\mathbb{R}^d$ such that

$$-\frac{1}{2}K(u) \le |\xi_0|^2 M(u) + \xi_0\cdot P(u) < 0,$$

then we can find another critical element $\widetilde{u}$ in $\mathcal{K}^+$ with

$$M(\widetilde{u}) = M(u) \le C, \quad E(\widetilde{u}) < E(u) = E_C^*, \quad \big\|\widetilde{u}\big\|_{ST(\mathbb{R})} = +\infty,$$

which is in contradiction with the definition of E_C^*. Hence $P(u)\equiv 0$. □

6.4 Negative regularity

In this subsection, we show that

Proposition 6.6. *For $d\ge 5$. Let u_c be the critical element as shown in Theorem 6.1, then $u_c \in L^\infty \dot{H}^{-\epsilon}$ for some $\epsilon > 0$.*

Proof. We drop the subscript c. Since we have $u\in L_t^\infty H^1$, then we have $u\in L_t^\infty L_x^p$, for any $p\in[2,\frac{2d}{d-2}]$,

$$\big\|f_1(u)+f_2(u)\big\|_{L^\infty L_x^r} \lesssim \big\|u\big\|_{L_t^\infty L_x^2}\big\|u\big\|_{L_t^\infty L_x^{p_1}}^{\frac{4}{d-2}} + \big\|u\big\|_{L_t^\infty L_x^2}\big\|u\big\|_{L_t^\infty L_x^{p_2}}^{\frac{4}{d-1}}$$

where $2\le p_1,p_2\le 2d/(d-2)$ and

$$\frac{1}{r} = \frac{1}{2} + \frac{4}{d-2}\times\frac{1}{p_1} = \frac{1}{2} + \frac{4}{d-1}\times\frac{1}{p_2}.$$

Therefore for any $r\in[\frac{2(d-1)}{d+3},\frac{2d}{d+4}]$, we have

$$f_1(u)+f_2(u)\in L_t^\infty L_x^r. \tag{6.5}$$

Now from (6.5), we claim that

$$u\in L_t^\infty \dot{B}_{2,\infty}^{-s_0}, \quad \text{for any } s_0 = \frac{d}{r} - \frac{d+4}{2} \in \left[0,\frac{2}{d-1}\right], \tag{6.6}$$

which implies the negative regularity of u by interpolation. Now we shows (6.6). By the time translation symmetry, it suffices to show that $u(0) \in \dot{B}^{-s_0}_{2,\infty}$. In fact, from Corollary 6.4, we have

$$\begin{aligned}
&\left\|u_N(0)\right\|^2_{L^2}\\
&= \Big\langle i\int_0^{\infty} e^{-it\Delta}P_N\Big(f_1(u(t))+f_2(u(t))\Big)\,dt,\\
&\quad -i\int_{-\infty}^{0} e^{-i\tau\Delta}P_N\Big(f_1(u(\tau))+f_2(u(\tau)\Big)\,d\tau\Big\rangle\\
&\le \int_0^{\infty}\int_{-\infty}^{0}\Big|\Big\langle P_N\Big(f_1(u(t))+f_2(u(t))\Big),\ e^{i(t-\tau)\Delta}P_N\Big(f_1(u(\tau))+f_2(u(\tau)\Big)\Big\rangle\Big|\,dtd\tau.
\end{aligned}$$

On one hand, by the dispersive estimate of $e^{it\Delta}$, we have

$$\begin{aligned}
&\Big|\Big\langle P_N\Big(f_1(u(t))+f_2(u(t))\Big),\ e^{i(t-\tau)\Delta}P_N\Big(f_1(u(\tau))+f_2(u(\tau)\Big)\Big\rangle\Big|\\
&\lesssim \Big\|P_N\Big(f_1(u(t))+f_2(u(t))\Big)\Big\|_{L^r_x}\Big\|e^{i(t-\tau)\Delta}P_N\Big(f_1(u(\tau))+f_2(u(\tau)\Big)\Big\|_{L^{r'}_x}\\
&\lesssim |t-\tau|^{d\left(\frac{1}{2}-\frac{1}{r}\right)}\big\|f_1(u(t))+f_2(u(t))\big\|_{L^r_x}\big\|f_1(u(\tau))+f_2(u(\tau))\big\|_{L^r_x}.
\end{aligned}$$

On the other hand, by Bernstein's inequality, we have

$$\begin{aligned}
&\Big|\Big\langle P_N\Big(f_1(u(t))+f_2(u(t))\Big),\ e^{i(t-\tau)\Delta}P_N\Big(f_1(u(\tau))+f_2(u(\tau)\Big)\Big\rangle\Big|\\
&\lesssim \Big\|P_N\Big(f_1(u(t))+f_2(u(t))\Big)\Big\|_{L^2_x}\Big\|P_N\Big(f_1(u(\tau))+f_2(u(\tau)\Big)\Big\|_{L^2_x}\\
&\lesssim N^{2d\left(\frac{1}{r}-\frac{1}{2}\right)}\big\|f_1(u(t))+f_2(u(t))\big\|_{L^r_x}\big\|f_1(u(\tau))+f_2(u(\tau))\big\|_{L^r_x}.
\end{aligned}$$

Therefore, for $d \ge 5$, we have

$$\begin{aligned}
\left\|u_N(0)\right\|^2_{L^2} &\lesssim \big\|f_1(u)+f_2(u)\big\|^2_{L^\infty_t L^r_x}\times\int_0^{\infty}\int_{-\infty}^{0}\min\left(|t-\tau|^{-1},N^2\right)^{d\left(\frac{1}{r}-\frac{1}{2}\right)}\,dtd\tau\\
&\lesssim N^{2s_0}\big\|f_1(u)+f_2(u)\big\|^2_{L^\infty_t L^r_x},
\end{aligned}$$

where $s_0=\frac{d}{r}-\frac{d+4}{2}$. This implies (6.6). □

Corollary 6.7. *For* $d \ge 5$. *Let* u_c *be the critical element as shown in Theorem* 6.1, *then for every* $\eta > 0$ *there exists* $C(\eta) > 0$ *such that*

$$\sup_{t\ge 0}\int_{|x-x(t)|\ge C(\eta)}\left|u(t,x)\right|^2\,dx \lesssim \eta.$$

Proof. The proof is the same as Lemma 8.2 in [30]. □

6.5 Control of the spatial center function of the critical element

Now we will use the virial argument to control the spatial center function of the critical element by acknowledge of the zero momentum and the compactness in H^1 of the critical element.

Proposition 6.8. *For $d \geq 5$. Let u_c be the critical element as shown in Theorem 6.1. Then*

$$|x(t)| = o(t), \quad \text{as } t \to +\infty.$$

Proof. We argue by contradiction. Suppose that there exist $\delta > 0$ and a sequence $t_n \to +\infty$ such that

$$|x(t_n)| > \delta t_n \quad \text{for all} \quad n \geq 1.$$

Let $\eta > 0$ be a small constant to be chosen later. By compactness and Corollary 6.3 and Corollary 6.7, there exist $x(t)$ and $C(\eta)$ such that for any $t \geq 0$,

$$\int_{|x-x(t)|>C(\eta)} \left(|\nabla u(t,x)|^2 + |u(t,x)|^2\right) \leq \eta. \tag{6.7}$$

Define

$$T_n := \inf\left\{t \in [0,t_n] : |x(t)| = |x(t_n)|\right\} \leq t_n, \quad R_n := C(\eta) + \sup_{t\in[0,T_n]} |x(t)|.$$

Now let ϕ be a smooth, radial function satisfying $0 \leq \phi \leq 1$, $\phi(x) = 1$ for $|x| \leq 1$, and $\phi(x) = 0$ for $|x| \geq 2$. Define the truncation "position" as following

$$X_R(t) = \int_{\mathbb{R}^d} x\phi\left(\frac{|x|}{R}\right) \cdot |u(t,x)|^2 \, dx,$$

then we have

$$\begin{aligned}\left|X_{R_n}(0)\right| &\leq \left|\int_{|x|\leq C(\eta)} x\phi\left(\frac{|x|}{R_n}\right) |u(0,x)|^2 \, dx\right| + \left|\int_{|x|\geq C(\eta)} x\phi\left(\frac{|x|}{R_n}\right) |u(0,x)|^2 \, dx\right| \\ &\leq C(\eta)M(u) + 2\eta R_n,\end{aligned}$$

and

$$\begin{aligned}\left|X_{R_n}(T_n)\right| \geq\ & |x(T_n)| \cdot M(u) - |x(T_n)| \cdot \left|\int_{\mathbb{R}^d} \left(1 - \phi\left(\frac{|x|}{R_n}\right)\right) |u(T_n,x)|^2 \, dx\right| \\ & - \left|\int_{|x-x(T_n)|\leq C(\eta)} (x - x(T_n))\, \phi\left(\frac{|x|}{R_n}\right) |u(T_n,x)|^2 \, dx\right| \\ & - \left|\int_{|x-x(T_n)|\geq C(\eta)} (x - x(T_n))\, \phi\left(\frac{|x|}{R_n}\right) |u(T_n,x)|^2 \, dx\right|\end{aligned}$$

$$\geq \big|x(T_n)\big| \cdot M(u) - \big|x(T_n)\big| \cdot \eta - C(\eta)M(u) - (R_n + |x(T_n)|) \cdot \eta$$
$$\geq \big|x(T_n)\big| \cdot (M(u) - \eta) - C(\eta) \cdot (M(u) + \eta).$$

Thus, taking $\eta > 0$ sufficiently small, we have

$$\Big|X_{R_n}(T_n) - X_{R_n}(0)\Big| \geq \frac{M(u)}{2}|x(T_n)| - 2M(u) \cdot C(\eta). \tag{6.8}$$

On the other hand, we have

$$\partial_t X_{R_n}(t) = 2\Im \int_{\mathbb{R}^d} \phi\left(\frac{|x|}{R_n}\right) \nabla u(t,x) \cdot \overline{u(t,x)}\, dx$$
$$+ 2\Im \int_{\mathbb{R}^d} \frac{x}{|x|R_n}\phi'\left(\frac{|x|}{R_n}\right) x \cdot \nabla u(t,x)\overline{u(t,x)}\, dx,$$

which together with Proposition 6.5 and Corollary 6.3 implies that for any $t \in [0, T_n]$,

$$\Big|\partial_t X_{R_n}(t)\Big| \leq \left|2\Im \int_{\mathbb{R}^d} \left(1 - \phi\left(\frac{|x|}{R_n}\right)\right) \nabla u(t,x) \cdot \overline{u(t,x)}\, dx\right|$$
$$+ \left|2\Im \int_{\mathbb{R}^d} \frac{x}{|x|R_n}\phi'\left(\frac{|x|}{R_n}\right) x \cdot \nabla u(t,x)\overline{u(t,x)}\, dx\right| \leq C\eta.$$

Hence, we have

$$\frac{M(u)}{2} \cdot \delta T_n - 2M(u) \cdot C(\eta) \leq \frac{M(u)}{2} \cdot \delta t_n - 2M(u) \cdot C(\eta)$$
$$< \Big|x(T_n)\Big| - C(\eta) \leq C\eta \cdot T_n.$$

Taking η sufficiently small such that $C\eta \leq \frac{M(u)}{4} \cdot \delta$, we obtain a contradiction with the fact $T_n \to +\infty$. □

6.6 Death of the critical element

We are in a position to preclude the soliton-like solution by a truncated virial identity.

Theorem 6.9. *For* $d \geq 5$. *The critical element* u_c *of* (1.1) *cannot be a soliton in the sense of Theorem* 6.1.

Proof. We still drop the subscript c. By Proposition 6.8, for any $\eta > 0$, there exists $T_0 = T_0(\eta) \in \mathbb{R}$ such that

$$|x(t)| \leq \eta t \ \text{ for all } \ t \geq T_0. \tag{6.9}$$

Now let ϕ be a smooth, radial function satisfying $0 \leq \phi \leq 1$, $\phi(x) = 1$ for $|x| \leq 1$, and $\phi(x) = 0$ for $|x| \geq 2$. For some R, we define

$$V_R(t) := \int_{\mathbb{R}^d} \phi_R(x)|u(t,x)|^2\, dx, \quad \phi_R(x) = R^2\phi\left(\frac{|x|^2}{R^2}\right).$$

On one hand, we have

$$\partial_t V_R(t) = 4\Im \int_{\mathbb{R}^d} \phi'\left(\frac{|x|^2}{R^2}\right) x \cdot \nabla u(t,x)\, \overline{u(t,x)}\, dx.$$

Therefore, we have

$$\left|\partial_t V_R(t)\right| \lesssim R \tag{6.10}$$

for all $t \geq 0$ and $R > 0$.

On the other hand, by Hölder's inequality, we have

$$\begin{aligned}
\partial_t^2 V_R(t) &= 4\int_{\mathbb{R}^d} \partial_{ij}(\phi_R) u_i(t,x)\bar{u}_j(t,x)\, dx - \int_{\mathbb{R}^d} (\Delta^2\phi_R)(x)|u(t,x)|^2\, dx \\
&\quad - \frac{4}{d}\int_{\mathbb{R}^d} (\Delta\phi_R)(x)|u(t,x)|^{2^*}\, dx + \frac{4}{d+1}\int_{\mathbb{R}^d} (\Delta\phi_R)(x)|u(t,x)|^{\frac{2d+2}{d-1}}\, dx \\
&= 4\int_{\mathbb{R}^d} \left(2|\nabla u(t,x)|^2 - 2|u(t,x)|^{2^*} + \frac{2d}{d+1}|u(t,x)|^{\frac{2d+2}{d-1}}\right) dx \\
&\quad + O\Bigg(\int_{|x|\geq R} \left(|\nabla u(t,x)|^2 + |u(t,x)|^{2^*} + |u(t,x)|^{\frac{2d+2}{d-1}}\right) dx \\
&\quad + \left(\int_{R\leq|x|\leq 2R} |u(t,x)|^{2^*}\, dx\right)^{(d-2)/d}\Bigg) \\
&= 4K(u(t)) + O\Bigg(\int_{|x|\geq R} \left(|\nabla u(t,x)|^2 + |u(t,x)|^{\frac{2d+2}{d-1}}\right) dx \\
&\quad + \left(\int_{R\leq|x|\leq 2R} |u(t,x)|^{2^*}\, dx\right)^{(d-2)/d}\Bigg).
\end{aligned}$$

By Lemma 3.9, we have

$$\begin{aligned}
&4K(u(t)) \\
&= 4\int_{\mathbb{R}^d} \left(2|\nabla u(t,x)|^2 - 2|u(t,x)|^{2^*} + \frac{2d}{d+1}|u(t,x)|^{\frac{2d+2}{d-1}}\right) dx \\
&\gtrsim \min\left(\bar{\mu}(m - E(u(t))),\ \frac{2}{2d-3}\left\|\nabla u(t)\right\|_{L^2}^2 + \frac{2d}{(d+1)(2d-3)}\left\|u(t)\right\|_{L^{\frac{2d+2}{d-1}}}^{\frac{2d+2}{d-1}}\right) \\
&\gtrsim E(u(t)),
\end{aligned}$$

Thus, choosing $\eta > 0$ sufficiently small and $R := C(\eta) + \sup_{t\in[T_0,T_1]} |x(t)|$ and by Corollary 6.3, we obtain

$$\partial_t^2 V_R(t) \gtrsim E(u(t)) = E(u_0),$$

which implies that for all $T_1 > T_0$

$$(T_1 - T_0)E(u_0) \lesssim R = C(\eta) + \sup_{t\in[T_0,T_1]} |x(t)| \leq C(\eta) + \eta T_1.$$

Taking η sufficiently small and T_1 sufficiently large, we obtain a contradiction unless $u \equiv 0$. But $u \equiv 0$ is not consistent with the fact that $\|u\|_{ST(\mathbb{R})} = \infty$. $\square$

Acknowledgements

The authors are partly supported by the NSF of China (No. 10901148, No. 11171033, No. 11231006). L. Zhao is supported by the Fundamental Research Funds for the Central Universities (WK0010000001, WK00100000013). The authors would like to thank Professors J. Holmer, K. Nakanishi and S. Roudenko for their valuable communications and comments.

References

[1] T. Akahori, S. Ibrahim, H. Kikuchi, and H. Nawa, Existence of a ground state and blow-up problem for a nonlinear Schrödinger equation with critical growth, to appear in Adv. Diff. Equat..

[2] T. Akahori, S. Ibrahim, H. Kikuchi, and H. Nawa, Existence of a ground state and scattering for a nonlinear Schrödinger equation with critical growth, preprint.

[3] T. Aubin, Problémes isopérimétriques et espaces de Sobolev, J. Diff. Geom. **11** (1976), 573–598.

[4] H. Bahouri and P. Gérard, High frequency approximation of solutions to critical nonlinear wave equations, Amer. J. Math. **121** (1999), 131–175.

[5] J. Bourgain, Global well-posedness of defocusing 3D critical NLS in the radial case, J. Amer. Math. Soc. **12** (1999), 145–171.

[6] J. Bourgain, Global solutions of nonlinear Schrödinger equations, Amer. Math. Soc. Colloq. Publ. **46**, Amer. Math. Soc., Providence, 1999.

[7] T. Cazenave, Semilinear Schrödinger equations, Courant Lecture Notes in Mathematics **10**, New York: New York University Courant Institute of Mathematical Sciences, 2003.

[8] J. Colliander, M. Keel, G. Staffilani, H. Takaoka, and T. Tao, Global existence and scattering for rough solution of a nonlinear Schrödinger equation on $\mathbb{R}^3$, Comm. Pure Appl. Math. **57** (2004), 987–1014.

[9] J. Colliander, M. Keel, G. Staffilani, H. Takaoka, and T. Tao, Global well-posedness and scattering for the energy-cirtical nonlinear Schrödinger equation in $\mathbb{R}^3$, Ann. of Math. **167** (2008), 767–865.

[10] B. Dodson, Global well-posedness and scattering for the defocusing, L^2-critical, nonlienar Schrödinger equation when $d \geq 3$, J. Amer. Math. Soc. **25** (2012), no. 2, 429–463.

[11] B. Dodson, Global well-posedness and scattering for the defocusing, L^2-critical, nonlienar Schrödinger equation when $d = 2$, arXiv:1006.1375.

[12] B. Dodson, Global well-posedness and scattering for the defocusing, L^2-critical, nonlienar Schrödinger equation when $d = 1$, arXiv:1010.0040.

[13] B. Dodson, Global well-posedness and scattering for the mass critical nonlienar Schrödinger equation with mass below the mass of the ground state, arXiv:1104.1114.

[14] T. Duyckaerts and F. Merle, Dynamic of threshold solutions for energy-critical NLS, GAFA **18** (2009), no. 6, 1787–1840.

[15] T. Duyckaerts, J. Holmer, and S. Roudenko, Scattering for the non-radial 3D cubic nonlinear Schrödinger equation, Math. Res. Lett. **15** (2008), no. 5–6, 1233–1250.

[16] T. Duyckaerts and S. Roudenko, Threshold solutions for the focusing 3D cubic Schrödinger equation. Revista. Math. Iber. **26** (2010), no. 1, 1–56.

[17] D. Foschi, Inhomogeneous Strichartz estimates, J. Hyper. Diff. Equat. **2** (2005), 1–24.

[18] J. Ginibre and G. Velo, Scattering theory in the energy space for a class of nonlinear Schrödinger equation, J. Math. Pures Appl. **64** (1985), 363–401.

[19] J. Ginibre and G. Velo, Time decay of finite energy solutions of the nonlinear Klein-Gordon and Schrödinger equation, Ann. Inst. Henri Poincaré, Physique théorique **43** (1985), no. 4, 399–442.

[20] R. T. Glassey, On the blowing up of solutions to the Cauchy problem for nonlinear Schrödinger equations, J. Math. Phys. **18** (1977), no. 9, 1794–1797.

[21] J. Holmer and S. Roudenko, A sharp condition for scattering of the radial 3d cubic nonlinear Schrödinger equation, Comm. Math. Phys. **282** (2008), no. 2, 435–467.

[22] J. Holmer and S. Roudenko, Divergence to infinite-variance nonradial solutions to the 3d NLS equation, Comm. PDEs **35** (2010), no. 5, 878–905.

[23] S. Ibrahim, N. Masmoudi, and K. Nakanishi, Scattering threshold for the focusing nonlinear Klein-Gordon equation, Analysis & PDE **4** (2011), no. 3, 405–460.

[24] M. Keel and T. Tao, Endpoint Strichartz estimates, Amer. J. Math. **120** (1998), no. 5, 955–980.

[25] C. E. Kenig and F. Merle, Global well-posedness, scattering and blow up for the energy-critical, focusing, nonlinear Schrödinger equation in the radial case, Invent. Math. **166** (2006), 645–675.

[26] C. E. Kenig and F. Merle, Global well-posedness, scattering and blow-up for the energy critical focusing non-linear wave equation, Acta Math. **201** (2008), no. 2, 147–212.

[27] S. Keraani, On the defect of compactness for the Strichartz estimates of the Schrödinger equations, J. Diff. Equat. **175** (2001), 353–392.

[28] S. Keraani, On the blow up phenomenon of the critical Schrödinger equation, J. Funct. Anal. **265** (2006), 171–192.

[29] R. Killip, T. Tao, and M. Visan, The cubic nonlinear Schrödinger equation in two dimensions with radial data, J. Eur. Math. Soc. **11** (2009), no. 6, 1203–1258.

[30] R. Killip and M. Visan, The focusing energy-critical nonlinear Schrödinger equation in dimensions five and higher, Amer. J. Math. **132** (2010), no. 2, 361–424.

[31] R. Killip and M. Visan, Nonlinear Schrödinger euqations at critical regularity, Proceedings of the Clay summer school "Evolution Equations", June 23–July 18, 2008, Eidgenössische technische Hochschule, Zürich.

[32] R. Killip, M. Visan, and X. Zhang, The mass-critical nonlinear Schrödinger equation with radial data in dimensions three and higher, Analysis & PDE **1** (2008), no. 2, 229–266.

[33] J. Krieger, K. Nakanishi, and W. Schlag, Global dynamics away from the ground state for the energy-critical nonlinear wave equation, Amer. J. Math. **135** (2013), no. 4, 935–965.

[34] J. Krieger, K. Nakanishi, and W. Schlag, Global dynamics above the ground state energy for the one-dimensional NLKG equation, Math. Z. **272** (2012), no. 1–2, 297–316.

[35] D. Li and X. Zhang, Dynamics for the energy critical nonlinear Schrödinger equation in high dimensions, J. Funct. Anal. **256** (2009), no. 6, 1928–1961.

[36] C. Miao, Y. Wu, and G. Xu, Dynamics for the focusing, energy-critical nonlinear Hartree equation, Forum Math. **27** (2015), no. 1, 373–447.

[37] C. Miao, G. Xu, and L. Zhao, Global well-posedness and scattering for the energy-critical, defocusing Hartree equation for radial data, J. Funct. Anal. **253** (2007), no. 2, 605–627.

[38] C. Miao, G. Xu, and L. Zhao, Global well-posedness, scattering and blow-up for the energy-critical, focusing Hartree equation in the radial case, Colloquium Mathematicum **114** (2009), 213–236.

[39] C. Miao, G. Xu, and L. Zhao, Global well-posedness and uniform bound for the defocusing $H^{1/2}$-subcritical Hartree equation in $\mathbb{R}^d$, Ann I. H. Poincaré, AN **26** (2009), no. 5, 1831–1852.

[40] C. Miao, G. Xu, and L. Zhao, Global well-posedness and scattering for the mass-critical Hartree equation with radial data, J. Math. Pures Appl. **91** (2009), 49–79.

[41] C. Miao, G. Xu, and L. Zhao, Global well-posedness and scattering for the energy-critical, defocusing Hartree equation in $\mathbb{R}^{1+n}$, Comm. PDEs **36** (2011), no. 5, 729–776.

[42] C. Miao, G. Xu, and L. Zhao, The dynamics of the 3D radial NLS with the combined terms, Comm. Math. Phys. **318** (2013), no. 3, 767–808.

[43] K. Nakanishi, Energy scattering for nonlinear Klein-Gordon and Schrödinger equations in spatial dimensions 1 and 2, J. Funct. Anal. **169** (1999), 201–225.

[44] K. Nakanishi, Remarks on the energy scattering for nonlinear Klein-Gordon and Schrödinger equations, Tohoku Math. J. **53** (2001), 285–303.

[45] K. Nakanishi and W. Schlag, Global dynamics above the ground state energy for the cubic NLS equation in $3D$, Calc. of Variations and PDE **44** (2012), no. 1–2, 1–45.

[46] T. Ogawa and Y. Tsutsumi, Blow-up of H^1 solution for the nonlinear Schrödinger equation, J. Diff. Equat. **92** (1991), 317–330.

[47] E. Ryckman and M. Visan, Global well-posedness and scattering for the defocusing energy-critical nonlinear Schrödinger equation in $\mathbb{R}^{1+4}$, Amer. J. Math. **129** (2007), 1–60.

[48] G. Talenti, Best constant in Sobolev inequality, Ann. Mat. Pura. Appl. **110** (1976), 353–372.

[49] T. Tao, Nonlinear dispersive equations, local and global analysis, CBMS. Regional Conference Series in Mathematics **106**. Published for the Conference Board of the Mathematical Science, Washington, DC; by the American Mathematical Society Providence, RI, 2006.

[50] T. Tao, M. Visan, and X. Zhang, The nonlinear Schrödinger equation with combined power-type nonlinearities, Comm. PDEs **32** (2007), 1281–1343.

[51] T. Tao, M. Visan, and X. Zhang, Minimal-mass blowup solutions of the mass-critical NLS, Forum Math. **20** (2008), no. 5, 881–919.

[52] M. Visan, The defocusing energy-critical nonlinear Schrödinger equation in higher dimensions, Duke Math. J. **138** (2007), no. 2, 281–374.

[53] J. Zhang, Sharp conditions of global existence for nonlinear Schrödinger and Klein-Gordon equations, Nonlinear Anal. T. M. A. **48** (2002), no. 2, 191–207.

[54] X. Zhang, On Cauchy problem of 3D energy critical Schrödinger equation with subcritical perturbations, J. Diff. Equat. **230** (2006), no. 2, 422–445.

Some Topics in Harmonic Analysis and Applications
ALM 34, pp. 299–329

Sharp Estimates for Bilinear Fourier Multiplier Operators

Akihiko Miyachi * and Naohito Tomita †

Abstract

We survey recent results on sharp estimates for bilinear Fourier multiplier operators in Lebesgue spaces and Hardy spaces. Proofs are given for some key results.

2000 Mathematics Subject Classification: 42B15, 42B20, 42B30.

Keywords and Phrases: bilinear Fourier multipliers, Hardy spaces.

1 Introduction

For $m \in L^\infty(\mathbb{R}^{2n})$, the bilinear Fourier multiplier operator T_m is defined by

$$T_m(f,g)(x) = \iint_{\mathbb{R}^n \times \mathbb{R}^n} e^{2\pi i x(\xi+\eta)} m(\xi,\eta) \widehat{f}(\xi)\widehat{g}(\eta)\, d\xi d\eta,$$

where f, $g \in \mathcal{S}(\mathbb{R}^n)$, $x \in \mathbb{R}^n$, and $\wedge$ denotes the Fourier transform

$$\widehat{f}(\xi) = \int_{\mathbb{R}^n} f(x) e^{-2\pi i x\xi}\, dx.$$

The function m is called the *multiplier*. If we use $K = \mathcal{F}^{-1}m$, the inverse Fourier transform of m on $\mathbb{R}^{2n}$, then we have

$$T_m(f,g)(x) = \iint_{\mathbb{R}^n \times \mathbb{R}^n} K(x-y, x-z) f(y) g(z)\, dy dz,$$

*Department of Mathematics, Tokyo Woman's Christian University, Zempukuji, Suginami-ku, Tokyo 167-8585, Japan. *E-mail:* `miyachi@lab.twcu.ac.jp`

†Department of Mathematics, Osaka University, Toyonaka, Osaka 560-0043, Japan. *E-mail:* `tomita@math.sci.osaka-u.ac.jp`

where the integral should be taken in an appropriate generalized sense if K is not integrable. The operator T_m satisfies the translation invariance

$$T_m\big(f(\cdot - a), g(\cdot - a)\big)(x) = T_m(f,g)(x-a) \tag{1.1}$$

and, for $t > 0$, the dilation formula

$$T_m(f,g)(x) = T_{m(t\cdot)}\big(f(t^{-1}\cdot), g(t^{-1}\cdot)\big)(tx). \tag{1.2}$$

The Fourier transform of $T_m(f,g)$ is given by the formula

$$\int_{\mathbb{R}^n} \big(T_m(f,g)\big)^{\wedge}(\zeta)\varphi(\zeta)\, d\zeta = \iint_{\mathbb{R}^n \times \mathbb{R}^n} \varphi(\xi+\eta) m(\xi,\eta) \widehat{f}(\xi)\widehat{g}(\eta)\, d\xi d\eta \tag{1.3}$$

for all $\varphi \in \mathcal{S}(\mathbb{R}^n)$.

The bilinear Fourier multipliers appear in many problems in analysis. There have been many important works and still there are many interesting problems. In this article, we shall survey the results of recent papers [7, 13, 15], which concern with boundedness of T_m in Lebesgue spaces L^p and Hardy spaces H^p.

For measurable functions f on $\mathbb{R}^n$, the L^p-quasinorm is defined by

$$\|f\|_p = \left(\int_{\mathbb{R}^n} |f(x)|^p\, dx \right)^{1/p}, \quad 0 < p \le \infty,$$

with the usual modification in the case $p = \infty$. The H^p-quasinorm is defined by

$$\|f\|_{H^p} = \|f^*\|_p, \quad 0 < p \le \infty,$$

for $f \in \mathcal{S}'(\mathbb{R}^n)$, where f^* is an appropriate grand maximal function associated with f (see, e.g., [16, Chapter III]). Notice that $\|f\|_{H^p} \approx \|f\|_p$ if $1 < p \le \infty$.

The problem we shall consider concerns with the estimate

$$\|T_m(f,g)\|_r \lesssim \|f\|_{H^p} \|g\|_{H^q}. \tag{1.4}$$

A prototype of T_m we are interested in is the one with homogeneous multipliers m such that $m(t\xi, t\eta) = m(\xi, \eta)$. For such m, the dilation formula (1.2) and a simple scaling argument show that the estimate (1.4) is possible only when

$$1/p + 1/q = 1/r. \tag{1.5}$$

Thus in this article we shall consider the estimate (1.4) only for p, q, r that satisfy (1.5). In the endpoint case $p = q = r = \infty$, it is convenient to replace (1.4) by the estimate

$$\|T_m(f,g)\|_{BMO} \lesssim \|f\|_\infty \|g\|_\infty. \tag{1.6}$$

In their pioneering works on bilinear Fourier multiplier operators, Coifman and Meyer [2, 3, 12] considered m satisfying

$$|\partial_\xi^\alpha \partial_\eta^\beta m(\xi,\eta)| \le C_{\alpha,\beta}(|\xi| + |\eta|)^{-|\alpha|-|\beta|} \tag{1.7}$$

and proved that the estimate (1.4) holds for $1 < p \le \infty$, $1 < q \le \infty$, and $1 < r < \infty$ satisfying (1.5). The estimate (1.6) is also implicitly given in [3, 12]. Kenig and Stein [11] proved weak type estimate for the case $p = q = 1$ and $r = 1/2$ and extended the results of Coifman and Meyer to the range $1/2 < r \le 1$. Grafakos and Torres [9] gave a general theory of multilinear Calderón-Zygmund operators and generalized the results of [2, 3, 12, 11]. Grafakos and Kalton [6] introduced Hardy spaces in the problem and extended the estimate to the full range $0 < p,\ q \le \infty$, $0 < r < \infty$.

Now, to ensure the estimate (1.4)–(1.5) or (1.6), it is sufficient to assume (1.7) for derivatives up to certain order. The problem we shall consider is to find the differentiability conditions generalizing (1.7) that are as small as possible to ensure (1.4)–(1.5) or (1.6).

We recall the case of linear Fourier multiplier operators. For $m \in L^\infty(\mathbb{R}^n)$, the linear Fourier multiplier operator $m(D)$ is defined by

$$m(D)f(x) = \int_{\mathbb{R}^n} e^{2\pi i x\xi} m(\xi)\widehat{f}(\xi)\, d\xi,$$

where $f \in \mathcal{S}(\mathbb{R}^n)$ and $x \in \mathbb{R}^n$. It is now well known that if

$$|\partial_\xi^\alpha m(\xi)| \le C_\alpha |\xi|^{-|\alpha|} \tag{1.8}$$

for all α, then the estimate

$$\|m(D)f\|_p \lesssim \|f\|_{H^p} \tag{1.9}$$

holds for all $0 < p < \infty$. We shall recall some works concerning weaker versions of the condition (1.8).

Hörmander [10] gave a condition that generalizes (1.8) for the case $1 < p < \infty$. To give Hörmander's result, we use some notations. For $s \in \mathbb{R}$, $1 < t < \infty$, and $f \in \mathcal{S}'(\mathbb{R}^n)$, we define the Sobolev norm by

$$\|f\|_{W_t^s(\mathbb{R}^n)} = \|(1-\Delta)^{s/2} f\|_t = \big\|\mathcal{F}^{-1}\big(\langle\xi\rangle^s \widehat{f}(\xi)\big)\big\|_t.$$

We take a function Ψ such that

$$\Psi \in C_0^\infty(\mathbb{R}^n), \quad \operatorname{supp}\Psi \subset \{\zeta \in \mathbb{R}^n : 1/2 \le |\zeta| \le 2\}, \tag{1.10}$$

and

$$\sum_{k\in\mathbb{Z}} \Psi(\zeta/2^k) = 1 \text{ for all } \zeta \in \mathbb{R}^n \setminus \{0\}. \tag{1.11}$$

The following theorem was essentially proved by Hörmander [10].

Theorem A ([10, Theorem 2.5]). *If $s > n/2$ and $1 < p < \infty$, then*

$$\|m(D)f\|_p \lesssim \sup_{j\in\mathbb{Z}} \big\{\|m(2^j\cdot)\Psi(\cdot)\|_{W_2^s(\mathbb{R}^n)}\big\}\|f\|_p.$$

Calderón and Torchinsky [1] extended Hörmander's theorem to the case $0 < p \le 1$.

Theorem B ([1, Theorem 4.6]). *If* $0 < p \le 1$ *and* $s > n/p - n/2$, *then*

$$\|m(D)f\|_p \lesssim \sup_{j\in\mathbb{Z}} \left\{ \|m(2^j\cdot)\Psi(\cdot)\|_{W_2^s(\mathbb{R}^n)} \right\} \|f\|_{H^p}.$$

For linear Fourier multiplier operators $m(D)$, Plancherel's theorem obviously implies that $m(D)$ is bounded in L^2 if $m \in L^\infty$, without any differentiability conditions. Interpolating this fact and Theorem A, Calderón and Torchinsky [1] proved the following theorem.

Theorem C ([1, Theorem 4.7]). *Let* $1 < p < 2$, $s > n/p - n/2$, *and* $1/t = 1/p - 1/2$. *Then*

$$\|m(D)f\|_p \lesssim \sup_{j\in\mathbb{Z}} \left\{ \|m(2^j\cdot)\Psi(\cdot)\|_{W_t^s(\mathbb{R}^n)} \right\} \|f\|_p.$$

It is known that Theorems A, B, and C are sharp in the sense that the numbers $n/2$ and $n/p - n/2$ in these theorems cannot be replaced by smaller numbers. The main subject of the papers [7, 13, 15] is to find similar critical orders for the case of bilinear operators T_m. Related results are also found in [18, 8, 5].

In Section 2, Theorem 2.2, we give the main results of [7] and [13]. Theorem 2.2 may be considered as a bilinear version of Theorems A and B. In Sections 3–8, we give proofs of some key parts of Theorem 2.2. In Section 9, we give the results of [15]. Theorem 9.2 of that section may be considered as a bilinear version of Theorem C.

Here we give a remark concerning the argument of this paper. Since we are interested in the estimate for operator norms, we usually tacitly assume that all functions, including multipliers, that appear in the argument are of the Schwartz class. In our proofs, we omit limiting argument that is necessary for rigorous proof.

Throughout this paper, we use the following notation. For $x \in \mathbb{R}^n$, we write $\langle x \rangle = (1 + |x|^2)^{1/2}$. We write $\mathbf{1}_E$ to denote the defining function of a set E. For measurable functions f on $\mathbb{R}^n$ and for measurable subsets $E \subset \mathbb{R}^n$ and $0 < p \le \infty$, we write $\|f\|_{p,E} = \|f(x)\|_{p,\, x\in E} = \|f\mathbf{1}_E\|_p$. For two nonnegative functions F and G on a set Ω, the notation $F(\omega) \lesssim G(\omega)$ means that there exists a constant $C > 0$ such that $F(\omega) \le CG(\omega)$ for all $\omega \in \Omega$. The notation $F(\omega) \approx G(\omega)$ means that $F(\omega) \lesssim G(\omega)$ and $G(\omega) \lesssim F(\omega)$. Usually we do not refer to the set Ω, which will be apparent from the context.

2 Product type Sobolev scale

In the main results of [7] and [13], product type Sobolev norm is used to measure smoothness of multipliers $m(\xi, \eta)$. The definition reads as follows.

Definition 2.1. For s_1, $s_2 \in \mathbb{R}$ and $F \in \mathcal{S}'(\mathbb{R}^{2n})$, we define

$$\begin{aligned}\|F\|_{W^{(s_1,s_2)}(\mathbb{R}^n\times\mathbb{R}^n)} &= \|(1-\Delta_\xi)^{s_1/2}(1-\Delta_\eta)^{s_2/2}F(\xi,\eta)\|_{2,\,(\xi,\eta)} \\ &= \|\langle x\rangle^{s_1}\langle y\rangle^{s_2}\mathcal{F}^{-1}F(x,y)\|_{2,\,(x,y)}.\end{aligned}$$

We take a function $\Psi \in C_0^\infty(\mathbb{R}^{2n})$ satisfying the conditions (1.10) and (1.11) with $\mathbb{R}^{2n}$ in place of $\mathbb{R}^n$. For $m \in L^\infty(\mathbb{R}^{2n})$, we define

$$A_{(s_1,s_2)}(m) = \sup_{j\in\mathbb{Z}} \|m(2^j\xi, 2^j\eta)\Psi(\xi,\eta)\|_{W^{(s_1,s_2)}(\mathbb{R}^n\times\mathbb{R}^n)}. \tag{2.1}$$

It can be shown that, up to the equivalence, the norm $A_{(s_1,s_2)}(m)$ does not depend on the choice of the function Ψ.

The following is the main results of [7, 13].

Theorem 2.2. *Let $0 < p,\ q,\ r \le \infty$ and $1/p+1/q = 1/r$. If $s_1 > \max\{n/2,\ n/p-n/2\}$, $s_2 > \max\{n/2,\ n/q - n/2\}$, and $s_1 + s_2 > n/r - n/2$, then the estimate*

$$\|T_m(f,g)\|_r \lesssim A_{(s_1,s_2)}(m)\|f\|_{H^p}\|g\|_{H^q} \tag{2.2}$$

holds, where the estimate should be replaced by

$$\|T_m(f,g)\|_{BMO} \lesssim A_{(s_1,s_2)}(m)\|f\|_\infty\|g\|_\infty \tag{2.3}$$

in the case $p = q = r = \infty$. Conversely, the estimate (2.2), with the same replacement (2.3) for $p = q = r = \infty$, holds only if $s_1 \ge \max\{n/2,\ n/p - n/2\}$, $s_2 \ge \max\{n/2,\ n/q - n/2\}$, and $s_1 + s_2 \ge n/r - n/2$.

To see easily the various conditions of Theorem 2.2, we divide the region of $(1/p, 1/q)$ into 7 regions $I_0, \ldots, I_6$ as in the following figure.

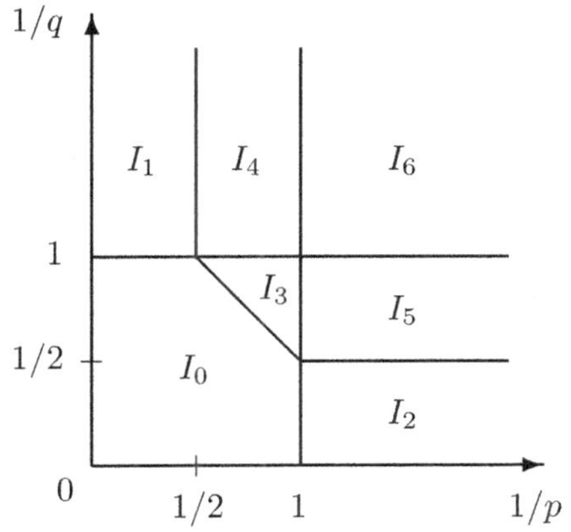

Then the assumptions on s_1 and s_2 in the if part of Theorem 2.2 are written as follows:

$$\begin{aligned}
&s_1 > n/2, \quad s_2 > n/2 && \text{if } (1/p,1/q) \in I_0;\\
&s_1 > n/2, \quad s_2 > n/q - n/2 && \text{if } (1/p,1/q) \in I_1;\\
&s_1 > n/p - n/2, \quad s_2 > n/2 && \text{if } (1/p,1/q) \in I_2;\\
&\begin{cases} s_1 > n/2, \quad s_2 > n/2, \\ s_1 + s_2 > n/p + n/q - n/2 \end{cases} && \text{if } (1/p,1/q) \in I_3;
\end{aligned}$$

$$\begin{cases} s_1 > n/2, \quad s_2 > n/q - n/2, \\ s_1 + s_2 > n/p + n/q - n/2 \end{cases} \quad \text{if} \quad (1/p, 1/q) \in I_4;$$

$$\begin{cases} s_1 > n/p - n/2, \quad s_2 > n/2, \\ s_1 + s_2 > n/p + n/q - n/2 \end{cases} \quad \text{if} \quad (1/p, 1/q) \in I_5;$$

$$\begin{cases} s_1 > n/p - n/2, \quad s_2 > n/q - n/2, \\ s_1 + s_2 > n/p + n/q - n/2 \end{cases} \quad \text{if} \quad (1/p, 1/q) \in I_6.$$

In Sections 3–7, we give proofs of some key assertions in the if part of Theorem 2.2. In Section 8, we give brief sketch of the proof of the only if part of Theorem 2.2.

Before we proceed to the proofs of the if part of Theorem 2.2, we observe that the claim of the if part follows if we prove the estimates for the following cases:

$$\begin{aligned} &0 < p \leq 1,\ q = \infty; \\ &p = q = \infty; \\ &0 < p \leq 1,\ q = 1; \\ &0 < p \leq 1,\ q = 2. \end{aligned}$$

In fact, if the estimates for the above cases are proved, then by the obvious symmetry between p and q and by interpolation, we can obtain the whole claim of the if part (see [13, §6]). In Sections 3–7, instead of giving proofs of all of the above cases, we shall give proofs for the following cases:

$$(p, q) = (2, \infty),\ (1, \infty),\ (\infty, \infty),\ (1, 1),\ (1, 2).$$

The estimate for $(p, q) = (2, \infty)$ is not necessary for interpolation but it is the key to our whole argument. The argument for the above cases will show important methods that can be applied to general cases.

3 Estimates for $L^2 \times L^\infty \to L^2$

In this section we prove the following proposition, which is the if part of Theorem 2.2 for $(p, q) = (2, \infty)$.

Proposition 3.1. *If $s > n/2$, then $\|T_m(f, g)\|_2 \lesssim A_{(s,s)}(m)\|f\|_2\|g\|_\infty$.*

Before we proceed to the proof, we shall give some basic ideas which will be used in the succeeding sections as well.

First, it is easy to see that if $\Theta(\xi, \eta)$ is a smooth function on $\mathbb{R}^{2n} \setminus \{(0,0)\}$ that is homogeneous of degree 0, then

$$A_{(s_1,s_2)}(m\Theta) \leq C_\Theta A_{(s_1,s_2)}(m) \tag{3.1}$$

(see [13, Lemma 2.4]). This implies that, in order to prove the estimate like (2.2), it is sufficient to prove it under the additional condition that m is supported in an cone. To be precise, if we prove that for each

$$(\xi_0, \eta_0) \in \Sigma = \{(\xi, \eta) \in \mathbb{R}^{2n} : |(\xi, \eta)| = 1\}$$

there exists an open neighborhood $\mathcal{O} \subset \Sigma$ of (ξ_0, η_0) such that the estimate (2.2) holds for m satisfying

$$\operatorname{supp} m \subset \Gamma(\mathcal{O}) = \{\zeta \in \mathbb{R}^{2n} \setminus \{0\} : \zeta/|\zeta| \in \mathcal{O}\},$$

then, using appropriate partition of unity in terms of smooth homogeneous functions of degree 0 and using (3.1), we can deduce the estimate for general m.

In our argument, we shall frequently decompose m as follows:

$$m(\xi,\eta) = \sum_{j\in\mathbb{Z}} m_j(\xi,\eta), \quad m_j(\xi,\eta) = m(\xi,\eta)\Psi(\xi/2^j, \eta/2^j). \tag{3.2}$$

Thus $T_m(f,g) = \sum_{j\in\mathbb{Z}} T_{m_j}(f,g)$. Observe that

$$\operatorname{supp} m_j(2^j\cdot) \subset \{\zeta \in \mathbb{R}^{2n} : 1/2 \le |\zeta| \le 2\} \tag{3.3}$$

and

$$\|m_j(2^j\cdot)\|_{W^{(s_1,s_2)}(\mathbb{R}^n\times\mathbb{R}^n)} \le A_{(s_1,s_2)}(m). \tag{3.4}$$

(Notice that this notation of m_j is different from the one used in [7, 13, 15].)

One of our tools is the pointwise estimate for $T_{m_j}(f,g)$, which is given in the following lemma.

Lemma 3.2. *Let $s > n/2$ and $\max\{1, n/s\} < \sigma < 2$. Suppose $j \in \mathbb{Z}$ and $\operatorname{supp} m \subset \{\zeta \in \mathbb{R}^{2n} : |\zeta| \le 2^{j+1}\}$. Then*

$$|T_m(f,g)(x)| \le C\|m(2^j\cdot)\|_{W^{(s,s)}(\mathbb{R}^n\times\mathbb{R}^n)}(\zeta_j * |f|^\sigma)(x)^{1/\sigma}(\zeta_j * |g|^\sigma)(x)^{1/\sigma}$$

for all $x \in \mathbb{R}^n$, where $\zeta_j(x) = 2^{jn}(1+2^j|x|)^{-s\sigma}$ and C is a constant depending only on n, s, and σ.

Proof. By the dilation formula (1.2), the proof can be reduced to the case $j = 0$. We write $K = \mathcal{F}^{-1}m$. Then Hölder's inequality gives

$$\begin{aligned}|T_m(f,g)(x)| &\le \iint |K(x-y,x-z)f(y)g(z)|\,dydz \\ &\le \left(\iint \langle x-y\rangle^{s\sigma'}\langle x-z\rangle^{s\sigma'}|K(x-y,x-z)|^{\sigma'}\,dydz\right)^{1/\sigma'} \\ &\quad\times\left(\iint \langle x-y\rangle^{-s\sigma}\langle x-z\rangle^{-s\sigma}|f(y)|^\sigma|g(z)|^\sigma\,dydz\right)^{1/\sigma}.\end{aligned}$$

The second factor on the last expression is $\approx (\zeta_0 * |f|^\sigma)(x)^{1/\sigma}(\zeta_0 * |g|^\sigma)(x)^{1/\sigma}$. For the first factor, Lemma 3.3 to be given below yields

$$\|\langle y\rangle^s\langle z\rangle^s K(y,z)\|_{\sigma',(y,z)} \lesssim \|\langle y\rangle^s\langle z\rangle^s K(y,z)\|_{2,(y,z)} = \|m\|_{W^{(s,s)}(\mathbb{R}^n\times\mathbb{R}^n)}$$

since $\operatorname{supp}\widehat{K} \subset \{|\zeta| \le 2\}$ and since $\sigma' > 2$. The proof is complete. □

The following lemma, which was already used in the above proof, will also be frequently used in our argument.

Lemma 3.3. *If $F \in \mathcal{S}'(\mathbb{R}^n)$ and $\operatorname{supp} \widehat{F} \subset \{\xi \in \mathbb{R}^n : |\xi| \le 10\}$, then, for each $2 \le q \le \infty$, $s \in \mathbb{R}$, and multi-index α, there exists a constant C depending only on n, q, s, and α such that*

$$\|\langle x \rangle^s \partial_x^\alpha F(x)\|_q \le C \|\langle x \rangle^s F(x)\|_2.$$

If $K \in \mathcal{S}'(\mathbb{R}^{2n})$ and $\operatorname{supp} \widehat{K} \subset \{\zeta \in \mathbb{R}^{2n} : |\zeta| \le 10\}$, then, for each $2 \le q \le \infty$, $s_1 \in \mathbb{R}$, $s_2 \in \mathbb{R}$, and multi-indices α and β, there exists a constant C depending only on n, q, s_1, s_2, α, and β such that

$$\|\langle x \rangle^{s_1} \langle y \rangle^{s_2} \partial_x^\alpha \partial_y^\beta K(x,y)\|_{q,\,(x,y)} \le C \|\langle x \rangle^{s_1} \langle y \rangle^{s_2} K(x,y)\|_{2,\,(x,y)}.$$

The inequalities in this lemma are, at least for the case $s = s_1 = s_2 = 0$, known as the Plancherel-Polya-Nikol'skij inequalities. Proofs can be found in, e.g., [19, Theorem 1.4.1], [5, Lemma A.1], or [7, Appendix A].

We explain notation that will be frequently used in our argument. For an even function $\phi \in C_0^\infty(\mathbb{R}^n)$ and for $j \in \mathbb{Z}$, we define the operator S_j by

$$S_j f = \phi(D/2^j) f = \mathcal{F}^{-1}(\phi(\xi/2^j)\widehat{f}(\xi))$$

and call them *S-operators.* For an even function $\psi \in C_0^\infty(\mathbb{R}^n)$ such that $\operatorname{supp} \psi \subset \mathbb{R}^n \setminus \{0\}$ and for $j \in \mathbb{Z}$, we define the operator Δ_j by

$$\Delta_j f = \psi(D/2^j) f = \mathcal{F}^{-1}(\psi(\xi/2^j)\widehat{f}(\xi))$$

and call them *D-operators.* Restricting ϕ and ψ to be even functions is not very important; it only gives the simple formula

$$\int S_j f(x) g(x)\, dx = \int f(x) S_j g(x)\, dx$$

and the similar one for Δ_j.

The proof of Proposition 3.1 is based on Lemma 3.4 to be given below. This lemma concerns with BMO functions and Carleson measures. Recall that a Borel measure μ on $\mathbb{R}^n \times (0,\infty)$ is called a *Carleson measure* if there exists a constant A such that

$$\mu\big(Q \times (0, \ell(Q))\big) \le A|Q|$$

for all cubes $Q \subset \mathbb{R}^n$, where $\ell(Q)$ denotes the sidelength of Q. The smallest value of A is called *Carleson norm* of μ and denoted by $\|\mu\|_C$. If μ is a Carleson measure, then the inequality

$$\iint_{\mathbb{R}^n \times (0,\infty)} F(x,t)\, d\mu(x,t) \le c_n \|\mu\|_C \int_{\mathbb{R}^n} NF(x)\, dx \tag{3.5}$$

holds with

$$NF(x) = \sup\{F(y,t) : y \in \mathbb{R}^n,\ t > 0,\ |x - y| < t\}, \quad x \in \mathbb{R}^n,$$

for all nonnegative Borel functions F on $\mathbb{R}^n \times (0,\infty)$ (see, e.g., [16, Chapter II, Section 2]).

Lemma 3.4. *Let $\epsilon > 0$ and*

$$\zeta_j(x) = 2^{jn}(1 + 2^j|x|)^{-n-\epsilon}, \quad x \in \mathbb{R}^n, \quad j \in \mathbb{Z}.$$

Let $g \in BMO$ and let Δ_j be D-operators. Then the measure μ on $\mathbb{R}^n \times (0, \infty)$ defined by

$$\mu = \sum_{j\in\mathbb{Z}} (\zeta_j * |\Delta_j g|^2)(x)\, dx \otimes \delta_{2^{-j}}(t) \tag{3.6}$$

is a Carleson measure with Carleson norm $\|\mu\|_C \le c_\epsilon \|g\|_{BMO}^2$.

The meaning of the notation (3.6) is that μ satisfies

$$\iint_{\mathbb{R}^n\times(0,\infty)} F(x,t)\, d\mu(x,t) = \sum_{j\in\mathbb{Z}} \int_{\mathbb{R}^n} F(x, 2^{-j})(\zeta_j * |\Delta_j g|^2)(x)\, dx$$

for all nonnegative Borel functions F on $\mathbb{R}^n \times (0, \infty)$. It is well known that for $g \in BMO$ the measure

$$\nu = \sum_{j\in\mathbb{Z}} |\Delta_j g(x)|^2\, dx \otimes \delta_{2^{-j}}(t)$$

is a Carleson measure with Carleson norm $\|\nu\|_C \lesssim \|g\|_{BMO}^2$ (see, e.g., [16, Chapter IV, Section 4.3]). Lemma 3.4 can be easily derived from this fact (see [7, Lemma 3.1]).

Finally, notice that the support of the Fourier transform of $T_m(f, g)$ is included in the closure of $\{\xi + \eta : m(\xi, \eta) \neq 0\}$. This can be immediately seen from (1.3).

Now we shall proceed to the proof of Proposition 3.1. We assume $s > n/2$ and $A_{(s,s)}(m) = 1$. By duality, it is sufficient to prove the estimate

$$\left| \int T_m(f, g)(x)h(x)\, dx \right| \lesssim \|f\|_2 \|g\|_\infty \|h\|_2. \tag{3.7}$$

As we have seen above, we may further assume that $\operatorname{supp} m \subset \Gamma(\mathcal{O})$, where $\mathcal{O} \subset \Sigma$ is a sufficiently small neighborhood of $(\xi_0, \eta_0) \in \Sigma$. If $\operatorname{supp} m \subset \Gamma(\mathcal{O})$ and if m_j is given by (3.2), then $\operatorname{supp} m_j \subset \Gamma(\mathcal{O}) \cap \{\zeta \in \mathbb{R}^{2n} : 2^{j-1} \le |\zeta| \le 2^{j+1}\}$. If $(\xi_0, \eta_0) \in \Sigma$, then at least two of ξ_0, η_0, $\xi_0 + \eta_0$ are not equal to 0. Thus we shall divide the proof into three cases.

Case 1: $\xi_0 \neq 0$ and $\eta_0 \neq 0$.

In this case, we can take the neighborhood $\mathcal{O}$ of (ξ_0, η_0) so that $|\xi| \approx |\eta| \approx 2^j$ for $(\xi, \eta) \in \operatorname{supp} m_j$. Hence we can take S-operators S_j and D-operators Δ_j such that $T_{m_j}(f, g) = S_j T_{m_j}(\Delta_j f, \Delta_j g)$. Thus

$$\int T_{m_j}(f, g)(x)h(x)\, dx = \int T_{m_j}(\Delta_j f, \Delta_j g)(x) S_j h(x)\, dx.$$

We take a real number σ such that $\max\{1, n/s\} < \sigma < 2$. Then using the pointwise inequality of Lemma 3.2 for $T_{m_j}(\Delta_j f, \Delta_j g)(x)$ and using Schwarz inequality, we obtain

$$\left| \int T_m(f, g)(x)h(x)\, dx \right| = \left| \sum_{j\in\mathbb{Z}} \int T_{m_j}(\Delta_j f, \Delta_j g)(x) S_j h(x)\, dx \right|$$

$$\lesssim \sum_j \int (\zeta_j * |\Delta_j f|^\sigma)(x)^{1/\sigma} (\zeta_j * |\Delta_j g|^\sigma)(x)^{1/\sigma} |S_j h(x)|\, dx$$
$$\lesssim \Big(\sum_j \int (\zeta_j * |\Delta_j f|^\sigma)(x)^{2/\sigma}\, dx \Big)^{1/2}$$
$$\times \Big(\sum_j \int (\zeta_j * |\Delta_j g|^\sigma)(x)^{2/\sigma} |S_j h(x)|^2\, dx \Big)^{1/2}.$$

For the first factor in the last expression, Young's inequality gives

$$\Big(\sum_j \int (\zeta_j * |\Delta_j f|^\sigma)(x)^{2/\sigma}\, dx \Big)^{1/2} \lesssim \Big(\sum_j \int |\Delta_j f(x)|^2\, dx \Big)^{1/2} \lesssim \|f\|_2.$$

For the second factor, notice that

$$(\zeta_j * |\Delta_j g|^\sigma)(x)^{2/\sigma} \lesssim (\zeta_j * |\Delta_j g|^2)(x)$$

by Hölder's inequality since $\sigma < 2$. Hence, Lemma 3.4 implies that the measure

$$\sum_{j\in\mathbb{Z}} (\zeta_j * |\Delta_j g|^\sigma)(x)^{2/\sigma}\, dx \otimes \delta_{2^{-j}}(t) \tag{3.8}$$

is a Carleson measure with Carleson norm $\lesssim \|g\|_{BMO}^2$. Hence the inequality (3.5) gives

$$\Big(\sum_j \int (\zeta_j * |\Delta_j g|^\sigma)(x)^{2/\sigma} |S_j h(x)|^2\, dx \Big)^{1/2}$$
$$\lesssim \|g\|_{BMO} \Big(\int \sup \big\{ |S_j h(y)|^2 \mid |y-x| < 2^{-j} \big\}\, dx \Big)^{1/2}$$
$$\lesssim \|g\|_{BMO} \|h\|_2 \lesssim \|g\|_\infty \|h\|_2,$$

where the second inequality holds because

$$\sup \big\{ |S_j h(y)|^2 : |y-x| < 2^{-j} \big\} \lesssim M(h)(x)^2$$

with M the Hardy-Littlewood maximal operator.

Case 2: $\xi_0 \neq 0$ and $\xi_0 + \eta_0 \neq 0$.

In this case, we can take the neighborhood $\mathcal{O}$ of (ξ_0, η_0) so that $|\xi| \approx |\xi+\eta| \approx 2^j$ for $(\xi, \eta) \in \operatorname{supp} m_j$. Hence we can take S-operators S_j and D-operators Δ_j such that $T_{m_j}(f,g) = \Delta_j T_{m_j}(\Delta_j f, S_j g)$. Thus, by Lemma 3.2 and Schwarz inequality, we obtain

$$\Big| \int T_m(f,g)(x) h(x)\, dx \Big| = \Big| \sum_{j\in\mathbb{Z}} \int T_{m_j}(\Delta_j f, S_j g)(x) \Delta_j h(x)\, dx \Big|$$
$$\lesssim \sum_j \int (\zeta_j * |\Delta_j f|^\sigma)(x)^{1/\sigma} (\zeta_j * |S_j g|^\sigma)(x)^{1/\sigma} |\Delta_j h(x)|\, dx$$

$$\lesssim \left(\sum_j \int (\zeta_j * |\Delta_j f|^\sigma)(x)^{2/\sigma}\, dx \right)^{1/2}$$

$$\times \left(\sum_j \int (\zeta_j * |S_j g|^\sigma)(x)^{2/\sigma} |\Delta_j h(x)|^2\, dx \right)^{1/2}.$$

The first factor in the last expression is $\lesssim \|f\|_2$. Since the integrand in the second factor is $\lesssim \|g\|_\infty^2 |\Delta_j h(x)|^2$, the second factor has the estimate

$$\lesssim \|g\|_\infty \left(\sum_j \int |\Delta_j h(x)|^2\, dx \right)^{1/2} \lesssim \|g\|_\infty \|h\|_2.$$

Case 3: $\eta_0 \neq 0$ and $\xi_0 + \eta_0 \neq 0$.

In this case, we can take the neighborhood $\mathcal{O}$ of (ξ_0, η_0) so that $|\eta| \approx |\xi+\eta| \approx 2^j$ for $(\xi, \eta) \in \operatorname{supp} m_j$. Hence we can take D-operators Δ_j and S-operators S_j such that $T_{m_j}(f,g) = \Delta_j T_{m_j}(S_j f, \Delta_j g)$. Thus, in the same way as in the above cases, we have

$$\left| \int T_m(f,g)(x) h(x)\, dx \right| = \left| \sum_j \int T_{m_j}(S_j f, \Delta_j g)(x) \Delta_j h(x)\, dx \right|$$

$$\lesssim \sum_j \int (\zeta_j * |S_j f|^\sigma)(x)^{1/\sigma} (\zeta_j * |\Delta_j g|^\sigma)(x)^{1/\sigma} |\Delta_j h(x)|\, dx$$

$$\lesssim \left(\sum_j \int (\zeta_j * |S_j f|^\sigma)(x)^{2/\sigma} (\zeta_j * |\Delta_j g|^\sigma)(x)^{2/\sigma}\, dx \right)^{1/2}$$

$$\times \left(\sum_j \int |\Delta_j h(x)|^2\, dx \right)^{1/2}.$$

The second factor in the last expression is $\lesssim \|h\|_2$. To estimate the first factor, we use Lemma 3.4 and the inequality (3.5). As mentioned in Case 1, Lemma 3.4 implies that the measure (3.8) is a Carleson measure with Carleson norm $\lesssim \|g\|_{BMO}^2$. We have

$$\sup \left\{ (\zeta_j * |S_j f|^\sigma)(y)^{2/\sigma} : |y - x| < 2^{-j} \right\} \lesssim M_\sigma(Mf)(x)^2,$$

where the operator M_σ is defined by $M_\sigma(F)(x) = M(|F|^\sigma)(x)^{1/\sigma}$. Thus, by (3.5),

$$\left(\sum_j \int (\zeta_j * |S_j f|^\sigma)(x)^{2/\sigma} (\zeta_j * |\Delta_j g|^\sigma)(x)^{2/\sigma}\, dx \right)^{1/2}$$

$$\lesssim \|g\|_{BMO} \|M_\sigma(Mf)\|_2 \lesssim \|g\|_\infty \|f\|_2.$$

Thus (3.7) is proved in all cases and the proof of Proposition 3.1 is complete.

4 Estimates for $H^1 \times L^\infty \to L^1$

In this section we prove the following proposition, which covers the if part of Theorem 2.2 for $(p,q)=(1,\infty)$.

Proposition 4.1. *If $s>n/2$, then $\|T_m(f,g)\|_1 \lesssim A_{(s,s)}(m)\|f\|_{H^1}\,|g\|_\infty$.*

We shall prove this proposition in a way almost parallel to the proof of the H^1-L^1-estimate for linear Fourier multiplier operators. Without loss of generality, we assume $s>n/2$ and $A_{(s,s)}(m)=1$. We freeze $g \in L^\infty$ with $\|g\|_\infty = 1$ and consider the linear operator $f \mapsto T_m(f,g)$. By virtue of the atomic decomposition theorem for H^1, it is sufficient to show the estimate $\|T_m(a,g)\|_1 \lesssim 1$ for all H^1-atoms a. By translation, we may assume that a satisfies

$$\operatorname{supp} a \subset \{x \in \mathbb{R}^n : |x| \le r\}, \quad \|a\|_\infty \le r^{-n}, \quad \int a(x)\,dx = 0$$

with $r = r_a > 0$.

By Proposition 3.1, we have

$$\|T_m(a,g)\|_2 \lesssim A_{(s,s)}(m)\|a\|_2\|g\|_\infty = \|a\|_2.$$

Hence, by Schwarz inequality,

$$\|T_m(a,g)(x)\|_{1,\,|x|\le 2r} \lesssim r^{n/2}\|a\|_2 \lesssim 1.$$

The rest of the proof is to show

$$\|T_m(a,g)(x)\|_{1,\,|x|>2r} \lesssim 1. \tag{4.1}$$

To prove this, we use the decomposition (3.2).

Consider first m_0 and $K_0 = \mathcal{F}^{-1}m_0$. We have

$$T_{m_0}(a,g)(x) = \iint_{|y|<r,\, z\in\mathbb{R}^n} K_0(x-y,x-z)a(y)g(z)\,dydz. \tag{4.2}$$

By the moment condition for a, we also have

$$\begin{aligned} T_{m_0}(a,g)(x) &= \iint \{K_0(x-y,x-z) - K_0(x,x-z)\}a(y)g(z)\,dydz \\ &= \iiint_{\substack{0<t<1\\ |y|<r,\, z\in\mathbb{R}^n}} \nabla_1 K_0(x-ty,x-z)\cdot(-y)a(y)g(z)\,dtdydz, \end{aligned} \tag{4.3}$$

where $\nabla_1 K_0(x,y)$ denotes the gradient of $K_0(x,y)$ with respect to x. Notice that $|x| \approx |x-y|$ for $|x|>2r$ and $|y|<r$. Hence, if $|x|>2r$, then (4.2) and Schwarz inequality give

$$|x|^s |T_{m_0}(a,g)(x)| \lesssim \iint_{|y|<r,\, z\in\mathbb{R}^n} |x-y|^s |K_0(x-y,x-z)| r^{-n}\,dydz$$

$$\lesssim r^{-n} \int_{|y|<r} \big\| |x-y|^s \langle z\rangle^s K_0(x-y,z) \big\|_{2,\,z}\, dy.$$

Hence, by Schwarz inequality, Minkowski's inequality for integrals, and (3.4), we obtain

$$\begin{aligned}
&\|T_{m_0}(a,g)(x)\|_{1,\,|x|>2r} \\
&\lesssim \big\| |x|^{-s} \big\|_{2,\,|x|>2r} \left\| r^{-n} \int_{|y|<r} \big\| |x-y|^s \langle z\rangle^s K_0(x-y,z) \big\|_{2,\,z}\, dy \right\|_{2,\,x} \\
&\lesssim r^{-s+n/2} r^{-n} \int_{|y|<r} \big\| |x-y|^s \langle z\rangle^s K_0(x-y,z) \big\|_{2,\,(z,x)}\, dy \\
&\lesssim r^{-s+n/2} \|m_0\|_{W^{(s,s)}} \le r^{-s+n/2}.
\end{aligned}$$

Similarly, from (4.3), we obtain

$$\begin{aligned}
&\langle x\rangle^s |T_{m_0}(a,g)(x)| \\
&\lesssim \iiint_{\substack{0<t<1\\ |y|<r,\, z\in\mathbb{R}^n}} \langle x-ty\rangle^s |\nabla_1 K_0(x-ty, x-z)| r^{1-n}\, dtdydz \\
&\lesssim r^{1-n} \iint_{0<t<1,\,|y|<r} \big\| \langle x-ty\rangle^s \langle z\rangle^s \nabla_1 K_0(x-ty, z) \big\|_{2,\,z}\, dtdy
\end{aligned}$$

for $|x|>2r$. Thus, by Schwarz inequality,

$$\begin{aligned}
&\|T_{m_0}(a,g)(x)\|_{1,\,|x|>2r} \\
&\lesssim \big\| \langle x\rangle^{-s} \big\|_2 \left\| r^{1-n} \iint_{0<t<1,\,|y|<r} \big\| \langle x-ty\rangle^s \langle z\rangle^s \nabla_1 K_0(x-ty,z) \big\|_{2,\,z}\, dtdy \right\|_{2,\,x} \\
&\lesssim r^{1-n} \iint_{0<t<1,\,|y|<r} \big\| \langle x-ty\rangle^s \langle z\rangle^s \nabla_1 K_0(x-ty,z) \big\|_{2,\,(z,x)}\, dtdy \\
&\lesssim r \|m_0\|_{W^{(s,s)}} \le r,
\end{aligned}$$

where the next to the last inequality follows from (3.3) and Lemma 3.3, and the last inequality from (3.4). Combining the two estimates, we obtain

$$\|T_{m_0}(a,g)(x)\|_{1,\,|x|>2r} \lesssim \min\{r^{-s+n/2},\, r\}. \tag{4.4}$$

Now consider general $j \in \mathbb{Z}$. Notice that the dilation formula (1.2) gives

$$T_{m_j}(a,g)(x) = T_{m_j(2^j\cdot)}\big(a(2^{-j}\cdot), g(2^{-j}\cdot)\big)(2^j x) \tag{4.5}$$

and that $2^{-jn}a(2^{-j}\cdot)$ is an H^1-atom supported on $\{|x| \le 2^j r\}$. Also recall (3.3) and (3.4). Hence, from (4.4), by changing variables, we obtain

$$\|T_{m_j}(a,g)(x)\|_{1,\,|x|>2r} \lesssim \min\{(2^j r)^{-s+n/2},\, 2^j r\}.$$

Thus

$$\|T_m(a,g)(x)\|_{1,\,|x|>2r} \lesssim \sum_{j\in\mathbb{Z}} \min\{(2^j r)^{-s+n/2},\, 2^j r\} \approx 1.$$

This completes the proof of Proposition 4.1.

5 Estimates for $L^\infty \times L^\infty \to BMO$

In this section we prove the following proposition, which covers the if part of Theorem 2.2 for $(p,q) = (\infty,\infty)$.

Proposition 5.1. *If $s > n/2$, then $\|T_m(f,g)\|_{BMO} \lesssim A_{(s,s)}(m)\|f\|_\infty\|g\|_\infty$.*

The proof is a slight modification of the proof of the L^∞-BMO-estimate of linear Fourier multiplier operators. We use the following lemma.

Lemma 5.2. *Let T_m be a bilinear Fourier multiplier operator and $K = \mathcal{F}^{-1}m$. Suppose*

$$\|T_m(f,g)\|_2 \le A\|f\|_2\|g\|_\infty, \quad \|T_m(f,g)\|_2 \le A\|f\|_\infty\|g\|_2, \tag{5.1}$$

and suppose

$$\iint_{\substack{|y|>2|x| \\ |z|>2|x|}} \big|K(x+y,x+z) - K(y,z)\big|\,dydz \le A \tag{5.2}$$

for all $x \in \mathbb{R}^n$. Then $\|T_m(f,g)\|_{BMO} \lesssim A\|f\|_\infty\|g\|_\infty$.

Proof. It is sufficient to prove that for each cube Q there exists a constant $a_Q \in \mathbb{C}$ such that

$$\frac{1}{|Q|}\int_Q |T_m(f,g)(x) - a_Q|\,dx \lesssim A\|f\|_\infty\|g\|_\infty.$$

By translation, we may assume the cube Q is centered at the origin. We may also assume $\|f\|_\infty = \|g\|_\infty = 1$. We take

$$a_Q = \iint_{\substack{y\in(Q^*)^c \\ z\in(Q^*)^c}} K(-y,-z)f(y)g(z)\,dydz,$$

where Q^* is the cube with the same center as Q and with the sidelength C times as large as Q, with C a sufficiently large constant. We decompose f and g as

$$f = f\mathbf{1}_{Q^*} + f\mathbf{1}_{(Q^*)^c} = f_0 + f_1,$$
$$g = g\mathbf{1}_{Q^*} + g\mathbf{1}_{(Q^*)^c} = g_0 + g_1.$$

Then

$$\begin{aligned} &|Q|^{-1}\int_Q |T_m(f,g)(x) - a_Q|\,dx \\ &\le |Q|^{-1}\int_Q |T_m(f_0,g_0)(x)|\,dx + |Q|^{-1}\int_Q |T_m(f_1,g_0)(x)|\,dx \\ &\quad + |Q|^{-1}\int_Q |T_m(f_0,g_1)(x)|\,dx + |Q|^{-1}\int_Q |T_m(f_1,g_1)(x) - a_Q|\,dx. \end{aligned} \tag{5.3}$$

The first three terms on the right hand side of (5.3) can be estimated by the L^2-estimate (5.1). In fact, for the third term,

$$|Q|^{-1}\int_Q |T_m(f_0,g_1)(x)|\,dx \le |Q|^{-1/2}\|T_m(f_0,g_1)\|_2 \le |Q|^{-1/2}A\|f_0\|_2\|g_1\|_\infty \lesssim A,$$

and the first and the second terms can be estimated in the same way. To estimate the last term of (5.3), notice that $|y| > 2|x|$ if $y \in (Q^*)^c$ and $x \in Q$. Hence we obtain

$$
\begin{aligned}
&|Q|^{-1} \int_Q |T_m(f_1, g_1)(x) - a_Q| \, dx \\
&= |Q|^{-1} \int_Q \left| \iint_{\substack{y \in (Q^*)^c \\ z \in (Q^*)^c}} (K(x-y, x-z) - K(-y,-z)) f(y) g(z) \, dydz \right| dx \\
&\leq |Q|^{-1} \int_Q \left(\iint_{\substack{|y|>2|x| \\ |z|>2|x|}} |K(x-y, x-z) - K(-y,-z)| \, dydz \right) dx \\
&\leq A
\end{aligned}
$$

by (5.2). This completes the proof of Lemma 5.2. □

Now we shall prove Proposition 5.1. We assume $s > n/2$ and $A_{(s,s)}(m) = 1$. By Proposition 3.1, we have the estimates (5.1) with $A \approx 1$. Thus, by Lemma 5.2, it is sufficient to show that $K = \mathcal{F}^{-1}m$ satisfies the estimate (5.2) with $A \approx 1$. We use the decomposition (3.2) and write $K_j = \mathcal{F}^{-1}m_j$.

Consider $K_0 = \mathcal{F}^{-1}m_0$. We have

$$
\begin{aligned}
&\iint_{\substack{|y|>2|x| \\ |z|>2|x|}} |K_0(x+y, x+z) - K_0(y,z)| \, dydz \\
&\leq \iint_{\substack{|y|>2|x| \\ z \in \mathbb{R}^n}} |K_0(x+y, x+z) - K_0(y, x+z)| \, dydz \\
&\quad + \iint_{\substack{y \in \mathbb{R}^n \\ |z|>2|x|}} |K_0(y, x+z) - K_0(y,z)| \, dydz.
\end{aligned} \tag{5.4}
$$

For the second integral on the right hand side of (5.4), we have

$$
\begin{aligned}
&\iint_{\substack{y \in \mathbb{R}^n \\ |z|>2|x|}} |K_0(y, x+z) - K_0(y,z)| \, dydz \\
&= \iint_{\substack{y \in \mathbb{R}^n \\ |z|>2|x|}} \left| \int_0^1 \nabla_1 K_0(y, tx+z) \cdot x \, dt \right| dydz \\
&\lesssim |x| \iint_{\mathbb{R}^{2n}} |\nabla_1 K_0(y,z)| \, dydz \lesssim |x| \left\| \langle y \rangle^s \langle z \rangle^s \nabla_1 K_0(y,z) \right\|_{2,\,(y,z)} \\
&\lesssim |x| \|m_0\|_{W^{(s,s)}} \leq |x|,
\end{aligned}
$$

where the second inequality follows from Schwarz inequality and the assumption $s > n/2$, the third inequality from (3.3) and Lemma 3.3, and the last inequality from (3.4). On the other hand,

$$
\iint_{\substack{y \in \mathbb{R}^n \\ |z|>2|x|}} |K_0(y, x+z) - K_0(y,z)| \, dydz \leq 2 \iint_{\substack{y \in \mathbb{R}^n \\ |z|>|x|}} |K_0(y,z)| \, dydz
$$

and, by Schwarz inequality,

$$\iint_{\substack{y\in\mathbb{R}^n\\|z|>|x|}} |K_0(y,z)|\,dydz$$
$$\le \left\| \langle y\rangle^{-s}|z|^{-s} \right\|_{2,\, y\in\mathbb{R}^n,\, |z|>|x|} \left\| \langle y\rangle^{s}|z|^{s}K_0(y,z) \right\|_{2,\,(y,z)\in\mathbb{R}^{2n}}$$
$$\lesssim |x|^{-s+n/2}\|m_0\|_{W^{(s,s)}} \le |x|^{-s+n/2}.$$

Thus

$$\iint_{\substack{y\in\mathbb{R}^n\\|z|>2|x|}} |K_0(y,x+z)-K_0(y,z)|\,dydz \lesssim |x|^{-s+n/2}.$$

By symmetry, the first integral on the right hand side of (5.4) has the same estimates. Thus, combining the above estimates, we obtain

$$\iint_{\substack{|y|>2|x|\\|z|>2|x|}} |K_0(x+y,x+z)-K_0(y,z)|\,dydz \lesssim \min\{|x|,\,|x|^{-s+n/2}\}. \tag{5.5}$$

For arbitrary $j\in\mathbb{Z}$, we have

$$K_j(y,z)=\mathcal{F}^{-1}m_j(y,z)=2^{2jn}\mathcal{F}^{-1}\big(m_j(2^j\cdot)\big)(2^jy,2^jz).$$

Hence (5.5) and a change of variables give

$$\iint_{\substack{|y|>2|x|\\|z|>2|x|}} |K_j(x+y,x+z)-K_j(y,z)|\,dydz \lesssim \min\{2^j|x|,\,(2^j|x|)^{-s+n/2}\}.$$

Thus

$$\iint_{\substack{|y|>2|x|\\|z|>2|x|}} |K(x+y,x+z)-K(y,z)|\,dydz \lesssim \sum_j \min\{2^j|x|,\,(2^j|x|)^{-s+n/2}\} \approx 1.$$

This completes the proof of Proposition 5.1.

6 Estimates for $H^1\times H^1\to L^{1/2}$

In this section we prove the following proposition.

Proposition 6.1. *If $s>3n/4$, then $\|T_m(f,g)\|_{1/2}\lesssim A_{(s,s)}(m)\|f\|_{H^1}\|g\|_{H^1}$.*

This proposition does not completely cover the if part of Theorem 2.2 for $(p,q)=(1,1)$ but the proof of the proposition will show the most essential ideas.

Our basic tool is the atomic decomposition for H^1. Here, however, observe that simple norm estimate

$$\|T_m(a,b)\|_{1/2}\lesssim A_{(s,s)}(m) \tag{6.1}$$

for H^1-atoms a and b does not directly imply the boundedness of T_m. For, if we decompose f, $g \in H^1$ as

$$f = \sum_i \lambda_i a_i, \quad g = \sum_k \mu_k b_k, \tag{6.2}$$

$$\sum_i \lambda_i \approx \|f\|_{H^1}, \quad \sum_k \mu_k \approx \|g\|_{H^1}, \tag{6.3}$$

where λ_i and μ_k are nonnegative real numbers and a_i and b_k are H^1-atoms, then by the subadditivity of $\|\cdot\|_{1/2}^{1/2}$ and by (6.1) we obtain

$$\begin{aligned}\|T_m(a,b)\|_{1/2}^{1/2} &\le \sum_i \sum_k \lambda_i^{1/2} \mu_k^{1/2} \|T_m(a_i, b_k)\|_{1/2}^{1/2} \\ &\lesssim \sum_i \sum_k \lambda_i^{1/2} \mu_k^{1/2} A_{(s,s)}(m)^{1/2}.\end{aligned}$$

But the last sums cannot be controlled by (6.3).

Thus we shall use a slightly stronger form of atomic decomposition, which reads as follows: If $f \in H^1$, then there exist functions $\widetilde{a}_i \in L^\infty$ and cubes Q_i such that

$$f = \sum_i \widetilde{a}_i, \quad \operatorname{supp} \widetilde{a}_i \subset Q_i, \quad \int \widetilde{a}_i(x)\, dx = 0,$$

$$\Big(\sum_i \|\widetilde{a}_i\|_\infty^\epsilon \mathbf{1}_{Q_i}(x) \Big)^{1/\epsilon} \le c_\epsilon f^*(x), \tag{6.4}$$

where f^* is an appropriate grand maximal function of f and the inequality (6.4) holds for every $\epsilon > 0$. (The $\widetilde{a}_i$ can be taken independent of ϵ.) For such a decomposition, see [17, Chapter VIII].

To prove Proposition 6.1, we assume $s > 3n/4$ and $A_{(s,s)}(m) = 1$. We decompose f as above and similarly decompose g as

$$g = \sum_k \widetilde{b}_k, \quad \operatorname{supp} \widetilde{b}_k \subset R_k, \quad \int \widetilde{b}_k(x)\, dx = 0,$$

$$\Big(\sum_k \|\widetilde{b}_k\|_\infty^\epsilon \mathbf{1}_{R_k}(x) \Big)^{1/\epsilon} \le c_\epsilon g^*(x),$$

where R_k are cubes. We have

$$T_m(f,g) = \sum_{i,k} T_m(\widetilde{a}_i, \widetilde{b}_k) \mathbf{1}_{Q_i^* \cap R_k^*} + \sum_{i,k} T_m(\widetilde{a}_i, \widetilde{b}_k) \mathbf{1}_{(Q_i^* \cap R_k^*)^c}. \tag{6.5}$$

For cubes Q, we write $Q^{**} = (Q^*)^*$.

The first sum on the right hand side of (6.5) is estimated as follows. By Proposition 3.1, T_m is bounded in $L^2 \times L^\infty \to L^2$ and $L^\infty \times L^2 \to L^2$. If $Q_i^* \cap R_k^* \neq \emptyset$ and $|Q_i| > |R_k|$, then by Hölder's inequality and by the $L^\infty \times L^2 \to L^2$ boundedness of T_m we have

$$\|T_m(\widetilde{a}_i, \widetilde{b}_k) \mathbf{1}_{Q_i^* \cap R_k^*}\|_{1/2} \le \|T_m(\widetilde{a}_i, \widetilde{b}_k)\|_2 |Q_i^* \cap R_k^*|^{3/2}$$

$$\lesssim \|\widetilde{a}_i\|_\infty \|\widetilde{b}_k\|_2 |R_k|^{3/2} \lesssim \|\widetilde{a}_i\|_\infty \|\widetilde{b}_k\|_\infty |R_k|^2 \le \|\widetilde{a}_i\|_\infty \|\widetilde{b}_k\|_\infty |Q_i^{**} \cap R_k^{**}|^2,$$

where the last inequality holds because $R_k \subset Q_i^{**} \cap R_k^{**}$ if $Q_i^* \cap R_k^* \neq \emptyset$ and $|Q_i| > |R_k|$. By symmetry the same estimate holds if $Q_i^* \cap R_k^* \neq \emptyset$ and $|Q_i| \le |R_k|$. If $Q_i^* \cap R_k^* = \emptyset$, then the estimate obviously holds. Hence, by subadditivity of $\|\cdot\|_{1/2}^{1/2}$ and by Hölder's inequality, we have

$$\begin{aligned}
&\Big\| \sum_{i,k} T_m(\widetilde{a}_i, \widetilde{b}_k) \mathbf{1}_{Q_i^* \cap R_k^*} \Big\|_{1/2}^{1/2} \le \sum_{i,k} \Big\| T_m(\widetilde{a}_i, \widetilde{b}_k) \mathbf{1}_{Q_i^* \cap R_k^*} \Big\|_{1/2}^{1/2} \\
&\lesssim \sum_{i,k} \|\widetilde{a}_i\|_\infty^{1/2} \|\widetilde{b}_k\|_\infty^{1/2} |Q_i^{**} \cap R_k^{**}| \\
&= \int \Big(\sum_i \|\widetilde{a}_i\|_\infty^{1/2} \mathbf{1}_{Q_i^{**}}(x) \Big) \Big(\sum_k \|\widetilde{b}_k\|_\infty^{1/2} \mathbf{1}_{R_k^{**}}(x) \Big) \, dx \\
&\le \Big\| \Big(\sum_i \|\widetilde{a}_i\|_\infty^{1/2} \mathbf{1}_{Q_i^{**}}(x) \Big)^2 \Big\|_1^{1/2} \Big\| \Big(\sum_k \|\widetilde{b}_k\|_\infty^{1/2} \mathbf{1}_{R_k^{**}}(x) \Big)^2 \Big\|_1^{1/2}.
\end{aligned}$$

If we take a $0 < \delta < 1/2$, then, since

$$M_\delta(\mathbf{1}_{Q_i})(x) \approx (1 + \ell(Q_i)^{-1}|x - c(Q_i)|)^{-n/\delta} \gtrsim \mathbf{1}_{Q_i^{**}}(x)$$

($c(Q_i)$ denotes the center of Q_i), Fefferman-Stein vector maximal inequality ([4]) and (6.4) give

$$\begin{aligned}
&\Big\| \Big(\sum_i \|\widetilde{a}_i\|_\infty^{1/2} \mathbf{1}_{Q_i^{**}}(x) \Big)^2 \Big\|_1 \lesssim \Big\| \Big(\sum_i M_\delta(\|\widetilde{a}_i\|_\infty \mathbf{1}_{Q_i})(x)^{1/2} \Big)^2 \Big\|_1 \\
&\lesssim \Big\| \Big(\sum_i \|\widetilde{a}_i\|_\infty^{1/2} \mathbf{1}_{Q_i}(x) \Big)^2 \Big\|_1 \lesssim \|f^*\|_1 \approx \|f\|_{H^1}.
\end{aligned}$$

Similarly we have

$$\Big\| \Big(\sum_k \|\widetilde{b}_k\|_\infty^{1/2} \mathbf{1}_{R_k^{**}}(x) \Big)^2 \Big\|_1 \lesssim \|g\|_{H^1}.$$

Combining the above inequalities, we obtain

$$\Big\| \sum_{i,k} T_m(\widetilde{a}_i, \widetilde{b}_k) \mathbf{1}_{Q_i^* \cap R_k^*} \Big\|_{1/2} \lesssim \|f\|_{H^1} \|g\|_{H^1}.$$

To estimate the second sum on the right hand side of (6.5), we write

$$\begin{aligned}
\widetilde{a}_i &= \lambda_1 a_i, \quad \lambda_i = \|\widetilde{a}_i\|_\infty |Q_i|, \\
\widetilde{b}_k &= \mu_k b_k, \quad \mu_k = \|\widetilde{b}_k\|_\infty |R_k|.
\end{aligned}$$

Then a_i and b_k are usual H^1-atoms and we have the usual atomic decompositions (6.2) and (6.3); in fact (6.3) holds because (6.4) yields

$$\sum_i \lambda_i = \int \sum_i \|\widetilde{a}_i\|_\infty \mathbf{1}_{Q_i}(x) \, dx \lesssim \int f^*(x) \, dx \approx \|f\|_{H^1}$$

and the similar inequality for μ_k and g. We shall prove that there exist functions u_i and v_k such that u_i depends only on m and a_i, v_k depends only on m and b_k, and

$$|T_m(a_i, b_k)(x)|\mathbf{1}_{(Q_i^*\cap R_k^*)^c}(x) \lesssim u_i(x)v_k(x), \qquad (6.6)$$
$$\|u_i\|_1 \lesssim 1, \quad \|v_k\|_1 \lesssim 1.$$

If this is proved, then the second sum on the right hand side of (6.5) can be estimated by Hölder's inequality as follows:

$$\Big\|\sum_{i,k} T_m(\widetilde{a}_i, \widetilde{b}_k)\mathbf{1}_{(Q_i^*\cap R_k^*)^c}\Big\|_{1/2} = \Big\|\sum_{i,k} \lambda_i\mu_k T_m(a_i, b_k)\mathbf{1}_{(Q_i^*\cap R_k^*)^c}\Big\|_{1/2}$$
$$\lesssim \Big\|\sum_{i,k} \lambda_i\mu_k u_i v_k\Big\|_{1/2} \le \Big\|\sum_{i} \lambda_i u_i\Big\|_{1}\Big\|\sum_{k} \mu_k v_k\Big\|_{1}$$
$$\lesssim \Big(\sum_i \lambda_i\Big)\Big(\sum_k \mu_k\Big) \lesssim \|f\|_{H^1}\|g\|_{H^1}.$$

(Notice that the estimate (6.6) should not be of the form

$$|T_m(a_i, b_k)(x)|\mathbf{1}_{(Q_i^*\cap R_k^*)^c}(x) \lesssim u_{i,k}(x)v_{i,k}(x),$$
$$\|u_{i,k}\|_1 \lesssim 1, \quad \|v_{i,k}\|_1 \lesssim 1,$$

which is in fact equivalent to the norm estimate $\|T_m(a_i, b_k)\|_{1/2,\,(Q_i^*\cap R_k^*)^c} \lesssim 1$.)

To prove the estimate (6.6), we simply write $a = a_i$, $b = b_k$, $Q = Q_i$, and $R = R_k$. We shall consider $x \in (Q^*)^c \cap (R^*)^c$, $x \in (Q^*)^c \cap R^*$, and $x \in Q^* \cap (R^*)^c$ separately.

First, assume $x \in (Q^*)^c \cap (R^*)^c$. We use the decomposition (3.2). Consider m_0 and write $K_0 = \mathcal{F}^{-1}m_0$. We have

$$T_{m_0}(a, b)(x) = \iint_{y\in Q,\, z\in R} K_0(x-y, x-z)a(y)b(z)\, dydz. \qquad (6.7)$$

By the moment condition of a, we also have

$$\begin{aligned} &T_{m_0}(a, b)(x) \\ &= \iint_{y\in Q,\, z\in R} \{K_0(x-y, x-z) - K_0(x-c_Q, x-z)\}a(y)b(z)\, dydz \\ &= \iiint_{\substack{0<t<1\\ y\in Q,\, z\in R}} \nabla_1 K_0(x-c_Q-t(y-c_Q), x-z)\cdot(-y+c_Q)a(y)b(z)\, dtdydz, \end{aligned} \qquad (6.8)$$

where c_Q denotes the center of the cube Q. Since $x \in (Q^*)^c \cap (R^*)^c$, we have $|x-c_Q| \approx |x-y|$ for $y \in Q$ and $|x-c_R| \approx |x-z|$ for $z \in R$. Hence, from (6.7), we obtain

$$\langle x-c_Q\rangle^s\langle x-c_R\rangle^s|T_{m_0}(a, b)(x)|$$

$$
\begin{aligned}
&\lesssim \iint_{y\in Q,\, z\in R} \langle x-y\rangle^{s}\langle x-z\rangle^{s}|K_0(x-y,x-z)|\,|Q|^{-1}|R|^{-1}\,dydz \\
&\lesssim |Q|^{-1}\int_Q \big\|\langle x-y\rangle^{s}\langle z\rangle^{s}K_0(x-y,z)\big\|_{\infty,\,z}\,dy \\
&\lesssim |Q|^{-1}\int_Q \big\|\langle x-y\rangle^{s}\langle z\rangle^{s}K_0(x-y,z)\big\|_{2,\,z}\,dy,
\end{aligned}
$$

where the last inequality holds by virtue of Lemma 3.3. We write the last expression as $g_0^Q(x)$:

$$
g_0^Q(x)=|Q|^{-1}\int_Q \big\|\langle x-y\rangle^{s}\langle z\rangle^{s}K_0(x-y,z)\big\|_{2,\,z}\,dy. \tag{6.9}
$$

By Minkowski's inequality for integrals, we have

$$
\begin{aligned}
\|g_0^Q\|_2 &\le |Q|^{-1}\int_Q \big\|\langle x-y\rangle^{s}\langle z\rangle^{s}K_0(x-y,z)\big\|_{2,\,(z,x)}\,dy \\
&= \big\|\langle x\rangle^{s}\langle z\rangle^{s}K_0(x,z)\big\|_{2,\,(z,x)} \approx \|m_0\|_{W^{(s,s)}} \le 1. \qquad (6.10)
\end{aligned}
$$

On the other hand, using the expression (6.8) and arguing in the same way as above, we obtain

$$
\langle x-c_Q\rangle^{s}\langle x-c_R\rangle^{s}|T_{m_0}(a,b)(x)| \lesssim \widetilde{g}_0^Q(x),
$$

where

$$
\begin{aligned}
\widetilde{g}_0^Q(x)=|Q|^{-1+1/n}\iint_{\substack{0<t<1\\ y\in Q}} &\big\|\langle x-c_Q-t(y-c_Q)\rangle^{s}\langle z\rangle^{s} \qquad (6.11)\\
&\cdot \nabla_1 K_0(x-c_Q-t(y-c_Q),z)\big\|_{2,\,z}\,dtdy.
\end{aligned}
$$

We also have

$$
\|\widetilde{g}_0^Q\|_2 \lesssim \ell(Q)\big\|\langle x\rangle^{s}\langle z\rangle^{s}\nabla_1 K_0(x,z)\big\|_{2,\,(z,x)} \lesssim \ell(Q)\|m_0\|_{W^{(s,s)}} \le \ell(Q), \tag{6.12}
$$

where the second inequality holds by virtue of Lemma 3.3. Combining the two estimates, we obtain

$$
\begin{aligned}
&|T_{m_0}(a,b)(x)|\mathbf{1}_{(Q^*)^c\cap(R^*)^c}(x) \\
&\lesssim \langle x-c_Q\rangle^{-s}\langle x-c_R\rangle^{-s}\min\{g_0^Q(x),\,\widetilde{g}_0^Q(x)\}. \qquad (6.13)
\end{aligned}
$$

Observe that g_0^Q and $\widetilde{g}_0^Q$ depend only on K_0 and Q.

By interchanging the rolls of y and z in the above argument, we obtain functions h_0^R and $\widetilde{h}_0^R$, depending only on K_0 and R, such that

$$
\begin{aligned}
&|T_{m_0}(a,b)(x)|\mathbf{1}_{(Q^*)^c\cap(R^*)^c}(x) \\
&\lesssim \langle x-c_Q\rangle^{-s}\langle x-c_R\rangle^{-s}\min\{h_0^R(x),\,\widetilde{h}_0^R(x)\} \qquad (6.14)
\end{aligned}
$$

and $\|h_0^R\|_2 \lesssim 1$, $\|\widetilde{h}_0^R\|_2 \lesssim \ell(R)$.

Combining the two estimates (6.13) and (6.14), we have

$$|T_{m_0}(a,b)(x)|\mathbf{1}_{(Q^*)^c\cap(R^*)^c}(x) \lesssim u_0(x)v_0(x)$$

with

$$u_0(x) = \langle x - c_Q\rangle^{-s}\mathbf{1}_{(Q^*)^c}(x)\sqrt{\min\{g_0^Q(x),\, \widetilde{g}_0^Q(x)\}},$$
$$v_0(x) = \langle x - c_R\rangle^{-s}\mathbf{1}_{(R^*)^c}(x)\sqrt{\min\{h_0^R(x),\, \widetilde{h}_0^R(x)\}}.$$

The function u_0 depends only on K_0 and Q, and the function v_0 depends only on K_0 and R. By Hölder's inequality and by (6.10) and (6.12),

$$\begin{aligned}\|u_0\|_1 &\le \left\|\langle x - c_Q\rangle^{-s}\mathbf{1}_{(Q^*)^c}(x)\right\|_{4/3}\left\|\sqrt{\min\{g_0^Q(x),\, \widetilde{g}_0^Q(x)\}}\right\|_4\\ &\lesssim \min\{1,\, \ell(Q)^{-s+3n/4}\}\cdot\min\left\{\|g_0^Q\|_2^{1/2},\, \|\widetilde{g}_0^Q\|_2^{1/2}\right\}\\ &\lesssim \min\{1,\, \ell(Q)^{-s+3n/4}\}\cdot\min\{1,\, \ell(Q)^{1/2}\}\\ &= \min\{\ell(Q)^{1/2},\, \ell(Q)^{-s+3n/4}\}.\end{aligned}$$

Similarly,

$$\|v_0\|_1 \lesssim \min\{\ell(R)^{1/2},\, \ell(R)^{-s+3n/4}\}.$$

For arbitrary $j \in \mathbb{Z}$, we use the dilation formula (4.5) and a change of variables to obtain functions u_j and v_j such that

$$\begin{aligned}&|T_{m_j}(a,b)(x)|\mathbf{1}_{(Q^*)^c\cap(R^*)^c}(x) \lesssim u_j(x)v_j(x),\\ &\|u_j\|_1 \lesssim \min\{(2^j\ell(Q))^{1/2},\, (2^j\ell(Q))^{-s+3n/4}\},\\ &\|v_j\|_1 \lesssim \min\{(2^j\ell(R))^{1/2},\, (2^j\ell(R))^{-s+3n/4}\},\end{aligned}$$

the function u_j depends only on m, j, and Q, and the function v_j depends only on m, j, and R.

Hence, for $T_m(a,b) = \sum_j T_{m_j}(a,b)$, we have

$$|T_m(a,b)(x)|\mathbf{1}_{(Q^*)^c\cap(R^*)^c}(x) \lesssim \sum_j u_j(x)v_j(x) \le \Big(\sum_j u_j(x)\Big)\Big(\sum_j v_j(x)\Big),$$

the function $\sum_j u_j$ depends only on m and Q, the function $\sum_j v_j$ depends only on m and R, and

$$\Big\|\sum_j u_j\Big\|_1 \lesssim \sum_j \min\{(2^j\ell(Q))^{1/2},\, (2^j\ell(Q))^{-s+3n/4}\} \approx 1,$$
$$\Big\|\sum_j v_j\Big\|_1 \lesssim \sum_j \min\{(2^j\ell(R))^{1/2},\, (2^j\ell(R))^{-s+3n/4}\} \approx 1.$$

Next, assume $x \in (Q^*)^c \cap R^*$. As above, we first consider $T_{m_0}(a,b)(x)$. Using (6.7), we obtain

$$\begin{aligned}
&\langle x - c_Q\rangle^s |T_{m_0}(a,b)(x)| \\
&\lesssim \iint_{y\in Q,\, z\in R} \langle x-y\rangle^s |K_0(x-y, x-z)|\, |Q|^{-1}|R|^{-1}\, dydz \\
&\lesssim |Q|^{-1}|R|^{-1} \int_Q \big\| \langle x-y\rangle^s \langle z\rangle^s K_0(x-y,z)\big\|_{2,\,z}\, dy,
\end{aligned}$$

where the latter $\lesssim$ follows from Schwarz inequality. Thus

$$|T_{m_0}(a,b)(x)|\mathbf{1}_{(Q^*)^c\cap R^*}(x) \lesssim \langle x - c_Q\rangle^{-s} g_0^Q(x)|R|^{-1},$$

where g_0^Q is the function of (6.9). Similarly, using (6.8), we obtain

$$|T_{m_0}(a,b)(x)|\mathbf{1}_{(Q^*)^c\cap R^*}(x) \lesssim \langle x - c_Q\rangle^{-s} \widetilde{g_0}^Q(x)|R|^{-1},$$

where $\widetilde{g_0}^Q$ is the function of (6.11). Combining the two estimates, we obtain

$$|T_{m_0}(a,b)(x)|\mathbf{1}_{(Q^*)^c\cap R^*}(x) \lesssim u_0(x)v(x)$$

with

$$u_0(x) = \langle x - c_Q\rangle^{-s}\mathbf{1}_{(Q^*)^c}(x)\min\big\{g_0^Q(x),\, \widetilde{g_0}^Q(x)\big\}, \tag{6.15}$$

$$v(x) = |R|^{-1}\mathbf{1}_{R^*}(x). \tag{6.16}$$

The function u_0 depends only on K_0 and Q. By Hölder's inequality and by (6.10) and (6.12),

$$\begin{aligned}
\|u_0\|_1 &\le \big\|\langle x - c_Q\rangle^{-s}\mathbf{1}_{(Q^*)^c}(x)\big\|_2 \Big\|\min\big\{g_0^Q(x),\, \widetilde{g_0}^Q(x)\big\}\Big\|_2 \\
&\lesssim \min\{1,\, \ell(Q)^{-s+n/2}\}\cdot \min\Big\{\|g_0^Q\|_2,\, \|\widetilde{g_0}^Q\|_2\Big\} \\
&\lesssim \min\{1,\, \ell(Q)^{-s+n/2}\}\cdot \min\{1,\, \ell(Q)\} \\
&= \min\{\ell(Q),\, \ell(Q)^{-s+n/2}\}.
\end{aligned} \tag{6.17}$$

For arbitrary $j \in \mathbb{Z}$, dilation gives

$$|T_{m_j}(a,b)(x)|\mathbf{1}_{(Q^*)^c\cap R^*}(x) \lesssim u_j(x)v(x),$$

where v is given by (6.16), the function u_j depends only on m, j, and Q, and

$$\|u_j\|_1 \lesssim \min\{2^j\ell(Q),\, (2^j\ell(Q))^{-s+n/2}\}.$$

Hence for $T_m(a,b) = \sum_j T_{m_j}(a,b)$ we have

$$|T_m(a,b)(x)|\mathbf{1}_{(Q^*)^c\cap R^*}(x) \lesssim \sum_j u_j(x)v(x),$$

the function $\sum_j u_j$ depends only on m and Q,

$$\Big\| \sum_j u_j \Big\|_1 \lesssim \sum_j \min\{2^j \ell(Q),\, (2^j \ell(Q))^{-s+n/2}\} \approx 1,$$

and obviously $\|v\|_1 \approx 1$.

Finally, estimates for $T_m(a, b)(x)$ for $x \in Q^* \cap (R^*)^c$ can be obtained in a way parallel to the second case. This completes the proof of Proposition 6.1.

7 Estimates for $H^1 \times L^2 \to L^{2/3}$

In this section we prove the following proposition, which covers the if part of Theorem 2.2 for $(p, q) = (1, 2)$.

Proposition 7.1. *If $s > n/2$, then $\|T_m(f, g)\|_{2/3} \lesssim A_{(s,s)}(m)\|f\|_{H^1}\|g\|_2$.*

We use the following lemma.

Lemma 7.2. *Suppose $\{F_j\} \subset \mathcal{S}'$ and suppose there exists a constant $B > 1$ such that $\operatorname{supp} \widehat{F_j} \subset \{\zeta \in \mathbb{R}^n : B^{-1}2^j \le |\zeta| \le B2^j\}$. Then, for each $0 < r < \infty$,*

$$\Big\| \sum_j F_j \Big\|_{H^r} \lesssim \Big\| \Big(\sum_j |F_j|^2 \Big)^{1/2} \Big\|_r .$$

This lemma is well known in the Littlewood-Paley theory. See, e.g., [19, 5.2.4].

To prove Proposition 7.1, assume $s > n/2$ and $A_{(s,s)}(m) = 1$. We take a real number σ such that $\max\{1, n/s\} < \sigma < 2$. By the same reason as given in Section 3, it is sufficient to consider m satisfying $\operatorname{supp} m \subset \Gamma(\mathcal{O})$, where $\mathcal{O}$ is a sufficiently small neighborhood of $(\xi_0, \eta_0) \in \Sigma$. We divide the proof into three cases.

Case 1: $\xi_0 \neq 0$ and $\eta_0 \neq 0$.

To prove the estimate of Proposition 7.1, it is sufficient to show that for each H^1-atom a and each $g \in L^2$ there exist functions u and v such that v depends only on g, and

$$|T_m(a, g)(x)| \lesssim u(x)v(x), \quad \|u\|_1 \lesssim 1, \quad \|v\|_2 \lesssim \|g\|_2. \tag{7.1}$$

In fact, suppose this is proved and take an arbitrary $f \in H^1$. We decompose f as

$$f = \sum_k \lambda_k a_k, \quad \sum_k \lambda_k \approx \|f\|_{H^1},$$

where λ_k are nonnegative real numbers and a_k are H^1-atoms. Let u_k and v be the functions corresponding to the H^1-atom a_k and $g \in L^2$. Then

$$|T_m(f, g)(x)| = \Big| \sum_k \lambda_k T_m(a_k, g)(x) \Big| \lesssim \sum_k \lambda_k u_k(x) v(x)$$

and hence Hölder's inequality gives

$$\|T_m(f,g)\|_{2/3} \le \Big\| \sum_k \lambda_k u_k \Big\|_1 \|v\|_2 \lesssim \sum_k \lambda_k \|v\|_2 \lesssim \|f\|_{H^1} \|g\|_2$$

as desired. Notice that v in (7.1) should be taken independent of a.

Suppose a is an H^1-atom and Q is a cube such that

$$\operatorname{supp} a \subset Q, \quad \|a\|_\infty \le |Q|^{-1}, \quad \int a(x)\, dx = 0. \tag{7.2}$$

To prove the estimates (7.1), we consider $x \in (Q^*)^c$ and $x \in Q^*$ separately.

First we estimate $T_m(a,g)(x)$ for $x \in (Q^*)^c$.

We use the decomposition (3.2). Consider m_0 and $K_0 = \mathcal{F}^{-1} m_0$. As in Sections 4 and 6, we have two expressions (6.7) and (6.8). For $x \in (Q^*)^c$, we first use (6.7) with $g(z)$ in place of $b(z)$ and use Hölder's inequality and Lemma 3.3 to obtain

$$\begin{aligned}
&\langle x - c_Q\rangle^s |T_{m_0}(a,g)(x)| \\
&\lesssim \iint_{y\in Q,\, z\in\mathbb{R}^n} \langle x-y\rangle^s |K_0(x-y, x-z)| |Q|^{-1} |g(z)|\, dydz \\
&\lesssim |Q|^{-1} \int_{y\in Q} \big\|\langle x-y\rangle^s \langle x-z\rangle^s K_0(x-y,x-z)\big\|_{\sigma',\, z} \big\|\langle x-z\rangle^s g(z)\big\|_{\sigma,\, z}\, dy \\
&\lesssim |Q|^{-1} \int_{y\in Q} \big\|\langle x-y\rangle^s \langle z\rangle^s K_0(x-y,z)\big\|_{2,\, z}\, dy\; M_\sigma(g)(x).
\end{aligned}$$

Thus

$$|T_{m_0}(a,g)(x)| \mathbf{1}_{(Q^*)^c}(x) \lesssim \langle x - c_Q\rangle^{-s} g_0^Q(x) M_\sigma(g)(x), \tag{7.3}$$

where g_0^Q is given by (6.9). Similarly, using (6.8) with $g(z)$ in place of $b(z)$, we obtain

$$|T_{m_0}(a,g)(x)| \mathbf{1}_{(Q^*)^c}(x) \lesssim \langle x - c_Q\rangle^{-s} \widetilde{g}_0^Q(x) M_\sigma(g)(x) \tag{7.4}$$

with $\widetilde{g}_0^Q$ given by (6.11). Combining (7.3) and (7.4), we obtain

$$|T_{m_0}(a,g)(x)| \mathbf{1}_{(Q^*)^c}(x) \lesssim u_0(x) M_\sigma(g)(x) \tag{7.5}$$

with u_0 given by (6.15). As we have seen in (6.17),

$$\|u_0\|_1 \lesssim \min\{\ell(Q)^{-s+n/2},\, \ell(Q)\}.$$

For arbitrary $j \in \mathbb{Z}$, using the dilation formula (4.5) and a change of variables, we obtain functions u_j such that

$$|T_{m_j}(a,g)(x)| \mathbf{1}_{(Q^*)^c}(x) \lesssim u_j(x) M_\sigma(g)(x) \tag{7.6}$$

and

$$\|u_j\|_1 \lesssim \min\{(2^j\ell(Q))^{-s+n/2},\, 2^j\ell(Q)\}. \tag{7.7}$$

Hence $|T_m(a,g)(x)|\mathbf{1}_{(Q^*)^c}(x) \lesssim u(x)v(x)$ with $u = \sum_j u_j$ and $v = M_\sigma(g)$. We have

$$\|u\|_1 \le \sum_j \|u_j\|_1 \lesssim \sum_j \min\{(2^j\ell(Q))^{-s+n/2},\, 2^j\ell(Q)\} \approx 1$$

and $\|v\|_2 = \|M_\sigma(g)\|_2 \lesssim \|g\|_2$ since $\sigma < 2$.

Next, we estimate $T_m(a,g)(x)$ for $x \in Q^*$. Recall that we are assuming $\operatorname{supp} m \subset \Gamma(\mathcal{O})$ with $\mathcal{O}$ a sufficiently small neighborhood of $(\xi_0,\eta_0) \in \Sigma$ and $\xi_0 \neq 0$ and $\eta_0 \neq 0$. In this case, we have $|\xi| \approx 2^j$ and $|\eta| \approx 2^j$ for $(\xi,\eta) \in \operatorname{supp} m_j$, and hence we can take D-operators Δ_j such that $T_{m_j}(a,g) = T_{m_j}(\Delta_j a, \Delta_j g)$. Hence, by Lemma 3.2,

$$\begin{aligned} |T_{m_j}(a,g)(x)| &= |T_{m_j}(\Delta_j a, \Delta_j g)| \\ &\lesssim \|m_j(2^j\cdot)\|_{W^{(s,s)}}(\zeta_j * |\Delta_j a|^\sigma)(x)^{1/\sigma}(\zeta_j * |\Delta_j g|^\sigma)(x)^{1/\sigma} \\ &\lesssim M_\sigma(\Delta_j a)(x) M_\sigma(\Delta_j g)(x). \end{aligned}$$

Thus, by Schwarz inequality for $\sum_{j\in\mathbb{Z}}$, we obtain $|T_m(a,g)(x)|\mathbf{1}_{Q^*}(x) \lesssim u(x)v(x)$ with

$$u(x) = \Big(\sum_j M_\sigma(\Delta_j a)(x)^2\Big)^{1/2}\mathbf{1}_{Q^*}(x), \quad v(x) = \Big(\sum_j M_\sigma(\Delta_j g)(x)^2\Big)^{1/2}.$$

By Schwarz inequality and by the L^2-boundedness of M_σ, $\sigma < 2$, we have $\|u\|_1 \lesssim \|a\|_2|Q|^{1/2} \le 1$ and $\|v\|_2 \lesssim \|g\|_2$. Thus the proof for Case 1 is complete.

Case 2: $\xi_0 \neq 0$ and $\xi_0 + \eta_0 \neq 0$.

In this case, for m satisfying $\operatorname{supp} m \subset \Gamma(\mathcal{O})$ with $\mathcal{O}$ a sufficiently small neighborhood of $(\xi_0,\eta_0) \in \Sigma$, we have $|\xi| \approx |\xi+\eta| \approx 2^j$ for $(\xi,\eta) \in \operatorname{supp} m_j$, $j \in \mathbb{Z}$. Hence the Fourier transform of $T_{m_j}(f,g)$ is supported on the annulus $\{\zeta \in \mathbb{R}^n : B^{-1}2^j \le |\zeta| \le B2^j\}$ with some constant $B > 1$. Hence, by Lemma 7.2, if we define

$$G_m(f,g) = \Big(\sum_{j\in\mathbb{Z}} |T_{m_j}(f,g)|^2\Big)^{1/2},$$

then we have

$$\|T_m(f,g)\|_{2/3} \lesssim \|T_m(f,g)\|_{H^{2/3}} \lesssim \|G_m(f,g)\|_{2/3}.$$

Thus it is sufficient to show the estimate $\|G_m(f,g)\|_{2/3} \lesssim \|f\|_{H^1}\|g\|_2$. By the same reason as in Case 1, this estimate follows if we prove that for each H^1-atom a and each $g \in L^2$ there exist functions u and v such that v depends only on g and

$$G_m(a,g)(x) \lesssim u(x)v(x), \quad \|u\|_1 \lesssim 1, \quad \|v\|_2 \lesssim \|g\|_2. \tag{7.8}$$

Suppose a is an H^1-atom satisfying (7.2) with a cube Q. We shall find the above $u(x)$ and $v(x)$ for $x \in (Q^*)^c$ and $x \in Q^*$ separately.

Firstly, for $x \in (Q^*)^c$, the argument in Case 1 applies to the present case as well and we obtain functions u_j satisfying (7.6) and (7.7). Hence

$$G_m(a,g)(x)\mathbf{1}_{(Q^*)^c}(x) \lesssim \Big(\sum_j u_j(x)^2\Big)^{1/2} M_\sigma(g)(x) \le \sum_j u_j(x) M_\sigma(g)(x)$$

and $\| \sum_j u_j \|_1 \lesssim 1$ and $\|M_\sigma(g)\|_2 \lesssim \|g\|_2$, which gives the desired estimate.

Next, consider $x \in Q^*$. In the present case, since $|\xi| \approx 2^j$ for $(\xi, \eta) \in \operatorname{supp} m_j$, we can take D-operators Δ_j such that $T_{m_j}(a, g) = T_{m_j}(\Delta_j a, g)$. Hence Lemma 3.2 gives

$$|T_{m_j}(a,g)(x)| \lesssim \|m_j(2^j\cdot)\|_{W^{(s,s)}} M_\sigma(\Delta_j a)(x) M_\sigma(g)(x) \le M_\sigma(\Delta_j a)(x) M_\sigma(g)(x).$$

Thus $G_m(a, g)(x)\mathbf{1}_{Q^*}(x) \lesssim u(x)v(x)$ with

$$u(x) = \Big(\sum_j M_\sigma(\Delta_j a)(x)^2\Big)^{1/2} \mathbf{1}_{Q^*}(x), \quad v(x) = M_\sigma(g)(x).$$

We have $\|u\|_1 \lesssim 1$ and $\|v\|_2 \lesssim \|g\|_2$ by the same reason as in the last part of Case 1. Thus the proof for Case 2 is complete.

Case 3: $\eta_0 \neq 0$ and $\xi_0 + \eta_0 \neq 0$.

By the same reason as in Case 2, it is sufficient to show the estimate (7.8) for H^1-atoms a and $g \in L^2$. Suppose a and Q satisfy (7.2).

The estimates for $x \in (Q^*)^c$ is obtained just in the same way as in Case 2.

Consider $x \in Q^*$. In the present case, we can take D-operators Δ_j such that $T_{m_j}(a, g) = T_{m_j}(a, \Delta_j g)$. Hence, by Lemma 3.2,

$$|T_{m_j}(a,g)(x)| \lesssim M_\sigma(a)(x) M_\sigma(\Delta_j g)(x).$$

Thus $G_m(a, g)(x)\mathbf{1}_{Q^*}(x) \lesssim u(x)v(x)$ with

$$u(x) = M_\sigma(a)(x)\mathbf{1}_{Q^*}(x), \quad v(x) = \Big(\sum_j M_\sigma(\Delta_j g)(x)^2\Big)^{1/2}.$$

Using the L^2-boundedness of M_σ, $\sigma < 2$, we easily see that $\|u\|_1 \lesssim 1$ and $\|v\|_2 \lesssim \|g\|_2$. This completes the proof of Proposition 7.1.

8 Proof of the only if part

In this section, we give propositions which will cover the only if part of Theorem 2.2 for typical (p, q)'s. We shall give only a brief sketch of the proof; for details, see [7, Propositions 7.1 and 7.2] and [13, §7].

Proposition 8.1. *The estimate* $\|T_m(f, g)\|_2 \lesssim A_{(s_1,s_2)}(m)\|f\|_2\|g\|_\infty$ *holds only if* $s_1 \ge n/2$ *and* $s_2 \ge n/2$.

Proof. To prove the necessity of the condition $s_2 \ge n/2$, we take functions ψ and φ in the class $C_0^\infty(\mathbb{R}^n)$ such that

$$\psi \neq 0, \quad \operatorname{supp}\psi \subset \{\xi \in \mathbb{R}^n : 9/10 \le |\xi| \le 11/10\},$$
$$\varphi(0) \neq 0, \quad \operatorname{supp}\varphi \subset \{\xi \in \mathbb{R}^n : |\xi| \le 1\}.$$

For sufficiently small $\epsilon > 0$, we set $m_\epsilon(\xi, \eta) = \psi(\xi)\varphi(\eta/\epsilon)$. For this m_ϵ, we have

$$T_{m_\epsilon}(f, g)(x) = \psi(D)f(x)\varphi(D/\epsilon)g(x),$$

where $\psi(D)$ and $\varphi(D/\epsilon)$ are linear Fourier multiplier operators. If we choose the function Ψ in the definition (2.1) appropriately, then, for sufficiently small $\epsilon > 0$, we have

$$m_\epsilon(2^j\xi, 2^j\eta)\Psi(\xi, \eta) = \begin{cases} m_\epsilon(\xi, \eta) & (j = 0), \\ 0 & (j \neq 0). \end{cases}$$

This implies

$$\begin{aligned} A_{(s_1,s_2)}(m_\epsilon) &= \|m_\epsilon\|_{W^{(s_1,s_2)}(\mathbb{R}^n \times \mathbb{R}^n)} \\ &= \|\psi(\xi)\|_{W_2^{s_1}(\mathbb{R}^n)}\|\varphi(\eta/\epsilon)\|_{W_2^{s_2}(\mathbb{R}^n)} \approx \epsilon^{-s_2+n/2}. \end{aligned}$$

Thus if the estimate of the proposition holds, then we must have

$$\|\psi(D)f(x)\varphi(D/\epsilon)g(x)\|_2 \lesssim \epsilon^{-s_2+n/2}\|f\|_2\|g\|_\infty. \tag{8.1}$$

If we test (8.1) for $\widehat{f} = \psi$ and $g = 1$, then we obtain $1 \lesssim \epsilon^{-s_2+n/2}$, which is possible only if $s_2 \geq n/2$.

The necessity of the condition $s_1 \geq n/2$ can be prove in a similar way by the use of the multipliers $m_\epsilon(\xi, \eta) = \varphi(\xi/\epsilon)\psi(\eta)$. □

Proposition 8.2. *For $0 < p \leq 1$, the estimate*

$$\|T_m(f, g)\|_p \lesssim A_{(s_1,s_2)}(m)\|f\|_{H^p}\|g\|_\infty$$

holds only if $s_1 \geq n/p - n/2$ and $s_2 \geq n/2$.

Proof. The idea is the same as in the proof of Proposition 8.1. The necessity of $s_2 \geq n/2$ can be proved in the same way as in the proof of Proposition 8.1. To prove the necessity of $s_1 \geq n/p - n/2$, we use the same functions φ and ψ as in the proof of Proposition 8.1, and define $m_\epsilon(\xi, \eta) = \varphi((\xi - \zeta_0)/\epsilon)\psi(\eta)$, where ζ_0 is a point in $\mathbb{R}^n$ with $|\zeta_0| = 1/10$ and $\epsilon > 0$ is sufficiently small. Then we have $A_{(s_1,s_2)}(m_\epsilon) \approx \epsilon^{-s_1+n/2}$. If the estimate of Proposition 8.2 holds, then we must have

$$\|\varphi((D - \zeta_0)/\epsilon)f(x)\psi(D)g(x)\|_p \lesssim \epsilon^{-s_1+n/2}\|f\|_{H^p}\|g\|_\infty. \tag{8.2}$$

We take an $f \in \mathcal{S}(\mathbb{R}^n)$ such that $\operatorname{supp}\widehat{f}$ is a compact subset of $\mathbb{R}^n \setminus \{0\}$ and $\widehat{f}(\xi) = 1$ in a neighborhood of ζ_0. We take an $\eta_0 \in \mathbb{R}^n$ such that $\psi(\eta_0) \neq 0$, and set $g(x) = e^{2\pi i \eta_0 x}$. Then (8.2) implies $\epsilon^{n-n/p} \lesssim \epsilon^{-s_1+n/2}$, which is possible only if $s_1 \geq n/p - n/2$. □

The next proposition covers the only if part of Theorem 2.2 for $(p, q) = (1, 1)$. The proof can be easily modified to the case of general (p, q) satisfying $0 < p \leq 1$ and $0 < q \leq 1$.

Proposition 8.3. *The estimate $\|T_m(f, g)\|_{1/2} \lesssim A_{(s_1,s_2)}(m)\|f\|_{H^1}\|g\|_{H^1}$ holds only if $s_1 \geq n/2$, $s_2 \geq n/2$, and $s_1 + s_2 \geq 3n/2$.*

Proof. The necessity of the conditions $s_1 \geq n/2$ and $s_2 \geq n/2$ can be proved in a similar way as in the proof of Proposition 8.2. We shall prove that $s_1 + s_2 \geq 3n/2$ is also necessary. Take a function $\varphi \in C_0^\infty(\mathbb{R}^n)$ be such that

$$\int \varphi(\xi)\, d\xi \neq 0, \quad \operatorname{supp} \varphi \subset \{|\xi| \leq 1/10\}, \quad \varphi(\xi) = 1 \text{ for } |\xi| \leq 1/20,$$

and take a $\zeta_0 \in \mathbb{R}^n$ satisfying $|\zeta_0| = \sqrt{2}$. For sufficiently small $\epsilon > 0$, we set

$$m_\epsilon(\xi, \eta) = \varphi\left(\frac{\xi + \eta - \zeta_0}{\epsilon}\right)\varphi(\xi - \eta).$$

This multiplier m_ϵ is supported on the narrow rectangle

$$R_\epsilon = \{|\xi + \eta - \zeta_0| \leq \epsilon/10,\, |\xi - \eta| \leq 1/10\},$$

which is included in the annulus $\{1 - \epsilon/10 - 1/10 \leq |(\xi, \eta)| \leq 1 + \epsilon/10 + 1/10\}$. By the same reason as in the proof of Proposition 8.1 and by some calculation, we have

$$A_{(s_1,s_2)}(m_\epsilon) = \|m_\epsilon\|_{W^{(s_1,s_2)}(\mathbb{R}^n \times \mathbb{R}^n)} \approx \epsilon^{n/2 - s_1 - s_2}.$$

The inverse Fourier transform $K_\epsilon = \mathcal{F}^{-1} m_\epsilon$ is given by

$$K_\epsilon(x, y) = c \exp\left(2\pi i \zeta_0 \frac{x+y}{2}\right) \epsilon^n \mathcal{F}^{-1}\varphi\left(\epsilon \frac{x+y}{2}\right) \mathcal{F}^{-1}\varphi\left(\frac{x-y}{2}\right),$$

where $\mathcal{F}^{-1}\varphi$ denotes the inverse Fourier transform of φ on $\mathbb{R}^n$. We take a function $f \in \mathcal{S}(\mathbb{R}^n)$ such that $\operatorname{supp} \widehat{f}$ is a compact subset of $\mathbb{R}^n \setminus \{0\}$ and $\widehat{f}(\xi)\widehat{f}(\eta) = 1$ for all $(\xi, \eta) \in \operatorname{supp} m_\epsilon$. Then

$$T_{m_\epsilon}(f, f)(x) = K_\epsilon(x, x) = c \exp(2\pi i \zeta_0 x) \epsilon^n \mathcal{F}^{-1}\varphi(\epsilon x) \mathcal{F}^{-1}\varphi(0)$$

and thus $\|T_{m^{(\epsilon)}}(f, f)\|_{1/2} \approx \epsilon^{-n}$. Hence, if the estimate of Proposition 8.3 holds, then we must have $\epsilon^{-n} \lesssim \epsilon^{n/2 - s_1 - s_2}$, which is possible only if $s_1 + s_2 \geq 3n/2$. □

9 Isotropic Besov scale

The result of Theorem 2.2 is essentially sharp as far as we use the product Sobolev norm to measure the smoothness of m, as in (2.1). In this section, we give another sharp estimate by using a different norm to measure the smoothness of m. The result may be considered as a bilinear version of Theorem C. We omit details of proofs. Details are given in [15].

For bilinear operators T_m basic L^2-estimate is not obvious at all. We have, however, the following theorem. We write Ψ to denote a function in $C_0^\infty(\mathbb{R}^{2n})$ satisfying (1.10) and (1.11) with $\mathbb{R}^{2n}$ in place of $\mathbb{R}^n$.

Theorem 9.1 ([15]). *Let $s > n/2$ and suppose $m \in L^\infty(\mathbb{R}^{2n})$ satisfies*

$$A = \sup_{j \in \mathbb{Z}} \left\{ \left\| \left\| (I - \Delta_\xi)^{s/2} \left[m(2^j \xi, 2^j \eta) \Psi(\xi, \eta) \right] \right\|_{2,\xi} \right\|_{\infty,\eta} \right.$$

$$+ \left\| \left\| (I - \Delta_\eta)^{s/2} \left[m(2^j \xi, 2^j \eta) \Psi(\xi, \eta) \right] \right\|_{2,\, \eta} \right\|_{\infty,\, \xi} \Bigg\} < \infty.$$

Then $\|T_m(f,g)\|_1 \lesssim A\|f\|_2\|g\|_2$.

This theorem improves the corresponding estimate given in Theorem 2.2. In particular, notice that the differentiability of m is assumed only up to $n/2+\epsilon$.

We shall derive an estimate by interpolating the results of Theorems 9.1 and 2.2. To do this, we use the Besov norms. We set

$$\Phi(\zeta) = 1 - \sum_{k=1}^{\infty} \Psi(\zeta/2^k), \quad \zeta \in \mathbb{R}^{2n}.$$

Thus $\Phi \in C_0^\infty(\mathbb{R}^{2n})$ and $\Phi \subset \{|\zeta| \le 2\}$. For $1 \le p,\ q \le \infty$, $s \in \mathbb{R}$, and $F \in \mathcal{S}'(\mathbb{R}^{2n})$, the Besov norm is defined by

$$\|F\|_{B^s_{p,q}(\mathbb{R}^{2n})} = \left(\|\Phi(D)F\|_p^q + \sum_{k=1}^{\infty} \|2^{ks}\Psi(D/2^k)F\|_p^q \right)^{1/q}$$

with the usual modification in the case $q=\infty$. For $m \in \mathcal{S}'(\mathbb{R}^{2n})$, we define

$$AB^s_{p,q}(m) = \sup_{j\in\mathbb{Z}} \|m(2^j\cdot)\Psi(\cdot)\|_{B^s_{p,q}(\mathbb{R}^{2n})}.$$

It is easy to see that if the support of $m \in \mathcal{S}'(\mathbb{R}^{2n})$ is included in $\{|\zeta| \le 10\}$ then

$$\begin{aligned} &\left\| \left\| (I-\Delta_\xi)^{s/2} m(\xi,\eta) \right\|_{2,\,\xi} \right\|_{\infty,\,\eta} + \left\| \left\| (I-\Delta_\eta)^{s/2} m(\xi,\eta) \right\|_{2,\,\eta} \right\|_{\infty,\,\xi} \\ &\lesssim \|m\|_{B^{s+\epsilon}_{\infty,\infty}(\mathbb{R}^{2n})} \end{aligned}$$

for every $\epsilon > 0$ (see [15, Lemma 4.1]). Hence Theorem 9.1 implies that if $s > n/2$ then the estimate

$$\|T_m(f,g)\|_r \lesssim AB^s_{\infty,\infty}(m)\|f\|_p\|g\|_q \tag{9.1}$$

holds for $(p,q,r) = (2,2,1)$. From this, by duality argument, we can deduce that the estimate (9.1) also holds for $(p,q,r) = (2,\infty,2)$ and $(\infty,2,2)$.

On the other hand, for $s>0$ and $\epsilon>0$, we have

$$\|m\|_{W^{(s/2,s/2)}(\mathbb{R}^n\times\mathbb{R}^n)} \lesssim \|m\|_{B^s_{2,2}(\mathbb{R}^{2n})} \lesssim \|m\|_{B^{s+\epsilon}_{2,\infty}(\mathbb{R}^{2n})}.$$

This inequality combined with Theorem 2.2 imply that if $s>n$ then the estimate

$$\|T_m(f,g)\|_r \lesssim AB^s_{2,\infty}(m)\|f\|_{H^p}\|g\|_{H^q}, \quad 1/r = 1/p + 1/q \tag{9.2}$$

holds if $(1/p, 1/q)$ is on the boundary of the region I_0 (see the figure in Section 2), where $\|\cdot\|_r$ should be replaced by $\|\cdot\|_{BMO}$ if $p=q=r=\infty$.

Interpolating (9.1) and (9.2), we obtain the former half of the following theorem.

Theorem 9.2 ([15]). *Let* $1 \le p, q \le \infty$, $2/3 \le r \le \infty$, *and* $1/p + 1/q = 1/r$. *If*

$$s > s(p,q) = \max\{n/2,\, n/p,\, n/q,\, n - n/r,\, n/r - n/2\}$$

and

$$\begin{aligned} 1/t &= s(p,q)/n - 1/2 \\ &= \max\{0,\, 1/p - 1/2,\, 1/q - 1/2,\, 1/2 - 1/r,\, 1/r - 1\}, \end{aligned}$$

then

$$\|T_m(f,g)\|_r \lesssim A B^s_{t,\infty}(m) \|f\|_{H^p} \|g\|_{H^q} \tag{9.3}$$

holds, where $\| \cdot \|_r$ *should be replaced by* $\| \cdot \|_{BMO}$ *if* $p = q = r = \infty$. *Conversely, if* $s > 0$ *and* (9.3), *with the same replacement in the case* $p = q = r = \infty$, *holds for some* $t \in [1, \infty]$, *then* $s \ge s(p,q)$.

The latter half of the above theorem (the necessity of the condition $s \ge s(p,q)$) can be proved with the aid of a result given in [14] concerning multipliers m satisfying the estimate

$$|\partial_\xi^\alpha \partial_\eta^\beta m(\xi, \eta)| \le C_{\alpha,\beta}(1 + |\xi| + |\eta|)^{-s},$$

where $s > 0$ is a fixed number. This is another interesting topic, but we shall not discuss it here.

References

[1] A. P. Calderón and A. Torchinsky, Parabolic maximal functions associated with a distribution II, Adv. Math. **24** (1977), 101–171.

[2] R. Coifman and Y. Meyer, Au delà des opérateurs pseudo-différentiels, Astérisque **57** (1978), 1–185.

[3] R. Coifman and Y. Meyer, Nonlinear harmonic analysis, operator theory and P.D.E., in Beijing lectures in harmonic analysis (Beijing, 1984), 3–45, Ann. of Math. Stud. **112**, Princeton University Press, Princeton, NJ, 1986.

[4] C. Fefferman and E. Stein, Some maximal inequalities, Amer. J. Math. **93** (1971), 107–115.

[5] M. Fujita and N. Tomita, Weighted norm inequalities for multilinear Fourier multipliers, Trans. Amer. Math. Soc. **364** (2012), 6335–6353.

[6] L. Grafakos and N. Kalton, Multilinear Calderón-Zygmund operators on Hardy spaces, Collect. Math. **52** (2001), 169–179.

[7] L. Grafakos, A. Miyachi, and N. Tomita, On multilinear Fourier multipliers of limited smoothness, Canad. J. Math. **65** (2013), 299–330.

[8] L. Grafakos and Z. Si, The Hörmander multiplier theorem for multilinear operators, J. Reine Angew. Math. **668** (2012), 133–147.

[9] L. Grafakos and R. Torres, Multilinear Calderón-Zygmund theory, Adv. Math. **165** (2002), 124–164.

[10] L. Hörmander, Estimates for translation invariant operators in L^p spaces, Acta Math. **104** (1960), 93–140.

[11] C. Kenig and E. M. Stein, Multilinear estimates and fractional integration, Math. Res. Lett. **6** (1999), 1–15.

[12] Y. Meyer and R. Coifman, Wavelets: Calderón-Zygmund and multilinear operators, Cambridge Stud. Adv. Math. **48**, Cambridge University Press, Cambridge, 1997.

[13] A. Miyachi and N. Tomita, Minimal smoothness conditions for bilinear Fourier multipliers, Rev. Mat. Iberoam. **29** (2013), 495–530.

[14] A. Miyachi and N. Tomita, Calderón-Vaillancourt type theorem for bilinear operators, Indiana Univ. Math. J. **62** (2013), 1165–1201.

[15] A. Miyachi and N. Tomita, Boundedness criterion for bilinear Fourier multiplier operators, Tohoku Math. J. **66** (2014), 55–76.

[16] E. M. Stein, Harmonic analysis: real-variable methods, orthogonality, and oscillatory integrals, Princeton Math. Ser. **43**, Princeton University Press, Princeton, 1993.

[17] J.-O. Strömberg and A. Torchinsky, Weighted Hardy Spaces, Springer Lecture Notes in Math. **1381**, Springer Verlag, 1989.

[18] N. Tomita, A Hörmander type multiplier theorem for multilinear operators, J. Funct. Anal. **259** (2010), 2028–2044.

[19] H. Triebel, Theory of Function Spaces, Birkhäuser Verlag, Basel, 1983.

Some Topics in Harmonic Analysis and Applications
ALM 34, pp. 331–367

Weighted Estimates for Fractional Type Marcinkiewicz Integral Operators Associated to Surfaces

Yoshihiro Sawano * and Kôzô Yabuta †

Abstract

In this paper, we shall discuss the weighted boundedness of the fractional type Marcinkiewicz integral operators associated to surfaces, and extend a result given by Chen, Fan and Ying in 2002 and the one given by authors. They showed that under certain conditions the fractional type Marcinkiewicz integral operators are bounded from the Triebel-Lizorkin spaces $\dot{F}^{\alpha}_{pq}(\mathbb{R}^n)$ to $L^p(\mathbb{R}^n)$. Recently one of the author, together with Xue and Yan, greatly weakened their assumptions, and we further extended their results to the case where the operators are associated to the surfaces of the form $\{x = \phi(|y|)y/|y|\} \subset \mathbb{R}^n \times (\mathbb{R}^n \setminus \{0\})$. In this paper, we investigate weighted estimates for these operators.

To prove our result, we discuss a characterization of the weighted homogeneous Triebel-Lizorkin spaces in terms of lacunary sequences.

2000 Mathematics Subject Classification: Primary 42B20; Secondary 42B25, 47G10.

Keywords and Phrases: L^p-boundedness, weighted L^p-boundedness, A_p, Marcinkiewicz integral, fractional integral operator, Triebel-Lizorkin spaces, Sobolev spaces.

1 Introduction

The fractional type Marcinkiewicz operator is defined by

*Department of Mathematics and Information Science, Tokyo Metropolitan University, Minami-Ohsawa 1-1, Hachioji 192-0397, Japan. *E-mail:* yoshihiro-sawano@celery.ocn.ne.jp

†Research Center for Mathematical Sciences, Kwansei Gakuin University, Gakuen 2-1, Sanda 669-1337, Japan. *E-mail:* kyabuta3@kwansei.ac.jp

This work was partially supported by Grant-in-Aid for Scientific Research (C) Nr. 23540228, Japan Society for the Promotion of Science.

$$\mu_{\Omega,\rho,\alpha}f(x) := \left(\int_0^\infty \left|\frac{1}{t^{\rho+\alpha}}\int_{B(t)} f(x-y)\frac{\Omega(y/|y|)}{|y|^{n-\rho}}dy\right|^2 \frac{dt}{t}\right)^{\frac{1}{2}}, \tag{1.1}$$

where we write $B(r) := \{|x| < r\} \subset \mathbb{R}^n$ for $r > 0$ here and below. This is a kind of singular integral operators. In this paper, we shall prove that this operator is bounded under a certain highly weak integrability assumption. To this end, we plan to employ a modified Littlewood-Paley decomposition adapted to our situation. It turns out that we can relax the integrability assumption on Ω and that the integral operator itself can be generalized to a large extent.

Let S^{n-1} be the unit sphere in the n-dimensional Euclidean space $\mathbb{R}^n$ $(n \geq 2)$, with the induced Lebesgue measure $d\sigma = d\sigma(x')$ and $\Omega \in L^1(S^{n-1})$. In the sequel, we often suppose that Ω satisfies the cancellation condition

$$\int_{S^{n-1}} \Omega(y')\,d\sigma(y') = 0. \tag{1.2}$$

Here, for the symbols x' and y', we adopt the following convention: sometimes they stand for points in S^{n-1}. But for $x \in \mathbb{R}^n \setminus \{0\}$, we abbreviate $x/|x|$ to x' in the present paper. We make this slight abuse of notation since no confusion is likely to occur.

In the present paper we deal with operators of Marcinkiewicz type. Define

$$\mu_{\Omega,\rho,\alpha,q}f(x) := \left(\int_0^\infty \left|\frac{1}{t^{\rho+\alpha}}\int_{B(t)} f(x-y)\frac{\Omega(y')}{|y|^{n-\rho}}dy\right|^q \frac{dt}{t}\right)^{\frac{1}{q}}, \tag{1.3}$$

where Ω is a suitable measurable function and ρ, α, q are constants. As a special case, by letting $\rho = 1$, $\alpha = 0$, $q = 2$, we can recapture Marcinkiewicz integral operator that Stein introduced in 1958 [32]. In 1960, Hörmander considered the parametric Marcinkiewicz integral operator $\mu_{\Omega,\rho,\alpha,2}$ [23]. Since then, about Marcinkiewicz type integral operators, many works appeared. A nice survey is given by Lu [26].

We fully formulate our results in the framework of Triebel-Lizorkin spaces of homogeneous type. For $\alpha \in \mathbb{R}$ and $p, q \in (1,\infty)$, we let $\dot{F}^\alpha_{pq}(\mathbb{R}^n)$ be the homogeneous Triebel-Lizorkin space defined in [34], and $\dot{F}^\alpha_{pq}(w)$ be the weighted homogeneous Triebel-Lizorkin space for a Muckenhoupt A_∞ class weight w (see e.g. [7]). Here by "a weight" we mean a measurable function w which satisfies $0 < w < \infty$ a.e.. Note that the space $\mathcal{S}_\infty(\mathbb{R}^n)$ given by

$$\mathcal{S}_\infty(\mathbb{R}^n) := \bigcap_{\alpha \in (\mathbb{N}\cup\{0\})^n} \left\{ f \in \mathcal{S}(\mathbb{R}^n) \,:\, \int_{\mathbb{R}^n} x^\alpha f(x)\,dx = 0 \right\}$$

is dense in $\dot{F}^\alpha_{pq}(w)$ as long as $\alpha \in \mathbb{R}$, $p, q \in (1,\infty)$ and $w \in A_\infty(\mathbb{R}^n)$. If $u \in (1,\infty)$, then define

$$u' := \frac{u}{u-1}, \quad \tilde{u} := \max\{u, u'\}.$$

Here and below a tacit understanding in the present paper is that the letter C is used for constants that may change from one occurrence to another, that is, the letter C will denote a positive constant which may vary from line to line but will remain independent of the relevant quantities.

About the fractional type Marcinkiewicz integral operator, in 2002, Chen, Fan and Ying obtained a result [10], which we recall now.

Theorem A. *Let $1 < p,\ q < \infty$ and $1 < r \le \infty$. Suppose $\Omega \in L^r(S^{n-1})$ satisfies the cancellation condition* (1.2). *If $|\alpha| < 2/(r'\tilde{p}\tilde{q})$ and $\rho = 1$, then*

$$\|\mu_{\Omega,\rho,\alpha,q} f\|_{L^p(\mathbb{R}^n)} \le C\|\Omega\|_{L^r(S^{n-1})}\|f\|_{\dot{F}^{\alpha}_{pq}(\mathbb{R}^n)}$$

for all $f \in \mathcal{S}_\infty(\mathbb{R}^n)$.

Si, Wang and Jiang discussed ones of somewhat different type [30]. We improved these results and we obtained the following result in [29, 36]:

Theorem 1. *Let $\rho > 0$, $1 < p,\ q < \infty$, and $\Omega \in L^1(S^{n-1})$.*

(i) *If $\alpha \in (0, 4/(\tilde{p}\tilde{q}))$ and Ω satisfies the cancellation condition* (1.2), *then*

$$\|\mu_{\Omega,\rho,\alpha,q} f\|_{L^p(\mathbb{R}^n)} \le C\|\Omega\|_{L^1(S^{n-1})}\|f\|_{\dot{F}^{\alpha}_{pq}(\mathbb{R}^n)}. \tag{1.4}$$

(ii) *If $\alpha \in (-\min\{4\beta/(\tilde{p}\tilde{q}),\ \rho\}, 0)$, and*

$$Z_\Omega := \sup_{\xi' \in S^{n-1}} \int_{S^{n-1}} \frac{|\Omega(y')|}{|y' \cdot \xi'|^\beta}\, d\sigma(y') < +\infty \tag{1.5}$$

for some $0 < \beta \le 1$, then

$$\|\mu_{\Omega,\rho,\alpha,q} f\|_{L^p(\mathbb{R}^n)} \le C Z_\Omega \|f\|_{\dot{F}^{\alpha}_{pq}(\mathbb{R}^n)}. \tag{1.6}$$

(iii) *If $\alpha = 0$ and $\Omega \in \mathrm{L}\log\mathrm{L}(S^{n-1})$ satisfies the cancellation condition* (1.2), *then*

$$\|\mu_{\Omega,\rho,\alpha,q} f\|_{L^p(\mathbb{R}^n)} \le C\|\Omega\|_{\mathrm{L}\log\mathrm{L}(S^{n-1})}\|f\|_{\dot{F}^{\alpha}_{pq}(\mathbb{R}^n)}. \tag{1.7}$$

Here $\mathrm{L}\log\mathrm{L}(S^{n-1})$ in the Orlicz space generated by $\Phi(t) = t\log(e+t)$.

To be precise, (1.4), (1.6) and (1.7) in the above hold for all $f \in \mathcal{S}_\infty(\mathbb{R}^n)$. In any case, by density we can extend them for all $f \in \dot{F}^{\alpha}_{pq}(\mathbb{R}^n)$.

About Theorems A and 1, a remark may be in order.

Remark 1. If $0 < \beta < 1$, $1/(1-\beta) < r \le \infty$ and $\Omega \in L^r(S^{n-1})$, it is easily seen that the condition (1.5) is satisfied. In this case

$$Z_\Omega \le C\|\Omega\|_{L^r(S^{n-1})}.$$

So, Theorem 1 includes completely Theorem A, where they assumed that $\Omega \in L^r(S^{n-1})$.

In the case $\alpha = 0$, $\rho = 1$ and $q = 2$, the conclusion in Theorem 1(iii) is shown to hold even when $\Omega \in \mathrm{L}\log\mathrm{L}^{1/2}(S^{n-1})$ in [3].

In [29] we considered in more generalized settings.

To describe the results there, by denoting $\mathbb{R}_+ = (0,\infty)$, we use one more function space $\Delta_\gamma(\mathbb{R}_+)$ with $\gamma \ge 1$. Regarding to Calderón-Zygmund singular integral and Marcinkiewicz integral operators, many authors discussed those operators with modified kernel $b(|\cdot|)\Omega(\cdot)$ in place of $\Omega(\cdot)$, where b belongs to the class $\Delta_\gamma(\mathbb{R}_+)$ of all measurable functions $h : [0,\infty) \to \mathbb{C}$ satisfying $\|h\|_{\Delta_\gamma} = \sup_{R>0}\left(R^{-1}\int_0^R |h(t)|^\gamma dt\right)^{1/\gamma} < \infty$ $(1 \le \gamma \le \infty)$, see [2, 15, 16, 17, 18], etc. We note that

$$\Delta_\infty(\mathbb{R}_+) = L^\infty(\mathbb{R}_+) \subset \Delta_\beta(\mathbb{R}_+) \subset \Delta_\alpha(\mathbb{R}_+) \quad \text{for } 1 \le \alpha < \beta,$$

and that all these inclusions are proper. We refer to [8, 25, 27] for extension and generalization of the space $\Delta_\gamma(\mathbb{R}_+)$.

Now, for $1 \le q < \infty$ and a positive function ϕ on $\mathbb{R}_+$, we define the operator $\mu_{\Omega,\rho,\phi,\alpha,q}$ and the modified one $\mu^{(b)}_{\Omega,\rho,\phi,\alpha,q}$ by

$$\mu_{\Omega,\rho,\phi,\alpha,q}f(x) := \left(\int_0^\infty \left|\frac{1}{t^\rho\phi(t)^\alpha}\int_{B(t)} f(x-\phi(|y|)y')\frac{\Omega(y)}{|y|^{n-\rho}}dy\right|^q \frac{dt}{t}\right)^{\frac{1}{q}}, \tag{1.8}$$

and

$$\mu^{(b)}_{\Omega,\rho,\phi,\alpha,q}f(x) := \left(\int_0^\infty \left|\frac{1}{t^\rho\phi(t)^\alpha}\int_{B(t)} f(x-\phi(|y|)y')\frac{b(|y|)\Omega(y)}{|y|^{n-\rho}}dy\right|^q \frac{dt}{t}\right)^{\frac{1}{q}}. \tag{1.9}$$

If $b \equiv 1$, $\phi \equiv \mathrm{id}_{(0,\infty)}$ and $q = 2$, then $\mu^{(b)}_{\Omega,\rho,\phi,\alpha,q} = \mu_{\Omega,\rho,\phi,\alpha,q} = \mu_{\Omega,\rho,\alpha,q}$.

In [29], we extended Theorem 1 to these operators. In this paper, we further investigate weighted estimates for fractional type Marcinkiewicz integrals associated to surfaces of the form $\{x = \phi(|y|)y/|y|\}$. For this purpose, next, we recall the definitions of classes of weights.

For $1 < p < \infty$, we say that a nonnegative function $\omega \in L^1_{\mathrm{loc}}(\mathbb{R}^n)$ is in $A_p(\mathbb{R}^n)$ if there is a constant $C > 0$ such that for any n-dimensional cubes Q with sides parallel to coordinate axes

$$\left(\frac{1}{|Q|}\int_Q \omega(x)\,dx\right)\left(\frac{1}{|Q|}\int_Q \omega(x)^{-1/(p-1)}\,dx\right)^{p-1} \le C < \infty. \tag{1.10}$$

If there is a constant $C > 0$ such that

$$M\omega(x) \le C\omega(x) \quad \text{for a.e. } x \in \mathbb{R}^n, \tag{1.11}$$

where $M\omega$ denotes the standard Hardy-Littlewood maximal function of ω on $\mathbb{R}^n$, then we say ω is in $A_1(\mathbb{R}^n)$. These are the usual Muckenhoupt weight classes on $\mathbb{R}^n$.

Next, we recall a class of weights on $\mathbb{R}_+$. Suppose that a nonnegative function ω is in $L^1_{\mathrm{loc}}(\mathbb{R}_+)$. For $1 < p < \infty$, we say that ω is in $A_p(\mathbb{R}_+)$ if there is a constant $C > 0$ such that for any interval $I \subset \mathbb{R}_+$,

$$\left(\frac{1}{|I|}\int_I \omega(r)\,dr\right)\left(\frac{1}{|I|}\int_I \omega(r)^{-1/(p-1)}\,dr\right)^{p-1} \le C < \infty. \tag{1.12}$$

If there is a constant $C > 0$ such that

$$\omega^*(r) \leq C\omega(r) \quad \text{for a.e. } r \in \mathbb{R}_+, \tag{1.13}$$

where ω^* denotes the standard Hardy-Littlewood maximal function of ω on $\mathbb{R}_+$, then we say ω is in $A_1(\mathbb{R}_+)$.

For $1 < p < \infty$, we define the weight classes $\widetilde{A}_p(\mathbb{R}^n)$ as follows:

$$\begin{aligned}\widetilde{A}_p(\mathbb{R}^n) := &\{\omega(x) = \nu_1(|x|)\nu_2(|x|)^{1-p} : \nu_1, \nu_2 \in A_1(\mathbb{R}_+) \text{ are decreasing }\}\\ &\cup\{\omega(x) = \nu_1(|x|)\nu_2(|x|)^{1-p} : \nu_1^2, \nu_2^2 \in A_1(\mathbb{R}_+)\}.\end{aligned}$$

For $p = 1$, we define

$$\widetilde{A}_1(\mathbb{R}^n) := \{\omega(x) = \nu_1(|x|) : \nu_1 \in A_1(\mathbb{R}_+) \text{ is decreasing or } {\nu_1}^2 \in A_1(\mathbb{R}_+)\}.$$

We know by [14] that the Hardy-Littlewood maximal operator M is bounded on $L^p(\omega)$ for $\omega \in \widetilde{A}_p(\mathbb{R}^n)$, and thus $\widetilde{A}_p(\mathbb{R}^n) \subset A_p(\mathbb{R}^n)$.

The space $L^p(\omega)$ denotes the weighted L^p-space associated to the weight ω defined by

$$L^p(\omega) = L^p(\mathbb{R}^n, \omega) := \left\{f : \|f\|_{L^p(\omega)} := \left(\int_{\mathbb{R}^n} |f(x)|^p \omega(x)\, dx\right)^{1/p} < \infty\right\}.$$

We formulate our main theorem.

Theorem 2. *Let $a > 0$, $\rho > 0$, $1 < p, q < \infty$, and $\Omega \in L^1(S^{n-1})$. Let w satisfy $w^{1+a} \in \widetilde{A}_2(\mathbb{R}^n)$ if $p \geq 2$, and let w satisfy $w^2 \in \widetilde{A}_1(\mathbb{R}^n)$ and $w^{1+a} \in \widetilde{A}_2(\mathbb{R}^n)$ if $1 < p < 2$. Let $c_0, c_1 > 1$. Suppose that $\phi : \mathbb{R}_+ \to \mathbb{R}_+$ is a nonnegative increasing C^1-function such that*

$$\phi(2t) \leq c_0\phi(t) \quad \text{for all } t \in \mathbb{R}_+, \tag{1.14}$$

$$\phi(t) \leq c_1 t\phi'(t) \quad \text{for all } t \in \mathbb{R}_+. \tag{1.15}$$

Define

$$\varphi(t) := \frac{\phi(t)}{t\phi'(t)} \quad \text{for all } t \in \mathbb{R}_+.$$

Then:

(i) *Let*

$$0 < \alpha < \frac{4a}{\tilde{p}\tilde{q}(1+a)c_1 \log_2 c_0}. \tag{1.16}$$

If Ω satisfies the cancellation condition (1.2), then for all $f \in \mathcal{S}_\infty(\mathbb{R}^n)$,

$$\|\mu_{\Omega,\rho,\phi,\alpha,q} f\|_{L^p(w)} \leq C\|\Omega\|_{L^1(S^{n-1})}\|f\|_{\dot{F}^\alpha_{pq}(w)}. \tag{1.17}$$

(ii) *Let $0 < \beta \leq 1$. Let*

$$-\min\left\{\frac{4a\beta}{\tilde{p}\tilde{q}(1+a)c_1 \log_2 c_0}, \frac{\rho}{\log_2 c_0}\right\} < \alpha < 0.$$

If ϕ satisfies the following additional condition

$$\varphi(t) \text{ or } t\phi'(t) \text{ is monotonic on } \mathbb{R}_+, \tag{1.18}$$

and Ω satisfies

$$Z_\Omega := \sup_{\xi' \in S^{n-1}} \int_{S^{n-1}} \frac{|\Omega(y')|}{|y' \cdot \xi'|^\beta} \, d\sigma(y') < +\infty, \tag{1.19}$$

then for all $f \in \mathcal{S}_\infty(\mathbb{R}^n)$,

$$\|\mu_{\Omega,\rho,\phi,\alpha,q} f\|_{L^p(w)} \le C Z_\Omega \|f\|_{\dot{F}^\alpha_{pq}(w)}. \tag{1.20}$$

(iii) *Let $\alpha = 0$. If $\Omega \in \mathrm{L}\log\mathrm{L}(S^{n-1})$ and it satisfies the cancellation condition* (1.2), *then for all $f \in \mathcal{S}_\infty(\mathbb{R}^n)$,*

$$\|\mu_{\Omega,\rho,\phi,\alpha,q} f\|_{L^p(w)} \le C \|\Omega\|_{\mathrm{L}\log\mathrm{L}(S^{n-1})} \|f\|_{\dot{F}^\alpha_{pq}(w)}. \tag{1.21}$$

In any case, by density we can extend (1.17), (1.20) *and* (1.21) *and have them for all $f \in \dot{F}^\alpha_{pq}(w)$.*

Note that (1.14) is referred to as the doubling condition.

We can prove Theorem 2 by modifying the proof of Theorem 3 in the case where $\gamma = \infty$, as in [29].

We state our main result in full generality. Theorem 2 is almost a direct consequence of the next theorem.

Theorem 3. *Suppose that we are given parameters*

$$(\rho, \gamma, \alpha, p, q, a) \in (0,\infty) \times (2,\infty) \times \mathbb{R} \times (1,\infty) \times (1,\infty) \times (0,\infty)$$

and the functions

$$(\Omega, b, w, \phi) \in L^1(S^{n-1}) \times \Delta_\gamma(\mathbb{R}_+) \times L^1_{\mathrm{loc}}(\mathbb{R}^n) \times C^1([0,\infty)).$$

First assume

$$\left(\frac{1}{p}, \frac{1}{q}\right) \in \text{Octagon } P_1P_2P_3P_4P_5P_6P_7P_8, \tag{1.22}$$

where $P_1 = (1/(2\gamma), 1/(2\gamma))$, $P_2 = (1-(1-1/(2\gamma))/\gamma, 1/(2\gamma))$, $P_3 = (1-(1-1/(2\gamma))/\gamma, 1/2)$, $P_4 = (1-1/(2\gamma), 1/2)$, $P_5 = (1-1/(2\gamma), 1-1/(2\gamma))$, $P_6 = ((1-1/(2\gamma))/\gamma, 1-1/(2\gamma))$, $P_7 = ((1-1/(2\gamma))/\gamma, 1/2)$, $P_8 = (1/(2\gamma), 1/2)$.

Consider three cases:

$$\text{Case } \alpha 1: \ \alpha \in \left(0, \frac{4a(1/\tilde{p} - 1/(2\gamma))(1/\tilde{q} - 1/(2\gamma))}{(1+a)(1-1/\gamma)^2 c_1 \log_2 c_0}\right), \tag{1.23}$$

$$\text{Case } \alpha 2: \ \alpha = 0, \tag{1.24}$$

$$\text{Case } \alpha 3: \ \alpha \in \left(-\min\left\{\frac{2a\beta(1/\tilde{p} - 1/(2\gamma))(1/\tilde{q} - 1/(2\gamma))}{(1+a)(1-1/\gamma)^2 c_1 \log_2 c_0}, \frac{\rho}{\log_2 c_0}\right\}, 0\right). \tag{1.25}$$

Assume in addition that Ω *satisfies the following conditions in the table:*

Case $\alpha 1$	*the cancellation condition* (1.2)
Case $\alpha 2$	(1.26) for some $\beta \in (0,1)$
Case $\alpha 3$	$\Omega \in \mathrm{L}\log\mathrm{L}(S^{n-1})$, (1.2)

Here (1.26) *is*

$$W_\Omega := \sup_{\xi' \in S^{n-1}} \int_{S^{n-1}\times S^{n-1}} \frac{|\Omega(y')\Omega(z')|}{|(y'-z')\cdot\xi'|^\beta}\, d\sigma(y')d\sigma(z') < +\infty. \tag{1.26}$$

Assume also that ϕ *satisfies* (1.14) *and* (1.15) *in Theorem* 2. *Finally assume that* w *is such that*

$$w^{1+a} \in \widetilde{A}_{2/\gamma'}(\mathbb{R}^n) \tag{1.27}$$

and that w *satisfies one of the following conditions in the table below:*

Case $W0$	$p = q = 2$	no other condition
Case $W1$	$2 < p, q < \infty$	(1.28), (1.29)
Case $W2$	$2 < p < \infty$ and $q = 2$	(1.28)
Case $W3$	$p = 2$ and $2 < q < \infty$	(1.29)
Case $W4$	$2 < p < \infty$ and $1 < q < 2$	(1.28), (1.30)
Case $W5$	$1 < p < 2$ and $2 < q < \infty$	(1.29), (1.31)
Case $W6$	$1 < p, q < 2$	(1.30), (1.31)
Case $W7$	$1 < p < 2$ and $q = 2$	(1.31)
Case $W8$	$p = 2$ and $1 < q < 2$	(1.30)

Here the conditions (1.28)–(1.31) *stand for*

$$w^{1-\gamma'} \in \widetilde{A}_1(\mathbb{R}^n), \tag{1.28}$$

$$w^{(1-p')\left(1-(\frac{p'}{(2\gamma)'})'\right)} \in \widetilde{A}_{\frac{(p'/(2\gamma)')'}{\gamma'}}(\mathbb{R}^n), \tag{1.29}$$

$$w^{1-(\frac{p'}{(2\gamma)'})'} \in \widetilde{A}_{\frac{(p/(2\gamma)')'}{\gamma'}}(\mathbb{R}^n), \tag{1.30}$$

$$w^{(1-2\gamma)(1-\gamma')} \in \widetilde{A}_1(\mathbb{R}^n), \tag{1.31}$$

respectively. Then:

Case $\alpha 1$: *for all* $f \in \mathcal{S}_\infty(\mathbb{R}^n)$,

$$\|\mu^{(b)}_{\Omega,\rho,\phi,\alpha,q} f\|_{L^p(w)} \le C\|b\|_{\Delta_\gamma}\|\Omega\|_{L^1(S^{n-1})}\|f\|_{\dot{F}^\alpha_{pq}(w)}. \tag{1.32}$$

Case $\alpha 2$: *for all* $f \in \mathcal{S}_\infty(\mathbb{R}^n)$,

$$\|\mu^{(b)}_{\Omega,\rho,\phi,\alpha,q} f\|_{L^p(w)} \le C\|b\|_{\Delta_\gamma} W_\Omega^{1/2}\|f\|_{\dot{F}^\alpha_{pq}(w)}. \tag{1.33}$$

Case $\alpha 3$: *for all* $f \in \mathcal{S}_\infty(\mathbb{R}^n)$,

$$\|\mu^{(b)}_{\Omega,\rho,\phi,\alpha,q} f\|_{L^p(w)} \le C\|b\|_{\Delta_{\max(2,\gamma)}} \|\Omega\|_{\mathrm{L}\log\mathrm{L}(S^{n-1})} \|f\|_{\dot{F}^\alpha_{pq}(w)}. \tag{1.34}$$

Remark 2. About the conditions (1.28)–(1.31), we adopt the following rules.

1. When $2 < p < \infty$, we postulate (1.28). In this case $\gamma > 1$ and hence the condition (1.28) makes sense.

2. When $2 < q < \infty$, we postulate (1.29).

3. When $1 < q < 2$, we postulate (1.30).

4. When $1 < p < 2$, we postulate (1.31). In this case $\gamma > 1$ and hence the condition (1.31) makes sense.

5. When $q = 2$, we can cover the results for homogeneous Sobolev spaces.

6. When $\alpha \ge 0$, we cannot remove the moment condition (1.2).

   Let h be an even $C^\infty(\mathbb{R})$-function satisfying $\chi_{[2,3]}(t) \le h(t) \le \chi_{[1,4]}(t)$ ($t \in \mathbb{R}$). Then $g(x) := \mathcal{F}^{-1}(h(|\cdot|))(x)$ belongs to $\mathcal{S}_\infty(\mathbb{R}^n)$ and $g(0) > 0$. So, choosing $\varepsilon > 0$ sufficiently small, and taking $A > 0$ sufficiently large, we see that $f(x) = Ag(\varepsilon x) \ge 1$ for $|x| \le 1 + \phi(1)$. That is, $f \in \mathcal{S}_\infty(\mathbb{R}^n)$ and $f(x) \ge 1$ for $|x| \le 1 + \phi(1)$.

   Now, for $x \in B(1)$ we have $|x - \phi(|y|)y'| \le 1 + \phi(1)$ if $|y| < 1$. Hence if $\Omega \equiv 1$ and $b \equiv 1$, we have for $0 < t < 1$,

$$\frac{1}{t^\rho \phi(t)^\alpha} \int_{B(t)} \frac{f(x - \phi(|y|)y')b(|y|)\Omega(y')}{|y|^{n-\rho}} dy \ge c \frac{1}{t^\rho \phi(t)^\alpha} \int_0^t \frac{r^{n-1}}{r^{n-\rho}} dr = \frac{c}{\rho \phi(t)^\alpha}.$$

   Thus, if $\alpha \ge 0$, we have

$$\mu^{(b)}_{\Omega,\rho,\phi,\alpha,q} f(x) = +\infty \quad \text{for } x \in B(1).$$

The case when $\gamma = \infty$ requires a simpler condition on weights.

Theorem 4. *Suppose that we are given parameters*

$$(\rho, \alpha, p, q) \in (0,\infty) \times \mathbb{R} \times (1,\infty) \times (1,\infty)$$

and the functions $\Omega \in L^1(S^{n-1})$, $w \in A_\infty(\mathbb{R}^n)$, *and* $b \in \Delta_\infty(\mathbb{R}_+)$. *Suppose* Ω *satisfies the same condition as in Theorem* 3. *Assume also that* ϕ *satisfies* (1.14) *and* (1.15) *in Theorem* 2. *Finally assume that* w *is such that*

$$w^{1+a} \in \widetilde{A}_2(\mathbb{R}^n) \tag{1.35}$$

for some $a > 0$. *In the case* $1 < p < 2$, *we assume furthermore* $w^2 \in \widetilde{A}_1(\mathbb{R}^n)$. *Then the same conclusions* (1.32)–(1.34) *with* $\gamma = \infty$ *in Theorem* 3 *hold.*

Theorem 2 Cases $\alpha 1$ and $\alpha 3$ are direct consequences of the corresponding assertions of Theorem 4. Indeed, to obtain Cases $\alpha 1$ and $\alpha 3$, we have only to apply Theorem 4 with $b \equiv 1$.

We rely upon the modified Littlewood-Paley decomposition for the proofs of Theorems 3 and 4, which we shall describe now. Let $\{a_k\}_{k\in\mathbb{Z}}$ be a lacunary sequence of positive numbers in the sense that $a_{k+1}/a_k \geq a > 1$ $(k \in \mathbb{Z})$. A sequence $\{\Phi_k\}_{k\in\mathbb{Z}}$ of $C^\infty(\mathbb{R}^n)$-functions is said to be a partition of unity adapted to $\{a_k\}_{k\in\mathbb{Z}}$ if

$$\operatorname{supp}\widehat{\Phi_k} \subset \{\xi \in \mathbb{R}^n : a_{k-1} \leq |\xi| \leq a_{k+1}\} \quad (k \in \mathbb{Z}),$$

$$\sum_{k\in\mathbb{Z}} \widehat{\Phi_k}(\xi) = 1 \quad (\xi \in \mathbb{R}^n \setminus \{0\}),$$

and

$$|\xi^\beta \partial^\beta \widehat{\Phi_k}(\xi)| \leq C_\beta \quad (\xi \in \mathbb{R}^n)$$

for any multiindex β.

Denote by $\mathcal{P}$ the set of all polynomials in $\mathbb{R}^n$. Let $1 < p, q < \infty$, $\alpha \in \mathbb{R}$ and $w \in A_\infty(\mathbb{R}^n)$. For $f \in \mathcal{S}'(\mathbb{R}^n)/\mathcal{P}$, we define the norm $\|f\|_{\dot{F}_{pq}^{\alpha,\{\Phi_k\}_{k\in\mathbb{Z}}}(w)}$ by

$$\|f\|_{\dot{F}_{pq}^{\alpha,\{\Phi_k\}_{k\in\mathbb{Z}}}(w)} := \left\| \Big(\sum_{k\in\mathbb{Z}} a_k^{\alpha q} |\Phi_k * f|^q \Big)^{1/q} \right\|_{L^p(w)}. \tag{1.36}$$

We admit that Proposition 1 below is true and we prove Theorem 3 first. We postpone the proof of Proposition 1 until the end of the paper.

Proposition 1. *Let $\alpha \neq 0$, $1 < p, q < \infty$ and $w \in A_p(\mathbb{R}^n)$. Let $\{a_k\}_{k\in\mathbb{Z}}$ be a lacunary sequence of positive numbers with $a_{k+1}/a_k \geq a > 1$ $(k \in \mathbb{Z})$. If $\|f\|_{\dot{F}_{pq}^{\alpha,\{\Phi_k\}_{k\in\mathbb{Z}}}(w)}$ and $\|f\|_{\dot{F}_{pq}^{\alpha,\{\Psi_k\}_{k\in\mathbb{Z}}}(w)}$ are equivalent for any two partitions of unity, $\{\Phi_k\}_{k\in\mathbb{Z}}$ and $\{\Psi_k\}_{k\in\mathbb{Z}}$, adapted to $\{a_k\}_{k\in\mathbb{Z}}$, then there exists $C_0 \geq a$ such that*

$$\frac{a_{k+1}}{a_k} \leq C_0 \qquad (k \in \mathbb{Z}),$$

and, in this case, $\|f\|_{\dot{F}_{pq}^{\alpha,\{\Phi_k\}_{k\in\mathbb{Z}}}(w)}$ is equivalent to the usual weighted homogeneous Triebel-Lizorkin space norm $\|f\|_{\dot{F}_{pq}^{\alpha}(w)}$.

We prepare one more lemma. See Section 5 for a generalization.

Lemma 1.1. *Let $1 \leq p_1, p_2, q_1, q_2 < \infty$, and $\alpha \in \mathbb{R}$. Define p and q by $1/p = (1-\theta)/p_1 + \theta/p_2$ and $1/q = (1-\theta)/q_1 + \theta/q_2$. Let $w_1 \in A_{p_1}(\mathbb{R}^n)$, $w_2 \in A_{p_2}(\mathbb{R}^n)$ and $v_1, v_2 \in L^1_{\mathrm{loc}}(\mathbb{R}^n)$ be nonnegative. Suppose T is a bounded sublinear map from $\dot{F}^\alpha_{p_j 2}(w_j)$ to $L^{q_j}(v_j)$ with bounds A_j $(j = 1, 2)$, then for any $0 < \theta < 1$, T is bounded from $\dot{F}^\alpha_{p2}(w_1^{(1-\theta)p/p_1} w_2^{\theta p/p_2})$ to $L^q(v_1^{(1-\theta)q/q_1} v_2^{\theta q/q_2})$ with bound $C A_1^{1-\theta} A_2^\theta$.*

Proof. Let I_α be the operator defined by $I_\alpha f(x) = |\cdot|^{-n+\alpha} * f(x)$. Then by the assumption the operator $T \circ I_\alpha$ is a bounded linear operator from $\dot{F}^0_{p_j 2}(w_j) = L^{p_j}(w_j)$

to $L^{q_j}(v_j)$ with bounds CA_j ($j = 1, 2$) (see, for example, [21] and [7, Theorem 1.4 (vi) and Remark 2.9]). Hence, by the Stein-Weiss theorem for interpolation of sublinear operators with change of measures [33, the comment after Theorem 2.11], we see that $T \circ I_\alpha$ is bounded from $L^p(w_1^{(1-\theta)p/p_1} w_2^{\theta p/p_2})$ to $L^q(v_1^{(1-\theta)q/q_1} v_2^{\theta q/q_2})$ with bound $CA_1^{1-\theta} A_2^\theta$, which in turn implies that T is bounded from $\dot{F}^\alpha_{p2}(w_1^{(1-\theta)p/p_1} w_2^{\theta p/p_2})$ to $L^q(v_1^{(1-\theta)q/q_1} v_2^{\theta q/q_2})$ with bound $CA_1^{1-\theta} A_2^\theta$. □

2 Preparation for the proof of Theorem 3

2.1 A setup

For $t > 0$, a function b on $\mathbb{R}_+$ and a homogeneous kernel Ω on $\mathbb{R}^n$, assume

$$\int_{B(t)\setminus B(t/2)} |b(|x|)\Omega(x')| \, dx < \infty.$$

For $\rho > 0$ and a nice function, we define the family $\{\sigma_t : t \in \mathbb{R}_+\}$ of measures and the maximal operator σ^* on $\mathbb{R}^n$ by

$$\int_{\mathbb{R}^n} f(x) \, d\sigma_t(x) = \frac{1}{t^\rho} \int_{B(t)\setminus B(t/2)} f(\phi(|x|)x') \frac{b(|x|)\Omega(x')}{|x|^{n-\rho}} \, dx, \tag{2.1}$$

$$\sigma^* f(x) = \sup_{t>0} \big| |\sigma_t| * f(x) \big|. \tag{2.2}$$

Note that

$$x \in \mathbb{R}^n \setminus \{0\} \mapsto \phi(|x|)x' \in \mathbb{R}^n \setminus \overline{B(\inf \phi)}$$

is a C^1-diffeomorphism, since $\phi \in C^1(\mathbb{R}_+, \mathbb{R}_+)$ satisfies (1.14) and (1.15). Therefore, if we consider the measure $\sigma_t^\dagger$ by

$$\int_{\mathbb{R}^n} f(x) \, d\sigma_t^\dagger(x) = \frac{1}{t^\rho} \int_{B(t)\setminus B(t/2)} f(x) \frac{b(|x|)\Omega(x')}{|x|^{n-\rho}} \, dx,$$

then the above diffeomorphism induces σ_t. So, about the absolute value of σ_t, we have

$$\int_{\mathbb{R}^n} f(x) \, d|\sigma_t|(x) = \frac{1}{t^\rho} \int_{B(t)\setminus B(t/2)} f(\phi(|x|)x') \frac{|b(|x|)\Omega(x')|}{|x|^{n-\rho}} \, dx.$$

We need the following lemma whose proof is in [29, Lemma 2.1].

Lemma 2.1. *Let $\Omega \in L^1(S^{n-1})$ and $b \in \Delta_1(\mathbb{R}_+)$.*

1. *For all admissible parameters,*

$$|\widehat{\sigma_t}(\xi)| \le 2^{n-\rho} \|\Omega\|_{L^1(S^{n-1})} \|b\|_{\Delta_1}. \tag{2.3}$$

2. *If in addition Ω satisfies* (1.2), *then we have*

$$|\widehat{\sigma_t}(\xi)| \le 2\|\Omega\|_{L^1(S^{n-1})} \|b\|_{\Delta_1} \phi(t)|\xi|. \tag{2.4}$$

Meanwhile, thanks to Lemma 2.2 below and the Minkowski inequality, for $p > \gamma'$ there exists $C > 0$ such that if $w \in \widetilde{A}_{p/\gamma'}(\mathbb{R}^n)$,

$$\|\sigma^*(f)\|_{L^p(w)} \le C\|b\|_{\Delta_\gamma}\|\Omega\|_{L^1(S^{n-1})}\|f\|_{L^p(w)}. \tag{2.5}$$

From the monotonicity, the doubling property (1.14) and (2.4) we get

$$\left(\int_{2^k}^{2^{k+1}} |\widehat{\sigma_t}(\xi)|^2 \frac{dt}{t\phi(t)^{2\alpha}}\right)^{1/2} \le 2\|\Omega\|_{L^1(S^{n-1})}\|b\|_{\Delta_1}|\xi|\phi(2^k)^{1-\alpha} \tag{2.6}$$

for $\alpha \in \mathbb{R}$ and $k \in \mathbb{Z}$.

As for the maximal operator σ^* given by (2.2), we invoke the following lemma in [12, Lemma 3.2]: we define the directional Hardy-Littlewood maximal function of F for a fixed vector $y' \in S^{n-1}$ by

$$M_{y'}F(x) := \sup_{r>0}\frac{1}{2r}\int_{-r}^{r} |f(x - ty')|\, dt, \quad x \in \mathbb{R}^n.$$

By the orthogonal decomposition $\mathbb{R}^n = H \oplus \mathbb{R}y'$, we can prove that $M_{y'}$ is bounded on $L^p(\mathbb{R}^n)$ for all $1 < p < \infty$ and that the bound is uniform over y'. By combining the Hölder inequality and the change of variables to polar coordinates, we can also prove

Lemma 2.2. *Let $\gamma > 1$. Then there exists $C > 0$ such that*

$$\sigma^*(f)(x) \le C\|b\|_{\Delta_\gamma}\|\Omega\|_{L^1(S^{n-1})}^{1/\gamma}\left(\int_{S^{n-1}} |\Omega(y')|M_{y'}(|f|^{\gamma'})(x)\, d\sigma(y')\right)^{1/\gamma'}. \tag{2.7}$$

The proof is in [12, Lemma 3.2].

We have the following estimate: for a precise proof, see [12, Lemma 2.4].

Lemma 2.3. *Using (1.14) and (1.15), we have for any $0 \le \beta < 1$,*

$$|\widehat{\sigma_t}(\xi)| \le CW_\Omega\|b\|_{\Delta_2}\frac{1}{(|\xi|\phi(t))^{\beta/2}}. \tag{2.8}$$

Note that this is where the number $\max\{2, \gamma\}$ appears in the right-hand side of (1.34).

2.2 Construction of partition of unity

For our purpose, we introduce a partition of unity and a characterization of the homogeneous Triebel-Lizorkin spaces associated to ϕ satisfying (1.14) and (1.15).

It is easily seen from (1.15) that $\{\phi(2^j)\}_{j\in\mathbb{Z}}$ is a lacunary sequence of positive numbers satisfying

$$\frac{\phi(2^{j+1})}{\phi(2^j)}\left(= \exp\left[\int_{2^j}^{2^{j+1}} (\log\phi(t))'\, dt\right]\right) \ge 2^{1/\|\varphi\|_{L^\infty(\mathbb{R}_+)}} > 1, \quad j \in \mathbb{Z} \tag{2.9}$$

(see e.g. [12, Lemma 2.8]). We denote $a_j := 1/\phi(2^{-j})$ and $a := 2^{1/\|\varphi\|_{L^\infty(\mathbb{R}_+)}}$.

Then $\{a_j\}_{j\in\mathbb{Z}}$ is also a lacunary sequence of the same lacunarity as $\{\phi(2^j)\}_{j\in\mathbb{Z}}$. Take a nonincreasing $C^\infty(\mathbb{R})$-function η such that $\chi_{[-1/a,1/a]}(t) \le \eta(t) \le \chi_{[-1,1]}(t)$ for all $t \in \mathbb{R}$.

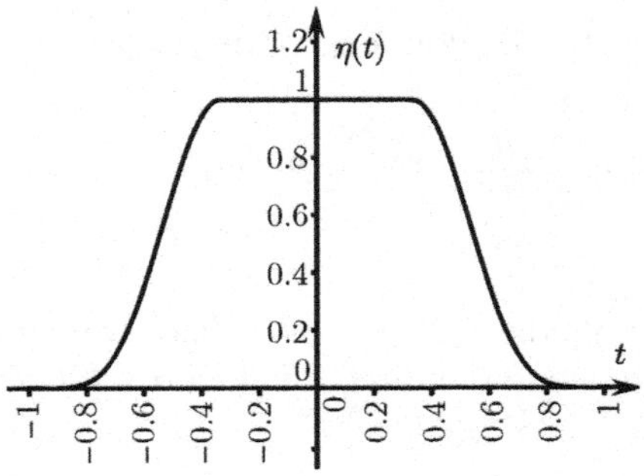

We define functions ψ_j on $\mathbb{R}^n$ by

$$\psi_j(\xi) = \eta\Big(\frac{|\xi|}{a_{j+1}}\Big) - \eta\Big(\frac{|\xi|}{a_j}\Big) \quad (\xi \in \mathbb{R}^n). \tag{2.10}$$

Then observe that

$$\psi_j(\xi) = \begin{cases} 0, & 0 \le |\xi| \le a_j/a, \ |\xi| \ge a\, a_{j+1}, \\ 1, & a_j \le |\xi| \le a_{j+1}, \end{cases} \tag{2.11}$$

and that

$$\operatorname{supp}\psi_j \subset \{a_j/a \le |\xi| \le a_{j+1}\}, \tag{2.12}$$

$$\operatorname{supp}\psi_j \cap \operatorname{supp}\psi_\ell = \emptyset, \ \text{ for } |j-\ell| \ge 2. \tag{2.13}$$

$$\sum_{j\in\mathbb{Z}} \psi_j(\xi) = 1 \quad (\mathbb{R}^n \setminus \{0\}). \tag{2.14}$$

That is, $\{\psi_j\}_{j\in\mathbb{Z}}$ is a smooth partition of unity adapted to $\{a_j\}_{j\in\mathbb{Z}}$.

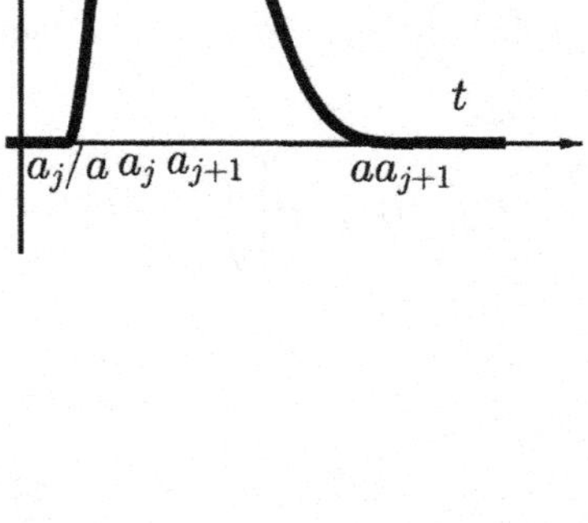

Let Ψ_j be defined on $\mathbb{R}^n$ by $\widehat{\Psi_j}(\xi) = \psi_j(\xi)$. By Proposition 1, we have for $w \in A_p(\mathbb{R}^n)$,

$$\left\| \left(\sum_{j=-\infty}^{\infty} |a_j^\alpha \Psi_j * f|^q \right)^{1/q} \right\|_{L^p(w)} \approx \|f\|_{\dot{F}^\alpha_{pq}(w)}. \tag{2.15}$$

if $\left\{ \frac{a_{j+1}}{a_j} \right\}_{j\in\mathbb{Z}}$ is bounded. Note that this condition is satisfied in our case, i.e. $a_{j+1}/a_j = \phi(2^{-j})/\phi(2^{-j-1}) \le c_1$.

2.3 A reduction by using the scaling invariance

As we did in [29, Section 2.4], for our purpose, it is sufficient to estimate the modified operator

$$\begin{aligned}&\tilde{\mu}^{(b)}_{\Omega,\rho,\phi,\alpha,q}(f)(x)\\ &:= \left(\int_0^\infty \left| \frac{1}{t^\rho \phi(t)^\alpha} \int_{B(t)\setminus B(t/2)} \frac{b(|y|)\Omega(y')}{|y|^{n-\rho}} f\left(x - \phi(|y|)y'\right) dy \right|^q \frac{dt}{t} \right)^{1/q}.\end{aligned} \tag{2.16}$$

If we use (2.1), then we have

$$\tilde{\mu}^{(b)}_{\Omega,\alpha,\rho,q}(f)(x) = \left(\int_0^\infty |\sigma_t * f(x)|^q \frac{dt}{t\phi(t)^{q\alpha}} \right)^{1/q}. \tag{2.17}$$

Now we proceed to the final preparation for the proof of Theorem 3. Let

$$\tilde{\mu}^{(b)}_{\Omega,\rho,\phi,\alpha,q,j} f(x) := \left(\sum_{k=-\infty}^{\infty} \int_{2^k}^{2^{k+1}} |\Psi_{j-k} * \sigma_t * f(x)|^q \frac{dt}{t\phi(t)^{q\alpha}} \right)^{1/q} \tag{2.18}$$

for each j. Using the partition of unity (2.10) and the triangle inequality as we did in [29, (2.24)], we then have

$$\tilde{\mu}^{(b)}_{\Omega,\rho,\phi,\alpha,q} f(x) \le \sum_{j\in\mathbb{Z}} \tilde{\mu}^{(b)}_{\Omega,\rho,\phi,\alpha,q,j} f(x), \quad x \in \mathbb{R}^n. \tag{2.19}$$

We need the $L^2(\mathbb{R}^n)$-estimate of $\tilde{\mu}^{(b)}_{\Omega,\rho,\phi,\alpha,q,j} f$. Note that we do not need the weighted estimate here.

Lemma 2.4 ([29, Lemma 2.5]). *Assume that* $\Omega \in L^1(S^{n-1})$ *satisfies* (1.2). *Then*

$$\|\tilde{\mu}^{(b)}_{\Omega,\rho,\phi,\alpha,2,j} f\|_{L^2(\mathbb{R}^n)} \le \begin{cases} C 2^{(\log_2 c_0 - \alpha/c_1)j} \|\Omega\|_{L^1(S^{n-1})} \|f\|_{\dot{F}^\alpha_{2,2}(\mathbb{R}^n)} & \text{if } j \ge 0, \\ C 2^{(1/c_1 - \alpha \log_2 c_0)j} \|\Omega\|_{L^1(S^{n-1})} \|f\|_{\dot{F}^\alpha_{2,2}(\mathbb{R}^n)} & \text{if } j \le 0. \end{cases} \tag{2.20}$$

Next, we treat the $L^p(w)$-estimate of $\tilde{\mu}^{(b)}_{\Omega,\rho,\phi,\alpha,q,j} f$.

In Section 3 we plan to distinguish three cases to prove

Lemma 2.5. *Assume that the parameters p, q, γ and the weight w satisfies*

$$p, q \in (1, \infty), \quad \gamma \in (1, \infty]$$

as well as either one of the following conditions in the table below:

$1 < q < r < \gamma q < \infty$	$w^{1-(r/q)'} \in \tilde{A}_{(r/q)'/\gamma'}(\mathbb{R}^n)$
$1 < q' < r' < \gamma q' < \infty$	$w^{(1-r')(1-(r'/q')')} \in \tilde{A}_{(r'/q')'/\gamma'}(\mathbb{R}^n)$
$\gamma' < q = r < \infty$	$w \in \tilde{A}_{r/\gamma'}(\mathbb{R}^n)$

If $\Omega \in L^1(S^{n-1})$, then we have

$$\|\tilde{\mu}^{(b)}_{\Omega,\rho,\phi,\alpha,q,j} f\|_{L^r(w)} \le \begin{cases} C 2^{-(\alpha/c_1)j} \|\Omega\|_{L^1(S^{n-1})} \|f\|_{\dot{F}^{\alpha}_{rq}(w)}, & j \ge 0, \\ C 2^{-(\alpha \log_2 c_0)j} \|\Omega\|_{L^1(S^{n-1})} \|f\|_{\dot{F}^{\alpha}_{rq}(w)}, & j < 0. \end{cases} \tag{2.21}$$

Let us set

$$\alpha(j) := \begin{cases} \alpha/c_1, & j \ge 0, \\ \alpha \log_2 c_0, & j < 0. \end{cases}$$

By using the strong decay of (2.21), interpolate (2.21) and (2.20) to have (2.21) again for any admissible p and q. Thus, in conclusion, (2.19) is summable over j by virtue of (2.21).

3 Proof of Theorem 3

3.1 Proof of Lemma 2.5

Here we do not need the cancellation property and hence we can consider its absolute value, which means that σ_t may be assumed to be positive.

(1) Let us consider the case that

$$1 < q < r < \gamma q.$$

Write

$$J := \|\tilde{\mu}^{(b)}_{\Omega,\rho,\phi,\alpha,q,j} f\|^q_{L^r(w)} = \left\| \left(\sum_{k \in \mathbb{Z}} \int_{2^k}^{2^{k+1}} \left| \Psi_{j-k} * \sigma_t * f(x) \right|^q \frac{dt}{t\phi(t)^{\alpha q}} \right)^{1/q} \right\|^q_{L^r(w)}.$$

Let us set $s = (r/q)' = r/(r-q)$. By the duality $L^{r/q}(w^{1-s})$-$L^s(w^{1-s})$, we can take a nonnegative function $h \in L^s(w^{1-s})$ with $\|h\|_{L^s(w^{1-s})} \le 2$ such that

$$J = \int_{\mathbb{R}^n} \sum_{k \in \mathbb{Z}} \left\{ \int_{2^k}^{2^{k+1}} \left| \Psi_{j-k} * \sigma_t * f(x) \right|^q \frac{dt}{t\phi(t)^{\alpha q}} \right\} h(x)\, dx.$$

Denote by $\|\sigma_t\|$ the total mass of σ_t. Hence, by the Hölder inequality

$$J = \sum_{k \in \mathbb{Z}} \int_{2^k}^{2^{k+1}} \left\{ \int_{\mathbb{R}^n} \left| \int_{\mathbb{R}^n} \Psi_{j-k} * f(x-y) d\sigma_t(y) \right|^q h(x)\, dx \right\} \frac{dt}{t\phi(t)^{\alpha q}}$$

$$\le \sum_{k\in\mathbb{Z}} \int_{2^k}^{2^{k+1}} \left\{ \int_{\mathbb{R}^n} \left[\int_{\mathbb{R}^n} |\Psi_{j-k} * f(x-y)|^q d\sigma_t(y) \right] \|\sigma_t\|^{q/q'} h(x)\, dx \right\} \frac{dt}{t\phi(t)^{\alpha q}}.$$

From (2.1), we deduce

$$\begin{aligned} J &\le C\|\Omega\|^q_{L^1(S^{n-1})} \|b\|_{\Delta_1}{}^{q-1} \\ &\quad \times \sum_{k\in\mathbb{Z}} \int_{2^k}^{2^{k+1}} \left\{ \int_{\mathbb{R}^n} \left[\int_{\mathbb{R}^n} |\Psi_{j-k} * f(y)|^q d|\sigma_t|(x-y) \right] h(x)\, dx \right\} \frac{dt}{t\phi(t)^{\alpha q}} \\ &= C\|\Omega\|^q_{L^1(S^{n-1})} \|b\|_{\Delta_1}{}^{q-1} \\ &\quad \times \sum_{k\in\mathbb{Z}} \int_{2^k}^{2^{k+1}} \left\{ \int_{\mathbb{R}^n} |\Psi_{j-k} * f(y)|^q \left(\int_{\mathbb{R}^n} h(x)\, d|\sigma_t|(x-y) \right) dy \right\} \frac{dt}{t\phi(t)^{\alpha q}}. \end{aligned}$$

Now, since $w^{1-s} \in \widetilde{A}_{s/\gamma'}(\mathbb{R}^n)$, by (2.5) and Hölder's inequality we get

$$\begin{aligned} J^{1/q} &\le C\|\Omega\|_{L^1(S^{n-1})} \|b\|_{\Delta_\gamma} \\ &\quad \times \left(\int_{\mathbb{R}^n} \sum_{k\in\mathbb{Z}} \left\{ \int_{2^k}^{2^{k+1}} |\Psi_{j-k} * f(y)|^q \sigma^*(h)(y) \frac{dt}{t\phi(t)^{\alpha q}} \right\} dy \right)^{1/q} \\ &\le C\|\Omega\|_{L^1(S^{n-1})} \|b\|_{\Delta_\gamma} \\ &\quad \times \left(\int_{\mathbb{R}^n} \sum_{k\in\mathbb{Z}} \frac{1}{\phi(2^k)^{\alpha q}} |\Psi_{j-k} * f(y)|^q \sigma^*(h)(y) w(y)^{1/s'} w(y)^{-1/s'} dy \right)^{1/q} \\ &\le C\|\Omega\|_{L^1(S^{n-1})} \|b\|_{\Delta_\gamma} \\ &\quad \times \left(\int_{\mathbb{R}^n} \left(\sum_{k\in\mathbb{Z}} \frac{1}{\phi(2^k)^{\alpha q}} |\Psi_{j-k} * f(y)|^q \right)^{s'} w(y) dy \right)^{1/(s'q)} \|h\|^{1/q}_{L^s(w^{1-s})} \\ &\le C\|\Omega\|_{L^1(S^{n-1})} \|b\|_{\Delta_\gamma} \left(\int_{\mathbb{R}^n} \left(\sum_{k\in\mathbb{Z}} \frac{1}{\phi(2^k)^{\alpha q}} |\Psi_{j-k} * f(y)|^q \right)^{r/q} w(y) dy \right)^{1/r}. \end{aligned}$$

Thus, by the doubling property of ϕ, we have

$$\|\tilde{\mu}^{(b)}_{\Omega,\rho,\phi,\alpha,q,j} f\|_{L^r(w)} \le \begin{cases} C2^{-(\alpha/c_1)j} \|f\|_{\dot{F}^\alpha_{rq}(w)}, & \text{for } j \ge 0, \\ C2^{-(\alpha \log_2 c_0)j} \|f\|_{\dot{F}^\alpha_{rq}(w)}, & \text{for } j < 0. \end{cases} \tag{3.1}$$

(2) Let us consider the case that

$$q' < r' < \gamma q'.$$

By the duality $L^r(w^{1-r'}; L^q(\frac{dt}{t}; \ell^q))$-$L^{r'}(w^{1-r'}; L^{q'}(\frac{dt}{t}; \ell^{q'}))$, there is a sequence of functions $g_k(x,t)$ such that

$$\left(\int_{\mathbb{R}^n} \left(\sum_{k\in\mathbb{Z}} \int_{2^k}^{2^{k+1}} |g_k(x,t)|^{q'} \frac{dt}{t} \right)^{r'/q'} w(x)^{1-r'} dx \right)^{1/r'} \le 2,$$

and that

$$\left\| \left(\sum_{k \in \mathbb{Z}} \int_{2^k}^{2^{k+1}} \left| \Psi_{j-k} * \sigma_t * f \right|^q \frac{dt}{t\phi(t)^{\alpha q}} \right)^{1/q} \right\|_{L^r(w)} = \int_{\mathbb{R}^n} \sum_{k \in \mathbb{Z}} \left\{ \int_{2^k}^{2^{k+1}} (\Psi_{j-k} * \sigma_t * f(x)) g_k(x,t) \frac{dt}{t\phi(t)^{\alpha}} \right\} dx.$$

Then we have

$$\begin{aligned}
&\int_{\mathbb{R}^n} \sum_{k \in \mathbb{Z}} \left\{ \int_{2^k}^{2^{k+1}} (\Psi_{j-k} * \sigma_t * f(x)) g_k(x,t) \frac{dt}{t\phi(t)^{\alpha}} \right\} dx \\
&\le \int_{\mathbb{R}^n} \sum_{k \in \mathbb{Z}} \left\{ \int_{2^k}^{2^{k+1}} \left(\int_{\mathbb{R}^n} |\Psi_{j-k} * f(y)| d|\sigma_t|(x-y) \right) |g_k(x,t)| \frac{dt}{t\phi(t)^{\alpha}} \right\} dx \\
&\le \int_{\mathbb{R}^n} \sum_{k \in \mathbb{Z}} \left\{ \int_{2^k}^{2^{k+1}} |\Psi_{j-k} * f(y)| \left(\int_{\mathbb{R}^n} |g_k(x,t)| d|\sigma_t|(x-y) \right) \frac{dt}{t\phi(t)^{\alpha}} \right\} dx \\
&\le C \int_{\mathbb{R}^n} \sum_{k \in \mathbb{Z}} \left\{ \int_{2^k}^{2^{k+1}} \frac{1}{\phi(2^k)^{\alpha}} |\Psi_{j-k} * f(y)| \left(\int_{\mathbb{R}^n} |g_k(x,t)| d|\sigma_t|(x-y) \right) \frac{dt}{t} \right\} dy.
\end{aligned}$$

By using the Hölder inequality for sequences, we have

$$\begin{aligned}
&\int_{\mathbb{R}^n} \sum_{k \in \mathbb{Z}} \left\{ \int_{2^k}^{2^{k+1}} (\Psi_{j-k} * \sigma_t * f(x)) g_k(x,t) \frac{dt}{t\phi(t)^{\alpha}} \right\} dx \\
&\le C \int_{\mathbb{R}^n} \left(\sum_{k \in \mathbb{Z}} \frac{1}{\phi(2^k)^{\alpha q}} |\Psi_{j-k} * f(y)|^q \right)^{1/q} w(y)^{1/r} \\
&\quad \times \left(\sum_{k \in \mathbb{Z}} \int_{2^k}^{2^{k+1}} \left(\int_{\mathbb{R}^n} |g_k(x,t)| d|\sigma_t|(x-y) \right)^{q'} \frac{dt}{t} \right)^{1/q'} w(y)^{-1/r} dy \\
&\le C 2^{-\alpha(j) j} \left(\int_{\mathbb{R}^n} \left(\sum_{k \in \mathbb{Z}} \frac{1}{\phi(2^k)^{\alpha q}} |\Psi_k * f(y)|^q \right)^{r/q} w(y)\, dy \right)^{1/r} \\
&\quad \times \left(\int_{\mathbb{R}^n} \left(\sum_{k \in \mathbb{Z}} \int_{2^k}^{2^{k+1}} \left(\int_{\mathbb{R}^n} |g_k(x,t)| d|\sigma_t|(x-y) \right)^{q'} \frac{dt}{t} \right)^{r'/q'} w(y)^{-r'/r}\, dy \right)^{1/r'}.
\end{aligned}$$

Write

$$K := \left(\int_{\mathbb{R}^n} \left(\sum_{k \in \mathbb{Z}} \int_{2^k}^{2^{k+1}} \left(\int_{\mathbb{R}^n} |g_k(x,t)| d|\sigma_t|(x-y) \right)^{q'} \frac{dt}{t} \right)^{r'/q'} w(y)^{-r'/r} dy \right)^{q'/r'}.$$

Set $s := (r'/q')'$ and $v := w^{-r'/r}$, so that we have $s' = r'/q'$. Hence

$$K = \left(\int_{\mathbb{R}^n} \left(\sum_{k \in \mathbb{Z}} \int_{2^k}^{2^{k+1}} \left(v(y)^{s/s'} \int_{\mathbb{R}^n} |g_k(x,t)| d|\sigma_t|(x-y) \right)^{q'} \frac{dt}{t} \right)^{s'} v(y)^{1-s} dy \right)^{1/s'}.$$

By duality, there exists $h \in L^s(v^{1-s})$ with $\|h\|_{L^s(v^{1-s})} = 1$ such that

$$\begin{aligned}
K &= \int_{\mathbb{R}^n} \left(\sum_{k\in\mathbb{Z}} \int_{2^k}^{2^{k+1}} \left(v(y)^{s/s'} \int_{\mathbb{R}^n} |g_k(x,t)| d|\sigma_t|(x-y) \right)^{q'} \frac{dt}{t} \right) h(y) v(y)^{1-s} dy \\
&= \int_{\mathbb{R}^n} \left(\sum_{k\in\mathbb{Z}} \int_{2^k}^{2^{k+1}} \left(\int_{\mathbb{R}^n} |g_k(x,t)| d|\sigma_t|(x-y) \right)^{q'} \frac{dt}{t} \right) h(y) dy \\
&= \int_{\mathbb{R}^n} \left(\sum_{k\in\mathbb{Z}} \int_{2^k}^{2^{k+1}} \left(\int_{\mathbb{R}^n} |g_k(z+y,t)| d|\sigma_t|(z) \right)^{q'} \frac{dt}{t} \right) h(y) dy.
\end{aligned}$$

Since the total measure $\|\sigma_t\|$ of σ_t is bounded by $2^{n-\rho}\|\Omega\|_{L^1(S^{n-1})}\|b\|_{\Delta_1}$, by using the Hölder inequality, we get

$$\begin{aligned}
K &\le C(\|\Omega\|_{L^1(S^{n-1})}\|b\|_{\Delta_1})^{q'/q} \sum_{k\in\mathbb{Z}} \int_{\mathbb{R}^n} \int_{2^k}^{2^{k+1}} \left(\int_{\mathbb{R}^n} |g_k(z+y,t)|^{q'} d|\sigma_t|(z) \frac{dt}{t} \right) h(y) dy \\
&= C(\|\Omega\|_{L^1(S^{n-1})}\|b\|_{\Delta_1})^{q'/q} \sum_{k\in\mathbb{Z}} \int_{2^k}^{2^{k+1}} \int_{\mathbb{R}^n} \left(\int_{\mathbb{R}^n} |g_k(z+y,t)|^{q'} h(y) dy \right) d|\sigma_t|(z) \frac{dt}{t} \\
&= C(\|\Omega\|_{L^1(S^{n-1})}\|b\|_{\Delta_1})^{q'/q} \sum_{k\in\mathbb{Z}} \int_{2^k}^{2^{k+1}} \int_{\mathbb{R}^n} \left(\int_{\mathbb{R}^n} |g_k(x,t)|^{q'} h(x-z) dx \right) d|\sigma_t|(z) \frac{dt}{t} \\
&= C(\|\Omega\|_{L^1(S^{n-1})}\|b\|_{\Delta_1})^{q'/q} \\
&\quad \times \sum_{k\in\mathbb{Z}} \int_{2^k}^{2^{k+1}} \left[\int_{\mathbb{R}^n} |g_k(x,t)|^{q'} \left(\int_{\mathbb{R}^n} h(x-z) d|\sigma_t|(z) \right) dx \right] \frac{dt}{t}.
\end{aligned}$$

By using the maximal operator σ^*, we obtain

$$\begin{aligned}
K &\le C(\|\Omega\|_{L^1(S^{n-1})}\|b\|_{\Delta_1})^{q'/q} \sum_{k\in\mathbb{Z}} \int_{2^k}^{2^{k+1}} \left(\int_{\mathbb{R}^n} |g_k(x,t)|^{q'} \sigma^*(h)(x) dx \right) \frac{dt}{t} \\
&= C(\|\Omega\|_{L^1(S^{n-1})}\|b\|_{\Delta_1})^{q'/q} \int_{\mathbb{R}^n} \left(\sum_{k\in\mathbb{Z}} \int_{2^k}^{2^{k+1}} |g_k(x,t)|^{q'} \frac{dt}{t} \right) \sigma^*(h)(x) dx \\
&= C(\|\Omega\|_{L^1(S^{n-1})}\|b\|_{\Delta_1})^{q'/q} \\
&\quad \times \int_{\mathbb{R}^n} \left(\sum_{k\in\mathbb{Z}} \int_{2^k}^{2^{k+1}} |g_k(x,t)|^{q'} \frac{dt}{t} \right) \sigma^*(h)(x) w(x)^{(1-r')/s'} w(x)^{(r'-1)/s'} dx \\
&\le C(\|\Omega\|_{L^1(S^{n-1})}\|b\|_{\Delta_1})^{q'/q} \left(\int_{\mathbb{R}^n} \left(\sum_{k\in\mathbb{Z}} \int_{2^k}^{2^{k+1}} |g_k(x,t)|^{q'} \frac{dt}{t} \right)^{s'} w(x)^{1-r'} dx \right)^{1/s'} \\
&\quad \times \left(\int_{\mathbb{R}^n} \sigma^*(h)(x)^s w(x)^{(1-r')(1-s)} dx \right)^{1/s}.
\end{aligned}$$

Assuming that $v^{1-s} = w^{(1-r')(1-s)} \in \widetilde{A}_{s/\gamma'}(\mathbb{R}^n)$ and that $s = (r'/q')' > \gamma'$, we deduce

$$\left(\int_{\mathbb{R}^n} \sigma^*(h)(x)^s w(x)^{(1-r')(1-s)}dx\right)^{1/s} \le C\|\Omega\|_{L^1(S^{n-1})}\|b\|_{\Delta_\gamma}\|h\|_{L^s(v^{1-s})}$$

from the estimate for $\sigma^*(h)(x)$ (see Lemma 2.2):

$$\sigma^*(h)(x) \le C\|b\|_{\Delta_\gamma}\|\Omega\|_{L^1(S^{n-1})}^{1/\gamma}\left(\int_{S^{n-1}} |\Omega(y')|M_{y'}(|h|^{\gamma'})(x)\,d\sigma(y')\right)^{1/\gamma'}.$$

Recall again that

$$s = \left(\frac{r'}{q'}\right)' > \gamma', \quad v^{1-s} = w^{(1-r')(1-s)} \in \widetilde{A}_{s/\gamma'}(\mathbb{R}^n).$$

Thus, we have

$$\begin{aligned}
&\left(\int_{\mathbb{R}^n}\left(\sum_{k\in\mathbb{Z}}\int_{2^k}^{2^{k+1}}\left(\int_{\mathbb{R}^n}|g_k(x,t)|d|\sigma_t|(x-y)\right)^{q'}\frac{dt}{t}\right)^{r'/q'} w(y)^{-r'/r}dy\right)^{1/r'}\\
&\le C\|\Omega\|_{L^1(S^{n-1})}\|b\|_{\Delta_\gamma}\left(\int_{\mathbb{R}^n}\left(\sum_{k\in\mathbb{Z}}\int_{2^k}^{2^{k+1}}|g_k(x,t)|^{q'}\frac{dt}{t}\right)^{r'/q'} w(x)^{1-r'}dx\right)^{1/r'}\\
&\le C\|\Omega\|_{L^1(S^{n-1})}\|b\|_{\Delta_\gamma}.
\end{aligned}$$

For the later use, we note that if $\gamma = \infty$, our assumption in this case means $v \in \widetilde{A}_{s'}(\mathbb{R}^n)$ i.e. $w^{(1-r')} \in \widetilde{A}_{r'/q'}(\mathbb{R}^n)$.

(3) Finally let us consider the case that

$$\gamma' < r = q.$$

In case $1 < r = q < \infty$, it is easy to see by using (2.5) that for $r > \gamma'$ and $w \in \widetilde{A}_{r/\gamma'}$,

$$\begin{aligned}
\|\tilde{\mu}^{(b)}_{\Omega,\alpha,\rho,q,j}f\|_{L^q(w)} &= \left(\int_{\mathbb{R}^n}\sum_{k\in\mathbb{Z}}\left(\int_{2^k}^{2^{k+1}}\left|\Psi_{j-k}*\sigma_t*f(x)\right|^q\frac{dt}{t\phi(t)^{\alpha q}}\right)w(x)dx\right)^{1/q}\\
&= \left(\sum_{k\in\mathbb{Z}}\int_{2^k}^{2^{k+1}}\left(\int_{\mathbb{R}^n}\left|\Psi_{j-k}*\sigma_t*f(x)\right|^q w(x)dx\right)\frac{dt}{t\phi(t)^{\alpha q}}\right)^{1/q}\\
&\le C\left(\sum_{k\in\mathbb{Z}}\int_{2^k}^{2^{k+1}}\left(\int_{\mathbb{R}^n}\left|\Psi_{j-k}*f(x)\right|^q w(x)dx\right)\frac{dt}{t\phi(t)^{\alpha q}}\right)^{1/q}\\
&\le C\left(\sum_{k\in\mathbb{Z}}\int_{\mathbb{R}^n}\frac{1}{\phi(2^k)^{\alpha q}}\left|\Psi_{j-k}*f(x)\right|^q w(x)dx\right)^{1/q}\\
&\le C2^{-\alpha(j)j}\|f\|_{\dot{F}^\alpha_{rq}(w)}.
\end{aligned}$$

3.2 The proof of Theorem 3 Case $\alpha 1$

Let

$$\alpha \in \left(0, \frac{a}{(1+a)c_1 \log_2 c_0} \cdot \frac{1/\tilde{p} - 1/(2\gamma)}{1/2 - 1/(2\gamma)} \cdot \frac{1/\tilde{q} - 1/(2\gamma)}{1/2 - 1/(2\gamma)}\right).$$

By interpolating (2.20) and (2.21), we claim that there exists $\delta > 0$ such that

$$\|\tilde{\mu}^{(b)}_{\Omega,\rho,\phi,\alpha,q,j} f\|_{L^p(w)} \le C2^{-\delta|j|} \|f\|_{\dot{F}^{\alpha}_{pq}(w)}. \tag{3.2}$$

Note that (3.2) is decomposed into two estimates:

$$\|\tilde{\mu}^{(b)}_{\Omega,\rho,\phi,\alpha,q,j} f\|_{L^p(w)} \le C2^{-\delta j} \|f\|_{\dot{F}^{\alpha}_{pq}(w)} \quad (j \ge 0), \tag{3.3}$$

$$\|\tilde{\mu}^{(b)}_{\Omega,\rho,\phi,\alpha,q,j} f\|_{L^p(w)} \le C2^{\delta j} \|f\|_{\dot{F}^{\alpha}_{pq}(w)} \quad (j \le 0). \tag{3.4}$$

However, (3.3) is (2.21) itself when $\delta = \alpha/c_1$. Therefore, we concentrate on (3.4). Here and below in this subsection, we let $j \le 0$.

Case $W0$. First, we shall obtain a general estimate which follows from (1.27). From (1.27), $w^{1+a} \in \widetilde{A}_{2/\gamma'}(\mathbb{R}^n)$. Observe also the condition $\gamma > 2$ allows us to use Lemma 2.5 with $\gamma' < q = r = 2$.

We apply Lemma 1.1 with $\theta = (1+a)^{-1}$.

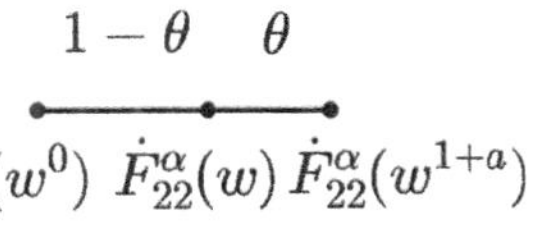

Hence, by (2.20) and (2.21), we get for $j \le 0$,

$$\|\tilde{\mu}^{(b)}_{\Omega,\rho,\phi,\alpha,2,j} f\|_{L^2(w)} \le C2^{(\frac{a}{c_1(1+a)} - \alpha \log_2 c_0)j} \|\Omega\|_{L^1(S^{n-1})} \|f\|_{\dot{F}^{\alpha}_{2,2}(w)}. \tag{3.5}$$

Thus, by (3.5) we obtain (3.4) with $\delta = \frac{a}{c_1(1+a)} - \alpha \log_2 c_0$.

Cases $W1$, $W4$, $W5$, $W6$. We place ourselves in the interior of the small square of the following $1/p - 1/q$ diagram:

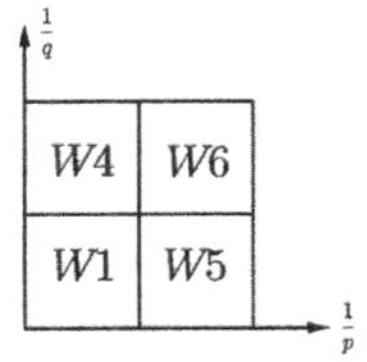

We shall begin with an arithmetic. The proof is in [29, p.18].

Lemma 3.1. *We can take* $1 < r_1, r_2 < \infty$ *and* $0 < \theta_1, \theta_2 < 1$ *satisfying*

$$\frac{1}{p} = \frac{\theta_1}{2} + \frac{1-\theta_1}{r_1}, \tag{3.6}$$

$$\frac{1}{q} = \frac{\theta_2}{2} + \frac{1-\theta_2}{r_2}, \tag{3.7}$$

$$\alpha < \frac{\theta_1\theta_2 a}{(1+a)c_1 \log_2 c_0} < \frac{a}{(1+a)c_1 \log_2 c_0} \cdot \frac{1/\tilde{p} - 1/(2\gamma)}{1/2 - 1/(2\gamma)} \cdot \frac{1/\tilde{q} - 1/(2\gamma)}{1/2 - 1/(2\gamma)}. \tag{3.8}$$

Our strategy of interpolations is as follows:
Step 1: We interpolate $\dot{F}^{\alpha}_{22}(w)$ and $\dot{F}^{\alpha}_{r_1 2}(w)$ to have an estimate for $\dot{F}^{\alpha}_{p2}(w)$.

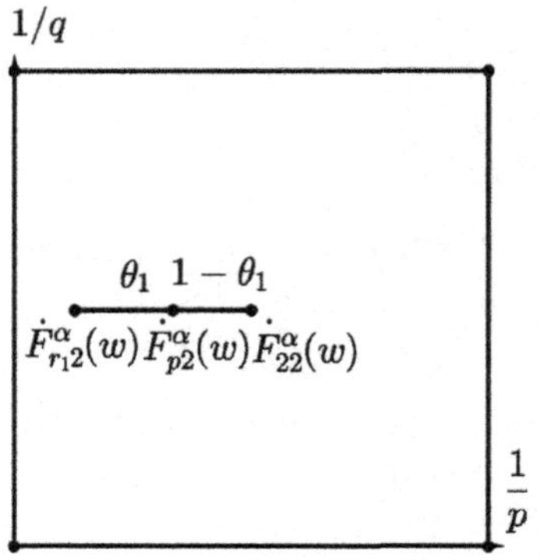

Step 2: We interpolate $\dot{F}^{\alpha}_{p2}(w)$ and $\dot{F}^{\alpha}_{pr_2}(w)$ to have an estimate for $\dot{F}^{\alpha}_{pq}(w)$.

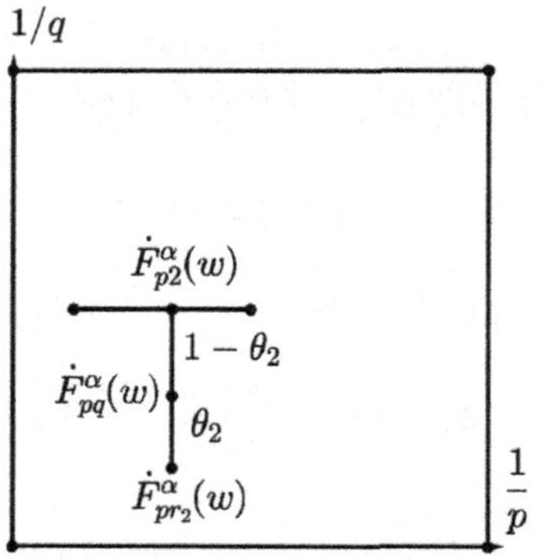

After interpolations in Steps 1 and 2, we are led to

$$\begin{aligned} &\|\tilde{\mu}^{(b)}_{\Omega,\rho,\phi,\alpha,q,j} f\|_{L^p(w)} \\ &\le C 2^{\{\theta_2\left(\theta_1\left(\frac{a}{c_1(1+a)} - \alpha\log_2 c_0\right) - (1-\theta_1)\alpha\log_2 c_0\right) - (1-\theta_2)\alpha\log_2 c_0\}j} \|f\|_{\dot{F}^{\alpha}_{pq}(w)}. \end{aligned} \tag{3.9}$$

An arithmetic shows that

$$\theta_2\left\{\theta_1\left(\frac{a}{c_1(1+a)} - \alpha\log_2 c_0\right) - (1-\theta_1)\alpha\log_2 c_0\right\} - (1-\theta_2)\alpha\log_2 c_0$$

$$= \frac{a\theta_1\theta_2}{c_1(1+a)} - \alpha \log_2 c_0 > 0.$$

Thus, taking

$$\delta := \min\left\{\frac{\alpha}{c_1}, \frac{a\theta_1\theta_2}{c_1(1+a)} - \alpha \log_2 c_0\right\},$$

we obtain the desired estimate (3.2).

What remains to do is to verify that our assumptions postulated on each domain are suitable to justify our Steps 1 and 2.

On the square $W1$

Step 1: Note that we can choose r_1 slightly less than 2γ in Lemma 3.1. According to [29, p.18], r_1 will do as long as $\widetilde{p} < r_1 < 2\gamma$. From (1.28) we conclude $w^{1-(r_1/2)'} \in \widetilde{A}_{(r_1/2)'/\gamma'}$, by virtue of the openness theorem. Hence, interpolating (2.21) with $r = r_1, q = 2$ and (3.5), we get

$$\|\tilde{\mu}^{(b)}_{\Omega,\rho,\phi,\alpha,2,j} f\|_{L^p(w)} \le C2^{\left(\theta_1(\frac{a}{c_1(1+a)} - \alpha \log_2 c_0) - (1-\theta_1)\alpha \log_2 c_0\right)j} \|f\|_{\dot{F}^{\alpha}_{p2}(w)}. \tag{3.10}$$

Step 2: Again choosing r_2 slightly less than 2γ in Lemma 3.1, as we did in [29, p.18], we see that from (1.29) it follows $w^{(1-p')(1-(p'/r_2')')} \in \widetilde{A}_{(p'/r_2')'/\gamma'}(\mathbb{R}^n)$. Again by virtue of the openness theorem, and hence (2.21) holds for $r = p, q = r_2$. We then interpolate (2.21) and (3.10) with $r = p, q = r_2$ to have (3.9).

On the square $W4$

Step 1: As we did on $W1$, we obtain (3.10).

Step 2: Choosing r_2 slightly larger than $(2\gamma)'$ in Lemma 3.1, we conclude that from (1.30) it follows $w^{1-(p/r_2)'} \in \widetilde{A}_{(p/r_2)'/\gamma'}(\mathbb{R}^n)$. Hence we get (2.21) for $r = p, q = r_2$. Interpolating (2.21) and (3.10), we have the desired estimate as in the last part of the case (a).

On the square $W5$

Step 1: Choosing r_1 slightly larger than $(2\gamma)'$ in Lemma 3.1, we see that from (1.31) it follows $w^{(1-r_1')(1-(r_1'/2)')} \in \widetilde{A}_{(r_1'/2)'/\gamma'}(\mathbb{R}^n)$. Thus, we can apply Lemma 2.5, and get (2.21) for $r = r_1, q = 2$. So, interpolating (3.5) and (2.21) with $r = r_1, q = 2$, we get (3.10).

Step 2: This is the same as the estimate on $W1$.

On the square $W6$

Step 1: This is the same as the estimate on $W5$.

Step 2: This is the same as the estimate on $W4$.

Cases $W2$, $W3$, $W7$, $W8$. This is a degenerate case: in the case $p=2$ or $q=2$, we can get the desired estimate (3.4) more simply, by applying interpolation once.

Thus by (2.18), (2.19) and (3.2) we obtain

$$\|\tilde{\mu}^{(b)}_{\Omega,\rho,\phi,\alpha,q}f\|_{L^p(w)} \le \sum_{j\in\mathbb{Z}} \|\tilde{\mu}^{(b)}_{\Omega,\rho,\phi,\alpha,q,j}f\|_{L^p(w)} \le C\|f\|_{\dot{F}^{\alpha}_{pq}(w)}. \tag{3.11}$$

This completes the proof of Theorem 3 Case $\alpha 1$.

Remark 3. The proof of Theorem 4 is similar to the above cases, so we omit the details of the proof.

3.3 The proof of Theorem 3 Case $\alpha 3$

Below we prove Theorem 3 Case $\alpha 3$. By the Schwarz inequality,

$$\begin{aligned}
|\widehat{\sigma_t}(\xi)| &= \frac{1}{t^\rho}\left|\int_{t/2<|y|\le t} e^{-i\phi(|y|)y'\cdot\xi}\frac{b(|y|)\Omega(y')}{|y|^{n-\rho}}\,dy\right| \\
&= \frac{1}{t^\rho}\left|\int_{t/2}^{t}\left(\int_{S^{n-1}}\Omega(y')e^{-i\phi(r)y'\cdot\xi}\,d\sigma(y')\right)b(r)r^{\rho-1}dr\right| \\
&\le \left(\int_{t/2}^{t}|b(r)|^2r^{-1}dr\right)^{1/2}\left(\int_{t/2}^{t}\left|\int_{S^{n-1}}\Omega(y')e^{-i\phi(r)y'\cdot\xi}\,d\sigma(y')\right|^2\frac{dr}{r}\right)^{1/2} \\
&\le C\|b\|_{\Delta_2}\left(\int_{t/2}^{t}\left|\int_{S^{n-1}}\Omega(y')e^{-i\phi(r)y'\cdot\xi}\,d\sigma(y')\right|^2\frac{dr}{r}\right)^{1/2}.
\end{aligned}$$

Recall that we have set

$$W_\Omega = \sup_{\xi\in\mathbb{R}^n\setminus\{0\}}\left(\int_{S^{n-1}\times S^{n-1}}\frac{|\Omega(y')\Omega(z')|}{|(y'-z')\cdot\xi'|^\beta}\,d\sigma(y')d\sigma(z')\right)^{1/2}. \tag{3.12}$$

Then, by (2.8), we have

$$\left(\int_{2^k}^{2^{k+1}}|\widehat{\sigma_t}(\xi)|^2\frac{dt}{t\phi(t)^{2\alpha}}\right)^{1/2} \le \frac{CW_\Omega}{|\xi|^{\beta/2}\phi(2^{k-1})^{\beta/2}\phi(2^{k+1})^{\alpha}}. \tag{3.13}$$

By (3.13) and the support property of ψ_{j-k}, we have

$$\begin{aligned}
&\|\tilde{\mu}^{(b)}_{\Omega,\rho,\phi,\alpha,2,j}f\|^2_{L^2(\mathbb{R}^n)} \\
&\le CW_\Omega^2\sum_{k=-\infty}^{\infty}\int_{\mathbb{R}^n}\frac{|\hat{f}(\xi)|^2\psi_{j-k}(|\xi|)^2}{|\xi|^\beta\phi(2^{k-1})^\beta\phi(2^{k+1})^{2\alpha}}\,d\xi \\
&\le CW_\Omega^2\sum_{\ell=-\infty}^{\infty}\int_{\mathbb{R}^n}\left(\frac{\phi(2^{-\ell+1})}{\phi(2^{j-\ell-1})}\right)^\beta\phi(2^{j-\ell+1})^{-2\alpha}|\hat{f}(\xi)|^2\psi_\ell(|\xi|)^2\,d\xi.
\end{aligned}$$

As in Case $\alpha 1$, we have that $\phi(2^{-\ell+1}) \le 2^{-(j-2)/c_1}\phi(2^{j-\ell-1})$ for $j \ge 2$, and $\phi(2^{-\ell+1}) \le 2^{-(j-2)\log_2 c_0}\phi(2^{j-\ell-1})$ for $j < 2$. Similarly, we have

$$\phi(2^{j-\ell-1}) \le 2^{(j-1)/c_1}\phi(2^{-\ell})$$

for $j \le 1$ and

$$\phi(2^{j-\ell-1}) \le c_0^{j-1}\phi(2^{-\ell}) = 2^{(j-1)\log_2 c_0}\phi(2^{-\ell})$$

for $j \ge 1$. So, as in the $L^2(\mathbb{R}^n)$-estimate in Case $\alpha 1$, we obtain

$$\|\tilde{\mu}^{(b)}_{\Omega,\rho,\phi,\alpha,2,j}f\|_{L^2(\mathbb{R}^n)} \le \begin{cases} C2^{-(\beta/(2c_1)+\alpha\log_2 c_0)j}\|f\|_{\dot{F}^{\alpha}_{2,2}} & \text{if } j \ge 1, \\ C2^{-((\beta\log_2 c_0)/2+\alpha/c_1)j}\|f\|_{\dot{F}^{\alpha}_{2,2}} & \text{if } j \le 0. \end{cases} \tag{3.14}$$

See [29, (2.26)] as well. As for the L^p-estimate, since $\alpha < 0$, it follows that

$$\phi(2^{j-\ell}) \le c_0^j\phi(2^{-\ell})$$

for $j \ge 0$ and that

$$\phi(2^{j-\ell}) \le 2^{j/c_1}\phi(2^{-\ell})$$

for $j \le 0$. Hence we get, as in the $L^p(w)$-estimate in Case $\alpha 1$, for any $1 < q, r < \infty$ with $\tilde{r} < \gamma\tilde{q}$, and $j \in \mathbb{Z}$,

$$\|\tilde{\mu}^{(b)}_{\Omega,\rho,\phi,\alpha,q,j}f\|_{L^r(w)} \le \begin{cases} C2^{-(\alpha/c_1)j}\|f\|_{\dot{F}^{\alpha}_{rq}(w)} \text{ for } j \le 0, \\ C2^{-(\alpha\log_2 c_0)j}\|f\|_{\dot{F}^{\alpha}_{rq}(w)} \text{ for } j \ge 0, \end{cases} \tag{3.15}$$

provided $w^{1-(r/q)'} \in \tilde{A}_{(r/q)'/\gamma'}(\mathbb{R}^n)$ in case $1 < q < r < \gamma q < \infty$, $w^{(1-r')(1-(r'/q')')} \in \tilde{A}_{(r'/q')'/\gamma'}(\mathbb{R}^n)$ in case $1 < q' < r' < \gamma q' < \infty$, and $w \in \tilde{A}_{r/\gamma'}(\mathbb{R}^n)$ in case $\gamma' < q = r < \infty$.

It follows that, for

$$\alpha \in \left(-\frac{2\beta}{c_1\log_2 c_0}\cdot\frac{1/\tilde{p} - 1/(2\gamma)}{1 - 1/\gamma}\cdot\frac{1/\tilde{q} - 1/(2\gamma)}{1 - 1/\gamma}, 0\right),$$

there still exists $\delta > 0$ such that

$$\|\tilde{\mu}^{(b)}_{\Omega,\rho,\phi,\alpha,q,j}f\|_{L^p(w)} \le C2^{-\delta|j|}\|f\|_{\dot{F}^{\alpha}_{pq}(w)}, \tag{3.16}$$

by using (3.15) in the case $j \le 0$, and interpolating (3.14) and (3.15) in the case $j > 0$, as in Case $\alpha 1$.

Thus by (2.18), (2.19) and (3.16) we obtain

$$\|\tilde{\mu}^{(b)}_{\Omega,\rho,\phi,\alpha,q}f\|_{L^p(w)} \le \sum_{j\in\mathbb{Z}}\|\tilde{\mu}^{(b)}_{\Omega,\rho,\phi,\alpha,q,j}f\|_{L^p(w)} \le C\|f\|_{\dot{F}^{\alpha}_{pq}(w)}.$$

This completes the proof of Theorem 3 Case $\alpha 3$.

3.4 The proof of Theorem 3 Case $\alpha 2$

This is close to the corresponding assertion in [29, Section 3.5] using the technique in [3]. □

4 Proof of Proposition 1

The part is an appendix of the present paper, where we prove Proposition 1. For the sake of completeness, we supply a whole proof. Let $\psi \in \mathcal{S}(\mathbb{R}^n)$ be chosen so that

$$\chi_{B(1)} \le \psi \le \chi_{B(a^{1/3})}.$$

Define

$$\varphi_k(\xi) := \psi(a_k^{-1}\xi) - \psi(a_{k-1}^{-1}\xi) \quad (\xi \in \mathbb{R}^n). \tag{4.1}$$

Notice that $\operatorname{supp}\varphi_k \subset \{\xi \in \mathbb{R}^n : a_{k-1} \le |\xi| \le a^{1/3}a_k\}$ $(k \in \mathbb{Z})$ and that $\varphi_k(\xi) = 1$ on $\{a^{1/3}a_{k-1} \le |\xi| \le a_k\}$. Let

$$\Phi_k := \mathcal{F}^{-1}\varphi_k. \tag{4.2}$$

Then, we see that $\{\Phi_k\}_{k\in\mathbb{Z}}$ is a partition of unity adapted to $\{a_k\}_{k\in\mathbb{Z}}$. Similarly, taking ψ so that

$$\chi_{B(a^{-1/3})} \le \psi \le \chi_{B(1)},$$

and setting

$$\varphi_k(\xi) := \psi(a_{k+1}^{-1}\xi) - \psi(a_k^{-1}\xi) \quad (\xi \in \mathbb{R}^n),$$

we obtain another partition of unity $\{\Psi_k\}_{k\in\mathbb{Z}}$ adapted to $\{a_k\}_{k\in\mathbb{Z}}$ satisfying $\operatorname{supp}\widehat{\Psi_k} \subset \{\xi \in \mathbb{R}^n : a_k/a^{1/3} \le |\xi| \le a_{k+1}\}$ $(k \in \mathbb{Z})$ and $\widehat{\Psi_k}(\xi) = 1$ on $\{a_k \le |\xi| \le a_{k+1}/a^{1/3}\}$. Note that $\{a_k \le |\xi| \le a^{2/3}a_k\} \subset \{a_k \le |\xi| \le a_{k+1}/a^{1/3}\}$. Let us take a function $\Theta \in \mathcal{S}$ so that $\operatorname{supp}(\mathcal{F}\Theta) \subset B(a^{1/3}/2 - 1/2)$. Consider

$$f_k(x) = f_k(x_1, x_2, \ldots, x_n) = \exp\left(i\frac{(a^{1/3}+a^{2/3})}{2}a_k x_1\right)\Theta(a_k x) \quad (x \in \mathbb{R}^n).$$

Then we have

$$\mathcal{F}f_k(\xi) = {a_k}^{-n}\mathcal{F}\Theta\left(\frac{\xi}{a_k} - \frac{(a^{1/3}+a^{2/3})}{2}\mathbf{e}_1\right) \quad (\xi \in \mathbb{R}^n),$$

where $\mathbf{e}_1 := (1, 0, \ldots, 0)$. It follows that $\operatorname{supp}\mathcal{F}f_k \subset \{a^{1/3}a_k \le |\xi| \le a^{2/3}a_k\}$. Hence we have

$$\Phi_j * f_k(x) = \delta_{(j-1)k}f_k(x) \quad \text{and} \quad \Psi_j * f_k(x) = \delta_{jk}f_k(x),$$

where

$$\delta_{jk} = \begin{cases} 1 & (j = k), \\ 0 & (j \ne k) \end{cases}$$

for $j, k \in \mathbb{Z}$.

$$\|f_k\|_{\dot{F}_{pq}^{\alpha,\{\Phi_k\}_{k\in\mathbb{Z}}}(w)} = {a_{k+1}}^{\alpha}\|f_k\|_{L^p(w)} = {a_{k+1}}^{\alpha}\|\Theta(a_k\cdot)\|_{L^p(w)}$$

and

$$\|f_k\|_{\dot{F}_{pq}^{\alpha,\{\Psi_k\}_{k\in\mathbb{Z}}}(w)} = {a_k}^{\alpha}\|f_k\|_{L^p(w)} = {a_k}^{\alpha}\|\Theta(a_k\cdot)\|_{L^p(w)}.$$

Since the two norms are assumed equivalent, we obtain

$$\frac{a_{k+1}}{a_k} \le C_0$$

for some $C_0 > 1$. Since $\frac{a_{k+1}}{a_k} \ge a$, we have $C_0 \ge a$.

Thus we have proved the first part of our proposition. We proceed to the second part. Let $\{a_k\}_{k\in\mathbb{Z}}$ be a lacunary sequence of positive numbers with $1 < a \le a_{k+1}/a_k \le C_0$ ($k \in \mathbb{Z}$), and $\{\Phi_k\}_{k\in\mathbb{Z}}$ be a partition of unity adapted to $\{a_k\}_{k\in\mathbb{Z}}$.

Now we can define the classical weighted homogeneous Triebel-Lizorkin spaces as follows: let $\psi \in \mathcal{S}(\mathbb{R}^n)$ be chosen so that $\chi_{B(a^{2/3})} \le \psi \le \chi_{B(a)}$. Define

$$\varphi_k(\xi) := \psi(a^{-k}\xi) - \psi(a^{-k+1}\xi) \quad (\xi \in \mathbb{R}^n).$$

Notice that $\varphi_k(\xi) \equiv 1$ on $\{a^k \le |\xi| \le a^{k+\frac{2}{3}}\}$.

Define

$$\|f\|_{\dot{F}_{pq}^{\alpha}(w)} = \left\| \left(\sum_{j=-\infty}^{\infty} |a^{\alpha j}\mathcal{F}^{-1}\varphi_j * f|^q \right)^{1/q} \right\|_{L^p(w)}.$$

Let us prove

$$\|f\|_{\dot{F}_{pq}^{\alpha,\{\Phi_k\}_{k\in\mathbb{Z}}}(w)} \le C\|f\|_{\dot{F}_{pq}^{\alpha}(w)}.$$

For each $k \in \mathbb{Z}$, we choose $m_k \in \mathbb{Z}$ so that

$$a^{m_k} \le a_k < a^{m_k+1}.$$

Combining with $a^{m_k+1} \le aa_k \le a_{k+1}$, we get $m_{k+1} \ge m_k + 1$. Meanwhile combining with $a_{k+1}/a_k \le C_0$, we have $m_{k+1} - m_k \le 1 + \log_a C_0$. Furthermore we have

$$\Phi_k = \Phi_k * \sum_{l=m_{k-1}}^{m_{k+1}+1} \varphi_l.$$

Consequently, we obtain

$$\begin{aligned}\|f\|_{\dot{F}_{pq}^{\alpha,\{\Phi_k\}_{k\in\mathbb{Z}}}(w)} &= \left\| \left(\sum_{k\in\mathbb{Z}} {a_k}^{\alpha q}|\Phi_k * f|^q \right)^{1/q} \right\|_{L^p(w)} \\ &= \left\| \left(\sum_{k\in\mathbb{Z}} {a_k}^{\alpha q}\left|\Phi_k * \sum_{l=m_{k-1}}^{m_{k+1}+1} \varphi_l * f\right|^q \right)^{1/q} \right\|_{L^p(w)}\end{aligned}$$

$$\leq \left\| \left(\sum_{k \in \mathbb{Z}} a_k{}^{\alpha q} \left| \Phi_k * \sum_{l=m_{k-1}}^{m_{k+1}+1} \varphi_l * f \right|^q \right)^{1/q} \right\|_{L^p(w)}.$$

We now invoke the Plancherel-Polya-Nikolskij inequality: we have

$$\left| \sum_{l=m_{k-1}}^{m_{k+1}+1} \varphi_l * f(x) \right| \leq C(1 + |a^{m_{k+1}+1}(x-y)|)^n M \left[\sum_{l=m_{k-1}}^{m_{k+1}+1} \varphi_l * f \right](y).$$

Using Plancherel's theorem, the assumption $|\xi^\beta \partial^\beta \widehat{\Phi_k}(\xi)| \leq C_\beta$ for all β and that $\operatorname{supp} \widehat{\Phi}_k \subset \{a_{k-1} \leq |\xi| \leq a_{k+1}\}$, we get

$$\int_{\mathbb{R}^n} (1 + |a^{m_{k+1}+1} x|)^n |\Phi_k(x)| \, dx \leq C. \tag{4.3}$$

See [29, pp.26–27] for details of (4.3). Hence, it follows that

$$\|f\|_{\dot{F}_{pq}^{\alpha, \{\Phi_k\}_{k \in \mathbb{Z}}}(w)} \leq C \left\| \left(\sum_{k \in \mathbb{Z}} a_k{}^{\alpha q} M \left[\left| \sum_{l=m_{k-1}}^{m_{k+1}+1} \varphi_l * f \right| \right]^q \right)^{1/q} \right\|_{L^p(w)}.$$

By the weighted Fefferman-Stein vector-valued maximal inequality (see [5, 19]), we obtain

$$\|f\|_{\dot{F}_{pq}^{\alpha, \{\Phi_k\}_{k \in \mathbb{Z}}}(w)} \leq C \left\| \left(\sum_{k \in \mathbb{Z}} a_k{}^{\alpha q} \left| \sum_{l=m_{k-1}}^{m_{k+1}+1} \varphi_l * f \right|^q \right)^{1/q} \right\|_{L^p(w)}.$$

If we use $(a_1 + a_2 + \cdots + a_N)^q \leq N^q (a_1{}^q + a_2{}^q + \cdots + a_N{}^q)$, then we obtain

$$\|f\|_{\dot{F}_{pq}^{\alpha, \{\Phi_k\}_{k \in \mathbb{Z}}}(w)} \leq C \left\| \left(\sum_{k \in \mathbb{Z}} a_k{}^{\alpha q} (m_{k+1} - m_{k-1} + 2)^q \sum_{l=m_{k-1}}^{m_{k+1}+1} |\varphi_l * f|^q \right)^{1/q} \right\|_{L^p(w)}.$$

Noting $m_{k+1} - m_k \geq 1$, $m_{k+1} - m_k \leq 1 + \log_a C_0$ and that $a^{m_{k-1}+1} \leq a^{m_k} \leq a_k < a^{m_k+1} \leq a^{m_{k-1}+1+[\log_a C_0]}$, we conclude

$$\begin{aligned} \|f\|_{\dot{F}_{pq}^{\alpha, \{\Phi_k\}_{k \in \mathbb{Z}}}(w)} &\leq C \left\| \left(\sum_{k \in \mathbb{Z}} a_k{}^{\alpha q} \sum_{l=m_{k-1}}^{m_{k-1}+2+[2\log_a C_0]} |\varphi_l * f|^q \right)^{1/q} \right\|_{L^p(w)} \\ &\leq C \|f\|_{F_{pq}^\alpha(w)}. \end{aligned}$$

Let us prove the reverse inequality. For each $k \in \mathbb{Z}$, we can choose $\ell_k \in \mathbb{Z}$ so that

$$a_{\ell_k} \leq a^{k-1} \leq a^{k+1} \leq a_{\ell_k+3}.$$

Then we have

$$\varphi_k \equiv \varphi_k (\Phi_{\ell_k} + \Phi_{\ell_k+1} + \Phi_{\ell_k+2} + \Phi_{\ell_k+3}).$$

As we did in [29, p.27], we have

$$\sup_{l\in\mathbb{Z}} \sharp\{k : \ell_k = l\} \le 3\log_a C_0$$

because $a_{l+3} \le {C_0}^3 a_l$. Thus, it follows that

$$\begin{aligned}\|f\|_{\dot F^\alpha_{pq}(w)} &= \left\| \left(\sum_{k=-\infty}^{\infty} |a^{\alpha k}\mathcal{F}^{-1}\varphi_k * f|^q \right)^{1/q} \right\|_{L^p(w)} \\ &= \left\| \left(\sum_{k=-\infty}^{\infty} |a^{\alpha k}\mathcal{F}^{-1}\varphi_k * (\mathcal{F}^{-1}\Phi_{\ell_k} + \cdots + \mathcal{F}^{-1}\Phi_{\ell_k+3}) * f|^q \right)^{1/q} \right\|_{L^p(w)} \\ &\le C\left\| \left(\sum_{k=-\infty}^{\infty} M[|a^{\alpha k}(\mathcal{F}^{-1}\Phi_{\ell_k} + \cdots + \mathcal{F}^{-1}\Phi_{\ell_k+3}) * f|]^q \right)^{1/q} \right\|_{L^p(w)}.\end{aligned}$$

Again by the weighted Fefferman-Stein vector-valued maximal inequality (see [5, 19]), we obtain

$$\begin{aligned}&\|f\|_{\dot F^\alpha_{pq}(w)} \\ &\le C\left\| \left(\sum_{k=-\infty}^{\infty} |a^{\alpha k}(\mathcal{F}^{-1}\Phi_{\ell_k} + \cdots + \mathcal{F}^{-1}\Phi_{\ell_k+3}) * f|^q \right)^{1/q} \right\|_{L^p(w)} \\ &\le C\left\| \left(\sum_{k=-\infty}^{\infty} \left(|{a_{\ell_k}}^\alpha \mathcal{F}^{-1}\Phi_{\ell_k} * f| + \cdots + |{a_{\ell_k+3}}^\alpha \mathcal{F}^{-1}\Phi_{\ell_k+3} * f| \right)^q \right)^{1/q} \right\|_{L^p(w)} \\ &\le C\|f\|_{\dot F^{\alpha,\{\Phi_k\}_{k\in\mathbb{Z}}}_{pq}(w)}.\end{aligned}$$

This completes the proof of our proposition.

5 Appendix: complex interpolation of homogeneous weighted Triebel-Lizorkin spaces

In this section, for a weight function w we use the following notations: $\dot F^{s,w}_{p,q} = \dot F^{s,w}_{p,q}(\mathbb{R}^n)$ and $L^{p,w} = L^{p,w}(\mathbb{R}^n)$ in place of $\dot F^s_{p,q}(w)$ and $L^p(w)$, respectively, because we use ℓ^q-valued L^p spaces $L^{p,w}(\ell^q)$ etc.

We shall prove the following result:

Theorem 5. *Let $w_0, w_1 \in A_\infty(\mathbb{R}^n)$, $1 < p_0, p_1 < \infty$, $1 \le q_0, q_1 \le \infty$, $0 < \theta < 1$ and $w = w_0^{(1-\theta)p/p_0} w_1^{\theta p/p_1}$. Then we have*

$$\left(\dot F^{s_0,w_0}_{p_0,q_0}(\mathbb{R}^n), \dot F^{s_1,w_1}_{p_1,q_1}(\mathbb{R}^n) \right)_\theta = \dot F^{s,w}_{p,q}(\mathbb{R}^n). \tag{5.1}$$

5.1 Atomic decomposition

Proposition 2. *Let $w \in A_\infty(\mathbb{R}^n)$. If $f \in \dot{F}^{s,w}_{p,q}(\mathbb{R}^n)$, then there exist a collection of functions $\{a_{\nu,m}\}_{(\nu,m)\in\mathbb{Z}\times\mathbb{Z}^n}$ and a sequence of coefficients $\{\lambda_{\nu,m}\}_{(\nu,m)\in\mathbb{Z}\times\mathbb{Z}^n}$ with the following conditions.*

1. *The coefficients $\{\lambda_{\nu,m}\}_{(\nu,m)\in\mathbb{Z}\times\mathbb{Z}^n}$ satisfy*

$$\left\| \left\{ 2^{\nu s} \sum_{m\in\mathbb{Z}^n} |\lambda_{\nu,m}\chi_{Q_{\nu,m}} \right\}_{\nu=-\infty}^{\infty} \right\|_{L^{p,w}(\ell^q)} \lesssim \|f\|_{\dot{F}^{s\ w}_{p,q}}.$$

   Here and below $Q_{\nu,m} = 2^{-\nu}m + [0, 2^{-\nu}]^n$ for $m \in \mathbb{Z}^n$ and $\chi_{Q_{\nu,m}}$ denotes the characteristic function on $Q_{\nu,m}$.

2. *The function $a_{\nu,m}$ is supported on $2^{-\nu}m + [-2^{-\nu+2}, 2^{-\nu+2}]^n$ and satisfies*

$$\|\partial^\alpha a_{\nu,m}\|_{L^\infty} \le 1$$

   for all α with $|\alpha| \le [s+1]_+ = \max\{[s+1], 0\}$, where $[\cdot]$ denotes the integer part.

3. *The function f can be decomposed as follows:*

$$f = \sum_{\nu=-\infty}^{\infty} \sum_{m\in\mathbb{Z}^n} \lambda_{\nu,m} a_{\nu,m},$$

   where the convergence is in $\mathcal{S}'(\mathbb{R}^n)$.

Denote by $\mathcal{S}'_\infty(\mathbb{R}^n)$ the topological dual. Then we know the following result.

Proposition 3. *Suppose we are given a collection of functions $\{a_{\nu,m}\}_{(\nu,m)\in\mathbb{Z}\times\mathbb{Z}^n}$ and a sequence of coefficients $\{\lambda_{\nu,m}\}_{(\nu,m)\in\mathbb{Z}\times\mathbb{Z}^n}$.*

1. *The coefficients $\{\lambda_{\nu,m}\}_{(\nu,m)\in\mathbb{Z}\times\mathbb{Z}^n}$ satisfy*

$$\left\| \left\{ 2^{\nu s} \sum_{m\in\mathbb{Z}^n} |\lambda_{\nu,m}|\chi_{Q_{\nu,m}} \right\}_{\nu=-\infty}^{\infty} \right\|_{L^{p,w}(\ell^q)} < \infty.$$

2. *The function $a_{\nu,m}$ is supported on $2^{-\nu}m + [-2^{-\nu+2}, 2^{-\nu+2}]^n$ and satisfies*

$$\|\partial^\alpha a_{\nu,m}\|_\infty \le 1$$

   for all α with $|\alpha| \le [s+1]_+$.

Then

$$f := \sum_{\nu=-\infty}^{\infty} \sum_{m\in\mathbb{Z}^n} \lambda_{\nu,m} a_{\nu,m}$$

converges in $\mathcal{S}'_\infty(\mathbb{R}^n)$ *and satisfies*

$$\|f\|_{\dot{F}^{s,w}_{p,q}} \lesssim \left\| \left\{ 2^{\nu s} \sum_{m\in\mathbb{Z}^n} |\lambda_{\nu,m}|\chi_{Q_{\nu,m}} \right\}_{\nu=-\infty}^{\infty} \right\|_{L^{p,w}(\ell^q)}.$$

See [20, 34, 24] for these propositions.

We let $\mathcal{D}$ denote the set of all dyadic cubes:

$$\mathcal{D} = \{2^{-\nu}m + [0, 2^{-\nu})^n\}_{\nu\in\mathbb{Z},\, m\in\mathbb{Z}^n}.$$

Let $\nu \in \mathbb{Z}$, $k \in \mathbb{Z}$, $l \in \mathbb{N}_0$, $m \in \mathbb{Z}^n$ and $M \geq n$. A function $m_{\nu,m}$ is said to be a (k, l, M)-smooth molecule for $Q_{\nu,m} = 2^{-\nu}m + [0, 2^{-\nu})^n$ if it satisfies the following conditions for some $N > M$:

(M1) $\displaystyle\int x^\gamma m_{\nu,m}(x)dx = 0$ for all $|\gamma| \leq k$; and

(M2) $|D^\gamma m_{\nu,m}(x)| \leq 2^{|\gamma|\nu}|Q_{\nu,m}|^{1/2}(1 + 2^\nu|x - 2^{-\nu}|)^{-N}$ for all multi-indices $\gamma \in \mathbb{N}_0^n$ with $|\gamma| \leq l$.

The conditions (M1) and (M2) are called the *moment* and *decay conditions*, respectively.

Let K, $L : \mathbb{R}^n \to \mathbb{R}$ and $M > n$. The family $\{m_{\nu,m}\}_{(\nu,m)\in\mathbb{Z}\times\mathbb{Z}^n}$ is said to be a *family of* (K, L, M)*-smooth molecules.*

We say that $\{m_{\nu,m}\}_{(\nu,m)\in\mathbb{Z}\times\mathbb{Z}^n}$ is a *family of smooth molecules for* $\dot{F}^{s,w}_{p,q}(\mathbb{R}^n)$ if it is a family of $(N + \epsilon, s + 1 + \epsilon, M)$-smooth molecules, where

$$N := \frac{n}{\min\{1, p, q\}} - n - s$$

for some constant $\epsilon > 0$, and M is a sufficiently large constant.

The number M needs to be chosen sufficiently large.

5.2 Some fundamental inequalities for complex interpolation

Finally, we need some fundamental inequalities for complex interpolation.

The first one concerns with the Poisson kernel, that is,

$$\mu_j(\theta, t) = \frac{\sin \pi\theta}{e^{\pi t} + e^{-\pi t} - 2\cos(j\pi + \pi\theta)}, \quad \theta \in [0, 1], \, t \in \mathbb{R},$$

for $j = 0, 1$.

Lemma 5.1 ([34]). *Let* $A = \{z : 0 < \operatorname{Re} z < 1\}$, $\bar{A} = \{z : 0 \leq \operatorname{Re} z \leq 1\}$ *and* $r \in (0, \infty)$. *Then there exist two positive functions* $\mu_0(\theta, t)$ *and* $\mu_1(\theta, t)$ *in* $(0, 1) \times \mathbb{R}$ *such that*

$$|g(z)|^r \leq \left(\frac{1}{1-\theta}\int_{\mathbb{R}} |g(\mathrm{it})|^r \mu_0(\theta, t)\, dt\right)^{1-\theta} \left(\frac{1}{\theta}\int_{\mathbb{R}} |g(1 + \mathrm{it})|^r \mu_1(\theta, t)\, dt\right)^{\theta}$$

with $\theta = \operatorname{Re} z$ *for any analytic function* $g(z)$ *on* A *which is uniformly continuous and bounded on* $\bar{A}$. *Furthermore, if* $\theta \in (0,1)$, *then*

$$\frac{1}{1-\theta}\int_{\mathbb{R}} \mu_0(\theta,t)\,dt = \frac{1}{\theta}\int_{\mathbb{R}} \mu_1(\theta,t)\,dt = 1.$$

Lemma 5.2 ([28, 34]). *Let* $\{\kappa_j\}_{j=-\infty}^{\infty}$ *be a positive sequence and* $\alpha > 0$. *Then we have*

$$\sum_{j=-\infty}^{\infty} \kappa_j \left(\sum_{k=-\infty}^{j} \kappa_k\right)^{\alpha-1} \le \frac{1}{\min\{\alpha,1\}}\left(\sum_{k=-\infty}^{\infty} \kappa_k\right)^{\alpha}.$$

The proof of a similar assertion can be found in [34, Lemma 1] and we essentially use this lemma in the form of [34, Lemma 1]. Here for the sake of convenience, we include the proof.

Proof. If $\alpha \ge 1$, then the inequality is trivial. If $\alpha < 1$, then we have

$$\begin{aligned}\sum_{j=-\infty}^{\infty} \kappa_j \left(\sum_{k=\infty}^{j} \kappa_k\right)^{\alpha-1} &\le \sum_{j=-\infty}^{\infty} \int_{\sum_{k=-\infty}^{j-1}\kappa_k}^{\sum_{k=0}^{j}\kappa_k} t^{\alpha-1}\,dt \\ &= \int_0^{\sum_{k=-\infty}^{\infty}\kappa_k} t^{\alpha-1}\,dt = \frac{1}{\alpha}\left(\sum_{k=-\infty}^{\infty}\kappa_k\right)^{\alpha}.\end{aligned}$$

The lemma is therefore proved. □

5.3 Proof of Theorem 5

First we prove that

$$\left(\dot{F}^{s_0,w_0}_{p_0,q_0}(\mathbb{R}^n), \dot{F}^{s_1,w_1}_{p_1,q_1}(\mathbb{R}^n)\right)_\theta \subset \dot{F}^{s,w}_{p,q}(\mathbb{R}^n) \tag{5.2}$$

in Step 1 and then

$$\left(\dot{F}^{s_0,w_0}_{p_0,q_0}(\mathbb{R}^n), \dot{F}^{s_1,w_1}_{p_1,q_1}(\mathbb{R}^n)\right)_\theta \supset \dot{F}^{s,w}_{p,q}(\mathbb{R}^n) \tag{5.3}$$

in Step 2.

Let us take a function $\varphi \in C^\infty_{\mathrm{c}}(\mathbb{R}^n)$ so that $\chi_{B(4)\setminus B(2)} \le \varphi \le \chi_{B(8)\setminus(1)}$. Define $\Phi_j(\xi) := 2^{jn}\mathcal{F}^{-1}\varphi(2^j\xi)$ for $\xi \in \mathbb{R}^n$.

Proof. Step 1. Let $f \in F(\dot{F}^{s_0,w_0}_{p_0,q_0}(\mathbb{R}^n), \dot{F}^{s_1,w_1}_{p_1,q_1}(\mathbb{R}^n))$. See [6] for the definition of $F(A,B)$, where A and B are Banach spaces. A normalization allows us to assume that

$$||f||_{F(\dot{F}^{s_0,w_0}_{p_0,q_0}(\mathbb{R}^n), \dot{F}^{s_1,w_1}_{p_1,q_1}(\mathbb{R}^n))} = 1.$$

We put $g_k(z) := \Phi_k * [f(\cdot,z)](x)$ for $k \in \mathbb{Z}$.

Fix an auxiliary parameter r so that $0 < r < \min\{p, q_0, q_1\}$. We abbreviate

$$G(x,\theta) := \left(\sum_{k=-\infty}^{\infty} |2^{ks} g_k(x,\theta)|^q \right)^{\frac{1}{q}}, \quad I := \left\{ \int_{\mathbb{R}^n} G(x,\theta)^p w(x)\,dx \right\}^{\frac{1}{p}}.$$

Then, we have

$$\|f(\theta)\|_{\dot{F}^{s,w}_{p,q}} = I = \left\{ \int_{\mathbb{R}^n} \left(\sum_{k=-\infty}^{\infty} (|2^{ks} g_k(x,\theta)|^r)^{\frac{q}{r}} \right)^{\frac{r}{q}\cdot\frac{p}{r}} w(x)\,dx \right\}^{\frac{1}{p}}. \tag{5.4}$$

Observe that the spaces involved $\ell^{q/r}$ and $L^{p,w}(\mathbb{R}^n)$ are Banach spaces. We use Lemma 5.1 with $2^{ks} g_k(z)$ instead of $g(z)$ with $x \in \mathbb{R}^n$ fixed. Let

$$a_k(x) := \frac{1}{1-\theta} \int_{\mathbb{R}} |g_k(x, \mathrm{i}t)|^r \mu_0(\theta,t)\,dt, \quad b_k(x) := \frac{1}{\theta} \int_{\mathbb{R}} |g_k(x, 1+\mathrm{i}t)|^r \mu_1(\theta,t)\,dt.$$

Then the Hölder inequality yields

$$\begin{aligned} &\left(\sum_{k=-\infty}^{\infty} (|2^{ks} g_k(x,\theta)|^r)^{\frac{q}{r}} \right)^{\frac{r}{q}} \\ &\le \left(\sum_{k=-\infty}^{\infty} 2^{ks_0 q_0} a_k(x)^{\frac{q_0}{r}} \right)^{\frac{r}{q_0}(1-\theta)} \left(\sum_{k=-\infty}^{\infty} 2^{ks_1 q_1} b_k(x)^{\frac{q_1}{r}} \right)^{\frac{r}{q_1}\theta}. \end{aligned} \tag{5.5}$$

By the Minkowski inequality we have

$$\left\| \{2^{ks_0 r} a_k(x)\}_{k=-\infty}^{\infty} \right\|_{\ell^{\frac{q_0}{r}}} \le \frac{1}{1-\theta} \int_{\mathbb{R}} \left(\sum_{k=-\infty}^{\infty} 2^{ks_0 q_0} |g_k(x,\mathrm{i}t)|^{q_0} \right)^{\frac{1}{q_0}\cdot r} \mu_0(\theta,t)\,dt.$$

A similar estimate holds for the second factor on the right-hand side of (5.5). Hence, (5.4) and (5.5) give

$$\begin{aligned} I \le \Bigg[\int_{\mathbb{R}^n} &\left(\frac{1}{1-\theta} \int_{\mathbb{R}} \left(\sum_{k=-\infty}^{\infty} 2^{ks_0 q_0} |g_k(x,\mathrm{i}t)|^{q_0} \right)^{\frac{1}{q_0}\cdot r} \mu_0(\theta,t)\,dt \right)^{(1-\theta)\frac{p}{r}} \\ &\times \left(\frac{1}{\theta} \int_{\mathbb{R}} \left(\sum_{k=-\infty}^{\infty} 2^{ks_1 q_1} |g_k(x,1+\mathrm{i}t)|^{q_1} \right)^{\frac{1}{q_1}\cdot r} \mu_1(\theta,t)\,dt \right)^{\theta\frac{p}{r}} w(x)\,dx \Bigg]^{\frac{1}{p}}. \end{aligned} \tag{5.6}$$

An arithmetic shows

$$1 = \frac{1}{\frac{p_0}{(1-\theta)p}} + \frac{1}{\frac{p_1}{\theta p}}.$$

If we use the Hölder inequality, then we have

$$I \lesssim \left\| \left(\int_{\mathbb{R}} \left(\sum_{k=-\infty}^{\infty} 2^{ks_0 q_0} |g_k(\cdot, it)|^{q_0} \right)^{\frac{1}{q_0} \cdot r} \frac{\mu_0(\theta, t)}{1-\theta} \, dt \right)^{(1-\theta)\frac{p}{r}} \right\|_{L^{\frac{p_0}{(1-\theta)p}, w_0}}^{\frac{1}{p}}$$

$$\times \left\| \left(\int_{\mathbb{R}} \left(\sum_{k=-\infty}^{\infty} 2^{ks_1 q_1} |g_k(\cdot, 1+it)|^{q_1} \right)^{\frac{1}{q_1} \cdot r} \frac{\mu_1(\theta, t)}{\theta} \, dt \right)^{\theta \frac{p}{r}} \right\|_{L^{\frac{p_1}{\theta p}, w_1}}^{\frac{1}{p}}.$$

Then we have

$$I \lesssim \left\| \int_{\mathbb{R}} \left(\sum_{k=-\infty}^{\infty} 2^{ks_0 q_0} |g_k(\cdot, it)|^{q_0} \right)^{\frac{1}{q_0} \cdot r} \frac{\mu_0(\theta, t)}{1-\theta} \, dt \right\|_{L^{\frac{p_0}{r}, w_0}}^{\frac{(1-\theta)}{r}} \tag{5.7}$$

$$\times \left\| \int_{\mathbb{R}} \left(\sum_{k=-\infty}^{\infty} 2^{ks_1 q_1} |g_k(\cdot, 1+it)|^{q_1} \right)^{\frac{1}{q_1} \cdot r} \frac{\mu_1(\theta, t)}{\theta} \, dt \right\|_{L^{\frac{p_1}{r}, w_1}}^{\frac{\theta}{r}}.$$

If we invoke the Minkowski inequality, then we have

$$I \lesssim \left[\int_{\mathbb{R}} \left\| \left(\sum_{k=-\infty}^{\infty} 2^{ks_0 q_0} |g_k(\cdot, it)|^{q_0} \right)^{\frac{1}{q_0}} \right\|_{L^{p_0, w_0}}^{r} \frac{\mu_0(\theta, t)}{1-\theta} \, dt \right]^{\frac{(1-\theta)}{r}}$$

$$\times \left[\int_{\mathbb{R}} \left\| \left(\sum_{k=-\infty}^{\infty} 2^{ks_1 q_1} |g_k(\cdot, 1+it)|^{q_1} \right)^{\frac{1}{q_1}} \right\|_{L^{p_1, w_1}}^{r} \frac{\mu_1(\theta, t)}{\theta} \, dt \right]^{\frac{\theta}{r}}$$

$$= \left[\int_{\mathbb{R}} \left\| 2^{ks_0} g_k(\cdot, it) \right\|_{L^{p_0, w_0}(\ell^{q_0})}^{r} \frac{\mu_0(\theta, t)}{1-\theta} \, dt \right]^{\frac{(1-\theta)}{r}}$$

$$\times \left[\int_{\mathbb{R}} \left\| 2^{ks_1} g_k(\cdot, 1+it) \right\|_{L^{p_1, w_1}(\ell^{q_1})}^{r} \frac{\mu_1(\theta, t)}{\theta} \, dt \right]^{\frac{\theta}{r}}.$$

If we recall that f is normalized, then we obtain

$$I \lesssim \sup_{t \in \mathbb{R}} \left\| 2^{ks_0} g_k(\cdot, it) \right\|_{L^{p_0, w_0}(\ell^{q_0})}^{1-\theta} \sup_{t \in \mathbb{R}} \left\| 2^{ks_1} g_k(\cdot, 1+it) \right\|_{L^{p_1, w_1}(\ell^{q_1})}^{\theta} \lesssim 1.$$

This proves (5.2).

Step 2. We may assume that $s \gg 1$, $l = 0, 1$. Assuming s large enough by virtue of Proposition 2, we can decompose f as follows: $f = \sum_{\nu \in \mathbb{Z}} \sum_{m \in \mathbb{Z}^n} \lambda_{\nu,m} a_{\nu,m}$, where each $a_{\nu,m}$ satisfies $\operatorname{supp}(a_{\nu,m}) \subset 3Q_{\nu,m}$, $\|\partial^\alpha a_{\nu,m}\|_\infty \leq 2^{|\alpha|\nu}$ and the coefficient $\lambda = \{\lambda_{\nu,m}\}_{\nu \in \mathbb{Z},\, m \in \mathbb{Z}^n}$ satisfies

$$\lambda_{\nu,m} \geq 0 \quad (\nu \in \mathbb{Z},\ m \in \mathbb{Z}^n), \qquad \left\| \left\{ 2^{\nu s} \sum_{m \in \mathbb{Z}^n} \lambda_{\nu,m} \chi_{Q_{\nu,m}} \right\}_{\nu=-\infty}^{\infty} \right\|_{L^{p,w}(\ell^q)} \lesssim \|f\|_{\dot{F}^{s,w}_{p,q}}.$$

It counts that we do not have to postulate the moment condition on $a_{\nu,m}$. By normalization, we assume that $\|f\|_{\dot{F}^{s,w}_{p,q}} = 1$.

For $\nu \in \mathbb{Z}$ and $x \in \mathbb{R}^n$, we define $\Lambda_\nu(x) := \sum_{m\in\mathbb{Z}^n} \lambda_{\nu,m}\chi_{Q_{\nu,m}}(x)$. Then note that

$$\left\| \left\{ 2^{\nu s} \sum_{m\in\mathbb{Z}^n} \lambda_{\nu,m}\chi_{Q_{\nu,m}} \right\}_{\nu=-\infty}^{\infty} \right\|_{L^{p,w}(\ell^q)} \lesssim 1.$$

With this in mind, we abbreviate by writing $R_{\nu,m} := 2^{\nu s}\lambda_{\nu,m}$ below.

For each $x \in \mathbb{R}^n$ and $z \in \mathbb{C}^n$ with $\mathrm{Re}(z) \in [0,1]$, we define

$$\begin{aligned}
\rho_1(z) &:= \frac{qs}{q_0}(1-z) + \frac{qs}{q_1}z - s_0(1-z) - s_1 z, \\
\rho_2(z) &:= \frac{q}{q_0}(1-z) + \frac{q}{q_1}z, \\
\rho_3(z) &:= \left(\frac{p}{qp_0} - \frac{1}{q_0}\right)(1-z) + \left(\frac{p}{qp_1} - \frac{1}{q_1}\right)z.
\end{aligned}$$

Also, we define a holomorphic function $\Lambda_{\nu,m}$ by

$$\begin{aligned}
&\Lambda_{\nu,m}(z) \\
&= \left\{ \frac{1}{|Q_{\nu,m}|} \int_{Q_{\nu,m}} 2^{\nu\frac{\rho_1(z)}{K}} \lambda_{\nu,m}{}^{\frac{\rho_2(z)}{K}} \left(\sum_{j=-\infty}^{\nu} \sum_{\tilde{m}\in\mathbb{Z}^n} R_{j,\tilde{m}}{}^q \chi_{Q_{j,\tilde{m}}}(y) \right)^{\frac{\rho_3(z)}{K}} dy \right\}^K
\end{aligned}$$

where K is a large integer. Abbreviate $\sum_{j=-\infty}^{\nu} \sum_{\tilde{m}\in\mathbb{Z}^n} R_{j,\tilde{m}}{}^q \chi_{Q_{j,\tilde{m}}}(y)$ to $S_\nu(y)$. Then, we have

$$\begin{aligned}
&|2^{\nu s_l}\Lambda_{\nu,m}(l+\mathrm{i}t)| \\
&\lesssim \left| \frac{1}{|Q_{\nu,m}|} \int_{Q_{\nu,m}} 2^{\frac{\nu}{K}\{s_l - s_0(1-l-\mathrm{i}t) - s_1(l+\mathrm{i}t)\}} R_{\nu,m}{}^{\frac{\rho_2(l+\mathrm{i}t)}{K}} S_\nu(y)^{\frac{\rho_3(l+\mathrm{i}t)}{K}} dy \right|^K \\
&\lesssim \left\{ \frac{1}{|Q_{\nu,m}|} \int_{Q_{\nu,m}} \left| 2^{\frac{\nu}{K}\{s_l - s_0(1-l-\mathrm{i}t) - s_1(l+\mathrm{i}t)\}} R_{\nu,m}{}^{\frac{\rho_2(l+\mathrm{i}t)}{K}} S_\nu(y)^{\frac{\rho_3(l+\mathrm{i}t)}{K}} \right| dy \right\}^K \\
&\lesssim \left\{ M\left[\sum_{m\in\mathbb{Z}^n} R_{\nu,m}{}^{\frac{q}{Kq_l}} S_\nu{}^{\frac{\rho_3(l)}{K}} \chi_{Q_{\nu,m}} \right](x) \right\}^K
\end{aligned}$$

for all $x \in Q_{\nu,m}$. Consequently, we obtain

$$\sum_{m\in\mathbb{Z}^n} |2^{\nu s_l}\Lambda_{\nu,m}(l+\mathrm{i}t)|\chi_{Q_{\nu,m}}(x) \lesssim \left\{ M\left[\sum_{m\in\mathbb{Z}^n} R_{\nu,m}{}^{\frac{q}{Kq_l}} S_\nu{}^{\frac{\rho_3(l)}{K}} \chi_{Q_{\nu,m}} \right](x) \right\}^K.$$

If we insert this pointwise estimate, then we have

$$\left\{\sum_{\nu=-\infty}^{\infty}\left(2^{\nu s_l}\sum_{m\in\mathbb{Z}^n}|\Lambda_{\nu,m}(l+\mathrm{it})|\chi_{Q_{\nu,m}}\right)^{q_l}\right\}^{\frac{1}{q_l}}$$

$$\leq\left\{\sum_{\nu=-\infty}^{\infty}\left(M\left[\sum_{m\in\mathbb{Z}^n}R_{\nu,m}{}^{\frac{q}{Kq_l}}S_\nu{}^{\frac{\rho_3(l)}{K}}\chi_{Q_{\nu,m}}\right]\right)^{Kq_l}\right\}^{\frac{1}{q_l}}$$

and hence

$$\left\|\left\{\sum_{\nu=-\infty}^{\infty}\left(2^{\nu s_l}\sum_{m\in\mathbb{Z}^n}|\Lambda_{\nu,m}(l+\mathrm{it})|\chi_{Q_{\nu,m}}\right)^{q_l}\right\}^{\frac{1}{q_l}}\right\|_{L^{p_l,w_l}}$$

$$\lesssim\left\|\left(\sum_{\nu=-\infty}^{\infty}\left[\sum_{m\in\mathbb{Z}^n}R_{\nu,m}{}^{\frac{q}{Kq_l}}S_\nu{}^{\frac{1}{Kq}\left(\frac{p}{p_l}-1\right)}\chi_{Q_{\nu,m}}\right]^{Kq_l}\right)^{\frac{1}{q_l}}\right\|_{L^{p_l,w_l}}$$

$$=\left\|\left(\sum_{\nu=-\infty}^{\infty}\sum_{m\in\mathbb{Z}^n}R_{\nu,m}{}^{q}\left(\sum_{j=-\infty}^{\nu}\sum_{\tilde{m}\in\mathbb{Z}^n}R_{j,\tilde{m}}{}^{q}\chi_{Q_{j,\tilde{m}}}\right)^{\frac{q_lp}{qp_l}-1}\chi_{Q_{\nu,m}}\right)^{\frac{1}{q_l}}\right\|_{L^{p_l,w_l}}.$$

If we write $S_\infty=\lim_{\nu\to\infty}S_\nu$ and use Lemma 5.2, then we have

$$\left\|\left\{\sum_{\nu=-\infty}^{\infty}\left(2^{\nu s_l}\sum_{m\in\mathbb{Z}^n}|\Lambda_{\nu,m}(l+\mathrm{it})|\chi_{Q_{\nu,m}}\right)^{q_l}\right\}^{\frac{1}{q_l}}\right\|_{L^{p_l,w_l}}\lesssim\left\|S_\infty{}^{\frac{p}{qp_l}}\right\|_{L^{p_l,w_l}}.$$

On the other hand, we have

$$\left\|S_\infty{}^{\frac{p}{qp_l}}\right\|_{L^{p_l,w_l}}=\left\|\left(\sum_{j=-\infty}^{\infty}\sum_{\tilde{m}\in\mathbb{Z}^n}R_{j,\tilde{m}}{}^{q}\chi_{Q_{j,\tilde{m}}}\right)^{\frac{p}{qp_l}}\right\|_{L^{p_l,w_l}}\lesssim 1.$$

Consequently, we obtain

$$\left\|\left\{\sum_{\nu=-\infty}^{\infty}\left(2^{\nu s_l}\sum_{m\in\mathbb{Z}^n}|\Lambda_{\nu,m}(l+\mathrm{it})|\chi_{Q_{\nu,m}}\right)^{q_l}\right\}^{\frac{1}{q_l}}\right\|_{L^{p_l,w_l}}\lesssim 1. \tag{5.8}$$

Now we define

$$F(z,x):=\sum_{\nu=-\infty}^{\infty}\sum_{m\in\mathbb{Z}^n}\Lambda_{\nu,m}(z)a_{\nu,m}(x),\quad \mathrm{Re}(z)\in[0,1],\ x\in\mathbb{R}^n.$$

Hence,

$$F(\theta)=f,\quad \|F(l+\mathrm{it})\|_{\dot{F}^{s_l,w_l}_{p_l,q_l}}\lesssim 1.$$

Hence it follows from the definition of interpolation spaces that

$$\|f\|_{(\dot{F}^{s_0,w_0}_{p_0,q_0},\dot{F}^{s_1,w_1}_{p_1,q_1})_\theta} \lesssim \|f\|_{\dot{F}^s_{p,q}}.$$

Therefore, the proof of interpolation is complete. □

Remark 4. A non-homogeneous counterpart is readily available.

References

[1] H. M. Al-Qassem, Weighted L^p boundedness for the function of Marcinkiewicz, Kyungpook Math. J. **46** (2006), 31–48.

[2] H. M. Al-Qassem and Y. Pan, On rough maximal operators and Marcinkiewicz integrals along submanifolds, Studia Math. **190** (2009), no. 1, 73–98.

[3] A. Al-Salman, H. Al-Qassem, L. Cheng, and Y. Pan, L^p bounds for the function of Marcinkiewicz, Math. Res. Lett. **9** (2002), 697–700.

[4] A. Al-Salman, On the L^2 boundedness of parametric Marcinkiewicz integral operator, J. Math. Anal. Appl. **375** (2011), 745–752.

[5] K. F. Andersen and R. T. John, Weighted inequalities for vecter-valued maximal functions and singular integrals, Studia Math. **69** (1980), 19–31.

[6] J. Bergh and J. Löfström, Interpolation spaces. An introduction. Grundlehren der Mathematischen Wissenschaften **223**, Springer-Verlag, Berlin-New York, 1976.

[7] H.-Q. Bui, Weighted Besov and Triebel spaces: Interpolation by the real method, Hiroshima Math. J. **12** (1982), 581–605.

[8] V. I. Burenkov, V. S. Guliyev, A. Serbetci, and T. V. Tararykova, Necessary and sufficient conditions for the boundedness of genuine singular integral operators in local Morrey-type spaces, Eurasian Math. J. **1** (2010), no. 1, 32–53.

[9] A. P. Calderón and A. Zygmund, A note on the interpolation of sublinear operations, Amer. J. Math. **78** (1956), 282–288.

[10] J. Chen, D. Fan, and Y. Ying, Singular integral operators on function spaces, J. Math. Anal. Appl. **276** (2002), 691–708.

[11] Y. Ding, D. Fan, and Y. Pan, On Littlewood-Paley functions and singular integrals, Hokkaido Math. J. **29** (2000), 537–552.

[12] Y. Ding, Q. Xue, and K. Yabuta, Boundedness of the Marcinkiewicz integrals with rough kernel associated to surfaces, Tohoku Math. J. **62** (2010), 233–262.

[13] Y. Ding, Q. Xue, and K. Yabuta, A remark to the L^2 boundedness of parametric Marcinkiewicz integral, J. Math. Anal. Appl. **378** (2012), 691–697.

[14] J. Duoandikoetxea, Weighted norm inequalities for homogeneous singular integrals, Trans. Amer. Math. Soc. **336** (1993), 869–880.

[15] J. Duoandikoetxea and J. L. Rubio de Francia, Maximal and singular integral operators via Fourier transform estimates, Invent. Math. **84** (1986), 541–561.

[16] D. Fan and Y. Pan, A singular integral operators with rough kernel, Proc. Amer. Math. Soc. **125** (1997), 3695–3703.

[17] D. Fan and Y. Pan, Singular integral operators with rough kernels supported by subvarieties, Amer. J. Math. **119** (1997), 799–839.

[18] R. Fefferman, A note on singular integrals, Proc. Amer. Math. Soc. **74** (1979), 266–270.

[19] C. Fefferman and E. M. Stein, Some maximal inequalities, Amer. J. Math. **93** (1971), 107–115.

[20] M. Frazier and B. Jawerth, A discrete transform and decompositions of distribution spaces, J. Funct. Anal. **93** (1990), no. 1, 34–170.

[21] J. García-Cuerva, Weighted H^p spaces, Dissertationes Math. **162** (1979).

[22] L. Grafakos, Modern Fourier Analysis, Second Edition, Springer-Verlag, 2009.

[23] L. Hörmander, Estimates for translation invariant operators in L^p spaces, Acta Math. **104** (1960), 93–140.

[24] M. Izuki and Y. Sawano, Atomic decomposition for weighted Besov and Triebel-Lizorkin spaces, Math. Nachr. **285** (2012), no. 1, 103–126.

[25] Y. Furuya-Komori, K. Matsuoka, E. Nakai, and Y. Sawano, Integral operators on B_σ-Morrey-Campanato spaces, Rev. Math. Complut. **26** (2013), no. 1, 1–32.

[26] S. Z. Lu, Marcinkiewicz integrals with rough kernels, Front. Math. China **3** (2008), 1–14.

[27] K. Matsuoka and E. Nakai, Fractional integral operators on $B^{p,\lambda}$ with Morrey-Campanato norms, Function spaces IX, 249–264, Banach Center Publ. **92**, Polish Acad. Sci. Inst. Math., Warsaw, 2011.

[28] T. Noi and Y. Sawano, Complex interpolation of Besov spaces and Triebel-Lizorkin spaces with variable exponents, J. Math. Anal. Appl. **387** (2012), no. 2, 676–690.

[29] Y. Sawano and K. Yabuta, Fractional type Marcinkiewicz integral operators associated to surfaces, J. Inequal. Appl. **232** (2014), 29 pp.

[30] Z. Y. Si, L. N. Wang, and Y. S. Jiang, Fractional type Marcinkiewicz integral on Hardy spaces, J. Math. Res. Exposition **31** (2011), 233–241.

[31] E. M. Stein, Interpolation of linear operators, Trans. Amer. Math. Soc. **83** (1956), 482–492.

[32] E. M. Stein, On the function of Littlewood-Paley, Lusin and Marcinkiewicz, Trans. Amer. Math. Soc. **88** (1958), 430–466.

[33] E. M. Stein and G. Weiss, Interpolation of operators with change of measures, Trans. Amer. Math. Soc. **87** (1958), 159–172.

[34] H. Triebel, Theory of Function Spaces, Birkhäuser, 1983.

[35] T. Walsh, On the function of Marcinkiewicz, Studia Math. **44** (1972), 203–217.

[36] Q. Xue, K. Yabuta, and J. Yan, Fractional type Marcinkiewicz integral operators on function spaces, Forum Math. DOI: 10.1515/forum-2013-0200, to appear.

[37] K. Yabuta, A remark on radial A_p weights, Hokkaido Math. J. **40** (2011), no. 2, 241–249.

Some Topics in Harmonic Analysis
and Applications
ALM 34, pp. 369–384

Commutator Estimates for the Dirichlet-to-Neumann Map in Lipschitz Domains

Zhongwei Shen*†

Abstract

We establish two commutator estimates for the Dirichlet-to-Neumann map associated with a second-order elliptic system in divergence form in Lipschitz domains. Our approach is based on Dahlberg's bilinear estimates.

2000 Mathematics Subject Classification: 35J25.

Keywords and Phrases: commutator, Dirichlet-to-Neumann map, Lipschitz domain.

1 Introduction

This paper concerns the L^2 boundedness of the commutator

$$[\Lambda, g]f = \Lambda(gf) - g\Lambda(f), \tag{1.1}$$

where Λ is the Dirichlet-to-Neumann map associated with the elliptic operator

$$\mathcal{L} = -\operatorname{div}\big(A(x)\nabla\big) = -\frac{\partial}{\partial x_i}\left\{a_{ij}^{\alpha\beta}(x)\frac{\partial}{\partial x_j}\right\} \quad \text{in } \Omega, \tag{1.2}$$

and Ω is a bounded Lipschitz domain in $\mathbb{R}^d$. Assume that $A = A(x) = \big(a_{ij}^{\alpha\beta}(x)\big)$, with $1 \le i,j \le d$ and $1 \le \alpha, \beta \le m$, is real, bounded measurable, and satisfies the ellipticity condition:

$$\mu|\xi|^2 \le a_{ij}^{\alpha\beta}(x)\xi_i^\alpha\xi_j^\beta \le \frac{1}{\mu}|\xi|^2, \quad \text{for any } \xi = (\xi_i^\alpha) \in \mathbb{R}^{d\times m} \text{ and } x \in \mathbb{R}^d, \tag{1.3}$$

*Department of Mathematics, University of Kentucky, Lexington, Kentucky 40506, USA, *E-mail:* zshen2@uky.edu

†Supported in part by NSF grant DMS-1161154.

where $\mu > 0$. It is well known that for any $f = (f^\alpha) \in H^{1/2}(\partial\Omega; \mathbb{R}^m)$, the Dirichlet problem

$$\mathcal{L}(u) = 0 \quad \text{in } \Omega \quad \text{and} \quad u = f \quad \text{on } \partial\Omega \tag{1.4}$$

has a unique solution in $H^1(\Omega; \mathbb{R}^m)$. The Dirichlet-to-Neumann map Λ is defined by

$$(\Lambda(f))^\alpha = \left(\frac{\partial u}{\partial \nu}\right)^\alpha = n_i a_{ij}^{\alpha\beta} \frac{\partial u^\beta}{\partial x_j}. \tag{1.5}$$

This gives a bounded linear map $\Lambda : H^{1/2}(\partial\Omega; \mathbb{R}^m) \to H^{-1/2}(\partial\Omega; \mathbb{R}^m)$. Under the additional symmetry condition $A^* = A$, i.e., $a_{ij}^{\alpha\beta} = a_{ji}^{\beta\alpha}$, as well as some smoothness condition on A, one may show that $\Lambda : H^1(\partial\Omega; \mathbb{R}^m) \to L^2(\partial\Omega; \mathbb{R}^m)$ is bounded (see [2, 12, 17, 6, 16, 15] and their references). The following are our main results of the paper.

Theorem 1.1. *Let Ω be a bounded Lipschitz domain in $\mathbb{R}^d$. Assume that A satisfies the ellipticity condition* (1.3) *and the symmetry condition $A^* = A$. Also assume that A is Hölder continuous. Then, for any $\mathbb{R}^m$-valued Lipschitz function f and any scalar Lipschitz function g on $\partial\Omega$,*

$$\|\Lambda(gf) - g\Lambda(f)\|_{L^2(\partial\Omega)} \le C\, \|g\|_{C^{0,1}(\partial\Omega)} \|f\|_{L^2(\partial\Omega)}, \tag{1.6}$$

where C depends only on A and Ω.

Theorem 1.2. *Assume that Ω and A satisfy the same conditions as in Theorem* 1.1. *Then, for any $\mathbb{R}^m$-valued Lipschitz function f and any scalar Lipschitz function g on $\partial\Omega$,*

$$\|\Lambda(gf) - g\Lambda(f)\|_{L^2(\partial\Omega)} \le C\, \|u\|_{L^\infty(\Omega)} \|g\|_{H^1(\partial\Omega)}, \tag{1.7}$$

where u is the solution of (1.4) *with boundary data f. The constant C in* (1.7) *depends only on A and Ω.*

A few remarks are in order.

Remark 1.3. If $\partial\Omega$ and A are sufficiently smooth, the operator Λ is a pseudo-differential operator of order one. It follows that

$$\|\Lambda(gf) - g\Lambda(f)\|_{L^p(\partial\Omega)} \le C \|g\|_{C^{0,1}(\partial\Omega)} \|f\|_{L^p(\partial\Omega)}$$

for any $1 < p < \infty$. This is a classical commutator estimate due to Calderón [1].

Remark 1.4. If $m = 1$, or $m > 1$ and $d = 2, 3$, the solutions of the Dirichlet problem (1.4) in Lipschitz domains satisfy the Agmon-Miranda maximum principle

$$\|u\|_{L^\infty(\Omega)} \le C\, \|f\|_{L^\infty(\partial\Omega)} \tag{1.8}$$

(it is the usual maximum principle for $m = 1$; see [7] for $m > 1$ and $d = 2, 3$). In this case the estimate (1.7) may be replaced by

$$\|\Lambda(gf) - g\Lambda(f)\|_{L^2(\partial\Omega)} \le C\, \|f\|_{L^\infty(\partial\Omega)} \|g\|_{H^1(\partial\Omega)}, \tag{1.9}$$

which, by duality, is equivalent to

$$\|\Lambda(gf) - g\Lambda(f)\|_{L^1(\partial\Omega)} \leq C\, \|f\|_{L^2(\partial\Omega)} \|g\|_{H^1(\partial\Omega)}. \tag{1.10}$$

We mention that (1.8) also holds in C^1 domains for $m > 1$ and $d \geq 2$. Whether the estimate (1.8) holds in Lipschitz domains for $m > 1$ and $d \geq 4$ remains a challenging open problem, even for second-order elliptic systems with constant coefficients.

Remark 1.5. Estimates in Theorems 1.1 and 1.2 were proved in [13] for the special case $\mathcal{L} = -\Delta$. They were used to analyze the asymptotic behavior of the Dirichlet-to-Neumann maps associated with elliptic systems with rapidly oscillating coefficients, arising in the theory of homogenization. We also point out that Theorems 1.1 and 1.2 continue to hold if $g(x) = \big(g^{\alpha\beta}(x)\big)$ is an $m \times m$ matrix that commutes with $A(x)$; i.e. $g^{\alpha\beta} a_{ij}^{\beta\gamma} = a_{ij}^{\alpha\beta} g^{\beta\gamma}$. The proof is the same.

Our approach to Theorems 1.1 and 1.2, which is similar to that used in [13] for $\mathcal{L} = -\Delta$, is based on Dahlberg's bilinear estimate

$$\begin{aligned}&\left| \int_\Omega \nabla u \cdot v\, dx \right| \\ &\leq C \left\{ \int_\Omega |\nabla u(x)|^2\, \delta(x)\, dx \right\}^{1/2} \left\{ \int_\Omega |\nabla v(x)|^2\, \delta(x)\, dx + \int_{\partial\Omega} |(v)^*|^2\, d\sigma \right\}^{1/2},\end{aligned} \tag{1.11}$$

where $\delta(x) = \operatorname{dist}(x, \partial\Omega)$, $u = (u^\alpha) \in H^1(\Omega; \mathbb{R}^m)$ is a weak solution of $\mathcal{L}(u) = 0$ in Ω, and $v = (v_i^\alpha) \in H^1(\Omega; \mathbb{R}^{dm})$. In (1.11) we have used $(v)^*$ to denote the nontangential maximal function of v. The bilinear estimate was proved in [5] for harmonic functions u in Lipschitz domains (see related work in [4, 9], where similar bilinear forms were used to solve L^p Dirichlet problems in Lipschitz domains). We also mention that the results in [5] were extended in [11] to a class of second-order elliptic operators in the upper half-space with time-independent complex coefficients. In Section 2 we provide a relative simple proof of (1.11), under the assumptions that A satisfies (1.3) and $|\nabla A(x)|^2\, \delta(x)\, dx$ is a Carleson measure on Ω. The proof, which is probably known to experts in the area, follows closely the basic argument in [8], where the equivalence in L^p norms between the square function and the nontangential maximal function was established for solutions of higher-order elliptic systems with constant coefficients in Lipschitz domains. In Section 3 we use a perturbation argument to prove (1.11) for elliptic operators with Hölder continuous coefficients.

The proof of Theorem 1.1 is given in Section 4, while Section 5 is devoted to the proof of Theorem 1.2. Let $h \in H^1(\Omega; \mathbb{R}^m)$ be a weak solution of $\mathcal{L}^*(h) = -\operatorname{div}(A^*(x)\nabla h) = 0$ in Ω. The connection between the bilinear estimate (1.11) and the commutator $[\Lambda, g]$ is made through the following identity

$$\begin{aligned}&\int_{\partial\Omega} \{\Lambda(gf) - g\Lambda(f)\} \cdot h\, d\sigma \\ &= \int_\Omega \frac{\partial v}{\partial x_i} \cdot u^\alpha \cdot a_{ji}^{\beta\alpha} \frac{\partial h^\beta}{\partial x_j}\, dx - \int_\Omega \frac{\partial v}{\partial x_i} \cdot a_{ij}^{\alpha\beta} \frac{\partial u^\beta}{\partial x_j} \cdot h^\alpha\, dx,\end{aligned} \tag{1.12}$$

where $v \in H^1(\Omega)$ is any extension of g to Ω. For the estimate (1.6) we construct v in such a way that $\|\nabla v\|_{L^\infty(\Omega)} \le C\|g\|_{C^{0,1}(\partial\Omega)}$, and $d\nu = |\nabla^2 v(x)|^2\,\delta(x)\,dx$ is a Carleson measure on Ω with norm $\|\nu\|_{\mathcal{C}}$ less than $C\|g\|^2_{C^{0,1}(\partial\Omega)}$. In the case of (1.7) we choose v to be the harmonic extension of g; i.e., $\Delta v = 0$ in Ω and $v = g$ on $\partial\Omega$.

Finally, we remark that although we will not pursue the approach in this paper, it seems possible, at least in the case of (1.6), to establish the commutator estimates for the Dirichlet-to-Neumann map, using the method of layer potentials. Indeed, let

$$\mathcal{S}(f)(x) = \int_{\partial\Omega} \Gamma(x,y) f(y)\, d\sigma(y)$$

denote the single layer potential for $\mathcal{L}$, where $\Gamma(x,y)$ is the matrix of fundamental solutions for $\mathcal{L}$ in $\mathbb{R}^d$ (one may suitably modify A outside Ω for the existence of $\Gamma(x,y)$). The conormal derivative of $u = \mathcal{S}(f)$ is given by $\big((1/2)I + \mathcal{K}\big)f$, where $\mathcal{K}$ is a singular integral operator on $\partial\Omega$. It follows that

$$\Lambda(f) = \big((1/2)I + \mathcal{K}\big)S^{-1}(f),$$

where $S = \mathcal{S}|_{\partial\Omega}$. This implies that

$$\begin{aligned}\big[\Lambda, g\big]f &= \big((1/2)I + \mathcal{K}\big)\big[S^{-1}, g\big]f + \big[\mathcal{K}, g\big]S^{-1}(f)\\ &= -\big((1/2)I + \mathcal{K}\big)S^{-1}\big[S, g\big]S^{-1}f + \big[\mathcal{K}, g\big]S^{-1}(f).\end{aligned}$$

Under the assumption that A is elliptic and Hölder continuous, it is known that for $1 < p < \infty$, the operator $\mathcal{K}$ is bounded on $L^p(\partial\Omega)$, and S is bounded from $L^p(\partial\Omega)$ to $W^{1,p}(\partial\Omega)$. If, in addition, $A^* = A$, then $\big((1/2)I + \mathcal{K}\big)$ is invertible on $L^2_0(\partial\Omega)$, the subspace of L^2 functions with mean value zero. Furthermore, if $d \ge 3$, S is an isomorphism from $L^2(\partial\Omega)$ to $W^{1,2}(\partial\Omega)$ (see [17, 16, 15]). As a result, we obtain

$$\begin{aligned}&\|\big[\Lambda, g\big]f\|_{L^2(\partial\Omega)}\\ &\le C\left\{\|\big[S, g\big]\|_{W^{-1,2}(\partial\Omega)\to W^{1,2}(\partial\Omega)} + \|\big[\mathcal{K}, g\big]\|_{W^{-1,2}(\partial\Omega)\to L^2(\partial\Omega)}\right\}\|f\|_{L^2(\partial\Omega)},\end{aligned} \tag{1.13}$$

where $W^{-1,2}(\partial\Omega) = H^{-1}(\partial\Omega)$ is the dual of $W^{1,2}(\partial\Omega) = H^1(\partial\Omega)$. This reduces the proof of the estimate (1.6) to that of estimates of commutators $[S, g]$ and $[\mathcal{K}, g]$.

2 Dahlberg's bilinear estimate, Part I

Let Ω be a bounded Lipschitz domain in $\mathbb{R}^d$ and $\delta(x) = \text{dist}(x, \partial\Omega)$. For a function u in Ω, the nontangential maximal function $(u)^*$ is defined by

$$(u)^*(Q) = \sup\big\{|u(x)| : \ x \in \Omega \text{ and } |x - Q| < \alpha_0\, \delta(x)\big\} \tag{2.1}$$

for $Q \in \partial\Omega$, where $\alpha_0 = \alpha_0(\Omega) > 1$ is a fixed large constant.

Let ν be a nonnegative measure on Ω. We call ν a Carleson measure if

$$\|\nu\|_{\mathcal{C}} := \sup\left\{\frac{\nu(B(Q,r)\cap\Omega)}{r^{d-1}} : Q \in \partial\Omega \text{ and } 0 < r < \text{diam}(\Omega)\right\}$$

is finite. The defining property of Carleson measures is that

$$\int_\Omega |u|\, d\nu \le C\|\nu\|_{\mathcal{C}} \int_{\partial\Omega} (u)^*\, d\sigma, \tag{2.2}$$

where C depends only on the Lipschitz character of Ω (see e.g. [10, Section 7.3] for the case of $\mathbb{R}^d_+$). The goal of this section is to establish the following theorem.

Theorem 2.1. *Let Ω be a bounded Lipschitz domain in $\mathbb{R}^d$. Let $\mathcal{L} = -\mathrm{div}(A(x)\nabla)$ with $A(x)$satisfying the ellipticity condition (1.3). Also assume that $A \in C^1(\Omega)$ and that*

$$d\nu = |\nabla A(x)|^2\, \delta(x)\, dx \text{ is a Carleson measure with norm } \|\nu\|_{\mathcal{C}} \text{ less than } C_0. \tag{2.3}$$

Let $u \in H^1(\Omega;\mathbb{R}^m)$ be a weak solution of $\mathcal{L}(u) = 0$ in Ω. Then, for any $v \in H^1(\Omega;\mathbb{R}^{d\times m})$,

$$\begin{aligned}&\left|\int_\Omega \nabla u\cdot v\, dx\right|\\ &\le C\left\{\int_\Omega |\nabla u(x)|^2\,\delta(x)\, dx\right\}^{1/2}\left\{\int_\Omega |\nabla v(x)|^2\,\delta(x)\, dx + \int_{\partial\Omega} |(v)^*|^2\, d\sigma\right\}^{1/2},\end{aligned} \tag{2.4}$$

where C depends only on d, m, μ, C_0, and Ω.

Proof. By a partition of unity it suffices to estimate the integral of $\nabla u\cdot v$ over $B(Q,r_0)\cap\Omega$ with $v \in H^1_0(B(Q,r_0))$, where $Q\in\partial\Omega$ and $r_0>0$ is small. Since the condition (2.3) is translation and rotation invariant, we may assume that $Q=0$ and

$$B(0,C_0r_0)\cap\Omega = B(0,C_0r_0)\cap\big\{(x',t)\in\mathbb{R}^d : x'\in\mathbb{R}^{d-1} \text{ and } t>\psi(x')\big\},$$

where ψ is a Lipschitz function on $\mathbb{R}^{d-1}$, $\psi(0)=0$, and $C_0 = 10(\|\nabla\psi\|_\infty + 1)$. Thus it is enough to establish the estimate (2.4) with Ω replaced by

$$D = \big\{(x',t)\in\mathbb{R}^d : x'\in\mathbb{R}^{d-1} \text{ and } t>\psi(x')\big\},$$

assuming that u is a weak solution of $\mathcal{L}(u)=0$ in $B(0,r_0)\cap D$ and $v\in H^1_0(B(0,r_0))$. Using a special change of variables invented by Kenig and Stein, we may further reduce the problem to the case of the upper half-space $\mathbb{R}^d_+ = \big\{(y',s)\in\mathbb{R}^d : y'\in\mathbb{R}^{d-1} \text{ and } s>0\big\}$. Indeed, consider the bi-Lipschitz map from $\mathbb{R}^d_+$ to D, defined by

$$(y',s)\to\Phi(y',s) = \big(y', cs+\eta_s*\psi(y')\big) = \big(y',F(y',s)\big),$$

where $\eta_s(y') = s^{1-d}\eta(y'/s)$ is a smooth compactly supported bump function and the constant $c = c(d,\|\nabla\psi\|_\infty)>0$ is so large that $\frac{\partial F}{\partial s}\ge 1$. The key observations here are (1) $c\le|\nabla\Phi(y',s)|\le C$; (2) $|\nabla^2\Phi(y',s)|^2\, s\, dy'ds$ is a Carleson measure on $\mathbb{R}^d_+$; and (3) all constants depend only on d and $\|\nabla\psi\|_\infty$ (see e.g. [5, 8]). It follows that the ellipticity condition (1.3) and the Carleson condition (2.3) are preserved under the change of variables $(x',t)=\Phi(y',s)$. As a result, we only need to show that

$$\left|\int_{\mathbb{R}^d_+} \nabla u\cdot v\, dx'dt\right| \le C\left\{\int_{\mathbb{R}^d_+} |\nabla u(x',t)|^2\, t\, dx'dt\right\}^{1/2}\left\{\int_{\mathbb{R}^d_+} |\nabla v(x',t)|^2\, t\, dx'dt+\int_{\mathbb{R}^{d-1}} |(v)^*|^2\, dx'\right\}^{1/2}, \tag{2.5}$$

where u is a weak solution of $\mathcal{L}(u)=0$ in $B(0,r_0)\cap\mathbb{R}^d_+$ and $v\in H^1_0(B(0,r_0))$. By suitably redefining A outside $B(0,3r_0)$, we may assume that $|\nabla A(x',t)|^2\, t\, dx'dt$ is a Carleson measure on $\mathbb{R}^d_+$ with a norm depending only on μ, C_0, and $\|\nabla\psi\|_\infty$.

We now proceed to prove the estimate (2.5), using an approach found in [8]. We begin by writing

$$\begin{aligned}\int_{\mathbb{R}^d_+} \nabla u\cdot v\, dx'dt &= \int_{\mathbb{R}^d_+} \nabla u\cdot v\cdot \frac{\partial t}{\partial t}\, dx'dt\\ &= -\int_{\mathbb{R}^d_+} \frac{\partial}{\partial t}\nabla u\cdot v\cdot t\, dx'dt-\int_{\mathbb{R}^d_+} \nabla u\cdot \frac{\partial v}{\partial t}\, t\, dx'dt,\end{aligned} \tag{2.6}$$

where we have used the integration by parts. By the Cauchy inequality, the second term in the right hand side of (2.6) is dominated in absolute value by the right hand side of (2.5). As for the first term, it is easy to see that an integration by parts in x' and the Cauchy inequality may be used to handle the integral of $\frac{\partial}{\partial t}\left(\frac{\partial u}{\partial x_i}\right)\cdot v\cdot t$ for $i=1,\dots,d-1$. As a result, it remains to bound the integral

$$\int_{\mathbb{R}^d_+} \frac{\partial^2 u}{\partial t^2}\cdot v\cdot t\, dx'dt \tag{2.7}$$

by the right hand side of (2.5). This will be done by using the assumption that u is a solution of a second-order elliptic system.

Indeed, using $\mathcal{L}(u)=0$ in $B(0,r_0)\cap\mathbb{R}^d_+$, we may write

$$a^{\alpha\beta}_{dd}\frac{\partial^2 u^\beta}{\partial t^2}=\nabla_{x'}\left\{F^\alpha(x',t)\right\}+G^\alpha(x',t), \tag{2.8}$$

where $\nabla_{x'}=(\frac{\partial}{\partial x_1},\dots,\frac{\partial}{\partial x_{d-1}})$, $F=(F^\alpha)$ and $G=(G^\alpha)$ satisfy $|F|\le C|\nabla u|$ and $|G|\le C|\nabla A|\,|\nabla u|$. Note that the $m\times m$ matrix $\left(a^{\alpha\beta}_{dd}\right)$ is invertible by (1.3), and its inverse $E=\left(b^{\alpha\beta}\right)$ satisfies the same type of ellipticity and Carleson conditions as A. This allows us to use the integration by parts in x' to obtain

$$\begin{aligned}\left|\int_{\mathbb{R}^d_+} \frac{\partial^2 u}{\partial t^2}\cdot v\cdot t\, dx'dt\right| &\le C\int_{\mathbb{R}^d_+}\Big\{|F|\,|\nabla\{E\cdot v\}|+|E|\,|\nabla A|\,|G|\,|v|\Big\}t\, dx'dt\\ &\le C\int_{\mathbb{R}^d_+}\Big\{|\nabla u|\,|\nabla E|\,|v|+|\nabla u|\,|\nabla v|+|\nabla A|\,|\nabla u|\,|v|\Big\}t\, dx'dt.\end{aligned} \tag{2.9}$$

It follows by the Cauchy inequality that the left hand side of (2.9) is bounded by

$$C\left\{\int_{\mathbb{R}^d_+} |\nabla u|^2\, t\, dx'dt\right\}^{1/2} \left\{\int_{\mathbb{R}^d_+} |\nabla v|^2\, t\, dx'dt + \int_{\mathbb{R}^d_+} |v|^2 \{|\nabla E|^2 + |\nabla A|^2\} t\, dx'dt\right\}^{1/2}. \tag{2.10}$$

Finally, since $\{|\nabla E|^2 + |\nabla A|^2\} t\, dx'dt$ is a Carleson measure on $\mathbb{R}^d_+$, we see that (2.10) is dominated by the right hand side of (2.5). This completes the proof. □

3 Dahlberg's bilinear estimate, Part II

In this section we show that the bilinear estimate (2.4) holds if A is elliptic, symmetric, and Hölder continuous. We mention that under these assumptions on A, the Dirichlet problem (1.4) is uniquely solvable for $f \in L^2(\partial\Omega; \mathbb{R}^m)$. Moreover, the solution u satisfies the nontangential maximal function and square function estimates:

$$\int_{\partial\Omega} |(u)^*|^2\, d\sigma + \int_\Omega |\nabla u(x)|^2\, \delta(x)\, dx \le C \int_{\partial\Omega} |f|^2\, d\sigma, \tag{3.1}$$

where C depends only on A and Ω (see e.g. [16, 15]).

Theorem 3.1. *Let Ω be a bounded Lipschitz domain in $\mathbb{R}^d$. Let $\mathcal{L} = -\mathrm{div}(A(x)\nabla)$ with $A(x)$ satisfying (1.3). Also assume that $A^* = A$ and that A is Hölder continuous in $\mathbb{R}^d$. Let $u \in H^1(\Omega; \mathbb{R}^m)$ be a weak solution of $\mathcal{L}(u) = 0$ in Ω. Then, for any $v \in H^1(\Omega; \mathbb{R}^{d\times m})$,*

$$\left|\int_\Omega \nabla u \cdot v\, dx\right| \le C\|u\|_{L^2(\partial\Omega)} \left\{\int_\Omega |\nabla v|^2\, \delta(x)\, dx + \int_{\partial\Omega} |(v)^*|^2\, d\sigma\right\}^{1/2}, \tag{3.2}$$

where the constant C depends at most on A and Ω.

Theorem 3.1 is a consequence of Theorem 2.1 and the following lemma.

Lemma 3.2. *Assume that Ω and A satisfy the same conditions as in Theorem 3.1. Let $u \in H^1_0(\Omega; \mathbb{R}^m)$ be a weak solution of $\mathcal{L}(u) = \mathrm{div}(f)$, where $f = (f_i^\alpha) \in L^2(\Omega; \mathbb{R}^{d\times m})$. Then*

$$\int_\Omega |\nabla u(x)|^2\, \delta(x) dx \le C \int_\Omega |f(x)|^2\, \delta(x)\, \phi(\delta(x))\, dx, \tag{3.3}$$

where $\phi(t) = \{|\ln(t)| + 1\}^2$, and C depends only on A and Ω.

Proof. By dilation and translation we may assume that $\Omega \subset [-1/4, 1/4]^d$. Without loss of generality we may further assume that $A(x)$ is periodic with respect to $\mathbb{Z}^d$. This would allow us to use results in [14] obtained for second-order elliptic systems in divergence form with periodic coefficients.

Let $w = (w^\alpha)$, where

$$w^\alpha(x) = -\int_\Omega \frac{\partial}{\partial y_i}\Big\{\Gamma^{\alpha\beta}(x,y)\Big\} f_i^\beta(y)\, dy,$$

and $\big(\Gamma^{\alpha\beta}(x,y)\big)$ denotes the matrix of fundamental solutions for $\mathcal{L}$ in $\mathbb{R}^d$, with pole at y. It follows from [14, Section 8] that

$$\int_\Omega |\nabla w(x)|^2\, \delta(x)\, dx \le C \int_\Omega |f(x)|^2\, \delta(x)\, \phi(\delta(x))\, dx, \tag{3.4}$$

and

$$\int_{\partial\Omega} |w|^2\, d\sigma \le C \int_\Omega |f(x)|^2\, \delta(x)\, \phi(\delta(x))\, dx. \tag{3.5}$$

Note that $\mathcal{L}(u-w)=0$ in Ω and $u-w=-w$ on $\partial\Omega$. Using (3.1), we obtain

$$\int_\Omega |\nabla(u-w)|^2\, \delta(x)\, dx \le C \int_{\partial\Omega} |w|^2\, d\sigma.$$

This, together with (3.5), gives

$$\int_\Omega |\nabla(u-w)|^2\, \delta(x)\, dx \le C \int_\Omega |f(x)|^2\, \delta(x)\, \phi(\delta(x))\, dx. \tag{3.6}$$

The desired estimate now follows from (3.4) and (3.6). □

Remark 3.3. Since $\phi(t) \le C_\varepsilon t^{-\varepsilon}$ for $0<t<1$ and $\varepsilon>0$, it follows from (3.3) that

$$\int_\Omega |\nabla u(x)|^2\, \delta(x)\, dx \le C_\varepsilon \int_\Omega |f(x)|^2\, \big[\delta(x)\big]^{1-\varepsilon}\, dx. \tag{3.7}$$

This, together with the energy estimate $\|\nabla u\|_{L^2(\Omega)} \le C\|f\|_{L^2(\Omega)}$, gives

$$\int_\Omega |\nabla u(x)|^2 \big[\delta(x)\big]^{\alpha_2}\, dx \le C \int_\Omega |f(x)|^2 \big[\delta(x)\big]^{\alpha_1}\, dx, \tag{3.8}$$

by complex interpolation, where $0 \le \alpha_1 < \alpha_2 \le 1$. Although the weighted norm inequality (3.8) for $0<\alpha_1<\alpha_2<1$ is sufficient for our purpose, it would be interesting to know if (3.8) holds for $0<\alpha_1=\alpha_2<1$.

Proof of Theorem 3.1. Without loss of generality we may assume that $\Omega \subset [-1/4, 1/4]^d$ and $A(x)$ is periodic with respect to $\mathbb{Z}^d$. Suppose that A is Hölder continuous of order η. We construct a matrix $B(x) = \big(b_{ij}^{\alpha\beta}(x)\big)$, where $b_{ij}^{\alpha\beta}$ is the solution of the Dirichlet problem

$$\Delta\big(b_{ij}^{\alpha\beta}\big) = 0 \quad \text{in } \Omega \quad \text{and} \quad b_{ij}^{\alpha\beta} = a_{ij}^{\alpha\beta} \quad \text{on } \partial\Omega.$$

It follows from the maximum principle that $B(x)$ satisfies the elliptic condition (1.3) with the same μ. Also note that $B \in C^\infty(\Omega)$, B is Hölder continuous of order η in $\overline{\Omega}$,

$$|\nabla B(x)| \le C\big[\delta(x)\big]^{\eta-1} \quad \text{for any } x \in \Omega, \tag{3.9}$$

and

$$|A(x) - B(x)| \le C\big[\delta(x)\big]^{\eta} \quad \text{for any } x \in \Omega \tag{3.10}$$

(here we have assumed that $0 < \eta < \eta_0(\Omega)$ is sufficiently small). In particular, the estimate (3.9) implies that $|\nabla B(x)|^2 \, \delta(x) \, dx$ is a Carleson measure on Ω.

Now, let $u \in H^1(\Omega; \mathbb{R}^m)$ be a weak solution of $\mathcal{L}(u) = 0$ in Ω and $v \in H^1(\Omega; \mathbb{R}^{d\times m})$. Write

$$\int_\Omega \nabla u \cdot v \, dx = \int_\Omega \nabla w \cdot v \, dx + \int_\Omega \nabla(u - w) \cdot v \, dx, \tag{3.11}$$

where $w \in H^1(\Omega; \mathbb{R}^m)$ is the solution of $\operatorname{div}\big(B(x)\nabla w\big) = 0$ in Ω and $w = u$ on $\partial\Omega$. By Theorem 2.1 and (3.1), the first term in the right hand side of (3.11) is bounded by

$$C\|u\|_{L^2(\partial\Omega)} \left\{ \int_\Omega |\nabla v|^2 \, \delta(x) \, dx + \int_{\partial\Omega} |(v)^*|^2 \, d\sigma \right\}^{1/2}.$$

The second term in the right hand side of (3.11) will be handled by the estimate (3.8). By the Cauchy inequality we see that

$$\begin{aligned} \left| \int_\Omega \nabla(w - u) \cdot v \, dx \right| &\le \left\{ \int_\Omega |\nabla(w-u)|^2 \, \big[\delta(x)\big]^{1-\eta} \, dx \right\}^{1/2} \left\{ \int_\Omega |v|^2 \, \big[\delta(x)\big]^{\eta - 1} \, dx \right\}^{1/2} \\ &\le C_\alpha \left\{ \int_\Omega |\nabla(w-u)|^2 \, \big[\delta(x)\big]^{1-\eta} \, dx \right\}^{1/2} \|(v)^*\|_{L^2(\partial\Omega)}, \end{aligned} \tag{3.12}$$

where we have used the fact that $\big[\delta(x)\big]^{\eta-1} dx$ is a Carleson measure on Ω for the second inequality. Note that $w - u \in H_0^1(\Omega; \mathbb{R}^m)$, and

$$\operatorname{div}\big(A\nabla(w-u)\big) = \operatorname{div}\big(A\nabla w\big) = \operatorname{div}\big((A - B)\nabla w\big).$$

It follows from the estimate (3.8) that

$$\begin{aligned} \int_\Omega |\nabla(w-u)|^2 \big[\delta(x)\big]^{1-\eta} \, dx &\le C \int_\Omega |(A-B)\nabla w|^2 \big[\delta(x)\big]^{1-2\eta} \, dx \\ &\le C \int_\Omega |\nabla w|^2 \, \delta(x) \, dx \\ &\le C \int_{\partial\Omega} |u|^2 \, d\sigma, \end{aligned} \tag{3.13}$$

where we have used (3.10) for the second inequality and (3.1) for the third. This, together with (3.12), completes the proof. □

4 Trilinear estimates and proof of Theorem 1.1

We begin with a lemma on extensions of Lipschitz functions.

Lemma 4.1. *Let Ω be a bounded Lipschitz domain. Let $g \in C^{0,1}(\partial\Omega)$ be a scalar Lipschitz function on $\partial\Omega$. Then there exists $v \in C(\overline{\Omega}) \cap C^\infty(\Omega)$ such that $v = g$ on $\partial\Omega$, $\|\nabla v\|_{L^\infty(\Omega)} \le C\|g\|_{C^{0,1}(\partial\Omega)}$, and $d\nu = |\nabla^2 v(x)|^2\,\delta(x)\,dx$ is a Carleson measure on Ω with norm $\|\nu\|_C \le C\|g\|^2_{C^{0,1}(\partial\Omega)}$, where C depends only on Ω.*

Proof. By a partition of unity we may assume that $\operatorname{supp}(g) \subset B(Q, r_0) \cap \partial\Omega$ for some $Q \in \partial\Omega$ and some small $r_0 > 0$. By translation and rotation we may assume that $Q = 0$ and

$$B(0, C_0 r_0) \cap \Omega = B(0, C_0 r_0) \cap \big\{(x', t) \in \mathbb{R}^d : \ x' \in \mathbb{R}^{d-1} \text{ and } t > \psi(x')\big\},$$

where ψ is a Lipschitz function on $\mathbb{R}^{d-1}$, $\psi(0) = 0$, and $C_0 = 10(1 + \|\nabla\psi\|_\infty)$. Let $D = \big\{(x', t) \in \mathbb{R}^d : x' \in \mathbb{R}^{d-1} \text{ and } t > \psi(x')\big\}$. Recall that there exists a bi-Lipschitz map $\Phi : \mathbb{R}^d_+ \to D$, such that $c \le |\nabla\Phi(x', t)| \le C$, and $|\nabla^2\Phi(x', t)|^2\, t\, dx'dt$ is a Carleson measure on $\mathbb{R}^d_+$.

Now, let $f(x') = g(\Phi^{-1}(x', 0))$. Let F be an extension of f to $\mathbb{R}^d_+$ so that $|F| \le \|f\|_\infty$, $|\nabla F| \le C\|\nabla_{x'} f\|_\infty$, and $|\nabla^2 F(x', t)|^2\, t\, dx'dt$ is a Carleson measure with norm less than $C\|\nabla_{x'} f\|^2_\infty$. It is not hard to verify that $v(x', t) = \varphi(x', t) F(\Phi^{-1}(x', t))$ satisfies all requirements in the lemma, if $\varphi \in C_0^\infty(B(0, C_0 r_0))$ and $\varphi = 1$ on $B(0, r_0)$.

Finally we remark that the extension F may be given by $F(x', t) = \eta_t * f(x')$, where $\eta_t(x') = t^{1-d}\eta(x'/t)$, $\eta \in C_0^\infty(\mathbb{R}^{d-1})$, and $\int_{\mathbb{R}^{d-1}} \eta = 1$. □

Let $f = (f^\alpha)$ be an $\mathbb{R}^m$-valued Lipschitz function on $\partial\Omega$ and g a scalar Lipschitz function on $\partial\Omega$. Let $u = (u^\alpha) \in H^1(\Omega; \mathbb{R}^m)$ be the weak solution of the Dirichlet problem

$$\mathcal{L}(u) = 0 \quad \text{in } \Omega \quad \text{and} \quad u = f \quad \text{on } \partial\Omega. \tag{4.1}$$

We will use v to denote the extension of g to Ω given by Lemma 4.1. Let $h \in H^1(\Omega; \mathbb{R}^m)$ be a weak solution of $\mathcal{L}^*(h) = \operatorname{div}\big(A^*(x)\nabla h\big) = 0$ in Ω. The proof of Theorem 1.1 relies on the following observation:

$$\begin{aligned}
&\int_{\partial\Omega} \big\{\Lambda(gf) - g\Lambda(f)\big\} \cdot h\, d\sigma \\
&= \int_{\partial\Omega} gf \cdot \Lambda^*(h)\, d\sigma - \int_{\partial\Omega} g\Lambda(f) \cdot h\, d\sigma \\
&= \int_{\partial\Omega} g \cdot u^\alpha \cdot n_i a_{ji}^{\beta\alpha} \frac{\partial h^\beta}{\partial x_j}\, d\sigma - \int_{\partial\Omega} g \cdot n_i a_{ij}^{\alpha\beta} \frac{\partial u^\beta}{\partial x_j} \cdot h^\alpha\, d\sigma \\
&= \int_\Omega \frac{\partial v}{\partial x_i} \cdot u^\alpha \cdot a_{ji}^{\beta\alpha} \frac{\partial h^\beta}{\partial x_j}\, dx - \int_\Omega \frac{\partial v}{\partial x_i} \cdot a_{ij}^{\alpha\beta} \frac{\partial u^\beta}{\partial x_j} \cdot h^\alpha\, dx,
\end{aligned} \tag{4.2}$$

where we have used $\mathcal{L}(u) = 0$ and $\mathcal{L}^*(h) = 0$ in Ω.

Lemma 4.2. *Let Ω be a bounded Lipschitz domain. Assume that A satisfies* (1.3) *and that $d\nu = |\nabla A(x)|^2\,\delta(x)\,dx$ is a Carleson measure on Ω with norm less than*

C_0. Let f, g, h, and u, v be given above. Then

$$\begin{aligned}
&\left| \int_{\partial\Omega} \{\Lambda(gf) - g\Lambda(f)\} \cdot h\, d\sigma \right| \\
&\le C\|g\|_{C^{0,1}(\partial\Omega)} \left\{ \int_\Omega |\nabla u|^2\, \delta(x)\, dx + \int_{\partial\Omega} |(u)^*|^2\, d\sigma \right\}^{1/2} \qquad (4.3)\\
&\cdot \left\{ \int_\Omega |\nabla h|^2\, \delta(x)\, dx + \int_{\partial\Omega} |(h)^*|^2\, d\sigma \right\}^{1/2},
\end{aligned}$$

where C depends only on d, m, μ, C_0, and Ω.

Proof. The lemma follows from (4.2) and Theorem 2.1. Indeed, since $\operatorname{div}(A^*\nabla h) = 0$ in Ω, by Theorem 2.1, the first term in the right hand side of (4.2) is bounded by

$$C \left\{ \int_\Omega |\nabla h|^2\, \delta(x)\, dx \right\}^{1/2} \left\{ \int_\Omega |\nabla F(x)|^2\, \delta(x)\, dx + \int_{\partial\Omega} |(F)^*|^2\, d\sigma \right\}^{1/2}, \qquad (4.4)$$

where $F = \left(\frac{\partial v}{\partial x_i} \cdot u^\alpha \cdot a_{ji}^{\beta\alpha} \right)$. Note that $|F| \le C|\nabla v|\, |u|$ and

$$|\nabla F| \le C\big\{ |\nabla^2 v|\, |u| + |\nabla v|\, |\nabla u| + |\nabla v|\, |u|\, |\nabla A| \big\}. \qquad (4.5)$$

It follows that

$$\int_{\partial\Omega} |(F)^*|^2\, d\sigma \le C\|g\|^2_{C^{0,1}(\partial\Omega)} \int_{\partial\Omega} |(u)^*|^2\, d\sigma, \qquad (4.6)$$

where we have used the estimate $\|\nabla v\|_{L^\infty(\Omega)} \le \|g\|_{C^{0,1}(\partial\Omega)}$ in Lemma 4.1. In view of (4.5) we obtain

$$\begin{aligned}
\int_{\partial\Omega} |\nabla F|^2\, \delta(x)\, dx &\le C\|g\|^2_{C^{0,1}(\partial\Omega)} \int_\Omega |\nabla u|^2\, \delta(x)\, dx + C \int_\Omega |u|^2 |\nabla^2 v|^2\, \delta(x)\, dx \\
&\quad + C\|g\|^2_{C^{0,1}(\partial\Omega)} \int_\Omega |u|^2 |\nabla A|^2\, \delta(x)\, dx \\
&\le C\|g\|^2_{C^{0,1}(\partial\Omega)} \left\{ \int_\Omega |\nabla u|^2\, \delta(x)\, dx + \int_{\partial\Omega} |(u)^*|^2\, d\sigma \right\}, \qquad (4.7)
\end{aligned}$$

where we have used the fact that $|\nabla^2 v(x)|^2\, \delta(x)\, dx$ is a Carleson measure on Ω with norm less than $C\|g\|^2_{C^{0,1}(\partial\Omega)}$ in Lemma 4.1, as well as the assumption that $|\nabla A(x)|^2\, \delta(x)\, dx$ is a Carleson measure on Ω. This, together with (4.4) and (4.6), shows that the first term in the right hand side of (4.2) is dominated by the right hand side of (4.3).

The second term in the right hand side of (4.2) may be handled in the same fashion. Since $\operatorname{div}(A\nabla u) = 0$ in Ω, by Theorem 2.1, this term is bounded by

$$C \left\{ \int_\Omega |\nabla u(x)|^2\, \delta(x) dx \right\}^{1/2} \left\{ \int_\Omega |\nabla G(x)|^2\, \delta(x)\, dx + \int_{\partial\Omega} |(G)^*|^2\, d\sigma \right\}^{1/2}, \qquad (4.8)$$

where $G = \left(\frac{\partial v}{\partial x_i} \cdot h^\alpha \cdot a_{ij}^{\alpha\beta}\right)$. Note that $|G| \le C|\nabla v|\,|h|$ and

$$|\nabla G| \le C\left\{|\nabla^2 v|\,|h| + |\nabla v|\,|\nabla h| + |\nabla v|\,|h|\,|\nabla A|\right\}.$$

The rest of the argument is identical to that for the first term (with the roles of u and h switched). We omit the details. □

Proof of Theorem 1.1. Let $f = (f^\alpha) \in C^{0,1}(\partial\Omega; \mathbb{R}^m)$ and $g \in C^{0,1}(\partial\Omega)$. Let $h \in H^1(\Omega; \mathbb{R}^m)$ be a weak solution of $\mathcal{L}(h) = 0$ in Ω. It follows from (4.2) that

$$\begin{aligned}
&\left|\int_{\partial\Omega} \{\Lambda(gf) - g\Lambda(f)\} \cdot h\, d\sigma\right| \\
&\le \left|\int_\Omega \frac{\partial v}{\partial x_i} \cdot u^\alpha \cdot a_{ji}^{\beta\alpha} \frac{\partial h^\beta}{\partial x_j}\, dx\right| + \left|\int_\Omega \frac{\partial v}{\partial x_i} \cdot a_{ij}^{\alpha\beta} \frac{\partial u^\beta}{\partial x_j} \cdot h^\alpha\, dx\right| = I_1 + I_2,
\end{aligned} \tag{4.9}$$

where v is the extension of g given by Lemma 4.1. We will show that

$$I_1 + I_2 \le C\|g\|_{C^{0,1}(\partial\Omega)} \|f\|_{L^2(\partial\Omega)} \|h\|_{L^2(\partial\Omega)}, \tag{4.10}$$

which yields the desired estimate by duality.

To estimate I_1, we construct an elliptic matrix $B(x) = (b_{ij}^{\alpha\beta})$, as in the proof of Theorem 3.1. Note that

$$\begin{aligned}
I_1 &\le \left|\int_\Omega \frac{\partial v}{\partial x_i} \cdot u^\alpha \cdot b_{ji}^{\beta\alpha} \frac{\partial h^\beta}{\partial x_j}\, dx\right| + \left|\int_\Omega \frac{\partial v}{\partial x_i} \cdot u^\alpha \cdot \left(a_{ji}^{\beta\alpha} - b_{ij}^{\alpha\beta}\right) \frac{\partial h^\beta}{\partial x_j}\, dx\right| \\
&= I_{11} + I_{12}.
\end{aligned} \tag{4.11}$$

Since $|\nabla^2 B(x)|^2\, \delta(x)\, dx$ is a Carleson measure on Ω, we may use Theorem 3.1 to show that

$$I_{11} \le C\|h\|_{L^2(\partial\Omega)} \|g\|_{C^{0,1}(\partial\Omega)} \left\{\int_\Omega |\nabla u|^2\, \delta(x)\, dx + \int_{\partial\Omega} |(u)^*|^2\, d\sigma\right\}^{1/2} \tag{4.12}$$

(see the proof of Lemma 4.2). This, together with (3.1), gives

$$|I_{11}| \le C\|h\|_{L^2(\partial\Omega)} \|g\|_{C^{0,1}(\partial\Omega)} \|f\|_{L^2(\partial\Omega)}. \tag{4.13}$$

To estimate I_{12}, we observe that

$$\begin{aligned}
I_{12} &\le \int_\Omega |\nabla v|\,|u|\,|A - B|\,|\nabla h|\, dx \\
&\le C\|g\|_{C^{0,1}(\partial\Omega)} \left\{\int_\Omega |\nabla h|^2\, \delta(x)\, dx\right\}^{1/2} \left\{\int_\Omega |u|^2 |A - B|^2 \left[\delta(x)\right]^{-1} dx\right\}^{1/2} \\
&\le C\|g\|_{C^{0,1}(\partial\Omega)} \|h\|_{L^2(\partial\Omega)} \left\{\int_\Omega |u|^2 |A - B|^2 \left[\delta(x)\right]^{-1} dx\right\}^{1/2},
\end{aligned}$$

where we have used the Cauchy inequality for the second inequality and (3.1) for the third. Since $|A(x) - B(x)| \le C\left[\delta(x)\right]^\eta$ for some $\eta > 0$, $|A - B|^2 \left[\delta(x)\right]^{-1} dx$ is a Carleson measure on Ω. It follows that

$$|I_{12}| \le C\|g\|_{C^{0,1}(\partial\Omega)} \|h\|_{L^2(\partial\Omega)} \|f\|_{L^2(\partial\Omega)}.$$

Finally, we point out that the term I_2 may be handled by the same manner as I_1, with the roles of u and h switched. We omit the details. □

5 Proof of Theorem 1.2

Throughout this section we will assume that A is elliptic, symmetric, and Hölder continuous. Let $f \in C^{0,1}(\partial\Omega;\mathbb{R}^m)$ and $u \in H^1(\Omega;\mathbb{R}^m)$ be the weak solution of $\mathcal{L}(u) = 0$ in Ω with boundary data $u = f$ on $\partial\Omega$. Let $h \in H^1(\Omega;\mathbb{R}^m)$ be a weak solution of $\mathcal{L}(h) = 0$ in Ω. In this section we shall use v to denote the harmonic extension of g to Ω; i.e., $\Delta v = 0$ in Ω and $v = g$ on $\partial\Omega$. It was proved in [3, 12] that

$$\int_{\partial\Omega} |(\nabla v)^*|^2 \, d\sigma + \int_\Omega |\nabla^2 v|^2 \, \delta(x) \, dx \le C \, \|g\|^2_{H^1(\partial\Omega)}, \tag{5.1}$$

where C depends only on the Lipschitz character of Ω. Our goal is to show that

$$\left| \int_{\partial\Omega} \{\Lambda(gf) - g\Lambda(f)\} \cdot h \, d\sigma \right| \le C \, \|g\|_{H^1(\partial\Omega)} \|u\|_{L^\infty(\Omega)} \|h\|_{L^2(\partial\Omega)}, \tag{5.2}$$

which yields the desired estimate in Theorem 1.2 by duality. We shall assume that $\|u\|_{L^\infty(\Omega)}$ is finite. Using the localized square function estimate, one may show that $|\nabla u(x)|^2 \, \delta(x) \, dx$ is a Carleson measure on Ω with norm less than $C\|u\|^2_{L^\infty(\Omega)}$, where C depends only on A and Ω.

To prove (5.2) we begin by recalling from (4.2) that

$$\begin{aligned} \int_{\partial\Omega} \{\Lambda(gf) - g\Lambda(f)\} \cdot h \, d\sigma &= \int_\Omega \frac{\partial v}{\partial x_i} \cdot u^\alpha \cdot a_{ji}^{\beta\alpha} \frac{\partial h^\beta}{\partial x_j} \, dx - \int_\Omega \frac{\partial v}{\partial x_i} \cdot a_{ij}^{\alpha\beta} \frac{\partial u^\beta}{\partial x_j} \cdot h^\alpha \, dx \\ &= I - J. \end{aligned} \tag{5.3}$$

To estimate I, we write

$$\begin{aligned} I &= \int_\Omega \frac{\partial v}{\partial x_i} \cdot u^\alpha \cdot b_{ji}^{\beta\alpha} \frac{\partial h^\beta}{\partial x_j} \, dx + \int_\Omega \frac{\partial v}{\partial x_i} \cdot u^\alpha \cdot \left(a_{ji}^{\beta\alpha} - b_{ji}^{\beta\alpha}\right) \frac{\partial h^\beta}{\partial x_j} \, dx \\ &= I_1 + I_2, \end{aligned} \tag{5.4}$$

where $B = (b_{ij}^{\alpha\beta})$ is the elliptic matrix constructed in the proof of Theorem 3.1. It follows from Theorem 3.1 that

$$|I_1| \le C\|h\|_{L^2(\partial\Omega)} \left\{ \int_\Omega |\nabla F|^2 \, \delta(x) \, dx + \int_{\partial\Omega} |(F)^*|^2 \, d\sigma \right\}^{1/2}, \tag{5.5}$$

where $F = \left(\frac{\partial v}{\partial x_i} \cdot u^\alpha \cdot b_{ji}^{\beta\alpha}\right)$. Using $(F)^* \le C \, \|u\|_{L^\infty(\Omega)} (\nabla v)^*$, we obtain

$$\int_{\partial\Omega} |(F)^*|^2 \, d\sigma \le C \, \|u\|^2_{L^\infty(\Omega)} \int_{\partial\Omega} |(\nabla v)^*|^2 \, d\sigma \le C \, \|u\|^2_{L^\infty(\Omega)} \|g\|^2_{H^1(\partial\Omega)}, \tag{5.6}$$

where we have used (5.1) for the second inequality. Since

$$|\nabla F| \le C\big\{ |\nabla^2 v| \, |u| + |\nabla v| \, |\nabla u| + |\nabla v| \, |u| \, |\nabla B| \big\},$$

we see that

$$\begin{aligned}\int_\Omega |\nabla F|^2\,\delta(x)\,dx &\le C\|u\|^2_{L^\infty(\Omega)}\int_\Omega |\nabla^2 v|^2\,\delta(x)\,dx + C\int_\Omega |\nabla v|^2|\nabla u|^2\,\delta(x)\,dx\\
&\quad + C\|u\|^2_{L^\infty(\Omega)}\int_\Omega |\nabla v|^2|\nabla B|^2\,\delta(x)\,dx\\
&\le C\|u\|^2_{L^\infty(\Omega)}\left\{\int_\Omega |\nabla^2 v|^2\,\delta(x)\,dx + \int_{\partial\Omega}|(\nabla v)^*|^2\,d\sigma\right\}\\
&\le C\|u\|^2_{L^\infty(\Omega)}\|g\|^2_{H^1(\partial\Omega)},\end{aligned}\tag{5.7}$$

where we have used the fact that $|\nabla u(x)|^2\delta(x)\,dx$ is a Carleson measure on Ω with norm less than $C\|u\|^2_{L^\infty(\Omega)}$, and that $|\nabla B(x)|^2\,\delta(x)\,dx$ is also a Carleson measure on Ω. This, together with (5.5) and (5.6), gives

$$|I_1| \le C\|g\|_{H^1(\partial\Omega)}\|u\|_{L^\infty(\Omega)}\|h\|_{L^2(\partial\Omega)}.\tag{5.8}$$

To bound I_2, we use

$$|I_2| \le C\|u\|_{L^\infty(\Omega)}\int_\Omega |\nabla v|\,|A-B|\,|\nabla h|\,dx.$$

It follows by the Cauchy inequality that

$$\begin{aligned}|I_2| &\le C\|u\|_{L^\infty(\Omega)}\left\{\int_\Omega |\nabla v|^2|A-B|^2\big[\delta(x)\big]^{-1}\,dx\right\}^{1/2}\left\{\int_\Omega |\nabla h|^2\,\delta(x)\,dx\right\}^{1/2}\\
&\le C\|u\|_{L^\infty(\Omega)}\|(\nabla v)^*\|_{L^2(\partial\Omega)}\|h\|_{L^2(\partial\Omega)}\\
&\le C\|u\|_{L^\infty(\Omega)}\|g\|_{H^1(\partial\Omega)}\|h\|_{L^2(\partial\Omega)},\end{aligned}$$

where we have used the fact that $|A-B|^2\big[\delta(x)\big]^{-1}\,dx$ is a Carleson measure on Ω. The estimate of $I = I_1 + I_2$ is now complete.

Next, we turn to the estimate of J in (5.3). Write

$$\begin{aligned}J &= \int_\Omega \frac{\partial v}{\partial x_i}\cdot b_{ij}^{\alpha\beta}\frac{\partial u^\beta}{\partial x_j}\cdot h^\alpha\,dx + \int_\Omega \frac{\partial v}{\partial x_i}\cdot\big(a_{ij}^{\alpha\beta} - b_{ij}^{\alpha\beta}\big)\frac{\partial u^\beta}{\partial x_j}\cdot h^\alpha\,dx\\
&= J_1 + J_2.\end{aligned}\tag{5.9}$$

For J_1, we use the integration by parts to obtain

$$\begin{aligned}J_1 &= -\int_\Omega \frac{\partial}{\partial x_j}\left\{b_{ij}^{\alpha\beta}\frac{\partial v}{\partial x_i}\right\}\cdot u^\beta\cdot h^\alpha\,dx - \int_\Omega \frac{\partial v}{\partial x_i}\cdot b_{ij}^{\alpha\beta}u^\beta\cdot\frac{\partial h^\alpha}{\partial x_i}\,dx\\
&\quad + \int_{\partial\Omega} n_j\frac{\partial v}{\partial x_i}\cdot b_{ij}^{\alpha\beta}u^\beta\cdot h^\alpha\,d\sigma\\
&= J_{11} + J_{12} + J_{13}.\end{aligned}\tag{5.10}$$

Note that $J_{12} = -I_1$ and by the Cauchy inequality,

$$|J_{13}| \le C\|u\|_{L^\infty(\Omega)}\|\nabla v\|_{L^2(\partial\Omega)}\|h\|_{L^2(\partial\Omega)} \le C\,\|u\|_{L^\infty(\Omega)}\|g\|_{H^1(\partial\Omega)}\|h\|_{L^2(\partial\Omega)}.$$

Also,

$$|J_{11}| \le \left|\int_\Omega b_{ij}^{\alpha\beta} \frac{\partial^2 v}{\partial x_i \partial x_j} \cdot u^\beta \cdot h^\alpha \, dx\right| + \left|\int_\Omega \frac{\partial b_{ij}^{\alpha\beta}}{\partial x_j} \cdot \frac{\partial v}{\partial x_i} \cdot u^\beta \cdot h^\alpha \, dx\right|. \tag{5.11}$$

Since ∇v is harmonic in Ω, we may use Theorem 3.1 to bound the first term in the right hand side of (5.11) by

$$C\|\nabla v\|_{L^2(\partial\Omega)} \left\{ \int_\Omega |\nabla G|^2 \, \delta(x) \, dx + \int_{\partial\Omega} |(G)^*|^2 \, d\sigma \right\}^{1/2}, \tag{5.12}$$

where $G = \left(b_{ij}^{\alpha\beta} u^\beta h^\alpha\right)$. As in the case of I_1, the term in (5.12) is dominated by the right hand side of (5.8), using the fact that $|\nabla B(x)|^2 \, \delta(x) \, dx$ and $|\nabla u(x)|^2 \, \delta(x) \, dx$ are Carleson measures on Ω. By the Cauchy inequality the second term in the right hand side of (5.11) is bounded by

$$\begin{aligned} &C\|u\|_{L^\infty(\Omega)} \left\{ \int_\Omega |\nabla v|^2 |\nabla B|^2 \left[\delta(x)\right]^{1-\eta} dx \right\}^{1/2} \left\{ \int_\Omega |h|^2 \left[\delta(x)\right]^{\eta-1} dx \right\}^{1/2} \\ &\le C\|u\|_{L^\infty(\Omega)} \|g\|_{H^1(\partial\Omega)} \|h\|_{L^2(\partial\Omega)}, \end{aligned}$$

where we have used the estimate $|\nabla B(x)| \le C\left[\delta(x)\right]^{\eta-1}$ as well as the fact that $\left[\delta(x)\right]^{\eta-1} dx$ is a Carleson measure on Ω.

Finally, it remains to estimate

$$J_2 = \int_\Omega \frac{\partial v}{\partial x_i} \cdot \left(a_{ij}^{\alpha\beta} - b_{ij}^{\alpha\beta}\right) \frac{\partial u^\beta}{\partial x_j} \cdot h^\alpha \, dx.$$

Recall that $|A(x) - B(x)| \le C\left[\delta(x)\right]^\eta$. By the Cauchy inequality we see that

$$\begin{aligned} J_2 &\le \left\{ \int_\Omega |\nabla v|^2 |A - B|^2 |\nabla u|^2 \left[\delta(x)\right]^{1-2\eta} dx \right\}^{1/2} \left\{ \int_\Omega |h|^2 \left[\delta(x)\right]^{2\eta-1} dx \right\}^{1/2} \\ &\le C\|h\|_{L^2(\partial\Omega)} \left\{ \int_\Omega |\nabla v|^2 |\nabla u|^2 \delta(x) \, dx \right\}^{1/2} \\ &\le C\|h\|_{L^2(\Omega)} \|u\|_{L^\infty(\Omega)} \|g\|_{H^1(\partial\Omega)}, \end{aligned}$$

where we have used the fact that $|\nabla u(x)|^2 \, \delta(x) \, dx$ is a Carleson measure with norm less than $C\|u\|^2_{L^\infty(\Omega)}$. This completes the proof of Theorem 1.2.

References

[1] A.P. Calderón, Algebra of singular integrals operators, Singular Integrals. Proc. Symp. Pure Math. **10** (1967), 18–55.

[2] B. Dahlberg, On the Poisson integral for Lipschitz and C^1-domains, Studia Math. **66** (1979), 13–24.

[3] B. Dahlberg, Weighted norm inequalities for the Lusin area integral and the nontangential maximal function for functions harmonic in a Lipschitz domain, Studia Math. **67** (1980), 297–314.

[4] B. Dahlberg, On the absolute continuity of elliptic measures, Amer. J. Math. **108** (1986), 1119–1138.

[5] B. Dahlberg, Poisson semigroup and singular integrals, Proc. Amer. Math. Soc. **97** (1986), 41–48.

[6] B. Dahlberg and C. Kenig, Hardy spaces and the Neumann problem in L^p for Laplace's equation in Lipschitz domains, Ann. of Math. **125** (1987), 437–466.

[7] B. Dahlberg and C. Kenig, L^p estimates for the three-dimensional system of elastostatics on Lipschitz domains, Lecture Notes in Pure and Applied Mathematics (Cora Sadoesky, ed.) **122**, Dekker, 1990, pp. 631–634.

[8] B. Dahlberg, C. Kenig, J. Pipher, and G. Verchota, Area integral estimates for higher order elliptic equations and systems, Ann. Inst. Fourier (Grenoble) **47** (1997), no. 5, 1425–1461.

[9] B. Dahlberg, C. Kenig, and G. Verchota, The Dirichlet problem for the biharmonic equation in a Lipschitz domain, Ann. Inst. Fourier (Grenoble) **36** (1986), no. 3, 109–135.

[10] L. Grafakos, Classical and Modern Fourier Analysis, Pearson Education, Inc., Upper Saddle River, New Jersey, 2004.

[11] S. Hofmann, Dahlberg's bilinear estimate for solutions of divergence form complex elliptic equations, Proc. Amer. Math. Soc. **136** (2008), 4223–4233.

[12] D. Jerison and C. Kenig, An identity with applications to harmonic measure, Bull. Amer. Math. Soc. **2** (1980), 447–451.

[13] C. Kenig, F. Lin, and Z. Shen, Periodic homogenization of Green and Neumann functions, Comm. Pure Appl. Math. **67** (2014), 1219–1262.

[14] C. Kenig, F. Lin, and Z. Shen, Convergence rates in L^2 for elliptic homogenization problems, Arch. Rational Mech. Anal. **203** (2012), no. 3, 1009–1036.

[15] C. Kenig and Z. Shen, Layer potential methods for elliptic homogenization problems, Comm. Pure Appl. Math. **64** (2011), 1–44.

[16] M. Mitrea and M. Taylor, Potential theory on Lipschitz domains in Riemannian manifolds: Hölder continuous metric tensors, Comm. PDEs **25** (2000), 1487–1536.

[17] G. Verchota, Layer potentials and regularity for the Dirichlet problem for Laplace's equation in Lipschitz domains, J. Funct. Anal. **59** (1984), 572–611.

Some Topics in Harmonic Analysis and Applications
ALM 34, pp. 385–397

A Note on L^p-norms of Quasi-modes*

Christopher D. Sogge† and Steve Zelditch‡

Abstract

In this note we show how improved L^p-estimates for certain types of quasi-modes are naturally equivalent to improved operator norms of spectral projection operators associated to shrinking spectral intervals of the appropriate scale. Using this, one can see that recent estimates that were stated for eigenfunctions also hold for the appropriate types of quasi-modes.

2010 Mathematics Subject Classification: 58J50 (35P20 42B10).

Keywords and Phrases: Eigenfunctions, quasi-modes.

1 Introduction and main results

Let (M, g) be a compact boundaryless Riemannian manifold of dimension $n \geq 2$. Let $0 = \lambda_0 < \lambda_1 \leq \lambda_2 \leq \cdots$ be the frequencies associated with Δ and let $e_j(x)$ be an orthonormal basis of eigenfunctions,

$$(\Delta^2 + \lambda_j^2)e_j(x) = 0 \quad \text{and} \quad \int_M |e_j|^2 \, dV = 1. \tag{1.1}$$

Here $dV = dV_g$ and $\Delta = \Delta_g$ denote the volume element and Laplace-Beltrami operator associated with (M, g), respectively. Also define the "unit-band" spectral projection operators (i.e. for frequency intervals of unit length) by

$$\chi_{[\lambda,\lambda+1]} f(x) = \sum_{\lambda_j \in [\lambda,\lambda+1]} E_j f, \tag{1.2}$$

*The research of the first author was partially supported by NSF grant # DMS-1069175 and the second by # DMS-1206527.

†Department of Mathematics, Johns Hopkins University, Baltimore, MD 21218, USA; *E-mail:* `sogge@jhu.edu`

‡Department of Mathematics, Northwestern University, Evanston, IL 60208, USA; *E-mail:* `s-zelditch@northwestern.ed`

where E_j denotes the projection onto the j-th eigenspace, i.e.,

$$E_j f(x) = \langle f, e_j \rangle \, e_j(x),$$

with $\langle \cdot , \cdot \rangle$ being the $L^2(dV)$-inner product.

Many recent works are concerned with L^p norms of eigenfunctions and their relations to the geometry of the geodesics of (M, g). The first author proved universal bounds on L^p norms of eigenfunctions [12]. For large p the bounds are saturated by zonal (rotationally invariant) spherical harmonics on the standard sphere S^n, which peak at the poles, and for low p they are saturated by elliptic Gaussian beams which concentrate on closed geodesics (see [11]). In general, a Riemannian manifold fails to have either kind of eigenfunction and that suggests that the L^p norms of eigenfunctions are strictly smaller on a typical (M, g) than in the case of the sphere. The theme of our works [17, 16, 20] is that for $p = \infty$ (or for "large p"), the L^p bounds can only be saturated if there exist (partial) poles or self-focal points. Less is known about necessary geometric conditions for existence of eigenfunctions saturating the low L^p norms. It is plausible that existence of a stable elliptic geodesic is necessary for saturation of low L^p norms. We refer to [14, 18, 19] for some recent results and to [15, 22, 23] for surveys of the problem.

Many of the techniques and results concerning L^p norms of eigenfunctions ("modes") apply equally well to quasi-modes or approximate eigenfunctions. Quasi-modes are linear combinations of eigenfunctions with eigenvalues concentrated in a short interval. Specifically, if λ ranges over a sequence $\{\mu_k\}$ tending to $+\infty$, then we say that $\{\psi_\lambda\}$ (with $\lambda = \mu_k$) is a quasi-mode of order s if

$$\int_M |\psi_\lambda|^2 \, dV = 1 \quad \text{and} \quad \|(\Delta + \lambda^2)\psi_\lambda\|_{L^2(M)} = O(\lambda^{-s}). \tag{1.3}$$

Note then that, by our convention of using λ^2 instead of λ in the second part of (1.7), the quasi-modes ψ_λ are associated with frequencies λ. Roughly speaking, the higher the s, the shorter the band of frequencies required to construct the quasi-mode. We would like to give bounds on L^p norms of quasi-modes of a given order s and relate them to the geometry of geodesics. For instance, we may ask if there exist quasi-modes of order s which saturate the L^p bounds and if they resemble zonal spherical harmonics for high p or Gaussian beams for low p. Of course, quasi-modes peaking at a point or concentrating on a closed geodesic exist much more often than actual eigenfunctions with these properties. But as in the case of modes, there are necessary geometric conditions on (M, g) for the construction of quasi-modes of order s which saturate the L^p bounds of [12]. It is not hard to see that there are no conditions if one lets $s = -1$, since the Schwartz kernel $\chi_{[\lambda,\lambda+1]}(\cdot, x)$ itself saturates the high L^p bounds and the spectral projection $\chi_{[\lambda,\lambda+1]}(\delta_\gamma)$ of the delta-function on a closed geodesic saturates the low L^p bounds. The question is whether one can find necessary geometric conditions under which quasi-modes with spectrum in just slightly shrinking intervals can saturate the L^p bounds. If so, Riemannian manifolds (M, g) which fail to satisfy the conditions are said to have "improved estimates" or to satisfy "sub-convexity bounds".

As this discussion indicates, closely related to estimates of modes and quasi-modes are mapping norm estimates of spectral projections for intervals. If

$$\sigma(p,n) = \begin{cases} n\left(\frac{1}{2}-\frac{1}{p}\right)-\frac{1}{2}, & \frac{2(n+1)}{n-1} \le p \le \infty, \\ \frac{n-1}{2}\left(\frac{1}{2}-\frac{1}{p}\right), & 2 \le p \le \frac{2(n+1)}{n-1}, \end{cases} \tag{1.4}$$

then it was shown in [12] that

$$\|\chi_{[\lambda,\lambda+1]} f\|_{L^p(M)} \le C\lambda^{\sigma(p,n)} \|f\|_{L^2(M)}, \quad \lambda \ge 1,\ p \ge 2. \tag{1.5}$$

These unit-band spectral projection estimates turn out to be sharp and cannot be improved on *any* compact boundaryless Riemannian manifold, as was shown in [13].

In this note we consider spectral projections for shrinking intervals. As with quasi-modes, the question is to find geometric conditions on (M,g) under which one can improve or "break convexity" for the mapping norm estimate. Breaking convexity bounds on mapping norms of spectral projections means finding a rate $\varepsilon(\lambda)$ of shrinking intervals $[\lambda, \lambda+\varepsilon(\lambda)]$ and a condition on (M,g) for which one can prove

$$\|\chi_{[\lambda,\lambda+\varepsilon(\lambda)]} f\|_{L^p(M)} \le C\sqrt{\varepsilon(\lambda)}\lambda^{\sigma(p,n)} \|f\|_{L^2(M)}, \tag{1.6}$$

assuming that $\varepsilon(\lambda) > 0$ and $\varepsilon(\lambda)$ tends monotonically to zero as $\lambda \to \infty$. Here, as before, $\sigma(p,n)$ is as in (1.4). The dependence of (1.6) on $\varepsilon(\lambda)$ is sharp in view of the fact that (1.5) cannot be improved. To see this one uses orthogonality and the dual version of (1.5).

In this note, we do not give new geometric conditions. Rather our purpose is to show that the conditions for improving mapping norm estimates or sub-convexity bounds for spectral projection operators associated to shrinking intervals are equivalent to the conditions improved bounds for all quasi-modes of the related order. The arguments are implicit in [5] and [16] but have not previously been made explicit. We shall focus on two different types of quasi-modes that naturally occur, although the methods will handle other types as well.

1.1 Quasi-modes of order $o(\lambda)$

The first type we call quasi-modes of order $o(\lambda)$ or "fat quasi-modes". In place of (1.7) they satisfy

$$\int_M |\psi_\lambda|^2 \, dV = 1 \quad \text{and} \quad \|(\Delta+\lambda^2)\psi_\lambda\|_{L^2(M)} = o(\lambda). \tag{1.7}$$

These occur at least implicitly in several recent works, including [2, 3, 4, 5, 6, 14, 16, 17, 20]. We would like to give conditions on (M,g) which break the convexity bounds, i.e. prohibit fat quasi-modes from saturating the L^p bounds of [12]. Our first result states that breaking convexity bounds for fat quasi-modes is equivalent to breaking convexity bounds for mapping norms of spectral projections for shrinking intervals. That is, we replace the unit intervals $[\lambda, \lambda+1]$ by shrinking intervals $[\lambda, \lambda+o(1)]$ and seek sufficient conditions on (M, g) so that if $\sigma(p, n)$ is as in (1.4), then

$$\|\chi_{[\lambda,\lambda+o(1)]}\|_{L^2(M)\to L^p(M)} = o(\lambda^{\sigma(p,n)}). \tag{1.8}$$

By this we of course mean that for every $\varepsilon > 0$ there is a $0 < \delta < 1$ and a $\Lambda < \infty$ (both depending on ε) so that

$$\|\chi_{[\lambda,\lambda+\delta]}\|_{L^2(M)\to L^p(M)} \le \varepsilon\lambda^{\sigma(p,n)}, \quad \lambda > \Lambda, \tag{1.9}$$

where

$$\chi_{[\lambda,\lambda+\delta]}f = \sum_{\lambda_j\in[\lambda,\lambda+\delta]} E_j f,$$

and

$$\|\chi_{[\lambda,\lambda+\delta]}\|_{L^2(M)\to L^p(M)}$$

denotes the $L^2(M) \to L^p(M)$ operator norm of $\chi_{[\lambda,\lambda+\delta]}$. Improved bounds (1.8) were shown to be generically true for large exponents p in [17] and [16], and in many cases in [2] and [19] for $2 < p < \frac{2(n+1)}{n-1}$.

The condition (1.7) says that in an L^2-sense the spectrum of the ψ_λ are concentrated in intervals of length $o(1)$ about $\lambda = \mu_k$ as $\mu_k \to \infty$. In view of this, the following result naturally links $o(\lambda^{\sigma(p,n)})$ estimates for these $o(\lambda)$ quasi-modes with the improved L^p-bounds in (1.9).

Theorem 1.1. *Suppose that $n \le 3$ and $p > 2$, $n = 4$ and $2 < p < \infty$ or $n \ge 5$ and $2 < p \le \frac{2n}{n-4}$. Then (1.8) is valid if and only if whenever λ ranges over a sequence $\mu_k \to \infty$ we have*

$$\|\psi_\lambda\|_{L^p(M)} = o(\lambda^{\sigma(p,n)}), \tag{1.10}$$

assuming that the ψ_λ satisfy the $o(\lambda)$ quasi-mode condition (1.7). If $n = 4$ and $p = \infty$ or $n \ge 5$ and $\frac{2n}{n-4} < p \le \infty$, the same conclusion is valid if we replace (1.7) by the condition that

$$\int_M |\psi_\lambda|^2\, dV = 1$$

and

$$\|(\Delta+\lambda^2)\psi_\lambda\|_{L^2(M)} + \lambda^{2-n(\frac{1}{2}-\frac{1}{p})}\|\chi_{[2\lambda,\infty)}\psi_\lambda\|_{L^p(M)} = o(\lambda), \tag{1.11}$$

where

$$\chi_{[2\lambda,\infty)}f = \sum_{\lambda_j\ge 2\lambda} E_j f$$

denotes the projection of f onto frequencies $\lambda_j \in [2\lambda, \infty)$.

Note that for all exponents $p > 2$ if $n \le 3$ or for $2 < p < \infty$ when $n = 4$ or $2 < p \le \frac{2n}{n-4}$ the second term in the last condition in (1.11) is dominated by the first term by Sobolev embeddings. Thus, for these exponents, the condition (1.11) is equivalent to the more succinct version (1.7). On the other hand for the other exceptional exponents $p > 2$ if $n \ge 4$, this term is needed for the conclusion as was argued in [16]. This is just due to Sobolev considerations.

1.2 Quasi-modes of order zero

The second type of quasi-mode we consider are those of order zero, which satisfy

$$\int_M |\psi_\lambda|^2 \, dV = 1 \quad \text{and} \quad \|(\Delta + \lambda^2)\psi_\lambda\|_{L^2(M)} = O(1), \tag{1.12}$$

assuming, as above, that λ ranges over a sequence $\mu_k \to \infty$. We refer to [24] for background and references.

Note that the second condition in (1.12) says that, in an L^2-sense, the spectrum of the ψ_λ are concentrated in intervals of length $1/\lambda$ about λ. Thus, the following result is the natural analog of Theorem 1.1.

Theorem 1.2. *Suppose that $n \le 3$ and $p > 2$, $n = 4$ and $2 < p < \infty$ or $n \ge 5$ and $2 < p \le \frac{2n}{n-4}$. Then*

$$\|\chi_{[\lambda,\lambda+1/\lambda]}\|_{L^2(M)\to L^p(M)} = o(\lambda^{\sigma(p,n)}), \tag{1.13}$$

where $\sigma(p,n)$ is as in (1.4) if and only if whenever λ ranges over a sequence $\mu_k \to \infty$ we have

$$\|\psi_\lambda\|_{L^p(M)} = o(\lambda^{\sigma(p,n)}), \tag{1.14}$$

assuming that the ψ_λ satisfy the zero order quasi-mode condition (1.13). If $n = 4$ and $p = \infty$ or $n \ge 5$ and $\frac{2n}{n-4} < p \le \infty$, the same conclusion is valid if we replace (1.13) by the condition that

$$\int_M |\psi_\lambda|^2 \, dV = 1,$$

and

$$\|(\Delta + \lambda^2)\psi_\lambda\|_{L^2(M)} + \lambda^{2-n(\frac{1}{2}-\frac{1}{p})}\|\chi_{[2\lambda,\infty)}\psi_\lambda\|_{L^p(M)} = O(1), \tag{1.15}$$

where $\chi_{[2\lambda,\infty)}$ is as above.

1.3 A unifying principle

As we shall see both of these theorems are a consequence of the following result, which is a variation of the ones in [5].

Proposition 1.3. *Suppose that $n \le 3$ and $p > 2$, $n = 4$ and $2 < p < \infty$ or $n \ge 5$ and $2 < p \le \frac{2n}{n-4}$. Then there is a uniform constant $C = C(M,g)$ depending only on our n-dimensional Riemannian manifold (M,g) so that for $0 < \delta < 1$ and sufficiently large λ we have*

$$\begin{aligned} &\|f\|_{L^p(M)} \\ &\le C \sup_{\mu\in[\lambda/2,2\lambda]} \|\chi_{[\mu,\mu+\delta]}\|_{L^2(M)\to L^p(M)} \times \big(\|f\|_{L^2(M)} + (\delta\lambda)^{-1}\|(\Delta+\lambda^2)f\|_{L^2(M)}\big). \end{aligned} \tag{1.16}$$

Similarly,

$$\|f\|_{L^p(M)} - \|\chi_{[2\lambda,\infty)}f\|_{L^p(M)}$$
$$\le C \sup_{\mu\in[\lambda/2,2\lambda]} \|\chi_{[\mu,\mu+\delta]}\|_{L^2(M)\to L^p(M)} \times \big(\|f\|_{L^2(M)} + (\delta\lambda)^{-1}\|(\Delta+\lambda^2)f\|_{L^2(M)}\big), \tag{1.17}$$

for the remaining exponents $p>2$, i.e., $p=\infty$ if $n=4$ and $\frac{2n}{n-4}<p\le\infty$ if $n\ge 5$.

2 Proof that Proposition 1.3 implies Theorems 1.1 and 1.2

We shall start by going over the simple argument which shows how Proposition 1.3 implies the two theorems.

Let us first suppose that (1.8) is valid. Thus, given $\varepsilon>0$ we can find a fixed $\delta=\delta(\varepsilon)\in(0,1)$ so that (1.9) is valid. In other words, we have that

$$\sup_{\mu\in[\lambda/2,2\lambda]} \|\chi_{[\mu,\mu+\delta]}\|_{L^2(M)\to L^p(M)} \le \varepsilon\lambda^{\sigma(p,n)},$$

if $\lambda\gg 1$ is large enough. By (1.16), we then have that

$$\|f\|_{L^p(M)} \le C\varepsilon\lambda^{\sigma(p,n)}\big(\|f\|_{L^2(M)} + (\delta\lambda)^{-1}\|(\Delta+\lambda^2)f\|_{L^2(M)}\big). \tag{2.1}$$

Since we have fixed $\delta\in(0,1)$, if we take $f=\psi_\lambda$, where the ψ_λ satisfy (1.7), we conclude that (2.1) yields

$$\|\psi_\lambda\|_{L^p(M)} \le 2C\varepsilon\lambda^{\sigma(p,n)} \tag{2.2}$$

for large enough λ. Thus, since $\varepsilon>0$ is arbitrary, we obtain (1.10) assuming that (1.8) is valid and $n\le 3$ or $2<p<\infty$ if $n=4$ or $2<p\le\frac{2n}{n-4}$ if $n\ge 5$. Since

$$n\left(\frac{1}{2}-\frac{1}{p}\right)-1<\sigma(p,n) \tag{2.3}$$

for all exponents $p>2$, we similarly obtain (1.10) for the remaining exponents from (1.17) assuming that (1.8) is valid and that the ψ_λ satisfy (1.11).

To prove the remaining half of Theorem 1.1, let

$$V_{[\lambda,\lambda+\delta]} = \{f\in L^2(M):\, E_jf=0,\ \lambda_j\notin[\lambda,\lambda+\delta]\} \tag{2.4}$$

denote the span of the eigenfunctions whose frequencies lie in $[\lambda,\lambda+\delta]$. It then follows that

$$\|\chi_{[\lambda,\lambda+\delta]}\|_{L^2(M)\to L^p(M)} = \sup_{f\in V_{[\lambda,\lambda+\delta]},\ \|f\|_2=1} \|f\|_{L^r(M)}, \tag{2.5}$$

and

$$\|(\Delta+\lambda^2)f\|_{L^2(M)} \le 3\delta\lambda, \quad \text{if } f \in V_{[\lambda,\lambda+\delta]} \text{ and } \|f\|_{L^2(M)} = 1, \tag{2.6}$$

assuming that $0 < \delta < 1$ and $\lambda \ge 1$. Note also that for such δ, λ, we have $\chi_{[2\lambda,\infty)}f = 0$ if $f \in V_{[\lambda,\lambda+\delta]}$. Thus if (1.10) is valid assuming either (1.7) or (1.11) as λ ranges over any sequence $\mu_k \to \infty$, we conclude that (1.8) must hold, completing the proof of Theorem 1.1.

The proof of Theorem 1.2 is very similar. First, if (1.13) is valid then by (2.1) with $\delta = 1/\lambda$, we conclude using (1.16) again that we must have (2.2) if the ψ_λ satisfy (1.12) and $n \le 3$ and $p > 2$, $2 < p < \infty$ if $n = 4$ or $2 < p \le \frac{2n}{n-4}$ if $n \ge 5$. We similarly obtain (1.14) assuming (1.15) from (1.17) for the remaining exponents $p > 2$, since for these we have

$$n\left(\frac{1}{2}-\frac{1}{p}\right) - 2 < \sigma(p,n).$$

Since the converse part of Theorem 1.2 follows from the corresponding proof of this part of Theorem 1.1 (using $\delta = 1/\lambda$ in (2.5) and (2.6)), we obtain the second theorem as well.

3 Proof of Proposition 1.3

We now prove Proposition 1.3. We first note that by the Sobolev estimate $\|u\|_p \le C\|u\|_{W^{k,2}}$ (with $k = \frac{n}{2}(\frac{1}{2}-\frac{1}{p})$) and orthogonality, if $2 < p < \infty$ and $n \le 4$ or $2 < p \le \frac{2n}{n-4}$ and $n \ge 5$, we have

$$\begin{aligned}\|\chi_{[2\lambda,\infty)}f\|_{L^p(M)} &\le C\|(-\Delta)^{\frac{n}{2}(\frac{1}{2}-\frac{1}{p})}\chi_{[2\lambda,\infty)}f\|_{L^2(M)}\\ &\le C\lambda^{n(\frac{1}{2}-\frac{1}{p})-2}\|(\Delta+\lambda^2)\chi_{[2\lambda,\infty)}f\|_{L^2(M)}\\ &\le C\lambda^{n(\frac{1}{2}-\frac{1}{p})-2}\|(\Delta+\lambda^2)f\|_{L^2(M)}.\end{aligned} \tag{3.1}$$

Similarly, for every $\varepsilon > 0$ we have

$$\|\chi_{[2\lambda,\infty)}f\|_{L^\infty(M)} \le C_\varepsilon\lambda^{\frac{n}{2}-2+\varepsilon}\|(\Delta+\lambda^2)f\|_{L^2(M)}, \quad \text{if } n \le 3. \tag{3.2}$$

Also, as we noted before, (1.5) cannot be improved, and, therefore, there must be a uniform constant $c > 0$, which is independent of $0 < \delta < 1$, so that if λ is large enough we have

$$c\delta\lambda^{\sigma(p,n)} \le \sup_{\mu\in[\lambda/2,2\lambda]} \|\chi_{[\mu,\mu+\delta]}\|_{L^2(M)\to L^p(M)},$$

and so

$$\begin{aligned}&\lambda^{\sigma(p,n)-1}\|(\Delta+\lambda^2)f\|_{L^2(M)}\\ &\le C\sup_{\mu\in[\lambda/2,2\lambda]}\|\chi_{[\mu,\mu+\delta]}\|_{L^2(M)\to L^p(M)} \times (\delta\lambda)^{-1}\|(\Delta+\lambda^2)f\|_{L^2(M)}.\end{aligned} \tag{3.3}$$

Therefore, if we recall (2.3) and combine (3.1)–(3.3), we conclude that if $2 < p \le \infty$ and $n \le 3$ or $2 < p < \infty$ and $n = 3$ or $2 < p \le \frac{2n}{n-4}$ and $n \ge 5$, then there is a uniform constant C, which is independent of $0 < \delta < 1$, so that for sufficiently large λ we have

$$\|\chi_{[2\lambda,\infty)} f\|_{L^p(M)} \le C(\delta\lambda)^{-1}\|(\Delta+\lambda^2)f\|_{L^2(M)} \times \sup_{\mu\in[\lambda/2,2\lambda]} \|\chi_{[\mu,\mu+\delta]}\|_{L^2(M)\to L^p(M)}.$$

From this, we deduce that we would have (1.16) and (1.17) if we could show that

$$\begin{aligned}&\|\chi_{[0,2\lambda)} f\|_{L^p(M)} \\ &\le C \sup_{\mu\in[\lambda/2,2\lambda]} \|\chi_{[\mu,\mu+\delta]}\|_{L^2(M)\to L^p(M)} \times \big(\|f\|_{L^2(M)} + (\delta\lambda)^{-1}\|(\Delta+\lambda^2)f\|_{L^2(M)}\big),\end{aligned} \tag{3.4}$$

for all exponents $2 < p \le \infty$ if $n \ge 2$ and δ and λ are as above.

To prove this we write

$$\chi_{[0,2\lambda)} f = \chi_{[\lambda-\delta,\lambda+\delta]} f + \sum_{k=1}^{\infty} \sum_{|\lambda-\lambda_j|\in(\delta k,\delta(k+1)]} \chi_{(0,2\lambda)} E_j f. \tag{3.5}$$

Clearly

$$\begin{aligned}\|\chi_{[\lambda-\delta,\lambda+\delta]} f\|_{L^p(M)} &\le \|\chi_{[\lambda-\delta,\lambda+\delta]}\|_{L^2(M)\to L^p(M)}\|f\|_{L^2(M)} \\ &\le 2 \sup_{\mu\in[\lambda/2,2\lambda]} \|\chi_{[\mu,\mu+\delta]}\|_{L^2(M)\to L^p(M)} \times \|f\|_{L^2(M)}.\end{aligned} \tag{3.6}$$

Similarly,

$$\begin{aligned}&\Big\|\sum_{|\lambda-\lambda_j|\in(\delta k,\delta(k+1)]} \chi_{(0,2\lambda)} E_j f\Big\|_{L^p(M)} \\ &\le \sup_{\mu\in[\lambda/2,2\lambda]} \|\chi_{[\mu,\mu+\delta]}\|_{L^2(M)\to L^p(M)} \times \Big\|\sum_{|\lambda-\lambda_j|\in(\delta k,\delta(k+1)]} \chi_{(0,2\lambda)} E_j f\Big\|_{L^2(M)} \\ &\le (\delta k\lambda)^{-1} \sup_{\mu\in[\lambda/2,2\lambda]} \|\chi_{[\mu,\mu+\delta]}\|_{L^2(M)\to L^p(M)} \\ &\quad \times \Big\|\sum_{|\lambda-\lambda_j|\in(\delta k,\delta(k+1)]} (\Delta+\lambda^2) E_j f\Big\|_{L^2(M)}.\end{aligned}$$

Therefore, by orthogonality and an application of the Cauchy-Schwarz inequality, we deduce that the $L^p(M)$-norm of the last term in (3.5) is

$$\le C \sup_{\mu\in[\lambda/2,2\lambda]} \|\chi_{[\mu,\mu+\delta]}\|_{L^2(M)\to L^p(M)} \times (\delta\lambda)^{-1}\|(\Delta+\lambda^2)f\|_{L^2(M)}.$$

From this and (3.6), we obtain (3.4), which completes the proof of Proposition 1.3.

4 Applications to breaking convexity bounds

In this section, we survey some recent results on breaking convexity bounds on mapping norms for $\chi_{[\lambda,\lambda+\varepsilon(\lambda)]}$ of the form (1.6). By Proposition 1.3, breaking such convexity bounds is equivalent to proving estimates

$$\|\psi_\lambda\|_{L^p(M)} \leq C\sqrt{\varepsilon(\lambda)}\lambda^{\sigma(p,n)}, \tag{4.1}$$

for quasi-modes satisfying

$$\|\psi_\lambda\|_{L^2(M)} \leq 1 \quad \text{and} \quad \|(\Delta+\lambda^2)\psi_\lambda\|_{L^2(M)} \leq \varepsilon(\lambda)\lambda, \tag{4.2}$$

assuming that $2 < p \leq \infty$ if $n \leq 3$, $2 < p < \infty$ if $n = 4$ and $2 < p \leq \frac{2n}{n-3}$ if $n \geq 5$. For the remaining exponents $p > 2$ if $n = 4$ or $n \geq 5$, we can obtain (4.1) if we also assume that

$$\lambda^{2-n(\frac{1}{2}-\frac{1}{p})}\|\chi_{[2\lambda,\infty)}\psi_\lambda\|_{L^p(M)} = O(1). \tag{4.3}$$

Note that the second condition in (4.2) is natural for (1.6) since it says that, in an L^2-sense, most of the spectrum of the ψ_λ are concentrated in intervals of size $O(\varepsilon(\lambda)\lambda)$ about λ.

4.1 Nondissipative focal points exist if convexity cannot be broken

It is proved in [17, Lemma 6.1] and [16, Theorem 2] that for generic (M,g) there exists a monotonically decreasing function $\varepsilon(\lambda) \to 0$ (the rate depending on (M,g)) so that (1.6) is valid. Thus, in this case, we also get (4.1) for $p = \infty$ and the above type of quasi-modes. However, $\varepsilon(\lambda)$ is not described explicitly.

If one *cannot* break the convexity bound on mapping norms, then one has

$$\|\chi_{[\lambda+o(1)]}\|_{L^2(M)\to L^\infty(M)} = \Omega(\lambda^{\frac{n-1}{2}}), \tag{4.4}$$

by which we mean that there is a uniform constant $c > 0$ so that

$$\limsup_{\lambda\to\infty} \lambda^{-\frac{n-1}{2}}\|\chi_{[\lambda,\lambda+\delta]}\|_{L^2(M)\to L^\infty(M)} \geq c, \tag{4.5}$$

whenever $\delta > 0$. In the real analytic case in [20] we found a necessary and sufficient condition for this. Specifically, we showed that (4.4) is valid if and only if our real analytic manifold (M,g) has a nondissipative self-focal point p. "Self-focal" means that there exists a time $T(p)$ so that, for all $\xi \in S_p^*M$ (the unit tangent space), the geodesic $\gamma_{p,\xi}(t)$ with initial data (p,ξ) returns to p at time $T(p)$. The geodesic is not assumed to smoothly close up (i.e. to be a closed geodesic), but just to loop back. The map $\Phi_p : S_p^*M \to S_p^*M$ taking the initial direction ξ to the terminal direction $\gamma'_{x,\xi}(T(p))$ is known as the first return map. It induces a unitary operator $U_p : L^2(S_p^*M) \to L^2(S_p^*M)$ defined by

$$U_p(f) = f(\Phi_p(\xi))\sqrt{J_p(\xi)},$$

where $\Phi_p^*|d\xi| = J_p(\xi)|d\xi|$ is the Jacobian with respect to the induced Riemannian volume density on S_p^*M. We say that a self-focal point is dissipative if U_p has no L^2 eigenfunction. Otherwise it is nondissipative.

To obtain (4.4) we showed that if there exists a nondissipative self-focal point x_0, then if $0 < \delta < 1$ we have

$$\limsup_{\lambda \to \infty} \lambda^{-(n-1)} \sum_{j=0}^{\infty} \rho(\delta^{-1}(\lambda - \lambda_j))|e_j(x_0)|^2 \geq c_0, \tag{4.6}$$

for some $c_0 > 0$. Here, $\rho \in \mathcal{S}(\mathbb{R})$ is any test function satisfying

$$\rho, \hat{\rho} \geq 0, \ \ \text{supp } \hat{\rho} \subset [-1, 1], \ \ \rho(s) \geq 1, \ \text{if } |s| \leq 1.$$

We prove (4.6) by using the von Neumann ergodic theorem for U_p together with arguments adapted from [7, 9, 10]. We obtain (4.5) from (4.6) due to the fact that

$$\|\chi_{[\lambda,\lambda+\delta]}\|^2_{L^2(M) \to L^\infty(M)} = \sup_{x \in M} \sum_{\lambda_j \in [\lambda, \lambda+\delta]} |e_j(x)|^2.$$

Conversely, if (4.5) is valid one of course gets that, for every $0 < \delta < 1$,

$$\limsup_{\lambda \to \infty} \Big(\sup_{x \in M} \lambda^{-(n-1)} \sum_{j=0}^{\infty} \rho(\delta^{-1}(\lambda - \lambda_j))|e_j(x)|^2 \Big) \geq c_0 > 0.$$

It follows from (4.4) and Proposition 1.3 that if our real analytic (M, g) has a nondissipative self-focal point then there is an $o(\lambda)$ sequence of quasi-modes satisfying

$$\|\psi_\lambda\|_{L^\infty(M)} = \Omega(\lambda^{\frac{n-1}{2}}). \tag{4.7}$$

We proved this in [20] by using a special case of Proposition 1.3.

4.2 Existence of a quasi-mode of order zero with maximal growth of L^∞ norms?

A natural question in [20] is whether (4.4) implies existence of a sequence of quasi-modes of order zero which saturate the L^∞ bounds. That is, if one has an L^2 eigenfunction of U_p at a self-focal point p, can one construct from it a quasi-mode of order zero with maximal sup-norm growth? This would definitely be the case if the L^2 eigenfunction were C^∞. In that case, we could use it as the symbol for a quasi-mode of order zero. In [16] such a construction was made in the special case of a self-focal point for which $\Phi_p = Id$, i.e. for which all of the loops at p were smoothly closed.

In view of Theorem 1.2, existence of a zeroth order quasi-mode of maximal sup norm growth would imply that

$$\limsup_{\lambda \to \infty} \lambda^{-\frac{n-1}{2}} \|\chi_{[\lambda, \lambda+1/\lambda]}\|_{L^2(M) \to L^\infty(M)} > 0. \tag{4.8}$$

4.3 Logarithmic shrinking of intervals for non-positively curved manifolds

In the case of non-positively curved manifolds we can be more specific about the shrinking rate of intervals and of L^p estimates.

Implicit in [1] is the fact that (1.6) is valid with $\varepsilon(\lambda) = 1/\ln(\lambda)$ if (M, g) has nonpositive sectional curvatures (see [15, §3.6]). This result was extended to all exponents $\frac{2(n+1)}{n-1} < p \le \infty$ in [8] (see also [5, §4]). Thus, for such manifolds we have

$$\|\psi_\lambda\|_{L^p(M)} \le C\lambda^{\sigma(p,n)}/\sqrt{\ln\lambda}, \quad \frac{2(n+1)}{n-1} < p \le \infty, \tag{4.9}$$

assuming that

$$\|\psi_\lambda\|_{L^2(M)} \le 1 \quad \text{and} \quad \|(\Delta+\lambda^2)\psi_\lambda\|_{L^2(M)} \le C\lambda/\ln\lambda,$$

as well as (4.3) if $n = 4$ or $n \ge 5$ and p is large.

The logarithmic estimate (4.9) reflects the hyperbolicity of the geodesic flow on negatively curved manifolds. It is obtained more generally in microlocal analysis of quasimodes around hyperbolic closed geodesics. It is plausible that one does not need global curvature assumptions but just the assumption that all closed geodesics of (M, g) are hyperbolic. In the case of *quantum integrable* Laplacians (such as a surface of revolution shaped like a peanut), it is possible to construct eigenfunctions which concentrate microlocally on a hyperbolic closed geodesic and whose L^p norms in tubes around the closed geodesic saturate the inequality (4.9). We refer to [23, §5.1] for further discussion and references.

It is possible to construct quasi-modes ψ_λ on negatively curved surfaces which partially concentrate on hyperbolic closed geodesics. A quasi-mode satisfying (4.2) with $\varepsilon(\lambda) = 1/\ln(\lambda)$ has been constructed recently by S. Brooks [4] on a compact hyperbolic surface whose microlocal lift (or defect measure) has mass $\ge \delta$ for a certain $\delta > 0$ along a closed geodesic γ. The estimate of (4.9) gives an upper bound on its L^p norms, and for $p = 6 + \varepsilon$ is precisely the order of magnitude of a quasi-mode associated to a hyperbolic geodesic. It would be interested to see if Brooks' quasi-mode saturates the bound (4.9).

4.4 $\varepsilon(\lambda) = \lambda^{-\varepsilon(p,n)}$ on a flat torus

In the case of the standard n-torus $\mathbb{T}^n = \mathbb{R}^n/\mathbb{Z}^n$, it was also shown in [5] that we have estimates of the form

$$\|\chi_{[\lambda,\lambda+\lambda^{-\varepsilon(p,n)}]}\|_{L^2(\mathbb{T}^n)\to L^p(\mathbb{T}^n)} \le \lambda^{\sigma(p,n)-\varepsilon(p,n)/2}, \quad \frac{2(n+1)}{n-1} < p \le \infty,$$

for certain powers $\varepsilon(p, n) > 0$. Therefore, by the above, we have

$$\|\psi_\lambda\|_{L^p(\mathbb{T}^n)} \le C\lambda^{\sigma(p,n)-\varepsilon(p,n)/2}, \quad \frac{2(n+1)}{n-1} < p \le \infty,$$

assuming that

$$\|\psi_\lambda\|_{L^2(\mathbb{T}^n)} \le 1 \quad \text{and} \quad \|(\Delta+\lambda^2)\psi_\lambda\|_{L^2(\mathbb{T}^n)} \le C\lambda^{1-\varepsilon(p,n)},$$

as well as (4.3) if p is large enough and $n \geq 4$.

References

[1] P. H. Bérard, On the wave equation on a compact manifold without conjugate points, Math. Z. **155** (1977), 249–276.

[2] M. Blair and C. D. Sogge, On Kakeya-Nikodym averages, L^p-norms and lower bounds for nodal sets of eigenfunctions in higher dimensions, J. European Math. Soc., to appear.

[3] J. Bourgain, Geodesic restrictions and L^p-estimates for eigenfunctions of Riemannian surfaces, Linear and complex analysis, Amer. Math. Soc. Transl. **226** (2009), Amer. Math. Soc., Providence, RI, 27–35.

[4] S. Brooks, Logarithmic-scale Quasimodes that do not Equidistribute, arXiv:1303.2484 (2013).

[5] J. Bourgain, P. Shao, C. D. Sogge, and X. Yao, On L^p-resolvent estimates and the density of eigenvalues for compact Riemannian manifolds, Comm. Math. Physics. **333** (2015), no. 3, 1483–1527.

[6] X. Chen and C. D. Sogge, A few endpoint geodesic restriction estimates for eigenfunctions, Comm. Math. Phys. **329** (2014), no. 2, 435–459.

[7] J. J. Duistermaat and V. W. Guillemin, The spectrum of positive elliptic operators and periodic bicharacteristics, Invent. Math. **29** (1975), 39–79.

[8] A. Hassell and M. Tacy, Improvement of eigenfunction estimates on manifolds of nonpositive curvature, Forum Math. **27** (2015), no. 3, 1435–1451.

[9] Y. G. Safarov, Asymptotics of a spectral function of a positive elliptic operator without a nontrapping condition, (Russian) Funktsional. Anal. i Prilozhen. **22** (1988), 53–65, 96; translation in Funct. Anal. Appl. **22** (1988), no. 3, 213–223 (1989).

[10] Y. G. Safarov and D. Vassiliev, The Asymptotic Distribution of Eigenvalues of Partial Differential Operators, translated from the Russian manuscript by the authors. Translations of Mathematical Monographs **155**, American Mathematical Society, Providence, RI, 1997.

[11] C. D. Sogge, Oscillatory integrals and spherical harmonics, Duke Math. J. **53** (1986), 43–65.

[12] C. D. Sogge, Concerning the L^p norm of spectral clusters for second-order elliptic operators on compact manifolds, J. Funct. Anal. **77** (1988), 123–138.

[13] C. D. Sogge, Fourier Integrals in Classical Analysis, Cambridge Tracts in Mathematics **105**, Cambridge University Press, Cambridge, 1993.

[14] C. D. Sogge, Kakeya-Nikodym averages and L^p-norms of eigenfunctions, Tohoku Math J. **63** (2011), 519–538.

[15] C. D. Sogge, Hangzhou Lectures on Eigenfunctions of the Laplacian, Annals of Math Studies **188**, Princeton University Press, 2014.

[16] C. D. Sogge, J. A. Toth, and S. Zelditch, About the blowup of quasimodes on Riemannian manifolds, J. Geom. Anal. **21** (2011), 150–173.

[17] C. D. Sogge and S. Zelditch, Riemannian manifolds with maximal eigenfunction growth, Duke Math. J. **114** (2002), 387–437.

[18] C. D. Sogge and S. Zelditch, Concerning the L^4 norms of typical eigenfunctions on compact surfaces, Recent Development in Geometry and Analysis, ALM 23, Int. Press, Beijing-Boston (2012), 407–423.

[19] C. D. Sogge and S. Zelditch, On eigenfunction restriction estimates and L^4-bounds for compact surfaces with nonpositive curvature, Advances in Analysis: The Legacy of Elias M. Stein, Princeton Mathematical Series, Princeton University Press, 2014, to appear.

[20] C. D. Sogge and S. Zelditch, Focal points and sup-norms of eigenfunctions, Rev. Mat. Iberoam, to appear.

[21] A. Weinstein, Asymptotics of eigenvalue clusters for the Laplacian plus a potential, Duke Math. J. **44** (1977), 883–892.

[22] S. Zelditch, CBMS Lectures, to appear.

[23] S. Zelditch, Recent developments in mathematical quantum chaos, Current developments in mathematics, 2009, 115–204, Int. Press, Somerville, MA, 2010.

[24] M. Zworski, Semiclassical Analysis, Graduate Studies in Mathematics **138**, Amer. Math. Soc., Providence, 2012.

Some Topics in Harmonic Analysis
and Applications
ALM 34, pp. 399–405

Astala's Conjecture from the Point of View of Singular Integrals on Metric Spaces

Alexander Volberg*†

Abstract

In the proof of Astala's conjecture on quasiconformal distortion obtained by Lacey-Sawyer-Uriarte-Tuero one of the key point is an estimate of the Ahlfors-Beurling operator in a certain weighted space. We show that the point of view of non-homogeneous Harmonic Analysis simplifies considerably this key point.

2010 Mathematics Subject Classification: 30C62, 35J15, 35J70, 42B20, 42B35.

Keywords and Phrases: Calderón-Zygmund operators on metric spaces, A_2 condition, weights, non-homogeneous harmonic analysis.

1 Introduction

Let us explain the setting. Let ϕ be a K-quasiconformal (QC) mapping, $0 < t < 2$. Astala [1] proved the following celebrated

$$\dim(E) = t \Rightarrow \dim(\phi(E)) \le t',$$

where

$$\frac{1}{t'} - \frac{1}{2} = \frac{1}{K}\left(\frac{1}{t} - \frac{1}{2}\right). \tag{1.1}$$

He asked whether for the borderline distortion one should have the absolute continuity of corresponding Hausdorff measures:

$$\mathcal{H}^t(E) = 0 \Rightarrow \mathcal{H}^{t'}(E) = 0\,? \tag{1.2}$$

*Supported by NSF grant DMS 0758552.

†Department of Mathematics, MSU, East Lansing, MI 48823, USA; *E-mail:* volberg@math.msu.edu

The answer is "yes", it was proved in [4], and one of the key point was the following theorem about weighted estimate of singular integral operators (SIO).

Theorem 1. *Let* $0 < d < 2$, *and* $\{Q_m\}_{m=1}^M$ *is a collection of dyadic squares on the plane such that it satisfy strong disjointness condition*

$$4Q_m \cap 4Q_{m'} = \emptyset\,,\ \forall m \neq m'\,, \tag{1.3}$$

and a packing condition, which says that for any dyadic square Q *of the same lattice*

$$\sum_{Q_m \subset Q} \ell(Q_m)^{2-d} \leq C\,\ell(Q)^{2-d}\,. \tag{1.4}$$

Consider measure $\mu := \sum_{m=1}^M \ell(Q_m)^{-d} \cdot m_2|Q_m$ *and the kernel* $t(x,y) := (x-y)^{-2}, x, y \in \cup_{m=1}^M Q_m$. *Then the operator with the kernel* $t(x,y)$ *is bounded from* $L^2(\mu)$ *to* $L^2(\mu)$ *in the sense that* ($X := \cup_{m=1}^M Q_m$)

$$\int_X \left| \int_X \frac{f(y)}{(x-y)^2}\, dm_2(y) \right| d\mu(x) \leq C_1 \int_X |f(x)|^2\, d\mu(x)\,. \tag{1.5}$$

Remarks. 1) The proof of this result in [4] is not too easy, and its mechanism is not obvious. Our goal in the present note is to clarify this mechanism.

2) We can write $d\mu = w\,dm_2$ with the obvious $w(x) := \sum_{m=1}^M \ell(Q_m)^{-d} \cdot \chi_{Q_m}$. It is very easy to check that

$$\frac{1}{|B|}\int_{B\cap X} w\, dm_2 \cdot \frac{1}{|B|}\int_{B\cap X} w^{-1}\, dm_2 \leq C_3 \tag{1.6}$$

independently of the disc B on the plane. This, of course, suggests that the theorem above is a particular case of a classical Hunt-Muckenhoupt-Wheeden weighted A_2 result. We warn the reader: this is **not so at all**. A very "tiny" difference is in integration over $B \cap X$ in (1.6). It cannot be replaced by the integration over B, which practically always equal to infinity for w^{-1}. This tiny difference is crucial. It gives the idea that we need to stick to X, namely, it seems at the first glance that we need the Hunt-Muckenhoupt-Wheeden weighted A_2 result on *metric space* X. However, at this moment there is no necessary and sufficient condition of the weighted type for singular integrals operators (SIO) on metric spaces. This problem is very close to a notoriously difficult two-weight problem for singular integrals. The difficulty is that we come into the realm of *non-homogeneous* Harmonic Analysis: meaning that neither $m_2|X$, nor $\mu|X$ is a doubling measure. There are partial A_2 type results in such situations, but only partial, see [6] for example.

3) This explains the subtlety of the proof in [4].

4) However, we show below that the point of view of non-homogeneous Harmonic Analysis on metric spaces is fruitful and allows us to simplify the proof of Theorem 1. But strangely enough, we need to look at it as **unweighted** non-homogeneous $T1$ theorem (in fact, it turns out that the mechanism is even simpler, but for that the reader should look at Section 2).

2 A simple proof of Theorem 1. The weighted estimate of Ahlfors-Beurling transform = unweighted estimate of a certain non-symmetric Calderón-Zygmund operator on a metric space

Let

$$\mu = \sum_i \ell(Q_i)^{-d} m_2 | Q_i \,.$$

Let us consider

$$K(x,y) := \begin{cases} 0, \text{ if } x, y \in \text{the same } Q_i, i = 1, \dots, M\,; \\ \dfrac{\ell(Q_i)^d}{(x-y)^2}\,, \text{ if } y \in Q_i, x \in Q_j, i \neq j. \end{cases}$$

It is obvious that the boundedness of this operator in $L^2(\mu)$ is exactly equivalent to proving Theorem 1.

We notice that it is enough to prove the boundedness for the formal adjoint operator T'_μ, whose kernel is

$$K'(x,y) := \begin{cases} 0, \text{ if } x, y \in \text{the same } Q_i, i = 1, \dots, M\,; \\ \dfrac{\ell(Q_j)^d}{(x-y)^2}\,, \text{ if } y \in Q_i, x \in Q_j, i \neq j\,. \end{cases}$$

Consider any metric space with any measure (we do not need even (X, d) to be a geometrically doubling metric space), and let

$$M_{\mu,3} f(x) := \sup_{R>0} \frac{1}{\mu(B(x,3R))} \int_{B(x,R)} |f(y)|\, d\mu(y)\,.$$

This maximal operator was widely used in [5]. It is an immediate consequence of Vitali's covering lemma that it is bounded in $L^2(\mu)$, no matter what is μ.

Now we have Theorem 1 as an immediate corollary of

Lemma 2. *The operator $f \to \int_X K'(x,y) f(y)\, d\mu(y)$ is pointwise majorized by $C\, M_{\mu,3} f(x)$.*

Proof. Let $f \geq 0$. Fix Q_j, its center c_j, $x \in Q_j$, and let us estimate

$$|T'_\mu f(x)| \leq \int_X |K'(x,y)| f(y)\, d\mu(y)\,.$$

Consider the centers c_i of squares Q_i such that $c_i \in 2^{a+1} Q_j \setminus 2^a Q_j$. We call this family of centers (and their squares too) family F_a. Notice that F_a can be non-empty only if $a \geq 2$ (see (1.3)).

If $x \in Q_j, y \in Q_i, Q_i \in F_a$ then

$$|K'(x,y)| \leq \frac{\ell_j^d}{|c_j - c_i|^2}\,, \text{where } \ell_j := \ell(Q_j)\,.$$

Therefore,

$$\begin{aligned}|T'_\mu f(x)| &\le \sum_{a=2}^{\infty}\Big(\sum_{c_i\in F_a}\frac{1}{|c_i-c_j|^2}\int_{Q_i} f\,d\mu\Big)\ell_j^d\\ &\le \sum_{a=2}^{\infty}2^{-ad}\frac{(2^a\ell_j)^d}{(2^a\ell_j)^2}\int_{\cup_{Q_i\in F_a}Q_i} f\,d\mu\\ &\le \sum_{a=2}^{\infty}2^{-ad}\frac{1}{(2^a\ell_j)^{2-d}}\int_{B(x,8\cdot 2^{a+1}\ell_j)} f\,d\mu := I\,.\end{aligned}$$

The last inequality holds because each square from the family F_a lie in the disc centered at $x \in Q_j$ of radius $R_a := 8\cdot 2^{a+1}\ell_j$. This is again obvious from (1.3). We continue:

$$\begin{aligned}I &\le C_1\sum_{a=2}^{\infty}2^{-ad}\frac{1}{(24\cdot 2^{a+1}\ell_j)^{2-d}}\int_{B(x,8\cdot 2^{a+1}\ell_j)} f\,d\mu\\ &\le C_2\sum_{a=2}^{\infty}2^{-ad}\frac{1}{\mu(B(x,3R_a))}\int_{B(x,R_a)} f\,d\mu \le C_3\,M_{\mu,3}f(x)\,.\end{aligned}$$

The last inequality is because

$$\mu(B(x,24\cdot 2^{a+1}\ell_j)) \le C_4\,(24\cdot 2^{a+1}\ell_j)^{2-d}\,. \tag{2.1}$$

In its turn this is easy: the disc $B(x,24\cdot 2^{a+1}\ell_j)$ is covered by a fixed number of dyadic squares of comparable size. Apply the packing condition (1.4) to each of these dyadic squares, and see (2.1) immediately. Theorem 1 is proved. □

3 $T1$ theorem for non-homogeneous metric measure spaces

The current section is not needed for the proof of Theorem 1 because the proof has been already given in the previous section.

Still we decided to include it here to explain the connection to other related questions. Also this section serves as a conceptual explanation of what is going on in Theorem 1. For the experts we want to emphasize a ceratin affinity to the problem considered in Volberg-Wick's paper [7]. But the situation in [7] is actually more singular so-to-speak.

Let us work in this section only with so-called geometrically doubling metric spaces (GDMS) (X,d) meaning that each ball of radius r can be covered by at most fixed number of balls of radius $r/2$. This guarantees (by Konyagin-Volberg's theorem [3]) that X is a metric measure space with measure σ, $\operatorname{supp}\sigma = X$, such that σ is a doubling measure ((X,d,σ) is sometimes called "homogeneous metric-measure space"). However, we need a non-homogeneous metric-measure space with the same underlying (X,d): namely, (X,d,μ), where μ is **not necessarily** doubling.

Given such (X, d, μ) with an extra condition

$$\mu(B(x,r)) \leq r^s \tag{3.1}$$

we can consider singular kernel of singularity s of Calderón-Zygmund type on X. It is $K(x,y), x, y \in X$ such that

I) $|K(x,y)| \leq d(x,y)^{-s}$.

II) $|K(x,y) - K(x',y)| \leq \frac{d(x,x')^\varepsilon}{d(x,y)^{s+\varepsilon}}$ for some $\varepsilon > 0$ and all x, x', y such that $d(x,x') \leq \frac{1}{2}d(x,y)$.

III) $|K(x,y) - K(x,y')| \leq \frac{d(x,x')^\varepsilon}{d(x,y)^{s+\varepsilon}}$ for the same $\varepsilon > 0$ and all y, y', x such that $d(y,y') \leq \frac{1}{2}d(x,y)$.

We can consider the (formal) operator T_μ.

$$f \in L^2(\mu) \to \int_X K(x,y)f(y)d\mu(y)\,.$$

Given all this, we have the following theorem called non-homogeneous $T1$ theorem. It was proved by Nazarov-Treil-Volberg if X is a Euclidean spaces, but the same proof can be adapted for GDMS (and actually this has been done by Hytönen-Martikainen [2]). Notice that the kernel actually can be allowed to be considerably worse (bigger) depending on μ than we list in I), II), III). See, for example, [2, 6]. But we do not need this here.

In the next result we assume that formal operator T_μ and its formal adjoint T'_μ can be correctly defined on characteristic functions of balls.

Theorem 3. *Operator T_μ with kernel K is bounded (can be extended to be a bounded operator from finite linear combinations of characteristic functions of balls) in $L^2(\mu)$ if and only if*

(i) $\|T_\mu \chi_B\|_\mu^2 \leq C\,\mu(B)$,

(ii) $\|T_\mu \chi_B\|_\mu^2 \leq C\,\mu(B)$,

where T'_μ is a (formal) operator with kernel $K(y,x)$.

Let us look at Lacey-Sawyer-Uriarte-Tuero's theorem 1 from the point of view of this theorem. The metric space is $X = \cup_{m=1}^M Q_m$, the metric d is the usual Euclidean metric, $\mu = \sum_m \ell(Q)^{-d} \cdot m_2|Q_m$. What is $K(x,y)$? Of course it is not $t(x,y) = (x-y)^{-2}$.

Let us consider

$$K(x,y) := \begin{cases} 0, & \text{if } x, y \in \text{the same } Q_i, i = 1, \ldots, M\,; \\ \dfrac{\ell(Q_i)^d}{(x-y)^2}\,, & \text{if } y \in Q_i, x \in Q_j, i \neq j. \end{cases}$$

It is really a simple calculation to see

Lemma 4. *Let $\tau \in (0,1)$ be an arbitrary number, $s := 2 - d$, $\varepsilon := \min(1, \tau d)$. Then thus defined K is a Calderón-Zygmund kernel on X with singularity s and Calderón-Zygmund parameter ε.*

Proof. It is trivial to see I) and II) by using disjointness condition (1.3). To check III) one consider cases:

III.1) y, y' are in the same square, x is in a different square. This is easy again by (1.3). If x joins y, y' in their square there is nothing to prove as K becomes zero.

III.2) x, y are in the same square but y' is in a different one. Then (1.3) and $|y-y'| \le \frac{1}{2}|x-y|$ would imply that such situation is impossible. If x, y' are in the same square but y is in a different one. Then again (1.3) and $|y-y'| \le \frac{1}{2}|x-y|$ would imply that such situation is impossible.

III.3) x, y, y' are all in different squares. Let $y \in Q_i$, $y' \in Q_{i'}$. As $|y-y'| \le \frac{1}{2}|x-y|$ and $\ell(Q_i) \le |y-y'|$, $\ell(Q_{i'}) \le |y-y'|$ (by (1.3)), we should think that squares $Q_i, Q_{i'}$ are small with respect to $|x-y|$. Here we do not estimate the difference, we estimate $|K(x,y)|$ and $|K(x,y')|$ separately:

$$
\begin{aligned}
|K(x,y) - K(x,y')| \le |K(x,y)| + |K(x,y')| &\le \frac{\ell(Q_i)^d + \ell(Q_{i'}^d}{|x-y|^2} \\
&\le C\frac{|y-y'|^{\tau d}|y-y'|^{(1-\tau)d}}{|x-y|^2} \le C\frac{|y-y'|^{\tau d}}{|x-y|^{2-d+\tau d}} \\
&= C\frac{|y-y'|^{\tau d}}{|x-y|^{s+\tau d}}.
\end{aligned}
$$

□

Remark. The integral operator with kernel $\int \frac{1}{(x-y)^2} \cdots dm_2(y)$ from Theorem 1 is exactly the sum of $\int K(x,y) \cdots d\mu(y)$ (we call attention of the reader to the change of measure!) plus the local operator $\int t_0(x,y) \cdots dm_2(y)$, where

$$
t_0(x,y) := \sum_{m=1}^{M} \frac{\chi_{Q_m}(x)\chi_{Q_m}(y)}{(x-y)^2}.
$$

The boundedness of t_0 in each $L^2(Q_m, \ell(Q_m)^{-d}\, dm_2)$ (and, thus, in the direct sum of these spaces, which is precisely $L^2(X,\mu)$) is obvious as the constant $\ell(Q_m)^{-d}$ just cancels out in norm estimate in the left and right hand sides, and operator with kernel $\frac{1}{(x-y)^2}$ is a classical (called Ahlfors–Beurling) operator, which is an isometry in $L^2(m_2)$.

So, in principle, we could have checked assumptions (i), (ii) of our $T1$ theorem 3 and then Theorem 1 would follow.

However, there is an even simpler way to see that the integral operator

$$
\int K(x,y) \cdots d\mu(y)
$$

is bounded in $L^2(\mu)$. This way was shown in Section 2.

Acknowledgements

The paper was partially supported by NSF grant DMS 1265549. The author is also thankful to the Hausdorff Institute for Mathematics, Bonn, Germany.

References

[1] K. Astala, Area distortion of quasiconformal mappings, Acta Math. **173** (1994), no. 1, 37–60.

[2] T. Hytönen and H. Martikainen, Non-homogeneous Tb theorem and random dyadic cubes on metric measure spaces, arXiv: 0911.4387v1, pp. 1–34.

[3] S. Konyagin and A. Volberg, On measures with the doubling condition, Izv. Akad. Nauk SSR, Ser. Mat. **51** (1987), no. 3, 666–675; translation in Math. USSR-Izv. **30** (1988), no. 3, 629–638.

[4] M. Lacey, E. Sawyer, and I. Uriarte-Tuero, Astala's conjecture on distortion of Hausdorff measures under quasiconformal maps in the plane, arXiv: 0805.4711v3, pp. 1–17.

[5] F. Nazarov, S. Treil, and A. Volberg, Cauchy integral and Calderón-Zygmund operators on nonhomogeneous spaces, IMRN, International Mathematics Research Notices (1997), no. 15, pp. 703–726.

[6] F. Nazarov, A. Reznikov, and A. Volberg, The proof of A_2 conjecture in a geometrically doubling metric space, preprint, 2011, pp. 1–24.

[7] A. Volberg and B. D. Wick, Bergman-type singular operators and the characterization of Carleson measures for Besov-Sobolev spaces on the complex ball, arXiv:0910.1142, to appear in Amer. J. Math.

Some Topics in Harmonic Analysis and Applications
ALM 34, pp. 407–419

C.S.I. for Besov Spaces $\dot{\Lambda}^{p,q}_{\alpha}(\mathbb{R}^n)$ with $\big(\alpha,(p,q)\big)\in(0,1)\times\big((0,1]\times(0,1]\setminus\{(1,1)\}\big)$

Jie Xiao* and Zhichun Zhai†

Abstract

Capacitary Strong-type Inequalities (C.S.I.) for the homogeneous Besov spaces $\dot{\Lambda}^{p,q}_{\alpha}(\mathbb{R}^n)$ with $\big(\alpha,(p,q)\big)\in(0,1)\times\big((0,1]\times(0,1]\setminus\{(1,1)\}\big)$ are established. Consequently, these inequalities are applied to an investigation of the embedding of a homogeneous Besov or fractional Sobolev space in its associated Lorentz or Lebesgue space on $\mathbb{R}^n$ or $\mathbb{R}^{1+n}_+$ via the fractional heat equation.

2000 Mathematics Subject Classification: 46E30, 47B38.

Keywords and Phrases: C.S.I., homogeneous Besov spaces/capacities, Lorentz spaces, embeddings.

1 Introduction

Capacitary Strong-type Inequalities (C.S.I.) for the homogeneous Besov spaces have ever been studied a lot and played an important role in the embedding theory of function spaces, partial differential equations and related areas. We refer the readers to [1, 5, 6, 8, 9, 13, 15] and the references therein.

Our main purpose here is to verify the following C.S.I. for the homogeneous Besov spaces/capacities $\dot{\Lambda}^{p,q}_{\alpha}(\mathbb{R}^n)/\ cap(\cdot;\dot{\Lambda}^{p,q}_{\alpha}(\mathbb{R}^n))$ with $\big(\alpha,(p,q)\big)\in(0,1)\times\big((0,1]\times(0,1]\setminus\{(1,1)\}\big)$:

$$\begin{aligned}&\int_0^\infty\Big(cap\Big(\{x\in\mathbb{R}^n:|f(x)|>t\};\dot{\Lambda}^{p,q}_{\alpha}(\mathbb{R}^n)\Big)\Big)^{1\vee\frac{q}{p}}\,dt^{p\vee q}\\&\lesssim\|f\|^{p\vee q}_{\dot{\Lambda}^{p,q}_{\alpha}(\mathbb{R}^n)},\quad\forall\ f\in\dot{\Lambda}^{p,q}_{\alpha}(\mathbb{R}^n).\end{aligned}\tag{1.1}$$

*Dept. of Math. and Stat., Memorial Univ., St. John's, NL A1C5S7, Canada; *E-mail:* jxiao@mun.ca

†Dept. of Math. and Stat., MacEwan Univ., Edmonton, AB T5J4S2, Canada; *E-mail:* zhaiz2@macewan.ca; zhichun1@ualberta.ca

Here and henceforth, $a \vee b = \max\{a, b\}$ and $a \wedge b = \min\{a, b\}$; $A \lesssim B$ means $A \le CB$ for some positive constant C and $A \approx B$ is used for both $A \lesssim B$ and $B \lesssim A$. The authors of [1, 5, 9, 13] proved (1.1) for $\alpha p < n$ and $(p, q) \in (1, \infty) \times (1, \infty)$. In [15], the first-named author of this paper established the endpoint case $(p, q) = (1, 1)$ and $\alpha \in (0, n)$ of (1.1). So, upon establishing (1.1) under $\big(\alpha, (p, q)\big) \in (0, 1) \times \big((0, 1] \times (0, 1] \setminus \{(1, 1)\}\big)$, we can see a relatively complete picture of the strong-type inequalities for the homogeneous Besov spaces/capacities. For two applications of (1.1), we consider the embedding of a homogeneous Besov or fractional Sobolev space in a Lorentz or Lebesgue space on $\mathbb{R}^n$ or $\mathbb{R}^{1+n}_+ := (0, \infty) \times \mathbb{R}^n$ via the fractional heat equation. More precisely, we not only decide the non-negative Radon measure $\mu(x)$ on $\mathbb{R}^n$ such that $\dot{\Lambda}^{p,q}_\alpha(\mathbb{R}^n)$ embeds in the Lorentz space $L^{p\wedge q, p\vee q}(\mathbb{R}^n, \mu)$, but also characterize the non-negative Radon measure $\mu(t, x)$ on $\mathbb{R}^{1+n}_+$ satisfying

$$\left(\int_{\mathbb{R}^{1+n}_+} \left|S_\beta(t^{2\beta}) f(x)\right|^p \, d\mu(t, x)\right)^{1/p} \lesssim \|f\|_{\dot{\Lambda}^{p,p}_\alpha(\mathbb{R}^n)}, \ \forall\, f \in \dot{\Lambda}^{p,p}_\alpha(\mathbb{R}^n).$$

Here $S_\beta(t) f(x)$ is the solution of the fractional heat equation (obtained by replacing the positive Laplacian $-\triangle$ in the classic heat equation by the fractional Laplacian $(-\triangle)^\beta$ with the initial data $f(x)$.

Before proceeding to the forthcoming two sections, let us agree to some standard notations concerning the homogeneous Besov spaces and their capacities. For $p, q \in (0, \infty)$ and $\alpha \in (0, \infty)$, $\dot{\Lambda}^{p,q}_\alpha(\mathbb{R}^n)$ is the homogeneous Besov space which is defined as the completion of all functions $f \in C^\infty_0(\mathbb{R}^n)$ (infinitely smooth functions with compact support) with the semi-norm

$$\|f\|_{\dot{\Lambda}^{p,q}_\alpha(\mathbb{R}^n)} = \left(\int_{\mathbb{R}^n} \|\triangle^k_h f\|^q_{L^p(\mathbb{R}^n)} |h|^{-(n+\alpha q)} \, dh\right)^{1/q} < \infty,$$

where $\|\cdot\|_{L^p(\mathbb{R}^n)}$ is the usual L^p-norm, k is an integer greater than α, and $\triangle^k_h$ denotes the k-difference: $\triangle^k_h = \triangle^1_h \triangle^{k-1}_h$; $\triangle^1_h f(x) = f(x+h) - f(x)$. In particular, if $\alpha \in (0, 1)$, then

$$\|f\|_{\dot{\Lambda}^{p,q}_\alpha(\mathbb{R}^n)} = \left(\int_{\mathbb{R}^n} \left(\int_{\mathbb{R}^n} |f(x+h) - f(x)|^p dx\right)^{q/p} |h|^{-(n+\alpha q)} dh\right)^{1/q}.$$

It is worth pointing out that $\dot{\Lambda}^{p,p}_\alpha(\mathbb{R}^n)$ is just the homogeneous Sobolev space $\dot{W}^{\alpha,p}(\mathbb{R}^n)$ with the fractional order α.

Meanwhile, $cap(\cdot, \dot{\Lambda}^{p,q}_\alpha(\mathbb{R}^n))$ is the homogeneous Besov capacity: for a compact set $K \subseteq \mathbb{R}^n$, one defines

$$cap(K; \dot{\Lambda}^{p,q}_\alpha(\mathbb{R}^n)) = \inf\left\{\|f\|^p_{\dot{\Lambda}^{p,q}_\alpha(\mathbb{R}^n)} : \ f \in C^\infty_0(\mathbb{R}^n) \ \& \ f \ge 1 \ \text{on} \ K\right\}.$$

For any set $E \subseteq \mathbb{R}^n$, one puts

$$cap(E; \dot{\Lambda}^{p,q}_\alpha(\mathbb{R}^n)) = \inf_{open\ O \supseteq E} \ \sup_{compact\ K \subseteq O} cap(K; \dot{\Lambda}^{p,q}_\alpha(\mathbb{R}^n)).$$

2 C.S.I.

From the definition of $cap(\cdot;\dot{\Lambda}^{p,q}_{\alpha}(\mathbb{R}^n))$ it readily follows that

$$cap\left(\{x\in\mathbb{R}^n : |f(x)|>t\};\dot{\Lambda}^{p,q}_{\alpha}(\mathbb{R}^n)\right) \lesssim t^{-p}\|f\|^p_{\dot{\Lambda}^{p,q}_{\alpha}(\mathbb{R}^n)},\ \forall\, f\in\dot{\Lambda}^{p,q}_{\alpha}(\mathbb{R}^n).$$

Such an inequality is usually referred to as a weak-type estimate, and can be actually strengthened through the following form.

Theorem 2.1. *Let* $p,q\in(0,1]$, $(p,q)\neq(1,1)$ *and* $\alpha\in(0,1)$. *Then* (1.1) *holds.*

Proof. In what follows, let us assume $s=1\vee\frac{q}{p}$. Let $\mathcal{G}$ be a C^∞-smooth function with

$$\mathcal{G}(t)=\begin{cases}0, & \forall\ t\le 0;\\ 1, & \forall\ t\ge 1.\end{cases}$$

For each integer j, define $\mathcal{G}_j(t)=2^j\mathcal{G}(2^{2-j}t-1)$. Note that for any $f\in C_0^\infty(\mathbb{R}^n)$ one has $\mathcal{G}_j(|f|)\in C_0^\infty(\mathbb{R}^n)$ and $2^{-j}\mathcal{G}_j(|f|)=1$ on the set $E_j=\{x\in\mathbb{R}^n : |f(x)|> 2^{j-1}\}$. According to the definition of $cap(\cdot;\dot{\Lambda}^{p,q}_{\alpha}(\mathbb{R}^n))$, one has $cap(E_j;\dot{\Lambda}^{p,q}_{\alpha}(\mathbb{R}^n))\le 2^{-ip}\|\mathcal{G}_j(|f|)\|^p_{\dot{\Lambda}^{p,q}_{\alpha}(\mathbb{R}^n)}$ and

$$\begin{aligned}
&\mathsf{I}(f;\alpha,p,q,s)\\
&:=\int_0^\infty\left(cap\left(\{x\in\mathbb{R}^n:|f(x)|>t\};\dot{\Lambda}^{p,q}_{\alpha}(\mathbb{R}^n)\right)\right)^s dt^{sp}\\
&\approx\sum_{j=-\infty}^{\infty}\left(cap\left(E_j;\dot{\Lambda}^{p,q}_{\alpha}(\mathbb{R}^n)\right)2^{pj}\right)^s\\
&\lesssim\sum_{j=-\infty}^{\infty}\|\mathcal{G}_j(|f|)\|^{ps}_{\dot{\Lambda}^{p,q}_{\alpha}(\mathbb{R}^n)}\\
&\approx\sum_{j=-\infty}^{\infty}\left(\int_{\mathbb{R}^n}\left(\int_{\mathbb{R}^n}|\mathcal{G}_j(|f|)(x+h)-\mathcal{G}_j(|f|)(x)|^p dx\right)^{q/p}|h|^{-(n+\alpha q)}dh\right)^{sp/q}.
\end{aligned}$$

It follows from the boundedness of $\mathcal{G}_j'(t)$ and mean-value theorem that (see [13])

$$\sum_{j=-\infty}^{\infty}|\mathcal{G}_j(|f|)(x+h)-\mathcal{G}_j(|f|)(x)|^p\lesssim|f(x+h)-f(x)|^p. \tag{2.1}$$

Two cases are taken into account.

Case 1: $0<p\le q\le 1$. In this case, $s=q/p$. It follows from the above

estimates for $\mathsf{I}(f;\alpha,p,q,s)$ and (2.1) that

$$\begin{aligned}
&\mathsf{I}(f;\alpha,p,q,s)\\
&\lesssim \sum_{j=-\infty}^{\infty}\left(\int_{\mathbb{R}^n}\left(\int_{\mathbb{R}^n}|\mathcal{G}_j(|f|)(x+h)-\mathcal{G}_j(|f|)(x)|^p dx\right)^{q/p}|h|^{-(n+\alpha q)}dh\right)^{sp/q}\\
&=\sum_{j=-\infty}^{\infty}\int_{\mathbb{R}^n}\left(\int_{\mathbb{R}^n}|\mathcal{G}_j(|f|)(x+h)-\mathcal{G}_j(|f|)(x)|^p dx\right)^{q/p}|h|^{-(n+\alpha q)}dh\\
&=\int_{\mathbb{R}^n}\sum_{j=-\infty}^{\infty}\left(\int_{\mathbb{R}^n}|\mathcal{G}_j(|f|)(x+h)-\mathcal{G}_j(|f|)(x)|^p dx\right)^{q/p}|h|^{-(n+\alpha q)}dh\\
&\lesssim \int_{\mathbb{R}^n}\left(\int_{\mathbb{R}^n}\sum_{j=-\infty}^{\infty}|\mathcal{G}_j(|f|)(x+h)-\mathcal{G}_j(|f|)(x)|^p dx\right)^{q/p}|h|^{-(n+\alpha q)}dh\\
&\lesssim \int_{\mathbb{R}^n}\left(\int_{\mathbb{R}^n}|f(x+h)-f(x)|^p dx\right)^{q/p}|h|^{-(n+\alpha q)}dh\\
&\approx \|f\|^{sp}_{\dot{\Lambda}^{p,q}_{\alpha}(\mathbb{R}^n)},
\end{aligned}$$

which implies (1.1).

Case 2: $0<q<p\le 1$. In this case, $s=1$. Then a combination of (2.1) and the Minkowski inequality derives

$$\begin{aligned}
&(\mathsf{I}(f;\alpha,p,q,s))^{q/p}\\
&\lesssim \left(\sum_{j=-\infty}^{\infty}\left(\int_{\mathbb{R}^n}\left(\int_{\mathbb{R}^n}|\mathcal{G}_j(|f|)(x+h)-\mathcal{G}_j(|f|)(x)|^p dx\right)^{q/p}|h|^{-(n+\alpha q)}dh\right)^{p/q}\right)^{q/p}\\
&\lesssim \int_{\mathbb{R}^n}\left(\sum_{j=-\infty}^{\infty}\int_{\mathbb{R}^n}|\mathcal{G}_j(|f|)(x+h)-\mathcal{G}_j(|f|)(x)|^p dx\right)^{q/p}|h|^{-(n+\alpha q)}dh\\
&\lesssim \int_{\mathbb{R}^n}\left(\int_{\mathbb{R}^n}\sum_{j=-\infty}^{\infty}|\mathcal{G}_j(|f|)(x+h)-\mathcal{G}_j(|f|)(x)|^p dx\right)^{q/p}|h|^{-(n-\alpha q)}dh\\
&\lesssim \int_{\mathbb{R}^n}\left(\int_{\mathbb{R}^n}|f(x+h)-f(x)|^p dx\right)^{q/p}|h|^{-(n+\alpha q)}dh\\
&\approx \|f\|^{sq}_{\dot{\Lambda}^{p,q}_{\alpha}(\mathbb{R}^n)},
\end{aligned}$$

and so (1.1). □

As an equivalent form of Theorem 2.1, the following corollary is true for the Netrusov capacity $H^{(\varepsilon)}_{d,\theta}(\cdot)$ with $(\varepsilon,d,\theta)\in(0,\infty]\times(0,\infty)\times(0,\infty)$ and has a root in [2] and [15]. Here $H_{d,\theta}^{(\varepsilon)}(\cdot)$ is determined via

$$H_{d,\theta}^{(\varepsilon)}(K) = \inf\left(\sum_{i=0}^{\infty}(m_i 2^{-id})^\theta\right)^{1/\theta},$$

where the infimum is taken over all countable coverings of $K \subseteq \mathbb{R}^n$ by balls whose radii r_j do not exceed ε, while m_i is the number of balls from this covering whose radii r_j belong to the interval $(2^{-i-1}, 2^{-i}]$, $i = 0, 1, 2, \ldots$. When $\theta = 1$ and $\varepsilon = \infty$, the Netrusov capacity $H_{d,1}^{(\infty)}(\cdot)$ coincides with the classical Hausdorff capacity $H_d^{(\infty)}(\cdot)$.

Corollary 2.2. *Let $p, q \in (0,1]$, $(p,q) \neq (1,1)$ and $\alpha \in (0,1)$. Then*

$$\int_0^\infty \left(H_{n-\alpha p, q/p}^{(\infty)}(\{x \in \mathbb{R}^n : |f(x)| > t\})\right)^{1\vee\frac{q}{p}} dt^{p\vee q} \lesssim \|f\|_{\dot{\Lambda}_\alpha^{p,q}(\mathbb{R}^n)}^{p\vee q}, \quad \forall\ f \in \dot{\Lambda}_\alpha^{p,q}(\mathbb{R}^n). \tag{2.2}$$

Proof. It follows from Theorem 2.1 and

$$cap(\cdot; \dot{\Lambda}_\alpha^{p,q}(\mathbb{R}^n)) \approx H_{n-\alpha p, q/p}^{(\infty)}(\cdot) \tag{2.3}$$

which can be seen from Netrusov [10, 11] and Adam [4]. □

From now on, $\mathcal{M}f$ represents the usual Hardy-Littlewood maximal operator of f defined as

$$\mathcal{M}f(x) = \sup_{r>0} r^{-n} \fint_{B(x,r)} |f(y)|dy, \quad \forall\ x \in \mathbb{R}^n,$$

where the barred integral denotes the usual integral average over the Euclidean ball $B(x,r)$ centered at x with radius r.

The following C.S.I. for the fractional Sobolev space $\dot{\Lambda}_\alpha^{p,p}(\mathbb{R}^n) = \dot{W}^{\alpha,p}(\mathbb{R}^n)$ will be used in Section 3.

Proposition 2.3. *Let $\frac{n}{n+\alpha} < p < 1$ and $\alpha \in (0,1)$. Then*

$$\int_0^\infty cap\left(\{x \in \mathbb{R}^n : |\mathcal{M}f(x)| > t\}; \dot{\Lambda}_\alpha^{p,p}(\mathbb{R}^n)\right) dt^p \lesssim \|f\|_{\dot{\Lambda}_\alpha^{p,p}(\mathbb{R}^n)}^p, \quad \forall\ f \in \dot{\Lambda}_\alpha^{p,p}(\mathbb{R}^n). \tag{2.4}$$

Proof. Suppose $f \in \dot{\Lambda}_\alpha^{p,p}(\mathbb{R}^n)$. It follows from [3, Theorem 7 (a)(i)] or [12, Theorem (i)] that for $\frac{n}{n+\alpha} < p < 1$ and $\alpha \in (0,1)$ one has

$$\int_0^\infty H_{n-\alpha p}^{(\infty)}(\{x \in \mathbb{R}^n : |\mathcal{M}f(x)| > t\})\, dt^p \lesssim \int_0^\infty H_{n-\alpha p}^{(\infty)}(\{x \in \mathbb{R}^n : |f(x)| > t\})\, dt^p.$$

This, along with (2.3), implies

$$\begin{aligned}&\int_0^\infty cap\left(\{x\in\mathbb{R}^n : |\mathcal{M}f(x)|>t\};\dot{\Lambda}^{p,p}_\alpha(\mathbb{R}^n)\right)dt^p\\ \approx&\int_0^\infty H^{(\infty)}_{n-\alpha p}\left(\{x\in\mathbb{R}^n : |\mathcal{M}f(x)|>t\}\right)dt^p\\ \lesssim&\int_0^\infty cap\left(\{x\in\mathbb{R}^n : |f(x)|>t\};\dot{\Lambda}^{p,p}_\alpha(\mathbb{R}^n)\right)dt^p\\ \lesssim&\|f\|^p_{\dot{\Lambda}^{p,p}_\alpha(\mathbb{R}^n)}\end{aligned}$$

and hence (2.4). □

3 Applications

For a non-negative Radon measure μ on $X=\mathbb{R}^n$ or $\mathbb{R}^{1+n}_+$, the symbols $L^{q,p}(X,\mu)$ and $L^q(X,\mu)$ denote the Lorentz space and the Lebesgue space of all functions f on X for which

$$\|f\|_{L^{q,p}(X,\mu)}=\left(\int_0^\infty(\mu(\{x\in X:|f(x)|>\lambda\}))^{p/q}d\lambda^p\right)^{1/p}<\infty$$

and

$$\|f\|_{L^q(X,\mu)}=\left(\int_X|f(x)|^q d\mu\right)^{1/q}<\infty$$

respectively. Moreover, we use $L^{q,\infty}(X,\mu)$ as the set of all μ-measurable functions f on X with

$$\|f\|_{L^{q,\infty}(X,\mu)}=\sup_{\lambda>0}\lambda(\mu(\{x\in X:|f(x)|>\lambda\}))^{1/q}<\infty.$$

Next, we write $\zeta^\alpha_p(\mu;t)$ for the (p,α)-variational capacity minimizing function associated with $\dot{\Lambda}^{p,p}_\alpha(\mathbb{R}^n)$ and μ on $\mathbb{R}^n$:

$$\zeta^\alpha_p(\mu;t)=\inf\{cap(O;\dot{\Lambda}^{p,p}_\alpha(\mathbb{R}^n)) : \text{bouded open } O\subseteq\mathbb{R}^n \ \ \& \ \ \mu(O)>t\}.$$

Similarly, $\eta^\alpha_p(\mu;t)$ is the (p,α)-variational capacity minimizing function associated with $\dot{\Lambda}^{p,p}_\alpha(\mathbb{R}^n)$ and μ on $\mathbb{R}^{1+n}_+$ defined by

$$\eta^\alpha_p(\mu;t)=\inf\{cap(O;\dot{\Lambda}^{p,p}_\alpha(\mathbb{R}^n)) : \text{bouded open } O\subseteq\mathbb{R}^n \ \ \& \ \ \mu(T(O))>t\},$$

where $T(O)$ is the tent based on O :

$$T(O)=\{(r,x)\in\mathbb{R}^{1+n}_+ : B(x,r)\subseteq O\}.$$

3.1 Embedding in Lorentz space on $\mathbb{R}^n$

As the first application of C.S.I. established in Section 2, we work out the structure of a non-negative Radon measure such that its induced Lorentz space contains a homogeneous Besov space.

Theorem 3.1. *Let μ be a non-negative Radon measure on $\mathbb{R}^n$ and $(\alpha,(p,q))\in(0,1)\times((0,1]\times(0,1]\setminus\ \{(1,1)\})$. Then*

$$\|f\|_{L^{p\wedge q,p\vee q}(\mathbb{R}^n,\mu)}\lesssim\|f\|_{\dot{\Lambda}^{p,q}_\alpha(\mathbb{R}^n)},\ \forall\ f\in\dot{\Lambda}^{p,q}_\alpha(\mathbb{R}^n) \tag{3.1}$$

if and only if

$$\mu(K)\lesssim\big(cap(K;\dot{\Lambda}^{p,q}_\alpha(\mathbb{R}^n))\big)^{\frac{p\wedge q}{p}},\ \forall\ compact\ K\subset\mathbb{R}^n. \tag{3.2}$$

In particular, if $0<p=q<1$ and $0<\alpha<1$, then the following five conditions are equivalent:

(a) $\|f\|_{L^p(\mathbb{R}^n,\mu)}\lesssim\|f\|_{\dot{\Lambda}^{p,p}_\alpha(\mathbb{R}^n)},\quad\forall\ \ f\in\dot{\Lambda}^{p,p}_\alpha(\mathbb{R}^n)$.

(b) $\|f\|_{L^{p,\infty}(\mathbb{R}^n,\mu)}\lesssim\|f\|_{\dot{\Lambda}^{p,p}_\alpha(\mathbb{R}^n)},\quad\forall\ \ f\in\dot{\Lambda}^{p,p}_\alpha(\mathbb{R}^n)$.

(c) $\sup_{t>0}t/\zeta^\alpha_p(\mu;t)<\infty$.

(d) $\mu(O)\lesssim cap(O;\dot{\Lambda}^{p,p}_\alpha(\mathbb{R}^n)),\ \forall\ $ *bounded open* $\ O\subseteq R^n$.

(e) $\|\mu\|_{p,\alpha}:=\sup_{r>0,x\in\mathbb{R}^n}r^{\alpha p-n}\mu(B(x,r))<\infty$.

Proof. (3.2) $\Longrightarrow$ (3.1) is a consequence of Theorem 2.1. For (3.1) $\Longrightarrow$ (3.2), let $K\subset\mathbb{R}^n$ be compact. Then, for any $\varepsilon>0$ there exists an $f\in\dot{\Lambda}^{p,q}_\alpha(\mathbb{R}^n)$ with $f\geq1$ on K and

$$cap(K;\dot{\Lambda}^{p,q}_\alpha(\mathbb{R}^n))+\varepsilon\geq\|f\|^p_{\dot{\Lambda}^{p,q}_\alpha(\mathbb{R}^n)}.$$

Theorem 2.1 implies

$$\mu(K)\lesssim\left(\int_0^\infty\mu\left(\{x\in\mathbb{R}^n:|f(x)|\geq t\}\right)^{\frac{p\vee q}{p\wedge q}}\,dt^{p\vee q}\right)^{\frac{p\wedge q}{p\vee q}}\lesssim\|f\|^{p\wedge q}_{\dot{\Lambda}^{p,q}_\alpha(\mathbb{R}^n)}$$

whence deriving (3.2) via the definition of $cap(K;\dot{\Lambda}^{p,q}_\alpha(\mathbb{R}^n))$.

Regarding the case $0<p=q<1$, we divide our argument into three parts.

Part 1: We show (a) $\Longrightarrow$ (b) $\Longrightarrow$ (d) $\Longrightarrow$ (a). In what follows, for $\lambda>0$ and $f\in\dot{\Lambda}^{p,p}_\alpha(\mathbb{R}^n)$, let

$$M_\lambda(f)=\{(t,x)\in\mathbb{R}^n:|f(x)|\geq\lambda\}.$$

(a) $\Longrightarrow$ (b). This follows from the following weak type estimate

$$t^p\mu(M_t(f))\leq p\int_0^\infty\mu(M_\lambda(f))\lambda^{p-1}d\lambda,\quad\forall\ \ t>0.$$

(b) $\Longrightarrow$ (d). Let (b) be true. For an given open set $O\subseteq\mathbb{R}^n$, and any function $f\in\dot{\Lambda}^{p,p}_\alpha(\mathbb{R}^n)$ with O being contained in the interior $\mathrm{Int}(\{x\in\mathbb{R}^n:|f(x)|\geq1\})$ of the level set $\{x\in\mathbb{R}^n:|f(x)|\geq1\}$, we have

$$\mu(O)\leq\mu(M_1(f))\lesssim\|f\|^p_{\dot{\Lambda}^{p,p}_\alpha(\mathbb{R}^n)},$$

thereby getting (d).

(d) $\Longrightarrow$ (a). If (d) is true, then for $f \in \dot{\Lambda}^{p,q}_{\alpha}(\mathbb{R}^n)$, $k = 1, 2, 3, \ldots$ and $B(0,k) \subseteq \mathbb{R}^n$, Theorem 2.1 implies

$$\begin{aligned}&\int_0^\infty (\mu(M_\lambda(f) \cap B(0,k)))d\lambda^p\\ &\lesssim \int_0^\infty cap(M_\lambda(f) \cap B(0,k)); \dot{\Lambda}^{p,p}_{\alpha}(\mathbb{R}^n) d\lambda^p\\ &\lesssim \int_0^\infty cap(M_\lambda(f); \dot{\Lambda}^{p,p}_{\alpha}(\mathbb{R}^n)) d\lambda^p\\ &\lesssim \|f\|^p_{\dot{\Lambda}^{p,p}_{\alpha}(\mathbb{R}^n)}.\end{aligned}$$

Letting $k \longrightarrow \infty$, we get (a).

Part 2: We verify (b) $\Longrightarrow$ (c) $\Longrightarrow$ (a).

If (b) holds, then

$$\mathsf{J}_p(\mu) := \sup_{0<\|f\|_{\dot{\Lambda}^{p,p}_{\alpha}(\mathbb{R}^n)}<\infty} \frac{\sup_{\lambda>0} \lambda \left(\mu\left(\{x \in \mathbb{R}^n : |f(x)| > \lambda\}\right)\right)^{\frac{1}{p}}}{\|f\|_{\dot{\Lambda}^{p,p}_{\alpha}(\mathbb{R}^n)}} < \infty.$$

For a given $f \in \dot{\Lambda}^{p,p}_{\alpha}(\mathbb{R}^n)$ and a bounded open set $O \subseteq \mathrm{Int}\left(\{x \in \mathbb{R}^n : |f(x)| \geq 1\}\right)$, we have

$$(\mu(O))^{1/p} \lesssim \mathsf{J}_p(\mu)\|f\|_{\dot{\Lambda}^{p,p}_{\alpha}(\mathbb{R}^n)}.$$

Note that

$$t \lesssim (\mathsf{J}_p(\mu))^p\, cap(O; \dot{\Lambda}^{p,p}_{\alpha}(\mathbb{R}^n)), \quad \forall\ 0 < t < \mu(O).$$

Hence

$$t \lesssim (\mathsf{J}_p(\mu))^p\, \zeta^{\alpha}_p(\mu; t),$$

which yields (c).

If (c) holds, then Theorem 2.1 implies that for each $k = 1, 2, 3, \ldots$,

$$\begin{aligned}&\int_0^\infty \left(\mu\left(M_\lambda(f) \cap B(0,k)\right)\right) d\lambda^p\\ &\leq \int_0^\infty \left(\frac{\left(\mu\left(M_\lambda(f) \cap B(0,k)\right)\right)}{\zeta^{\beta}_p\left(\mu; \mu\left(M_\lambda(f) \cap B(0,k)\right)\right)}\right) cap\left(M_\lambda(f) \cap B(0,k); \dot{\Lambda}^{p,p}_{\alpha}(\mathbb{R}^n)\right) d\lambda^p\\ &\lesssim \left(\sup_{t>0} \frac{t}{\zeta^{\alpha}_p(\mu; t)}\right) \int_0^\infty cap\left(\{x \in \mathbb{R}^n : |f(x)| > \lambda\} \cap B(0,k); \dot{\Lambda}^{p,p}_{\alpha}(\mathbb{R}^n)\right) d\lambda^p\\ &\lesssim \left(\sup_{t>0} \frac{t}{\zeta^{\alpha}_p(\mu; t)}\right) \|f\|^p_{\dot{\Lambda}^{p,p}_{\alpha}(\mathbb{R}^n)}.\end{aligned}$$

Letting $k \longrightarrow \infty$ in the previous inequalities we have

$$\int_0^\infty \left(\mu\left(M_\lambda(f)\right)\right) d\lambda^p \lesssim \left(\sup_{t>0} \frac{t}{\zeta^{\beta}_p(\mu; t)}\right) \|f\|^p_{\dot{\Lambda}^{p,p}_{\alpha}(\mathbb{R}^n)}.$$

This validates (a).

Part 3: We verify (d) $\Longleftrightarrow$ (e).

The implication (d) $\Longrightarrow$ (e) follows from the facts $cap(\cdot;\dot{\Lambda}^{p,p}_{\alpha}(\mathbb{R}^n)) \approx H^{(\infty)}_{n-\alpha p}(\cdot)$ and $H^{(\infty)}_{n-\alpha p}(B(x,r)) \approx r^{n-\alpha p}$. Conversely, if (e) holds, then for a bounded open $O \subset \mathbb{R}^n$, we can find a sequence of open balls $\{B_i\}$ with radii $\{r_i\}$ to cover O, whence getting

$$\mu(O) \le \sum_i \mu(B_i) \lesssim \|\mu\|_{p,\alpha} \sum_i r_i^{n-\alpha p},$$

which implies

$$\mu(O) \lesssim \|\mu\|_{p,\alpha} H^{(\infty)}_{n-\alpha p}(O) \lesssim \|\mu\|_{p,\alpha} cap(O;\dot{\Lambda}^{p,p}_{\alpha}(\mathbb{R}^n)).$$

Thus, (d) holds. □

3.2 Embedding in Lebesgue space on $\mathbb{R}^{1+n}_{+}$ via $(\partial_t + (-\triangle)^{\beta})v = 0$

As the second application of C.S.I. obtained in Section 2, we discuss the embedding in a diagonal Lorentz space (i.e. Lebesgue space) induced by a non-negative Radon measure on $\mathbb{R}^{1+n}_{+}$ of the extension of a given fractional Sobolev space $\dot{\Lambda}^{p,p}_{\alpha}(\mathbb{R}^n)$ via the fractional heat equation

$$\begin{cases} \partial_t v(t,x) + (-\triangle)^{\beta} v(t,x) = 0, & \forall\ (t,x) \in \mathbb{R}^{1+n}_{+}; \\ v(0,x) = f(x), & \forall\ x \in \mathbb{R}^n, \end{cases} \tag{3.3}$$

where $-\triangle$ is the positive Laplacian with respect to x and

$$(-\triangle)^{\beta} u(t,x) = \mathcal{F}^{-1}(|\xi|^{2\beta}\mathcal{F}(u(t,\xi)))(x)$$

with $\mathcal{F}$ and $\mathcal{F}^{-1}$ being the Fourier transform and the inverse Fourier transform. Here

$$v(t,x) = S_{\beta}(t)f(x) := K^{\beta}_t * f(x)$$

solves (3.3),

$$K^{\beta}_t(x) := (2\pi)^{-\frac{n}{2}} \int_{\mathbb{R}^n} e^{ix\cdot\xi} e^{-t|\xi|^{2\beta}} d\xi, \quad \forall \quad (t,x) \in \mathbb{R}^{1+n}_{+},$$

and $g * h$ means the convolution between g and h on $\mathbb{R}^n$.

The following lemma comes from [14, Lemma 2.4] and [16, Lemma 2.4].

Lemma 3.2. *Let $p \in (0,1)$, $\beta \in (0,1]$ and $\alpha \in (0,1)$. Given $f \in \dot{\Lambda}^{p,p}_{\alpha}(\mathbb{R}^n)$, $\lambda > 0$, and a non-negative Radon measure μ on $\mathbb{R}^{1+n}_{+}$, let*

$$\begin{cases} E^{\alpha,\beta}_{\lambda}(f) = \{(t,x) \in \mathbb{R}^{1+n}_{+} : |S_{\beta}(t^{2\beta})f(x)| > \lambda\}; \\ O^{\alpha,\beta}_{\lambda}(f) = \{y \in \mathbb{R}^n : \sup_{x\in B(y,t)} |S_{\beta}(t^{2\beta})f(x)| > \lambda\}. \end{cases}$$

(i) *For any natural number* k, *one has*

$$\begin{cases} \mu\left(E_{\lambda}^{\alpha,\beta}(f)\cap T(B(0,k))\right)\le \mu\left(T(O_{\lambda}^{\alpha,\beta}(f)\cap B(0,k))\right); \\ cap\left(O_{\lambda}^{\alpha,\beta}(f)\cap B(0,k);\dot{\Lambda}_{\alpha}^{p,p}(\mathbb{R}^n)\right)\ge \eta_p^{\alpha}\left(\mu;\mu\left(T(O_{\lambda}^{\alpha,\beta}(f)\cap B(0,k))\right)\right). \end{cases}$$

(ii) *There exist two constants* $\theta_1,\theta_2>0$ *such that*

$$\begin{cases} \sup_{y\in B(x,t)}|S_\beta(t^{2\beta})f(y)|\le \theta_1\mathcal{M}f(x),\ \forall\, x\in\mathbb{R}^n; \\ (t,x)\in T(O)\Longrightarrow S_\beta(t^{2\beta})|f|(x)\ge\theta_2,\ \forall\ \textit{bounded open} \\ \qquad O\subset Int(\{x\in\mathbb{R}^n: f(x)\ge 1\}). \end{cases}$$

Below is a resolution to the above described embedding problem.

Theorem 3.3. *Let* μ *be a non-negative Radon measure on* $\mathbb{R}_+^{1+n}$, $\beta\in(0,1]$, $\frac{n}{n+\alpha}<p<1$ *and* $\alpha\in(0,1)$. *Then the following five conditions are equivalent:*

(a) $\|v(t^{2\beta},x)\|_{L^p(\mathbb{R}_+^{1+n},\mu)}\lesssim\|f\|_{\dot{\Lambda}_{\alpha}^{p,p}(\mathbb{R}^n)},\ \forall\ f\in\dot{\Lambda}_{\alpha}^{p,p}(\mathbb{R}^n)$.

(b) $\|v(t^{2\beta},x)\|_{L^{p,\infty}(\mathbb{R}_+^{1+n},\mu)}\lesssim\|f\|_{\dot{\Lambda}_{\alpha}^{p,p}(\mathbb{R}^n)},\ \forall\ f\in\dot{\Lambda}_{\alpha}^{p,p}(\mathbb{R}^n)$.

(c) $\sup_{t>0} t/\eta_p^{\alpha}(\mu;t)<\infty$.

(d) $\mu(T(O))\lesssim cap(O;\dot{\Lambda}_{\alpha}^{p,p}(\mathbb{R}^n)),\ \ \forall$ *bounded open* $O\subseteq\mathbb{R}^n$.

(e) $|||\mu|||_{p,\alpha}:=\sup_{(r,x)\in\mathbb{R}_+^{1+n}} r^{\alpha p-n}\mu\left(T(B(x,r))\right)<\infty$.

Proof. The argument consists of three parts.

Part 1: We prove (a) $\Longrightarrow$ (b) $\Longrightarrow$ (d) $\Longrightarrow$ (a).

(a) $\Longrightarrow$ (b). This follows from the following weak-type estimate

$$t^p\mu(E_t^{\alpha,\beta}(f))\le p\int_0^\infty \mu(E_\lambda^{\alpha,\beta}(f))\lambda^{p-1}d\lambda,\quad \forall\, t>0.$$

(b) $\Longrightarrow$ (d). If (b) is true, then

$$\mathsf{K}_p(\mu):=\sup_{0<\|f\|_{\dot{\Lambda}_{\alpha}^{p,p}(\mathbb{R}^n)}<\infty}\frac{\sup_{\lambda>0}\lambda\left(\mu\left(\{(t,x)\in\mathbb{R}_+^{1+n}:|v(t^{2\beta},x)|>\lambda\}\right)\right)^{\frac{1}{p}}}{\|f\|_{\dot{\Lambda}_{\alpha}^{p,p}(\mathbb{R}^n)}}<\infty.$$

For a given $f\in\dot{\Lambda}_{\alpha}^{p,p}(\mathbb{R}^n)$ and a bounded open set $O\subseteq\mathrm{Int}\left(\{x\in\mathbb{R}^n: f(x)\ge 1\}\right)$, then Lemma 3.2 (ii) implies

$$\left(\mu\left(T(O)\right)\right)^{1/p}\lesssim\mathsf{K}_p(\mu)\|f\|_{\dot{\Lambda}_{\alpha}^{p,p}(\mathbb{R}^n)}$$

and hence (d) follows from the definition of $cap(O;\dot{\Lambda}_{\alpha}^{p,p}(\mathbb{R}^n))$.

(d) $\Longrightarrow$ (a). To see this, we assume that (d) is valid. Then

$$\mathsf{L}_p(\mu):=\sup\left\{\frac{\mu\left(T(O)\right)}{cap(O;\dot{\Lambda}_{\alpha}^{p,p}(\mathbb{R}^n))}:\ \text{bounded open}\ \ O\subseteq\mathbb{R}^n\right\}<\infty.$$

If $f \in \dot{\Lambda}_{\alpha}^{p,p}(\mathbb{R}^n)$ and $k = 1,2,3,\dots$, then Lemma 3.2 (i)–(ii) and Proposition 2.3 imply

$$\begin{aligned}
&\int_0^\infty \left(\mu\left(E_\lambda^{\alpha,\beta}(f)\cap T(B(0,k))\right)\right) d\lambda^p \\
&\le \int_0^\infty \left(\mu\left(T(O_\lambda^{\alpha,\beta}(f)\cap B(0,k))\right)\right) d\lambda^p \\
&\le \int_0^\infty \left(\mu_k\left(T(\{x\in\mathbb{R}^n : \theta_1\mathcal{M}f(x)>\lambda\}\cap B(0,k))\right)\right) d\lambda^p \\
&\le \int_0^\infty \left(\mu\left(T(\{x\in\mathbb{R}^n : \theta_1\mathcal{M}f(x)>\lambda\}\cap B(0,k))\right)\right) d\lambda^p \\
&\lesssim \mathsf{L}_p(\mu)\int_0^\infty cap\left(\{x\in\mathbb{R}^n : \theta_1\mathcal{M}f(x)>\lambda\}\cap B(0,k);\dot{\Lambda}_{\alpha}^{p,p}(\mathbb{R}^n)\right) d\lambda^p \\
&\lesssim \mathsf{L}_p(\mu)\int_0^\infty cap\left(\{x\in\mathbb{R}^n : \theta_1\mathcal{M}f(x)>\lambda\};\dot{\Lambda}_{\alpha}^{p,p}(\mathbb{R}^n)\right) d\lambda^p \\
&\lesssim \mathsf{L}_p(\mu)\|f\|^p_{\dot{\Lambda}_{\alpha}^{p,p}(\mathbb{R}^n)}.
\end{aligned}$$

Letting $k \longrightarrow \infty$ in the above inequality we have

$$\int_0^\infty \mu\left(E_\lambda^{\alpha,\beta}(f)\right) d\lambda^p \lesssim \mathsf{L}_p(\mu)\|f\|^p_{\dot{\Lambda}_{\alpha}^{p,p}(\mathbb{R}^n)}.$$

This derives (a).

Part 2: We verify (b) $\Longrightarrow$ (c) $\Longrightarrow$ (a).

If (b) holds, then for any bounded open set $O \subseteq \mathbb{R}^n$, we have

$$\mu\left(T(O)\right)^{1/p} \lesssim \mathsf{K}_p(\mu)\|f\|_{\dot{\Lambda}_{\alpha}^{p,p}(\mathbb{R}^n)}.$$

Note that

$$t \lesssim (\mathsf{K}_p(\mu))^p\, cap(O;\dot{\Lambda}_{\alpha}^{p,p}(\mathbb{R}^n)),\quad \forall\ 0<t<\mu\left(T(O)\right).$$

Hence

$$t \lesssim (\mathsf{K}_p(\mu))^p\,\eta_p^\alpha(\mu;t).$$

Therefore (c) holds.

If (c) holds, then Lemma 3.2 (i)–(ii) and Proposition 2.3 yield that for each $k = 1,2,3,\dots$,

$$\begin{aligned}
&\int_0^\infty \left(\mu\left(E_\lambda^{\alpha,\beta}(f)\cap T(B(0,k))\right)\right) d\lambda^p \\
&\le \int_0^\infty \left(\frac{\left(\mu\left(E_\lambda^{\alpha,\beta}(f)\cap T\left(B(0,k)\right)\right)\right)}{\eta_p^\beta\left(\mu;\mu\left(E_\lambda^{\alpha,\beta}(f)\cap T\left(B(0,k)\right)\right)\right)}\right) cap\left(O_\lambda(f)\cap B(0,k);\dot{\Lambda}_{\alpha}^{p,p}(\mathbb{R}^n)\right) d\lambda^p \\
&\lesssim \left(\sup_{t>0}\frac{t}{\eta_p^\alpha(\mu;t)}\right)\int_0^\infty cap\left(\{x\in\mathbb{R}^n : \theta_1\mathcal{M}f(x)>\lambda\}\cap B(0,k);\dot{\Lambda}_{\alpha}^{p,p}(\mathbb{R}^n)\right) d\lambda^p \\
&\lesssim \left(\sup_{t>0}\frac{t}{\eta_p^\alpha(\mu;t)}\right)\|f\|^p_{\dot{\Lambda}_{\alpha}^{p,p}(\mathbb{R}^n)}.
\end{aligned}$$

Letting $k \longrightarrow \infty$ in the previous inequality we have

$$\int_0^\infty \left(\mu\left(E_\lambda^{\alpha,\beta}(f)\right)\right) d\lambda^p \lesssim \left(\sup_{t>0} \frac{t}{\eta_p^\beta(\mu;t)}\right) \|f\|^p_{\dot{\Lambda}_\alpha^{p,p}(\mathbb{R}^n)}.$$

This gives (a).

Part 3: We show (d) $\Longleftrightarrow$ (e).

It suffices to verify (e) $\Longrightarrow$ (d). Suppose that $O \subseteq \mathbb{R}^n$ is a bounded open set and is covered by a sequence of dyadic cubes $\{I_j\}$ in $\mathbb{R}^n$ with $\sum_j |I_j|^{\frac{n-\alpha p}{n}} < \infty$ and $|I_j|$ being the volume of I_j. According to [7, Lemma 4.1] there exists another sequence of dyadic cubes $\{J_j\}$ in $\mathbb{R}^n$ such that

$$\begin{cases} \mathrm{Int}(J_j) \cap \mathrm{Int}(J_k) = \emptyset \text{ for } j \neq k; \\ \bigcup_j J_j = \bigcup_k I_k; \\ \sum_j |J_j|^{\frac{n-\alpha p}{n}} \leq \sum_k |I_k|^{\frac{n-\alpha p}{n}}; \\ T(O) \subseteq \bigcup_j T(\mathrm{Int}(5\sqrt{n}J_j)), \end{cases}$$

where $5\sqrt{n}J_j$ stands for the cube with side-length $5\sqrt{n}$ times side-length of J_j and the same centre as J_j's. Then

$$\mu(T(O)) \lesssim |||\mu|||_{p,\alpha} \sum_j |5\sqrt{n}J_j|^{\frac{n-\alpha p}{n}} \lesssim |||\mu|||_{p,\alpha} \sum_j |J_j|^{\frac{n-\alpha p}{n}} \lesssim |||\mu|||_{p,\alpha} \sum_j |I_j|^{\frac{n-\alpha p}{n}}.$$

The fact $cap(\cdot;\dot{\Lambda}_\alpha^{p,p}(\mathbb{R}^n)) \approx H^\infty_{n-\alpha p}(\cdot)$ (a special form of (2.3)) and the definition of $H^\infty_{n-\alpha p}(O)$ imply

$$\mu(T(O)) \lesssim |||\mu|||_{p,\alpha} cap(O;\dot{\Lambda}_\alpha^{p,p}(\mathbb{R}^n)).$$

□

Acknowledgements

This project was in part supported by NSERC of Canada and URP of Memorial University.

References

[1] D. R. Adams, On the existence of capacitary strong type estimates in $\mathbb{R}^n$, Ark. Mat. **14** (1976), 125–140.

[2] D. R. Adams, A note on Choquet integral with respect to Hausdorff capacity, in: Function Spaces and Applications, Lund, 1986, in: Lecture Notes in Math. **1302**, Springer, 1988, pp.115–124.

[3] D. R. Adams, Choquet integrals in potential theory, Publ. Mat. **42** (1998), 3–66.

[4] D. R. Adams, Besov capacity redux, J. Math. Sci. **162** (2009), 307–318.

[5] D. R. Adams and J. Xiao, Strong type estimates for homogeneous Besov capacities, Math. Ann. **325** (2003), 695–709.

[6] G. Dafni, G. E. Karadzhov, and J. Xiao, Classes of Carleson-type measures generated by capacities, Math. Z. **258** (2008), 827–844.

[7] G. Dafni and J. Xiao, Some new tent spaces and duality theorems for fractional Carleson measures and $Q_\alpha(\mathbb{R}^n)$, J. Funct. Anal. **208** (2004), 377–422.

[8] G. E. Karadzhov and J. Xiao, Carleson type theorems for certain convolution operators, Integr. Equ. Oper. Theory **55** (2006), 429–438.

[9] V. G. Maz'ya, On capacitary strong type estiamtes for fractional norms, Zap. Nauchn. Sem. Leningrad Otdel. Math. Inst. steklov. (LOMI) **70** (1977), 161–168 (in Russian).

[10] Yu. V. Netrusov, Metric estimates of the capacities of sets in Besov spaces (in Russian), Tr. Mat. Inst. Steklova **190** (1989), 159–185; English transl.: Proc. Steklov Inst. Math. **190** (1992), 167–192.

[11] Yu. V. Netrusov, Estimates of capacities associated with Besov spaces (in Russian), Zap. Nauchn. Semin. POMI **201** (1992), 124–156; English transl.: J. Math. Sci. **78** (1996), 199–217.

[12] J. Orobitg and J. Verdera, Choquet integrals, Hausdorff content and the Hardy-Littlewood maximal operator, Bull. London Math. Soc. **30** (1998), 145–150.

[13] Z. Wu, Strong type estiamtes and Carleson measures for Lipschitz spaces, Proc. Amer. Math. Soc. **127** (1999), 3243–3249.

[14] J. Xiao, Carleson embeddings for Sobolev spaces via heat equation, J. Differ. Equations **224** (2006), 277–295.

[15] J. Xiao, Homogeneous endpoint Besov space embeddings by Hausdorff capacity and heat equation, Adv. Math. **207** (2006), 828–846.

[16] Z. Zhai, Carleson measure problems for parabolic Bergman spaces and homogeneous Sobolev spaces, Nonlinear Anal. TMA **73** (2010), 2611–2630.

A List of Ph.D. Students, Post-doctors and Visiting Scholars Supervised by Professor Shanzhen Lu and Foreign Collaborators Who Worked with Professor Shanzhen Lu

1. Ph.D. Students

Zhixin Liu (1986—1989)
Heping Liu (1987—1990)
Yan Zhang (1987—1990)
Haiyang Huang (1988—1991)
Sigang Qiu (1989—1990)
Dachun Yang (1989—1992)
Yong Ding (1992—1995)
Bolin Ma (1992—1995)
Wengu Chen (1993—1996)
Guoquan Li (1996—1999)
Xiaochun Li (1996—1997)
Yiqing Gui (1997—2000)
Lin Tang (1998—2001)
Jingshi Xu (1998—2001)
Dunyan Yan (1998—2001)
Pu Zhang (1998—2001)
Huoxiong Wu (2000—2003)
Qiang Wu (2000—2003)
Junfeng Li (2000—2005)
Mingju Liu (2001—2004)
Qingying Xue (2001—2004)
Huixia Mo (2003—2006)
Yan Lin (2003—2008)
Junyang Gao (2004—2007)
Xia Xia (2004—2007)
Zunwei Fu (2005—2008)
Fayou Zhao (2007—2010)
Zengyan Si (2008—2011)
Shaoguang Shi (2009—2012)

2. Post-doctors

Dinghui Guo (1991—1993)
Guoen Hu (1993—1995)
Rong Yuan (1994—1996)
Zongguang Liu (1999—2001)
Shuangping Tao (2004—2006)
Dongxiang Chen (2006—2008)

3. Visiting Scholars

Yinsheng Jiang (1987)
Changmei Tan (1993)
Weiquan Liu (1994)
Jiacheng Lan (2000)
Senhua Lan (2000)
Lanzhe Liu (2000)
Xiao Yu (2012)

4. Foreign Collaborators

Guido Weiss
Mitchell H Taibleson

Loukas Grafakos
Dashan Fan
Yibiao Pan
Yongsheng Han
Guozhen Lu
Fernando Soria
Kozo Yabuta
Shuichi Sato

Photographs

Shanzhen Lu was on the rostrum of the satellite conference (Harmonic Analysis and Its Applications) of International Congress of Mathematicians, 2002

Shanzhen Lu attended the International Conference on Harmonic Analysis in Changsha, 2006

Shanzhen Lu attended the International Conference on Applicable Harmonic Analysis in Beijing, 2006

Shanzhen Lu attended the International Conference on PDEs and Harmonic Analysis in Beijing, 2007

Shanzhen Lu attended the International Conference on Harmonic Analysis at Hokkaido of Japan, 2007

Shanzhen Lu attended the International Workshop on Fourier Analysis and PDEs in Beijing, 2008

Shanzhen Lu attended the International Conference on Harmonic Analysis and Partial Differential Equations with Applications in Beijing, 2009

Shanzhen Lu attended the International Conference on Harmonic Analysis and Applications in Mudanjiang, 2010

Shanzhen Lu attended the International Workshop on Harmonic Analysis and PDE in Hangzhou, 2011.

Shanzhen Lu attended the International Conference on Harmonic Analysis and Its Applications in Nanjing, 2012

Shanzhen Lu attended the International Conference on Harmonic Analysis and Its Applications in Jinhua, 2012

Shanzhen Lu attended the International Conference on Harmonic Analysis and Its Applications in Haikou, 2013

Shanzhen Lu attended the Workshop on Harmonic Analysis in Linyi, 2013

Shanzhen Lu was at Washington University in St. Louis, 1980

Shanzhen Lu and Guido Weiss were in St. Louis, 1981

Shanzhen Lu and Alan McIntosh were in suburban Canberra, 1987

Shanzhen Lu and R. R. Coifman were at Shinkansen trains in Japan, 1990

Shanzhen Lu and C. E. Kenig were in Tohoku University in Sendai, 1990

As President of Beijing Normal University, Shanzhen Lu was at his office in 1995

Shanzhen Lu was in Fuzhou with J. C. Chen, D. C. Yang, D. S. Fan, Y. Ding and L. X. Yan, 2008

Shanzhen Lu was in Beijing Normal University with D. C. Yang, Z. J. Wu, F. Soria, C. Perez, C. C. Lin, Y. S. Han and D. S. Fan, 2009

Shanzhen Lu was in Nanjing train station with G. Z. Lu, S. Chanillo, M. Lacey and J. -L. Menaldi, 2012

Shanzhen Lu was appointed with Chancellor of Wenzhou-Kean University (China- US cooperation university) , 2012

As Chancellor of Wenzhou-Kean University, Shanzhen Lu was at his office in Wenzhou, 2013

Shanzhen Lu was in his office with Guozhen Lu (left) and Fanghua Lin (right) in Beijing, 2013

Made in the USA
Monee, IL
01 March 2026

45072374R00254